Y0-ACA-452

Advanced Heat Transfer

Lectures from the Advanced Heat Transfer
Short Course held at the University of Illinois,
Urbana, April 24-28, 1967, in cooperation with the
Program of the State Technical Services Act and
the American Society of Mechanical Engineers

Advanced Heat Transfer

B.T. CHAO, Editor

Published in cooperation with
Engineering Publications Office,
College of Engineering

University of Illinois Press

Urbana, Chicago, London 1969

Manufactured in the United States of America
Library of Congress Catalog Card No. 68-24626

252 78412 x

PREFACE

Technological developments associated with nuclear power generation, high-speed flight, and space exploration have provided unprecedented impetus to the need of predicting heat transfer rates with higher precision and greater detail. As part of the activities commemorating the Centennial Year of the University of Illinois, a one-week concentrated course titled "Advanced Heat Transfer" was offered to the public during April of 1967. The lecture notes were preprinted and distributed to the participants. A large number of requests for copies of these notes were subsequently received and could not be accommodated due to a shortage of supply. This book is a collection of the revised course lectures.

The first chapter presents the physical concepts and the mathematical formulation of the basic laws governing the convective heat transfer process. Significant parameters of fluid flow and heat transfer and possible simplifications of a problem are shown to be deducible from dimensional considerations. A full section is devoted to the methodology of similarity analysis, and another to the discussion and evaluation of Meksyn's asymptotic method of integrating boundary layer equations. This chapter includes a description of a simple coordinate transformation technique not previously given in textbooks.

Fundamental concepts and equations of radiative heat transfer are described in the second chapter. The prediction of reflection and emission properties of certain classes of surfaces from electromagnetic theory is explained and illustrated. The calculation of radiative transfer among surfaces, taking into account their directional and spectral properties, is thoroughly discussed. This chapter culminates with a treatment of the important but complex subject of simultaneous conduction, convection, and radiation transfer processes.

The third chapter discusses and compares a number of numerical methods suitable for solving steady and transient heat conduction problems using digital computers. Truncation error, consistency, stability, and the rate of convergence of various difference schemes are examined. The importance and utility of recognizing the physical consistency in writing the difference equation for the boundary conditions are emphasized and demonstrated.

The next three chapters discuss separately the reason for the unsatisfactory state of affairs for nucleate pool boiling, the physics and chemistry of dropwise condensation, and the current status of understanding of the choking phenomenon in vapor-liquid flow. Subsequent chapters treat heat and mass transfer in laminar boundary layers, in turbulent separated flows, and in chemically reacting gases. Another treats combined free and forced convection flows in parallel channels. Recent work on the influence of surface curvature on radiative transfer in the presence of a gray heat generating gas, on the heat transfer processes in particulate suspensions, and on the use of digital techniques for statistically analyzing turbulence data appear in three separate chapters. A final chapter surveys the heat transfer problem in bioengineering.

It is our hope that the present volume would not only be useful to engineers responsible for heat transfer research and development work, but also that it may serve as a subsidiary text for advanced courses in heat transfer.

My thanks are due to all authors for their cooperation in making the course a success and in undertaking the arduous task of preparing and revising the manuscripts; to Miss Ann Riggins, editor, Engineering Publications Office, for her supervision in preparing the camera-ready copy; and to the staffs of the Engineering Publications Office and the University of Illinois Press for their assistance and courtesy in printing this book. Professor Ross J. Martin and Dr. Marvin E. Krasnow were instrumental in organizing the course. Professor John C. Chato served as the conference co-chairman. In addition to helping organize the technical aspects of the course, he also ably arranged the social events which delightfully provided some refreshing moments. To all these people, I offer my sincere gratitude.

March 1968 B. T. CHAO

ACKNOWLEDGMENTS

The editor and the authors wish to record their indebtedness to the various reviewers, whose names appear below, for their thoughtful comments and valuable suggestions.

Professor Richard J. Grosh, Dean of the College of Engineering, Purdue University, West Lafayette, Indiana.

Dr. Frank Kreith, Professor of Mechanical Engineering, University of Colorado, Boulder, Colorado.

Professor Helmut H. Korst, Head of the Department of Mechanical and Industrial Engineering, University of Illinois, Urbana, Illinois.

Dr. Thomas J. Hanratty, Professor of Chemical Engineering, University of Illinois, Urbana, Illinois.

Dr. John L. Hudson, Assistant Professor of Chemical Engineering, University of Illinois, Urbana, Illinois.

CONTENTS

Selected Topics on Convective Heat Transfer 1
B. T. CHAO

Selected Topics on Radiative Heat Transfer 75
R. G. HERING

Numerical Methods in Heat Transfer 157
A. M. CLAUSING

Nucleate Pool Boiling 217
J. W. WESTWATER

Dropwise Condensation 233
J. W. WESTWATER

Critical Two-Phase Flow 263
H. K. FAUSKE

Heat and Mass Transfer in Separated Flows 277
R. H. PAGE

Mass Transfer Cooling in Laminar Boundary Layers 301
J. P. HARTNETT

Heat Transfer in Chemically Reacting Gases 319
P. M. CHUNG

Recent Developments in Turbulence Measuring
and Analyzing Techniques 339
B. G. JONES

Radiative Transfer Between Two Concentric Gray Spheres
Separated by Radiating Gas 373
R. VISKANTA

Heat Transfer in Bioengineering 395
J. C. CHATO

Heat Transfer Processes of Particulate Suspensions 415
S. L. SOO

Combined Free and Forced Convection Flows in Channels 439
J. C. CHATO

CONTENTS

SELECTED TOPICS ON CONVECTIVE HEAT TRANSFER
by B. T. Chao

1. CONSERVATION EQUATIONS AND CONSTITUTIVE RELATIONSHIPS. 1

1.1 Integral Formulation of the Conservation Principles and the Differential Equations Deduced from Them. 1
1.2 Constitutive Relations and the Transport Property for Momentum and Heat 8
1.3 Summary 9

2. COORDINATE TRANSFORMATION. 15

2.1 Development of the Transformation Formula 15
2.2 Illustrations 16

3. SIMILITUDE AND BOUNDARY LAYER APPROXIMATIONS 25

3.1 Flow and Heat Transfer Parameters 25
3.2 Boundary Layer Approximations 31

4. SIMILARITY ANALYSIS OF BOUNDARY LAYER EQUATIONS. 40

4.1 Transient Temperature Field in an Isotropic Infinite Solid Due to an Infinite Line Source of Constant Strength 40
4.2 Steady, Incompressible, Two-dimensional, Laminar Boundary Layer with Pressure Gradient. 42
4.3 Steady, Laminar Forced Convection in Incompressible Wedge Flow with Nonuniform Wall Temperature. 44
4.4 Steady, Laminar Free Convection Over a Vertical Plate with Uniform Heat Flux 46
4.5 Steady, Compressible, Two-dimensional Laminar Boundary Layer with Pressure Gradient and Heat Transfer. 48
4.6 Two-dimensional, Incompressible Jet (Laminar or Turbulent). 50
4.7 Concluding Remarks. 52

5. MEKSYN'S METHOD OF LAMINAR BOUNDARY LAYER ANALYSIS 54

5.1 Transformed Boundary Layer Equations. 54
5.2 Series Inversion. 56
5.3 Euler's Transformation and the Sum of Nonconvergent Series. 57
5.4 Meksyn's Method of Integrating Boundary Layer Equations 58
5.5 Forced Convection Over Flat Plate at Uniform Wall Temperature 61
5.6 Stagnation Point Heat Transfer. 63
5.7 Supersonic Flow Past a Flat Plate with and Without Heat Transfer. 65
5.8 Concluding Remarks. 71

SELECTED TOPICS ON CONVECTIVE HEAT TRANSFER*

B. T. Chao**

1. CONSERVATION EQUATIONS AND CONSTITUTIVE RELATIONSHIPS

Engineers engaged in the design and development of heat transfer equipment in which flow of fluids takes place often need information on temperatures, pressures, velocities, and so forth. Such information may be obtained from experimentation or analysis or both. The purpose of this chapter is to develop the fundamental equations which govern the underlying physical processes. A knowledge and understanding of them are necessary for any analytical endeavor. To keep the discussion reasonably simple, only single phase, single component media are considered, although the procedure used has been generalized to treat multiphase and multicomponent systems.

It will be assumed that the readers have some familiarity with the Cartesian tensor notation which is adopted for convenience.***

1.1 INTEGRAL FORMULATION OF THE CONSERVATION PRINCIPLES AND THE DIFFERENTIAL EQUATIONS DEDUCED FROM THEM

1.1.1 The Continuity Equation

We consider a fixed region in space. Let its volume be V and its enclosing surface be S. The region may or may not be simply connected. Let n_i be the unit, outdrawn vector as shown in Figure 1.1; the index i takes on the values 1, 2, and 3 and thus its three components are n_1, n_2, and n_3. The principle of mass conservation states: the rate of increase of mass in V equals the mass influx through its surface. Thus

$$\frac{\partial}{\partial t}\int_V \rho dV = -\oint \rho v_i n_i dS, \qquad (1.1)^{****}$$

where t is the time, ρ the density, and v_i the vector velocity at dS. The symbol $\oint$

*This paper contains five sections and was given as four separate lectures at the conference.

**Professor of Mechanical and of Nuclear Engineering, University of Illinois, Urbana, Illinois.

***An excellent exposition of the use of Cartesian tensors in elasticity, fluid dynamics, electromagnetism, etc., may be found in H. J. Jeffreys and B. S. Jeffreys, Methods of Mathematical Physics, Cambridge University Press, 1950, Ch. 3.

****The summation convention is used throughout this discussion, i.e., whenever the index or subscript is repeated in one term, that term is summed over that index. Thus $v_i n_i = v_1 n_1 + v_2 n_2 + v_3 n_3$.

designates a surface integral over S. Applying the divergence theorem to the right side of Equation (1.1) leads to

$$\int_V \frac{\partial \rho}{\partial t}\, dV + \int_V \frac{\partial(\rho v_i)}{\partial x_i}\, dV = 0, \tag{1.2}$$

where x_i denotes the vector coordinate. We note that $\partial(\rho v_i)/\partial x_i$ is simply $\mathrm{div}(\rho\vec{v})$ in vector notation. Since V is arbitrary, it follows from Equation (1.2) that

$$\frac{\partial \rho}{\partial t} + \frac{\partial(\rho v_i)}{\partial x_i} = 0, \tag{1.3}$$

which is the familiar Eulerian form of the continuity equation. An alternate form of Equation (1.3) is

$$\frac{D\rho}{Dt} + \rho \frac{\partial v_i}{\partial x_i} = 0, \tag{1.4}$$

where

$$\frac{D}{Dt} \equiv \frac{\partial}{\partial t} + v_i \frac{\partial}{\partial x_i}$$

is the substantial derivative. It is a time differentiation following the fluid particles. To see this, we consider any quantity, $F(x_i,t)$, be it a scalar or vector, at space coordinate x_i and time t. To an observer who travels with the fluid, the coordinate becomes $x_i + v_i\delta t$ at the later instant $t + \delta t$. Hence the time rate of change of the quantity F as seen by this observer is

$$\lim_{\delta t \to 0} \frac{F(x_i + v_i\delta t,\ t + \delta t) - F(x_i,t)}{\delta t}$$

$$= \left(\frac{\partial}{\partial t} + v_i \frac{\partial}{\partial x_i}\right) F \, . \tag{1.5}$$

Clearly, if the flow is incompressible, the continuity equation reduces to

$$\frac{\partial v_i}{\partial x_i} = 0, \tag{1.6}$$

irrespective of whether the motion is steady or not.

1.1.2 The Linear Momentum Equation

A basic principle of the dynamics of fluid motion is that of the conservation of linear momentum: the rate of increase of linear momentum in V plus its outflow rate through S equals the resultant external forces acting on the mass system inside V. The external forces are usually of two types: (1) the body force which is distributed through V and (2) the surface force. The foregoing principle may be restated in mathematical language as follows:

$$\frac{\partial}{\partial t} \int_V \rho v_i dV + \oint v_i \rho v_j n_j dS$$

$$= \int_V \rho X_i dV + \oint t_i dS, \tag{1.7}$$

where X_i is the body force per unit mass and t_i is the stress vector. The latter is the surface force per unit area. The index, j, like i, takes on the values 1, 2, and 3. Using the divergence theorem, the left side of Equation (1.7) may be transformed to become

$$\int_V \frac{\partial}{\partial t} (\rho v_i) dV + \int_V \frac{\partial}{\partial x_j} (v_i \rho v_j) dV$$

which is identical to

$$\int_V \rho \left(\frac{\partial v_i}{\partial t} + v_j \frac{\partial v_i}{\partial x_j}\right) dV \tag{1.8}$$

in view of the continuity equation. We may thus rewrite Equation (1.7) as

$$\int_V \rho \left(\frac{\partial v_i}{\partial t} + v_j \frac{\partial v_i}{\partial x_j}\right) dV$$

$$= \int_V \rho X_i dV + \oint t_i dS. \tag{1.9}$$

Let the arbitrary volume, V, be of the order ℓ^3, where ℓ is its typical length. If we divide each term of Equation (1.9) by ℓ^2 and examine the limiting condition when $V \to 0$, we arrive at the important result:

$$\lim_{V \to 0} \frac{1}{\ell^2} \oint t_i dS = 0, \tag{1.10}$$

which implies that the stress forces are in local equilibrium in spite of the fact that the fluid may be in acceleration.

We now proceed to explore the nature of the stress vector t_i. To this end, let us consider a tetrahedron of fluid as illustrated in Figure 1.2, with vertex at x_i and with three of its faces parallel to the coordinate planes. The fourth is slanted and has an area δA. If $\hat{i}$, $\hat{j}$, and $\hat{k}$ denote unit vectors parallel to the coordinate axes as shown, then the outdrawn unit normals to the above three faces are $-\hat{i}$, $-\hat{j}$, and $-\hat{k}$. We designate the outdrawn unit normal to δA by n_i. Clearly, the stress vector acting on δA depends not only on x_i and t but also on its orientation which is completely prescribed by n_i. Hence

$$t_i = t_i(x_i, n_i, t).$$

Upon applying Equation (1.10) to the stress tetrahedron of Figure 1.2, we obtain

$$t_i(x_i, n_i, t) + n_1 t_i(x_i, -\hat{i}, t)$$
$$+ n_2 t_i(x_i, -\hat{j}, t) + n_3 t_i(x_i, -\hat{k}, t) = 0. \quad (1.11)$$

If we rotate the slanted face such that $n_i = \hat{i}$, then $n_1 = 1$ and $n_2 = n_3 = 0$. Under this condition, Equation (1.11) becomes

$$t_i(x_i, \hat{i}, t) = -t_i(x_i, -\hat{i}, t),$$

which is a result to be expected. Likewise, we recognize that

$$t_i(x_i, \hat{j}, t) = -t_i(x_i, -\hat{j}, t)$$

and a similar result for $\hat{k}$. Consequently, we have

$$t_i(x_i, n_i, t) = n_1 t_i(x_i, \hat{i}, t) + n_2 t_i(x_i, \hat{j}, t)$$
$$+ n_3 t_i(x_i, \hat{k}, t), \quad (1.12)$$

which is of the form:

$$\left.\begin{aligned} t_1 &= n_1\pi_{11} + n_2\pi_{21} + n_3\pi_{31} \\ t_2 &= n_1\pi_{12} + n_2\pi_{22} + n_3\pi_{32} \\ t_3 &= n_1\pi_{13} + n_2\pi_{23} + n_3\pi_{33}, \end{aligned}\right\} \quad (1.13)$$

where

$$\pi_{11} = t_1(x_i, \hat{i}, t),$$

$$\pi_{21} = t_1(x_i, \hat{j}, t),$$

$$\pi_{31} = t_1(x_i, \hat{k}, t),$$

$$\pi_{12} = t_2(x_i, \hat{i}, t) \text{ etc.}$$

Either Equation (1.12) or Equations (1.13) shows that the stress vector is expressible as a linear function of the components of n_i.

Using the summation convention, the three equations of (1.13) may be conveniently written as

$$t_i = \pi_{ji} n_j. \quad (1.14)$$

The nine quantities π_{ji} are the components of the stress tensor. Often, one merely speaks of the tensor π_{ji}. From the foregoing, it is clear that π_{ji} has a simple physical meaning. It represents the i-th component of the stress acting on a surface element whose outdrawn normal is in the j-direction. In view of Equation (1.14), the second term on the right side of Equation (1.9) may be written as $\oint \pi_{ji} n_j dS$, which can then be transformed to the volume integral $\int_V (\partial\pi_{ji}/\partial x_j) dV$ according to the divergence theorem. The linear momentum equation now becomes

$$\int_V \rho\left(\frac{\partial v_i}{\partial t} + v_j \frac{\partial v_i}{\partial x_j}\right) dV$$

$$= \int_V \left(\rho X_i + \frac{\partial \pi_{ji}}{\partial x_j}\right) dV. \quad (1.15)$$

It follows that

$$\rho\left(\frac{\partial v_i}{\partial t} + v_j \frac{\partial v_i}{\partial x_j}\right) = \rho X_i + \frac{\partial \pi_{ji}}{\partial x_j} \quad (1.16a)$$

or, equivalently,

$$\frac{Dv_i}{Dt} = X_i + \frac{1}{\rho}\frac{\partial \pi_{ji}}{\partial x_j}. \quad (1.16b)$$

An alternate form is

$$\frac{\partial}{\partial t}(\rho v_i) + \frac{\partial}{\partial x_j}(\rho v_i v_j) = \rho X_i + \frac{\partial \pi_{ji}}{\partial x_j}. \quad (1.16c)$$

Note that Dv_i/Dt is the acceleration computed by following the fluid motion and $(\partial/\partial x_j)(\rho v_i v_j)$ is the outflow rate of linear momentum per unit volume.

If one takes an elementary cube and considers all the forces acting on it (body as well as surface forces), one finds, upon invoking the principle of conservation of angular momentum and passing to the limit of a differential element, that

$$\pi_{ji} = \pi_{ij} \quad , \qquad (1.17)^*$$

which states that the stress tensor is symmetrical. Because of this, it has only six, rather than nine, different components in general. The result given in Equation (1.17) follows from the fact that, as the size of the cube diminishes, the body forces decrease at a rate faster than the surface forces by an order of magnitude.

It is customary to define a thermodynamic pressure, p, such that the equation of state of the fluid is

$$p = p(\rho, T),$$

where T is the temperature. We may now define a viscous stress tensor τ_{ij} according to

$$\pi_{ij} = -p\delta_{ij} + \tau_{ij} \quad , \qquad (1.18)$$

where δ_{ij} is the Kronecker delta. The physical reason for this separation into a static pressure, p, and a viscous stress tensor, τ_{ij}, will be given in Section 1.2.

Using the relation (1.18), the divergence of the stress tensor π_{ji} may be written as

$$\frac{\partial \pi_{ji}}{\partial x_j} = -\frac{\partial p}{\partial x_i} + \frac{\partial \tau_{ji}}{\partial x_j} \quad . \qquad (1.19)$$

1.1.3 The Kinetic Energy Equation

If we take the scalar product of v_i and the momentum equation, Equation (1.16a), we find

$$\rho\left(v_i \frac{\partial v_i}{\partial t} + v_i v_j \frac{\partial v_i}{\partial x_j}\right)$$

$$= \rho v_i X_i + v_i \frac{\partial \pi_{ji}}{\partial x_j} \quad . \qquad (1.20)$$

*This excludes the consideration of extraneous couples which generally arise in polarized media.

If we let $q^2 = v_i v_i$, which is the square of the speed, we recognize that the left side equals

$$\rho \frac{\partial}{\partial t}\left(\frac{q^2}{2}\right) + \rho v_j \frac{\partial}{\partial x_j}\left(\frac{q^2}{2}\right)$$

or simply

$$\rho \frac{D}{Dt}\left(\frac{q^2}{2}\right) .$$

Hence Equation (1.20) may be alternately written as

$$\rho \frac{D}{Dt}\left(\frac{q^2}{2}\right) = \rho v_i X_i + v_i \frac{\partial \pi_{ji}}{\partial x_j}$$

$$= \rho v_i X_i + \frac{\partial}{\partial x_j}(v_i \pi_{ji})$$

$$- \pi_{ji} \frac{\partial v_i}{\partial x_j} \quad , \qquad (1.21)$$

which is the kinetic energy equation. In view of the continuity requirement,

$$\rho \frac{D}{Dt}\left(\frac{q^2}{2}\right)$$

also equals

$$\frac{\partial}{\partial t}\left(\frac{\rho q^2}{2}\right) + \frac{\partial}{\partial x_j}\left(v_j \frac{\rho q^2}{2}\right).$$

Thus the left side of Equation (1.21) is the time rate of increase of the kinetic energy of the fluid in a unit, fixed volume in space plus the rate of outflow of kinetic energy through its surface. The right side gives the causes of such changes. The first and second terms represent, respectively, the rate of work done by the body and surface forces per unit volume. The third term needs further explanation; this will be done in the next section. Here, we wish to note that the kinetic energy equation is deduced from the momentum equation; it does not represent a separate principle.

1.1.4 Physical Interpretation of the Scalar Product $\pi_{ji}(\partial v_i/\partial x_j)$

We begin by examining the simple case of one-dimensional flow in the x_1-direction. Let us consider a fluid element AB of unit cross section and of length dx, at time, t, as shown in Figure 1.3a. At a later instant, t + dt, A would move through a distance $v_1 dt$ and B would move through

$$\left[v_1 + \left(\frac{\partial v_1}{\partial x_1}\right) dx_1\right] dt.$$

Hence $\partial v_1/\partial x_1$ is the elongation per unit length of the fluid element per unit time; it is the rate of normal strain. We also observe that the original volume is $dx_1 \cdot 1$ and the new volume is

$$\left[dx_1 + \left(\frac{\partial v_1}{\partial x_1}\right) dx_1\, dt\right] \cdot 1,$$

giving rise to a change

$$\left(\frac{\partial v_1}{\partial x_1}\right) dx_1\, dt.$$

Thus the work done by the normal stress π_{11} per unit volume per unit time is

$$\pi_{11} \frac{\partial v_1}{\partial x_1}.$$

By a similar argument, we find that the rate of volumetric strain, for the general three-dimensional case, is

$$\frac{\partial v_1}{\partial x_1} + \frac{\partial v_2}{\partial x_2} + \frac{\partial v_3}{\partial x_3},$$

which we recognize to be the divergence of the velocity vector, $\partial v_i/\partial x_i$. This quantity is also known as the dilatation. The work done against volume change by the normal stress on the fluid per unit volume and time is

$$\pi_{11} \frac{\partial v_1}{\partial x_1} + \pi_{22} \frac{\partial v_2}{\partial x_2} + \pi_{33} \frac{\partial v_3}{\partial x_3}.$$

Next, let us consider two perpendicular fluid elements, OA and OB, at time, t, in a two-dimensional flow field (see Figure 1.3b). At a later instant, t + dt, they would assume a new configuration O'A' and O'B'. Besides elongation which has just been described, the elements are also subjected to shear deformation and bodily rotation. Referring to that figure, it is seen that O'A' has rotated through an angle $(\partial v_2/\partial x_1)dt$ in the counterclockwise direction, while O'B' has rotated through an angle $(\partial v_1/\partial x_2)dt$ in the clockwise direction. Hence the angle between the two lines of fluid particles originally along x_1- and x_2-axis is decreasing at a rate

$$\frac{\partial v_2}{\partial x_1} + \frac{\partial v_1}{\partial x_2}.$$

If, for instance, a section of the fluid element is initially rectangular, like OACB, it will be deformed into a shape like O'A'C'B'. It can be shown that a circular fluid element will be deformed into an ellipse.

Simultaneously with the deformation, the fluid element also experiences a rotation. Since OA rotates at an angular speed $\partial v_2/\partial x_1$ (counterclockwise) while OB rotates at a speed $\partial v_1/\partial x_2$ (clockwise), the fluid mass has, at that instant, a rotational speed

$$\frac{1}{2}\left(\frac{\partial v_2}{\partial x_1} - \frac{\partial v_1}{\partial x_2}\right), \text{ counterclockwise.}$$

To further explore the physical nature of the quantity $\pi_{ji}(\partial v_i/\partial x_j)$, we decompose the tensor, $\partial v_i/\partial x_j$, into a symmetrical and an antisymmetrical part as follows:

$$\frac{\partial v_i}{\partial x_j} \equiv \underbrace{\frac{1}{2}\left(\frac{\partial v_i}{\partial x_j} + \frac{\partial v_j}{\partial x_i}\right)}_{\text{symmetrical}} + \underbrace{\frac{1}{2}\left(\frac{\partial v_i}{\partial x_j} - \frac{\partial v_j}{\partial x_i}\right)}_{\text{antisymmetrical}}, \quad (1.22)$$

which is merely an identity.

The symmetrical part has the following components:

$$\frac{1}{2}\begin{bmatrix} 2\frac{\partial v_1}{\partial x_1} & \frac{\partial v_1}{\partial x_2} + \frac{\partial v_2}{\partial x_1} & \frac{\partial v_1}{\partial x_3} + \frac{\partial v_3}{\partial x_1} \\ \frac{\partial v_2}{\partial x_1} + \frac{\partial v_1}{\partial x_2} & 2\frac{\partial v_2}{\partial x_2} & \frac{\partial v_2}{\partial x_3} + \frac{\partial v_3}{\partial x_2} \\ \frac{\partial v_3}{\partial x_1} + \frac{\partial v_1}{\partial x_3} & \frac{\partial v_3}{\partial x_2} + \frac{\partial v_2}{\partial x_3} & 2\frac{\partial v_3}{\partial x_3} \end{bmatrix}. \quad (1.22a)$$

The diagonal elements are the rate of normal strain, the remaining are the rate of shear strain. In formulating the product

$$\frac{1}{2}\pi_{ji}\left(\frac{\partial v_i}{\partial x_j}+\frac{\partial v_j}{\partial x_i}\right) ,$$

each component of the stress tensor π_{ji} is multiplied by the corresponding component of the symmetrical tensor defined by (1.22a). Since π_{ji} is also symmetrical, the product contains six instead of nine terms. Clearly, such a product represents the work done by surface forces against deformation (dilatation and shear) per unit volume per unit time.

The components of the antisymmetrical part are:

$$\frac{1}{2}\begin{bmatrix} 0 & \frac{\partial v_1}{\partial x_2}-\frac{\partial v_2}{\partial x_1} & \frac{\partial v_1}{\partial x_3}-\frac{\partial v_3}{\partial x_1} \\ \frac{\partial v_2}{\partial x_1}-\frac{\partial v_1}{\partial x_2} & 0 & \frac{\partial v_2}{\partial x_3}-\frac{\partial v_3}{\partial x_2} \\ \frac{\partial v_3}{\partial x_1}-\frac{\partial v_1}{\partial x_3} & \frac{\partial v_3}{\partial x_2}-\frac{\partial v_2}{\partial x_3} & 0 \end{bmatrix} \tag{1.22b}$$

which, as has been explained, represent the rates of angular rotation of the fluid element. Clearly,

$$\pi_{ji}\left(\frac{\partial v_i}{\partial x_j}-\frac{\partial v_j}{\partial x_i}\right) = 0.$$

We therefore conclude that $\pi_{ji}(\partial v_i/\partial x_j)$ is the work done by surface forces against deformation per unit volume per unit time. Since

$$\pi_{ji} = -p\delta_{ji} + \tau_{ji},$$

$$\pi_{ji}\frac{\partial v_i}{\partial x_j} = -p\frac{\partial v_i}{\partial x_i} + \tau_{ji}\frac{\partial v_i}{\partial x_j} . \tag{1.23}$$

The first term on the right side is the thermodynamic work done by pressure forces against volume change. The second term represents the irreversible conversion of mechanical energy into heat as a result of viscosity.

Since the components of the symmetrical tensor

$$\frac{1}{2}\left(\frac{\partial v_i}{\partial x_j}+\frac{\partial v_j}{\partial x_i}\right)$$

represent rates of strain, it is called the rate of deformation tensor, to be denoted by D_{ij}. From the foregoing discussion, it is seen that

$$\tau_{ji}\frac{\partial v_i}{\partial x_j} = \tau_{ij}D_{ij} .$$

1.1.5 The Energy Equation

The third basic principle of mechanics is that of the conservation of energy: the rate of increase of the sum of internal and kinetic energy of the fluid inside V plus their outflow rate through S equals the rate of work done by the body and surface forces on the fluid inside V, plus the influx of heat across S and any extraneous internal heat source such as that due to chemical reaction, passage of electric current, radioactive decay, etc. Let e denote the internal energy per unit mass, q_i the heat flux vector, and Q the rate of internal heat generation per unit mass. Thus

$$\frac{\partial}{\partial t}\int_V \rho\left(e+\frac{q^2}{2}\right)dV + \oint \rho\left(e+\frac{q^2}{2}\right)v_i n_i dS = \int_V \rho v_i X_i dV + \oint v_i \pi_{ji} n_j dS - \oint q_i n_i dS + \int_V \rho Q dV . \tag{1.24}$$

Again, upon invoking the divergence theorem and recognizing that V is arbitrary, there is obtained

$$\frac{\partial}{\partial t}\left[\rho\left(e + \frac{q^2}{2}\right)\right] + \frac{\partial}{\partial x_i}\left[\rho v_i\left(e + \frac{q^2}{2}\right)\right]$$

$$= \rho v_i X_i + \frac{\partial}{\partial x_j}(v_i \tau_{ji}) - \frac{\partial q_i}{\partial x_i} + \rho Q. \qquad (1.25)$$

By using the continuity equation, one sees that the left side is equivalent to

$$\rho \frac{D}{Dt}\left(e + \frac{q^2}{2}\right).$$

If one subtracts Equation (1.21) from Equation (1.25), the result is

$$\rho \frac{De}{Dt} = \rho Q - \frac{\partial q_i}{\partial x_i} - p\frac{\partial v_i}{\partial x_i} + \tau_{ji}\frac{\partial v_i}{\partial x_j}, \qquad (1.26)$$

which is the equation for the conservation of energy. As we have previously noted,

$$\tau_{ji}\frac{\partial v_i}{\partial x_j} = \tau_{ij}D_{ij}.$$

Since

$$h = e + \frac{p}{\rho},$$

where h is the enthalpy per unit mass, it follows that

$$\frac{De}{Dt} = \frac{Dh}{Dt} - \frac{1}{\rho}\frac{Dp}{Dt} + \frac{p}{\rho^2}\frac{D\rho}{Dt}$$

$$= \frac{Dh}{Dt} - \frac{1}{\rho}\frac{Dp}{Dt} - \frac{p}{\rho}\frac{\partial v_i}{\partial x_i}.$$

Hence the energy equation can be written in terms of the enthalpy as

$$\rho\frac{Dh}{Dt} = \rho Q - \frac{\partial q_i}{\partial x_i} + \frac{Dp}{Dt} + \tau_{ij}D_{ij}. \qquad (1.27)$$

We conclude this section by presenting two alternate forms of the energy equation. If we regard the internal energy as a function of T and ρ, then

$$de = \left.\frac{\partial e}{\partial T}\right|_\rho dT + \left.\frac{\partial e}{\partial \rho}\right|_T d\rho. \qquad (1.28)$$

It is known that

$$\left.\frac{\partial e}{\partial T}\right|_\rho = c_v,$$

the specific heat at constant volume, and

$$\left.\frac{\partial e}{\partial \rho}\right|_T = \frac{p}{\rho^2} - \frac{T}{\rho^2}\left.\frac{\partial p}{\partial T}\right|_\rho.$$

Thus

$$\frac{De}{Dt} = c_v\frac{DT}{Dt} + \frac{1}{\rho}\left(T\left.\frac{\partial p}{\partial T}\right|_\rho - p\right)\frac{\partial v_i}{\partial x_i}. \qquad (1.29)$$

Consequently, Equation (1.26) may be written as

$$c_v\rho\frac{DT}{Dt} = \rho Q - \frac{\partial q_i}{\partial x_i}$$

$$+ \tau_{ij}D_{ij} - T\left.\frac{\partial p}{\partial T}\right|_\rho \frac{\partial v_i}{\partial x_i}. \qquad (1.30)$$

For an ideal gas, the last term reduces to $-p(\partial v_i/\partial x_i)$.

On the other hand, if we regard the enthalpy as a function of T and p, then

$$dh = \left.\frac{\partial h}{\partial T}\right|_p dT + \left.\frac{\partial h}{\partial p}\right|_T dp.$$

We recognize that

$$\left.\frac{\partial h}{\partial T}\right|_p = c_p,$$

which is the specific heat at constant pressure and

$$\left.\frac{\partial h}{\partial p}\right|_T = \frac{T}{\rho^2}\left.\frac{\partial \rho}{\partial T}\right|_p + \frac{1}{\rho}.$$

Thence

$$\frac{Dh}{Dt} = c_p\frac{DT}{Dt} + \left(\frac{T}{\rho^2}\left.\frac{\partial \rho}{\partial T}\right|_p + \frac{1}{\rho}\right)\frac{Dp}{Dt} \qquad (1.31)$$

and Equation (1.27) becomes

$$c_p\rho\frac{DT}{Dt} = \rho Q - \frac{\partial q_i}{\partial x_i} + \tau_{ij}D_{ij}$$

$$- \frac{T}{\rho}\left.\frac{\partial \rho}{\partial T}\right|_p \frac{Dp}{Dt}. \qquad (1.32)$$

We note that for an ideal gas

$$\frac{T}{\rho}\frac{\partial\rho}{\partial T}\bigg|_p = -1.$$

In the above, we consider that all thermodynamic functions like temperature, enthalpy, internal energy, etc., retain their classical meaning even though the fluid is in motion. In the case of a gas, support of this contention has been provided, at least in part, by the kinetic theory. In general, experimental evidence has in no instance contradicted the consequences of such assertion and this we regard as the ultimate justification.

1.2 CONSTITUTIVE RELATIONS AND THE TRANSPORT PROPERTY FOR MOMENTUM AND HEAT

In the general case of compressible fluid flow, the unknowns in the problem are p, ρ, e, v_i, τ_{ji}, and q_i, totaling fifteen.* The conservation principles lead to the continuity equation, the momentum equation which has three components, and the energy equation. These, together with the equation of state of the flow medium, provide a total of six equations and are thus insufficient to determine the problem. The missing equations are those which relate the components of viscous stress tensor with the rates of strain and the components of the heat flux vector with the rate of spatial change of temperature.

For gases, the fundamental approach to seek such relations is via the discipline of statistical mechanics. The book by Chapman and Cowling(1)** gives a detailed account of the method. It is beyond the scope of the present notes to dwell on this subject. Suffice it to state that, excluding conditions of extreme temperature and pressure, the linear relations prescribed below are valid.

$$\tau_{ij} = (\mu' - \tfrac{2}{3}\mu)\frac{\partial v_k}{\partial x_k}\delta_{ij} + \mu\left(\frac{\partial v_i}{\partial x_j} + \frac{\partial v_j}{\partial x_i}\right) \quad (1.33)$$

*Each of the vectors v_i and q_i has three components. The symmetrical tensor, τ_{ji}, has six components.

**Superscript numbers in parentheses refer to References at the end of each section.

$$q_i = -k\frac{\partial T}{\partial x_i} \quad . \quad (1.34)$$

In Equations (1.33) and (1.34), k is the thermal conductivity, μ the ordinary dynamic viscosity, and μ' the dilational or bulk viscosity. They are the transport properties of the fluid. μ' is also called the second viscosity coefficient; it is associated with the bulk expansion or compression of the gas and its effect would disappear if the divergence vanishes.

In the literature, Equation (1.33) is oftentimes referred to as the constitutive relation and Equation (1.34) as the Fourier law of heat conduction for isotropic media. They show that viscous stresses would appear when there is spatial variation of velocity, and heat flow would occur when there is spatial variation of temperature. As a fluid particle moves from one point to another, the presence of such spatial variation would cause it to continuously adapt or adjust itself to the new surroundings. This is, of course, the physical reason why velocity and temperature differences give rise to, respectively, viscous stresses and heat flow.

In a gas, it is the collision of the molecules which renders feasible the adaptation process described above. Clearly, such a process must take place at a finite rate and thus local thermodynamic equilibrium can exist only when heat flow and viscous stress are vanishingly small. This observation suggests the natural separation of π_{ij} into p and τ_{ij} as suggested in Equation (1.18), since the thermodynamic pressure p is calculable from the equation of state which is strictly for thermodynamic equilibrium.

The three principal forms of energy of a gas molecule are associated with the translational, rotational, and vibrational modes of freedom. Of the three, the vibrational mode is the most difficult to excite. Hence, if there is rapid change in the state of a gas and if the gas molecules have appreciable vibrational energy (e.g., gases at high temperatures), there will be a lag behind the equilibrium value. In such cases, the main contribution to μ' comes from the lag or vibrational energy. For monatomic gases at ordinary pressures, the energy of the molecules is translational, thus $\mu' = 0$. For polyatomic gases, μ' is generally small but not zero. At the present time, there is a dearth of information on μ' and in a great majority of engineering analysis it is simply ignored. Further discussion on the subject may be found in a book by Hirschfelder, Curtiss, and Bird.[2]

The constitutive relation, Equation (1.33), could be derived from a different point of view, without the aid of kinetic theory. One may begin by defining a Newtonian fluid which is isotropic and whose stresses are linearly related to the corresponding rates of strain. By observing certain invariant behavior under coordinate transformation, one is led to the identical result with the magnitude of μ' undetermined. One also finds that if the thermodynamic pressure is taken to equal the arithmetic mean of the normal stresses, then $\mu' = 0$. Such a phenomenological approach is the usual treatment given in engineering textbooks on fluid dynamics (e.g., Fluid Dynamics, by V. L. Streeter). For incompressible flow, the divergence of the velocity vanishes. Consequently, the question concerning μ' does not arise.

When the system temperatures are such that radiation plays a role, the first effect is its contribution to the heat flux vector, q_i. Denoting the radiant flux vector by q_i^r, we may modify Equation (1.34) to read:

$$q_i = -k \frac{\partial T}{\partial x_i} + q_i^r . \tag{1.34a}$$

The nature of q_i^r is discussed at some length in Professor R. G. Hering's lecture on "Simultaneous Conduction, Convection, and Radiation Transfer"* and, thus we shall refrain from considering it here. In what follows, we regard that molecular conduction is the only mechanism contributing to q_i.

When the linear constitutive equations are inserted into the momentum equation, Equation (1.16), the resulting system is called the Navier-Stokes equations of motion (commonly with μ' set to zero).

Using Equations (1.33) and (1.34), the terms $\partial q_i/\partial x_i$ and $\tau_{ij}D_{ij}$ in the energy equation, respectively, become:

$$\frac{\partial q_i}{\partial x_i} = - \frac{\partial}{\partial x_i}\left(k \frac{\partial T}{\partial x_i}\right) \tag{1.35}$$

and

$$\tau_{ij}D_{ij} = (\mu' - \frac{2}{3}\mu)\left(\frac{\partial v_i}{\partial x_i}\right)^2 + \frac{\mu}{2}\left(\frac{\partial v_i}{\partial x_j} + \frac{\partial v_j}{\partial x_i}\right)^2 = \Phi, \tag{1.36}$$

which is called the dissipation function. It can be shown to be always positive for $\mu' \geq 0$. It represents the irreversible conversion into heat of mechanical work against deformation as a result of viscosity.

As stated at the outset of this discussion, only single phase, single component fluid is considered. If the fluid has more than one component and is heterogeneous with possible chemical reaction among components, the situation becomes complicated. The derivation of the conservation equations under such circumstances may be found in a recent paper by Hayday.[3] Furthermore, a multicomponent gas could be initially uniform but the uniformity could be destroyed in regions of steep temperature gradient as a result of thermal diffusion. We shall refrain from discussing such complicated effects in this review.

The conservation equations described herein are presumed to be applicable to laminar as well as turbulent flows. In arriving at the differential form, the continuity of the derivatives has been tacitly assumed. A question naturally arises: Can the random, seemingly discontinuous, turbulent motion of a fluid be meaningfully described by them? It is known that the scale of turbulence is macroscopic. The smallest length may be characterized by the Kolmogorov scale which is several orders of magnitude larger than the mean free path of air under the usual conditions valid for a continuum. Hence the answer to the foregoing question is affirmative. While we shall limit ourselves to the discussion of laminar flow in this review, it is pertinent that the conservation equations have been used in the time-averaged form in the development of semi-empirical theories of turbulent heat transfer.

1.3 SUMMARY

We conclude this section by listing the conservation and the constitutive equations in Cartesian tensor notation as well as in vector-tensor form for the readers' reference. The latter will be used in Section 2, in which a technique of

*This lecture is included in this book beginning on page 142

coordinate transformation for vectors and tensors will be presented.

1.3.1 The Continuity Equation

$$\frac{\partial \rho}{\partial t} + \frac{\partial}{\partial x_i}(\rho v_i) = 0$$

$$\frac{\partial \rho}{\partial t} + \nabla \cdot (\rho \vec{v}) = 0$$

or

$$\frac{D\rho}{Dt} + \rho \frac{\partial v_i}{\partial x_i} = 0 ,$$

$$\frac{D\rho}{Dt} + \rho \nabla \cdot \vec{v} = 0$$

where

$$\frac{D}{Dt} \equiv \frac{\partial}{\partial t} + v_i \frac{\partial}{\partial x_i}$$

$$\frac{D}{Dt} \equiv \frac{\partial}{\partial t} + \vec{v} \cdot \nabla$$

∇ denotes the vector differential operator. Thus $\nabla \cdot \vec{v}$ is the divergence of $\vec{v}$.

1.3.2 The Linear Momentum Equation

$$\rho \frac{Dv_i}{Dt} = \rho X_i - \frac{\partial p}{\partial x_i} + \frac{\partial \tau_{ji}}{\partial x_j}$$

$$\rho \frac{D\vec{v}}{Dt} = \rho \vec{X} - \nabla p + \nabla \cdot \overline{\overline{\tau}}$$

An equivalent form for the left side is

$$\frac{\partial}{\partial t}(\rho v_i) + \frac{\partial}{\partial x_j}(\rho v_i v_j)$$

$$\frac{\partial}{\partial t}(\rho \vec{v}) + \nabla \cdot (\rho \vec{v}, \vec{v})$$

$\overline{\overline{\tau}}$ denotes the viscous stress tensor. The comma denotes a dyadic product.

1.3.3 The Kinetic Energy Equation

$$\rho \frac{D}{Dt}\left(\frac{q^2}{2}\right) = \rho v_i X_i - v_i \frac{\partial p}{\partial x_i} + v_i \frac{\partial \tau_{ji}}{\partial x_j}$$

$$\rho \frac{D}{Dt}\left(\frac{q^2}{2}\right) = \rho \vec{v} \cdot \vec{X} - \vec{v} \cdot \nabla p + \vec{v} \cdot (\nabla \cdot \overline{\overline{\tau}})$$

An equivalent form for the left side is

$$\frac{\partial}{\partial t}\left(\frac{\rho q^2}{2}\right) + \frac{\partial}{\partial x_i}\left(v_i \frac{\rho q^2}{2}\right)$$

$$\frac{\partial}{\partial t}\left(\frac{\rho q^2}{2}\right) + \nabla \cdot \left(\vec{v} \frac{\rho q^2}{2}\right)$$

1.3.4 The Energy Equation

$$\rho \frac{De}{Dt} = \rho Q - \frac{\partial q_i}{\partial x_i} - p \frac{\partial v_i}{\partial x_i} + \Phi$$

$$\rho \frac{Dh}{Dt} = \rho Q - \frac{\partial q_i}{\partial x_i} + \frac{Dp}{Dt} + \Phi$$

$$c_v \rho \frac{DT}{Dt} = \rho Q - \frac{\partial q_i}{\partial x_i} - T \left.\frac{\partial p}{\partial T}\right|_\rho \frac{\partial v_i}{\partial x_i} + \Phi$$

$$c_p \rho \frac{DT}{Dt} = \rho Q - \frac{\partial q_i}{\partial x_i} - \frac{T}{\rho} \left.\frac{\partial \rho}{\partial T}\right|_p \frac{Dp}{Dt} + \Phi$$

where

$$\Phi = \tau_{ij} D_{ij}, \quad D_{ij} = \frac{1}{2}\left(\frac{\partial v_i}{\partial x_j} + \frac{\partial v_j}{\partial x_i}\right)$$

Write:

$\nabla \cdot \vec{q}$ for $\frac{\partial q_i}{\partial x_i}$

and

$\nabla \cdot \vec{v}$ for $\frac{\partial v_i}{\partial x_i}$

$$\Phi = \bar{\bar{\tau}} : \bar{\bar{D}}, \quad \bar{\bar{D}} = \frac{1}{2}\left[\nabla, \vec{v} + (\nabla, \vec{v})_c\right]$$

The double dot (:) denotes scalar multiplication and the subscript c refers to conjugate.

(Note: For ideal gas $T \left.\frac{\partial p}{\partial T}\right|_\rho = p$ and $\frac{T}{\rho} \left.\frac{\partial \rho}{\partial T}\right|_p = -1$)

1.3.5 Constitutive Equations for Newtonian Fluids

$$\tau_{ij} = \left(\mu' - \frac{2}{3}\mu\right) \frac{\partial v_k}{\partial x_k} \delta_{ij} + \mu \left(\frac{\partial v_i}{\partial x_j} + \frac{\partial v_j}{\partial x_i}\right)$$

$$\bar{\bar{\tau}} = \left(\mu' - \frac{2}{3}\mu\right) \nabla \cdot \vec{v}\, \bar{\bar{I}} + \mu \left[\nabla, \vec{v} + (\nabla, \vec{v})_c\right]$$

$\bar{\bar{I}}$ is the identity tensor; it plays the same role as δ_{ij} in the Cartesian tensor system.

$$q_i = -k \frac{\partial T}{\partial x_i}$$

$$\vec{q} = -k \nabla T$$

REFERENCES

1. Chapman, S. and T. G. Cowling. The Mathematical Theory of Non-Uniform Gases. Cambridge, England: Cambridge University Press, 1952, Ch. 3.

2. Hirschfelder, J. O., C. F. Curtiss, and R. B. Bird. The Molecular Theory of Gases and Liquids. New York: John Wiley & Sons, 1965, pp. 503, 521, 711.

3. Hayday, A. A., "On Balance Equations for Heterogeneous Continua," Applied Scientific Research, 16 (1966), p. 65.

General

4. Tsien, H. S., "The Equations of Gas Dynamics," in Fundamentals of Gas Dynamics, H. W. Emmons, ed., Part A. Princeton, N. J.: Princeton University Press, 1958.

5. Howarth, L., "The Equations of Flow in Gases," in Modern Developments in Fluid Dynamics -- High Speed Flow. L. Howarth, ed., Vol. I, Ch. 2. Oxford, England: Oxford University Press, 1956.

6. Schlichting, H. Boundary-Layer Theory. 6th ed. New York: The McGraw-Hill Book Co., 1968, Chs. 3, 12.

7. Bird, R. B., W. E. Stewart, and E. N. Lightfoot. Transport Phenomena. New York: John Wiley & Sons, 1960, Chs. 3, 10. Appendix A of this reference contains a brief review of vector and tensor formalism, dyadic product, double dot multiplication, etc.

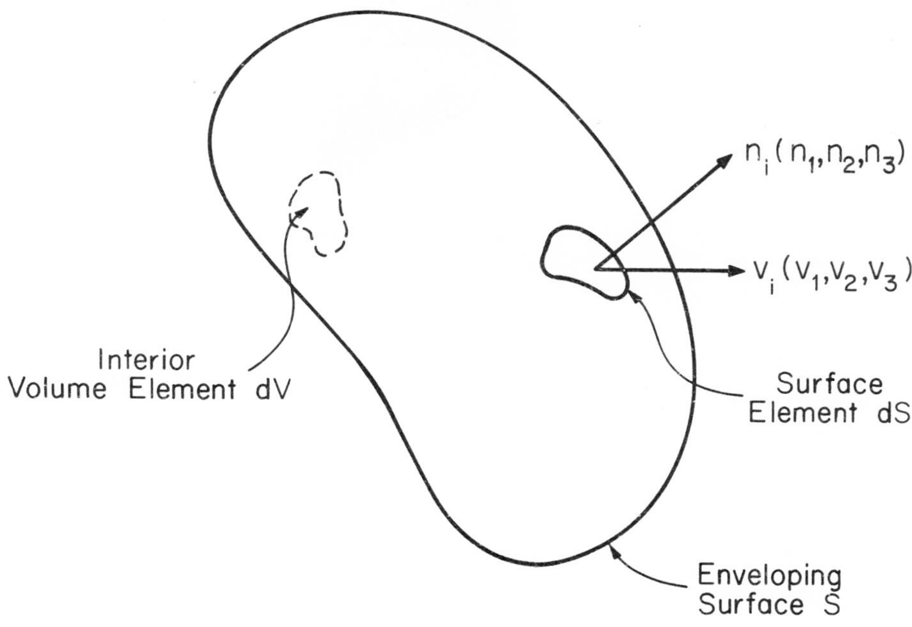

FIGURE 1.1 FLOW THROUGH A CONTROL VOLUME.

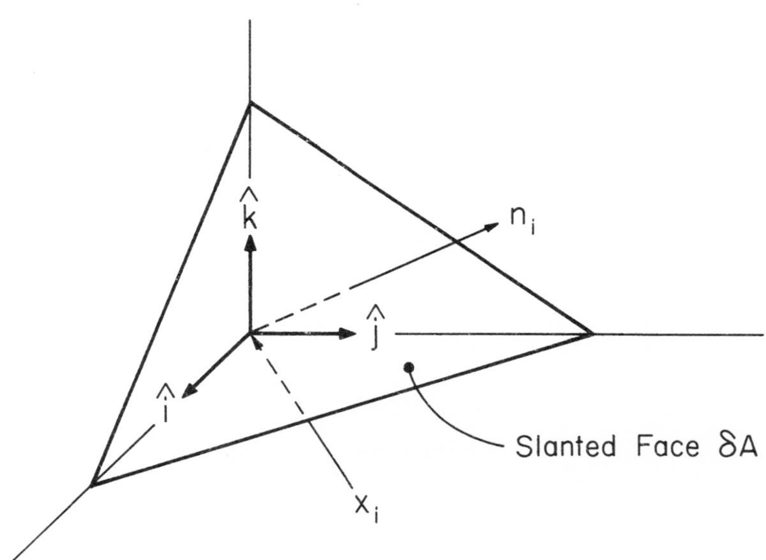

FIGURE 1.2 STRESS TETRAHEDRON.

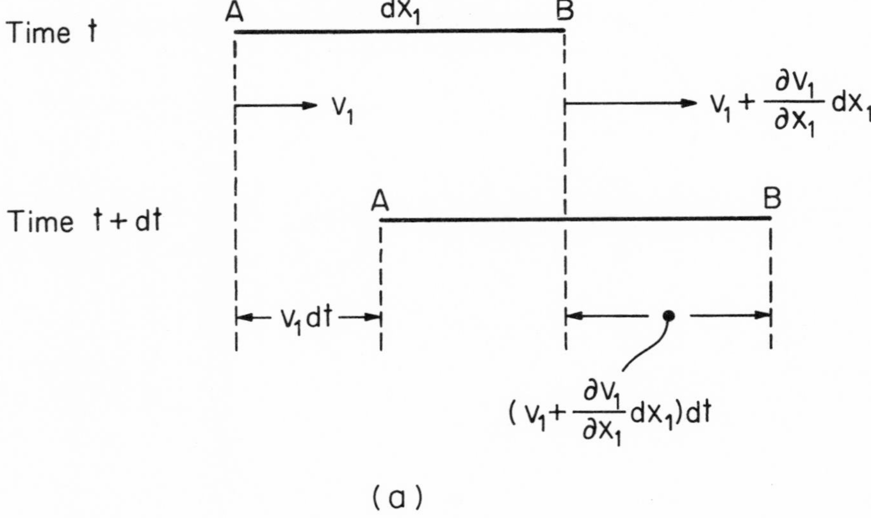

(a)

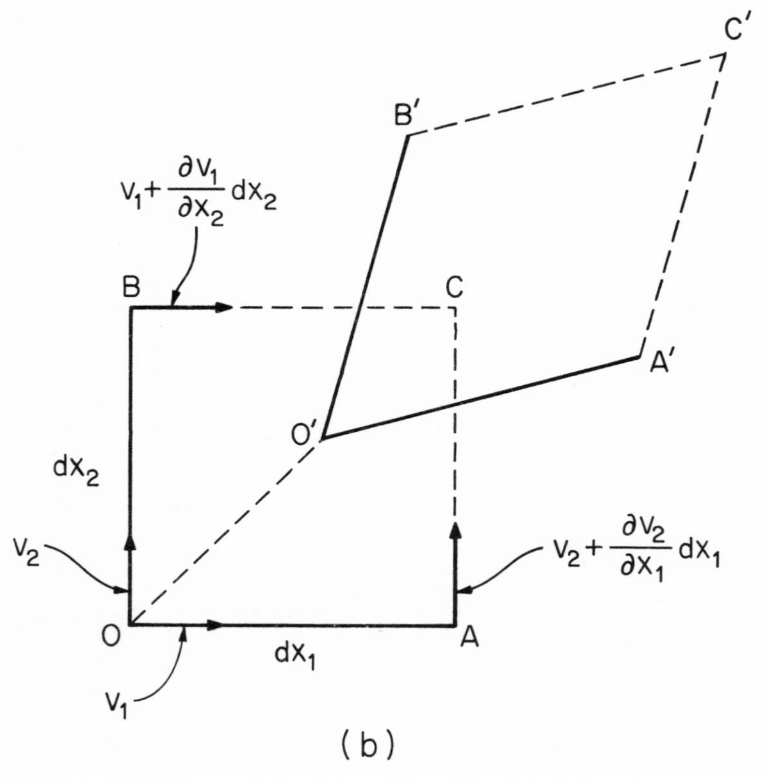

(b)

FIGURE 1.3 DEFORMATION OF FLUID ELEMENTS.

2. COORDINATE TRANSFORMATION

The basic conservation and constitutive equations given in the summary of the previous section in the invariant vector-tensor form are valid for any coordinate system. In order to expand them in component form for the cylindrical, spherical, or any other orthogonal curvilinear system of coordinates, we present a general, yet relatively simple, formula for the del-operator.

The problem of coordinate transformation has been extensively treated in textbooks on continuum mechanics. A brief description of the procedure with emphasis on fluid dynamic equations can be found in Reference 1. Others merely tabulate the results for cylindrical and spherical coordinates without indicating how they were obtained.*

2.1 DEVELOPMENT OF THE TRANSFORMATION FORMULA

Referring to Figure 2.1, $\vec{x}$ is the position vector of a point P in three-dimensional space and u_1, u_2, u_3 are the orthogonal curvilinear coordinates. Thus

$$\vec{x} = \vec{x}(u_1, u_2, u_3).$$

The partial derivative $\partial\vec{x}/\partial u_1$ is a vector tangent to the u_1 curve at P. We define a unit vector $\hat{u}_1$ parallel to $\partial\vec{x}/\partial u_1$. Hence

$$\hat{u}_1 = \frac{\partial\vec{x}/\partial u_1}{|\partial\vec{x}/\partial u_1|},$$

where $|\ |$ denotes magnitude. If we further let

$$h_1 = \left|\frac{\partial\vec{x}}{\partial u_1}\right| ,$$

then

$$\frac{\partial\vec{x}}{\partial u_1} = h_1 \hat{u}_1 . \tag{2.1a}$$

Likewise,

$$\frac{\partial\vec{x}}{\partial u_2} = h_2 \hat{u}_2 \tag{2.1b}$$

and

$$\frac{\partial\vec{x}}{\partial u_3} = h_3 \hat{u}_3 , \tag{2.1c}$$

where $\hat{u}_2$ and $\hat{u}_3$ are unit vectors, respectively, parallel to $\partial\vec{x}/\partial u_2$ and $\partial\vec{x}/\partial u_3$. We note that h_1, h_2, and h_3 are, in essence, scaling factors. They are constants in the Cartesian coordinate system but may vary with location in the curvilinear system.

By chain rule of differentiation, we have

$$d\vec{x} = \frac{\partial\vec{x}}{\partial u_1} du_1 + \frac{\partial\vec{x}}{\partial u_2} du_2 + \frac{\partial\vec{x}}{\partial u_3} du_3$$

$$= \hat{u}_1 h_1 du_1 + \hat{u}_2 h_2 du_2 + \hat{u}_3 h_3 du_3 , \tag{2.2}$$

which implies that $h_1 du_1$, $h_2 du_2$, and $h_3 du_3$ are, respectively, the projection of $d\vec{x}$ on the three coordinate axes. Clearly, they represent the differential arc lengths. Thus we write

$$ds_1 = h_1 du_1 ,$$

$$ds_2 = h_2 du_2 ,$$

and

$$ds_3 = h_3 du_3 . \tag{2.3}$$

We now proceed to expand the gradient of a scalar, $\nabla\varphi$, into its components in the general orthogonal curvilinear system. From the definition of $\nabla\varphi$, we have

$$\nabla\varphi = \hat{u}_1 \frac{\partial\varphi}{\partial s_1} + \hat{u}_2 \frac{\partial\varphi}{\partial s_2} + \hat{u}_3 \frac{\partial\varphi}{\partial s_3}$$

$$= \frac{\hat{u}_1}{h_1}\frac{\partial\varphi}{\partial u_1} + \frac{\hat{u}_2}{h_2}\frac{\partial\varphi}{\partial u_2} + \frac{\hat{u}_3}{h_3}\frac{\partial\varphi}{\partial u_3} , \tag{2.4}$$

which is valid for any scalar φ. In particular, if $\varphi = u_1$, then,

$$\nabla u_1 = \frac{\hat{u}_1}{h_1} , \tag{2.5a}$$

*See, e.g., Transport Phenomena, by R. B. Bird, W. E. Stewart, and E. N. Lightfoot, John Wiley & Sons, 1960, pp. 82, 317.

since u_1, u_2, and u_3 form an orthogonal set. Likewise,

$$\nabla u_2 = \frac{\hat{u}_2}{h} \qquad (2.5b)$$

and

$$\nabla u_3 = \frac{\hat{u}_3}{h_3} . \qquad (2.5c)$$

Thus Equation (2.4) may be written as

$$\nabla\varphi = (\nabla u_1)\frac{\partial\varphi}{\partial u_1} + (\nabla u_2)\frac{\partial\varphi}{\partial u_2} + (\nabla u_3)\frac{\partial\varphi}{\partial u_3}$$

$$= \sum_{i=1}^{3} (\nabla u_i)\frac{\partial\varphi}{\partial u_i} . \qquad (2.6)$$

From Equations (2.2) and (2.4), we find that

$$(\nabla\varphi)\cdot d\vec{x} = \frac{\partial\varphi}{\partial u_1} du_1 + \frac{\partial\varphi}{\partial u_2} du_2 + \frac{\partial\varphi}{\partial u_3} du_3 = d\varphi . \qquad (2.7)$$

In particular, if we set $\varphi = u_1$, there results

$$(\nabla u_1)\cdot\frac{\partial\vec{x}}{\partial u_1} du_1 = du_1$$

in view of the orthogonality of ∇u_1 with $\partial\vec{x}/\partial u_2$ and $\partial\vec{x}/\partial u_3$. Consequently, we have

$$(\nabla u_1)\cdot\frac{\partial\vec{x}}{\partial u_1} = 1 \qquad (2.8)$$

and similar expressions for u_2 and u_3.

Following precisely the same procedure, we obtain for the divergence of a vector:

$$\nabla\cdot\vec{a} = \hat{u}_1\cdot\frac{\partial\vec{a}}{\partial s_1} + \hat{u}_2\cdot\frac{\partial\vec{a}}{\partial s_2} + \hat{u}_3\cdot\frac{\partial\vec{a}}{\partial s_3}$$

$$= \sum_{i=1}^{3} (\nabla u_i)\cdot\frac{\partial\vec{a}}{\partial u_i} , \qquad (2.9)$$

where ∇u_i is related to $\partial\vec{x}/\partial u_i$ according to Equation (2.8). Clearly, other operations, like the curl or the dyadic product, would lead to similar expressions. Thus we arrive at the general transformation formula for the del-operator:

$$\nabla_{(\cdot)} G = \sum_{i=1}^{3} (\nabla u_i)_{(\cdot)}\frac{\partial G}{\partial u_i} \qquad (2.10a)$$

with

$$(\nabla u_i)\cdot\frac{\partial\vec{x}}{\partial u_i} = 1 , \qquad (2.10b)$$

where $\vec{x}$ is the position vector. The function G may be a scalar, a vector, or a tensor. The symbol $(\cdot)$ is interpreted to mean a gradient, a dot, a cross, or a comma operation.

2.2 ILLUSTRATIONS

We begin by considering the cylindrical coordinate system and demonstrate that all relevant terms in the conservation and constitutive equations can be systematically derived with ease.

Let the coordinates be denoted by r, θ, and z and the corresponding unit vectors by $\hat{r}$, $\hat{\theta}$, and $\hat{z}$ (we may regard $u_1 = r$, $u_2 = \theta$, and $u_3 = z$). From the geometry of Figure 2.2, one sees that

$$d\vec{x} = \hat{r}\,dr + \hat{\theta}r\,d\theta + \hat{z}\,dz. \qquad (2.11)$$

Since

$$d\vec{x} = \frac{\partial\vec{x}}{\partial r}dr + \frac{\partial\vec{x}}{\partial\theta}d\theta + \frac{\partial\vec{x}}{\partial z}dz,$$

it follows then that

$$\frac{\partial\vec{x}}{\partial r} = \hat{r},$$

$$\frac{\partial\vec{x}}{\partial\theta} = r\hat{\theta},$$

and

$$\frac{\partial\vec{x}}{\partial z} = \hat{z}.$$

Hence

$$\nabla r = \hat{r}, \qquad (2.12)$$

$$\nabla\theta = \frac{\hat{\theta}}{r} ,$$

and

$$\nabla z = \hat{z} , \tag{2.12}$$

which are the immediate results of Equation (2.10b). We shall need the partial derivatives of the unit vectors; they can be deduced from elementary geometric considerations and are tabulated in Table 2.1. We note that in the Cartesian system, all elements in a corresponding tabulation are zero.

TABLE 2.1

PARTIAL DERIVATIVES OF UNIT VECTORS IN CYLINDRICAL COORDINATE SYSTEM

	$\hat{r}$	$\hat{\theta}$	$\hat{z}$
$\frac{\partial}{\partial r}$	0	0	0
$\frac{\partial}{\partial \theta}$	$\hat{\theta}$	$-\hat{r}$	0
$\frac{\partial}{\partial z}$	0	0	0

The foregoing completes the information necessary to obtain the required results. The remaining work is straightforward and quite mechanical. What is needed is some careful accounting.

(1) Gradient of a Scalar: ∇T.

According to Equation (2.10a), one has

$$\begin{aligned} \nabla T &= \sum_{i=1}^{3} (\nabla u_i) \frac{\partial T}{\partial u_i} \\ &= \nabla r \frac{\partial T}{\partial r} + \nabla\theta \frac{\partial T}{\partial \theta} + \nabla z \frac{\partial T}{\partial z} \\ &= \hat{r} \frac{\partial T}{\partial r} + \frac{\hat{\theta}}{r} \frac{\partial T}{\partial \theta} + \hat{z} \frac{\partial T}{\partial z} . \end{aligned} \tag{2.13}$$

(2) Divergence of a Vector: $\nabla \cdot \vec{v}$

From the preceding result, one may immediately write

$$\nabla \cdot \vec{v} = \hat{r} \cdot \frac{\partial \vec{v}}{\partial r} + \frac{\hat{\theta}}{r} \cdot \frac{\partial \vec{v}}{\partial \theta} + \hat{z} \cdot \frac{\partial \vec{v}}{\partial z} . \tag{2.14}$$

If we let the components of $\vec{v}$ in the r-, θ-, and z-directions be denoted by v_r, v_θ, and v_z, then

$$\vec{v} = \hat{r}\, v_r + \hat{\theta}\, v_\theta + \hat{z}\, v_z . \tag{2.15}$$

It follows that

$$\left.\begin{aligned} \frac{\partial \vec{v}}{\partial r} &= \hat{r} \frac{\partial v_r}{\partial r} + \hat{\theta} \frac{\partial v_\theta}{\partial r} + \hat{z} \frac{\partial v_z}{\partial r} \\ \frac{\partial \vec{v}}{\partial \theta} &= \hat{r} \left(\frac{\partial v_r}{\partial \theta} - v_\theta \right) + \hat{\theta} \left(\frac{\partial v_\theta}{\partial \theta} + v_r \right) \\ &\quad + \hat{z} \frac{\partial v_z}{\partial \theta} \\ \frac{\partial \vec{v}}{\partial z} &= \hat{r} \frac{\partial v_r}{\partial z} + \hat{\theta} \frac{\partial v_\theta}{\partial z} + \hat{z} \frac{\partial v_z}{\partial z} \end{aligned}\right\} \tag{2.16}$$

by carrying out the indicated differentiation and using the information in Table 2.1. Since

$$\hat{r} \cdot \hat{r} = 1,$$

and

$$\hat{r} \cdot \hat{\theta} = \hat{r} \cdot \hat{z} = 0,$$

we obtain from Equation (2.14)

$$\begin{aligned} \nabla \cdot \vec{v} &= \frac{\partial v_r}{\partial r} + \frac{1}{r} \left(\frac{\partial v_\theta}{\partial \theta} + v_r \right) + \frac{\partial v_z}{\partial z} \\ &= \frac{1}{r} \frac{\partial}{\partial r}(r\, v_r) + \frac{1}{r} \frac{\partial v_\theta}{\partial \theta} + \frac{\partial v_z}{\partial z} . \end{aligned} \tag{2.17}$$

Thus the continuity equation,

$$\frac{\partial \rho}{\partial t} + \nabla \cdot (\rho \vec{v}) = 0,$$

in cylindrical coordinates becomes

$$\frac{\partial \rho}{\partial t} + \frac{1}{r}\frac{\partial}{\partial r}(r\,\rho\,v_r) + \frac{1}{r}\frac{\partial}{\partial \theta}(\rho\,v_\theta) + \frac{\partial}{\partial z}(\rho\,v_z) = 0\,. \tag{2.18}$$

Using Equation (2.13) and by analogy, one may readily write

$$\nabla \cdot (k\nabla T) = \frac{1}{r}\frac{\partial}{\partial r}\left(kr\frac{\partial T}{\partial r}\right) + \frac{1}{r^2}\frac{\partial}{\partial \theta}\left(k\frac{\partial T}{\partial \theta}\right) + \frac{\partial}{\partial z}\left(k\frac{\partial T}{\partial z}\right)\,. \tag{2.19}$$

(3) Tensor: $\nabla,\vec{v}$

First, we note

$$\nabla,\vec{v} = \hat{r},\frac{\partial \vec{v}}{\partial r} + \frac{\hat{\theta}}{r},\frac{\partial \vec{v}}{\partial \theta} + \hat{z},\frac{\partial \vec{v}}{\partial z}\,. \tag{2.20}$$

Using Equation (2.16), one sees immediately that the components of the tensor are

$$\nabla,\vec{v} = \begin{bmatrix} \hat{r},\hat{r}\,\dfrac{\partial v_r}{\partial r} & \hat{r},\hat{\theta}\,\dfrac{\partial v_\theta}{\partial r} & \hat{r},\hat{z}\,\dfrac{\partial v_z}{\partial z} \\ \hat{\theta},\hat{r}\,\dfrac{1}{r}\left(\dfrac{\partial v_r}{\partial \theta} - v_\theta\right) & \hat{\theta},\hat{\theta}\,\dfrac{1}{r}\left(\dfrac{\partial v_\theta}{\partial \theta} + v_r\right) & \hat{\theta},\hat{z}\,\dfrac{1}{r}\dfrac{\partial v_z}{\partial \theta} \\ \hat{z},\hat{r}\,\dfrac{\partial v_r}{\partial z} & \hat{z},\hat{\theta}\,\dfrac{\partial v_\theta}{\partial z} & \hat{z},\hat{z}\,\dfrac{\partial v_z}{\partial z} \end{bmatrix} \tag{2.21}$$

the arrangement of which may be compared with the same tensor in the Cartesian system.

(4) Vector: $\vec{v}\cdot\nabla,\vec{v}$

Using Equations (2.15) and (2.21), carrying out the indicated multiplication, and observing that

$$\hat{r}\cdot\hat{r},\hat{r} = \hat{r}, \quad \hat{r}\cdot\hat{\theta},\hat{r} = 0,$$

$$\hat{r}\cdot\hat{z},\hat{r} = 0 \quad \hat{r}\cdot\hat{r},\hat{\theta} = \hat{\theta},$$

etc., one needs little effort to find:

$$\vec{v}\cdot\nabla,\vec{v} = \hat{r}\left[v_r\frac{\partial v_r}{\partial r} + \frac{v_\theta}{r}\left(\frac{\partial v_r}{\partial \theta} - v_\theta\right) + v_z\frac{\partial v_r}{\partial z}\right] + \hat{\theta}\left[v_r\frac{\partial v_\theta}{\partial r} + \frac{v_\theta}{r}\left(\frac{\partial v_\theta}{\partial \theta} + v_r\right) + v_z\frac{\partial v_\theta}{\partial z}\right] + \hat{z}\left[v_r\frac{\partial v_z}{\partial r} + \frac{v_\theta}{r}\frac{\partial v_z}{\partial \theta} + v_z\frac{\partial v_z}{\partial z}\right]. \tag{2.22}$$

(5) Divergence of a Tensor: $\nabla\cdot\bar{\bar{\tau}}$

Again, we note

$$\nabla\cdot\bar{\bar{\tau}} = \hat{r}\cdot\frac{\partial \bar{\bar{\tau}}}{\partial r} + \frac{\hat{\theta}}{r}\cdot\frac{\partial \bar{\bar{\tau}}}{\partial \theta} + \hat{z}\,\frac{\partial \bar{\bar{\tau}}}{\partial z} \tag{2.23}$$

in accordance with Equation (2.10a). We designate the various components of the viscous stress tensor as follows:

$$\bar{\bar{\tau}} = \begin{bmatrix} \hat{r},\hat{r}\;\tau_{rr} & \hat{r},\hat{\theta}\;\tau_{r\theta} & \hat{r},\hat{z}\;\tau_{rz} \\ \hat{\theta},\hat{r}\;\tau_{\theta r} & \hat{\theta},\hat{\theta}\;\tau_{\theta\theta} & \hat{\theta},\hat{z}\;\tau_{\theta z} \\ \hat{z},\hat{r}\;\tau_{zr} & \hat{z},\hat{\theta}\;\tau_{z\theta} & \hat{z},\hat{z}\;\tau_{zz} \end{bmatrix} \tag{2.24}$$

The significance of the double subscript has been explained in Section 1.1.2. For instance, $\tau_{r\theta}$ represents shear stress in the θ-direction, acting on a surface whose normal is in the r-direction. Since $\bar{\bar{\tau}}$ is symmetrical,

$$\tau_{r\theta} = \tau_{\theta r},$$

$$\tau_{rz} = \tau_{zr},$$

and

$$\tau_{\theta z} = \tau_{z\theta}.$$

To carry out the indicated differentiation, we need to evaluate terms like

$$\frac{\partial}{\partial r}(\hat{r},\hat{r}\;\tau_{rr}) = \frac{\partial}{\partial r}(\hat{r},\hat{r})\;\tau_{rr} + \hat{r},\hat{r}\frac{\partial \tau_{rr}}{\partial r}. \quad (2.25)$$

Now, if $\vec{a}$ and $\vec{b}$ denote any two vectors and u_i the coordinate: r, θ, or z, it is known that

$$\frac{\partial}{\partial u_i}(\vec{a},\vec{b}) = \frac{\partial \vec{a}}{\partial u_i},\vec{b} + \vec{a},\frac{\partial \vec{b}}{\partial u_i} \quad (2.26)$$

which is a form of the chain rule of differentiation. From Table 2.1, it is clear that all derivatives of the comma product of the unit vectors in Equation (2.24) with respect to r or z will vanish. On the other hand, corresponding derivatives with respect to θ will contribute. They are:

$$\frac{\partial}{\partial \theta}(\hat{r},\hat{r}) = \hat{\theta},\hat{r} + \hat{r},\hat{\theta};$$

$$\frac{\partial}{\partial \theta}(\hat{\theta},\hat{r}) = \frac{\partial}{\partial \theta}(\hat{r},\hat{\theta}) = -\hat{r},\hat{r} + \hat{\theta},\hat{\theta};$$

$$\frac{\partial}{\partial \theta}(\hat{z},\hat{r}) = \hat{z},\hat{\theta};$$

$$\frac{\partial}{\partial \theta}(\hat{r},\hat{z}) = \hat{\theta},\hat{z};$$

$$\frac{\partial}{\partial \theta}(\hat{\theta},\hat{\theta}) = -\hat{r},\hat{\theta} - \hat{\theta},\hat{r};\ \text{etc.} \quad (2.27)$$

With the foregoing results and observing that $\hat{r}\cdot\hat{\theta} = \hat{r}\cdot\hat{z} = \hat{\theta}\cdot\hat{z} = 0$, one finds from Equation (2.23):

$$\nabla\cdot\bar{\bar{\tau}} = \hat{r}\left[\frac{1}{r}\frac{\partial}{\partial r}(r\;\tau_{rr}) + \frac{1}{r}\frac{\partial \tau_{r\theta}}{\partial \theta} - \frac{\tau_{\theta\theta}}{r} + \frac{\partial \tau_{rz}}{\partial z}\right] + \hat{\theta}\left[\frac{1}{r^2}\frac{\partial}{\partial r}(r^2\;\tau_{r\theta}) + \frac{1}{r}\frac{\partial \tau_{\theta\theta}}{\partial \theta} + \frac{\partial \tau_{\theta z}}{\partial z}\right] + \hat{z}\left[\frac{1}{r}\frac{\partial}{\partial r}(r\;\tau_{rz}) + \frac{1}{r}\frac{\partial \tau_{\theta z}}{\partial \theta}\;\frac{\partial \tau_{zz}}{\partial z}\right]. \quad (2.28)$$

Using Equations (2.22) and (2.28), the three components of the momentum equation can be written down immediately.

(6) Stress-Strain Rate Relation for Newtonian Fluids

As has been given in Section 1.3.5,

$$\bar{\bar{\tau}} = (\mu' - \frac{2}{3}\mu)\nabla\cdot\vec{v}\;\bar{\bar{I}} + \mu\left[\nabla,\vec{v} + (\nabla,\vec{v})_c\right]. \quad (2.29)$$

The components of the identity tensor $\bar{\bar{I}}$ are as follows:

$$\bar{\bar{I}} = \begin{bmatrix} \hat{r},\hat{r}\;1 & 0 & 0 \\ 0 & \hat{\theta},\hat{\theta}\;1 & 0 \\ 0 & 0 & \hat{z},\hat{z}\;1 \end{bmatrix}. \quad (2.30)$$

In Equation (2.21), if we interchange the elements in rows with those in columns, we obtain $(\nabla,\vec{v})_c$. Thus

$$(\nabla,\vec{v})_c = \begin{bmatrix} \hat{r},\hat{r}\frac{\partial v_r}{\partial r} & \hat{r},\hat{\theta}\frac{1}{r}\left(\frac{\partial v_r}{\partial \theta} - v_\theta\right) & \hat{r},\hat{z}\frac{\partial v_r}{\partial z} \\ \hat{\theta},\hat{r}\frac{\partial v_\theta}{\partial r} & \hat{\theta},\hat{\theta}\frac{1}{r}\left(\frac{\partial v_\theta}{\partial \theta} + v_r\right) & \hat{\theta},\hat{z}\frac{\partial v_\theta}{\partial z} \\ \hat{z},\hat{r}\frac{\partial v_z}{\partial r} & \hat{z},\hat{\theta}\frac{1}{r}\frac{\partial v_z}{\partial \theta} & \hat{z},\hat{z}\frac{\partial v_z}{\partial z} \end{bmatrix} \quad (2.31)$$

Using Equations (2.24), (2.30), (2.21), and (2.31), we are directly led to the following desired results:

(a) Normal Stresses

$$\tau_{rr} = 2\mu\frac{\partial v_r}{\partial r} + (\mu' - \frac{2}{3}\mu)\nabla\cdot\vec{v} \quad (2.32a)$$

$$\left.\begin{aligned} \tau_{\theta\theta} &= 2\mu\;\frac{1}{r}\left(\frac{\partial v_\theta}{\partial \theta} + v_r\right) + (\mu' - \frac{2}{3}\mu)\;\nabla\cdot\vec{v} \\ \tau_{zz} &= 2\mu\;\frac{\partial v_z}{\partial z} + (\mu' - \frac{2}{3}\mu)\;\nabla\cdot\vec{v}\;, \end{aligned}\right\} \quad (2.32a)$$

where $\nabla\cdot\vec{v}$ is given by Equation (2.17).

(b) Shear Stresses

$$\left.\begin{aligned} \tau_{r\theta} &= \mu\left(\frac{1}{r}\frac{\partial v_r}{\partial\theta} - \frac{v_\theta}{r} + \frac{\partial v_\theta}{\partial r}\right) \\ \tau_{rz} &= \mu\left(\frac{\partial v_z}{\partial r} + \frac{\partial v_r}{\partial z}\right) \\ \tau_{\theta z} &= \mu\left(\frac{1}{r}\frac{\partial v_z}{\partial\theta} + \frac{\partial v_\theta}{\partial z}\right) . \end{aligned}\right\} \quad (2.32b)$$

(7) Dissipation Function, $\Phi = \overline{\overline{T}}:\overline{\overline{D}}$

It has been shown in Section 1.14 that

$$\overline{\overline{T}}:\overline{\overline{D}} = \overline{\overline{T}}:\nabla,\vec{v} \quad ,$$

which is also equal to $\overline{\overline{T}}:(\nabla,v)_c$ in view of the symmetry of $\overline{\overline{T}}$. The scalar multiplication of the two tensors (also called the double dot multiplication) is performed by simply forming the product of the corresponding components and summing the results. Thus

$$\Phi = \tau_{rr}\frac{\partial v_r}{\partial r} + \tau_{\theta\theta}\frac{1}{r}\left(\frac{\partial v_\theta}{\partial\theta} + v_r\right) + \tau_{zz}\frac{\partial v_z}{\partial z} + \tau_{r\theta}\left(\frac{1}{r}\frac{\partial v_r}{\partial\theta} - \frac{v_\theta}{r} + \frac{\partial v_\theta}{\partial r}\right)$$
$$+ \tau_{rz}\left(\frac{\partial v_z}{\partial r} + \frac{\partial v_r}{\partial z}\right) + \tau_{\theta z}\left(\frac{1}{r}\frac{\partial v_z}{\partial\theta} + \frac{\partial v_\theta}{\partial z}\right) \quad (2.33a)$$

which becomes

$$\Phi = 2\mu\left[\left(\frac{\partial v_r}{\partial r}\right)^2 + \frac{1}{r^2}\left(\frac{\partial v_\theta}{\partial\theta} + v_r\right)^2 + \left(\frac{\partial v_z}{\partial z}\right)^2\right] + \mu\left(\frac{1}{r}\frac{\partial v_r}{\partial\theta} - \frac{v_\theta}{r} + \frac{\partial v_\theta}{\partial r}\right)^2$$
$$+ \mu\left(\frac{\partial v_z}{\partial r} + \frac{\partial v_r}{\partial z}\right)^2 + \mu\left(\frac{1}{r}\frac{\partial v_z}{\partial\theta} + \frac{\partial v_\theta}{\partial z}\right)^2 + (\mu' - \frac{2}{3}\mu)(\nabla\cdot\vec{v})^2 \quad (2.33b)$$

by using the results given in Equations (2.32a,b).

The energy equation in any of the forms listed in Section 1.3.4 can now be written down by noting that

$$\frac{D(\cdot)}{Dt} = \frac{\partial(\cdot)}{\partial t} + v_r\frac{\partial(\cdot)}{\partial r} + \frac{v_\theta}{r}\frac{\partial(\cdot)}{\partial\theta} + v_z\frac{\partial(\cdot)}{\partial z} ,$$

where $(\cdot)$ stands for the scalar e, h, or T.

To deduce the equations in spherical coordinates, we refer to Figure 2.3. The u_i's are the radial coordinate r, the polar angle θ, and the azimuthal angle φ. Hence the position vector

$$\vec{x} = \vec{x}(r,\theta,\varphi) .$$

From the figure, one sees that

$$d\vec{x} = \hat{r}\,dr + \hat{\theta}r d\theta + \hat{\varphi}\,r\sin\theta\,d\varphi, \quad (2.34)$$

where $\hat{r}$, $\hat{\theta}$, and $\hat{\varphi}$ denote, as before, unit vectors. Since

$$d\vec{x} = \frac{\partial\vec{x}}{\partial r}dr + \frac{\partial\vec{x}}{\partial\theta}d\theta + \frac{\partial\vec{x}}{\partial\varphi}d\varphi ,$$

it follows then that

$$\frac{\partial\vec{x}}{\partial r} = \hat{r},$$

$$\frac{\partial\vec{x}}{\partial\theta} = \hat{\theta}r,$$

and

$$\frac{\partial\vec{x}}{\partial\varphi} = \hat{\varphi}r\sin\theta .$$

In view of Equation (2.10b), one concludes that

$$\left.\begin{aligned} \nabla r &= \hat{r}, \\ \nabla\theta &= \frac{\hat{\theta}}{r}, \\ \text{and} \quad \nabla\varphi &= \frac{\hat{\varphi}}{r\sin\theta} . \end{aligned}\right\} \quad (2.35)$$

The partial derivatives of the unit vectors are given in the following table.

TABLE 2.2

PARTIAL DERIVATIVES OF UNIT VECTORS IN SPHERICAL COORDINATE SYSTEM

	$\hat{r}$	$\hat{\theta}$	$\hat{\varphi}$
$\frac{\partial}{\partial r}$	0	0	0
$\frac{\partial}{\partial \theta}$	$\hat{\theta}$	$-\hat{r}$	0
$\frac{\partial}{\partial \varphi}$	$\hat{\varphi}\sin\theta$	$\hat{\varphi}\cos\theta$	$-(\hat{r}\sin\theta + \hat{\theta}\cos\theta)$

The foregoing completes the information necessary for the problem. The modus operandi is precisely the same as that we have shown in detail for the cylindrical coordinate system. The only difference is that the arithmetic becomes somewhat lengthy.

For two-dimensional flow along a curved wall, an orthogonal system of coordinates as shown in Figure 2.4 may be used. The x-axis is along the wall and the y-axis is perpendicular to it. Thus the curvilinear net will consist of curves parallel to the wall and of straight lines normal to them. The radius of curvature at a point P(x,o) is R(x). Tollmien[2] gave the Navier-Stokes equations of motion for incompressible flow in such a system of coordinates without showing the derivation. Using Equations (2.10a,b) and the information on the partial derivatives of the unit vectors $\hat{x}$ and $\hat{y}$ with respect to the coordinates (see Table 2.3), one may demonstrate that all of Tollmien's results can be conveniently deduced. In fact, the corresponding equations for compressible flow can likewise be obtained with ease.

TABLE 2.3

PARTIAL DERIVATIVES OF UNIT VECTORS IN TWO-DIMENSIONAL CURVILINEAR ORTHOGONAL COORDINATE SYSTEM

	$\hat{x}$	$\hat{y}$
$\frac{\partial}{\partial x}$	$-\hat{y}\frac{1}{R}$	$\hat{x}\frac{1}{R}$
$\frac{\partial}{\partial y}$	0	0

We conclude this section by listing the following results relevant to the energy equation, with which Tollmien was not concerned.

$$\frac{De}{Dt} = \frac{\partial e}{\partial t} + \frac{R}{R + y} u \frac{\partial T}{\partial x} + v \frac{\partial T}{\partial y} \tag{2.36}$$

$$\nabla\cdot(k\nabla T) = \left(\frac{R}{R + y}\right)^2 \frac{\partial}{\partial x}\left(k\frac{\partial T}{\partial x}\right) + \frac{Ry}{(R + y)^3}\frac{dR}{dx} k \frac{\partial T}{\partial x} + \frac{1}{R + y} k \frac{\partial T}{\partial y} + \frac{\partial}{\partial y}\left(k\frac{\partial T}{\partial y}\right) \tag{2.37}$$

$$\Phi = 2\mu \left[\left(\frac{R}{R+y}\right)^2\left(\frac{\partial u}{\partial x} + \frac{v}{R}\right)^2 + \left(\frac{\partial v}{\partial y}\right)^2\right] + \mu\left[\frac{R}{R+y}\left(\frac{\partial v}{\partial x} - \frac{u}{R}\right) + \frac{\partial u}{\partial y}\right]^2 + \left(\mu' - \frac{2}{3}\mu\right)\left(\nabla\cdot\vec{v}\right)^2 . \tag{2.38}$$

REFERENCES

1. Goldstein, S., ed. Modern Developments in Fluid Dynamics. Vol. I. Oxford, England: Oxford University Press, 1952, p. 101.

2. Tollmien, W., "Grenzschichttheorie," in Handbuch der Experimentalphysik, L. Schiller, ed., Vol. IV, Part I, 1931, p. 248.

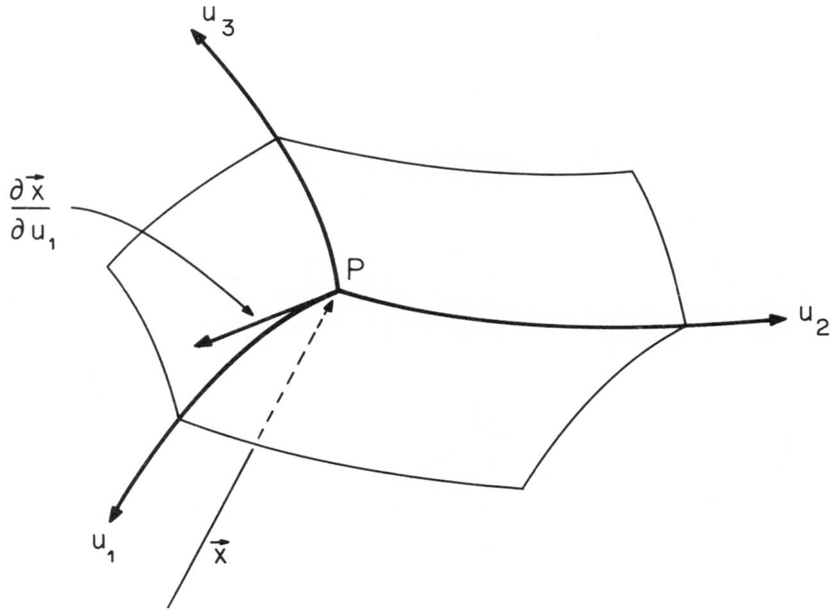

FIGURE 2.1 GENERAL ORTHOGONAL CURVILINEAR COORDINATES.

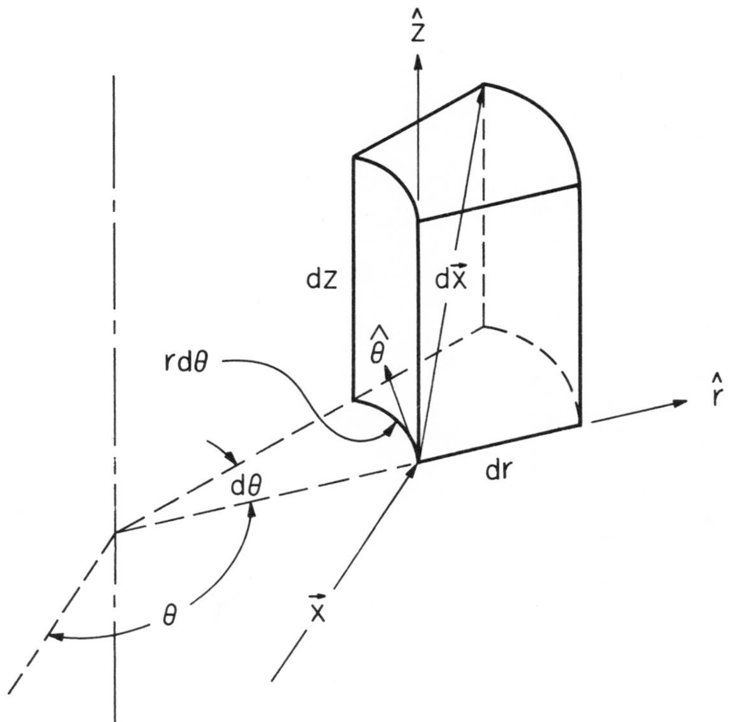

FIGURE 2.2 CYLINDRICAL COORDINATE SYSTEM.

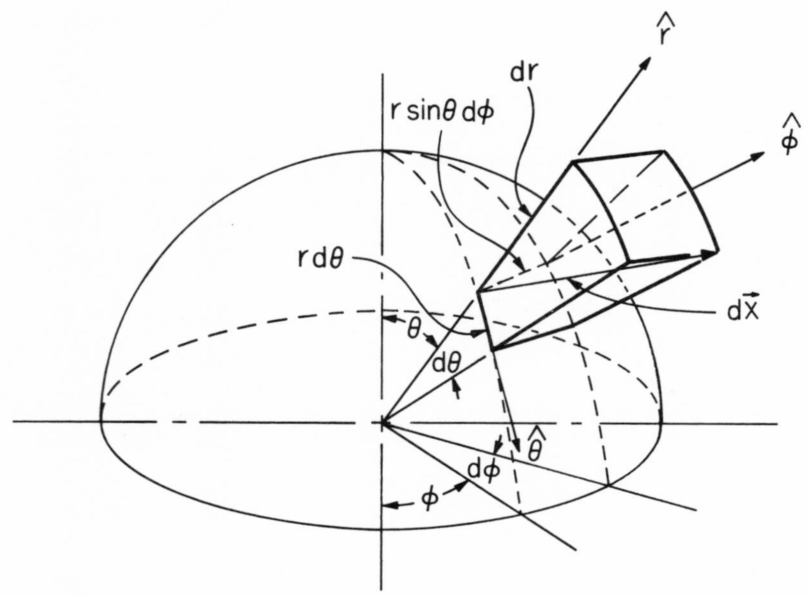

FIGURE 2.3 SPHERICAL COORDINATE SYSTEM.

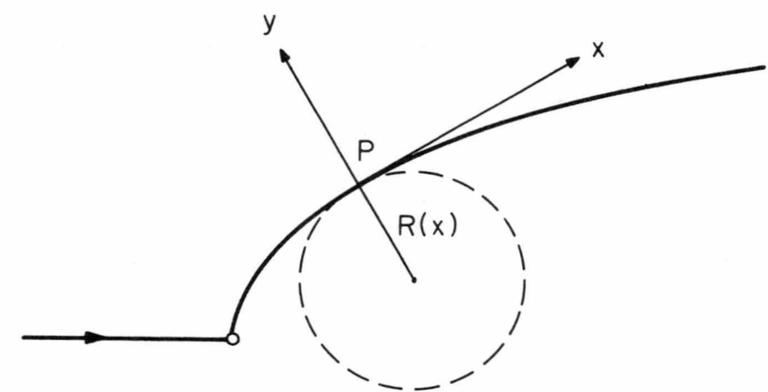

FIGURE 2.4 TWO-DIMENSIONAL CURVILINEAR ORTHOGONAL COORDINATE SYSTEM.

3. SIMILITUDE AND BOUNDARY LAYER APPROXIMATIONS

Only under rare circumstances could information on heat transfer be obtained by directly solving the governing conservation equations satisfying the initial and boundary conditions appropriate for the problem. Examples may be found in current textbooks. (See References 1 and 2.) In a great majority of applications, however, the equations are so complicated that they defy mathematical solutions. One turns to seek approximate, but realistic, answers to the problem. A basic tool, intimately connected with the latter procedure, is the method of dimensional analysis.

To the engineer, dimensional analysis is often associated with the idea of simulation. It is based on a single premise that any relation describing a physical phenomenon must be dimensionally homogeneous. Different approaches have been advocated by different authors. It is the opinion of this writer that, if one's understanding of the problem is of such extent that the governing equations could be written, an effective and reliable approach is to deduce the relevant dimensionless parameters from these equations together with the initial and boundary conditions. The development which follows is based on this viewpoint.

While the relevant nondimensional parameters of a problem can be correctly found in this manner, the assessment of their relative importance requires a qualitative understanding of the physical phenomena involved. This qualitative understanding often plays a central role in engineering analysis and its importance cannot be overemphasized.

3.1 FLOW AND HEAT TRANSFER PARAMETERS

3.1.1 Combined Forced and Free Convection in Low-Speed Flow

We first examine the relatively simple problem of steady transfer of heat between a solid and its surrounding fluid by combined forced and free convection. For definiteness, imagine that the body is held fixed. It has a typical linear dimension, L, and a uniform surface temperature, T_w. The fluid is of infinite extent; its free stream pressure, temperature, and velocity are, respectively, p_∞, T_∞, and U_∞, all of which are uniform. The flow is regarded as incompressible and all properties are assumed to be constants except the density change due to temperature variations. The latter consideration is, of course, mandatory since free convection is of interest here.

The buoyancy force per unit mass of the fluid is $-g_i\beta(T - T_\infty)$, where g_i is the component of gravitational acceleration along the coordinate axis in question, T is the local fluid temperature, and β is the thermal expansion coefficient defined by:

$$\beta = -\frac{1}{\rho}\left(\frac{\partial \rho}{\partial T}\right)_p ,$$

where ρ is the density. The expression given above for the buoyancy force is valid when the density difference (due to temperature difference) is small as compared with the density itself.

We now proceed to deduce the dimensionless form of the governing equations by referring (a) all lengths to L, (b) all velocities to U_∞,[**] (c) pressure difference to ρU_∞^2, and (d) temperature difference to $T_w - T_\infty$. It is worth noting that, in incompressible flow, pressure is the dynamic variable (in contrast to compressible flow wherein pressure is the thermodynamic variable) and hence ρU_∞^2 is an appropriate reference quantity. Also, in view of the constant property assumption made, only pressure and temperature differences enter into the problem.

Designating the nondimensional quantities by an asterisk, we have

$$\left.\begin{aligned} x_i^* &= \frac{x_i}{L}, \quad v_i^* = \frac{v_i}{U_\infty}, \\ p^* &= \frac{p - p_\infty}{\rho U_\infty^2}, \\ \text{and} \quad T^* &= \frac{T - T_\infty}{T_w - T_\infty}. \end{aligned}\right\} \qquad (3.1)$$

In terms of the foregoing nondimensional variables, the continuity equation becomes

$$\frac{\partial v_i^*}{\partial x_i^*} = 0 . \qquad (3.2)$$

[**] Here we assume that significant forced flow prevails and thus U_∞ ia an appropriate reference velocity.

The constitutive relation for a Newtonian fluid, under the condition of incompressible flow, is

$$\tau^*_{ij} = \frac{\partial v^*_i}{\partial x^*_j} + \frac{\partial v^*_j}{\partial x^*_i} , \qquad (3.3)$$

where

$$\tau^*_{ij} = \frac{\tau_{ij}}{\mu\, U_\infty / L} ,$$

and the Fourier law of conduction is

$$q^*_i = - \frac{\partial T^*}{\partial x^*_i} , \qquad (3.4)$$

where

$$q^*_i = \frac{q_i}{k(T_w - T_\infty)/L} .$$

The momentum equation becomes

$$v^*_j \frac{\partial v^*_i}{\partial x^*_j} = - \underbrace{\left[\frac{g\,\beta(T_w - T_\infty)L}{U^2_\infty}\right]}_{\frac{Gr}{Re^2}} \frac{g_i}{g} T^* - \frac{\partial p^*}{\partial x^*_i} + \underbrace{\left[\frac{\mu}{\rho U_\infty L}\right]}_{\frac{1}{Re}} \frac{\partial^2 v^*_i}{\partial x^*_j \partial x^*_j} \qquad (3.5)$$

Recalling that the rate of deformation tensor is

$$D_{ij} = \frac{1}{2}\left(\frac{\partial v_i}{\partial x_j} + \frac{\partial v_j}{\partial x_i}\right),$$

we obtain its nondimensional counterpart:

$$D^*_{ij} = \frac{1}{2}\left(\frac{\partial v^*_i}{\partial x^*_j} + \frac{\partial v^*_j}{\partial x^*_i}\right) = \frac{D_{ij}}{U_\infty/L} .$$

It follows that the dimensionless dissipation function is

$$\Phi^* = \tau^*_{ij}\, D^*_{ij} = \frac{\Phi}{\mu(U_\infty/L)^2} , \qquad (3.6)$$

and the energy equation becomes

$$v^*_i \frac{\partial T^*}{\partial x^*_i} = Q^* + \underbrace{\left[\frac{k}{c_p \rho U_\infty L}\right]}_{\frac{1}{Pr \cdot Re}} \frac{\partial^2 T^*}{\partial x^*_i \partial x^*_i} + \underbrace{\left[\frac{\mu\, U_\infty}{c_p \rho (T_w - T_\infty) L}\right]}_{\frac{E}{Re}} \Phi^* , \qquad (3.7)$$

where

$$Q^* = \frac{Q}{c_p U_\infty (T_w - T_\infty)/L} . \qquad (3.8)$$

The various quantities in the square brackets of the momentum and energy equation are all dimensionless. They may be conveniently taken as the parameters of the problem. We recognize that:

$$\frac{\rho U_\infty L}{\mu} = Re, \text{ the Reynolds number}, \qquad (3.9)$$

$$\frac{g\beta(T_w - T_\infty)L}{U^2_\infty} = \frac{g\beta(T_w - T_\infty)L^3\rho^2}{\mu^2}\left(\frac{\mu}{\rho U_\infty L}\right)^2 = \frac{Gr}{Re^2} , \qquad (3.10)$$

where Gr is the Grashof number,

$$\frac{k}{c_p \rho U_\infty L} = \frac{k}{c_p \mu}\, \frac{\mu}{\rho U_\infty L} = \frac{1}{Pr \cdot Re} , \qquad (3.11)$$

where Pr is the Prandtl number (the product Pr · Re is also known as the Peclet number), and

$$\frac{\mu U_\infty}{c_p \rho (T_w - T_\infty) L} = \frac{U^2_\infty}{c_p (T_w - T_\infty)}\, \frac{\mu}{\rho U_\infty L} = \frac{E}{Re} , \qquad (3.12)$$

where E is the Eckert number. The velocity boundary conditions are:

$$v_i^* = 0 \text{ at the body surface,}$$

$$v_1^* = 1,\ v_2^* = v_3^* = 0 \text{ in the free stream,}^{**} \tag{3.13}$$

and the temperature boundary conditions are:

$$T^* = 1 \text{ at the body surface,}$$

$$T^* = 0 \text{ in the free stream.} \tag{3.14}$$

A problem of great engineering significance is that of similarity. A knowledge of it is not only of value in analysis, but also is essential in the testing of scaled models, planning of experiments, and correlation of data. In particular, we are concerned with geometrically similar objects, similarly oriented with respect to the local gravitation vector. The question is: How should one choose the free stream velocity, U_∞, fluid properties ρ, μ, β, k, and c_p, temperature difference, $T_w - T_\infty$, body length, L, and heat source, Q, such that the velocity and temperature fields will be similar? From Equations (3.2), (3.5), (3.7), and the boundary conditions (3.13) and (3.14), it is apparent that the solution for v_i^* and T^*, in terms of the independent variables, x_i^*, will be identical for two systems if the bracketed quantities in Equations (3.5) and (3.7) are identical and the strength of the similarly distributed heat source is proportional to $c_p U_\infty (T_w - T_\infty)/L$. This implies that Re, Gr, Pr, and E are the same for the two systems. For the problem under consideration, the boundary conditions in Equations (3.13) and (3.14) do not give rise to any additional similarity requirements.

The local heat flux at the surface of the body is

$$q_w = -k \left(\frac{\partial T}{\partial n}\right)_w ,$$

where the subscript w refers to solid surface and n denotes the outdrawn normal. Its nondimensional form is

$$q_w^* = \frac{q_w}{k(T_w - T_\infty)/L} = -\left(\frac{\partial T^*}{\partial n^*}\right)_w . \tag{3.15}$$

Hence the local Nusselt number is

$$Nu \equiv \frac{q_w L}{(T_w - T_\infty)k} = q_w^* , \tag{3.16}$$ ***

which is a function of $(x_i^*)_w$, Q^*, Re, Gr, Pr, and E. The average Nusselt number for the entire surface is thus expressible by a relation of the form:

$$\overline{Nu} = f(Q^*, \text{Re}, \text{Gr}, \text{Pr}, \text{E}), \tag{3.17}$$

which holds irrespective of whether the flow separates or reattaches to the surface. If there is no extraneous heat source in the fluid, $Q^* = 0$. In a great majority of forced convection problems, in which free convection has a significant effect, the dissipation term in the energy equation can be neglected. Under such circumstances, Equation (3.17) simplifies to

$$\overline{Nu} = f(\text{Re}, \text{Gr}, \text{Pr}) . \tag{3.18}$$

It is also customary to express the heat transfer coefficient in terms of the Stanton number. It is defined by

$$St \equiv \frac{q_w}{(T_w - T_\infty) c_p \rho U_\infty} = \frac{q_w^*}{\text{Pr} \cdot \text{Re}} , \tag{3.19}$$

which is clearly also a function of $(x_i^*)_w$, Q^*, Re, Gr, Pr, and E. The discussion given above for $\overline{Nu}$ is equally applicable to the average Stanton number.

In the case of pure free convection, U_∞ vanishes and thus is no longer feasible for use as a reference. If one considers a fluid mass being accelerated from rest by the buoyancy force over a distance, L, and if all the work done is converted into kinetic energy, one finds that it will attain a velocity of the order:

$$\sqrt{g\beta (T_w - T_\infty) L} ,$$

which may be used as the reference velocity. Ignoring dissipation and repeating the analysis similar to what we have done before, we find that the continuity equation assumes

** This implies that the direction of U_∞ coincides with x_1-axis.

*** The definition for the local Nusselt number is not unique; the reference length, L, could be replaced by an appropriate local distance coordinate.

the same form as Equation (3.2), but the momentum equation becomes,

$$v_j^* \frac{\partial v_i^*}{\partial x_j^*} = -\frac{g_i}{g} T^* + \frac{1}{\sqrt{Gr}} \frac{\partial^2 v_i^*}{\partial x_j^* \partial x_j^*}, \quad (3.20)$$

and the energy equation is

$$v_i^* \frac{\partial T^*}{\partial x_i^*} = Q^* + \frac{1}{Pr\sqrt{Gr}} \frac{\partial^2 T^*}{\partial x_i^* \partial x_i^*}, \quad (3.21)$$

where

$$Q^* = \frac{Q}{c_p\sqrt{g\beta(T_w - T_\infty)^3/L}} . \quad (3.22)$$

Hence, to preserve similarity, the systems must have identical Pr and Gr and the heat source in the fluid, if present, must be similarly distributed with strength proportional to

$$c_p\sqrt{\frac{g\beta(T_w - T_\infty)^3}{L}} .$$

3.1.2 Forced Convection in High-Speed Gas Flow

Again, for definiteness, we consider a solid body of typical length dimension, L, in a Newtonian fluid moving with uniform, undisturbed velocity, U_∞. The body may execute an oscillatory or some unsteady motion of characteristic time, t_c. As before, we designate all free stream quantities by the subscript ∞ and all nondimensional quantities by an asterisk. Thus

$$\left.\begin{aligned} t^* &= \frac{t}{t_c}, & x_i^* &= \frac{x_i}{L}, \\ v_i^* &= \frac{v_i}{U_\infty}, & p^* &= \frac{p}{p_\infty}, \\ T^* &= \frac{T}{T_\infty}, & h^* &= \frac{h}{h_\infty}, \\ \mu^* &= \frac{\mu}{\mu_\infty}, & \rho^* &= \frac{\rho}{\rho_\infty}, \\ \text{and} \\ k^* &= \frac{k}{k_\infty} . \end{aligned}\right\} \quad (3.23)$$

As noted earlier, pressure is now a thermodynamic variable. The reference quantities p_∞, ρ_∞, and T_∞ are, of course, related by the equation of state and <u>only two of them can be independently specified</u>. Furthermore, in contrast to the case considered in Section 3.1.1, the problem now depends on the absolute level of temperature and pressure (or density). Using the nondimensional variables defined in Equation (3.23), we find that

$$\tau_{ij}^* = (\mu'^* - \frac{2}{3}\mu^*) \frac{\partial v_k^*}{\partial x_k^*} \delta_{ij} + \mu^* \left(\frac{\partial v_i^*}{\partial x_j^*} + \frac{\partial v_j^*}{\partial x_i^*} \right), \quad (3.24)$$

where

$$\tau_{ij}^* = \frac{\tau_{ij}}{\mu_\infty U_\infty / L},$$

and that

$$q_i^* = -k^* \frac{\partial T^*}{\partial x_i^*}, \quad (3.25)$$

where

$$q_i^* = \frac{q_i}{k_\infty T_\infty / L} .$$

The continuity equation becomes

$$\left[\frac{L}{U_\infty t_c}\right] \frac{\partial \rho^*}{\partial t^*} + \frac{\partial(\rho^* v_i^*)}{\partial x_i^*} = 0, \quad (3.26)$$

and the momentum equation is

$$\left[\frac{L}{U_\infty t_c}\right] \rho^* \frac{\partial v_i^*}{\partial t^*} + \rho^* v_j^* \frac{\partial v_i^*}{\partial x_j^*}$$

$$= -\left[\frac{p_\infty}{\rho_\infty U_\infty^2}\right] \frac{\partial p^*}{\partial x_i^*} + \underbrace{\left[\frac{\mu_\infty}{\rho_\infty U_\infty L}\right]}_{\frac{1}{Re_\infty}} \frac{\partial \tau_{ji}^*}{\partial x_j^*}, \quad (3.27)$$

where the body force has been ignored. The energy equation, Equation (1.27), becomes

$$\left[\frac{L}{U_\infty t_c}\right]\left\{\rho^* \frac{\partial h^*}{\partial t^*} - \left[\frac{p_\infty}{\rho_\infty h_\infty}\right]\frac{\partial p^*}{\partial t^*}\right\} + \rho^* v_i^* \frac{\partial h^*}{\partial x_i^*}$$

$$= \rho^* Q^* + \underbrace{\left[\frac{k_\infty T_\infty}{\rho_\infty U_\infty L h_\infty}\right]}_{\frac{1}{Pr_\infty Re_\infty}\,\frac{c_{p\infty} T_\infty}{h_\infty}} \frac{\partial}{\partial x_i^*}\left(k^* \frac{\partial T^*}{\partial x_i^*}\right) + \left[\frac{p_\infty}{\rho_\infty h_\infty}\right] v_i^* \frac{\partial p^*}{\partial x_i^*} + \underbrace{\left[\frac{\mu_\infty U_\infty}{\rho_\infty L h_\infty}\right]}_{\frac{1}{Re_\infty}\,\frac{U_\infty^2}{h_\infty}} \Phi^* , \qquad (3.28)$$

with

$$Q^* = \frac{Q}{U_\infty h_\infty / L} \qquad (3.29)$$

and

$$\Phi^* = \frac{\Phi}{\mu_\infty (U_\infty/L)^2} = \tau_{ij}^* D_{ij}^* \qquad (3.30)$$

where τ_{ij}^* is given by Equation (3.24) and

$$D_{ij}^* = \frac{1}{2}\left(\frac{\partial v_i^*}{\partial x_j^*} + \frac{\partial v_j^*}{\partial x_i^*}\right).$$

The various quantities in the square brackets have the following physical interpretation. The parameter, $L/U_\infty t_c$, is the ratio of the body length to the distance which a fluid particle will traverse during the time characteristic of the unsteady motion of the solid body. The unsteady effect will become less significant if this ratio is small. The quantity p_∞/ρ_∞ has the dimension of velocity squared. In fact, $\sqrt{p_\infty/\rho_\infty}$ is of the order of the mean speed of random molecular motion and is also roughly the acoustic speed in the free stream. Hence

$$\frac{p_\infty}{\rho_\infty U_\infty^2} \sim \frac{1}{M_\infty^2} ,$$

where M_∞ is the free stream Mach number. As has been noted previously in Section 1, the internal energy of a gas has its origin mainly by virtue of the translational and rotational motion of its molecules. For air, the average translational energy per molecule is $3/2\ \tilde{k}T$, $\tilde{k}$ being the Boltzmann constant. Its rotational energy is $\tilde{k}T$ per molecule. Thus e_∞ has a magnitude of the order of p_∞/ρ_∞, and $p_\infty/\rho_\infty h_\infty$ is of the order of unity since

$$h_\infty = e_\infty + \frac{p_\infty}{\rho_\infty} .$$

The ratio $c_{p\infty}T_\infty/h_\infty$ is a number close to 1; it is, in fact, equal to 1 if c_p is a constant. Clearly,

$$\frac{U_\infty^2}{h_\infty} \sim M_\infty^2 .$$

In order to have two similar flow fields, it is necessary that not only the quantities in square brackets of Equations (3.27) and (3.28) are identical but also the nondimensional state equations,

$$p^* = p^*(\rho^*, T^*)$$

and

$$h^* = h^*(\rho^*, T^*) ,$$

are the same for both flows. In addition, one also requires that the fluid properties, μ'/μ_∞, μ/μ_∞, and k/k_∞ are unique functions of ρ^* and T^*. These conditions cannot all be satisfied in general.

In the majority of practical situations, the gas can be assumed to be "ideal." Its equation of state in nondimensional form is

$$p^* = \rho^* T^* , \qquad (3.31)$$

and the enthalpy is a function of temperature only. Thus

$$h = \int_0^T c_p dT, \tag{3.32}$$

and

$$h^* = \frac{\int_0^{T^*} (c_p/c_{p\infty})\, dT^*}{\int_0^1 (c_p/c_{p\infty})\, dT^*} . \tag{3.33}$$

Denoting the ratio of the specific heats, c_p/c_v by γ, we find

$$\left.\begin{aligned} \frac{p_\infty}{\rho_\infty U_\infty^2} &= \frac{1}{\gamma_\infty M_\infty^2} , \\ \frac{c_{p\infty} T_\infty}{h_\infty} &= \frac{1}{\int_0^1 (c_p/c_{p\infty})\, dT^*} , \end{aligned}\right\} \tag{3.34}$$

$$\left.\begin{aligned} \frac{p_\infty}{\rho_\infty h_\infty^2} &= \frac{\gamma_\infty - 1}{\gamma_\infty} \frac{c_{p\infty} T_\infty}{h_\infty} , \\ &\text{and} \\ \frac{U_\infty^2}{h_\infty} &= (\gamma_\infty - 1)\, M_\infty^2 \frac{c_{p\infty} T_\infty}{h_\infty} . \end{aligned}\right\} \tag{3.34}$$

If Ω denotes the ratio $L/U_\infty t_c$, the governing conservation equations for the flow of ideal gases are as follows:

Continuity:

$$\Omega \frac{\partial \rho^*}{\partial t^*} + \frac{\partial(\rho^* v_i^*)}{\partial x_i^*} = 0 \tag{3.35}$$

Momentum:

$$\Omega \rho^* \frac{\partial v_i^*}{\partial t^*} + \rho^* v_j^* \frac{\partial v_i^*}{\partial x_j^*} = - \frac{1}{\gamma_\infty M_\infty^2} \frac{\partial p^*}{\partial x_i^*} + \frac{1}{Re_\infty} \frac{\partial \tau_{ji}^*}{\partial x_j^*} \tag{3.36}$$

Energy:

$$\Omega \rho^* \frac{\partial h^*}{\partial t^*} - \Omega \frac{\gamma_\infty - 1}{\gamma_\infty} \left(\frac{c_{p\infty} T_\infty}{h_\infty}\right) \frac{\partial p^*}{\partial t^*} + \rho^* v_i^* \frac{\partial h^*}{\partial x_i^*} = \rho^* Q^* + \frac{1}{Pr_\infty Re_\infty} \left(\frac{c_{p\infty} T_\infty}{h_\infty}\right) \frac{\partial}{\partial x_i^*} \left(k^* \frac{\partial T^*}{\partial x_i^*}\right)$$

$$+ \frac{\gamma_\infty - 1}{\gamma_\infty} \left(\frac{c_{p\infty} T_\infty}{h_\infty}\right) v_i^* \frac{\partial p^*}{\partial x_i^*} + \frac{(\gamma_\infty - 1) M_\infty^2}{Re_\infty} \left(\frac{c_{p\infty} T_\infty}{h_\infty}\right) \Phi^* . \tag{3.37}$$

The state equation is

$$p^* = \rho^* T^* . \tag{3.38}$$

If the boundary conditions are the same as those considered in Section 3.1.1 and the initial conditions are also similar, one concludes by examining Equations (3.35) to (3.38) together with Equations (3.24), (3.29), (3.33), and (3.34) that if (a) T_w is made proportional to T_∞, (b) Q is made proportional to $U_\infty h_\infty/L$, (c) Ω, M_∞, Re_∞, Pr_∞, and γ_∞ are the same for both flows, and (d) μ'/μ_∞, μ/μ_∞, k/k_∞, and $c_p/c_{p\infty}$ depend only on T^*, then the velocity, temperature, and density fields will be similar. The conditions listed in (d) are met if all these properties vary with powers of T, although the variation may be different from one another. Under the foregoing conditions, the local heat flux at the body surface will obviously be similar. In nondimensional form, it is

$$q_w^* = \frac{q_w}{k_\infty T_\infty/L} = - \left(k^* \frac{\partial T^*}{\partial n^*}\right)_w . \tag{3.39}$$

The local Stanton number based on enthalpy difference may be defined by

$$St \equiv \frac{q_w}{(h_w - h_{aw})\rho_\infty U_\infty} , \tag{3.40}$$

where h_{aw} is the adiabatic wall enthalpy. In terms of dimensionless variables, Equation (3.40) may be written as

$$St = \frac{q_w^*}{h_w^* - h_{aw}^*} \frac{1}{Pr_\infty Re_\infty} \frac{1}{\int_0^1 (c_p/c_{p\infty})\, dT^*} . \tag{3.41}$$

If we denote

$$\frac{c_p}{c_{p\infty}} = \left(\frac{T}{T_\infty}\right)^{\alpha_{c_p}} ,$$

$$\frac{k}{k_\infty} = \left(\frac{T}{T_\infty}\right)^{\alpha_k} ,$$

$$\frac{\mu'}{\mu_\infty} = \left(\frac{T}{T_\infty}\right)^{\alpha_{\mu'}} ,$$

and

$$\frac{\mu}{\mu_\infty} = \left(\frac{T}{T_\infty}\right)^{\alpha_\mu} ,$$

then, we are led to the conclusion that, for high-speed flow of an ideal gas, the Stanton number depends, in general, on

$$(x_i^*)_w,\ t^*,\ T_w^*,\ Q^*,\ \Omega,\ M_\infty,\ Re_\infty,\ Pr_\infty,\ \gamma_\infty,$$

and the four α's -- a formidable list! Fortunately, in many applications, μ' can be ignored, and the specific heat and the Prandtl number can be taken as constants. Then the viscosity and conductivity will vary with temperature in the same manner, i.e.,

$$\frac{\mu}{\mu_\infty} = \frac{k}{k_\infty} = \left(\frac{T}{T_\infty}\right)^{\alpha} .$$

If, in addition, the motion is steady and there is no extraneous heat source, the average Stanton number would then depend on

$$T_w^*,\ M_\infty,\ Re_\infty,\ Pr_\infty,\ \gamma,\ \text{and}\ \alpha.$$

The nondimensional governing equations deduced and discussed in this section are also useful for making order of magnitude estimates of the various terms when they are applied to boundary layers. This is briefly considered in the section which follows.

3.2 BOUNDARY LAYER APPROXIMATIONS

Let us consider a fixed, streamlined solid body immersed in a fluid of small viscosity which flows past the body at a moderate speed. At the solid surface, the velocity is zero. At small distances from the surface the fluid velocity becomes close to that of the main stream. The recognition of the existence of such a boundary layer is one of the cornerstones of modern fluid dynamics. It was first brought to attention by Prandtl in 1904. He noted that the action of viscosity on a flow past a streamlined body was often confined to a thin layer adjacent to it, and to a thin wake behind it. Outside this layer, the flow is inviscid.

When the fluid moves into a region of rising pressure, it is slowed down. The slower-moving fluid in the boundary layer may not have enough momentum to overcome the combined action of viscous and pressure forces with the consequence that it is brought to rest. The forward flow then separates from the solid wall, and farther downstream there is a slow reversed flow. If the region of the separated flow is extensive, it can produce an appreciable influence on the external flow which is no longer predictable by the inviscid fluid theory without the boundary layer.

When the external flow is supersonic, there is an additional mechanism by which the boundary layer can interact with the external stream. Ideally, in a purely supersonic field a disturbance such as that produced by an obstacle cannot be propagated upstream. However, the fluid in the boundary layer sufficiently close to the wall will be subsonic, thus providing a mechanism which makes possible a modification of the upstream condition.

In spite of the foregoing complications, it is still useful to consider separately the boundary layer approximation and the inviscid approximation for the external flow as has been commonly done.

Quite analogous to the velocity boundary layer described above is the temperature field around a hot streamlined body in a fast flowing fluid, particularly when the thermal conductivity of the fluid is not excessively high. This narrow zone, adjacent to the surface, in which the temperature gradient is large is referred to as the thermal boundary layer. Thus both the velocity and thermal boundary layers are seen to be essentially "skin" effects and, as such, the momentum and energy equations simplify considerably. For plane flow, as well as for flows past smooth curved walls of moderate curvature such as a cylindrical obstacle with axis normal to the flow, one finds that to a very great extent these equations do not depend on the shape of the solid body. This provides enormous simplifications to the solution of the problem. While three-dimensional boundary layers have been studied, we shall restrict ourselves to the discussion of two-dimensional cases.

3.2.1 Boundary Layer Equations for Combined Forced and Free Convection in Low-Speed Flow

Let us now re-examine, within the context of boundary layer theory, the problem of combined forced and free convection in low-speed flow described in Section 3.1.1. For definiteness, we consider a vertical plate with main fluid stream flowing upward and parallel to it.

We adopt Cartesian coordinates (x,y), x being measured along the surface in the direction of the main stream, and y perpendicular to it. The associated velocity components are (u,v). At a certain x, the velocity and thermal boundary layers have thicknesses δ and δ_t, respectively. We assume that both are much less than a typical length, L, of the body. Somewhere in the velocity boundary layer, the viscous force becomes of the same order of magnitude as the inertia force. Thus

$$\mu \frac{\partial^2 u}{\partial y^2} \sim \rho u \frac{\partial u}{\partial x}$$

and, crudely, we may write

$$\mu \frac{U_\infty}{\delta^2} \sim \rho \frac{U_\infty^2}{L} ,$$

where we have taken x to be of order L, i.e., we have excluded consideration of the region close to the leading edge. Clearly,

$$\delta \sim \sqrt{\frac{\nu L}{U_\infty}} ,$$

where $\nu = \mu/\rho$, the kinematic viscosity. In nondimensional form, we have

$$\delta^* \left(= \frac{\delta}{L}\right) \sim \frac{1}{\sqrt{Re}} . \qquad (3.42)$$

This important result may be arrived at from a different viewpoint. If we were to seek an inviscid solution to the problem, there would be a finite slip velocity at the plate surface and thereby the existence of a vortex sheet. In reality, this vorticity would diffuse away from the surface at a rate controlled by the fluid kinematic viscosity. In time, t, the diffusing distance is of the order $\sqrt{\nu t}$. Now a fluid particle starting at the leading edge will, during that time, travel through a distance,

$$L = U_\infty t$$

along the length of the plate. Hence we again conclude that

$$\delta \sim \sqrt{\frac{\nu L}{U_\infty}} .$$

The continuity equation may be written as

$$\frac{\partial v}{\partial y} = - \frac{\partial u}{\partial x} ,$$

which is of the order U_∞/L. Upon integrating across the layer from y = 0 to y = δ, we immediately find that v is of the order $\delta U_\infty/L$. That is,

$$v^* \left(= \frac{v}{U_\infty}\right) \sim \delta^* , \qquad (3.43)$$

which is $\ll 1$ as we originally assumed.

Inside the thermal boundary layer, due to the large temperature gradient $\partial T/\partial y$, there exists a region where conduction is of the same order of magnitude as convection. Hence

$$k \frac{\partial^2 T}{\partial y^2} \sim c_p \rho u \frac{\partial T}{\partial x}$$

or

$$k \frac{T_w - T_\infty}{\delta_t^2} \sim c_p \rho U_\infty \frac{T_w - T_\infty}{L} .$$

Then

$$\delta_t \sim \sqrt{\frac{\kappa L}{U_\infty}} ,$$

where $\kappa = k/c_p\rho$ is the thermal diffusivity of the fluid. If one considers the diffusion of heat from a plane source at the surface of the plate and follows a reasoning completely analogous to the diffusion of vorticity, one arrives at a similar result. Thus

$$\delta_t^* \left(= \frac{\delta_t}{L}\right) \sim \frac{1}{\sqrt{Pr \cdot Re}} \qquad (3.44)$$

and

$$\frac{\delta_t}{\delta} \sim \frac{1}{\sqrt{Pr}} . \qquad (3.45)$$

Hence, in forced flows of gases whose Prandtl numbers do not differ much from unity, the thermal boundary layer is expected to have approximately the same thickness as the velocity boundary layer. In viscous oils, $\delta_t \ll \delta$ since $Pr \gg 1$. The Prandtl number of most liquid metals is of the order of 10^{-2} or less, hence $\delta_t \gg \delta$. It may be worth noting that, in liquid metals, the assumption of thin thermal boundary layer may not be valid, particularly when the velocity is low.

The nondimensional momentum equation in the x-direction, according to Equation (3.5), is

$$u^* \frac{\partial u^*}{\partial x^*} + v^* \frac{\partial u^*}{\partial y^*} = \frac{Gr}{Re^2} T^* - \frac{dp^*}{dx^*} + \frac{1}{Re}\left(\frac{\partial^2 u^*}{\partial x^{*2}} + \frac{\partial^2 u^*}{\partial y^{*2}}\right), \qquad (3.46)^{**}$$

$$1 \quad 1 \quad \delta^* \frac{1}{\delta^*} \qquad 1 \qquad \delta^{*2}\left(1 \quad \frac{1}{\delta^{*2}}\right)$$

where

$$\frac{dp^*}{dx^*} = -u_1^* \frac{du_1^*}{dx^*}$$

$$1 \qquad 1 \quad 1$$

and subscript 1 refers to the edge of the boundary layer. If the fluid is unconfined,

$$\frac{du_1^*}{dx^*} = 0 ,$$

and the pressure term in Equation (3.46) drops out.

The order of magnitude of the various terms in Equation (3.46) can be estimated using the information already obtained and the results are as shown. Of the two viscous force terms, only one need be retained. Since T^* is of the order of 1, we conclude that the natural convection effect will become significant when

$$\frac{Gr}{Re^2} \sim 1 \text{ or greater.} \qquad (3.47)$$

If one examines the momentum equation in the y-direction, one finds that all terms are of the order of δ^* or less. The pressure gradient $\partial p^*/\partial y^*$ is, at most, of $O(\delta^*)$. Hence the pressure difference across the boundary layer can be neglected and

$$p(x,y) = p(x).$$

For this reason, the pressure term in Equation (3.46) is written as dp^*/dx^* instead of $\partial p^*/\partial x^*$.

We now turn to the energy equation. According to Equation (3.7), we have

$$u^* \frac{\partial T^*}{\partial x^*} + v^* \frac{\partial T^*}{\partial y^*} = Q^* + \frac{1}{Pr\ Re}\left(\frac{\partial^2 T^*}{\partial x^{*2}} + \frac{\partial^2 T^*}{\partial y^{*2}}\right) + E \frac{1}{Re} \Phi^*, \qquad (3.48)$$

$$1 \quad 1 \quad \delta^* \frac{1}{\delta_t^*} \qquad \delta_t^{*2}\left(1 \quad \frac{1}{\delta_t^{*2}}\right) \qquad \delta^{*2} \frac{1}{\delta^{*2}}$$

where

$$\Phi^* = 2\left[\left(\frac{\partial u^*}{\partial x^*}\right)^2 + \left(\frac{\partial v^*}{\partial y^*}\right)^2\right] + \left(\frac{\partial u^*}{\partial y^*} + \frac{\partial v^*}{\partial x^*}\right)^2 .$$

$$1 \qquad 1 \qquad \left(\frac{1}{\delta^*} + \delta^*\right)^2$$

It is thus seen that conduction in the x-direction can be ignored in the boundary layer and the frictional dissipation would become significant when the Eckert number

$$E \sim 1 \text{ or larger.} \qquad (3.49)$$

Also, of the many terms in Φ^*, only one, $(\partial u^*/\partial y^*)^2$ need be retained. The extraneous source will be important if

$$Q^* \left[= \frac{QL}{c_p U_\infty (T_w - T_\infty)}\right]$$

attains a magnitude of the order of unity.

** Since the x-axis is vertically upward, the x-component of the local gravitational acceleration vector is -g.

With the foregoing simplifications and reverting to dimensional quantities, we obtain the appropriate boundary layer equations for the problem as follows.

Continuity:

$$\frac{\partial u}{\partial x} + \frac{\partial v}{\partial y} = 0 \tag{3.50}$$

Momentum:

$$\rho\left(u\frac{\partial u}{\partial x} + v\frac{\partial u}{\partial y}\right) = \rho g\,\beta(T - T_\infty) - \frac{dp}{dx} + \mu\frac{\partial^2 u}{\partial y^2} \tag{3.51}$$

Energy:

$$c_p\rho\left(u\frac{\partial T}{\partial x} + v\frac{\partial T}{\partial y}\right) = \rho Q + k\frac{\partial^2 T}{\partial y^2} + \mu\left(\frac{\partial u}{\partial y}\right)^2 . \tag{3.52}$$

If the variation of viscosity and conductivity is large, the viscous force term in Equation (3.51) should be written as

$$\frac{\partial}{\partial y}\left[\mu\left(\frac{\partial u}{\partial y}\right)\right]$$

and the conduction term in Equation (3.52) as

$$\frac{\partial}{\partial y}\left[k\left(\frac{\partial T}{\partial y}\right)\right].$$

To further ascertain if all the terms in Equations (3.51) and (3.52) need be retained in specific situations, we consider the following cases:

Case (1):

A vertical plate 6 in. high is suspended in water at 70° F. The water flows vertically upward at a velocity of 1 ft/sec and the plate surface is heated to a temperature of 100° F. Using available property tables, one finds

$$Gr \simeq 3 \times 10^8$$

and

$$Re \simeq 6 \times 10^4 .$$

Hence

$$\frac{Gr}{Re^2} \simeq \frac{1}{12} ,$$

indicating that the influence of free convection is small. One could delete the term $\rho g\beta(T - T_\infty)$ in Equation (3.51) without introducing large errors.

The Eckert number has been found to be approximately

$$1.3 \times 10^{-6} ,$$

indicating that the frictional heating effect is completely negligible, which is what one would expect. Hence the last term of the right side of Equation (3.52) should be deleted.

If the water velocity is reduced to 1/4 ft/sec, the ratio Gr/Re^2 becomes approximately 1.3. Under this condition, the buoyancy force due to density variation must be taken into account. Needless to say, the frictional dissipative effect is even smaller due to the reduced velocity.

Case (2):

The same plate as in Case (1), but it is suspended in atmospheric air flowing past the plate at 300 ft/sec. Calculations show that

$$Gr \simeq 7.6 \times 10^6 ,$$

$$Re \simeq 8.8 \times 10^5 ,$$

and hence

$$\frac{Gr}{Re^2} \sim 10^{-5} ,$$

indicating that the buoyancy force in the momentum equation can be dropped. The Eckert number is approximately 0.5, suggesting that the frictional dissipation term in the energy equation should probably be retained.

3.2.2 Boundary Layer Equations in High-Speed Gas Flow

The general problem of unsteady, compressible boundary layer flow has not received the attention it deserves. The physical process involved depends on the nature of the unsteadiness, e.g., one may consider a body moving impulsively from rest or the periodic motion of an oscillatory body. In the former, the dimensionless parameters governing the "small" time behavior are different from those for "long" time; in the latter, the high frequency behavior is different from that for low frequency. Unsteadiness in the velocity field could also be induced by a temperature change of the solid body situated in an otherwise steady flow as a consequence of changes in fluid properties. It is not difficult to see that the simple conclusion concerning the relative thickness of the velocity and thermal boundary layers in steady flow would no longer hold.

We shall begin by considering the steady flow case. As before, we restrict ourselves to the consideration of two-dimensional problems.

Using assumptions and arguments similar to those described in Section 3.2.1, one finds for steady compressible flow:

$$\left.\begin{aligned} \delta^* &\sim \frac{1}{\sqrt{Re}}\ , \\ \delta_t^* &\sim \frac{1}{\sqrt{Pr\ Re}}\ , \end{aligned}\right\} \qquad (3.53)$$

and

$$v^* \sim \delta^* \ll 1\ , \qquad (3.53)$$

which are identical to those found previously for incompressible flow. As noted before, the Prandtl number of gases does not differ much from unity. Hence we conclude that $\delta^* \sim \delta_t^*$. If large temperature variation occurs across the boundary layer, it is logical to evaluate the relevant properties in relationships (3.53) corresponding to a temperature somewhere between the wall and free stream value.

We now proceed to make an order of magnitude estimate of the various terms in Equations (3.35), (3.36), and (3.37) with the unsteady terms deleted.

Continuity:

$$\frac{\partial(\rho^* u^*)}{\partial x^*} + \frac{\partial(\rho^* v^*)}{\partial y^*} = 0\ , \qquad (3.54)$$

$$1 \qquad\qquad 1$$

since $\rho^* = 0(1)$.

Momentum (x-component):

$$\rho^* u^* \frac{\partial u^*}{\partial x^*} + \rho^* v^* \frac{\partial u^*}{\partial y^*} = -\frac{1}{\gamma_\infty M_\infty^2}\frac{dp^*}{dx^*} + \frac{1}{Re_\infty}\frac{\partial \tau_{ji}^*}{\partial x_j^*}\ , \qquad (3.55)^{**}$$

$$1\quad 1\quad 1 \qquad 1\quad \delta^* \quad \frac{1}{\delta^*} \qquad\qquad\qquad \delta^{*2}\quad \frac{1}{\delta^{*2}}$$

where

$$-\frac{1}{\gamma_\infty M_\infty^2}\frac{dp^*}{dx^*} = \rho_1^*\, u_1^*\, \frac{du_1^*}{dx^*}\ . \qquad (3.55a)$$

$$1 \quad 1 \quad 1$$

Again, subscript 1 refers to the edge of the boundary layer, and

$$\frac{\partial \tau_{ji}^*}{\partial x_j^*} = \frac{\partial}{\partial x^*}\left[\left(\mu'^* - \frac{2}{3}\mu^*\right)\left(\frac{\partial u^*}{\partial x^*} + \frac{\partial v^*}{\partial y^*}\right)\right] + 2\frac{\partial}{\partial x^*}\left(\mu^* \frac{\partial u^*}{\partial x^*}\right) + \frac{\partial}{\partial y^*}\left[\mu^*\left(\frac{\partial u^*}{\partial y^*} + \frac{\partial v^*}{\partial x^*}\right)\right]. \qquad (3.55b)$$

$$1 \qquad 1 \qquad 1 \qquad 1 \qquad\qquad 1 \quad 1 \quad 1 \qquad \frac{1}{\delta^*}\left[1\left(\frac{1}{\delta^*} \qquad \delta^*\right)\right]$$

** The quantity $\partial\tau_{ji}^*/\partial x_j^*$ in Equation (3.55) and the same in Equations (3.55b) and (3.59) are to be interpreted as the x-component of the vector.

It is seen that there is only one term in $\partial\tau_{ji}^*/\partial x_j^*$ which need be retained, i.e.,

$$\frac{\partial}{\partial y^*}\left[\mu^*\left(\frac{\partial u^*}{\partial y^*}\right)\right].$$

It is large compared to all other terms by a factor $1/\delta^{*2}$. This conclusion holds irrespective of whether the flow is steady or unsteady. It is also clear that, for boundary layer flow, our lack of knowledge on the bulk viscosity coefficient is immaterial.

A detailed examination of the y-momentum equation would lead us to the same conclusion obtained for the incompressible case.

Energy:

$$\rho^* u^* \frac{\partial h^*}{\partial x^*} + \rho^* v^* \frac{\partial h^*}{\partial y^*} = \rho^* Q^* + \frac{1}{Pr_\infty Re_\infty}\left(\frac{c_{p\infty}T_\infty}{h_\infty}\right)\left[\frac{\partial}{\partial x^*}\left(k^* \frac{\partial T^*}{\partial x^*}\right) + \frac{\partial}{\partial y^*}\left(k^* \frac{\partial T^*}{\partial y^*}\right)\right]$$

$$1 \quad 1 \qquad 1 \qquad 1 \quad \delta^* \quad \frac{1}{\delta^*} \qquad\qquad \delta^{*2} \qquad 1 \qquad\qquad 1 \qquad\qquad \frac{1}{\delta^{*2}}$$

$$+ \frac{\gamma_\infty - 1}{\gamma_\infty}\left(\frac{c_{p\infty}T_\infty}{h_\infty}\right)\left(u^* \frac{\partial p^*}{\partial x^*} + v^* \frac{\partial p^*}{\partial y^*}\right)$$

$$1 \qquad 1 \qquad\qquad 1 \qquad \delta^* \cdot \delta^*$$

$$+ (\gamma_\infty - 1) M_\infty^2 \frac{1}{Re_\infty}\left(\frac{c_{p\infty}T_\infty}{h_\infty}\right) \Phi^* , \tag{3.56}$$

$$1 \qquad \delta^{*2} \qquad 1 \qquad \frac{1}{\delta^{*2}}$$

where

$$\Phi^* = 2\mu^*\left[\left(\frac{\partial u^*}{\partial x^*}\right)^2 + \left(\frac{\partial v^*}{\partial y^*}\right)^2\right] + \mu^*\left(\frac{\partial u^*}{\partial y^*} + \frac{\partial v^*}{\partial x^*}\right)^2 + \left(\mu'^* - \frac{2}{3}\mu^*\right)\left(\frac{\partial u^*}{\partial x^*} + \frac{\partial v^*}{\partial y^*}\right)^2 .$$

$$1 \qquad 1 \qquad\qquad 1 \qquad\qquad 1 \left(\frac{1}{\delta^*} \qquad \delta^*\right)^2 \qquad\qquad 1 \qquad 1 \qquad 1$$

Quantities which can be ignored are shown crossed by dashed arrows. Again, we note that only one term in Φ^*, namely,

$$\mu^*\left(\frac{\partial u^*}{\partial y^*}\right)^2 ,$$

need be retained. It is larger than other quantities by a factor of $1/\delta^{*2}$. It is also seen that dissipative effects will become significant when

$$(\gamma_\infty - 1)\, M_\infty^2 \sim 1 \text{ or larger}, \tag{3.57}$$

and the influence of extraneous heat source in the fluid must be taken into account when

$$\frac{QL}{U_\infty h_\infty} \sim 1 \text{ or larger}. \tag{3.58}$$

We now turn to the case of unsteady flow. The analysis of Section 3.1.2 shows that an additional parameter $\Omega(=L/U_\infty t_c)$ appears. The question is: What should one choose for the characteristic time, t_c, in boundary layer flow? As it turns out, the answer depends on the nature of the unsteadiness. It is to a discussion of this aspect of the problem that the following four paragraphs are devoted.

We first consider a streamlined object moving from rest in a quiescent mass of fluid. The vorticity formed at the surface will diffuse away from it by virtue of molecular transport and be convected by the flow relative to the object. At time, t, later, the diffusion distance is of the order $\sqrt{\nu t}$, which is also the boundary layer thickness. Alternatively, one may speak of the diffusion time which is of the order δ^2/ν. The concept of the diffusion of vorticity has been proved most fruitful in the analysis of unsteady velocity boundary layers. It is analogous to the diffusion of heat in thermal boundary layers.

If we take the following variables: a reference length in the direction of motion, L, a reference velocity, u_o, a reference time, t_c, and kinematic viscosity, ν, which is the fluid property governing the rate of diffusion of vorticity, we could form two nondimensional parameters, namely, $\nu t_c/L^2$ and $L/u_o t_c$, or alternatively, $u_o L/\nu$ and $L/u_o t_c$. Studies conducted by Stewartson(3) demonstrated that the former pair are relevant for "small" times and the latter for "large" times. At small times, the boundary layer growth is controlled by the diffusion process and, hence δ^2/ν would be the appropriate value for t_c. Thus

$$\frac{\nu t_c}{L^2} = \left(\frac{\delta}{L}\right)^2$$

and it represents the ratio of the rate of diffusion through L to that through δ. It is a quantity much less than 1. The parameter

$$\frac{L}{u_o t_c} = \frac{(\nu/\delta^2)}{(u_o/L)}$$

and it represents the ratio of the rate of diffusion through δ to the rate of convection through L. Thus, at small times,

$$\frac{L}{u_o t_c} \gg 1.$$

This suggests that the convective contribution in the momentum equation may be neglected and the resulting equation becomes <u>linear</u> if the fluid properties are taken as constants. This observation has been utilized by several investigators in the formulation of power series solutions for the problem. In passing, we note that the diffusion rate ν/δ^2 and the convection rate u_o/L will be of similar magnitude when the flow approaches steadiness. Under the latter condition,

$$\frac{\nu}{\delta^2} \sim \frac{u_o}{L}$$

or

$$\frac{\delta}{L} \sim \frac{1}{\sqrt{Re}},$$

a result which we noted previously for steady boundary layers. At the early stages of the boundary layer growth,

$$\frac{\delta}{L} \ll \left(\frac{u_o L}{\nu}\right)^{-1/2}.$$

At "large" times from the start of motion, i.e., as the growth enters its final stage, the characteristic time is the actual time. Under this condition,

$$\delta^* \sim \frac{1}{\sqrt{Re}}.$$

For the two-dimensional flow considered here, the x-component of the momentum equation, Equation (3.36), becomes

$$\left(\frac{L}{u_o t}\right) \rho^* \frac{\partial u^*}{\partial t^*} + \rho^* u^* \frac{\partial u^*}{\partial x^*} + \ldots = -\frac{1}{\gamma_\infty M_\infty^2} \frac{\partial p^*}{\partial x^*} + \frac{1}{Re_\infty} \frac{\partial \tau_{ji}^*}{\partial x_j^*}. \quad (3.59)$$

$$1 \quad 1 \qquad 1\ 1 \quad 1 \qquad\qquad \delta^{*2} \left(\frac{1}{\delta^{*2}}, \ldots 1\right)$$

It is obvious that, at sufficiently large times,

$$\frac{L}{u_o t} \ll 1.$$

In fact, $L/u_o t$ could be of the same order as δ^{*2}, or even less. When this occurs, it may not be justified to delete terms in

$$\frac{\partial \tau^*_{ji}}{\partial x^*_j} ,$$

which are small compared to

$$\frac{\partial}{\partial y^*}\left[\mu^*\left(\frac{\partial u^*}{\partial y^*}\right)\right] ,$$

since they may be of the same order as the unsteady term. For this reason, it is not strictly valid to use the usual boundary layer equations given below by Equations (3.61) and (3.62) for the analysis of its behavior approaching steady state.

Similar precautions should be made for oscillatory boundary layers either in the presence or in the absence of a mean flow. A limit needs to be imposed on the frequency, f, such that Lf/u_o remains large as compared to $\nu/u_o L$, in order that the boundary layer approximations are uniformly valid. It is not difficult to realize that completely analogous problems arise in connection with the analysis of unsteady thermal boundary layer equations.

With these precautions, we conclude this section by tabulating the boundary layer equations for unsteady, two-dimensional, high-speed, ideal gas flow:

$$\frac{\partial \rho}{\partial t} + \frac{\partial(\rho u)}{\partial x} + \frac{\partial(\rho v)}{\partial y} = 0, \tag{3.60}$$

$$\rho\left(\frac{\partial u}{\partial t} + u\frac{\partial u}{\partial x} + v\frac{\partial u}{\partial y}\right) = -\frac{\partial p}{\partial x} + \frac{\partial}{\partial y}\left(\mu\frac{\partial u}{\partial y}\right), \tag{3.61}$$

where

$$-\frac{\partial p}{\partial x} = \rho_1\frac{\partial u_1}{\partial t} + \rho_1 u_1\frac{\partial u_1}{\partial x}$$

and

$$c_p\rho\left(\frac{\partial T}{\partial t} + u\frac{\partial T}{\partial x} + v\frac{\partial T}{\partial y}\right) = \rho Q + \frac{\partial p}{\partial t} + u\frac{\partial p}{\partial x} + \frac{\partial}{\partial y}\left(k\frac{\partial T}{\partial y}\right) + \mu\left(\frac{\partial u}{\partial y}\right)^2 . \tag{3.62}$$

In addition, we have

$$p = \rho RT, \tag{3.63}$$

R being the gas constant and

$$k = k(T), \tag{3.64}$$

$$\mu = \mu(T),$$

$$c_p = c_p(T). \tag{3.64}$$

If we denote the stagnation enthalpy by h_o, which is $h + u^2/2$ since $v^2 \ll u^2$, the energy equation (3.62) can be rewritten as

$$\rho\left(\frac{\partial h_o}{\partial t} + u\frac{\partial h_o}{\partial x} + v\frac{\partial h_o}{\partial y}\right) = \rho Q + \frac{\partial p}{\partial t} + \frac{\partial}{\partial y}\left(\mu\frac{\partial h_o}{\partial y}\right) + \frac{\partial}{\partial y}\left[\mu\left(\frac{1}{Pr} - 1\right)\frac{\partial h}{\partial y}\right]. \tag{3.65}$$

In the context of boundary layer theory, the pressure is determined by the external flow and thus is regarded as known. The system of seven equations, (3.60) to (3.64) contain seven unknowns: u, v, ρ, T, μ, k, and c_p. In principle, solutions can be found when appropriate initial and boundary conditions are prescribed.

Finally, we remark that the method of similitude analysis as expounded in this section can be used for turbulent flow. It was employed by von Kármán in his derivation of the similarity law for turbulent shear layers. The boundary layer equations given in this section are commonly taken as the starting point in the development of semi-rational theories of turbulent transport of momentum and heat in such layers.

REFERENCES

1. Jakob, M. Heat Transfer, Vol. I. New York: John Wiley & Sons, 1950, p. 451.

2. Bird, R. B., W. E. Stewart, and E. N. Lightfoot. Transport Phenomena. New York: John Wiley & Sons, 1960, Chs. 10, 11.

3. Stewartson, K., "On the Impulsive Motion of a Flat Plate in a Viscous Liquid," Quarterly Journal of Mechanics, 4 (1951), p. 182.

General

4. Gukhman, A. A. Introduction to the Theory of Similarity. New York: Academic Press, 1965.

5. Tsien, H. S., "The Equations of Gas Dynamics," in Fundamentals of Gas Dynamics. H. W. Emmons, ed., Part A. Princeton, N. J.: Princeton University Press, 1958.

6. Schlichting, H. Boundary Layer Theory. 6th ed. New York: The McGraw-Hill Book Co., 1968, Chs. 7, 12.

4. SIMILARITY ANALYSIS OF BOUNDARY LAYER EQUATIONS

Closely coupled with the development of boundary layer theory has been the advancement of a mathematical technique which reduces the governing partial differential equations to ordinary differential equations. Generally speaking, it involves a transformation of both the dependent and independent variables. The procedure is called similarity analysis for reasons that will soon become apparent. If the stated objective is achieved, the solutions are designated as "similar." In this section, we shall briefly explore the idea, the methodology involved, and the condition under which such similar solution would exist. No mathematical rigor is attempted. While we are primarily interested in analyzing the boundary layer equations, it is pertinent to note that similar solutions are by no means limited to them. Recently, two books[1,2] written on the subject appeared. They should be consulted for further information.

Instead of previewing the broad aspects of similarity analysis, we propose to introduce the subject by examining specific examples. In Section 4.1, we consider a well-known unsteady heat conduction problem, the solution of which has been given by Carslaw and Jaeger.[3] We shall demonstrate that identical result can be obtained using similarity analysis.

4.1 TRANSIENT TEMPERATURE FIELD IN AN ISOTROPIC INFINITE SOLID DUE TO AN INFINITE LINE SOURCE OF CONSTANT STRENGTH

We adopt the following nomenclature: k = thermal conductivity, q_L = rate of heat liberation per unit length of the line source, r = radial distance from the line source, T = temperature, t = time, and κ = thermal diffusivity. Initially, the solid is at a uniform temperature which we conveniently take as a datum. The thermal properties are regarded as constants. The governing equation is:

$$\frac{1}{\kappa}\frac{\partial T}{\partial t} = \frac{\partial^2 T}{\partial r^2} + \frac{1}{r}\frac{\partial T}{\partial r} ; \qquad (4.1)$$

$$r > 0, \quad t > 0.$$

The initial condition is

$$T(r,0) = 0 \qquad (4.2)$$

and the boundary conditions are

$$\lim_{r\to 0} r\frac{\partial T}{\partial r} = -\frac{q_L}{2\pi k} \qquad (4.3)$$

and

$$T(\infty, t) = 0 . \qquad (4.4)$$

We seek the possibility of expressing the solution in the form:

$$T(r,t) = \Phi(t)\ f(\eta), \qquad (4.5)$$

with

$$\eta = r\ g(t) . \qquad (4.6)$$

The two functions $\Phi(t)$ and $g(t)$ are to be determined such that Equation (4.1) would reduce to an ordinary differential equation in f with compatible boundary conditions. It is pertinent that if we start by taking

$$\eta = r^m g(t)$$

where $m > 0$, the end result remains identical. The only difference is the amount of calculation involved. For the problem under consideration, the arithmetic becomes simpler if one selects $m = 2$. However, we shall proceed with Equation (4.6), since an analogous expression is often used in boundary layer analyses.

One has, from Equation (4.6),

$$\left.\begin{aligned} \frac{\partial \eta}{\partial t} &= \frac{\eta}{g}\frac{dg}{dt} , \\ \text{and} \\ \frac{\partial \eta}{\partial r} &= g . \end{aligned}\right\} \qquad (4.7)$$

Hence

$$r\left(\frac{\partial T}{\partial r}\right) = \Phi(t)\ \eta f' ,$$

where the prime denotes differentiation with respect to η. The boundary condition (4.3) can be satisfied if we set

$$\Phi(t) = \frac{q_L}{2\pi k} ,$$

which is a constant, and further demand that

$$\lim_{\eta\to 0} \eta f' = -1 . \qquad (4.8)$$

Thus, from now on, we replace Equation (4.5) by

$$T(r,t) = \frac{q_L}{2\pi k} f(\eta) \ . \tag{4.9}$$

Needless to say, at this stage of the analysis, there is no assurance that our objective can be achieved. Using Equations (4.6), (4.7), and (4.9), one finds that Equation (4.1) could be written as

$$\frac{1}{\kappa g^3}\frac{dg}{dt} = \frac{f''}{\eta f'} + \frac{1}{\eta^2} = -a, \tag{4.10}$$

where a is a separation constant. From the first order differential equation for g, one obtains

$$g = \frac{1}{\sqrt{2a\kappa t + C_1}} \ ,$$

C_1 being an integration constant. It is clear that in order to satisfy the initial condition (4.2) and the boundary condition (4.4), C_1 must be zero. It follows that

$$\eta = \frac{r}{\sqrt{2a\kappa t}} \tag{4.11}$$

and

$$f(\infty) = 0 \ . \tag{4.12}$$

The coalescence of the two conditions into one is necessary since, otherwise, we would have conditions incompatible to f. Experience shows that this requirement usually limits the existence of similar solutions to problems in which there is no characteristic length. For example, if we were to consider a cylindrical source of radius r_o or if the solid is bounded by a cylinder of finite radius, no similar solutions can be found. It is also known that if the physical parameters in a problem provide a characteristic time, similar solutions may not exist.

The differential equation for f is

$$f'' + \left(\frac{1}{\eta} + a\eta\right) f' = 0 \ . \tag{4.13}$$

Integrating once gives

$$\eta f' = C_2 e^{-a\eta^2/2} \ .$$

To satisfy the boundary condition (4.8), one requires $C_2 = -1$. We now transform the independent variable η to ζ according to

$$\zeta = \frac{a\eta^2}{2} = \frac{r^2}{4\kappa t} \ . \tag{4.14}$$

It follows that

$$\frac{df}{d\zeta} = -\frac{1}{2\zeta} e^{-\zeta}$$

which, upon integration, gives

$$f = \frac{1}{2}\int_{\zeta}^{\infty} \frac{e^{-u}}{u}\, du \ , \tag{4.15}$$

since $f(\infty) = 0$. We note that f does not involve a, suggesting that the latter can be arbitrarily chosen. The required solution for the temperature is

$$T = \frac{q_L}{4\pi k}\int_{\frac{r^2}{4\kappa t}}^{\infty} \frac{e^{-u}}{u}\, du$$

$$= -\frac{q_L}{4\pi k} \mathrm{Ei}\left(-\frac{r^2}{4\kappa t}\right) \ , \tag{4.16}$$

where

$$\mathrm{Ei}(-x) = -\int_x^{\infty} \frac{e^{-u}}{u}\, du$$

is the exponential integral. The dimensionless temperature, kT/q_L, is a unique function of $r^2/4\kappa t$. In other words, the transient temperature distribution in the solid, at various times, can be represented by a single curve by scaling the radial distance inversely with the square root of time. This is, then, the reason that such a solution is described as "similar."

If q_L is time dependent, similar solution does not generally exist. However, if it varies as powers of time, i.e., $q_L \sim t^n$, then similar solutions can again be found. Retaining Equation (4.9) as the defining equation for $f(\eta)$ and letting

$$\eta = \frac{r}{2\sqrt{\kappa t}} \ ,$$

one may easily show that f satisfies

$$f'' + \frac{f'}{\eta} + 2\eta f' - 4nf = 0 \tag{4.17}$$

with

$$\lim_{\eta\to 0} \eta f' = -1 \tag{4.18a}$$

and

$$f(\infty) = 0 \,. \tag{4.18b}$$

4.2 STEADY, INCOMPRESSIBLE, TWO-DIMENSIONAL, LAMINAR BOUNDARY LAYER WITH PRESSURE GRADIENT

As usual, the coordinates (x,y) are taken along and normal to the solid surface respectively, $y = 0$ being at the solid surface. The corresponding velocity components are (u,v). Following from the discussion presented in Section 3, the boundary layer equations are:

$$\frac{\partial u}{\partial x} + \frac{\partial v}{\partial y} = 0 \tag{4.19}$$

$$u\frac{\partial u}{\partial x} + v\frac{\partial u}{\partial y} = u_1 \frac{du_1}{dx} + \nu \frac{\partial^2 u}{\partial y^2} \,, \tag{4.20}$$

where $u_1(x)$ is the velocity at the edge of the boundary layer and ν is the kinematic viscosity. The boundary conditions are:

$$y = 0, \qquad u = v = 0, \tag{4.21}$$

$$\lim_{y\to\infty} u = u_1(x) \,. \tag{4.22}$$

At the entrance, $x = x_o$, let $u = u_2(y)$. (4.23)

The two velocity functions $u_1(x)$ and $u_2(y)$ are solely determined by conditions in the external flow. Here, we exclude the possibility of having flow separation. The foregoing system of equations can be solved for any physically meaningful expressions of $u_1(x)$ and $u_2(y)$. However, similar solutions exist only when $u_1(x)$ and $u_2(y)$ are of special forms. In what follows, we present a procedure by which they may be determined.

We introduce a stream function $\psi(x,y)$ defined by:

$$u = \frac{\partial\psi}{\partial y}, \qquad v = -\frac{\partial\psi}{\partial x} \,. \tag{4.24}$$

Then the continuity equation is identically satisfied and the momentum equation becomes:

$$\frac{\partial\psi}{\partial y}\frac{\partial^2\psi}{\partial x\partial y} - \frac{\partial\psi}{\partial x}\frac{\partial^2\psi}{\partial y^2} = u_1\frac{du_1}{dx} + \nu\frac{\partial^3\psi}{\partial y^3} \,. \tag{4.25}$$

We seek the possibility of expressing ψ in the form:

$$\psi(x,y) = \Phi(x)\; f(\eta) \,, \tag{4.26a}$$

with

$$\eta = y\, g(x) \,. \tag{4.27}$$

A remark is perhaps in order at this point. One may wonder why the particular form of η is selected. The reasoning which led to the selection could be stated as follows: If we start by writing it in the most general form, $\eta = \eta(x,y)$, then in view of the boundary condition (4.22), we would require

$$\lim_{y\to\infty} \Phi(x)\frac{\partial\eta}{\partial y} f' = u_1(x) \,, \tag{4.28}$$

where

$$f' \equiv \frac{df}{d\eta} \,.$$

Since, by supposition, f is a function of η only,

$$(f')_{\eta=\eta(x,\infty)}$$

cannot be dependent on x and must be a constant. For convenience and without losing generality, we may assign its value to be unity. It follows that, as $y\to\infty$, $\partial\eta/\partial y$ can only be a function of x. A simple choice is

$$\eta = yg(x) + g_1(x) \,.$$

However, on the boundary $y = 0$, η must reduce to a numeral. This requires that $g_1(x)$ can only be a constant. It is natural to locate the origin of the transformed coordinate $\eta = 0$ on the boundary; hence

$$g_1(x) = 0.$$

Obviously, η defined by (4.27) is not the only form which satisfies the foregoing requirement. Consider, for instance,

$$\eta = y(1 + e^{-by})g(x) \,,$$

where $b > 0$. We may easily verify that it satisfies both conditions, namely, (a)

$$\left(\frac{\partial\eta}{\partial y}\right)_{y\to\infty}$$

does not contain y, and (b) $\eta = 0$ on the boundary. It has been found, however, that the form of η is important when one attempts to reduce the partial differential equation

(4.25) to an ordinary differential equation for f with compatible boundary conditions.

Using Equation (4.27) and with $f'(\infty) = 1$, we see from Equation (4.28) that

$$\Phi(x)\ g(x) = u_1(x).$$

Hence we rewrite Equation (4.26a) as

$$\psi(x,y) = \frac{u_1(x)}{g(x)}\ f(\eta). \qquad (4.26b)$$

Differentiating Equation (4.26b) as indicated by the various terms in Equation (4.25), and substituting the results into it, one obtains after some simplification and rearrangement

$$(f')^2 - \left(1 - \frac{u_1}{u_1'}\frac{g'}{g}\right) ff''$$

$$= 1 + \nu \frac{g^2}{u_1'} f''', \qquad (4.29)$$

where

$$g' \equiv \frac{dg}{dx},$$

$$u_1' \equiv \frac{du_1}{dx},$$

and the primes associated with f denote differentiation with respect to η. It is seen that Equation (4.29) will reduce to an ordinary differential equation if

(a) $\nu(g^2/u_1')$ is independent of x and hence is a constant, a, and

(b) $1 - (u_1/u_1')(g'/g)$ is also independent of x and thus is another constant, b.

From (a), one immediately has

$$g = \sqrt{\frac{au_1'}{\nu}}\ . \qquad (4.30)$$

Thence,

$$\frac{g'}{g} = \frac{1}{2}\left(\frac{u_1''}{u_1'}\right),$$

and condition (b) becomes

$$1 - \frac{1}{2}\frac{u_1 u_1''}{(u_1')^2} = b$$

or,

$$\frac{u_1''}{u_1'} = 2(1-b)\left(\frac{u_1'}{u_1}\right)$$

which, upon integrating twice, yields two results depending on whether b equals, or does not equal, 1/2.

Case (1):

$$b \neq \frac{1}{2}$$

$$u_1 = (C_1 x + C_2)^m, \qquad (4.31a)$$

with

$$m = \frac{1}{2b-1}\ .$$

Case (2):

$$b = \frac{1}{2}$$

$$u_1 = C_3 e^{C_4 x}\ . \qquad (4.31b)$$

In Equation (4.31), C_1 to C_4 are integration constants. Thus, if $u_1(x)$ varies either as powers of x or as an exponential of x, similar solutions are possible. Clearly, the external flow governs the value of b. On the other hand, the choice of the constant a remains arbitrary. Different values of a will only alter the numerical scale of the coordinate η. We thus set a = b and obtain from Equation (4.29)

$$f''' + ff'' + \beta\left[1 - (f')^2\right] = 0, \qquad (4.32)$$

where

$$\beta = \frac{1}{b} = \frac{2m}{m+1}\ .$$

Equation (4.32) was first given by Falkner and Skan.[(4)]

The precise form of the similarity variable η and other relevant quantities are listed below for both cases.

Case (1):

$$\left.\begin{aligned} u_1 &= (C_1 x + C_2)^m, \qquad \beta = \frac{2m}{m+1} \\ \eta &= y\sqrt{\frac{bu_1'}{\nu}} = y\sqrt{\frac{m+1}{2}\frac{C_1}{\nu}u_1^{(m-1)/m}} \end{aligned}\right\} (4.33)$$

$$\left.\begin{aligned} \psi &= \sqrt{\frac{2}{m+1}\,\frac{\nu}{C_1}\,u_1^{(m+1)/m}}\; f \\ u &= u_1 f' \\ \text{and} \\ v &= -\sqrt{\frac{m+1}{2}\,\nu C_1 u_1^{(m-1)/m}}\left(f + \frac{m-1}{m+1}\,\eta f'\right) \end{aligned}\right\} \quad (4.33)$$

The boundary conditions (4.21) and (4.22) transform to

$$f(0) = 0,$$

$$f'(0) = 0,$$

and

$$f'(\infty) = 1.$$

Since Equation (4.32) is a third order ordinary differential equation, the function, f, is thus completely determined.

We now turn to examine if the entrance condition (4.23) can be satisfied. First, it must be compatible with the boundary condition on f'. This is possible if $u_1(x_o) = 0$, since then, the condition $x = x_o$ transform to $\eta \to \infty$ for $0 < m < 1$ and to $\eta = 0$ for $m < 0$ or $m > 1$. Clearly, $u_2(y)$ must necessarily vanish.

For flow past a wedge of included angle $\beta\pi$, placed symmetrically in a uniform main stream,

$$u_1 = Cx^m,$$

$$m = \frac{\beta}{2 - \beta}, \quad (4.34)$$

where x is measured from the front stagnation located at $x_o = 0$ and m may assume any value between 0 and 1. In view of the finding just described, we conclude that a similar solution exists for wedge flow. Readers are referred to published literature(5,6) for further information.

Case (2):

$$\left.\begin{aligned} u_1 &= C_3 e^{C_4 x}, \qquad \beta = 2 \\ \eta &= y\sqrt{\frac{C_4 u_1}{2\nu}} \\ \psi &= \sqrt{\frac{2\nu u_1}{C_4}}\; f \\ u &= u_1 f' \\ \text{and} \\ v &= -\sqrt{\frac{\nu C_4 u_1}{2}}\,(f + \eta f')\,. \end{aligned}\right\} \quad (4.35)$$

In this case, the entrance condition at $x = x_o$ cannot be made to merge into the boundary condition on f' for finite x_o.

4.3 STEADY, LAMINAR FORCED CONVECTION IN INCOMPRESSIBLE WEDGE FLOW WITH NONUNIFORM WALL TEMPERATURE (Figure 4.1)

Here we are concerned with the problem of finding the type of temperature variation over the wedge surface which admits similar solutions for the temperature field. We adopt the same coordinate system used in Section 4.2; the origin is located at the front stagnation. Assuming constant properties, the energy equation for the thermal boundary layer is:

$$u\frac{\partial T}{\partial x} + v\frac{\partial T}{\partial y} = \kappa\frac{\partial^2 T}{\partial y^2} + \frac{\nu}{c_p}\left(\frac{\partial u}{\partial y}\right)^2, \quad (4.36)$$

where T is the temperature, c_p is the specific heat, and κ is the thermal diffusivity. In many physical situations, the assumption of constant properties may not be warranted, particularly when the flow is of such speed that dissipation needs to be considered. Nevertheless, we shall proceed with this assumption and examine how the inclusion of the dissipative effect limits the possibility of obtaining similar solutions.

The boundary conditions to be considered are:

$$y = 0, \quad T = T_w(x) \quad (4.37)$$

$$\lim_{y\to\infty} T = T_\infty \quad (4.38)$$

and the entrance condition is

$$x = 0, \quad T = T_\infty\,. \quad (4.39)$$

In the above, T_∞ is the undisturbed uniform, free stream temperature.

For wedge flow, the velocity at the edge of the boundary layer, as has been previously stated, is

$$u_1 = Cx^m , \qquad (4.40)$$

where

$$m = \frac{\beta}{2 - \beta}$$

and $\beta\pi$ is the included angle of the wedge. The similarity variable is

$$\eta = y\sqrt{\frac{m+1}{2}\frac{C}{\nu}x^{(m-1)}} \qquad (4.41)$$

according to Equation (4.33), but rewritten to conform with relationship (4.40). The velocity components are:

$$u = u_1 f' \qquad (4.42a)$$

$$v = -u_1\sqrt{\frac{m+1}{2}\frac{\nu}{Cx^{m+1}}}\left(f + \frac{m-1}{m+1}\eta f'\right). \qquad (4.42b)$$

To examine the possibility of obtaining similar temperature fields, we consider

$$T - T_\infty = X(x)\ \theta(\eta) . \qquad (4.43)$$

It is seen that the boundary condition (4.37) can be satisfied if we set $\theta(0) = 1$ and let

$$T_w(x) = T_\infty + X(x) . \qquad (4.44)$$

The boundary condition (4.38) and the entrance condition (4.39) are satisfied if $\theta(\infty) = 0$, since $\eta\to\infty$ for either $y\to\infty$ or $x = 0$ except when $m = 1$. The latter corresponds to the two-dimensional stagnation flow for which the boundary layer has a constant thickness and the entrance condition becomes physically identical to Equation (4.38).

Using Equations (4.41), (4.42), and (4.43), one finds, after a straightforward calculation, that Equation (4.36) becomes

$$\theta'' + Pr\ f\theta' - \frac{2}{m+1}\frac{xX'}{X} Pr\ f'\theta$$

$$+ \frac{u_1^2}{X}\frac{Pr}{c_p}(f'')^2 = 0. \qquad (4.45)$$

Our objective of finding an ordinary differential equation for θ is achieved if

(a) xX'/X does not depend on x and hence is a constant, n, and

(b) u_1^2/X does not depend on x.

From (a) we readily obtain

$$X = Ax^n, \qquad (4.46)$$

where A is an integration constant. Since u_1 is given by (4.40), condition (b) can be satisfied only if

$$n = 2m . \qquad (4.47)$$

Hence we conclude that, in the presence of dissipation, the temperature field is similar if

$$T_w(x) = T_\infty + Ax^{2m}, \quad m = \frac{\beta}{2-\beta} , \qquad (4.48)$$

and, in this case, θ satisfies

$$\theta'' + Pr\ f\theta' - \frac{4m}{m+1} Pr\ f'\theta$$

$$+ \frac{C^2}{A}\frac{Pr}{c_p}(f'')^2 = 0 , \qquad (4.49)$$

with

$$\theta(0) = 1 \qquad (4.50a)$$

and

$$\theta(\infty) = 0. \qquad (4.50b)$$

On the other hand, if dissipation is negligible, the temperature field remains similar when the wall temperature varies according to

$$T_w(x) = T_\infty + Ax^n, \qquad (4.51)$$

where n may be chosen independent of m. Here, the temperature function θ satisfies

$$\theta'' + Pr\ f\theta' - \frac{2n}{m+1} Pr\ f'\theta = 0 . \qquad (4.52)$$

The boundary conditions (4.50a,b) remain unaltered.

4.4 STEADY, LAMINAR FREE CONVECTION OVER A VERTICAL PLATE WITH UNIFORM HEAT FLUX (Figure 4.2)

We take coordinates (x,y) along and normal to the plate with their origin located at its bottom edge which is horizontal. Thus the x-axis is vertically upward. We denote the corresponding velocity components by (u,v). The governing boundary layer equations are:

$$\frac{\partial u}{\partial x} + \frac{\partial v}{\partial y} = 0, \tag{4.53}$$

$$u\frac{\partial u}{\partial x} + v\frac{\partial u}{\partial y} = g\beta(T - T_\infty) + \nu\frac{\partial^2 u}{\partial y^2}, \tag{4.54}$$

and

$$u\frac{\partial T}{\partial x} + v\frac{\partial T}{\partial y} = \kappa\frac{\partial^2 T}{\partial y^2}, \tag{4.55}$$

where g is the gravitational acceleration, β is the volumetric thermal expansion coefficient, T is the temperature, ν is the kinematic viscosity, and κ is the thermal diffusivity of the fluid whose undisturbed temperature is T_∞.

The appropriate boundary conditions are:

$$y = 0; \quad u = v = 0, \tag{4.56a}$$

$$\frac{\partial T}{\partial y} = -\frac{q_w}{k} \tag{4.56b}$$

$$\lim_{y\to\infty} u = 0, \tag{4.57a}$$

$$\lim_{y\to\infty} T = T_\infty, \tag{4.57b}$$

and the entrance condition is

$$x = 0; \quad u = 0, \tag{4.58a}$$

$$T = T_\infty . \tag{4.58b}$$

In (4.56b), q_w is the uniform wall flux and k is the thermal conductivity.

We again introduce a stream function $\psi(x,y)$ defined by:

$$u = \frac{\partial\psi}{\partial y}, \quad v = -\frac{\partial\psi}{\partial x} \tag{4.59}$$

and consider

$$\psi(x,y) = \Phi(x)\, f(\eta), \text{ and} \tag{4.60}$$

$$T - T_\infty = \frac{q_w}{k}\frac{\theta(\eta)}{G(x)} \tag{4.60}$$

with

$$\eta = y\, G(x) . \tag{4.61}$$

The form chosen for $T - T_\infty$ in Equation (4.60) follows from a consideration of the boundary condition (4.56b). If one starts with a more general form,

$$T - T_\infty = X(x)\theta(\eta),$$

one finds that

$$X(x)\; G(x)\; \theta'(0) = -\frac{q_w}{k}$$

as required by Equation (4.56b). Since $\theta'(0)$ must be a numerical constant which we conveniently choose to be -1, we are led to the result that

$$X(x) = \frac{q_w}{kG(x)} .$$

Using Equations (4.60) and (4.61), one obtains

$$u = \Phi\, Gf' \tag{4.62}$$

and

$$-v = \Phi' f + \Phi\frac{G'}{G}\eta f', \tag{4.63}$$

where the prime associated with f denotes differentiation with respect to η,

$$\Phi' \equiv \frac{d\Phi}{dx},$$

and

$$G' \equiv \frac{dG}{dx} .$$

By straightforward differentiation, one may express $\partial u/\partial x$, $\partial u/\partial y$, $\partial T/\partial x$, $\partial T/\partial y$, ..., etc., in terms of f, θ, Φ, G, and their derivatives. Upon substituting these results into Equations (4.54) and (4.55) one finds, after some rearrangement, that the momentum equation becomes

$$f''' + \frac{\Phi'}{\nu G} ff'' - \left(\frac{\Phi'}{\nu G} + \frac{\Phi G'}{\nu G^2}\right)(f')^2 + \frac{q_w g\beta}{k\nu}\,\frac{1}{\Phi G^4}\,\theta = 0 \tag{4.64}$$

and the energy equation is

$$\theta'' + \frac{\Phi'}{\nu G} \Pr f\theta' + \frac{\Phi G'}{\nu G^2} \Pr f'\theta = 0, \quad (4.65)$$

where Pr is the Prandtl number. Clearly, Equations (4.64) and (4.65) will become ordinary differential equations for f and θ if

$$\text{(a)} \quad \frac{q_w g\beta}{k\nu} \frac{1}{\Phi G^4} = a_1 ,$$

$$\text{(b)} \quad \frac{\Phi G'}{\nu G^2} = a_2 ,$$

and

$$\text{(c)} \quad \frac{\Phi'}{\nu G} = a_3 ,$$

where a_1, a_2, and a_3 are constants. Since there are only two unknown functions, they are not all arbitrary. One may, without losing generality, select $a_1 = 1$ and obtain

$$\Phi = \frac{q_w g\beta}{k\nu} \frac{1}{G^4} . \quad (4.66)$$

With this, condition (b) becomes

$$\frac{1}{G^6} \frac{dG}{dx} = a_2 \frac{k\nu^2}{q_w g\beta}$$

which, upon integration, yields

$$\frac{1}{G^5} = -5a_2 \frac{k\nu^2 x}{q_w g\beta} + C_1 .$$

Since both G(x) and x are positive, a_2 must be negative. For convenience, we select $a_2 = -1$. The entrance conditions (4.58a,b) will merge into the boundary conditions (4.57a,b) if we set the integration constant $C_1 = 0$. Hence

$$G(x) = \left(\frac{q_w g\beta}{5k\nu^2 x}\right)^{1/5} \quad (4.67)$$

and

$$\eta = y \left(\frac{q_w g\beta}{5k\nu^2 x}\right)^{1/5} . \quad (4.68)$$

From Equations (4.66) and (4.67), one finds

$$\Phi(x) = 5\nu \left(\frac{q_w g\beta}{5k\nu^2}\right)^{1/5} x^{4/5} \quad (4.69)$$

and

$$a_3 = 4 .$$

This completes the similarity analysis. We summarize the results as follows:

Momentum:

$$f''' + 4ff'' - 3(f')^2 + \theta = 0 \quad (4.70)$$

Energy:

$$\theta'' + 4\Pr f\theta' - \Pr f'\theta = 0, \quad (4.71)$$

with

$$f(0) = 0, \; f'(0) = 0; \quad (4.72a)$$

$$\theta'(0) = 1 \quad (4.72b)$$

and

$$f'(\infty) = 0; \quad (4.73a)$$

$$\theta(\infty) = 0 . \quad (4.73b)$$

The velocity components are:

$$u = \Phi G f'$$

$$= 5^{3/5} \nu^{1/5} \left(\frac{q_w g\beta}{k}\right)^{2/5} x^{3/5} f' \quad (4.74)$$

$$v = -\Phi' f - \Phi \frac{G'}{G} \eta f'$$

$$= \nu^{3/5} \left(\frac{q_w g\beta}{k}\right)^{1/5} \frac{1}{(5x)^{1/5}} (\eta f' - 4f), \quad (4.75)$$

and the temperature is

$$T - T_\infty = \frac{q_w}{k} \left(\frac{5k\nu^2}{q_w g\beta}\right)^{1/5} \theta . \quad (4.76)$$

The existence of similar solutions for free convection over a heated vertical plate at constant temperature was first found by E. Pohlhausen according to Schmidt and Beckmann.[7] It is now known that similar solutions also exist if the plate surface temperature or heat flux varies as powers of x or as exponentials of x. Further discussion including unsteady flow may be found in a comprehensive paper by Yang.[8]

4.5 STEADY, COMPRESSIBLE, TWO-DIMENSIONAL, LAMINAR BOUNDARY LAYER WITH PRESSURE GRADIENT AND HEAT TRANSFER (constant T_w)

We consider an ideal gas of constant c_p, Pr = 1, and having viscosity varying linearly with temperature according to $\mu = CT$. We shall use the nomenclature of Sections 4.2 and 4.3; additional symbols will be explained as they arise. All quantities with subscript o refer to the stagnation condition and those with subscript 1 refer to the edge of the boundary layer. The similarity analysis presented below follows essentially that given by Li and Nagamatsu,[9] but modified to conform to the procedure suggested in the present notes.

The governing equations for the problem are:

$$\frac{\partial(\rho u)}{\partial x} + \frac{\partial(\rho v)}{\partial y} = 0 \qquad (4.77)$$

$$\rho\left(u\frac{\partial u}{\partial x} + v\frac{\partial u}{\partial y}\right) = -\frac{dp}{dx} + \frac{\partial}{\partial y}\left(\mu\frac{\partial u}{\partial y}\right) \qquad (4.78)$$

$$\rho\left(u\frac{\partial h_o}{\partial x} + v\frac{\partial h_o}{\partial y}\right) = \frac{\partial}{\partial y}\left(\frac{\partial h_o}{\partial y}\right), \qquad (4.79)$$

where the stagnation enthalpy

$$h_o = h + \frac{u^2}{2} .$$

Denoting the gas constant by R; the equation of state is

$$p = \rho RT. \qquad (4.80)$$

Hence

$$\rho T = \rho_1 T_1$$

as the pressure is constant across the boundary layer. In the main stream, the effects of viscosity and heat conduction can be neglected. Thus

$$\frac{dp}{dx} = -\rho_1 u_1 \frac{du_1}{dx} = c_p \rho_1 \frac{dT_1}{dx} \qquad (4.81)$$

and

$$a_1^2 + \frac{\gamma-1}{2} u_1^2 = a_{o1}^2, \text{ a constant,} \qquad (4.82)$$

where a is the acoustic speed and γ is the specific heat ratio. Equation (4.82) follows from the fact that the total temperature of the main stream, T_{o1}, is a constant. The boundary conditions are:

$$y = 0; \quad u = v = 0, \qquad (4.83a)$$

$$h_o = h_w = c_p T_w \qquad (4.83b)$$

$$y \to \infty; \quad u = u_1(x), \qquad (4.84a)$$

$$h_o = h_{o1} = c_p T_{o1} . \qquad (4.84b)$$

The entrance condition is

$$x = x_o; \quad u = u_1(x_o), \qquad (4.85a)$$

$$h_o = h_{o1} . \qquad (4.85b)$$

Introducing a stream function defined by

$$\psi(x,y) = \int_o^y \frac{u}{T}\, dy \qquad (4.86)$$

and hence

$$u = T\frac{\partial\psi}{\partial y} ,$$

one finds from Equation (4.77), after integrating once with respect to y, that

$$-\frac{v}{T} = \frac{\partial\psi}{\partial x} + \frac{c_p}{RT_1}\frac{dT_1}{dx}\psi . \qquad (4.87)$$

In Equations (4.78) and (4.79), we replace ρ by p/RT, μ by $(\mu_1/T_1)T$, $-dp/dx$ by $c_p\rho_1(dT_1/dx)$, and eliminate v by using Equation (4.87). The results are:

$$\frac{\partial\psi}{\partial y}\frac{\partial u}{\partial x} - \frac{\partial u}{\partial y}\left(\frac{c_p}{RT_1}\frac{dT_1}{dx}\psi + \frac{\partial\psi}{\partial x}\right)$$

$$= -\frac{c_p}{T_1}\frac{dT_1}{dx} + \frac{R}{p}\frac{\mu_1}{T_1}\frac{\partial}{\partial y}\left(T\frac{\partial u}{\partial y}\right) \qquad (4.88)$$

$$\frac{\partial\psi}{\partial y}\frac{\partial h_o}{\partial x} - \frac{\partial h_o}{\partial y}\left(\frac{c_p}{RT_1}\frac{dT_1}{dx}\psi + \frac{\partial\psi}{\partial x}\right)$$

$$= \frac{R}{p}\frac{\mu_1}{T_1}\frac{\partial}{\partial y}\left(T\frac{\partial h_o}{\partial y}\right). \qquad (4.89)$$

To examine the possibility of obtaining similar solutions, we again consider

$$\psi(x,y) = \Phi(x)\ f(\eta) \qquad (4.90)$$

and

$$h_o(x,y) = h_{o1}\ H(\eta)\ , \qquad (4.91)$$

with $\eta = \eta(x,y)$. Expression (4.91) follows from a consideration of the boundary conditions on h_o: (4.83b) and (4.84b). Clearly, they can be satisfied by simply requiring

$$H(0) = \frac{T_w}{T_{o1}}\ , \text{ a constant,}$$

and

$$H(\infty) = 1\ , \qquad (4.92)$$

where we anticipate that $\eta = 0$ corresponds to $y = 0$ and $\eta \to \infty$ corresponds to $y \to \infty$. To satisfy the boundary condition on u as $y \to \infty$, we require

$$\Phi(x)\left[T(x,y)\ f'\ \frac{\partial\eta}{\partial y}\right]_{y\to\infty} = u_1(x). \qquad (4.93)$$

Now $f'(\infty)$ must be a constant which we conveniently select to be 1. Hence Equation (4.93) can be satisfied if we let

$$\frac{\partial\eta}{\partial y} = \frac{u_1(x)}{\Phi(x)\ T(x,y)}\ .$$

Or,

$$\eta = \frac{u_1(x)}{\Phi(x)}\int_0^y \frac{dy}{T(x,y)}\ . \qquad (4.94)$$

It follows that

$$u = u_1 f' \qquad (4.95)$$

and

$$-\frac{v}{T} = \Phi' f + \Phi f'\frac{\partial\eta}{\partial x} - \frac{c_p}{RT_1}\frac{dT_1}{dx}\Phi f, \qquad (4.96)$$

where $\Phi' \equiv d\Phi/dx$. Hence the boundary condition (4.83a) becomes

$$f(0) = f'(0) = 0 \qquad (4.97)$$

and the boundary condition (4.84a) becomes

$$f'(\infty) = 1\ . \qquad (4.98)$$

Substituting Equations (4.90), (4.91), (4.94), and (4.95) into Equations (4.88) and (4.89), and making use of the following relation

$$\frac{T}{T_1} = \left(1 + \frac{\gamma-1}{2}M_1^2\right)H - \frac{\gamma-1}{2}M_1^2(f')^2, \qquad (4.99)$$

where M is the Mach number, one finds that the momentum equation becomes

$$f''' + b\ ff'' = \frac{1}{m}\left[(f')^2 - H\right] \qquad (4.100)$$

and the energy equation is

$$H'' + bf\ H' = 0\ . \qquad (4.101)$$

In the above, b and $1/m$ represent, respectively,

$$b \equiv \frac{pT_1}{R\mu_1}\frac{\Phi^2}{u_1}\left(\frac{c_p}{RT_1}\frac{dT_1}{dx} + \frac{\Phi'}{\Phi}\right) \qquad (4.102)$$

$$\frac{1}{m} \equiv \frac{pT_1}{R\mu_1}\frac{u_1'}{u_1^2}\Phi^2\left(1 + \frac{\gamma-1}{2}M_1^2\right). \qquad (4.103)$$

In order that Equations (4.100) and (4.101) become ordinary differential equations, both b and $1/m$ must be independent of x, and hence they are constants. Also, no generality will be lost if we assign one of them to be unity. We set $b = 1$ and obtain

$$\frac{c_p}{RT_1}\frac{dT_1}{dx} + \frac{\Phi'}{\Phi}$$

$$= \frac{m}{u_1}\frac{du_1}{dx_1}\left(1 + \frac{\gamma-1}{2}M_1^2\right). \qquad (4.104)$$

Since

$$\frac{\gamma-1}{2}M_1^2 = \frac{u_1^2}{2c_pT_1}$$

and u_1 and T_1 are related according to Equation (4.81), one may integrate Equation (4.104) to obtain

$$\Phi(x) = Au_1^m\frac{1}{a_1^{m+[2\gamma/(\gamma-1)]}}\ , \qquad (4.105)$$

where A is an integration constant. Having determined Φ, Equation (4.103) can be integrated to obtain the required form of $u_1(x)$ which would result in similar solutions. The general case is very complicated and cannot be expressed as explicit functions of x. However, Li and Nagamatsu found that when

$$m = -\frac{2\gamma-1}{\gamma-1},$$

the mainstream velocity again turned out to be powers of x. Readers are referred to the original paper[9] for details.

4.6 TWO-DIMENSIONAL, INCOMPRESSIBLE JET (LAMINAR OR TURBULENT)

The problem of a jet emerging from a narrow slit and undergoing constant pressure mixing with the surrounding fluid has been discussed by Schlichting.[10] It is the intent of this section to demonstrate that all pertinent results given in the said reference can be systematically deduced by using the procedure expounded in the present chapter.

A system of coordinates and the physical configuration of the jet are shown in Figure 4.3. The slit is assumed to be infinitely small, and hence the jet velocity there is infinitely large in order that the mass flow rate is finite. The total x-momentum per unit width of the jet, J, is independent of the distance, x, from the slit and is a constant. Thus

$$J = \rho \int_{-\infty}^{\infty} u^2\, dy = \text{constant}. \qquad (4.106)$$

The governing conservation equations are:

$$\frac{\partial u}{\partial x} + \frac{\partial v}{\partial y} = 0 \qquad (4.107)$$

and

$$u\frac{\partial u}{\partial x} + v\frac{\partial u}{\partial y} = \epsilon(x)\frac{\partial^2 u}{\partial y^2}, \qquad (4.108)$$

where $\epsilon(x)$ is the eddy diffusivity for momentum transfer and is assumed to be a function of x only.* For a laminar jet, $\epsilon(x)$ should be replaced by ν, the fluid kinematic viscosity. The appropriate boundary conditions are

$$y = 0; \quad v = \frac{\partial u}{\partial y} = 0 \qquad (4.109a)$$

$$\lim_{y\to\infty} u = 0, \qquad (4.109b)$$

and the entrance condition is

$$x = 0, \quad u = 0. \qquad (4.110)$$

Following a procedure identical to that used in the previous examples, we introduce a stream function $\psi(x,y)$ defined by Equation (4.24) and consider

$$\psi(x,y) = \Phi(x)\, F(\eta)$$

with $\eta = yg(x)$. After a straightforward calculation, we find that

$$u = \Phi g F', \qquad (4.111a)$$

$$-v = \Phi' F + \Phi\frac{g'}{g}\eta F', \qquad (4.111b)$$

and that Equation (4.108) becomes

$$\left(1 + \frac{\Phi}{\Phi'}\frac{g'}{g}\right)(F')^2 - FF'' = \epsilon(x)\frac{g}{\Phi'}F'''. \qquad (4.112)$$

In Equations (4.111a,b) and (4.112), the prime on Φ or g denotes differentiation with respect to x, while those on F denote differentiation with respect to η. Equation (4.112) will become an ordinary differential equation for F if:

(a) $\frac{\Phi}{\Phi'}\frac{g'}{g} = a_1$

and

(b) $\epsilon(x)\frac{g}{\Phi'} = a_2$,

where a_1 and a_2 are constants. An integration of (a) yields $g = C_1\Phi^{a_1}$, C_1 being an integration constant. Following Schlichting, we introduce the kinematic momentum $K(=J/\rho)$ and rewrite Equation (4.106) as

$$K = \Phi^2 g \int_{-\infty}^{\infty} (F')^2 d\eta, \qquad (4.113)$$

*Recent measurements made by Professor B. G. Jones at the University of Illinois indicate that this assumption does not strictly hold.

which immediately leads to the requirement that $\Phi^2 g$ must be a constant. It follows then $a_1 = -2$ and $\Phi^2 g = C_1$. In contrast to a_1, the constants C_1 and a_2 may be arbitrarily chosen. For convenience, we set $C_1 = 1$ and $a_2 = 1/2$. Consequently,

$$\Phi = \left[6 \int_0^x \varepsilon(x)\, dx\right]^{1/3}, \qquad (4.114)$$

$$g = \left[6 \int_0^x \varepsilon(x)\, dx\right]^{-2/3}; \qquad (4.115)$$

and Equation (4.112) becomes

$$F''' + 2FF'' + 2(F')^2 = 0 \qquad (4.116)$$

with $F(0) = F''(0) = 0$ and $F'(\infty) = 0$. The first two conditions follow from (4.109a), while the last follows from (4.109b). Since $\eta = yg(x)$, the entrance condition (4.110) also transforms to $F'(\infty) = 0$. It is pertinent to note that the required solution for F must also satisfy:

$$\int_{-\infty}^{\infty} (F')^2 d\eta = K . \qquad (4.117)$$

Equation (4.116) can be directly integrated once to give

$$F'' + 2FF' = 0 \qquad (4.118)$$

in which the constant of integration has been set to zero in view of the two boundary conditions at $\eta = 0$. Upon further integration, we obtain from Equation (4.118)

$$F = C_2 \tanh (C_2\eta) , \qquad (4.119)$$

which satisfies the condition $F'(\infty) = 0$. The constant C_2 must be so chosen that Equation (4.117) is satisfied. Accordingly,

$$C_2 = \left(\frac{3}{4} K\right)^{1/3} . \qquad (4.120)$$

To facilitate the presentation of results, we define:

$$I \equiv 6 \int_0^x \varepsilon(x)\, dx, \qquad (4.121)$$

and

$$\xi \equiv C_2\eta = C_2 y\, I^{-2/3}. \qquad (4.122)$$

Then,

$$\psi = C_2 I^{1/3} \tanh \xi , \qquad (4.123)$$

$$u = C_2^2 I^{-1/3} \operatorname{sech}^2\xi , \qquad (4.124)$$

$$v = 2C_2\varepsilon(x) I^{-2/3} (2\xi \operatorname{sech}^2 \xi - \tanh\xi). \qquad (4.125)$$

The transverse velocity at the jet boundary is

$$v_\infty = -2C_2\varepsilon(x) I^{-2/3} \qquad (4.126)$$

and the volume rate of discharge per unit width of the slit is

$$Q = 2C_2 I^{1/3} . \qquad (4.127)$$

For a laminar jet, $\varepsilon(x) = \nu$, hence $I = 6\nu x$ and

$$\xi = \frac{1}{48^{1/3}} \left(\frac{K}{\nu^2}\right)^{1/3} \frac{y}{x^{2/3}} .$$

When these are substituted into Equations (4.124) to (4.127), all corresponding expressions given in Reference 10 are obtained. For a turbulent jet, Schlichting considered

$$\varepsilon(x) = \varepsilon_s \left(\frac{x}{s}\right)^{1/2} ,$$

s being a characteristic distance from the slit. Thence,

$$I = 4\varepsilon_s s^{-1/2} x^{3/2}$$

and

$$\xi = \sigma \frac{y}{x}$$

with

$$\sigma = \frac{1}{4}\left(\frac{3Ks}{\epsilon_s^2}\right)^{1/3} .$$

Again, when the latter expressions for I and ξ are introduced into Equations (4.124) to (4.127), we recover Schlichting's results. It goes without saying that the existence of the similar solution for the turbulent jet is not restricted to the particular form of the eddy diffusivity considered by Schlichting, namely, $\epsilon(x) \sim x^{1/2}$. Similar solutions for the circular jet, either laminar or turbulent, for wake flow behind a flat plate, etc., can be readily obtained by using the present procedure.

4.7 CONCLUDING REMARKS

In this section we have presented a procedure by which the possibility of obtaining similar solutions of partial differential equations with well-posed boundary and entrance (or initial) conditions may be examined. From the outset, emphasis is placed on the consideration of the boundary conditions in the selection of transformation variables, both dependent and independent. An important aspect of the procedure is the recognition that certain constants arising during the analysis are arbitrary while others are not. Other methods have also been suggested and a brief account of them can be found in Reference 1.

Historically, the existence of similar solutions was recognized from physical considerations. We have noted earlier that when the problem under consideration provides a characteristic length or a characteristic time, in all probability one would not be able to find a similar solution. On the other hand, this does not mean that if a problem provides no such fundamental length or time, a similar solution can always be found. A reader can be easily convinced of this fact by simply reviewing the many problems in fluid dynamics and heat transfer whose solutions are known and found to be non-similar.

In all six examples given, the similarity variable, η, assumes the same form (the fifth has a superficial difference). One naturally wonders if such choice should always be made. There is no theoretical reason why this must be the case. Without doubt there are many problems, old and new, remaining to be solved and a joint effort among the engineers and the applied mathematicians may prove most worthwhile.

REFERENCES

1. Hansen, A. G. Similarity Analysis of Boundary Value Problems in Engineering. New York: Prentice-Hall, 1964.

2. Kline, S. J. Similitude and Approximation Theory. New York: McGraw-Hill Book Co., 1965, Ch. 4, Sec. 10c.

3. Carslaw, H. S. and J. C. Jaeger. Conduction of Heat in Solids. Oxford, England: Oxford University Press, 1959, p. 261.

4. Falkner, V. M. and S. W. Skan, "Some Approximate Solutions of the Boundary Layer Equations," Philosophical Magazine, 12 (1931), p. 865.

5. Rosenhead, L., ed. Laminar Boundary Layers. Oxford, England: Oxford University Press, 1963, Ch. 5, Sec. 21.

6. Schlichting, H. Boundary Layer Theory. 6th ed. New York: The McGraw-Hill Book Co., 1968, p. 150.

7. Schmidt, E. and W. Beckmann, "Das Temperatur -und Geschwindigkeitsfeld von einer Wärme abgebenden senkrechten Platte bei natürlicher Konvektion," Technische Mechanik und Thermodynamik, (1930), pp. 341, 391.

8. Yang, K. T., "Possible Similarity Solutions for Laminar Free Convection on Vertical Plates and Cylinders," Journal of Applied Mechanics (1960), p. 230.

9. Li, T. Y. and H. T. Nagamatsu, "Similar Solutions of Compressible Boundary-Layer Equations," Journal of the Aeronautical Sciences, 22 (1955), p. 607.

10. Schlichting, H. Boundary Layer Theory, 6th ed. New York: The McGraw-Hill Book Co., 1968, p. 170 (for laminar jet) and p. 696 (for turbulent jet).

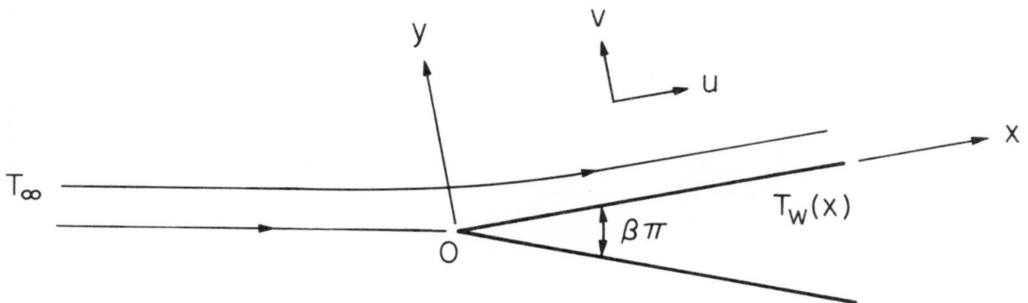

FIGURE 4.1 LAMINAR FORCED CONVECTION IN WEDGE FLOW.

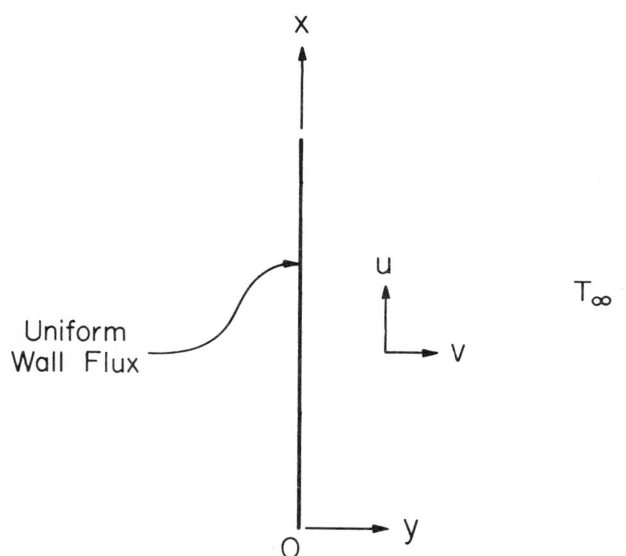

FIGURE 4.2 LAMINAR FREE CONVECTION OVER A VERTICAL PLATE.

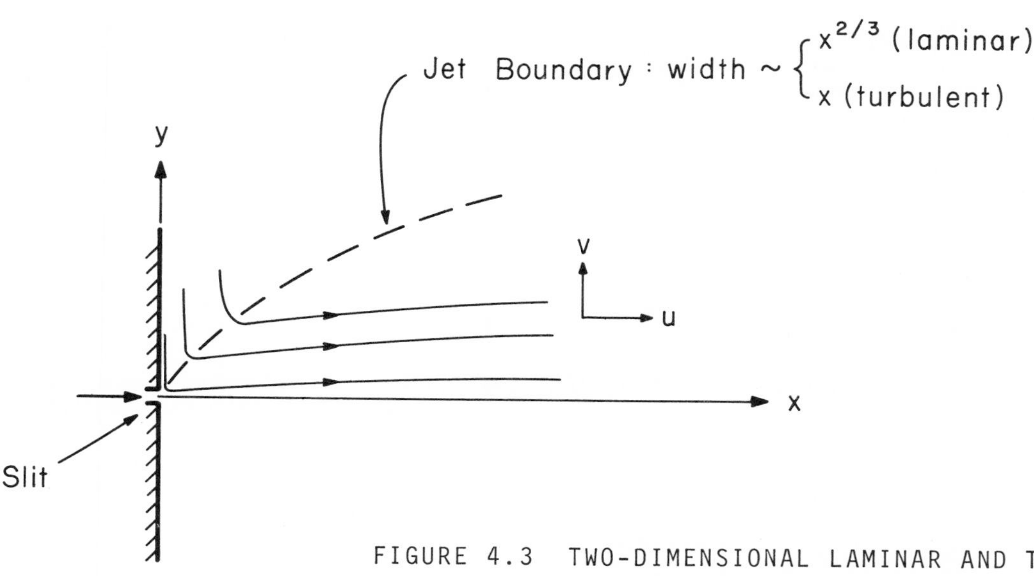

FIGURE 4.3 TWO-DIMENSIONAL LAMINAR AND TURBULENT JET.

5. MEKSYN'S METHOD OF LAMINAR BOUNDARY LAYER ANALYSIS*

In this section, we briefly describe an analytical technique devised by D. Meksyn for solving laminar boundary layer equations. It was reported in a series of papers which appeared in the Proceedings of the Royal Society of London, beginning in 1948. His asymptotic method of solution makes full use of the physical nature of the flow within the boundary layer. As we shall soon see, good approximate solutions can often be obtained with only a moderate amount of calculation. The asymptotic method has also been found to be quite effective in the calculation of heat transfer through thermal boundary layers. However, there are important questions concerning the technique which remain unanswered at the present time. They will be brought out in the following discussion.

A comprehensive account of the method may be found in Meksyn's excellent book.(1)

5.1 TRANSFORMED BOUNDARY LAYER EQUATIONS

For simplicity, we first consider steady, incompressible, two-dimensional boundary layer flow in the presence of pressure gradient and heat transfer. The appropriate boundary layer equations for constant properties are given by (4.19), (4.20), and (4.36) of the previous section. They are reproduced below for convenient reference,

$$\frac{\partial u}{\partial x} + \frac{\partial v}{\partial y} = 0 \qquad (5.1)$$

$$u\frac{\partial u}{\partial x} + v\frac{\partial u}{\partial y} = u_1\frac{du_1}{dx} + \nu\frac{\partial^2 u}{\partial y^2} \qquad (5.2)$$

$$u\frac{\partial T}{\partial x} + v\frac{\partial T}{\partial y} = \kappa\frac{\partial^2 T}{\partial y^2} + \frac{\nu}{c_p}\left(\frac{\partial u}{\partial y}\right)^2 . \qquad (5.3)$$

The usual boundary conditions will be considered. The continuity equation can be identically satisfied by introducing the stream function ψ defined by:

$$u = \frac{\partial \psi}{\partial y}, \qquad v = -\frac{\partial \psi}{\partial x} . \qquad (5.4)$$

*Unless stated otherwise, the nomenclature employed in Section 4 is used throughout in this section.

The boundary layer is a skin effect, while the physical coordinates (x,y) pertain to the general case of plane motion and, as such, they are not well suited for the very special nature of the boundary layer flow. Following Piercy, Preston, and Whitehead,(2) Meksyn used coordinates based on the velocity potential and stream function of the inviscid flow outside the boundary layer. Thus the coordinates (x,y) are first transformed to (ξ,χ) according to:

$$\left.\begin{matrix} x \\ \\ y \end{matrix}\right\} \Rightarrow \left\{\begin{matrix} \xi = \displaystyle\int_0^x \frac{u_1}{U_\infty}\,dx \\ \\ \chi = \dfrac{u_1}{U_\infty}\,y \end{matrix}\right. , \qquad (5.5)$$

where U_∞ is the uniform free stream velocity. Its use in Equation (5.5) is for convenience only; any other appropriate reference velocity may be adopted. A second transformation which is essentially identical to that used by Blasius:

$$\left.\begin{matrix} \xi \\ \\ \chi \end{matrix}\right\} \Rightarrow \left\{\begin{matrix} \xi \\ \\ \eta = \chi\sqrt{\dfrac{U_\infty}{2\nu\xi}} \end{matrix}\right.$$

$$= y\,u_1(x)\sqrt{\frac{1}{2\nu U_\infty \xi}} , \qquad (5.6)$$

and a velocity function, $f(\xi,\eta)$, are then introduced. The latter is related to the stream function according to

$$\psi = \sqrt{2\nu U_\infty \xi}\; f(\xi,\eta) . \qquad (5.7)$$

It follows that the velocity components are

$$u = \frac{\partial \psi}{\partial y} = u_1 f' \qquad (5.8a)$$

$$v = -\frac{\partial \psi}{\partial x} = -u_1\sqrt{\frac{\nu}{2\xi U_\infty}}\left[f + 2\xi\frac{\partial f}{\partial \xi} - (1-\beta)\eta f'\right] , \qquad (5.8b)$$

where the prime denotes partial differentiation with respect to η, and the nondimensional parameter β is defined by

$$\beta = 2\xi U_\infty \frac{1}{u_1^2}\frac{du_1}{dx} = 2\xi\frac{d(\ell n\, u_1)}{d\xi} . \qquad (5.9)$$

When these results are substituted into Equation (5.2), one finds that the momentum equation transforms to

$$f''' + ff'' + \beta\left[1 - (f')^2\right] = 2\xi\left(f'\frac{\partial f'}{\partial \xi} - f''\frac{\partial f}{\partial \xi}\right), \tag{5.10}$$

and the boundary conditions are:

$$\eta = 0; \quad f = 0, \quad f' = 0 \tag{5.11a}$$

$$\eta = \infty; \quad f' = 1. \tag{5.11b}$$

We assume that the entrance condition at $\xi = 0$ coincides with the boundary condition at $\eta = \infty$. The energy equation (5.3) transforms to

$$\frac{\partial^2 T}{\partial \eta^2} + \Pr f\frac{\partial T}{\partial \eta} + \Pr\frac{u_1^2}{c_p}(f'')^2 = 2\Pr\xi\left(f'\frac{\partial T}{\partial \xi} - \frac{\partial f}{\partial \xi}\frac{\partial T}{\partial \eta}\right), \tag{5.12}$$

where T is the temperature, Pr is the Prandtl number, and c_p is the specific heat. Appropriate dimensionless temperatures may be selected depending on the free stream and boundary conditions. A common situation is that the free stream temperature, T_∞, is uniform and the wall temperature has some prescribed variation, $T_w(x)$. In this case, it is convenient to put

$$\frac{T - T_\infty}{T_w(x) - T_\infty} = \theta(\xi,\eta) \tag{5.13}$$

and obtain from Equation (5.12) an equation for θ:

$$\frac{\partial^2\theta}{\partial \eta^2} + \Pr f\frac{\partial \theta}{\partial \eta} + \Pr\frac{u_1^2}{c_p(T_w - T_\infty)}(f'')^2 = 2\Pr\xi\left(f'\frac{\partial\theta}{\partial \xi} + f'\theta\frac{U_\infty}{u_1}\frac{dT_w/dx}{T_w - T_\infty} - \frac{\partial f}{\partial \xi}\frac{\partial \theta}{\partial \eta}\right), \tag{5.14}$$

with

$$\theta(\xi,0) = 1, \quad \theta(\xi,\infty) = 0 \tag{5.15a}$$

and

$$\theta(0,\eta) = 0\,. \tag{5.15b}$$

For an adiabatic wall, the appropriate form for θ is

$$\frac{T - T_\infty}{u_1^2/2c_p} = \theta(\xi,\eta), \tag{5.16}$$

thence Equation (5.12) becomes

$$\frac{\partial^2\theta}{\partial \eta^2} + \Pr f\frac{\partial \theta}{\partial \eta} + 2\Pr(f'')^2 = 2\Pr\xi\left(f'\frac{\partial\theta}{\partial \xi} + 2f'\frac{\theta}{u_1^2}\frac{du_1}{dx}U_\infty - \frac{\partial f}{\partial \xi}\frac{\partial \theta}{\partial \eta}\right), \tag{5.17}$$

with

$$\frac{\partial\theta}{\partial\eta}(\xi,0) = 0, \quad \theta(\xi,\infty) = 0 \tag{5.18a}$$

and

$$\theta(0,\eta) = 0\,. \tag{5.18b}$$

For axisymmetrical boundary layers over a body of revolution, the momentum and energy equations may be transformed to a form identical to Equations (5.10) and (5.12) when suitable expressions for ξ and η are used. Meksyn[3] has also demonstrated that the equations for two-dimensional compressible boundary layer can be likewise transformed.

At the forward portion of the boundary layer, the right side of Equation (5.10) is small and can often be neglected. It then reduces to the ordinary differential equation of Falkner and Skan, with β considered as the local value. If we seek the form of $u_1(x)$ which will lead to a constant value of β in Equation (5.9), we find that u_1 should either be powers or exponentials of x -- a result which we have already obtained in Section 4.2.

It may be of interest to re-examine the problem considered in Section 4.3 in the light of Equation (5.14). Here

$$u_1 = Cx^m$$

and f if a function of η only. If θ were to be independent of ξ, examination of Equation (5.14) shows that both

$$\text{(a)} \qquad \xi \frac{U_\infty}{u_1} \frac{1}{T_w - T_\infty} \frac{dT_w}{dx}$$

and

$$\text{(b)} \qquad \frac{u_1^2}{T_w - T_\infty}$$

must not depend on x. These considerations lead immediately to the result that

$$T_w = T_\infty + Ax^{2m}$$

and, at the same time, Equation (5.14) reduces to Equation (4.49).

In Equation (5.17), we see that if θ were a function of η alone, $\xi(1/u_1^2)(du_1/dx)$ must be independent of x. This is precisely the condition that β is a constant. We can therefore conclude that, for the Falkner-Skan flow past a perfectly insulated object, the temperature field is similar.

Intimately associated with Meksyn's method of integrating the boundary layer equations are the subjects of series inversion and Euler's transformation for summing semi-divergent series. These are considered in the following two sections.

5.2 SERIES INVERSION

The problem under consideration may be described as follows. Given the series

$$y = x^p \sum_{n=0}^{\infty} c_n x^n , \qquad \text{(a)}$$

where p is an integer equal to or greater than one, it is required to determine the coefficients in the inverted series,

$$x = y^{1/p} \sum_{m=0}^{\infty} b_m y^{m/p} \qquad \text{(b)}$$

which is valid for sufficiently small y. We may demonstrate that Series (b) is the correct form for the inverted series by writing Series (a) as

$$x = c_0^{-1/p} y^{1/p} \left(1 + \frac{c_1}{c_0} x + \frac{c_2}{c_0} x^2 + \dots\right)^{-1/p}$$

followed by expanding the term $(\cdot)^{-1/p}$ and repeatedly substituting x inside the parenthesis by itself.

For convenience, we rewrite the coefficients in Series (b) as $A_m/(m+1)$ and obtain

$$x = y^{1/p} \sum_{m=0}^{\infty} \frac{A_m}{m+1} y^{m/p} ; \qquad \text{(c)}$$

thence

$$dx = \frac{1}{p} \sum_{m=0}^{\infty} A_m y^{[(m+1)/p]-1} dy . \qquad \text{(d)}$$

Let us now regard x as a complex variable and consider the line integral:

$$\oint_{(\cdot)} y^{-1/p} dx$$

along a closed circuit once in the counterclockwise direction around x = 0. In view of Series (d), we have

$$\oint_{(\cdot)} y^{-1/p} dx = \frac{1}{p} \oint_{(p)} \left[A_0 y^{-1} + A_1 y^{-1+(1/p)} + A_2 y^{-1+(2/p)} + \dots\right] dy$$

$$= 2\pi i A_0 ,$$

where $i = \sqrt{-1}$. The symbol ⓟ denotes a p-fold circuit around y = 0. This is necessary because, in Series (c), y occurs with a fractional power 1/p. In a similar manner, we find

$$\oint_{(\cdot)} y^{-2/p} dx = 2\pi i A_1 ,$$

$$\oint_{(\cdot)} y^{-3/p} dx = 2\pi i A_2 ,$$

and, in general,

$$A_m = \frac{1}{2\pi i} \oint y^{-(m+1)/p}\, dx \,. \qquad \text{(e)}$$

Therefore, we conclude from the well-known theorem of residues that A_m is the coefficient of x^{-1} in the series expansion of $y^{-(m+1)/p}$ in ascending powers of x. Since

$$y^{-(m+1)/p} = x^{-(m+1)} \left(\sum_{n=0}^{\infty} c_n x^n \right)^{-(m+1)/p} ,$$

the coefficient of x^{-1} is also the coefficient of x^m in the expansion

$$\left(\sum_{n=0}^{\infty} c_n x^n \right)^{-(m+1)/p} .$$

The above result has been obtained by Sim[(4)] through a somewhat different approach.

We summarize the result in this section as follows: If

$$y = x^p \sum_{n=0}^{\infty} c_n x^n , \qquad (5.19a)$$

p being an integer ≥ 1, then

$$x = \sum_{m=0}^{\infty} \frac{A_m}{m+1} y^{(m+1)/p} , \qquad (5.19b)$$

where the A_m's are the coefficient of x^m in the expansion

$$\left(\sum_{n=0}^{\infty} c_n x^n \right)^{-(m+1)/p} \qquad (5.20)$$

5.3 EULER'S TRANSFORMATION AND THE SUM OF NONCONVERGENT SERIES

Meksyn's method often leads to nonconvergent series which are developed from presumably convergent expressions and are asymptotic in nature (although no general proof can be given). To find their sum, Euler's transformation is used. In this section, we briefly present the idea without attempting any mathematical rigor.

Let us consider the series

$$S(x) = \sum_{n=0}^{\infty} a_n x^{n+1}, \quad x > 0 \qquad \text{(f)}$$

which is convergent for small x. Let

$$x = \frac{y}{1-y}, \quad \text{thus} \quad y = \frac{x}{1+x} . \qquad \text{(g)}$$

Since $x > 0$, y is positive and is always less than 1. Substituting expression (g) into Series (f) yields

$$S(x) = \sum_{p=0}^{\infty} a_p y^{p+1} (1-y)^{-p-1}$$

$$= \sum_{p=0}^{\infty} a_p y^{p+1} \sum_{m=0}^{\infty} \binom{p+m}{m} y^m , \qquad \text{(h)}$$

where the binomial coefficient $\binom{p+m}{m}$ stands for

$$\frac{(p+m)!}{p!\; m!} ,$$

which is also equal to $\binom{p+m}{p}$.

If, in Series (h), we write $n = p + m$ and invert the order of summation, we have

$$S(x) = \sum_{p=0}^{\infty} a_p \sum_{n=p}^{\infty} \binom{n}{n-p} y^{n+1}$$

$$= \sum_{n=0}^{\infty} y^{n+1} \sum_{p=0}^{\infty} \binom{n}{p} a_p = \sum_{n=0}^{\infty} b_n y^{n+1}, \qquad \text{(i)}$$

where

$$b_n = \sum_{p=0}^{\infty} \binom{n}{p} a_p = \sum_{p=0}^{n} \binom{n}{p} a_p . \qquad \text{(j)}$$

Thus

$$b_o = a_o, \; b_1 = \binom{1}{o} a_o + \binom{1}{1} a_1 = a_o + a_1$$

$$b_2 = \binom{2}{0} a_0 + \binom{2}{1} a_1 + \binom{2}{2} a_2$$

$$= a_0 + 2! \; a_1 + a_2, \text{ etc.}$$

The coefficients b_n may be conveniently computed from a_n by using a tabulation scheme shown below.

$$\begin{array}{ccccccc} a_0 & & a_1 & & a_2 & & a_3 \\ & a_0 + a_1 & & a_1 + a_2 & & a_2 + a_3 & \\ & & a_0 + 2a_1 + a_2 & & a_1 + 2a_2 + a_3 & & \\ & & & a_0 + 3a_1 + 3a_2 + a_3 & & & \end{array}$$

The first term in the array is b_n.

Series (f) may become divergent for large x, but Series (i) could remain convergent for the corresponding y. If such were the case, Series (i) represents the sum S(x). Otherwise, Euler's transformation may be repeatedly applied until a convergent series is obtained. If we set x = 1, then y = 1/2, and it follows from Series (f) and (i) that

$$\sum_{n=0}^{\infty} a_n = \sum_{n=0}^{\infty} \frac{b_n}{2^{n+1}} , \qquad (5.21)$$

which is Euler's transformation formula. A discussion of Euler's method from a much broader viewpoint has been given by Hardy.[5]

As an illustration, let us consider a series constructed from the convergent expression:

$$S(x) = \frac{x}{1 + 2x} = x - 2x^2 + 4x^3 - 8x^4 + 16x^5 - \ldots \text{(k)}$$

For x = 1, the series is obviously nonconvergent. Using the tabulation scheme, one may readily find the coefficients of the transformed series. Thus

$$\begin{array}{ccccccccc} 1 & & -2 & & 4 & & -8 & & 16 \\ & -1 & & 2 & & -4 & & 8 & \\ & & 1 & & -2 & & 4 & & \\ & & & -1 & & 2 & & & \\ & & & & 1 & & & & \end{array}$$

Thence $b_0 = 1$, $b_1 = -1$, $b_2 = 1$, $b_3 = -1$, ..., and according to Equation (5.21)

$$S(1) = \frac{1}{2} - \frac{1}{2^2} + \frac{1}{2^3} - \frac{1}{2^4} + \frac{1}{2^5} - \frac{1}{2^6} + \frac{1}{2^7} - \frac{1}{2^8} + \ldots \qquad (l)$$

Six terms of Series (l) lead to S(1) = 0.3281; eight terms lead to 0.3320, which is quite close to the exact value 1/3.

Euler's method can be used to sum asymptotic series. In such series, it often happens that the absolute values of the successive terms decrease initially, reach a minimum, and eventually increase. For best results, Euler's transformation should be applied to the non-convergent terms only and with the process repeated if necessary. Meksyn's book[1] listed a set of rules for using Euler's method and it is important that these rules be followed.

5.4 MEKSYN'S METHOD OF INTEGRATING BOUNDARY LAYER EQUATIONS

A comprehensive preview of Meksyn's procedure is given in the introductory section of his book.[1] Here, we illustrate his method in some detail by applying it to the well-known Blasius problem. We shall also briefly comment on his success in integrating the Falkner-Skan equation, and the general equation, Equation (5.10). Further examples illustrating its application to the calculation of thermal boundary layers will be given in later sections.

Blasius was concerned with the development of a laminar boundary layer over a semi-infinite flat plate in a parallel main stream. From the discussion of Section 4.2, it is known that the pressure gradient in the direction of main flow is zero, the velocity field is similar, and the nondimensional velocity function, $f(\eta)$, satisfies the equation

$$f''' + ff'' = 0 , \tag{5.22}$$

with

$$f(0) = 0, \quad f'(0) = 0 \tag{5.23a}$$

and

$$f'(\infty) = 1 . \tag{5.23b}$$

As before, the prime denotes differentiation with respect to

$$\eta \left(= y\sqrt{\frac{U_\infty}{2\nu x}} \right).$$

The Maclaurin series expansion which satisfies Equations (5.22) and (5.23a) is

$$f = \sum_{n=2}^{\infty} \frac{a_n \eta^n}{n!} , \tag{5.24}$$

where

$$a_2 = a,\ a_3 = a_4 = 0,\ a_5 = -a^2,$$

$$a_6 = a_7 = 0,\ a_8 = 11a^3,\ a_9 = a_{10} = 0,$$

$$a_{11} = -375a^4, \text{ etc.}$$

The determination of the unknown parameter a in the coefficients is the heart of the problem. It is in this respect that Meksyn's method differs from that of Blasius and others. The main difficulty associated with the power series (5.24) is that it has only a finite radius of convergence. We shall soon see how Meksyn has devised a means which avoids, to a great extent, such difficulty.

Within the boundary layer, $f''(\sim \partial u/\partial y)$ is a rapidly decreasing function of η and, as such, it can be regarded as an asymptotic solution of the boundary layer equation. Substituting Series (5.24) into Equation (5.22) for f only, we formally transform the latter into a linear, ordinary differential equation for f''. Upon integration, we find

$$f'' = Ae^{-F(\eta)} , \tag{5.25}$$

where

$$F(\eta) = \int_0^{\eta} f d\eta = \eta^3 \sum_{n=0}^{\infty} c_n \eta^n \tag{5.26}$$

with

$$c_0 = \frac{a}{3!} ,\quad c_1 = c_2 = 0, \quad c_3 = -\frac{a^2}{6!} ,$$

$$c_4 = c_5 = 0, \quad c_6 = \frac{11a^3}{9!} ,\quad c_7 = c_8 = 0,$$

$$c_9 = -\frac{375a^4}{12!} , \text{ etc.} \tag{5.27}$$

The integration constant A is seen to equal a since $f''(0) = a$. To satisfy the boundary condition (5.23b), we require

$$f'(\infty) = a \int_0^{\infty} e^{-F} d\eta = 1 . \tag{5.28}$$

To evaluate the integral, Meksyn uses what is virtually the method of steepest descent. The function $F(\eta)$ is always positive and is a monotonically increasing function of η. Its stationary point is at $\eta = 0$. Accordingly, the main contribution to the integral in Equation (5.28) comes from the region close to $\eta = 0$. In his earlier work, Meksyn simply took one term in $F(\eta)$ and carried out the integration to obtain the first approximation. Later, he suggested an improved procedure by changing the variable of integration from η to $F(\eta)$ which he denoted by τ. Thus

$$\tau \equiv F(\eta) = \eta^3 \sum_{n=0}^{\infty} c_n \eta^n , \tag{5.29}$$

and its inverse is

$$\eta = \sum_{m=0}^{\infty} \frac{A_m}{m+1} \tau^{(m+1)/3} \qquad (5.30)$$

in which the coefficients A_m can be evaluated by using the method described in Section 5.2. It was found that

$$A_0 = \left(\frac{a}{3!}\right)^{-1/3}, \qquad A_1 = A_2 = 0,$$

$$A_3 = \frac{1}{15}\left(\frac{a}{3!}\right)^{-1/3}, \qquad A_4 = A_5 = 0,$$

$$A_6 = -\frac{1}{180}\left(\frac{a}{3!}\right)^{-1/3}, \qquad A_7 = A_8 = 0,$$

$$A_9 = \frac{23}{89,100}\left(\frac{a}{3!}\right)^{-1/3}, \text{ etc.} \qquad (5.31)$$

Differentiating Equation (5.30), inserting the values of A_m given above, and factoring out the first term, we obtain

$$d\eta = \left(\frac{3!}{a}\right)^{1/3} \frac{\tau^{-2/3}}{3} \left(1 + \frac{1}{15}\tau - \frac{1}{180}\tau^2 + \frac{23}{89,100}\tau^3 - \ldots\right) d\tau. \qquad (5.32)$$

Using Equation (5.32), we integrate Equation (5.28) in Gamma functions and obtain

$$\frac{1}{3}(3!\ a^2)^{1/3}\left\{\Gamma\left(\frac{1}{3}\right) + \frac{1}{15}\Gamma\left(\frac{4}{3}\right) - \frac{1}{180}\Gamma\left(\frac{7}{3}\right) + \frac{23}{89,100}\Gamma\left(\frac{10}{3}\right) - \ldots\right\} = 1$$

or

$$\frac{\Gamma\left(\frac{1}{3}\right)}{3}(6a^2)^{1/3}(1 + 0.02222 - 0.00247 + 0.00027 - \ldots) = 1. \qquad (5.33)$$

Thence a = 0.4696 which is precisely Hartree's result(6) who numerically integrated Equation (5.22) with boundary conditions (5.23a,b). Since the series in (5.33) are rapidly convergent, there is no need to use Euler's transformation.

It is instructive to examine the correspondence between the new variable τ and the Blasius variable η. The results of calculation based on four nonvanishing terms in Equation (5.30) are shown in Table 5.1. It is seen that $\tau \ll 1$ for $\eta \leq 1$. When τ reaches unity, $\eta = 2.375$, at which point the u-component of velocity in the boundary layer already attains approximately 90 per cent of the free stream value. For this reason, the use of τ-series has significant advantage over the η-series in the evaluation of boundary layer quantities.

TABLE 5.1

CORRESPONDENCE BETWEEN τ and η

τ	.001	.005	.01	.05	.1	.5	1.0
η	.2338	.3998	.5037	.8620	1.0869	1.8706	2.3750

The velocity field, (u,v), can be expressed in terms of the incomplete Gamma function as follows:

$$\frac{u}{U_\infty} = f' = \int_0^\eta f''\,d\eta = \frac{a}{3}\left(\frac{3!}{a}\right)^{1/3}\int_0^\tau e^{-\tau}\tau^{-2/3}\left(1 + \frac{\tau}{15} - \frac{\tau^2}{180} + \frac{23\tau^3}{89,100} - \ldots\right)d\tau$$

$$= 0.3659\left\{\Gamma\left(\frac{1}{3}, \tau\right) + \frac{1}{15}\Gamma\left(\frac{4}{3},\tau\right) - \frac{1}{180}\Gamma\left(\frac{7}{3},\tau\right) + \frac{23}{89,100}\Gamma\left(\frac{10}{3},\tau\right) + \ldots\right\} \qquad (5.34a)$$

$$\frac{v}{U_\infty}\sqrt{Re_x} = \frac{1}{\sqrt{2}}(\eta f' - f) = \frac{1}{\sqrt{2}}\int_0^\eta \eta f''\, d\eta$$

$$= \frac{a}{3\sqrt{2}}\left(\frac{3!}{a}\right)^{2/3}\int_0^\tau e^{-\tau}\tau^{-1/3}\left(1 + \frac{\tau}{12} - \frac{11\tau^2}{2,100} + \frac{157\tau^3}{1,134,000} - \dots\right) d\tau$$

$$= 0.6049\left\{\Gamma\left(\frac{2}{3},\tau\right) + \frac{1}{12}\Gamma\left(\frac{5}{3},\tau\right) - \frac{11}{2,100}\Gamma\left(\frac{8}{3},\tau\right) + \frac{157}{1,134,000}\Gamma\left(\frac{11}{3},\tau\right) - \dots\right\} . \quad (5.34b)$$

As $\tau \to \infty$, Equation (5.34b) gives

$$\frac{v}{U_\infty}\sqrt{Re_x} = 0.860 . \quad (5.34c)$$

Expressions for the displacement thickness, momentum thickness, and friction coefficient obtained by the method are all in good agreement with known results.

Meksyn integrated the Falkner-Skan equation, Equation (4.32), using essentially the same procedure. In this case, the boundary condition for $\eta = \infty$ requires that

$$\int_0^\infty e^{-F(\eta)}\,\Phi(\eta)\, d\eta = 1 , \quad (5.35)$$

where $F(\eta)$ is defined by an expression same as Equation (5.26), and $\Phi(\eta)$ is a slowly varying function. To evaluate the integral, Meksyn again transformed the variable η to τ and obtained

$$\sum_{m=0}^{\infty} d_m\,\Gamma\left(\frac{m+1}{3}\right) = 1 . \quad (5.36)$$

In Equation (5.36), the coefficients d_m contain various powers of the unknown parameter $a[=f''(0)]$ and, in contrast to Equation (5.33), it cannot be factored out but may be evaluated by successive approximation. Meksyn verified Equation (5.36) for several β's by using numerical values of 'a' reported by Hartree(6) and showed that the sum of the left side of Equation (5.36) was close to unity. The series was found to be divergent and use was made of Euler's transformation.

For the integration of the general equation (5.10), Meksyn used similar procedures. He considered the boundary layer over an elliptic cylinder and demonstrated good agreement between his results and Schubauer's measurements for velocities, almost down to separation, and for the position of the separation point. His results were also satisfactory downstream of separation. It should be mentioned that, in this case, a is a function of ξ and its derivatives appear in the coefficients d_m of Equation (5.36). Fortunately, the first three, d_o, d_1, and d_2, do not contain such derivatives and they have a major contribution to the sum. This makes possible considerable simplification to the problem.

With these brief comments, we now turn to examples in which Meksyn's idea is applied to the integration of thermal boundary layer equations.

5.5 FORCED CONVECTION OVER FLAT PLATE AT UNIFORM WALL TEMPERATURE

We begin by considering a simple problem in convective heat transfer, namely, the prediction of heat transfer rate over a flat plate of uniform temperature in parallel, incompressible flow. Historically, the problem was solved by Pohlhausen in 1921. It is instructive to see how Pohlhausen's main result can be effectively obtained by using Meksyn's procedure and with ease.

Since the fluid properties are taken as constants, the velocity field is known and is given by the Blasius solution (see Section 5.4). From the analysis of Section 4.3, we know that the temperature field is also similar. If we denote the uniform wall temperature by T_w, the uniform free stream temperature by T_∞, and define

$$\frac{T - T_\infty}{T_w - T_\infty} = \theta(\eta) , \tag{5.37}$$

then the energy equation is

$$\theta'' + Pr\, f\, \theta' = 0 , \tag{5.38}$$

where the prime denotes differentiation with respect to η. The boundary conditions are

$$\theta(0) = 1 \tag{5.39a}$$

and

$$\theta(\infty) = 0 . \tag{5.39b}$$

Integrating Equation (5.38) and making use of the boundary conditions (5.39a,b) leads to

$$\theta(\eta) = 1 + \theta'(0) \int_0^\eta e^{-PrF}\, d\eta , \tag{5.40}$$

where F is again defined by Equation (5.26). The unknown nondimensional temperature derivative $\theta'(0)$ is given by

$$\theta'(0) = - \left[\int_0^\infty e^{-PrF}\, d\eta \right]^{-1} . \tag{5.41}$$

To evaluate the integral in Equation (5.41), we again transform η to the new variable $\tau[\equiv F(\eta)]$ and integrate in Gamma functions. The result is

$$\int_0^\infty e^{-PrF}\, d\eta = \frac{C(Pr)}{0.4790\, Pr^{1/3}} , \tag{5.42}$$

where

$$C(Pr) = 1 + 2.222 \times 10^{-2}\, Pr^{-1} - 2.469 \times 10^{-3}\, Pr^{-2} + 2.677 \times 10^{-4}\, Pr^{-3} + 2.473 \times 10^{-4}\, Pr^{-4} - 1.189 \times 10^{-4}\, Pr^{-5} + \ldots . \tag{5.43}$$

The numerical constants in Equation (5.43) were calculated by the writer. They differ slightly from those originally given by Meksyn who reported the first four terms. C(Pr) is a slowly varying function and, for $Pr > 0.72$, it is practically a constant and close to unity. This is shown in Table 5.2.

TABLE 5.2

VALUES OF C(Pr) IN EQUATION (5.43)

Pr:	0.01	0.1	0.72	1	10	100
C(Pr):	1.389	1.156	1.027	1.020	1.002	1.0002

Pohlhausen calculated the numerical values of $-\theta'(0)$ for Pr ranging from 0.6 to 15. His results are shown in Table 5.3. Included in the table are our results calculated from the following expression:

$$-\theta'(0) = \frac{0.4790\, Pr^{1/3}}{C(Pr)} , \tag{5.44}$$

with C(Pr) given by Equation (5.43). It is seen that the agreement is very satisfactory. The local wall flux is given by

$$q_w = -k\left(\frac{\partial T}{\partial y}\right)_0 = -k(T_w - T_\infty)\sqrt{\frac{U_\infty}{2\nu x}}\, \theta'(0)$$

$$= 0.3387\, \frac{k(T_w - T_\infty)}{x}\, \frac{Pr^{1/3}}{C(Pr)}\, Re_x^{1/2} , \tag{5.45}$$

and the corresponding Nusselt number is

$$Nu_x = \frac{q_w x}{(T_w - T_\infty)k}$$

$$= 0.3387 \frac{Pr^{1/3}}{C(Pr)} Re_x^{1/2} . \qquad (5.46)$$

The average Nusselt number over a length, L, of the plate is

$$\overline{Nu} = 0.677 \frac{Pr^{1/3}}{C(Pr)} Re_L^{1/2} . \qquad (5.47)$$

For $0.6 < Pr < 15$, Pohlhausen recommended the use of the following approximate formula:

$$\overline{Nu} = 0.664 \, Pr^{1/3} \, Re_L^{1/2} , \qquad (5.48)$$

which may be directly compared with Equation (5.47).

TABLE 5.3

COMPARISON OF POHLHAUSEN'S RESULTS WITH THOSE EVALUATED FROM EQUATION (5.44)

	$-\theta'(0)$	
Pr	Pohlhausen	From (5.44)
0.6	0.390	0.3915
0.7	0.414	0.4139
0.8	0.434	0.4341
1.0	0.470	0.4695
7.0	0.912	0.9134
10.0	1.032	1.030
15.0	1.181	1.180

Note: Pohlhausen's definition of η differs from that used in the present analysis by a factor of $\sqrt{2}$. This has been incorporated in the above comparison.

5.6 STAGNATION POINT HEAT TRANSFER

A knowledge of the local heat transfer at the forward stagnation point of a blunt-nosed object moving through a fluid is of technological interest because of the high rate of heat transfer which may occur. In this section, we briefly present an analysis for the two-dimensional front stagnation in steady, incompressible flow. The wall is either at a uniform temperature, T_w, or having a uniform flux, q_w. Results for the axisymmetrical front stagnation will also be given.

Let us consider the two-dimensional case with uniform wall temperature. The coordinates (x,y) are along and normal to the solid surface, respectively; x = 0 being the front stagnation. The corresponding velocity components are (u,v). At the vicinity of the front stagnation, the main stream velocity at the edge of the boundary layer is given by

$$u_1 = Cx . \qquad (5.49)$$

If a cylinder of radius R is placed in a uniform stream of undisturbed velocity, U_∞, with its axis normal to the flow, $C = 2(U_\infty/R)$. The form of the external stream velocity described by Equation (5.49) holds also for axisymmetrical bodies if the x-coordinate is measured along the meridian.

From the analysis presented in Section 4.2, it is known that the velocity and temperature field are similar. Referring to Equation (4.33), we have for m = 1 and $C_2 = 0$,

$$\eta = y\sqrt{\frac{C}{\nu}} \tag{5.50}$$

which does not contain x. The velocity function $f(\eta)$ satisfies

$$f''' + ff'' + 1 - (f')^2 = 0, \tag{5.51}$$

since $\beta = 1$. The boundary conditions are

$$f(0) = 0, \quad f'(0) = 0, \tag{5.52a}$$

and

$$f'(\infty) = 1 . \tag{5.52b}$$

The series solution which satisfies Equation (5.51) and the boundary condition (5.52a) is

$$f(\eta) = \sum_{n=2}^{\infty} \frac{a_n \eta^n}{n!} \tag{5.53}$$

in which

$$a_2 = a, \quad a_3 = -1, \quad a_4 = 0,$$

$$a_5 = a^2, \quad a_6 = -2a, \quad a_7 = 2,$$

$$a_8 = -a^3, \quad a_9 = 4a^2, \text{ etc.}$$

The unknown constant a is determined from the boundary condition (5.52b) which leads to an integral of the same form as Equation (5.35). Some comments have already been made in Section 5.4 concerning the evaluation of the integral by Meksyn's procedure. The result is

$$a = f''(0) = 1.2326$$

The temperature function,

$$\theta(\eta) = \frac{T - T_\infty}{T_w - T_\infty}, \tag{5.54}$$

satisfies

$$\theta'' + Pr\, f\, \theta' = 0 \tag{5.55}$$

with

$$\theta(0) = 1 \tag{5.56a}$$

and

$$\theta(\infty) = 0 . \tag{5.56b}$$

It is seen that Equation (5.55) is formally identical to Equation (5.38) with a different velocity function. We may conveniently use an integration procedure same as that explained in the previous section and obtain

$$-\theta'(0) = \frac{0.6608\, Pr^{1/3}}{C(Pr)}, \tag{5.57}$$

where

$$C(Pr) = 1 + 1.1583 \times 10^{-1}\, Pr^{-1/3} + 4.41 \times 10^{-2}\, Pr^{-2/3} + 1.18 \times 10^{-3}\, Pr^{-1} + 2.00 \times 10^{-4}\, Pr^{-4/3} - 3.11 \times 10^{-3}\, Pr^{-5/3} + \ldots . \tag{5.58}$$

The local wall flux is

$$q_w = -k(T_w - T_\infty)\sqrt{\frac{C}{\nu}}\, \theta'(0)$$

$$= 0.6608\, k(T_w - T_\infty)\frac{Pr^{1/3}}{C(Pr)}\left(\frac{C}{\nu}\right)^{1/2}, \tag{5.59}$$

which is independent of x. Conversely, if a uniform wall flux is prescribed, the wall temperature rise at the stagnation is also uniform and is given by

$$T_w - T_\infty = \frac{q_w}{k}\left(\frac{\nu}{C}\right)^{1/2} \frac{C(Pr)}{0.6608\, Pr^{1/3}} . \tag{5.60}$$

For heat transfer at the axisymmetrical front stagnation, similar analysis can be made. The result is

$$-\theta'(0) = \frac{0.8501\, Pr^{1/3}}{C_a(Pr)}, \tag{5.61}$$

where

$$C_a(Pr) = 1 + 8.460 \times 10^{-2} Pr^{-1/3} + 2.352 \times 10^{-2} Pr^{-2/3} + 9.11 \times 10^{-3} Pr^{-1} + 2.1 \times 10^{-4} Pr^{-4/3} + 8.9 \times 10^{-4} Pr^{-5/3} + \ldots \quad (5.62)$$

Heat transfer near the front stagnation in axisymmetrical flow was earlier studied by Sibulkin.[7] A comparison of his results with those computed from Equation (5.61) is made in Table 5.4. The local wall flux is

$$q_w = 0.8501\, k(T_w - T_\infty) \frac{Pr^{1/3}}{C_a(Pr)} \left(\frac{C}{\nu}\right)^{1/2} \quad (5.63)$$

and, conversely, the wall temperature rise due to a prescribed uniform wall flux is

$$T_w - T_\infty = \frac{q_w}{k} \left(\frac{\nu}{C}\right)^{1/2} \frac{C_a(Pr)}{0.8501\, Pr^{1/3}} \cdot \quad (5.64)$$

For a sphere of radius, R, placed in a uniform stream of undisturbed velocity, U_∞, the constant, C, in both Equations (5.63) and (5.64) becomes

$$\frac{3}{2} \frac{U_\infty}{R} .$$

Meksyn's procedure of integrating boundary layer equations has also been successfully used in the analysis of unsteady stagnation point heat transfer. Readers are referred to Reference 8 for details.

TABLE 5.4

COMPARISON WITH SIBULKIN'S RESULTS

Pr	$-\theta'(0)$ Sibulkin	$-\theta'(0)$ From (5.61)
0.6	0.625	0.6241
0.8	0.700	0.6983
1.0	0.763	0.7607
2.0	0.988	1.002
10.0	1.76	1.752

5.7 SUPERSONIC FLOW PAST A FLAT PLATE WITH AND WITHOUT HEAT TRANSFER

Meksyn first integrated the compressible boundary layer equations over a flat plate in parallel flow in 1948[9] employing his asymptotic method. He considered an adiabatic plate and used only one leading term in the power series expansion for velocity and for temperature. The integration was carried out for a particular Prandtl number of 0.733. The results of his calculation were in satisfactory agreement with those reported by Emmons and Brainerd.[10,11] The problem was re-examined by Meksyn in 1955,[12] when he considered the more general case of heat transfer and arbitrary Prandtl number close to unity. Again, he used only the first term in the series expansion for velocity and for temperature. The results were found less satisfactory. Recently, a third paper[13] on the same subject appeared in which Meksyn improved the method of integration. In what follows, we present a brief account of his procedure as described in Reference 13.

5.7.1 Transformed Boundary Layer Equations for the Flat Plate

In all three papers mentioned above, the starting point of Meksyn's analysis is the transformed boundary layer equations due to Emmons and Brainerd. It is assumed that Pr and c_p are constants and that the viscosity (hence the conductivity) varies with temperature according to

$$\frac{\mu}{\mu_\infty} = \frac{k}{k_\infty} = \theta^{\omega} \equiv \varphi , \tag{5.65}$$

where $\theta = T/T_\infty$ and ω is a constant. For air, ω is close to 0.5 at high temperatures but increases to approximately 1 at low temperatures. For the reader's convenient reference, we recapitulate below the governing equations for the problem:

$$\frac{\partial(\rho u)}{\partial x} + \frac{\partial(\rho v)}{\partial y} = 0 \tag{5.66}$$

$$\rho\left(u\frac{\partial u}{\partial x} + v\frac{\partial u}{\partial y}\right) = \frac{\partial}{\partial y}\left(\mu\frac{\partial u}{\partial y}\right) \tag{5.67}$$

$$c_v\rho\left(u\frac{\partial T}{\partial x} + v\frac{\partial T}{\partial y}\right) = -p\left(\frac{\partial u}{\partial x} + \frac{\partial v}{\partial y}\right) + \frac{\partial}{\partial y}\left(k\frac{\partial T}{\partial y}\right) + \mu\left(\frac{\partial u}{\partial y}\right)^2 \tag{5.68}$$

and the equation of state

$$p = \rho RT . \tag{5.69}$$

Using Equation (5.69), and since p is a constant, we may rewrite Equations (5.66), (5.67), and (5.68) as

$$T\left(\frac{\partial u}{\partial x} + \frac{\partial v}{\partial y}\right) = u\frac{\partial T}{\partial x} + v\frac{\partial T}{\partial y} \tag{5.70}$$

$$u\frac{\partial u}{\partial x} + v\frac{\partial u}{\partial y} = \nu_\infty\frac{T}{T_\infty}\frac{\partial}{\partial y}\left(\varphi\frac{\partial u}{\partial y}\right) \tag{5.71}$$

$$p\frac{\gamma}{\gamma-1}\left(\frac{\partial u}{\partial x} + \frac{\partial u}{\partial y}\right) = k_\infty\frac{\partial}{\partial y}\left(\varphi\frac{\partial T}{\partial y}\right) + \mu\left(\frac{\partial u}{\partial y}\right)^2 . \tag{5.72}$$

The appropriate boundary conditions are:

$$y = 0; \quad u = 0, \quad v = 0 , \tag{5.73a}$$

$$\text{either } T = T_w \quad \text{or} \quad \frac{\partial T}{\partial y} = 0 , \tag{5.73b}$$

$$y = \infty; \quad u = U_\infty, \quad T = T_\infty , \tag{5.73c,d}$$

and the entrance condition is

$$x = 0; \quad u = U_\infty, \quad T = T_\infty . \tag{5.74a,b}$$

Using the procedure expounded in Section 4, one may readily demonstrate that similar solutions exist for both the velocity and the temperature field. The plate may be at uniform temperature or it may be perfectly insulated.

Besides $\theta(=T/T_\infty)$, Emmons and Brainerd introduced the following nondimensional variables:

$$\left.\begin{aligned} U &= \frac{u}{U_\infty} , \\ V &= v\sqrt{\frac{x}{\nu_\infty U_\infty}} , \\ \eta &= y\sqrt{\frac{U_\infty}{\nu_\infty x}} , \\ \xi &= \eta U - 2V . \end{aligned}\right\} \tag{5.75*}$$

In terms of these variables, the governing equations become

$$\theta(\xi' - U) = \theta'\xi \tag{5.76}$$

$$(\varphi U')' + \frac{\xi}{2\theta} U' = 0 \tag{5.77}$$

$$(\varphi\theta')' + \frac{Pr}{2}(\xi' - U) + b\,Pr\,\varphi(U')^2 = 0 , \tag{5.78}$$

where the primes denote differentiation with respect to η, and

$$b = (\gamma - 1)\, M_\infty^2 .$$

The transformed boundary conditions are

$$U(0) = 0, \quad \xi(0) = 0 , \tag{5.79a}$$

either

$$\theta(0) = \theta_o\left(\equiv \frac{T_w}{T_\infty}\right) \quad \text{or} \quad \theta'(0) = 0 \tag{5.79b}$$

---

*The dimensionless coordinate η defined here differs from that used in Section 5.4 by a factor of $\sqrt{2}$.

and

$$U(\infty) = 1, \quad \theta(\infty) = 1 . \qquad (5.79c,d)$$

5.7.2 Series Expansions for U, ξ, and θ

The Maclaurin series expansions for U, ξ, and θ which satisfy Equations (5.76), (5.77), (5.78), and the boundary conditions (5.79a,b) can be obtained in the usual way. They are

$$U(\eta) = \sum_{n=1}^{\infty} \frac{U^{(n)}(0)}{n!} \eta^n , \qquad (5.80)$$

with

$$U^{(1)}(0) \equiv \alpha, \qquad U^{(2)}(0) = -\frac{\omega\alpha\theta'_o}{\theta_o} ,$$

$$U^{(3)}(0) = \omega(2\omega+1)\alpha \frac{(\theta'_o)^2}{\theta_o^2} + \omega\alpha^3\beta \frac{1}{\theta_o}, \text{ etc.},$$

where

$$\theta'_o \equiv \theta^{(1)}(0)$$

and

$$\beta = b \text{ Pr} = (\gamma - 1) M_\infty^2 \text{ Pr} .$$

$$\xi(\eta) = \sum_{n=2}^{\infty} \frac{\xi^{(n)}(0)}{n!} \eta^n , \qquad (5.81)$$

with

$$\xi^{(2)}(0) = \alpha ,$$

$$\xi^{(3)}(0) = (1 - \omega)\alpha \frac{\theta'_o}{\theta_o} , \text{ etc.},$$

and

$$\theta(\eta) = \sum_{n=0}^{\infty} \frac{\theta^{(n)}(0)}{n!} \eta^n , \qquad (5.82)$$

with

$$\theta^{(0)}(0) \equiv \theta_o , \qquad \theta^{(1)}(0) \equiv \theta'_o ,$$

$$\theta^{(2)}(0) = -\alpha^2\beta - \omega \frac{(\theta'_o)^2}{\theta_o} ,$$

$$\theta^{(3)}(0) = \omega(2\omega+1) \frac{(\theta'_o)^3}{\theta_o^2} + 4\omega\alpha^2\beta \frac{\theta'_o}{\theta_o}, \text{ etc.}$$

For a plate with a given uniform surface temperature, the unknowns in the above three series are α and θ'_o; for an adiabatic plate, the unknowns are α and θ_o. The coefficients of the series simplify considerably in the latter case. These unknowns are determined from the boundary conditions (5.79c,d). However, when Pr is close to unity, Meksyn presented an approximate integral for $\theta(\eta)$ which he first introduced in his 1955 paper. That integral is seemingly able to give very good results.

5.7.3 Meksyn's Approximate Integration of the Energy Equation

Using Equations (5.76) and (5.77), the energy equation (5.78) may be recast into the following form:

$$\theta'' + \left[(1 - \text{Pr}) \frac{\varphi'}{\varphi} - \text{Pr} \frac{U''}{U'}\right] \theta' + b \text{ Pr}(U')^2 = 0 , \qquad (5.83)$$

which, upon integrating once, gives

$$\theta' = \frac{\theta'_o}{\alpha^{\text{Pr}}} \left(\frac{\varphi_o}{\varphi}\right)^{1-\text{Pr}} (U')^{\text{Pr}} - b \text{ Pr} \frac{(U')^{\text{Pr}}}{\varphi^{1-\text{Pr}}} \int_0^\eta \varphi^{1-\text{Pr}} (U')^{2-\text{Pr}} d\eta. \quad (5.84)$$

Since Pr $\sim$ 1, $\varphi^{1-\text{Pr}}$ is a slowly varying function in the boundary layer. This is particularly true for an adiabatic plate. Consequently, Meksyn regarded $\varphi^{1-\text{Pr}}$ as a constant in integrating Equation (5.84) from $\eta = \eta$ to $\eta = \infty$. The result is

$$\theta = 1 - \underbrace{\frac{\theta'_o}{\alpha^{\text{Pr}}} \int_\eta^\infty (U')^{\text{Pr}} d\eta}_{I_1} + \underbrace{b \text{ Pr} \int_\eta^\infty (U')^{\text{Pr}} d\eta \int_0^\eta (U')^{2-\text{Pr}} d\eta}_{I_2} . \qquad (5.85)$$

In Equation (5.85), if we set $\theta_o' = 0$ and $Pr = 1$, we are led to the well-known result:

$$\theta = 1 + \frac{\gamma - 1}{2} M_\infty^2 (1 - U^2) . \qquad (5.86)$$

To evaluate the first integral, I_1, we rewrite Equation (5.77) as

$$\frac{(\varphi U')'}{\varphi U'} = - \frac{\xi}{2\varphi\theta} \simeq - \frac{\alpha \eta^2}{4\theta_o^{\omega+1}} , \qquad (5.87)$$

where only the first term in the series expansion for ξ and θ has been used. Integrating Equation (5.87) once leads to

$$U' = \alpha \exp\left(- \frac{\alpha \eta^3}{12\theta_o^{\omega+1}}\right) ; \qquad (5.88)$$

thence

$$(U')^{Pr} = \alpha^{Pr} \exp\left(- \frac{\alpha \, Pr \, \eta^3}{12\theta_o^{\omega+1}}\right) . \qquad (5.89)$$

Using Equation (5.89) and recognizing that

$$\int_\eta^\infty (U')^{Pr} \, d\eta$$

$$= \alpha^{Pr} Pr^{-1/3} \int_{\eta Pr^{1/3}}^\infty \exp\left(- \frac{\alpha \, t^3}{12\theta_o^{\omega+1}}\right) dt, \qquad (5.90)$$

we may immediately establish

$$I_1 = \frac{\theta_o'}{\alpha} \left[1 - U(\eta \, Pr^{1/3})\right] Pr^{-1/3} . \qquad (5.91)$$

To evaluate the second integral, I_2, Meksyn, following Kuerti,[14] set

$$Pr = 1 + \epsilon \qquad (5.92)$$

and considered $\epsilon \ll 1$. The inner integral is integrated by parts to give

$$U(U')^{-\epsilon} + \epsilon \int_0^\eta U(U')^{-\epsilon} \frac{U''}{U'} d\eta ,$$

which is then inserted into I_2. A further partial integration gives

$$I_2 = b \, Pr \frac{1 - U^2}{2} + b \, Pr \, \epsilon \int_\eta^\infty (U')^{1+\epsilon} d\eta \int_0^\eta U(U')^{-\epsilon} \frac{U''}{U'} d\eta . \qquad (5.93)$$

From Equation (5.76), we readily find that

$$\frac{\xi}{\theta} = \int_0^\eta \frac{U}{\theta} d\eta . \qquad (5.94)$$

Using Equation (5.94), Equation (5.77) becomes

$$(\varphi U')' + \frac{1}{2} U' \int_0^\eta \frac{U}{\theta} d\eta = 0 . \qquad (5.95)$$

We again introduce the approximation

$$\varphi \equiv \theta^\omega \simeq \theta_o^\omega$$

and, in the integrand of Equation (5.95), we let $\theta \simeq \theta_o$. Introducing a new function, $\zeta(\eta)$, defined by

$$\zeta = \int_0^\eta U d\eta , \qquad (5.96)$$

we may rewrite Equation (5.95) in the approximate form

$$\zeta''' + \frac{\zeta\zeta''}{2\theta_o^{\omega+1}} = 0 . \qquad (5.97)$$

The inner integral in Equation (5.93) can then be transformed into

$$- \frac{1}{2\theta_o^{\omega+1}} \int_0^\eta \zeta\zeta' (\zeta'')^{-\epsilon} d\eta$$

and, with this, Equation (5.93) becomes

$$I_2 = b\ Pr\ \frac{1 - U^2}{2} + b\ Pr\ \epsilon\ I(\eta;\epsilon)\ , \qquad (5.98)$$

where

$$I(\eta;\epsilon) = -\frac{1}{2\theta_o^{\omega+1}} \int_\eta^\infty (\zeta'')^{1+\epsilon}\, d\eta \int_0^\eta \zeta\zeta'\,(\zeta'')^{-\epsilon}\, d\eta\ .$$

Since $\epsilon \ll 1$, we consider $I(\eta;0)$ in lieu of $I(\eta;\epsilon)$. Thus

$$I(\eta;0) = -\frac{1}{4}(1 - U^2) - \frac{1}{2}\ U' \int_0^\eta U d\eta \qquad (5.99)$$

and

$$I_2 = b\ \frac{1 - U^2}{2}\ Pr^{1/2} + \frac{b\ Pr(1-Pr)}{2}\, U' \int_0^\eta U d\eta,$$

in which the approximation $Pr^{-1/2} \simeq 1 - \epsilon/2$ has been made.

Inserting the foregoing results for I_1 and I_2 into Equation (5.85), we finally obtain

$$\theta(\eta) = 1 - \frac{\theta'_o}{\alpha}\left[1 - U\ (\eta\ Pr^{1/3})\right] Pr^{-1/3} + b\left[\frac{1-U^2}{2}\ Pr^{1/2} + \frac{Pr(1-Pr)}{2}\ U' \int_0^\eta U d\eta\right], \quad (5.100)$$

which is valid for Pr close to unity.

Equation (5.100) was first given by Meksyn in Reference 12 and cited again in Reference 13 without indicating, in either reference, the intervening steps in arriving at the result. The development given above is due to this writer and he should be responsible for any inaccuracies. Since several approximations are introduced in the derivation, it is thought worthwhile to have the details shown.

Several immediate conclusions may be drawn from Equation (5.100):

(a) For a gas of $Pr = 1$, Equation (5.100) simplifies to

$$\theta(\eta) = 1 - \frac{\theta'_o}{\alpha}(1 - U) + \frac{\gamma-1}{2}\, M_\infty^2\, (1 - U^2). \qquad (5.101)$$

Thence

$$\theta_o = 1 - \frac{\theta'_o}{\alpha} + \frac{\gamma-1}{2}\, M_\infty^2\ . \qquad (5.102)$$

It follows that

$$\theta(\eta) = \theta_o + \left[\frac{\gamma-1}{2}\, M_\infty^2 + (1 - \theta_o)\right] U - \frac{\gamma-1}{2}\, M_\infty^2\, U^2\ , \qquad (5.103)$$

which is the well-known Crocco's integral.

(b) Setting $\eta = 0$ in Equation (5.100) leads to

$$\frac{\theta'_o}{\alpha} = (1 - \theta_o)\ Pr^{1/3} + \frac{\gamma-1}{2}\, M_\infty^2\, Pr^{5/6}\ , \qquad (5.104)$$

which agrees well with Crocco's numerical result for $Pr = 0.725$.

(c) For adiabatic plate, $\theta'_o = 0$, we have

$$\theta(\eta) = 1 + \frac{\gamma-1}{2}\, M_\infty^2 \left[(1 - U^2)\ Pr^{1/2} + Pr(1-Pr)\ U' \int_0^\eta U d\eta\right], \quad (5.105)$$

which, upon putting $\eta = 0$, becomes

$$\theta_o = 1 + \frac{\gamma-1}{2} M_\infty^2 \, Pr^{1/2} . \tag{5.106}$$

The temperature assumed by the adiabatic plate is commonly called the recovery temperature, T_r, and the ratio

$$\frac{T_r - T_\infty}{U_\infty^2/2c_p}$$

is called the temperature recovery factor, r. Hence it follows from Equation (5.106) that

$$r = Pr^{1/2} , \tag{5.107}$$

which is a well-known result.

5.7.4 Drag Coefficient for the Adiabatic Plate

As stated in Section 5.7.2, the two unknowns of the problem are α and θ_o. The latter is given by Equation (5.106); it only remains to evaluate α.

Equation (5.77) can be integrated to give

$$\varphi U' = \varphi_o \alpha \, e^{-F(\eta)} , \tag{5.108}$$

where

$$F(\eta) = \frac{1}{2} \int_0^\eta \frac{\xi}{\theta^{\omega+1}} \, d\eta . \tag{5.109}$$

Dividing Equation (5.108) by φ and integrating from $\eta = 0$ to $\eta = \infty$, we obtain

$$1 = \varphi_o \alpha \int_0^\infty e^{-F} \varphi^{-1} \, d\eta, \tag{5.110}$$

which determines α.

Using the series expansion for ξ and θ, we find from Equation (5.109) that

$$F = \eta^3 \sum_{n=0}^\infty c_n \eta^n \equiv \tau , \tag{5.111}$$

with

$$c_o = \frac{\alpha}{12\theta_o^{\omega+1}} ,$$

$$c_1 = 0 ,$$

$$c_2 = \frac{(7\omega+3)\alpha^3\beta}{240\theta_o^{\omega+2}} ,$$

$$c_3 = - \frac{\alpha^2}{2880 \, \theta_o^{2(\omega+1)}} , \text{ etc.}$$

We transform Equation (5.110) to the τ variable and write it as

$$1 = \alpha \int_0^\infty e^{-\tau} \frac{\varphi_o}{\varphi} \frac{d\eta}{d\tau} \, d\tau , \tag{5.112}$$

where

$$\frac{\varphi_o}{\varphi} \frac{d\eta}{d\tau} = \tau^{-2/3} \sum_{m=0}^\infty d_m \tau^{m/3} . \tag{5.113}$$

The first four coefficients in the series expansion of Equation (5.113) are

$$d_o = \frac{1}{3} \left(\frac{12\theta_o^{\omega+1}}{\alpha} \right)^{1/3} ,$$

$$d_1 = 0 ,$$

$$d_2 = - \frac{3}{5} (1-\omega) \alpha \beta \theta_o^\omega ,$$

and

$$d_3 = \frac{1}{45} \left(\frac{12\theta_o^{\omega+1}}{\alpha} \right)^{1/3} .$$

Substituting the foregoing into Equation (5.112) and integrating, we find, after some rearrangement,

$$1 = \frac{\Gamma\left(\frac{1}{3}\right)}{3} \left(1 + \frac{1}{45} + \ldots \right) X - \frac{1}{20} \frac{(1-\omega)\beta}{\theta_o} X^3 + \ldots , \tag{5.114}$$

where

$$X = \left(12\theta_o^{\omega+1}\ \alpha^2\right)^{1/3} . \qquad (5.115)$$

Meksyn solved Equation (5.114) by successive approximations. He pointed out that if only the first term on the right side of Equation (5.114) were retained, the result became Blasius' solution. It is

$$X^{-1} = \frac{\Gamma(1/3)}{3}\left(1 + \frac{1}{45} + \ldots\right) , \qquad (5.116)$$

which is identical to Equation (5.33) if $\theta_o = 1$, i.e., $T_w = T_\infty$, and if proper account is taken for the difference in the definition of η.

A second approximation is obtained by inserting the value of X given by Equation (5.116) into the term involving X^3 in Equation (5.114). The result is

$$X = 1.096\left[1 + \frac{(1.096)^3}{20}\ \frac{\beta(1-\omega)}{\theta_o}\right] \qquad (5.117)$$

or,

$$\alpha = 0.331\left[1 + \frac{0.0986\beta(1-\omega)}{\theta_o}\right]\frac{1}{\theta_o^{(\omega+1)/2}} . \qquad (5.118)^*$$

The local wall shear is

$$\tau_w = \left(\mu\frac{\partial u}{\partial y}\right)_o = \mu_\infty\ \theta_o^\omega\ \alpha\ U_\infty\sqrt{\frac{U_\infty}{\nu_\infty x}} \qquad (5.119)$$

and thence the average drag coefficient for one side of the plate over length, L, is

$$C_D = \frac{4\theta_o^\omega\ \alpha}{\sqrt{Re_L}} , \qquad (5.120)$$

where

$$Re_L = \frac{U_\infty L}{\nu_\infty} .$$

Using Equation (5.118) for α, we find

$$C_D\sqrt{Re_L} = 1.324\left[1 + \frac{0.0986\ \beta(1-\omega)}{\theta_o}\right]\theta_o^{(\omega-1)/2} . \qquad (5.121)^*$$

Meksyn demonstrated that results calculated from Equation (5.121) showed very good agreement with those of Crocco.

Meksyn also analyzed the case with heat transfer. The calculation becomes somewhat laborious although the procedure is unchanged and remains simple. The corresponding integral in Equation (5.110) leads to a divergent expression and use is made of Euler's transformation for its evaluation. Interested readers should consult Reference 13 for details.

5.8 CONCLUDING REMARKS

Meksyn's asymptotic method of integration leads to very satisfactory results in all examples considered in this section. It has also been successfully applied to the calculation of free convection flows over a heated vertical wall, as well as flows involving mixed free and forced convection.(15) Speaking in general terms, his method usually works well for incompressible, similar flows. However, if the physical nature of the problem excludes the existence of similar solutions and if the boundary conditions are such that the derivatives of the unknown parameter a with respect to $x(\sim\xi)$ coordinate occurs in leading terms of the series (5.36), the evaluation of the unknown parameter a becomes a major problem, and this writer is unaware of any effective means for coping with such a situation.

For compressible flow, the momentum and the energy equation are, in general, coupled. Even for the relatively simple case for which similar solutions exist,

---

*There is a misprint in Meksyn's equation (3.6) of Reference 13; the quantity $\theta_o^{\omega+1}$ should be replaced by $\theta_o^{(\omega+1)/2}$. Also, in his equation (3.7), $\theta_o^{\omega-(1/2)}$ should read $\theta_o^{(\omega-1)/2}$.

Meksyn's procedure leads to the requirement that two series of the form of Equation (5.36) need be simultaneously satisfied. The two unknowns, for example, U'(0) and θ'(0), would appear in the coefficients d_m of both series. The evaluation of these unknowns could be met with difficulties. Conceivably, one could use the method of successive approximations. However, the procedure may be slowly converging and thus is not useful. For the flat plate in supersonic flow considered by Meksyn and described in Section 5.7, it is only fortunate that the energy equation can be approximately integrated and thus one of the unknowns θ'(0) (or, θ(0) for the adiabatic plate) can be evaluated independent of the other. This is, indeed, a very special case and can only be regarded as an exception.

Meksyn's asymptotic method of integration has, in this writer's opinion, opened a new avenue in analyzing boundary layer flow. However, it seems also clear, from the discussion presented above, that the procedure is not entirely free from difficulty. Considerable ingenuity is often needed to improve the approximations. To broaden its application, more research and development appear necessary.

REFERENCES

1. Meksyn, D. New Methods in Laminar Boundary-Layer Theory. New York: Pergamon Press, 1961.

2. Piercy, N. A. V., J. H. Preston, and L. G. Whitehead, "The Approximate Prediction of Skin-Friction and Lift," Philosophical Magazine, 26 (1938), p. 791.

3. Meksyn, D., "The Boundary Layer Equations of Compressible Flow. Separation," Zeitschrift für augewandte Mathematik und Mechanik, 38 (1958), p. 372.

4. Sim, A. C., "A Generalization of Reversion Formulae with Their Application to Non-Linear Differential Equations," Philosophical Magazine, 42 (1951), p. 228.

5. Hardy, G. H. Divergent Series. Oxford, England: Oxford University Press, 1949, Ch. 8.

6. Hartree, D. R., "On an Equation Occurring in Falkner and Skan's Approximate Treatment of the Equations of the Boundary Layer," Proceedings, Cambridge Philosophical Society, 33 (1937), p. 223.

7. Sibulkin, M., "Heat Transfer Near the Forward Stagnation Point of a Body of Revolution," Journal of the Aeronautical Sciences, 19 (1952), p. 570.

8. Chao, B. T. and D. R. Jeng, "Unsteady Stagnation Point Heat Transfer," Journal of Heat Transfer, 87 (1965), p. 221.

9. Meksyn, D., "Integration of the Boundary Layer Equations for a Plane in a Compressible Fluid," Proceedings, Royal Society (London), A, 195 (1948), p. 180.

10. Emmons, H. W. and J. G. Brainerd, "Temperature Effects in a Laminar Compressible Fluid Boundary Layer Along a Flat Plate," Journal of Applied Mechanics, A-105, 8 (1941).

11. Brainerd, J. G. and H. W. Emmons, "Effect of Variable Viscosity on Boundary Layers, with a Discussion of Drag Measurement," Journal of Applied Mechanics, A-1, 9 (1942).

12. Meksyn, D., "Integration of the Boundary Layer Equations for a Plane in Compressible Flow with Heat Transfer," Proceedings, Royal Society (London), A, 231 (1955), p. 274.

13. ———, "Supersonic Flow Past a Semi-Infinite Plane," Zeitschrift für augewandte Mathematik und Physik, 16 (1965), p. 344.

14. Kuerti, G., "The Laminar Boundary Layer in Compressible Flow," Advances in Applied Mechanics, 2 (1951), p. 21.

15. Brindley, J., "An Approximate Technique for Natural Convection in a Boundary Layer," International Journal of Heat and Mass Transfer, 6 (1963), p. 1035.

CONTENTS

SELECTED TOPICS ON RADIATIVE HEAT TRANSFER
by R. G. Hering

1. FUNDAMENTAL CONCEPTS OF RADIATIVE TRANSFER 75

1.1 Introduction. 75
1.2 Spectral Intensity of Radiation 75
1.3 Radiation Flux, Energy, and Stress Tensor 76
1.4 Interaction of Radiation with Matter. 80
1.5 Equation of Transfer. 83
1.6 Conservation of Radiant Energy. 85

2. ELECTROMAGNETIC THEORY OF RADIATION. 89

2.1 Introduction. 89
2.2 Fundamentals of Electrodynamics 89
2.3 Plane Electromagnetic Waves in Unbounded Media. 96
2.4 Reflection and Refraction at a Plane Surface. 101
2.5 Reflectivity and Emissivity 105

3. RADIATIVE TRANSFER BETWEEN SURFACES. 121

3.1 Introduction. 121
3.2 Radiation Properties of Surfaces. 121
3.3 Radiant Transfer in an Enclosure. 124
3.4 Radiant Transfer in an Enclosure of Specular-Diffuse Surfaces 129

4. SIMULTANEOUS CONDUCTION, CONVECTION, AND RADIATION 142

4.1 Introduction. 142
4.2 One-dimensional Radiation Transfer. 142
4.3 Approximate Radiation Flux Expressions. 145
4.4 Radiative Equilibrium 146
4.5 Combined Conduction and Radiation 148
4.6 Combined Conduction, Convection, and Radiation. 149

SELECTED TOPICS ON RADIATIVE HEAT TRANSFER *

R. G. Hering**

1. FUNDAMENTAL CONCEPTS OF RADIATIVE TRANSFER

1.1 INTRODUCTION

Certain basic definitions and fundamental concepts are necessary to provide a foundation for the analysis of energy transport by thermal radiation. These are presented and discussed here using a macroscopic point of view. From this viewpoint, the basic quantity describing radiation phenomena is the spectral radiant intensity. Using this concept, the radiation heat flux, radiation energy density, and radiation stress tensor follow directly. Certain coefficients are then introduced to describe the interaction of radiation with matter. In keeping with the macroscopic viewpoint, these coefficients are assumed to be specified or independently determined. Generally, the evaluation of these coefficients requires a microscopic approach employing classical or quantum mechanics. With the interaction quantities defined, the equation governing the variation of intensity within a participating medium is formulated. The discussion concludes with the conservation equation for radiant energy.

Consideration in the development to follow is limited to systems within which local thermodynamic equilibrium may be assumed to exist. All material velocities are taken as small relative to the speed of light so that relativistic effects may be ignored. The effects of polarization are neglected. Only coherent scattering is considered; that is, no account is made for changes in frequency of the radiation due to scattering.

1.2 SPECTRAL INTENSITY OF RADIATION

The analysis of a radiation field requires consideration of the transport of energy. For this purpose the ray concept of geometrical optics is inadequate, and it is customary to introduce the concept of spectral radiant intensity.(1)*** This fundamental quantity may be defined with the aid of Figure 1.1 as follows. Let $\Delta^4 \mathcal{E}_\lambda$ be the radiant energy in the wavelength interval $(\lambda, \lambda + \Delta\lambda)$ transported during the time interval, Δt, across an arbitrarily oriented area element, ΔA, located at point $\vec{r}$, and confined to the elemental solid angle, $\Delta\omega$, about the direction defined by the unit vector, $\vec{s}$. The apex of the elementary cone is at $\vec{r}$ and the unit normal $\vec{n}$ to the surface makes an angle θ with the direction $\vec{s}$. The <u>spectral radiant intensity</u>

---

*The four sections of this paper were presented as separate lectures at the conference.

**Professor of Mechanical Engineering, University of Illinois, Urbana, Illinois.

***Superscript numbers in parentheses refer to References at the end of each section.

at $\vec{r}$ in direction $\vec{s}$ is defined by the following limit:

$$I_\lambda(\vec{r},\vec{s},t) = \lim_{\substack{\Delta t,\Delta A \\ \Delta\omega,\Delta\lambda}\} \to 0} \left| \frac{\Delta^4 \mathcal{E}_\lambda}{\cos\theta\ \Delta A \Delta\omega \Delta\lambda \Delta t} \right| = \frac{d^4 \mathcal{E}_\lambda}{\cos\theta\ dA d\omega d\lambda dt} . \tag{1.1}$$

As a consequence of this definition, the spectral radiant intensity may be interpreted as the instantaneous rate at which radiant energy within the spectral interval $(\lambda, \lambda + d\lambda)$ is transported at $\vec{r}$ in the direction $\vec{s}$ per unit solid angle, per unit wavelength interval, and per unit area normal to the $\vec{s}$-direction. In a medium which absorbs, emits, and scatters radiation, the spectral intensity is, in general, a function of position, $\vec{r}$, direction, $\vec{s}$, wavelength, λ, and time, t. A radiation field is said to be isotropic at a point if the intensity is independent of direction at that point. If, in addition, the intensity at any given instant is everywhere identical in the domain under consideration, the radiation field is called homogeneous and isotropic.

For certain applications, it is useful to express the spectral intensity in terms of two components. The first, I_λ^+, corresponds to direction $\vec{s}$ such that $\vec{n}\cdot\vec{s} > 0$; while the other, I_λ^-, corresponds to directions for which $\vec{n}\cdot\vec{s} < 0$. The direction $\vec{s}$ may be specified by the angles θ and φ of a local spherical polar coordinate system at $\vec{r}$ with polar axis aligned along $\vec{n}$. The angle θ is the polar angle measured relative to the surface normal $\vec{n}$, and φ is the azimuthal angle measured relative to some fixed line in the plane of dA. In terms of this coordinate system, I_λ^+ corresponds to polar angles θ between 0 and $\pi/2$ and I_λ^- to θ values between $\pi/2$ and π.

$$I_\lambda(\vec{r},\vec{s},t) = \begin{cases} I_\lambda^+(\vec{r},\theta,\varphi,t), & 0 \le \theta \le \frac{\pi}{2} \\ I_\lambda^-(\vec{r},\theta,\varphi,t), & \frac{\pi}{2} \le \theta \le \pi \end{cases} . \tag{1.2}$$

Frequently, the intensity is based on frequency rather than wavelength. The advantage of a frequency definition is that, unlike the wavelength, the frequency remains constant when radiation passes from one medium into another. Although the frequency definition is not employed here, the intensity based on frequency is related to that based on wavelength by the following relation:

$$I_\nu = I_\lambda \left| \frac{d\lambda}{d\nu} \right| = \frac{c}{\nu^2} I_{\lambda = c/\nu} . \tag{1.3}$$

In the above ν is the frequency corresponding to wavelength $\lambda (= c/\nu)$ and c is the speed of light in the medium.

Often interest lies in the radiant energy transported over all wavelengths. The corresponding intensity shall be differentiated from the spectral intensity by the absence of the subscript λ, and simply called the intensity or total intensity. The intensity is given by the integral of the spectral intensity over all wavelengths.

$$I(\vec{r},\vec{s},t) = \int_0^\infty I_\lambda(\vec{r},\vec{s},t)\ d\lambda . \tag{1.4}$$

1.3 RADIATION FLUX, ENERGY, AND STRESS TENSOR

With the foregoing definition of intensity, three quantities of substantial interest may be defined. These are the radiation flux, radiation energy density, and the radiation stress tensor.

1.3.1 Spectral Radiant Flux and Flux Vector

It follows from the definition of spectral intensity that the radiant energy in the wavelength interval $(\lambda, \lambda + d\lambda)$ traversing an element of surface area dA per unit time and within $d\omega$ about the direction $\vec{s}$ is

$$\frac{d^4\mathcal{E}_\lambda}{dt} = I_\lambda \cos\theta\ dA d\lambda d\omega \tag{1.5}$$

Hence the monochromatic energy crossing the surface per unit area and per unit time within $d\omega$ about the direction $\vec{s}$ is

$$I_\lambda \cos\theta \, d\omega. \quad (1.6)$$

The net amount of monochromatic energy passing through dA per unit time and per unit area is called the spectral radiant flux, q^r_λ, and is given by

$$q^r_\lambda = \int_{\omega=4\pi} I_\lambda \cos\theta \, d\omega \,. \quad (1.7)$$

The solid angle, $d\omega$, may be expressed in terms of a local polar coordinate system at $\vec{r}$ as

$$d\omega = \sin\theta \, d\theta d\varphi \,. \quad (1.8)$$

With this, q^r_λ is written as

$$q^r_\lambda = \int_0^{2\pi} \int_0^{\pi} I_\lambda \cos\theta \sin\theta \, d\theta d\varphi. \quad (1.9)$$

It follows that the spectral radiant flux is zero in an isotropic radiation field.

Introducing the intensity components I^+_λ and I^-_λ, q^r_λ may be expressed as

$$q^r_\lambda = q^{r+}_\lambda - q^{r-}_\lambda \,, \quad (1.10)$$

where

$$\left.\begin{aligned} q^{r+}_\lambda &= \int_0^{2\pi} \int_0^{\pi/2} I^+_\lambda \cos\theta \sin\theta \, d\theta d\varphi \,, \\ q^{r-}_\lambda &= - \int_0^{2\pi} \int_{\pi/2}^{\pi} I^-_\lambda \cos\theta \sin\theta \, d\theta d\varphi. \end{aligned}\right\} \quad (1.11)$$

The one-sided fluxes q^{r+}_λ and q^{r-}_λ represent the flux in the direction $\vec{n}$ and opposite to $\vec{n}$, respectively. The minus sign is included in q^{r-}_λ to account for the fact that by definition both one-sided fluxes are positive quantities.

It is evident from the expression for q^r_λ of Equation (1.7) that this quantity depends on the orientation of the element at the selected point. A more general concept eliminating the need for specifying the orientation may be introduced by noting that $\cos\theta$ is simply the scalar product of the unit normal to dA and the direction $\vec{s}$.

$$\cos\theta = \vec{n} \cdot \vec{s} \,. \quad (1.12)$$

Then

$$q^r_\lambda = \int_{\omega=4\pi} I_\lambda \cos\theta \, d\omega = \int_{4\pi} I_\lambda \vec{n} \cdot \vec{s} d\omega = \vec{n} \cdot \left[\int_{4\pi} I_\lambda \vec{s} d\omega \right] . \quad (1.13)$$

The vector quantity within the bracket is known as the spectral radiant flux vector, $\vec{q}^{\,r}_\lambda$.

$$\vec{q}^{\,r}_\lambda = \int_{4\pi} I_\lambda \vec{s} d\omega \,. \quad (1.14)$$

In terms of indicial notation, the component of the flux vector along the x_i-axis, say $q^r_{\lambda,i}$, is

$$q^r_{\lambda,i} = \int_{4\pi} I_\lambda \ell_i d\omega, \quad (i = 1,2,3) \quad (1.15)$$

where ℓ_i denotes the direction cosine of the direction $\vec{s}$ with respect to the x_i-axis. Once having evaluated the spectral radiant flux vector at $\vec{r}$, the net radiant flux across a unit area surface element with normal $\vec{n}$ (direction cosines n_i) is

$$q_\lambda^r = \vec{n}\cdot\vec{q}_\lambda^r = n_i q_{\lambda,i}^r = \int_{4\pi} I_\lambda n_i \ell_i d\omega \ . \quad (1.16)$$

The summation convention is implied in the above.

The total radiation flux quantities are simply the integral of the spectral values over all wavelengths. To differentiate these from their spectral counterparts, the subscript λ is suppressed. Thus, for example, the radiation flux vector is

$$\vec{q}^r = \int_0^\infty \vec{q}_\lambda^r \, d\lambda \ . \quad (1.17)$$

It may be noted that $\vec{q}^r$ gives the net flux of radiant energy across a surface element in a fashion completely analogous to that of the heat flux vector in conductive heat transfer.

1.3.2 Radiation Energy Density

Although radiation is propagated at a very large velocity -- the speed of light -- the velocity of propagation is finite. Therefore a finite volume must contain a finite amount of energy in the form of radiation. The monochromatic radiation energy density at a given point, u_λ^r, is the amount of radiant energy within the spectral interval $(\lambda, \lambda + d\lambda)$ in transit in the neighborhood of the point per unit volume and per unit wavelength interval. The relationship between energy density and intensity may be obtained[(2)] by considering an infinitesimal volume, υ, with convex surface area, a, enclosing an arbitrary point, P, in a radiation field (Figure 1.2). About υ construct a locally concave enveloping surface of area A whose linear dimensions are large in comparison to those of υ. Now the radiation traversing υ must previously have crossed some element of A. Consider a beam crossing the element dA of A and passing through the element da of a. Let θ_A and θ_a denote the angles which unit normals to dA and da subtend with the line joining the area elements. The radiant energy in the wavelength interval $(\lambda, \lambda + d\lambda)$ flowing across dA and passing through da in time dt is

$$I_\lambda \cos\theta_A \, dA d\omega d\lambda dt = I_\lambda \cos\theta_A \, dA \left(\frac{\cos\theta_a \, da}{L^2}\right) d\lambda dt \ . \quad (1.18)$$

This energy remains in υ for the time required to traverse the length ℓ, that is, $dt = \ell/c$, where c is the velocity of light in the medium. The contribution of this beam to the spectral radiant energy contained in υ may be written as

$$\frac{I_\lambda}{c}\left(\frac{dA\cos\theta_A}{L^2}\right)(\ell \, da \cos\theta_a) d\lambda = \frac{I_\lambda}{c} d\Omega d\upsilon d\lambda \ , \quad (1.19)$$

since $(\ell \, da \cos\theta_a)$ and $(dA\cos\theta_A/L^2)$ are just the element of volume $d\upsilon$ of υ and the solid angle subtended by dA at P, respectively. To account for the contributions of all elements of A to the energy in υ requires integration of Equation (1.19) over υ and Ω.

$$\frac{d\lambda}{c}\int_\upsilon \int_{\Omega=4\pi} I_\lambda d\Omega d\upsilon = \frac{\upsilon d\lambda}{c}\int_{\Omega=4\pi} I_\lambda d\Omega . \quad (1.20)$$

Hence the <u>monochromatic radiant energy density</u> follows as

$$u_\lambda^r = \frac{1}{c}\int_{4\pi} I_\lambda d\Omega \ . \quad (1.21)$$

For an isotropic radiation field with intensity I_λ^o, the monochromatic energy density, $u_\lambda^{r,o}$, is

$$u_\lambda^{r,o} = \frac{4\pi}{c} I_\lambda^o \ . \tag{1.22}$$

1.3.3 Radiation Pressure and the Radiation Stress Tensor

The existence of radiation pressure follows from thermodynamic reasoning as well as the wave and quantum theories of radiation. According to the latter, a quantum of energy, $h\nu$, is associated with a momentum transfer, $h\nu/c$, in the direction of propagation, where h is Planck's constant. Consequently, radiant energy in the amount $\mathcal{E}$ traversing a medium in a specific direction results in a momentum transfer, $\mathcal{E}/c$, in that direction. It therefore follows that there is a stress field associated with a radiation field.

Consider a pencil of radiation with vertex at point $\vec{r}$ of an arbitrarily oriented element of surface dA in a radiation field. Let θ denote the angle between the surface normal to dA and the direction of propagation of the considered pencil. If $d\omega$ is the solid angle of the beam, the rate at which radiant energy in the wavelength interval $(\lambda, \lambda + d\lambda)$ traverses dA is

$$I_\lambda \cos\theta \, d\omega d\lambda dA \ . \tag{1.23}$$

Associated with this energy transfer is the momentum transfer

$$\frac{I_\lambda \cos\theta \, d\omega d\lambda dA}{c} \tag{1.24}$$

in the direction of propagation. The transport of the normal component of the momentum across dA is

$$\frac{I_\lambda \cos^2\theta \, d\omega d\lambda dA}{c} \ . \tag{1.25}$$

Accounting for all wavelengths and directions of traversal yields the net rate of transport of normal momentum across dA as

$$\frac{dA}{c} \int_{4\pi} I \cos^2\theta \, d\omega \tag{1.26}$$

or a radiation pressure, p^r, as

$$p^r = \frac{1}{c} \int_{4\pi} I \cos^2\theta \, d\omega \ . \tag{1.27}$$

The foregoing considerations may be extended to yield the radiation stress tensor. For this purpose consider a unit area surface element at $\vec{r}$ with its normal along the x_i-axis. Let the intensity of radiation at $\vec{r}$ in direction $\vec{s}$ with direction cosines ℓ_i (i = 1,2,3) be I. Then

$$\frac{I \, \ell_i d\omega}{c} \ , \qquad (i = 1,2,3) \tag{1.28}$$

is the momentum transported across the unit area by the radiation in the direction of propagation. The components of this momentum in the x_j-directions (j = 1,2,3) are

$$\frac{I \, \ell_i \ell_j d\omega}{c} \ , \qquad (i,j = 1,2,3) \ . \tag{1.29}$$

Therefore the net flux of the x_j directed component of radiant momentum across a surface whose normal is in the x_i-coordinate direction is

$$\frac{1}{c} \int_{4\pi} I \, \ell_i \ell_j d\omega \ . \tag{1.30}$$

A radiation stress tensor, P_{ij}^r, may be defined as

$$P_{ij}^r = \frac{1}{c} \int_{4\pi} I \, \ell_i \ell_j d\omega \ , \quad (i,j = 1,2,3). \tag{1.31}$$

It may be noted that P_{ij}^r is symmetric ($P_{ij}^r = P_{ji}^r$) and each diagonal term P_{11}^r, P_{22}^r, and P_{33}^r is the radiation pressure on a surface normal to the x_1-, x_2-, and x_3-axis, respectively. The sum of the diagonal terms is

$$P_{ii}^r = P_{11}^r + P_{22}^r + P_{33}^r = \frac{1}{c} \int_{4\pi} I \, \ell_i \ell_i d\omega = \frac{1}{c} \int_{4\pi} I d\omega = u^r , \tag{1.32}$$

since the sum of the squares of the direction cosines is unity.

For an isotropic radiation field, the intensity is independent of direction, and then

$$P_{11}^{r,o} = P_{22}^{r,o} = P_{33}^{r,o} = \frac{u^{r,o}}{3}$$

$$P_{ij}^{r,o} = 0 \qquad (i \neq j),$$

which may be more compactly written with indicial notation as

$$P_{ij}^{r,o} = \frac{u^{r,o}}{3}\,\delta_{ij}\,, \tag{1.33}$$

where δ_{ij} is the Kronecker delta which has the value 1 when i = j and zero otherwise.

The radiation stress tensor may be put in a form which is analogous to the mechanical stress tensor in deformable media.(3) For this purpose the tensor P_{ij}^{r} is divided into an isotropic part and an anisotropic part corresponding to a subdivision of the intensity into such elements. Let

$$I = I' + I''\,, \tag{1.34}$$

where the isotropic part I' is the average of I over all directions.

$$4\pi I' = \int_{4\pi} I\, d\omega\ .$$

Then the isotropic part of the radiation stress tensor is

$$\frac{1}{c}\int_{4\pi} I'\,\ell_i \ell_j\, d\omega = \frac{I'}{c}\int_{4\pi} \ell_i \ell_j\, d\omega = \frac{I'}{c}\,\frac{4\pi}{3}\,\delta_{ij} = \frac{\delta_{ij}}{3c}\int_{4\pi} I\, d\omega = \frac{u^r}{3}\,\delta_{ij} = \bar{p}^r\,\delta_{ij}\ ,$$

and the anisotropic part is given by

$$\frac{1}{c}\int_{4\pi} I''\ell_i \ell_j\, d\omega = -\frac{1}{c}\int_{4\pi} (I' - I)\,\ell_i \ell_j\, d\omega = -\,\tau_{ij}^r\ .$$

Thus the radiation pressure tensor can be written as

$$\pi_{ij}^r = -\,P_{ij}^r = -\,\bar{p}^r\,\delta_{ij} + \tau_{ij}^r\,, \tag{1.35}$$

and the analogy is therefore complete. The minus sign has been introduced to conform to the customary notation in fluid dynamics.

1.4 INTERACTION OF RADIATION WITH MATTER

It is customary to introduce certain coefficients which describe on a macroscopic basis the interaction of radiation with matter. Consideration here is limited to those necessary to discuss the propagation of radiation within a medium which absorbs, scatters, and emits thermal radiation. The analogous quantities necessary to describe the interaction of radiation with surfaces are introduced in Section 3.2.

1.4.1 Absorption Coefficient

The energy of a beam of radiation is generally reduced as it propagates through a material by the process of absorption. By absorption reference is made to the transformation of radiant energy into other forms of energy within the material. This process is described by the volumetric absorption coefficient, which may be defined as follows. Consider a monochromatic beam of thermal radiation with spectral intensity, I_λ, traveling in the $\vec{s}$-direction. In traveling a distance ds along its path, it will undergo an attenuation due to absorption in the intervening medium. This absorption is taken proportional to the distance traversed and the magnitude of the intensity such that the change in intensity due to absorption, $dI_{\lambda,a}$, is

$$dI_{\lambda,a} = -\kappa_\lambda I_\lambda ds, \qquad (1.36)$$

where κ_λ is the volumetric absorption coefficient for radiation of wavelength, λ. As a consequence of this definition, the energy lost by absorption from a pencil of monochromatic radiation incident on an elemental volume of matter with cross section dA and length ds per unit time is

$$\frac{d^4\mathcal{E}_{\lambda,a}}{dt d\lambda} = (-dI_{\lambda,a}) \cos\theta \, dAd\omega = \kappa_\lambda I_\lambda (dA \cos\theta \, ds) d\omega = \kappa_\lambda I_\lambda dVd\omega \, . \qquad (1.37)$$

Thus $\kappa_\lambda I_\lambda$ represents the monochromatic energy absorbed by the medium from the considered beam per unit time, per unit volume, and per unit solid angle. The absorption coefficient is usually taken independent of the direction of the incident beam. Under such conditions, the monochromatic radiation absorbed per unit time and volume is

$$\kappa_\lambda \int_{4\pi} I_\lambda d\omega \, . \qquad (1.38)$$

Sometimes a mass absorption coefficient rather than a volume absorption coefficient is used. The mass absorption coefficient, $\kappa_{\lambda,m}$, is related to the volumetric coefficient by the relation

$$\kappa_\lambda = \rho_m \kappa_{\lambda,m} \quad , \qquad (1.39)$$

where ρ_m is the mass density of the material.

1.4.2 Scattering Coefficient and Scattering Function

As a monochromatic beam of radiation traverses a medium, it may also be weakened as a result of a redistribution of its energy in wavelength and space. When no changes occur in wavelength of the redirected or scattered energy, the process is referred to as coherent scattering. Consideration here is limited to coherent scattering. Note that unlike the absorption process, the energy scattered out of a beam is not lost to the radiation field but appears in beams traveling in other directions. The scattering phenomena is characterized by a monochromatic volumetric scattering coefficient, σ_λ, defined by

$$dI_{\lambda,s} = -\sigma_\lambda I_\lambda ds \, , \qquad (1.40)$$

where $dI_{\lambda,s}$ denotes the change in spectral intensity attributed to scattering when a beam of intensity, I_λ, travels a distance, ds, along its path. Consequently, the monochromatic radiation lost by scattering from a pencil incident on a volume of matter with cross section dA and length ds per unit time is

$$\frac{d^4\mathcal{E}_{\lambda,s^-}}{dt d\lambda} = (-dI_{\lambda,s}) \cos\theta \, dAd\omega = \sigma_\lambda I_\lambda (dA \cos\theta \, ds) d\omega = \sigma_\lambda I_\lambda dVd\omega \, . \qquad (1.41)$$

Thus $\sigma_\lambda I_\lambda$ represents the monochromatic energy scattered out of a single beam per unit time, per unit volume, and per unit solid angle. When σ_λ is independent of direction, the monochromatic energy scattered per unit time and per unit volume is

$$\sigma_\lambda \int_{4\pi} I_\lambda d\omega \, . \qquad (1.42)$$

Since the scattered energy of a beam reappears in other directions, the energy of a beam traversing in direction $\vec{s}$ is enhanced by the scattering of radiation from other directions, $\vec{s}'$, into direction $\vec{s}$. To describe this phenomena, we introduce the scattering or phase function, $p_\lambda(\vec{s}' \rightarrow \vec{s})$, defined such that

$$\left[\frac{p_\lambda\,(\vec{s}\,' \rightarrow \vec{s})\,d\omega}{4\pi}\right]\left[\sigma_\lambda\, I_\lambda\,(\vec{s}\,')\,d\omega'\,dV\right] \qquad (1.43)$$

is the fraction of the monochromatic radiation incident on dV from within $d\omega'$ about the $\vec{s}\,'$-direction, which is scattered into $d\omega$ about the $\vec{s}$-direction. Since the total amount of monochromatic energy scattered out of the incident beam is $\sigma_\lambda I_\lambda(\vec{s}\,')\,d\omega'\,dV$, the integral over all scattering directions $\vec{s}$ of

$$\frac{p_\lambda\,(\vec{s}\,' \rightarrow \vec{s})\,d\omega}{4\pi}$$

must be unity.

$$\frac{1}{4\pi}\int_{\omega=4\pi} p_\lambda(\vec{s}\,' \rightarrow \vec{s})\,d\omega = 1. \qquad (1.44)$$

For an isotropically scattering medium, the phase function has the value unity.

According to Equation (1.43), the increase in the monochromatic energy of a beam propagating in the $\vec{s}$-direction and confined to the elemental solid angle, $d\omega$, due to scattering of radiation from all directions per unit time is

$$\frac{d^4\mathcal{E}_{\lambda,s+}}{dt\,d\lambda} = \left[\frac{\sigma_\lambda}{4\pi}\int_{\omega'=4\pi} p_\lambda(\vec{s}\,' \rightarrow \vec{s})\,I_\lambda(\vec{s}\,')\,d\omega'\right] dV\,d\omega\ . \qquad (1.45)$$

A scattering coefficient defined on a unit mass basis, $\sigma_{\lambda,m}$, is related to that defined here by the relation

$$\sigma_\lambda = \rho_m \sigma_{\lambda,m}\ .$$

1.4.3 Extinction Coefficient

The attenuation of a beam of radiation by both absorption and scattering is often represented by the monochromatic volumetric extinction coefficient, β_λ, defined as

$$\beta_\lambda = \kappa_\lambda + \sigma_\lambda\ .$$

A similarly defined mass extinction coefficient, $\beta_{\lambda,m}$, may be used through the definition

$$\beta_\lambda = \rho_m\,\beta_{\lambda,m}\ .$$

1.4.4 Emission Coefficient

The process of emission is the creation of thermal radiation at the expense of other forms of energy. The energy emitted per unit time by a substance of volume, dV, into the solid angle, $d\omega$, about the direction $\vec{s}$ between wavelengths $(\lambda,\ \lambda + d\lambda)$ per unit wavelength interval is

$$\frac{d^4\mathcal{E}_{\lambda,e}}{dt\,d\lambda} = \eta_\lambda\ dV\,d\omega\ . \qquad (1.46)$$

The proportionality factor, η_λ, is the monochromatic volumetric emission coefficient and represents the monochromatic thermal radiation emitted by matter per unit time, volume, and solid angle. Often a mass emission coefficient, j_λ, is used. This coefficient is related to the volumetric emission coefficient by $\eta_\lambda = \rho_m j_\lambda$.

Assuming local thermodynamic equilibrium, Kirchhoff's law may be used to relate the emission coefficient to the absorption coefficient.

$$\eta_\lambda = \kappa_\lambda\, I_{b,\lambda}(T) = \frac{\kappa_\lambda\, e_{b,\lambda}(T)}{\pi}\ . \qquad (1.47)$$

In the above, $I_{b,\lambda}(T)$ and $e_{b,\lambda}(T)$ are the local black body spectral intensity and spectral emissive power, respectively, given by Planck's law as

$$e_{b,\lambda}(T) = \pi\, I_{b,\lambda}(T)$$

$$= \frac{c_1}{n^2\lambda^5\left(e^{c_2/n\lambda T} - 1\right)}\ , \qquad (1.48)$$

where c_1 and c_2 are the first and second Planck radiation constants, n is the refractive index, and T the absolute temperature.

Since thermal emission is independent of direction, the local monochromatic emission per unit time and volume is

$$\frac{d^3\mathcal{E}_{\lambda,e}}{dt d\lambda dV} = \int_{4\pi} \eta_\lambda d\omega = \frac{\kappa_\lambda e_{b,\lambda}}{\pi} \int_{4\pi} d\omega = 4\kappa_\lambda e_{b,\lambda} \, . \tag{1.49}$$

1.5 EQUATION OF TRANSFER

The equation of transfer expresses the variation of the spectral radiant intensity within a material in terms of the interaction coefficients. It may be derived by writing a radiant energy balance about an arbitrary volume fixed in space. For this purpose, consider the volume, V, with its enclosing surface, A, within a medium capable of absorbing, emitting, and scattering thermal radiation (Figure 1.3). According to Equation (1.21), the quantity

$$\frac{I_\lambda}{c} d\omega d\lambda \tag{1.50}$$

may be interpreted as the radiant energy per unit volume with wavelengths between λ and $\lambda + d\lambda$ which is confined to the differential solid angle $d\omega$ about the $\vec{s}$-direction. Then the time rate of increase of this energy contained in the volume, V, is

$$\frac{\partial}{\partial t} \int_V \frac{I_\lambda}{c} dV d\omega d\lambda \, . \tag{1.51}$$

This change in energy content within V is attributed to the net inflow of such energy across the enclosing surface, A, to the augmentation of the energy as a consequence of emission and the scattering of energy from other directions into the $\vec{s}$-direction, and to the depletion resulting from absorption and scattering out of the selected direction.

The net transport of the considered energy across the enclosing surface may be evaluated as follows. In time, dt, the energy transported across a typical surface area element, dA, with outdrawn normal $\vec{n}$ is that contained in a cylinder with slant height $c\vec{s}dt$ and base dA. Then on a unit time basis, the radiant energy within the wavelength interval λ to $\lambda + d\lambda$ and confined to $d\omega$ about the $\vec{s}$-direction which is transported into V across dA is

$$- \frac{(I_\lambda/c) \; c\vec{s} \cdot \vec{n} \; dA dt d\omega d\lambda}{dt} \, . \tag{1.52}$$

Accounting for all elements of A yields

$$- \int_A I_\lambda \vec{s} \cdot \vec{n} \; dA d\omega d\lambda = - \int_V \vec{s} \cdot \nabla I_\lambda dV d\omega d\lambda \, , \tag{1.53}$$

for the net inflow across the boundary of V. The right side follows from the divergence theorem.

The increase of the considered energy within V due to emission follows from Equation (1.49) as

$$\int_V \eta_\lambda dV d\omega d\lambda \, , \tag{1.54}$$

and due to scattering into the $\vec{s}$-direction from Equation (1.45) as

$$\int_V \left[\frac{\sigma_\lambda}{4\pi} \int_{\omega'=4\pi} p_\lambda(\vec{s}' \to \vec{s}) \; I_\lambda(\vec{s}') d\omega' \right] dV d\omega d\lambda \, . \tag{1.55}$$

The depletion due to absorption and to scattering out of the considered direction follows from Equations (1.37) and (1.41) as

$$\int_V (\kappa_\lambda + \sigma_\lambda) I_\lambda dV d\omega d\lambda \tag{1.56}$$

Equating the change in energy content to the sum of the net transport across the boundary and the gains and losses due to emission, absorption, and scattering yields

$$\frac{\partial}{\partial t}\int_V \frac{I_\lambda}{c}\,dV = -\int_V \vec{s}\cdot\nabla I_\lambda\,dV + \int_V \eta_\lambda\,dV - \int_V (\kappa_\lambda + \sigma_\lambda) I_\lambda\,dV + \int_V \left[\frac{\sigma_\lambda}{4\pi}\int_{\omega'=4\pi} p_\lambda(\vec{s}\,' \to \vec{s}) I_\lambda(\vec{s}\,')\,d\omega'\right] dV \,. \quad (1.57)$$

Since V is arbitrary, it follows that

$$\frac{1}{c}\frac{\partial I_\lambda}{\partial t} + \vec{s}\cdot\nabla I_\lambda = \eta_\lambda - (\kappa_\lambda + \sigma_\lambda) I_\lambda + \frac{\sigma_\lambda}{4\pi}\int_{\omega'=4\pi} p_\lambda(\vec{s}\,' \to \vec{s}) I_\lambda(\vec{s}\,')\,d\omega' \,. \quad (1.58)$$

This integro-differential equation is the <u>equation of transfer</u>. Because of the factor 1/c, the first term in Equation (1.58) is usually negligible in comparison to the remaining terms for most engineering applications. The resulting equation is sometimes referred to as the quasi-steady equation of transfer. Note, however, that time variations in I_λ may still enter through the temperature, T, in the emission coefficient. For convenience, the first and last terms are often[1] combined and called the effective emission coefficient, $\eta_{e,\lambda}$. Following this procedure the equation of transfer becomes

$$\frac{1}{c}\frac{\partial I_\lambda}{\partial t} + \vec{s}\cdot\nabla I_\lambda = \eta_{e,\lambda} - \beta_\lambda I_\lambda \,, \quad (1.59)$$

where

$$\eta_{e,\lambda} = \eta_\lambda + \frac{\sigma_\lambda}{4\pi}\int_{\omega'=4\pi} p_\lambda(\vec{s}\,' \to \vec{s}) I_\lambda(\vec{s}\,')\,d\omega' \,. \quad (1.60)$$

A formal solution to the quasi-steady equation of transfer may be readily obtained by treating the effective emission coefficient as a specified function. For this purpose, let the coordinate s be in the direction of the unit vector $\vec{s}$. The directional derivative $(\vec{s}\cdot\nabla)$ then reduces to d/ds and the quasi-steady form of Equation (1.59) is

$$\frac{dI_\lambda}{ds} + \beta_\lambda I_\lambda = \eta_{e,\lambda} \,. \quad (1.61)$$

Equation (1.61) may be integrated from an initial point, s_o, where the intensity is $I_\lambda(s_o)$ to an arbitrary point, s, along the ray. The result is

$$I_\lambda(s) = I_\lambda(s_o)\exp\left[-\tau_\lambda(s,s_o)\right] + \int_{s'=s_o}^{s'=s} \eta_{e,\lambda}(s')\exp\left[-\tau_\lambda(s,s')\right] ds' \,, \quad (1.62)$$

where $\tau_\lambda(s,s_o)$ is the optical distance along s from s_o to s, and is defined as follows:

$$\tau_\lambda(s,s_o) = \int_{s=s_o}^{s'=s} \beta_\lambda ds' \ . \qquad (1.63)$$

The factor $\tau_\lambda(s,s')$ is the optical distance from s' to s.

$$\tau_\lambda(s,s') = \int_{s''=s'}^{s''=s} \beta_\lambda ds'' \ . \qquad (1.64)$$

The physical interpretation of Equation (1.62) is clearly evident. The intensity at a point s in a given direction $\vec{s}$ is that at s_o attenuated by the factor $\exp[-\tau_\lambda(s,s_o)]$ and enhanced by energy contributions from $\eta_{e,\lambda}$ at all anterior points between s_o and s, each attenuated by the factor $\exp[-\tau_\lambda(s,s')]$ to account for the attenuation by the intervening media between s' and s.

1.6 CONSERVATION OF RADIANT ENERGY

An equation expressing the conservation of radiant energy may be derived by integrating the equation of transfer over all directions.(1) The result obtained by multiplying Equation (1.59) by $d\omega$ and integrating over the 4π solid angles is

$$\int_{4\pi} \frac{1}{c}\frac{\partial I_\lambda}{\partial t} d\omega + \int_{4\pi} (\vec{s}\cdot\nabla) I_\lambda d\omega = \int_{4\pi} \eta_{e,\lambda} d\omega - \int_{4\pi} \beta_\lambda I_\lambda d\omega \ . \qquad (1.65)$$

Each of the integrals is related to quantities previously defined.

$$\int_{4\pi} \frac{1}{c}\frac{\partial I_\lambda}{\partial t} d\omega = \frac{\partial}{\partial t}\left[\frac{1}{c}\int_{4\pi} I_\lambda d\omega\right] = \frac{\partial u_\lambda^r}{\partial t} \ . \qquad (1.66)$$

$$\int_{4\pi} (\vec{s}\cdot\nabla) I_\lambda d\omega = \int_{4\pi} \left[\nabla\cdot(\vec{s}I_\lambda)\right] d\omega = \nabla\cdot \int_{4\pi} I_\lambda \vec{s} d\omega = \nabla\cdot\vec{q}_\lambda^r \ . \qquad (1.67)$$

$$\int_{4\pi} \beta_\lambda I_\lambda d\omega = \beta_\lambda \int_{4\pi} I_\lambda d\omega = \beta_\lambda c u_\lambda^r = \beta_\lambda \mathcal{G}_\lambda \ . \qquad (1.68)$$

$$\int_{4\pi} \eta_{e,\lambda} d\omega = \int_{\omega=4\pi} \left[\eta_\lambda + \frac{\sigma_\lambda}{4\pi}\int_{\omega'=4\pi} p_\lambda(\vec{s}'\to\vec{s}) I(\vec{s}') d\omega'\right] d\omega = 4\pi\eta_\lambda + \sigma_\lambda c u_\lambda^r = 4\pi\eta_\lambda + \sigma_\lambda \mathcal{G}_\lambda . \quad (1.69)$$

The quantity $\mathcal{G}_\lambda$ introduced above is the monochromatic radiant energy incident on an elemental volume from all directions per unit time and per unit area. In terms of intensity and energy density, $\mathcal{G}_\lambda$ is

$$\mathcal{G}_\lambda = c u_\lambda^r = \int_{4\pi} I_\lambda d\omega \ . \qquad (1.70)$$

Substituting Equations (1.66) to (1.69) into Equation (1.65) gives the result

$$\frac{\partial u_\lambda^r}{\partial t} + \nabla \cdot \vec{q}_\lambda^r = - (\kappa_\lambda + \sigma_\lambda) \mathcal{G}_\lambda + 4\pi\eta_\lambda + \sigma_\lambda \mathcal{G}_\lambda , \quad (1.71)$$

or

$$\frac{\partial u_\lambda^r}{\partial t} + \nabla \cdot \vec{q}_\lambda^r = 4\pi\eta_\lambda - \kappa_\lambda \mathcal{G}_\lambda . \quad (1.72)$$

Equation (1.72) expresses the conservation principle for radiant energy in the spectral interval $(\lambda, \lambda + d\lambda)$. Note that the scattering terms are absent from Equation (1.72). This result confirms the physical fact that all incident energy which is scattered emerges from the volume and is not stored. The conservation equation of total radiant energy is obtained by integrating Equation (1.72) over all wavelengths. Thus

$$\frac{\partial u^r}{\partial t} + \nabla \cdot \vec{q}^r = 4\pi\eta - \kappa \mathcal{G} , \quad (1.73)$$

where the total emission coefficient (η), total absorption coefficient (κ), and total irradiation $(\mathcal{G})$ are defined as

$$\left. \begin{aligned} \eta &= \int_0^\infty \eta_\lambda d\lambda , \\ \kappa \mathcal{G} &= \int_0^\infty \kappa_\lambda \mathcal{G}_\lambda d\lambda , \\ \mathcal{G} &= \int_0^\infty \mathcal{G}_\lambda d\lambda . \end{aligned} \right\} \quad (1.74)$$

The conservation equations of radiant energy must be modified for relativistic conditions.[1]

REFERENCES

1. Viskanta, R., "Radiation Transfer and Interaction of Convection with Radiation Heat Transfer," Advances in Heat Transfer, Vol. III. New York: Academic Press, 1966.

2. Milne, E. A., "Thermodynamics of the Stars," in Handbuch der Astrophysik, G. Eberhard, et al., eds., Vol. III, Part I, pp. 65-255, Springer, Berlin, 1930.

3. Rosseland, S. Theoretical Astrophysics. London: Oxford University Press (Clarendon), 1936.

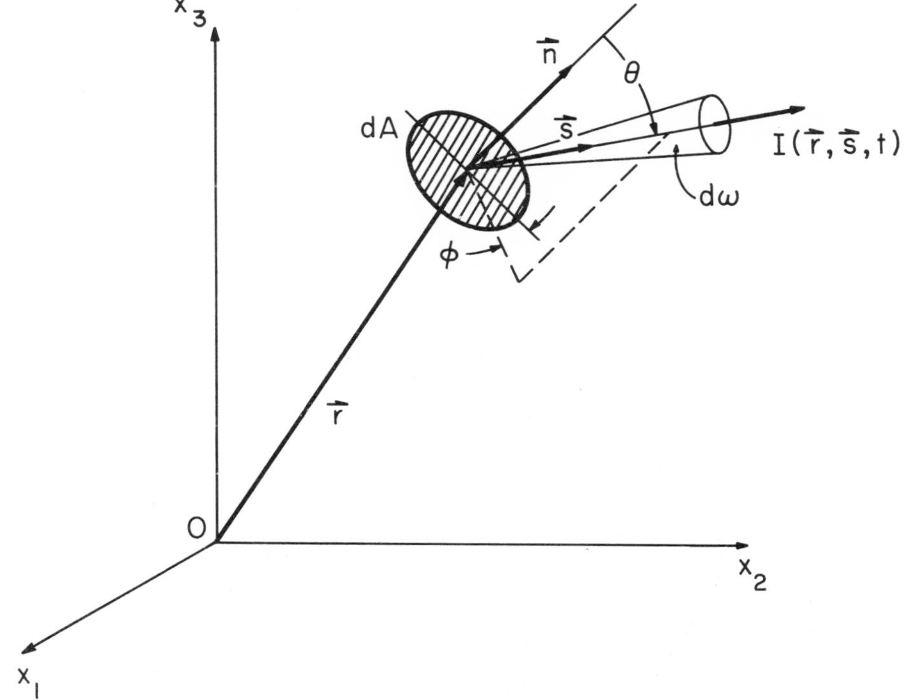

FIGURE 1.1 DEFINITION OF INTENSITY.

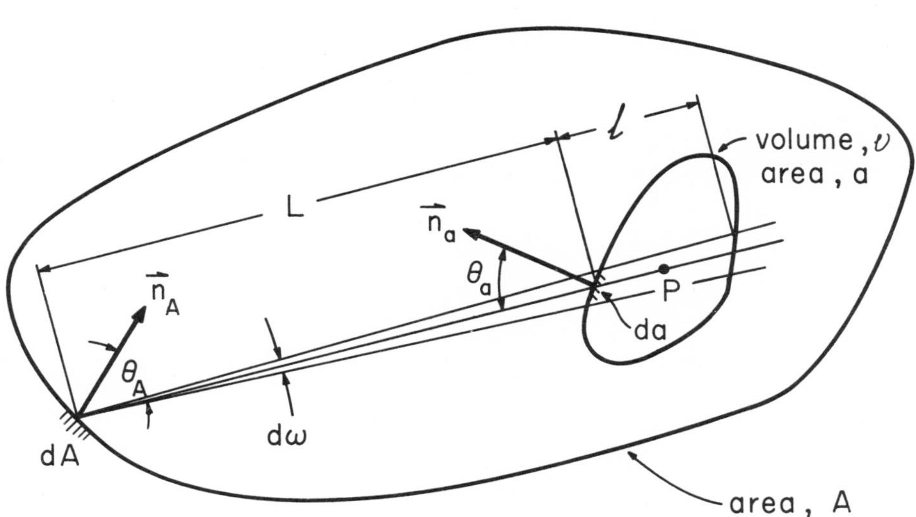

FIGURE 1.2 RADIANT ENERGY DENSITY.

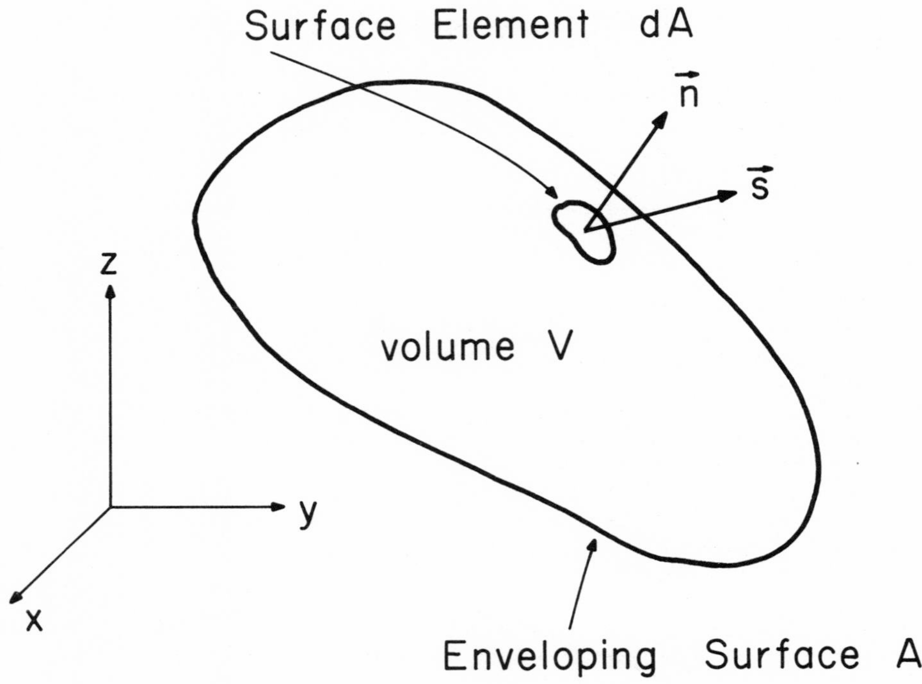

FIGURE 1.3 EQUATION OF TRANSFER.

2. ELECTROMAGNETIC THEORY OF RADIATION

2.1 INTRODUCTION

Thermal radiation differs from other forms of electromagnetic radiation only in the method of generation and the observable effects it produces when absorbed. Just as the bombardment of metal targets by electrons yields X rays, the temperature of a system causes emission of energy in the form of electromagnetic radiation. Consequently, the classical laws of electrodynamics are available for studying the transport of thermal radiation. It is the purpose here to present only a minute portion of the vast field which electrodynamics encompasses. In order to provide a self-contained approach, the study is initiated from the basic experimental laws of physics concerning the forces between electrical charges and between current elements. These action-at-a-distance phenomena are interpreted in terms of certain field quantities, namely, the electric field and the magnetic induction. The introduction of these concepts enables the experimental laws to be concisely given in terms of a system of partial differential equations. These equations predict the transport of energy by electromagnetic waves. After a brief study of some of the characteristics of this energy transport mechanism in unbounded media, attention is directed to the interaction of electromagnetic radiation with surfaces. This study culminates with the prediction of the reflection and emission properties of certain classes of surfaces.

2.2 FUNDAMENTALS OF ELECTRODYNAMICS

Experiments on forces between electrically charged bodies have led to the observation that charges experience two types of electrical forces. One, which depends only on the relative position of the charges, is called the electrostatic force, while the other, which depends on their relative velocities, is termed the magnetic force. The vector equation

$$\vec{F} = q(\vec{E} + \vec{V} \times \vec{B})$$

summarizes these results for the mechanical force, $\vec{F}$, on a charge, q, with velocity, $\vec{V}$. The quantities $\vec{E}$ and $\vec{B}$ are the electric field and magnetic induction, respectively. The fundamental problem of electrodynamics is the evaluation of the electric field and magnetic induction in space and time in terms of known system parameters. These fundamental quantities satisfy a set of partial differential equations which express the fundamental laws of electricity and magnetism determined from experiment. These laws and the subsequent development of the equations of electrodynamics are briefly reviewed in this section. Rationalized mks units are used throughout.

2.2.1 Coulomb's Law

Coulomb established experimentally that in a vacuum the force $\vec{F}$ acting on a point charge q_1 located at $\vec{r}_1$, due to another point charge q_2 located at $\vec{r}_2$, may be expressed as

$$\vec{F} = \frac{q_1 q_2}{4\pi\epsilon_o} \frac{(\vec{r}_1 - \vec{r}_2)}{r_{12}^3} , \tag{2.1}$$

where

$$r_{12} = |\vec{r}_1 - \vec{r}_2|$$

and

$$\epsilon_o = 8.854 \times 10^{-12} \frac{\text{farad}}{\text{meter}}$$

is a constant called the permittivity of free space. It follows that the electric field at an arbitrary point, $\vec{r}$ (Figure 2.1), due to a point charge, q_1, is

$$\vec{E}(\vec{r}) = \frac{q_1}{4\pi\epsilon_o} \frac{(\vec{r} - \vec{r}_1)}{r^3} , \tag{2.2}$$

with

$$r = |\vec{r} - \vec{r}_1| .$$

Experiment confirms that the electric field at point $\vec{r}$ due to a number of point charges is the vector sum of that due to each charge. Should the charges be sufficient in number to be described by a charge density per unit volume, q_υ, the resultant field due to a finite volume, V, of charges is

$$\vec{E}(\vec{r}) = \frac{1}{4\pi\epsilon_o} \int_V q_\upsilon(\vec{r}') \frac{(\vec{r} - \vec{r}')}{r^3} d\upsilon' . \tag{2.3}$$

In the above $d\upsilon'$ is a volume element about the point $\vec{r}'$, where the charge density is

$q_\upsilon(\vec{r}')$. Since the vector factor in the integrand is the negative gradient of the scalar

$$\frac{1}{r} = \frac{1}{|\vec{r} - \vec{r}'|}$$

with respect to $\vec{r}$, Equation (2.3) may be written as

$$\vec{E}(\vec{r}) = -\frac{1}{4\pi\varepsilon_o}\nabla \int_V \frac{q_\upsilon(\vec{r}')}{|\vec{r} - \vec{r}'|} d\upsilon' \quad . \tag{2.4}$$

This integral expression is not the most convenient form for the evaluation of the field. Another integral result, Gauss' law, leads to a differential equation for $\vec{E}(\vec{r})$. To obtain Gauss' law, consider a point charge, q, and a closed surface of area, A, as illustrated in Figure 2.2. Let r denote the distance from the charge q to a point on A with $\vec{n}$ the outward unit normal to the surface element dA at that point. If the electric field and the unit normal at the point make an angle θ, the product of the normal component of $\vec{E}$ and the area element is

$$\vec{E} \cdot \vec{n}\, dA = \frac{q}{4\pi\varepsilon_o} \frac{\cos\theta}{r^2} dA = \frac{q}{4\pi\varepsilon_o} d\omega ,$$

where $d\omega$ is the solid angle subtended by dA at the position of the charge. The integral of the normal component of $\vec{E}$ over the closed surface A is

$$\int_A \vec{E} \cdot \vec{n}\, dA = \frac{q}{\varepsilon_o} ,$$

if q lies within A and zero otherwise. This result is Gauss' law for a single point charge. The analog to the above result for a charge density q_υ is

$$\int_A \vec{E} \cdot \vec{n}\, dA = \int_V \frac{q_\upsilon}{\varepsilon_o} d\upsilon ,$$

when V is the volume enclosed by A. The divergence theorem may be employed to transform the area integral to a volume integral yielding the result

$$\int_V \left(\nabla \cdot \vec{E} - \frac{q_\upsilon}{\varepsilon_o}\right) d\upsilon = 0 .$$

Since the volume, V, is arbitrary and the integrand assumed continuous, the integral has a zero value in general only if the integrand vanishes. Thus

$$\nabla \cdot \vec{E} = \frac{q_\upsilon}{\varepsilon_o} . \tag{2.5}$$

Equation (2.5) is one of the basic equations of electrodynamics. It is later modified to account for the presence of material media. The derivation of this equation depends on the inverse square law for the force between charges and the linear superposition of the effects of different charges.

2.2.2 Ampere's Law

The modern point of view of magnetic phenomena is that all magnetic effects are associated with the motion of charges, i.e., current. Ampere established the basic experimental laws relating the magnetic induction, $\vec{B}$, to the current, as well as the law of force between one current element and another. In vacuum, Ampere's law for the force on a current element, $I_1 \vec{d\ell}_1$, in the presence of a current element, $I\vec{d\ell}$, is

$$\vec{dF} = \frac{\mu_o}{4\pi} \frac{I I_1 \vec{d\ell}_1 \times (\vec{d\ell} \times \vec{r})}{r^3} , \tag{2.6}$$

where $\vec{d\ell}$ and $\vec{d\ell}_1$ represent length elements in the directions of current flow and $\vec{r}$ is the coordinate vector from the element $\vec{d\ell}$ to $\vec{d\ell}_1$ with r its scalar length. The constant

$$\mu_o = 4\pi \times 10^{-7} \frac{\text{henry}}{\text{meter}}$$

is known as the permeability of free space. The view may be taken that the current element, $I\vec{d\ell}$, produces a magnetic induction at the point, P, where $I_1\vec{d\ell}_1$ is located (Figure 2.3) so that Equation (2.6) may be written as

$$\vec{dF} = I_1 \vec{d\ell}_1 \times \vec{dB} ,$$

where

$$\vec{dB} = \frac{\mu_o}{4\pi} \frac{I(\vec{d\ell} \times \vec{r})}{r^3} .$$

Now a current corresponds to charges in motion and may be described in terms of a current density, $\vec{J}$, measured in units of positive charge crossing unit area per unit time, the direction of motion of the charges defining the direction of $\vec{J}$. Thus employing the superposition principle for a current density, $\vec{J}$, distributed over a volume, V, the basic definition of magnetic induction is

$$\vec{B}(\vec{r}) = \frac{\mu_o}{4\pi} \int_V \vec{J}(\vec{r}') \times \frac{(\vec{r} - \vec{r}')}{r^3} \, d\upsilon' \quad . \qquad (2.7)$$

This expression is the magnetic analog of Equation (2.3) for the electric field. As in the situation for the electric field, Equation (2.7) is not in its most useful form for many applications. Again, since

$$\frac{(\vec{r} - \vec{r}')}{r^3} = -\nabla \left(\frac{1}{r}\right) ,$$

the result for $\vec{B}$ may be transformed into the form

$$\vec{B}(\vec{r}) = \frac{\mu_o}{4\pi} \nabla \times \int_V \frac{\vec{J}(\vec{r}')}{|\vec{r} - \vec{r}'|} \, d\upsilon' \quad . \qquad (2.8)$$

Since the divergence of a curl operation is identically zero, it follows that

$$\nabla \cdot \vec{B} = 0 \; . \qquad (2.9)$$

This is another of the equations governing the field vectors of electrodynamics. Still another equation governing the steady $\vec{B}$ field may be obtained by evaluating the curl of $\vec{B}$ and using the result that for steady fields, the conservation of charge principle requires $\nabla \cdot \vec{J} = 0$. Since this derivation is somewhat lengthy[1] we simply state the result:

$$\nabla \times \vec{B} = \mu_o \vec{J} \; . \qquad (2.10)$$

This fundamental equation of electrodynamics is later modified to include the effects of material media and extended to include time dependent fields.

2.2.3 Faraday's Law

Faraday in about 1831 observed that a transient current is induced in a circuit if (a) the current in an adjacent stationary circuit is varied, (b) the adjacent circuit with steady current is moved relative to the first, and (c) a permanent magnet is moved into or out of the circuit. He interpreted the transient current flow as being due to a changing magnetic flux linked by the circuit with the changing flux inducing an electric field around the circuit. The line integral of this field around the circuit is called the electromotive force, $\mathcal{E}$. In order to put these observations in mathematical form, let the circuit,* C, bound an open surface, A, with local unit normal to dA denoted $\vec{n}$. The magnetic induction in the neighborhood of the circuit is $\vec{B}$. The surface integral of $\vec{B}$ over the area A is termed the magnetic flux, Φ, linking the circuit and is defined by

$$\Phi = \int_A \vec{B} \cdot d\vec{A} \; . \qquad (2.11)$$

The electromotive force, $\mathcal{E}$, around the circuit is

$$\mathcal{E} = \oint_C \vec{E} \cdot d\vec{\ell} \; . \qquad (2.12)$$

With these, Faraday's law may be expressed as follows:

$$\mathcal{E} = -\frac{\partial \Phi}{\partial t} \; . \qquad (2.13)$$

Inserting the definitions of Φ and $\mathcal{E}$ yields

$$\oint_C \vec{E} \cdot d\vec{\ell} = -\frac{\partial}{\partial t} \int_A \vec{B} \cdot d\vec{A} \; ,$$

which after using Stokes' theorem becomes

$$\int_A \left(\nabla \times \vec{E} + \frac{\partial \vec{B}}{\partial t} \right) \cdot d\vec{A} = 0 \; . \qquad (2.14)$$

---

*As is later shown [Equation (2.37)], the tangential component of $\vec{E}$ is continuous across the boundary between the wire and space just outside it. Thus the circuit need not be a physical circuit. All that is required is that Φ be the flux passing through the surface enclosed by the integration path.

By the usual arguments we find

$$\nabla \times \vec{E} + \frac{\partial \vec{B}}{\partial t} = 0 . \qquad (2.15)$$

Equation (2.15) is the last of the electrodynamic equations.

2.2.4 Constitutive Relations

The basic equations of electrodynamics in vacua must be modified to account for induced charges and currents which arise when materials are subjected to fields. Allowance is made for these phenomena by the introduction of auxiliary vectors, the electric displacement, $\vec{D}$, and the magnetic field, $\vec{H}$.

The charge density in Equation (2.5) represents the total charge density per unit volume, that placed throughout space, q_υ, as well as that induced in the material, q_υ^i. The induced charge is described in terms of a polarization vector, $\vec{P}$, representing the electric moment per unit volume and the negative divergence of which yields the induced charge density per unit volume. Then Equation (2.5) may be written as

$$\nabla \cdot \vec{E} = \frac{q_\upsilon + q_\upsilon^i}{\varepsilon_o} = \frac{q_\upsilon}{\varepsilon_o} - \frac{1}{\varepsilon_o} \nabla \cdot \vec{P}$$

or

$$\nabla \cdot \vec{D} = q_\upsilon , \qquad (2.16)$$

where

$$\vec{D} = \varepsilon_o \vec{E} + \vec{P} . \qquad (2.17)$$

The vector $\vec{D}$ defined by Equation (2.17) is the electric displacement. Many substances may be treated as homogeneous and isotropic with the polarization proportional to the applied electric field.

$$\vec{P} = \chi_e \varepsilon_o \vec{E} . \qquad (2.18)$$

The constant χ_e is called the electric susceptibility of the medium. For such materials, the electric displacement and field are related as follows:

$$\vec{D} = \varepsilon_o \vec{E} + \vec{P} = \varepsilon_o \vec{E} + \chi_e \varepsilon_o \vec{E} = \varepsilon_o (1 + \chi_e) \vec{E} = \varepsilon_o \kappa_e \vec{E} = \varepsilon \vec{E} . \qquad (2.19)$$

In the above κ_e is the dielectric constant $(= \varepsilon/\varepsilon_o)$ and ε is the absolute permittivity of the dielectric.

In a material medium, the current density appearing in Equation (2.10) must account for the conduction current, $\vec{J}$, representing the actual transport of charge and the effective current, $\vec{J}_M$, resulting from the circulating currents inside atoms and molecules. The effective atomic current density is specified in terms of the magnetization or magnetic moment per unit volume, $\vec{M}$, as

$$\vec{J}_M = \nabla \times \vec{M} . \qquad (2.20)$$

Employing this relation in Equation (2.10) and subsequent rearrangement yields

$$\nabla \times \vec{H} = \vec{J} , \qquad (2.21)$$

where

$$\vec{H} = \frac{\vec{B}}{\mu_o} - \vec{M} . \qquad (2.22)$$

The vector $\vec{H}$ is the magnetic field. For materials which may be considered homogeneous and isotropic with the magnetization proportional to the magnetic field, $\vec{M}$ is related to $\vec{H}$ through the magnetic susceptibility, χ_m.

$$\vec{M} = \chi_m \mu_o \vec{H} . \qquad (2.23)$$

Introducing Equation (2.23) into Equation (2.22) gives

$$\vec{B} = \mu_o \vec{H} + \vec{M} = \mu_o \vec{H} + \chi_m \mu_o \vec{H} = \mu_o (1 + \chi_m) \vec{H} = \mu_o \kappa_m \vec{H} = \mu \vec{H} , \qquad (2.24)$$

where $\kappa_m = (1 + \chi_m)$ is the relative permeability and μ is the absolute permeability of the medium. Except for ferromagnetic substances, κ_m differs from unity by only a few parts in 10^5, so that in the absence of magnetic materials, μ is commonly taken equal to the permeability of free space.

If an isotropic medium is characterized by a scalar electrical conductivity, σ, the relation between the conduction current and electric field is

$$\vec{J} = \sigma \vec{E} . \tag{2.25}$$

Equation (2.25) is known as Ohm's law.

2.2.5 Displacement Current

An inconsistency exists between the equations of electrodynamics presented earlier and the charge conservation equation for unsteady fields. The conservation of charge principle leads to the result

$$\frac{\partial q_\upsilon}{\partial t} + \nabla \cdot \vec{J} = 0 , \tag{2.26}$$

while the divergence of Equation (2.21) yields

$$\nabla \cdot (\nabla \times \vec{H}) = \nabla \cdot \vec{J} . \tag{2.27}$$

The divergence of a curl operation is identically zero, while according to Equation (2.26) the divergence of the current density is zero only in the absence of charge accumulation. This difficulty was resolved by Maxwell who made the assumption that the conduction current, $\vec{J}$, in Equation (2.27) must be augmented by an additional term to obtain a total effective current the divergence of which vanishes. Introducing this term, $\vec{J}_D$, in Equation (2.21) we have

$$\nabla \times \vec{H} = \vec{J} + \vec{J}_D . \tag{2.28}$$

Imposing the condition that the divergence of $(\vec{J} + \vec{J}_D)$ vanish gives

$$-\frac{\partial q_\upsilon}{\partial t} + \nabla \cdot \vec{J}_D = 0 , \tag{2.29}$$

where Equation (2.26) has been used to eliminate $\nabla \cdot \vec{J}$. Differentiation of Equation (2.16) with respect to time and subsequent use in Equation (2.29) yields

$$\vec{J}_D = \frac{\partial \vec{D}}{\partial t} . \tag{2.30}$$

$\vec{J}_D$ is known as the displacement current, and its introduction renders the electrodynamical equations consistent. The concept of displacement current has important consequences particularly in dielectrics where the conduction current is zero. Its existence is verified by experiment.

2.2.6 Maxwell Equations

Throughout the previous sections, certain relationships were emphasized as fundamental equations of electrodynamics. Coulomb's law led to Equation (2.5) which was modified to account for material media and ultimately resulted in Equation (2.16). Ampere's law led to the development of Equations (2.9) and (2.10). The latter equation was modified for the presence of material media and corrected for time-dependent fields to give Equation (2.28). Finally Faraday's observations were expressed by Equation (2.15). This set of four equations,

$$\nabla \cdot \vec{D} = q_\upsilon , \tag{2.31}$$

$$\nabla \cdot \vec{B} = 0 , \tag{2.32}$$

$$\nabla \times \vec{H} - \frac{\partial \vec{D}}{\partial t} = \vec{J} , \tag{2.33}$$

$$\nabla \times \vec{E} + \frac{\partial \vec{B}}{\partial t} = 0 , \tag{2.34}$$

known as the Maxwell equations, form the basis of all electromagnetic phenomena. They relate the four field vectors $\vec{E}$, $\vec{D}$, $\vec{B}$, and $\vec{H}$ to the current and charge distribution.

2.2.7 Boundary Conditions

It remains to establish the conditions satisfied by $\vec{D}$, $\vec{E}$, $\vec{B}$, and $\vec{H}$ at the interface between media of different electric and magnetic properties. For this purpose consider a small cylinder oriented so that its faces are in regions 1 and 2 and its sides are normal to the surface boundary, S, as shown in Figure 2.4a. The integral over this volume, V, of Equations (2.31) and (2.32) are the equations

$$\int_V \nabla \cdot \vec{D} d\upsilon = \int_A \vec{D} \cdot \vec{n}\, dA = \int_V q_\upsilon\, d\upsilon$$

and

$$\int_V \nabla \cdot \vec{B} d\upsilon = \int_A \vec{B} \cdot \vec{n}\, dA = 0 ,$$

where A is the surface area of the cylinder and $\vec{n}$ is the local unit normal to the cylinder. Since the ends of the cylinder are taken as small and the fields continuous over these surfaces, the integrals over the surface area of the cylinder may be subdivided into three contributions.

$$\vec{D}_1 \cdot \vec{n}_1\, \delta A_1 + \vec{D}_2 \cdot \vec{n}_2\, \delta A_2 + \text{(contributions by walls)} = q_s \delta A ,$$

$$\vec{B}_1 \cdot \vec{n}_1\, \delta A_1 + \vec{B}_2 \cdot \vec{n}_2\, \delta A_2 + \text{(contributions by walls)} = 0 ,$$

where q_s is the surface charge per unit area. In the limit of vanishing cylinder height, the contribution of the cylinder walls tends to zero. Consequently, in this limit

$$(\vec{D}_2 - \vec{D}_1) \cdot \vec{n} = q_s , \tag{2.35}$$

$$(\vec{B}_2 - \vec{B}_1) \cdot \vec{n} = 0 , \tag{2.36}$$

in which the relation between unit normals $\vec{n} = -\vec{n}_1 = \vec{n}_2$ has been used. Thus the normal component of $\vec{B}$ is continuous across the boundary, while the normal component of $\vec{D}$ is discontinuous by an amount equal to the surface charge density.

Now consider a small, narrow circuit, C, with sides parallel and perpendicular to the interface, S, and enclosing the area, A, such as shown in Figure 2.4b. Let $\vec{a}$ denote a unit vector perpendicular to the plane of the rectangle. Integrating Equations (2.34) and (2.33) over the area, A, and using Stokes' theorem gives

$$\int_A \nabla \times \vec{E} \cdot \vec{a}\, dA = \oint_C \vec{E} \cdot d\vec{\ell} = -\frac{\partial}{\partial t} \int_A \vec{B} \cdot \vec{a}\, dA$$

and

$$\int_A \nabla \times \vec{H} \cdot \vec{a}\, dA = \oint_C \vec{H} \cdot d\vec{\ell} = \int_A \vec{J} \cdot \vec{a}\, dA + \frac{\partial}{\partial t} \int_A \vec{D} \cdot \vec{a}\, dA ,$$

where $d\vec{\ell}$ is an element of the circuit, C. For a sufficiently small circuit, the above may be written as

$$-\vec{E}_1 \cdot \vec{t}\, \delta s + \vec{E}_2 \cdot \vec{t}\, \delta s + \text{(contributions by ends)} = -\frac{\partial \vec{B}}{\partial t} \cdot \vec{a}\, \delta s \delta h ,$$

$$-\vec{H}_1 \cdot \vec{t}\, \delta s + \vec{H}_2 \cdot \vec{t}\, \delta s + \text{(contributions by ends)} = \vec{J}_s \cdot \vec{a}\, \delta s + \frac{\partial \vec{D}}{\partial t} \cdot \vec{a}\, \delta s \delta h ,$$

where $\vec{J}_s$ is surface current per unit area. In the limit as $\delta h \to 0$, the above reduce to

$$\vec{n} \times (\vec{E}_2 - \vec{E}_1) = 0 , \tag{2.37}$$

$$\vec{n} \times (\vec{H}_2 - \vec{H}_1) = \vec{J}_s , \tag{2.38}$$

assuming $\partial\vec{B}/\partial t$ and $\partial\vec{D}/\partial t$ are finite. The relation $\vec{t} = \vec{a} \times \vec{n}$ and the fact that the orientation of the circuit, hence $\vec{a}$, is arbitrary have been utilized to arrive at the above. Thus the tangential component of the electric field is continuous across the interface, whereas the tangential component of the magnetic field is discontinuous by an amount equal to the surface current density.

2.2.8 Energy Density and Poynting Vector

For a single charge, q, the rate of doing work by external electromagnetic fields $\vec{E}$ and $\vec{B}$ is $q\vec{V}\cdot\vec{E}$, where $\vec{V}$ is the velocity of the charge. No contribution occurs from the $\vec{B}$ field, since the magnetic force is normal to the velocity. If there exists a continuous distribution of charge and current, the total rate of doing work by the fields in a finite volume, V, is

$$\int_V \vec{J}\cdot\vec{E}\, d\upsilon .$$

This integral represents the conversion of electromagnetic energy into other forms of energy such as mechanical and thermal. Eliminating $\vec{J}$ with Equation (2.33) yields

$$\int_V \vec{J}\cdot\vec{E}\, d\upsilon = \int_V \left[\vec{E}\cdot\nabla\times\vec{H} - \vec{E}\cdot\frac{\partial\vec{D}}{\partial t}\right] d\upsilon . \tag{2.39}$$

Employing the vector identity,

$$\nabla\cdot(\vec{E}\times\vec{H}) = \vec{H}\cdot(\nabla\times\vec{E}) - \vec{E}\cdot(\nabla\times\vec{H}),$$

and Faraday's law, Equation (2.34), transforms Equation (2.39) to

$$\int_V \vec{J}\cdot\vec{E}\, d\upsilon = -\int_V \nabla\cdot(\vec{E}\times\vec{H})\, d\upsilon - \frac{\partial}{\partial t}\int_V \left[\frac{\vec{E}\cdot\vec{D}}{2} + \frac{\vec{H}\cdot\vec{B}}{2}\right] d\upsilon . \tag{2.40}$$

If the divergence theorem is used to transform the last integral to the right, there results

$$-\frac{\partial}{\partial t}\int \left[\frac{\vec{E}\cdot\vec{D}}{2} + \frac{\vec{H}\cdot\vec{B}}{2}\right] d\upsilon = \int_V \vec{J}\cdot\vec{E}\, d\upsilon + \int_A (\vec{E}\times\vec{H})\cdot d\vec{A} , \tag{2.41}$$

where A is the surface area enclosing V. The left side is interpreted as the rate of decrease of the sum of the electric and magnetic field energies within the volume, V. The first term on the right represents the Joule heat losses in the absence of impressed electromotive forces. The interpretation of the remaining term is a decrease of energy within the volume by an outward flow of electromagnetic energy through the surface bounding the volume. As a consequence of these interpretations, let the electromagnetic energy per unit volume be denoted by u, where

$$u = u_e + u_m = \frac{\vec{E}\cdot\vec{D}}{2} + \frac{\vec{H}\cdot\vec{B}}{2} , \tag{2.42}$$

with u_e and u_m the corresponding electric and magnetic contributions. The vector $\vec{N}$ given by

$$\vec{N} = \vec{E}\times\vec{H} \tag{2.43}$$

is called the Poynting vector and represents the instantaneous flow rate of electromagnetic energy normal to a unit area in the field. In the applications with which we are later concerned, $\vec{N}$ is a rapidly varying periodic function of time whose instantaneous value cannot be observed experimentally. The time average taken over a time interval large compared to the period, however, is a measurable quantity, $\langle\vec{N}\rangle$. For a harmonic electromagnetic field, the mean energy flow rate is

$$\langle\vec{N}\rangle = \frac{1}{2} \operatorname{Re} (\vec{E} \times \vec{H}^*) , \qquad (2.44)$$

where Re denotes "real part of" and H* is the complex conjugate of $\vec{H}$.

2.3 PLANE ELECTROMAGNETIC WAVES IN UNBOUNDED MEDIA

Some of the simplest solutions to the equations of electrodynamics are those that depend upon time and a single spatial coordinate. The factors characterizing these one-dimensional fields also determine in general the features possessed by more complex fields. Consequently, we shall study the properties of plane waves in unbounded media. Attention is restricted to an isotropic and homogeneous medium obeying the linear constitutive relations of Equations (2.19), (2.24), and (2.25). These material relations may be employed to eliminate $\vec{D}$, $\vec{B}$, and $\vec{J}$ from the Maxwell equations yielding their following special form for the materials considered.

$$\nabla \cdot \vec{E} = 0 \qquad (I)$$

$$\nabla \cdot \vec{H} = 0 \qquad (II)$$

$$\nabla \times \vec{H} - \sigma\vec{E} - \epsilon \frac{\partial\vec{E}}{\partial t} = 0 \qquad (III)$$

$$\nabla \times \vec{E} + \mu \frac{\partial\vec{H}}{\partial t} = 0 \qquad (IV)$$

In writing these equations we have assumed the absence of free charges. Surface charges and currents will also be excluded from consideration.

2.3.1 Plane Waves

The Maxwell equations are four simultaneous differential equations relating the two field vectors, $\vec{E}$ and $\vec{H}$. By elimination, differential equations which each of the vectors must satisfy may be determined. Applying the vector operator curl to (IV) and substituting (III) in the resulting expression gives

$$\nabla \times (\nabla \times \vec{E}) = - \epsilon\mu \frac{\partial^2\vec{E}}{\partial t^2} - \sigma\mu \frac{\partial\vec{E}}{\partial t} . \qquad (2.45)$$

Employing the vector identity

$$\nabla \times (\nabla \times \vec{E}) = \nabla(\nabla \cdot \vec{E}) - \nabla^2\vec{E}$$

with the first term to the right zero by (I), Equation (2.45) transforms to

$$\nabla^2\vec{E} = \epsilon\mu \frac{\partial^2\vec{E}}{\partial t^2} + \sigma\mu \frac{\partial\vec{E}}{\partial t} . \qquad (2.46)$$

A similar equation may be derived for the magnetic field vector by an analogous procedure.

$$\nabla^2\vec{H} = \epsilon\mu \frac{\partial^2\vec{H}}{\partial t^2} + \sigma\mu \frac{\partial\vec{H}}{\partial t} . \qquad (2.47)$$

Thus $\vec{E}$ and $\vec{H}$ satisfy the same differential equation. Both equations are vector equations, which means that each of the three components of $\vec{E}$ and $\vec{H}$ separately satisfies the same scalar equation; namely,

$$\nabla^2 w = \epsilon\mu \frac{\partial^2 w}{\partial t^2} + \sigma\mu \frac{\partial w}{\partial t} . \qquad (2.48)$$

The scalar function, w, may be any one of the components of $\vec{E}$ or $\vec{H}$. Consider a possible solution to Equation (2.48) which depends on only one spatial coordinate, say z, and time. When $\sigma = 0$, that is, electrical nonconductors, Equation (2.48) is satisfied by

$$w(z,t) = f(z - \upsilon t) + g(z + \upsilon t) , \qquad (2.49)$$

where f and g are arbitrary functions of their respective arguments and

$$\upsilon = \frac{1}{\sqrt{\epsilon\mu}} . \qquad (2.50)$$

Since the arguments $(z \mp \upsilon t)$ are not altered when t and z are replaced with $(t + \tau)$ and $(z \pm \upsilon\tau)$, respectively, each of the functions represents the propagation of a disturbance in the z-direction with a velocity given by Equation (2.50). The function $f(z - \upsilon t)$ represents propagation in the positive z-coordinate direction and the function $g(z + \upsilon t)$ propagation in the negative z-direction. For many applications, periodic disturbances are of considerable importance. Since the Maxwell equations are linear, superposition principles apply and essentially no loss of generality is incurred by using a harmonic particular solution. Other wave forms may be constructed from the harmonic solutions by the methods of Fourier series. It is convenient to use the exponential form of the harmonic solution where it is understood that physical quantities are obtained by taking the real parts of complex quantities. Also, since there is no essential difference in character between the waves propagating in the opposite coordinate directions, discussion is limited to those propagating in the positive coordinate direction. Then, solutions to Equations (2.46) and (2.47) for a nonconducting medium are the following harmonic plane waves of angular frequency, ω, propagating in the positive z-coordinate direction.

$$\vec{E} = \vec{E}_o \exp\left[i\omega \left(t - \frac{z}{\upsilon} \right) \right]. \qquad (2.51)$$

$$\vec{H} = \vec{H}_o \exp\left[i\omega \left(t - \frac{z}{\upsilon} \right) \right]. \qquad (2.52)$$

$\vec{E}_o$ and $\vec{H}_o$ are complex amplitudes constant in time and space.

In free space, $\mu = \mu_o$, $\epsilon = \epsilon_o$, and the Maxwell equations predict the propagation of electromagnetic waves at a velocity, c, given by

$$c = \frac{1}{\sqrt{\epsilon_o \mu_o}}. \qquad (2.53)$$

Insertion of the numerical values for the constants ϵ_o and μ_o yields a value for c which is within about one part in 30,000 of the measured value of the velocity of light. This prediction of the velocity of light was one of the early successes of Maxwell's theory for the electromagnetic nature of light. The ratio of c to the velocity of propagation in a medium is its index of refraction, n. Hence using Equations (2.50) and (2.53),

$$n = \frac{c}{\upsilon} = \sqrt{\frac{\epsilon\mu}{\epsilon_o \mu_o}} = \sqrt{\kappa_e \kappa_m} . \qquad (2.54)$$

This relation is often referred to as Maxwell's formula. Except for ferromagnetic materials, κ_m is usually within one part in 10^5 of unity. Then the square of the refractive index should be equal to the dielectric constant, κ_e, which has been assumed to be a constant of the material. However, simply passing natural light through a prism demonstrates that the refractive index is dependent on the frequency (or wavelength) of the light. To retain Maxwell's formula, it is assumed that κ_e is not a constant of the material but rather a function of the frequency of the field. This frequency dependence can only be treated by accounting for the atomic structure of matter and is outside the scope of the present discussion.

Before passing on to conducting media, we generalize the plane wave solution to arbitrary direction of propagation. Note first that plane electromagnetic waves are characterized by the condition that $\vec{E}$ and $\vec{H}$ are functions only of time and the distance from a fixed plane. For waves traveling in the direction of the z-axis, the fixed plane is the xy-plane, and the distance from this plane is the z-coordinate. If $\vec{s}$ is the unit vector in the direction of propagation of a plane wave, all points, P, with position vector $\vec{r}$ such that $\vec{r} \cdot \vec{s}$ = constant lie on a plane perpendicular to $\vec{s}$. Thus a monochromatic plane harmonic wave of angular frequency, ω, propagating in the $\vec{s}$-direction may be expressed as

$$\vec{E} = \vec{E}_o \exp\left[i\omega \left(t - n\frac{\vec{s} \cdot \vec{r}}{c} \right) \right], \qquad (2.55)$$

$$\vec{H} = \vec{H}_o \exp\left[i\omega \left(t - n\frac{\vec{s} \cdot \vec{r}}{c} \right) \right], \qquad (2.56)$$

where Equation (2.54) has been used to replace υ.

Turning now to conducting media, we write in analogy to Equations (2.55) and (2.56):

$$\vec{E} = \vec{E}_o \exp\left[i\omega \left(t - \tilde{n}\frac{\vec{s} \cdot \vec{r}}{c} \right) \right], \qquad (2.57)$$

$$\vec{H} = \vec{H}_o \exp\left[i\omega \left(t - \tilde{n}\frac{\vec{s} \cdot \vec{r}}{c} \right) \right], \qquad (2.58)$$

where the refractive index, n, has been replaced with the parameter $\tilde{n}$, independent of time and space. Direct substitution into Equations (2.46) and (2.47) verifies that these are solutions, provided

$$\tilde{n}^2 = \frac{\epsilon\mu}{\epsilon_o \mu_o} \left[1 - i \left(\frac{\sigma}{\epsilon\omega} \right) \right]$$

$$= \kappa_e \kappa_m \left[1 - i \left(\frac{\sigma}{\epsilon\omega} \right) \right]. \qquad (2.59)$$

Since $\tilde{n}^2$ is complex, so must be $\tilde{n}$. Let

$$\tilde{n} = n(1 - ik) , \tag{2.60}$$

where n and k are positive real quantities. Upon squaring and equating real and imaginary parts of the result to those of Equation (2.59), one obtains

$$\left.\begin{aligned} n^2 &= \frac{\kappa_e \kappa_m}{2}\left[\sqrt{1 + \left(\frac{\sigma}{\epsilon\omega}\right)^2} + 1\right] , \\ n^2k^2 &= \frac{\kappa_e \kappa_m}{2}\left[\sqrt{1 + \left(\frac{\sigma}{\epsilon\omega}\right)^2} - 1\right] . \end{aligned}\right\} \tag{2.61}$$

The parameters $\tilde{n}$, n, and (nk) are called the complex refractive index, the refractive index, and the attenuation index, respectively. Utilizing Equation (2.60), Equation (2.57) may be written as

$$\vec{E} = \vec{E}_o \exp\left[i\omega\left(t - n\frac{\vec{s}\cdot\vec{r}}{c}\right)\right] \exp\left[-\frac{nk\omega}{c}(\vec{s}\cdot\vec{r})\right] . \tag{2.62}$$

Inspection of this equation shows that the significance of the real part of the complex refractive index, n, is the same as that in a dielectric. That is, it is the factor by which the speed of light is reduced in the medium compared with its value in a vacuum. Also, it is evident that the attenuation index, nk, is a measure of the reduction of the field strength with distance in the direction of propagation. Thus, unlike in a perfect dielectric, the waves experience a damping as they progress in conductors.

Further insight into the nature of the electromagnetic fields associated with the propagation of radiation may be obtained by subjecting the wave solutions to the Maxwell equations. Since the fields given by Equations (2.57) and (2.58) in the limit of vanishing conductivity yield those appropriate to nonconductors, only these expressions need be considered. Making use of the relations

$$\nabla \equiv -\left(\frac{i\omega\tilde{n}}{c}\right)\vec{s} \quad , \quad \frac{\partial}{\partial t} \equiv i\omega , \tag{2.63}$$

the Maxwell equations (I) and (II) yield

$$\begin{aligned} \vec{s}\cdot\vec{E} &= 0 , \\ \vec{s}\cdot\vec{H} &= 0 . \end{aligned} \tag{2.64}$$

Hence both the electric and magnetic field vectors are perpendicular to the direction of wave propagation. Waves with $\vec{E}$ and $\vec{H}$ in the plane of the wavefront are called transverse waves. The fourth Maxwell equation gives

$$\vec{H} = \frac{\tilde{n}}{\mu c}\vec{s}\times\vec{E} . \tag{2.65}$$

Thus the magnetic field is completely specified in terms of the electric field. It is so directed that the vectors $\vec{E}$, $\vec{H}$, and $\vec{s}$ form a right-handed triad of orthogonal vectors. The implication of the complex factor $\tilde{n}$ can be found by writing it in the form $re^{i\varphi}$, where

$$r = \sqrt{n^2(1 + k^2)} = \sqrt{\kappa_e\kappa_m}\left[1 + \left(\frac{\sigma}{\epsilon\omega}\right)^2\right]^{1/4} , \tag{2.66}$$

$$\tan\varphi = \left[\frac{1 - \sqrt{1 + (\sigma/\epsilon\omega)^2}}{1 + \sqrt{1 + (\sigma/\epsilon\omega)^2}}\right]^{1/2} . \tag{2.67}$$

With this, the relation for $\vec{H}$ may be written as

$$\vec{H} = \frac{\sqrt{n^2(1 + k^2)}}{\mu c} e^{i\varphi}\,\vec{s}\times\vec{E} . \tag{2.68}$$

It follows from the above that in nonconductors the magnetic and electric fields are in phase, while in conductors the fields differ in phase by φ. Figure 2.5 illustrates the relationship between the field vectors for a harmonic wave propagating in the positive z-coordinate direction in a dielectric.

The rate at which energy traverses a surface in an electromagnetic field is measured by the Poynting vector, $\vec{N} = \vec{E} \times \vec{H}$, defined in Equation (2.43). In view of Equation (2.65), it is apparent that the energy flow is in the direction of wave propagation. As noted earlier, the mean energy flow rate is of primary interest. According to Equation (2.44), the mean energy flow rate per unit area normal to the direction of propagation is given by

$$\langle \vec{N} \rangle = \frac{1}{2} \operatorname{Re}(\vec{E} \times \vec{H}^*) \tag{2.69}$$

for harmonic fields. Inserting the expression for $\vec{E}$ from (2.57) and forming the complex conjugate of $\vec{H}$ after use of (2.68), the mean energy flow rate is found as

$$\langle \vec{N} \rangle = \frac{n\vec{s}}{2\mu c} \left| \vec{E}_o \right|^2 \exp\left[-\frac{2nk\omega}{c} (\vec{r} \cdot \vec{s}) \right] . \tag{2.70}$$

The coefficient of $(\vec{r} \cdot \vec{s})$ in this expression is a measure of the reduction in the radiation intensity in passing a distance $\vec{r} \cdot \vec{s}$ from the origin in the direction of wave propagation. It is related to the absorption coefficient, κ_λ, by the relation

$$\kappa_\lambda = \frac{2nk\omega}{c} = \frac{4\pi\nu nk}{c} = \frac{4\pi nk}{\lambda_o} = \frac{4\pi k}{\lambda} , \tag{2.71}$$

where ν is the frequency of the radiation and λ_o and λ are the wavelengths in vacuum and in the medium, respectively. The reciprocal of the absorption coefficient is the distance in which the radiation intensity is reduced by the factor e; that is, the skin depth, δ. It follows that high-frequency fields in conductors are attenuated to a much greater degree than those of low frequency.

The effects of conductivity and frequency on the propagation of plane waves are most easily studied by considering two limiting situations. According to Equation (2.61), the behavior of the parameters n and nk is essentially determined by the parameter $(\sigma/\epsilon\omega)^2$.

First consider the case when $(\sigma/\epsilon\omega)^2 \ll 1$. This is generally the situation for poorly conducting materials and high-frequency radiation. Expansion of Equation (2.61) in powers of $(\sigma/\epsilon\omega)^2$ yields

$$n = \left\{ \frac{\kappa_e \kappa_m}{2} \left[1 + \frac{1}{2}\left(\frac{\sigma}{\epsilon\omega}\right)^2 + \dots + 1 \right] \right\}^{1/2} \cong \sqrt{\kappa_e \kappa_m} , \tag{2.72}$$

$$nk = \left\{ \frac{\kappa_e \kappa_m}{2} \left[1 + \frac{1}{2}\left(\frac{\sigma}{\epsilon\omega}\right)^2 + \dots - 1 \right] \right\}^{1/2} \cong \frac{1}{2}\sqrt{\kappa_e \kappa_m}\left(\frac{\sigma}{\epsilon\omega}\right)$$

$$= \frac{1}{2\epsilon_o} \sqrt{\frac{\kappa_e}{\kappa_m}} \left(\frac{\sigma}{\omega}\right) = \sqrt{\frac{\kappa_m}{\kappa_e}} \left(\frac{30\lambda_o}{\rho_e}\right) , \tag{2.73}$$

where ρ_e is the electrical resistivity (ohm meter) and λ_o the wavelength (meter) in vacuum. Within this approximation, the absorption coefficient, κ_λ, is

$$\kappa_\lambda = \frac{2nk\omega}{c} \cong \frac{1}{c\epsilon_o} \sqrt{\frac{\kappa_m}{\kappa_e}}\, \sigma$$

$$= 377 \sqrt{\frac{\kappa_m}{\kappa_e}} \frac{1}{\rho_e} . \tag{2.74}$$

Thus, under these conditions, the behavior is similar to that in dielectrics characterized by an index of refraction, $\sqrt{\kappa_e \kappa_m}$, and a small frequency independent absorption coefficient.

Consider the case when $(\sigma/\epsilon\omega)^2 \gg 1$. This approximation corresponds to the typical situation for metals at frequencies less than that of visible light. To this approximation, unity may be neglected in Equation (2.61), thereby giving the result

$$n \cong \sqrt{\frac{\kappa_e \kappa_m \sigma}{2\varepsilon\omega}} = \frac{1}{2}\frac{\sqrt{\kappa_m}}{\sqrt{\varepsilon_o}}\sqrt{\frac{\sigma}{\omega}} = \sqrt{\kappa_m}\sqrt{\frac{30\lambda_o}{\rho_e}}, \tag{2.75}$$

$$k \cong 1 .$$

For nonmagnetic substances κ_m may be taken as unity.

From Equation (2.68), the ratio of the amplitude of $\vec{H}$ to $\vec{E}$ is

$$\frac{|\vec{H}|}{|\vec{E}|} = \sqrt{\frac{\varepsilon}{\mu}}\left[1 + \left(\frac{\sigma}{\varepsilon\omega}\right)^2\right]^{1/4} = \sqrt{\frac{\varepsilon_o}{\mu_o}}\sqrt{\frac{\kappa_e}{\kappa_m}}\left[1 + \left(\frac{\sigma}{\varepsilon\omega}\right)^2\right]^{1/4}$$

$$= 2.654 \times 10^{-3}\sqrt{\frac{\kappa_e}{\kappa_m}}\left[1 + \left(\frac{\sigma}{\varepsilon\omega}\right)^2\right]^{1/4} . \tag{2.76}$$

Therefore the magnetic field in poorly conducting materials is generally much less than the electric field, whereas for good conductors it usually dominates. It also follows from Equation (2.67) that while the magnetic field is nearly in phase with the electric field in poor conductors, it lags the electric field by almost 45 degrees in metals. Finally, according to Equation (2.42), the electromagnetic energy density is equally divided between the magnetic and electric forms in dielectrics and almost entirely in the magnetic component for good conductors.

2.3.2 Polarization

Another aspect of the propagation of electromagnetic radiation is polarization. As shown earlier, the electric field has no component in the direction of propagation but, in general, can be resolved into two mutually perpendicular components lying in the plane of the wave front. Let a fixed Cartesian reference system be aligned with the z-axis in the direction of wave propagation. Denote by E_x and E_y the electric field components in this system. In general, these components are not in phase so that the trace of the resultant field with time in the xy-plane is some curve. The polarization of the radiation is defined in terms of this curve.

Consider a monochromatic harmonic wave, the components of which are given in the coordinate system described above as

$$E_x = E_{x,o} \sin\left[\omega\left(t - \frac{nz}{c}\right)\right], \tag{2.77}$$

$$E_y = E_{y,o} \sin\left[\omega\left(t - \frac{nz}{c}\right) + \varphi\right], \tag{2.78}$$

where $E_{x,o}$ and $E_{y,o}$ are unequal constants and φ is the phase difference in the components. Let us examine the trace of the resultant field with time at the point $z = 0$ in space. To determine the curve traced out by the resultant field at the selected point, the parameter t must be eliminated between Equations (2.77) and (2.78) when $z = 0$. This can be accomplished by employing the first to eliminate $\sin \omega t$ and $\cos \omega t$ from the expanded form of the second and then rationalizing the result. One obtains

$$\left(\frac{E_x}{E_{x,o}}\right)^2 + \left(\frac{E_y}{E_{y,o}}\right)^2 - 2\left(\frac{E_x}{E_{x,o}}\right)\left(\frac{E_y}{E_{y,o}}\right)\cos\varphi = \sin^2\varphi . \tag{2.79}$$

This is the equation of an ellipse and therefore the resultant electric field vector traces out an ellipse in the plane perpendicular to the direction of propagation. The radiation is said to be elliptically polarized. If the amplitudes of both components are equal ($E_{x,o} = E_{y,o}$) and the phase difference is $m\pi/2$ with m an odd integer, Equation (2.79) degenerates to the equation of a circle. The corresponding radiation is called circularly polarized. Should the phase difference be zero or an integral multiple of π, or either amplitude be zero, the ellipse reduces to a straight line. The radiation is then called linear or plane polarized. A linearly polarized harmonic wave is shown in Figure 2.5.

Radiation which is the result of generation by a very large number of microscopic sources comprising a macroscopic source, as, for example, that attributed to emission within a heated body, has the property that the instantaneous field has no preferential direction. Such radiation, referred to as natural or unpolarized radiation, may be regarded as the superposition of two linearly polarized waves of identical average amplitude vibrating in mutually perpendicular planes with their phase difference changing in a completely random fashion.

2.4 REFLECTION AND REFRACTION AT A PLANE SURFACE

A great deal of interesting and significant information can be obtained from the analysis of the interaction of a plane electromagnetic wave with the interface between media of different electric and magnetic properties. In the analysis to follow, the boundary is taken perfectly plane with its linear dimensions very large in comparison to the wavelength of the incident radiation.

2.4.1 Optical Laws

Consider two homogeneous and isotropic materials of infinite extent with a common plane boundary. Let the medium below and the medium above the plane z = 0 have permeabilities, permittivities, and conductivities μ_1, ϵ_1, σ_1, and μ_2, ϵ_2, σ_2, respectively. An electromagnetic plane wave traveling in the $\vec{s}^i$-direction in the medium designated 2 is incident on the boundary as illustrated in Figure 2.6. To satisfy the electromagnetic boundary conditions in general, a reflected and refracted wave must exist. Let the reflected and refracted beams propagate in the $\vec{s}^r$- and $\vec{s}^t$-directions. The incident, reflected, and refracted radiation fields will, in general, have components along the three Cartesian axes and may be written as:

Incident:

$$\vec{E}^i = \vec{E}_o^i \exp\left\{ i\omega^i \left[t - \frac{\tilde{n}_2(\vec{r}\cdot\vec{s}^i)}{c} \right] \right\},$$

Reflected:

$$\vec{E}^r = \vec{E}_o^r \exp\left\{ i\omega^r \left[t - \frac{\tilde{n}_2(\vec{r}\cdot\vec{s}^r)}{c} \right] \right\}, \qquad (2.80)$$

Refracted:

$$\vec{E}^t = \vec{E}_o^t \exp\left\{ i\omega^t \left[t - \frac{\tilde{n}_1(\vec{r}\cdot\vec{s}^t)}{c} \right] \right\}.$$

In the above, $\vec{r}$ is a radius vector from the origin located on the interface and is given by

$$\vec{r} = \vec{i}x + \vec{j}y + \vec{k}z.$$

Note that allowances have been made for possible frequency changes in the reflected and refracted waves. The propagation vectors may be expressed in terms of the coordinates as follows:

$$
\left.
\begin{aligned}
&\text{Incident:} \\
&\qquad \vec{s}^{\,i} = \vec{i}\,\sin\theta^i - \vec{k}\,\cos\theta^i\,, \\
&\text{Reflected:} \\
&\qquad \vec{s}^{\,r} = \vec{i}\,\sin\theta^r\,\cos\varphi^r + \vec{j}\,\sin\theta^r\,\sin\varphi^r + \vec{k}\,\cos\theta^r\,, \\
&\text{Refracted:} \\
&\qquad \vec{s}^{\,t} = \vec{i}\,\sin\theta^t\,\cos\varphi^t + \vec{j}\,\sin\theta^t\,\sin\varphi^t - \vec{k}\,\cos\theta^t\,.
\end{aligned}
\right\} \tag{2.81}
$$

The existence of boundary conditions at $z = 0$ which must be satisfied at all points on the plane at all times implies that the spatial and time variation of all fields must be identical at $z = 0$. Consequently, the arguments of the exponential functions of the fields in Equation (2.80) must all be equal at $z = 0$.

$$
\begin{aligned}
\omega^i\left[t - \frac{\tilde{n}_2}{c}(x\,\sin\theta^i)\right] &= \omega^r\left[t - \frac{\tilde{n}_2}{c}(x\,\sin\theta^r\,\cos\varphi^r + y\,\sin\theta^r\,\sin\varphi^r)\right] \\
&= \omega^t\left[t - \frac{\tilde{n}_1}{c}(x\,\sin\theta^t\,\cos\varphi^t + y\,\sin\theta^t\,\sin\varphi^t)\right].
\end{aligned} \tag{2.82}
$$

It follows that $\omega^i = \omega^r = \omega^t$; that is, the frequency of the reflected and refracted radiation is identical to that of the incident beam. Turning now to the geometrical factors, since a y-component is absent from the incident beam, the same component must vanish from the reflected and refracted ray arguments. Thus $\varphi^r = \varphi^t = 0$ or, in other words, the incident, reflected, and refracted beams all lie in a plane. The plane determined by the boundary normal and the incident ray is called the plane of incidence. For equality of the x-component of the incident and reflected field arguments, we must have

$$\sin\theta^i = \sin\theta^r, \text{ or } \theta^i = \theta^r\,. \tag{2.83}$$

Hence the angle of reflection, θ^r, equals the angle of incidence, θ^i. Finally, for the x-component of the incident and refracted field arguments to agree, we demand that

$$\tilde{n}_2\,\sin\theta^i = \tilde{n}_1\,\sin\theta^t\,. \tag{2.84}$$

When both $\tilde{n}_1$ and $\tilde{n}_2$ are real (electrical nonconductors), Equation (2.84) is recognized as Snell's law of refraction. Interested readers are referred to the literature (see Reference 2) for the implications of the complex form of Snell's law. Thus the electromagnetic waves obey all the experimental laws of reflection and refraction at the smooth interface separating two isotropic media.

2.4.2 Fresnel's Equations

The application of the boundary conditions is greatly simplified by employing the optical laws. Since the incident, reflected, and refracted rays all lie in a plane, geometrical difficulties are reduced to a minimum. Also, one need only deal with the complex field amplitudes since the exponential factors are identical at the interface. The boundary conditions which must be satisfied are given in Equations (2.35) through (2.38). These simplify for the media under consideration in the absence of surface charges and currents to the statement that the normal components of $\vec{D}(=\varepsilon\vec{E})$ and $\vec{B}(=\mu\vec{H})$, as well as the tangential components of $\vec{E}$ and $\vec{H}$, must be continuous across the interface. To apply the boundary conditions, it is convenient to resolve the electric fields and the associated magnetic fields into components parallel to the plane of incidence (subscript $\shortparallel$) and normal to the plane of incidence (subscript $\perp$). Let the incident, reflected, and refracted fields be denoted E^i, E^r, and E^t, respectively; their respective complex amplitudes in the plane of incidence as $E^i_{\shortparallel}$,

$E^r_{\parallel}$, $E^t_{\parallel}$, and normal to the plane of incidence as $E^i_{\perp}$, $E^r_{\perp}$, $E^t_{\perp}$.

Consider the application of the boundary conditions to the electric field component which is parallel to the plane of incidence. Figure 2.7a illustrates the situation in which the directions of the electric fields have been arbitrarily directed subject to the requirement that they be normal to the direction of propagation. The orientation of the magnetic fields has been chosen to give a positive flow of energy in the direction of propagation. The requirements that the tangential component of $\vec{E}$ and $\vec{H}$ must be continuous across the boundary yields

$$\left(E^i_{\parallel} - E^r_{\parallel}\right)\cos\theta - E^t_{\parallel}\cos\theta^t = 0\ , \tag{2.85}$$

$$\frac{\tilde{n}_2}{\mu_2}\left(E^i_{\parallel} + E^r_{\parallel}\right) - \frac{\tilde{n}_1}{\mu_1}E^t_{\parallel} = 0\ . \tag{2.86}$$

In Equation (2.86), the components of the magnetic vector have been replaced with the associated electric field components through Equation (2.65). The requirement that the normal component of $\vec{B}$ is continuous across the boundary is automatically satisfied, since the magnetic field component normal to the interface is zero. The continuity condition on the normal component of $\vec{D}$ simply yields Equation (2.86) when Snell's law is used. The above relations give the reflected and transmitted fields in terms of the incident field as

$$E^r_{\parallel} = \left[\frac{\mu_2\tilde{n}_1^2\cos\theta - \mu_1\tilde{n}_2\sqrt{\tilde{n}_1^2 - \tilde{n}_2^2\sin^2\theta}}{\mu_2\tilde{n}_1^2\cos\theta + \mu_1\tilde{n}_2\sqrt{\tilde{n}_1^2 - \tilde{n}_2^2\sin^2\theta}}\right]E^i_{\parallel}\ , \tag{2.87}$$

$$E^t_{\parallel} = \left[\frac{2\mu_1\tilde{n}_1\tilde{n}_2\cos\theta}{\mu_2\tilde{n}_1^2\cos\theta + \mu_1\tilde{n}_2\sqrt{\tilde{n}_1^2 - \tilde{n}_2^2\sin^2\theta}}\right]E^i_{\parallel}\ . \tag{2.88}$$

The complex coefficients of the incident field components imply that the reflected and transmitted waves differ in phase from the incident wave.

For the electric field component normal to the plane of incidence (Figure 2.7b), the boundary condition requiring continuity of the normal component of $\vec{D}$ is automatically satisfied, while the requirement for continuity of $\vec{E}$ and $\vec{H}$ across the interface yields

$$E^i_{\perp} + E^r_{\perp} - E^t_{\perp} = 0\ , \tag{2.89}$$

$$\frac{\tilde{n}_2}{\mu_2}\left(E^i_{\perp} - E^r_{\perp}\right)\cos\theta - \frac{\tilde{n}_1}{\mu_1}\cos\theta^t E^t_{\perp} = 0. \tag{2.90}$$

Again the magnetic field components have been eliminated with Equation (2.65). The condition that the normal component of $\vec{B}$ is continuous across the interface duplicates Equation (2.89). Simultaneous solution yields the reflected and transmitted fields as

$$E^r_{\perp} = \left[\frac{\mu_1\tilde{n}_2\cos\theta - \mu_2\sqrt{\tilde{n}_1^2 - \tilde{n}_2^2\sin^2\theta}}{\mu_1\tilde{n}_2\cos\theta + \mu_2\sqrt{\tilde{n}_1^2 - \tilde{n}_2^2\sin^2\theta}}\right]E^i_{\perp}\ , \tag{2.91}$$

$$E^t_{\perp} = \left[\frac{2\mu_1\tilde{n}_2\cos\theta}{\mu_1\tilde{n}_2\cos\theta + \mu_2\sqrt{\tilde{n}_1^2 - \tilde{n}_2^2\sin^2\theta}}\right]E^i_{\perp}\ . \tag{2.92}$$

Again, the coefficients are in general complex. Equations (2.87), (2.88), (2.91), and (2.92) are known as the Fresnel amplitude relations. Two special cases are of particular importance and these are considered next.

When both materials are dielectrics, the conductivities σ_1 and σ_2 are zero and the complex refractive indices reduce to the refractive indices. In this case the ratio of the field amplitudes simplify to the following expressions:

$$\frac{E^r_{11}}{E^i_{11}} = \frac{\mu_2 n^2_{12}\cos\theta - \mu_1\sqrt{n^2_{12} - \sin^2\theta}}{\mu_2 n^2_{12}\cos\theta + \mu_1\sqrt{n^2_{12} - \sin^2\theta}} \rightarrow \frac{n^2_{12}\cos\theta - \sqrt{n^2_{12} - \sin^2\theta}}{n^2_{12}\cos\theta + \sqrt{n^2_{12} - \sin^2\theta}} = \frac{\tan(\theta - \theta^t)}{\tan(\theta + \theta^t)} \tag{2.93}$$

$$\frac{E^t_{11}}{E^i_{11}} = \frac{2\mu_1 n_{12}\cos\theta}{\mu_2 n^2_{12}\cos\theta + \mu_1\sqrt{n^2_{12} - \sin^2\theta}} \rightarrow \frac{2 n_{12}\cos\theta}{n^2_{12}\cos\theta + \sqrt{n^2_{12} - \sin^2\theta}} = \frac{2\sin\theta^t\cos\theta}{\sin(\theta + \theta^t)\cos(\theta - \theta^t)} \tag{2.94}$$

$$\frac{E^r_\perp}{E^i_\perp} = \frac{\mu_1\cos\theta - \mu_2\sqrt{n^2_{12} - \sin^2\theta}}{\mu_1\cos\theta + \mu_2\sqrt{n^2_{12} - \sin^2\theta}} \rightarrow \frac{\cos\theta - \sqrt{n^2_{12} - \sin^2\theta}}{\cos\theta + \sqrt{n^2_{12} - \sin^2\theta}} = -\frac{\sin(\theta - \theta^t)}{\sin(\theta + \theta^t)} \tag{2.95}$$

$$\frac{E^t_\perp}{E^i_\perp} = \frac{2\mu_1\cos\theta}{\mu_1\cos\theta + \mu_2\sqrt{n^2_{12} - \sin^2\theta}} \rightarrow \frac{2\cos\theta}{\cos\theta + \sqrt{n^2_{12} - \sin^2\theta}} = \frac{2\sin\theta^t\cos\theta}{\sin(\theta + \theta^t)} \tag{2.96}$$

In the above n_{12} is the index of refraction of medium 1 relative to medium 2 (n_1/n_2). The equivalent expressions to the right in each case are the results appropriate for $\mu_1 = \mu_2 = \mu_o$, which is generally true for nonmagnetic materials. In this case the relative refractive index is

$$\sqrt{\frac{\kappa_{e,1}}{\kappa_{e,2}}}\ .$$

We shall exclude from consideration the phenomena of total reflection which occurs when $n_{12}\sin\theta^i > 1$.

The second situation of particular interest is the interface between a dielectric and a conductor. Let medium 2 be the dielectric and the refracting medium 1 conducting. Then

$$\begin{aligned} \tilde{n}_2 &= n_2\ , \\ \tilde{n}_1 &= n_1(1 - i\,k_1)\ . \end{aligned} \tag{2.97}$$

The presence of the complex refractive index for the conducting medium introduces considerable complexity into the field coefficients. In view of later considerations, attention is restricted to the reflected fields for nonmagnetic materials with $\mu_1 = \mu_2 = \mu_o$. After a lengthy process of rationalization of Equations (2.87) and (2.91), the magnitude of the reflected field in terms of the incident is found as

$$|E^r_{11}| = r_{11}\,|E^i_{11}|\ , \tag{2.98}$$

$$|E^r_\perp| = r_\perp\,|E^i_\perp|\ , \tag{2.99}$$

where

$$\left.\begin{aligned} r^2_\perp &= \frac{(a - \cos\theta)^2 + b^2}{(a + \cos\theta)^2 + b^2}\ , \\ r^2_{11} &= r^2_\perp \cdot \frac{(a - \sin\theta\tan\theta)^2 + b^2}{(a + \sin\theta\tan\theta)^2 + b^2}\ , \end{aligned}\right\} \tag{2.100}$$

and

$$\left.\begin{aligned} 2a^2 &= n_2^2\left[\sqrt{4n_{12}^4 k_1^2 + \left[n_{12}^2\left(1 - k_1^2\right) - \sin^2\theta\right]^2} + n_{12}^2\left(1 - k_1^2\right) - \sin^2\theta\right], \\ 2b^2 &= n_2^2\left[\sqrt{4n_{12}^4 k_1^2 + \left[n_{12}^2\left(1 - k_1^2\right) - \sin^2\theta\right]^2} - n_{12}^2\left(1 - k_1^2\right) + \sin^2\theta\right]. \end{aligned}\right\} \quad (2.101)$$

The factor n_{12} denotes the ratio of the real part of the complex refractive index of medium 1 to the refractive index of the dielectric medium.

2.5 REFLECTIVITY AND EMISSIVITY

Fresnel's equations provide the necessary information from which the monochromatic specular reflectivity of optically smooth surfaces may be evaluated. The monochromatic specular reflectivity is defined as the ratio of the monochromatic intensity of the reflected beam to that of the incident beam. From the reflectivity, the directional emissivity follows and finally the hemispherical properties. In order to reduce the complexity of the relationships, the medium through which the radiation propagates before striking the interface is taken to have a refractive index of unity and an extinction coefficient of zero. These conditions correspond to the important situation encountered by a surface in a vacuum or gaseous environment. Finally, the results are limited to nonmagnetic substances ($\mu_1 = \mu_2 = \mu_o$).

2.5.1 Dielectrics

In order to determine the monochromatic specular reflectivity, the mean energy flow in the reflected and incident beams must be evaluated. The energy incident on a unit area of the interface per unit time, $\mathcal{E}^i$, is given by the scalar product of the Poynting vector, $\langle\vec{N}\rangle$, and the unit normal to the surface, $\vec{n}$. Then

$$\mathcal{E}^i = \langle\vec{N}^i\rangle \cdot \vec{n} = \frac{1}{2}\sqrt{\frac{\varepsilon_o}{\mu_o}}\,|\vec{E}_o^i|^2\cos\theta = \frac{1}{2}\sqrt{\frac{\varepsilon_o}{\mu_o}}\left[\left(E_{11}^i\right)^2 + \left(E_\perp^i\right)^2\right]\cos\theta\,. \quad (2.102)$$

Similarly, the energy of the reflected radiation leaving the unit area in the specular direction per unit time is

$$\mathcal{E}^r = \langle\vec{N}^r\rangle \cdot \vec{n} = \frac{1}{2}\sqrt{\frac{\varepsilon_o}{\mu_o}}\,|\vec{E}_o^r|^2\cos\theta = \frac{1}{2}\sqrt{\frac{\varepsilon_o}{\mu_o}}\left[\left(E_{11}^r\right)^2 + \left(E_\perp^r\right)^2\right]\cos\theta\,. \quad (2.103)$$

The monochromatic specular reflectivity, $\rho_{s,\lambda}(\theta)$, is therefore

$$\rho_{s,\lambda}(\theta) = \frac{\left(E_{11}^r\right)^2 + \left(E_\perp^r\right)^2}{\left(E_{11}^i\right)^2 + \left(E_\perp^i\right)^2}\,. \quad (2.104)$$

If γ is the angle which the $\vec{E}$ vector of the incident wave makes with the plane of incidence, then

$$E_{11}^i = E\cos\gamma,\quad E_\perp^i = E\sin\gamma, \quad (2.105)$$

where

$$E^2 = |\vec{E}_o^i|^2 = \left(E_{11}^i\right)^2 + \left(E_\perp^i\right)^2\,. \quad (2.106)$$

With this, Equation (2.104) may be written as

$$\rho_{s,\lambda}(\theta) = \rho^{11}_{s,\lambda}(\theta)\cos^2\gamma + \rho^{\perp}_{s,\lambda}(\theta)\sin^2\gamma , \qquad (2.107)$$

with

$$\rho^{11}_{s,\lambda}(\theta) = \left(\frac{E^r_{11}}{E^i_{11}}\right)^2 , \qquad (2.108)$$

$$\rho^{\perp}_{s,\lambda}(\theta) = \left(\frac{E^r_{\perp}}{E^i_{\perp}}\right)^2 . \qquad (2.109)$$

The factors $\rho^{11}_{s,\lambda}(\theta)$ and $\rho^{\perp}_{s,\lambda}(\theta)$ are simply the monochromatic specular reflectivities for linearly polarized radiation parallel and normal to the plane of incidence, respectively. The specular reflectivity, therefore, depends on the polarization of the incident radiation. According to Equations (2.93) and (2.95), the plane polarized components of the reflectivity are given by

$$\rho^{11}_{s,\lambda}(\theta) = \left[\frac{n^2\cos\theta - \sqrt{n^2 - \sin^2\theta}}{n^2\cos\theta + \sqrt{n^2 - \sin^2\theta}}\right]^2 , \qquad (2.110)$$

$$\rho^{\perp}_{s,\lambda}(\theta) = \left[\frac{\cos\theta - \sqrt{n^2 - \sin^2\theta}}{\cos\theta + \sqrt{n^2 - \sin^2\theta}}\right]^2 , \qquad (2.111)$$

where n is the refractive index of the medium and θ is the polar angle of incidence. The wavelength dependence of the reflectivities enters through the refractive index. Should the refracting medium not be in the presence of a gas or vacuum, n is replaced with n_{12}, the ratio of the refractive index of the refracting medium to that in which the radiation is propagating.

Of particular significance is the monochromatic specular reflectivity for unpolarized or natural radiation. In this case, the mean amplitudes of the two polarized components in the incident beam are equal and their phase differences completely random. The corresponding reflectivity, $\bar{\rho}_{s,\lambda}$, may be obtained by averaging over all directions. Since the averages of $\sin^2\gamma$ and $\cos^2\gamma$ are 1/2,

$$\bar{\rho}_{s,\lambda}(\theta) = \frac{\rho^{11}_{s,\lambda}(\theta) + \rho^{\perp}_{s,\lambda}(\theta)}{2} , \qquad (2.112)$$

or

$$\bar{\rho}_{s,\lambda}(\theta) = \frac{1}{2}\left\{\left[\frac{n^2\cos\theta - \sqrt{n^2 - \sin^2\theta}}{n^2\cos\theta + \sqrt{n^2 - \sin^2\theta}}\right]^2 + \left[\frac{\cos\theta - \sqrt{n^2 - \sin^2\theta}}{\cos\theta + \sqrt{n^2 - \sin^2\theta}}\right]^2\right\} . \qquad (2.113)$$

The monochromatic specular reflectivity for unpolarized radiation, as well as the reflectivities for linearly polarized radiation parallel and normal to the plane of incidence, are illustrated in Figure 2.8 for selected values of the refractive index. One significant trend that may be observed is the vanishing value for $\rho^{11}_{s,\lambda}$; the implication being that if radiation is incident at this critical angle, the reflected wave is polarized entirely in the direction normal to the plane of incidence. Another interesting result is the relatively constant value for the unpolarized reflectivity, $\bar{\rho}_{s,\lambda}$, when the angle of incidence is within about 50 degrees from the normal. A representative comparison of experiment and theory is shown in Figure 2.9 for black glass.[4] The agreement of the measurements with the theoretical relationships is excellent.

The expressions for the specular reflectivity are particularly simple if the incidence is normal; that is, $\theta = 0$. In this case, there is no distinction between $\rho^{11}_{s,\lambda}$, $\rho^{\perp}_{s,\lambda}$, and $\bar{\rho}_{s,\lambda}$. The result for normal incidence, usually denoted $\rho_{N,\lambda}$ and called the monochromatic normal specular reflectance, is

$$\rho_{N,\lambda} = \left[\frac{n-1}{n+1}\right]^2 . \qquad (2.114)$$

Equation (2.114) is particularly useful for the evaluation of the effective refractive index of dielectrics from reflection measurements at near normal incidence.

The monochromatic directional emissivity, $\epsilon_{\lambda}(\theta)$, of an optically smooth dielectric follows from the reflectivity through the relation

$$\epsilon_\lambda(\theta) = 1 - \overline{\rho}_{s,\lambda}(\theta) \ . \qquad (2.115)$$

The use of this relation is restricted to materials of sufficient thickness such that the energy transmitted through the interface is characteristic of the material's temperature and not the energy from external sources which may have been transmitted completely through the substance. Inserting the value for $\overline{\rho}_{s,\lambda}(\theta)$ from Equation (2.103) yields

$$\epsilon_\lambda(\theta) = 1 - \frac{1}{2}\left[\left(\frac{n^2\cos\theta - \sqrt{n^2 - \sin^2\theta}}{n^2\cos\theta + \sqrt{n^2 - \sin^2\theta}}\right)^2 + \left(\frac{\cos\theta - \sqrt{n^2 - \sin^2\theta}}{\cos\theta + \sqrt{n^2 - \sin^2\theta}}\right)^2\right] ; \qquad (2.116)$$

whereas the monochromatic normal emissivity, $\epsilon_{N,\lambda}$, follows from Equation (2.114) as

$$\epsilon_{N,\lambda} = \epsilon_\lambda(0) = \frac{4n}{(n+1)^2} \ . \qquad (2.117)$$

Directional emissivity distributions for selected values of the refractive index are illustrated in Figure 2.10. In general, the directional emissivity is almost constant with the angle measured from the normal until a value of about 50 degrees is reached. Thereafter, there is a gradual decrease to zero at grazing angles. These trends are of course expected in view of the variation of the reflectivity with direction of incidence. Since the refractive indices of poorly conducting materials do not generally exceed a value of about three, dielectrics are characterized by large values of emissivity and small values of reflectivity.

Monochromatic hemispherical emissivity may be evaluated from the expression for the directional emissivity. The hemispherical emissivity, $\epsilon_{H,\lambda}$, is obtained from the directional emissivity by the relation

$$\epsilon_{H,\lambda} = \frac{1}{\pi}\int_{2\pi} \epsilon_\lambda(\theta)\cos\theta \, d\omega = \int_0^{\pi/2} \epsilon_\lambda(\theta)\sin 2\theta \, d\theta \ .$$

Substitution of Equation (2.116) for $\epsilon_\lambda(\theta)$, and subsequent integration yields the hemispherical emissivity for a nonconductor in terms of the refractive index as

$$\frac{\epsilon_{H,\lambda}}{\epsilon_{N,\lambda}} = \frac{1}{2}\left[\frac{2}{3} + \frac{1}{3n} + \frac{n^2(n+1)(n^2+2n-1)}{(n^2+1)^2(n-1)} - \frac{4n^3(n^4+1)}{(n^2+1)^3(n-1)^2}\,\ell n\, n + \frac{n(n+1)^2(n^2-1)^2}{2(n^2+1)^3}\,\ell n\left(\frac{n+1}{n-1}\right)\right],$$

where $\epsilon_{N,\lambda}$ is given by Equation (2.117). Figure 2.11 illustrates the hemispherical and normal emissivity expressions as a function of the refractive index. In Figure 2.12 the ratio $\epsilon_{H,\lambda}/\epsilon_{N,\lambda}$ is plotted against $\epsilon_{N,\lambda}$. It may be observed that for materials with refractive indices below about three, the hemispherical emissivity has a value approximately 95 per cent of that for the normal emissivity.

2.5.2 Conductors

The development of expressions for monochromatic specular reflectivity for conducting materials proceeds in an analogous fashion as that for dielectrics. In particular, Equation (2.107) continues to apply, but now the factors $\rho^{11}_{s,\lambda}(\theta)$ and $\rho^{\perp}_{s,\lambda}(\theta)$ are evaluated with r^2_{11} and $r^2_{\perp}$ given by Equation (2.100). The result for unpolarized radiation is

$$\bar{\rho}_{s,\lambda}(\theta) = \frac{1}{2}\left[\frac{(a - \cos\theta)^2 + b^2}{(a + \cos\theta)^2 + b^2}\right]\left[1 + \frac{(a - \sin\theta\ \tan\theta)^2 + b^2}{(a + \sin\theta\ \tan\theta)^2 + b^2}\right], \tag{2.118}$$

with a and b given by Equation (2.101). For normal incidence, the factors a and b take on the values n and nk, respectively. Then $r_{11} = r_{\perp}$ and the reflectivity for normal incidence reduces to the simple expression

$$\rho_{N,\lambda} = \frac{(n - 1)^2 + n^2k^2}{(n + 1)^2 + n^2k^2}. \tag{2.119}$$

The directional emissivity follows from the relation $\epsilon_{\lambda}(\theta) = 1 - \bar{\rho}_{s,\lambda}(\theta)$ as

$$\epsilon_{\lambda}(\theta) = \frac{1}{2}\left\{\frac{4a\cos\theta}{\left[(a + \cos\theta)^2 + b^2\right]}\right\}\left\{1 + \frac{\left[a^2 + b^2 + \sin^2\theta\right]\left[1/\cos^2\theta\right]}{\left[(a + \sin\theta\ \tan\theta)^2 + b^2\right]}\right\}. \tag{2.120}$$

The complexity of this expression is evident when one recalls that both quantities a and b are complicated functions of the angle of incidence, as well as the properties of the conductor. Typical directional emissivity distributions for good electrical conductors are illustrated in Figure 2.13 for selected values of the optical parameters n and k. Again, $\epsilon_{\lambda}(\theta)$ is nearly constant with increasing polar angle to about 45 degrees from the surface normal. In contrast to dielectrics, however, the directional emissivity for good conductors has values which are large relative to that at the normal for near grazing incidence. Also, except at large polar angles, the directional emissivity is small compared to the values for nonconductors shown in Figure 2.10.

The emissivity in the direction of the surface normal, $\epsilon_{N,\lambda}$, follows from Equation (2.120) as

$$\epsilon_{N,\lambda} = \frac{4n}{(n + 1)^2 + n^2k^2}. \tag{2.121}$$

Metals constitute a class of materials of considerable importance. For these substances the ratio $(\sigma/\varepsilon\omega)^2 \gg 1$ and the approximations cited earlier are appropriate. Then, neglecting $\sin^2\theta$ with respect to $n^2(1 + k^2)$, the expression for a and b in Equation (2.101) yields

$$a \simeq n, \quad b \simeq nk.$$

The resulting expression for reflectivity may be written as

$$\begin{aligned}\bar{\rho}_{s,\lambda}(\theta) &= \frac{1}{2}\left\{\left[\frac{(n - \cos\theta)^2 + n^2k^2}{(n + \cos\theta)^2 + n^2k^2}\right] + \left[\frac{\left[n - (1/\cos\theta)\right]^2 + n^2k^2}{\left[n + (1/\cos\theta)\right]^2 + n^2k^2}\right]\right\} \\ &= \frac{1}{2}\left\{\left[\frac{\cos^2\theta - 2n\cos\theta + n^2(1 + k^2)}{\cos^2\theta + 2n\cos\theta + n^2(1 + k^2)}\right] + \left[\frac{n^2(1 + k^2)\cos^2\theta - 2n\cos\theta + 1}{n^2(1 + k^2)\cos^2\theta + 2n\cos\theta + 1}\right]\right\}.\end{aligned} \tag{2.122}$$

In Figure 2.9, experimental data for platinum at 2μ is compared to the theoretical distribution calculated from Equation (2.122) using reported values for the optical parameters.(4) The experimental values are consistently lower than the predicted values by up to 5 per cent. This decrease may be a result of impurities in the platinum.

The directional emissivity in this approximation is

$$\epsilon_{\lambda}(\theta) = \frac{2n}{\left[(n + \cos\theta)^2 + n^2k^2\right]}\left\{\cos\theta + \frac{1}{\cos\theta}\left[\frac{(n + \cos\theta)^2 + n^2k^2}{\left[n + (1/\cos\theta)^2 + n^2k^2\right]}\right]\right\}. \tag{2.123}$$

The integration of this expression for directional emissivity may be carried out and the resulting expression for the hemispherical emissivity is[3]

$$\epsilon_{H,\lambda} = 4n - 4n^2 \ln\left[\frac{1+2n+n^2(1+k^2)}{n^2(1+k^2)}\right] + \frac{4n^2(1-k^2)}{k}\tan^{-1}\left[\frac{k}{1+n(1+k^2)}\right] + \frac{4}{n(1+k^2)}$$

$$- \frac{4}{n^2(1+k^2)^2}\ln\left[1+2n+n^2(1+k^2)\right] + \frac{4(1-k^2)}{n^2k(1+k^2)^2}\tan^{-1}\left[\frac{nk}{1+n}\right] . \qquad (2.124)$$

The monochromatic hemispherical emissivity is illustrated in Figure 2.14 as a function of the refractive index for selected values of the parameter k. The solid curves were calculated by numerical integration of Equation (2.120) and the dashed by evaluation of Equation (2.124). For refractive index values greater than unity, Equation (2.124) is exceptionally accurate yielding values for $\epsilon_{H,\lambda}$, which are imperceptible from those obtained from the exact relation except for very small values of k. Such is not the case, however, for $n < 1$ except for large values of k.[5] In Figure 2.15, the ratio of monochromatic hemispherical to monochromatic normal emissivity is shown.

A number of investigators have used the theoretical expressions for monochromatic reflectivity to evaluate the total emissivity of metals. In general the different expressions reported may be attributed to the approximations introduced, as well as to the different atomic models employed to describe the optical parameters n and k. For example, Davisson and Weeks[6] took the complex refractive index as

$$\tilde{n}^2 = 1 - i\left(\frac{60\,\lambda}{\rho_e}\right), \qquad (2.125)$$

where ρ_e is the electrical resistivity in units of ohm-cm and λ is the wavelength in cm. The atomic model which is the basis for this relation ignores the effects of bound electrons and assumes that the free electrons oscillate at the frequency of and in phase with the impressed field. It should be expected to yield reasonable results for metals, provided the frequency is not too high. The value of n and nk corresponding to Equation (2.125) are

$$\left.\begin{aligned} n &= \left\{\frac{1}{2}\left[\sqrt{1+\left(\frac{60\,\lambda}{\rho_e}\right)^2}+1\right]\right\}^{1/2}, \\ nk &= \left\{\frac{1}{2}\left[\sqrt{1+\left(\frac{60}{\rho_e}\right)^2}-1\right]\right\}^{1/2}. \end{aligned}\right\} \qquad (2.126)$$

These investigators graphically integrated the equivalent of the monochromatic directional emissivity expression given by Equation (2.120) and fit their results within a few per cent with the equation

$$\epsilon_{H,\lambda} = \epsilon_{N,\lambda}\left[1 + 0.305\exp\left(-1.3368\sqrt{\frac{\rho_e}{\lambda}}\right)\right]. \qquad (2.127)$$

Also, $\epsilon_{N,\lambda}$ was approximated with a series expansion of Equation (2.119).

$$\epsilon_{N,\lambda} = 0.365\left(\frac{\rho_e}{\lambda}\right)^{1/2} + 0.0667\left(\frac{\rho_e}{\lambda}\right) - 0.00912\left(\frac{\rho_e}{\lambda}\right)^{3/2} + \ldots . \qquad (2.128)$$

The first term of Equation (2.128) may be recognized as the Hagen-Ruben relation.

Expanding the exponential in Equation (2.127) and combining with Equation (2.128) gives

$$\epsilon_{H,\lambda} = 0.476\left(\frac{\rho_e}{\lambda}\right)^{1/2} - 0.236\left(\frac{\rho_e}{\lambda}\right) + 0.139\left(\frac{\rho_e}{\lambda}\right)^{3/2} - 0.0662\left(\frac{\rho_e}{\lambda}\right)^2 + \dots \quad (2.129)$$

The total hemispherical emissivity follows from

$$\epsilon_H = \frac{1}{\sigma T^4}\int_0^\infty \epsilon_{H,\lambda}\, e_{b,\lambda}\, d\lambda \,,$$

where σ, T, and $e_{b,\lambda}$ denote the Stefan-Boltzmann constant, absolute temperature, and black body monochromatic emissive power, respectively. Substitution of the expression for $\epsilon_{H,\lambda}$ of Equation (2.129) and subsequent integration yields

$$\epsilon_H = 0.751(\rho_e T)^{1/2} - 0.632(\rho_e T) + 0.670(\rho_e T)^{3/2} - 0.607(\rho_e T)^2 + \dots , \quad (2.130)$$

where the temperature is in degrees Kelvin and ρ_e in units of ohm-cm.

Schmidt and Eckert[7] followed a similar procedure replacing the infinite series expression of Equation (2.128) with the polynomial

$$\epsilon_{N,\lambda} = 0.365\left(\frac{\rho_e}{\lambda}\right)^{1/2} - 0.0464\left(\frac{\rho_e}{\lambda}\right). \quad (2.131)$$

Note that the coefficient of the second term does not agree with the corresponding term of Equation (2.128). This alteration was an effort by the authors to correct for the remaining terms in the series. Equation (2.131) was integrated over wavelength in the customary manner to give the total emissivity in the normal direction.

$$\epsilon_N = 0.576(\rho_e T)^{1/2} - 0.124(\rho_e T) \,. \quad (2.132)$$

These investigators also graphically integrated the expression for $\epsilon_\lambda(\theta)$ and fit their results with the following expressions.

$$\epsilon_{H,\lambda} = \begin{cases} 0.476\sqrt{\frac{\rho_e}{\lambda}} - 0.148\left(\frac{\rho_e}{\lambda}\right) ; & 0 < \frac{\rho_e}{\lambda} < 0.5 \,, \\ 0.442\sqrt{\frac{\rho_e}{\lambda}} - 0.0995\left(\frac{\rho_e}{\lambda}\right) ; & 0.5 < \frac{\rho_e}{\lambda} < 2.5 \,. \end{cases} \quad (2.133)$$

Subsequent integration over wavelength gave the following results for total hemispherical emissivity.

$$\epsilon_H = \begin{cases} 0.751\sqrt{\rho_e T} - 0.396(\rho_e T) \,, & 0 < \rho_e T < 0.2 \,. \\ 0.698\sqrt{\rho_e T} - 0.266(\rho_e T) \,, & 0.2 < \rho_e T < 0.5 \,. \end{cases} \quad (2.134)$$

More recently, Parker and Abbott[8] have followed similar procedures while accounting for the electronic relaxation time. In the limit of zero relaxation time, these investigators obtained the following result for total hemispherical emittance.

$$\epsilon_H = 0.766(\rho_e T)^{1/2} - (0.309 - 0.0887 \,\ell n\, \rho_e T)(\rho_e T) - 0.0175(\rho_e T)^{3/2} \,. \quad (2.135)$$

A comparison of the theoretical expressions for total hemispherical emittance to measurements for various metals is illustrated in Figure 2.16. Agreement of the data with the theory is good. The expressions developed by Schmidt and Eckert [Equation (2.134)] appear to agree with the data at large values of $\sqrt{\rho_e T}$ better than either of the other relations.

REFERENCES

1. Jackson, J. D. Classical Electrodynamics. New York: John Wiley & Sons, 1962.

2. Stratton, J. A. Electromagnetic Theory. New York: McGraw-Hill Book Co., 1941.

3. Dunkle, R. V., "Emissivity and Inter-Reflection Relationships for Infinite Parallel Specular Surfaces," Symposium on Thermal Radiation of Solids, S. Katzoff, ed., NASA SP-55, 1965, pp. 39-44.

4. Brandenberg, W. M., "The Reflectivity of Solids at Grazing Angles," Measurement of Thermal Radiation Properties of Solids, J. Richmond, ed., NASA SP-31, 1963, pp. 75-82.

5. Hering, R. G. and T. F. Smith, "Surface Radiation Properties from Electromagnetic Theory," to be published in International Journal of Heat and Mass Transfer.

6. Davison, C. D. and J. R. Weeks, Jr., "The Relation Between the Total Thermal Emissive Power of a Metal and Its Electrical Resistivity," Journal of the Optical Society of America, 8:5 (May, 1924), pp. 581-605.

7. Schmidt, E. and E. Eckert, "Directional Distribution of Heat Radiation from Surfaces," Forschung auf dem Gebiete des Ingenieurwesens, 6 (1935), pp. 175-183.

8. Parker, W. J. and G. L. Abbott, "Theoretical and Experimental Studies of the Total Emittance of Metals," Symposium on Thermal Radiation of Solids, S. Katzoff, ed., NASA SP-55, 1964, pp. 11-20.

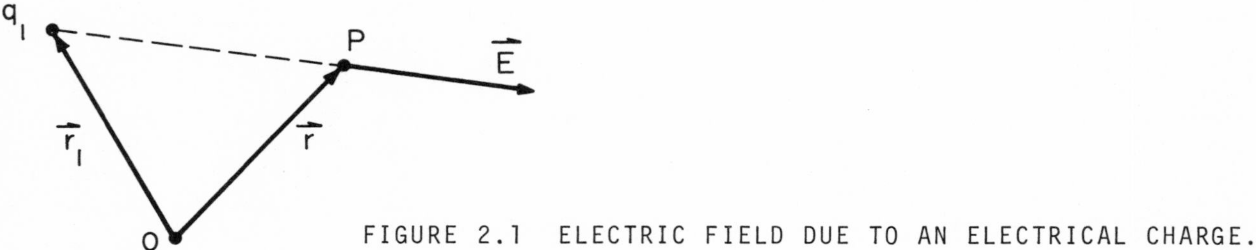

FIGURE 2.1 ELECTRIC FIELD DUE TO AN ELECTRICAL CHARGE.

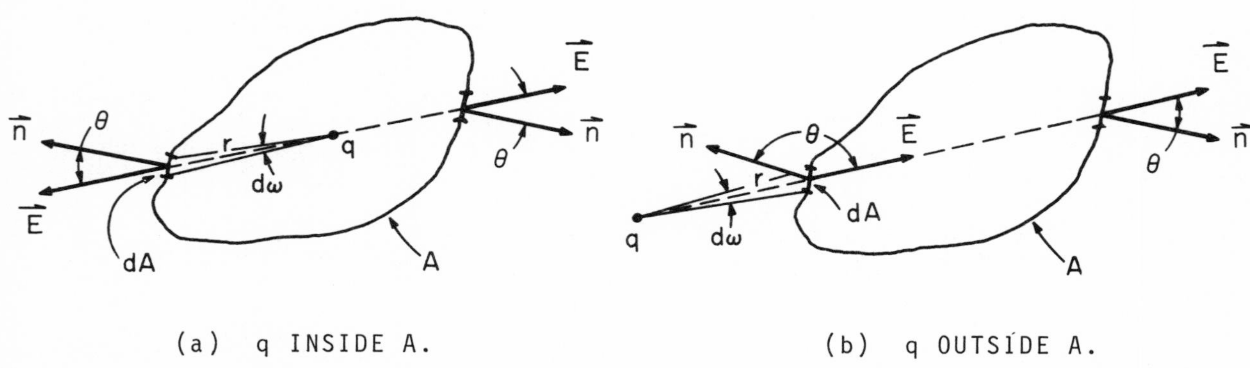

(a) q INSIDE A.

(b) q OUTSIDE A.

FIGURE 2.2 GAUSS' LAW.

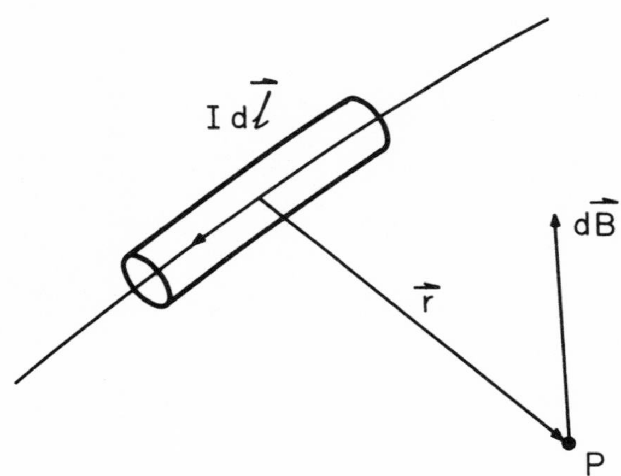

FIGURE 2.3 MAGNETIC INDUCTION DUE TO A CURRENT ELEMENT.

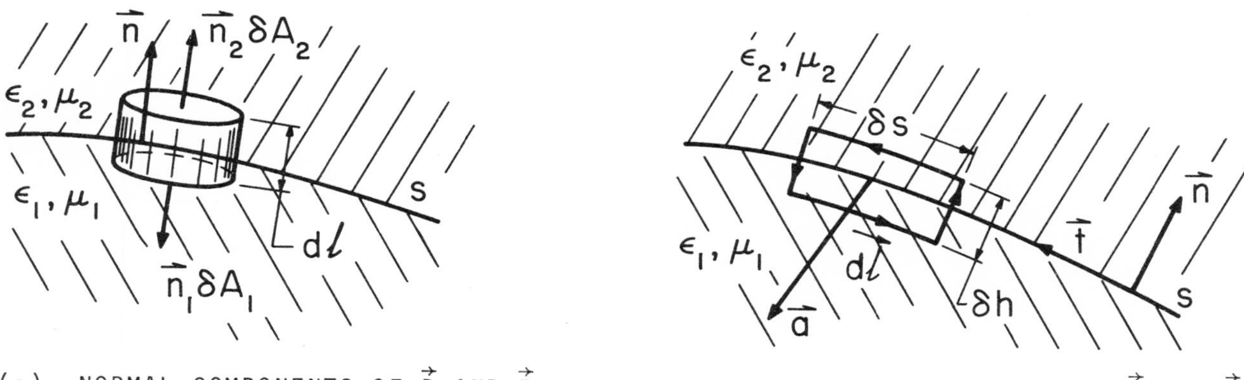

(a) NORMAL COMPONENTS OF $\vec{D}$ AND $\vec{B}$. (b) TANGENTIAL COMPONENTS OF $\vec{E}$ AND $\vec{H}$.

FIGURE 2.4 ELECTROMAGNETIC BOUNDARY CONDITIONS.

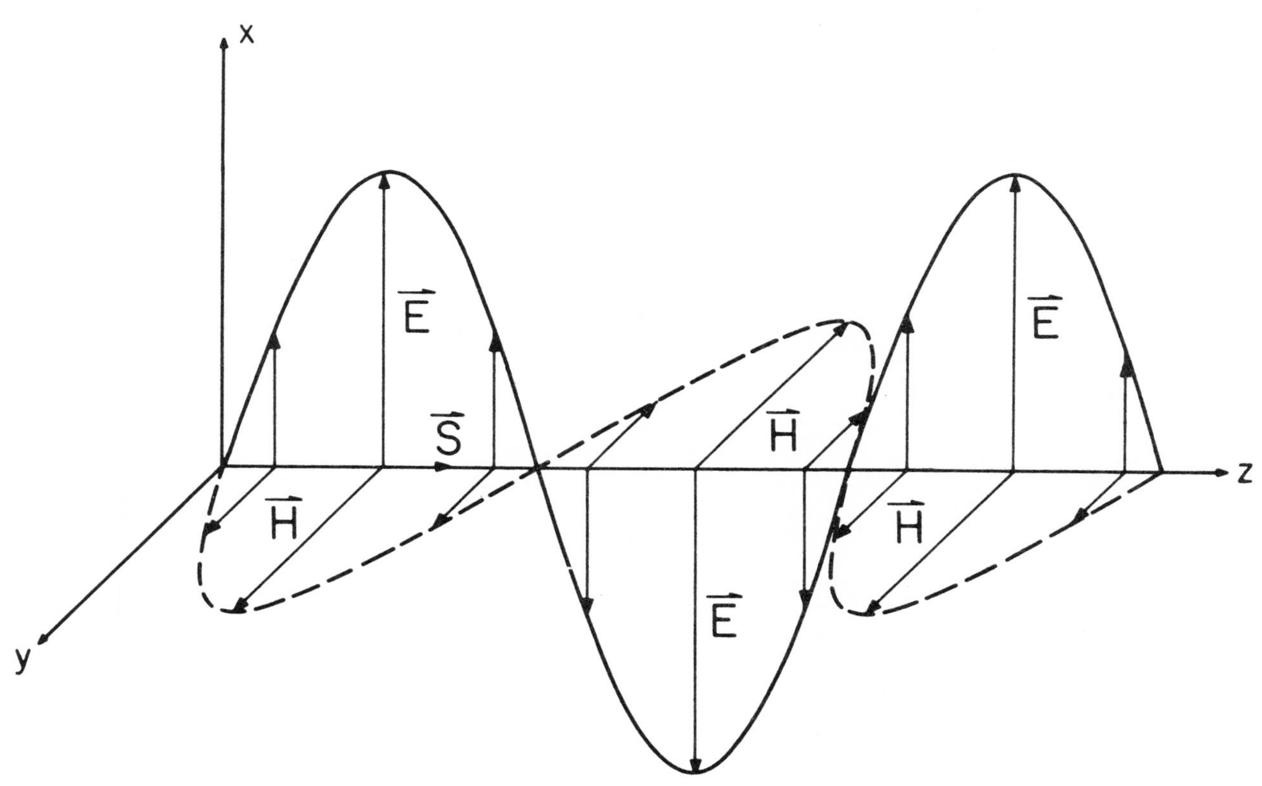

FIGURE 2.5 LINEARLY POLARIZED HARMONIC WAVE.

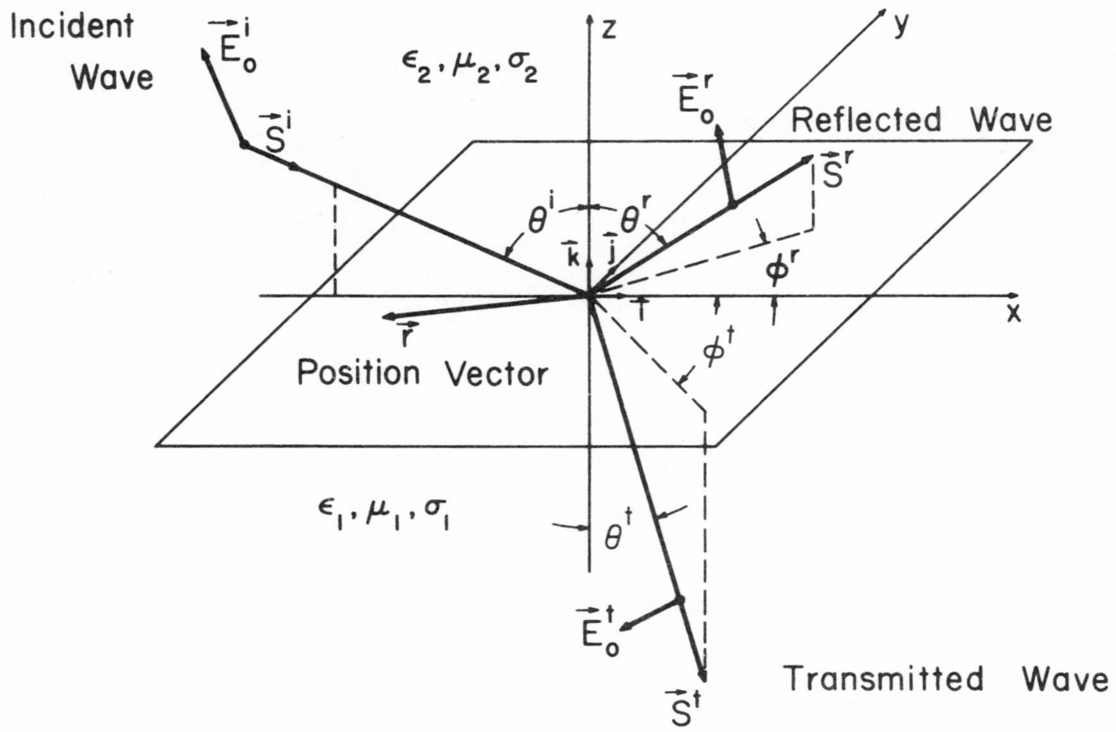

FIGURE 2.6 INTERACTION OF AN ELECTROMAGNETIC WAVE WITH A PLANE INTERFACE.

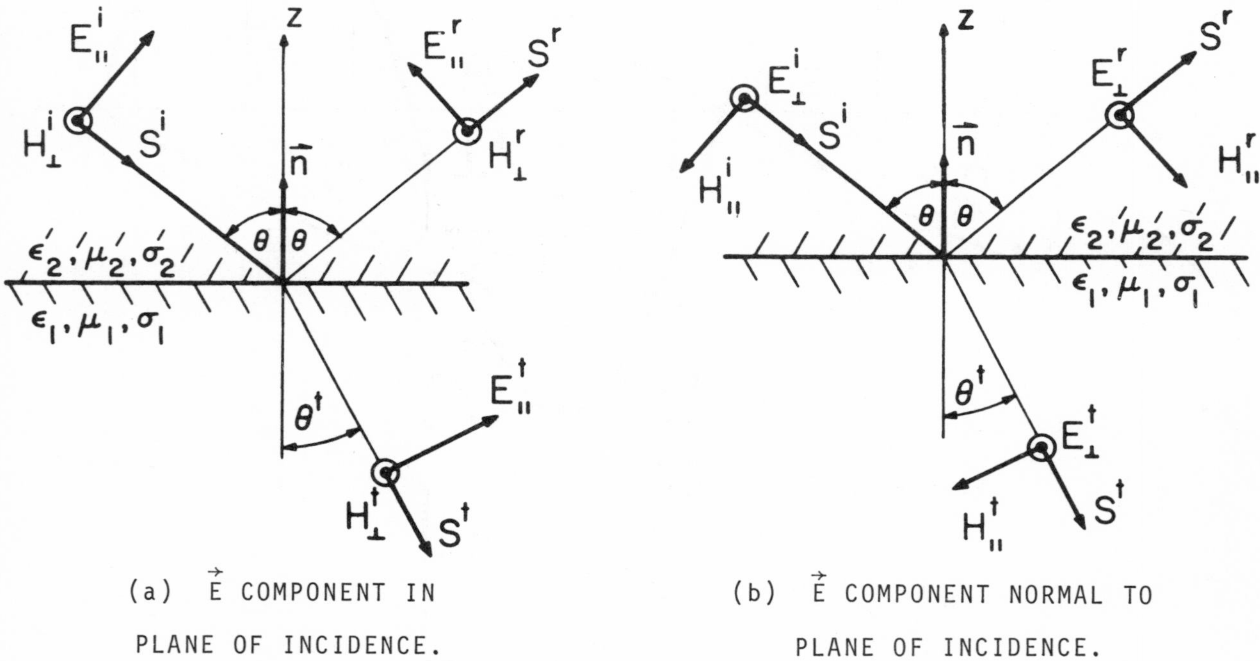

(a) $\vec{E}$ COMPONENT IN PLANE OF INCIDENCE.

(b) $\vec{E}$ COMPONENT NORMAL TO PLANE OF INCIDENCE.

FIGURE 2.7 REFLECTION AND REFRACTION AT A PLANE INTERFACE.

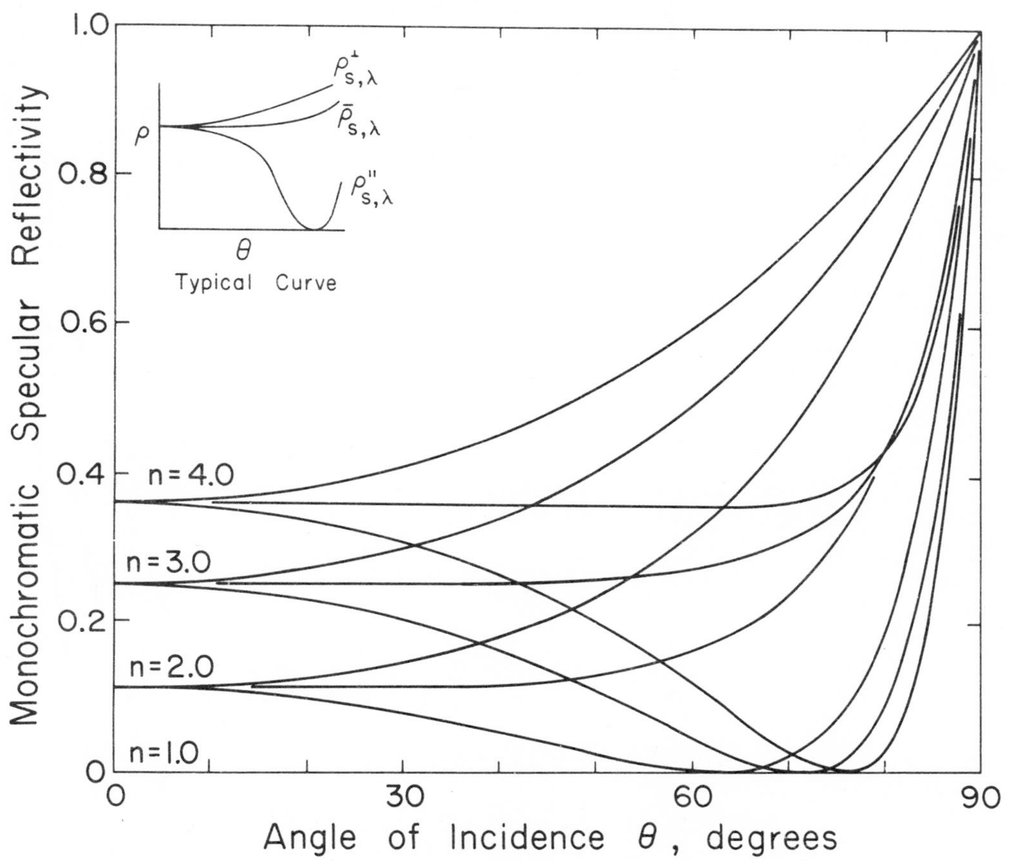

FIGURE 2.8 SPECULAR REFLECTIVITY OF ELECTRICAL NONCONDUCTORS.

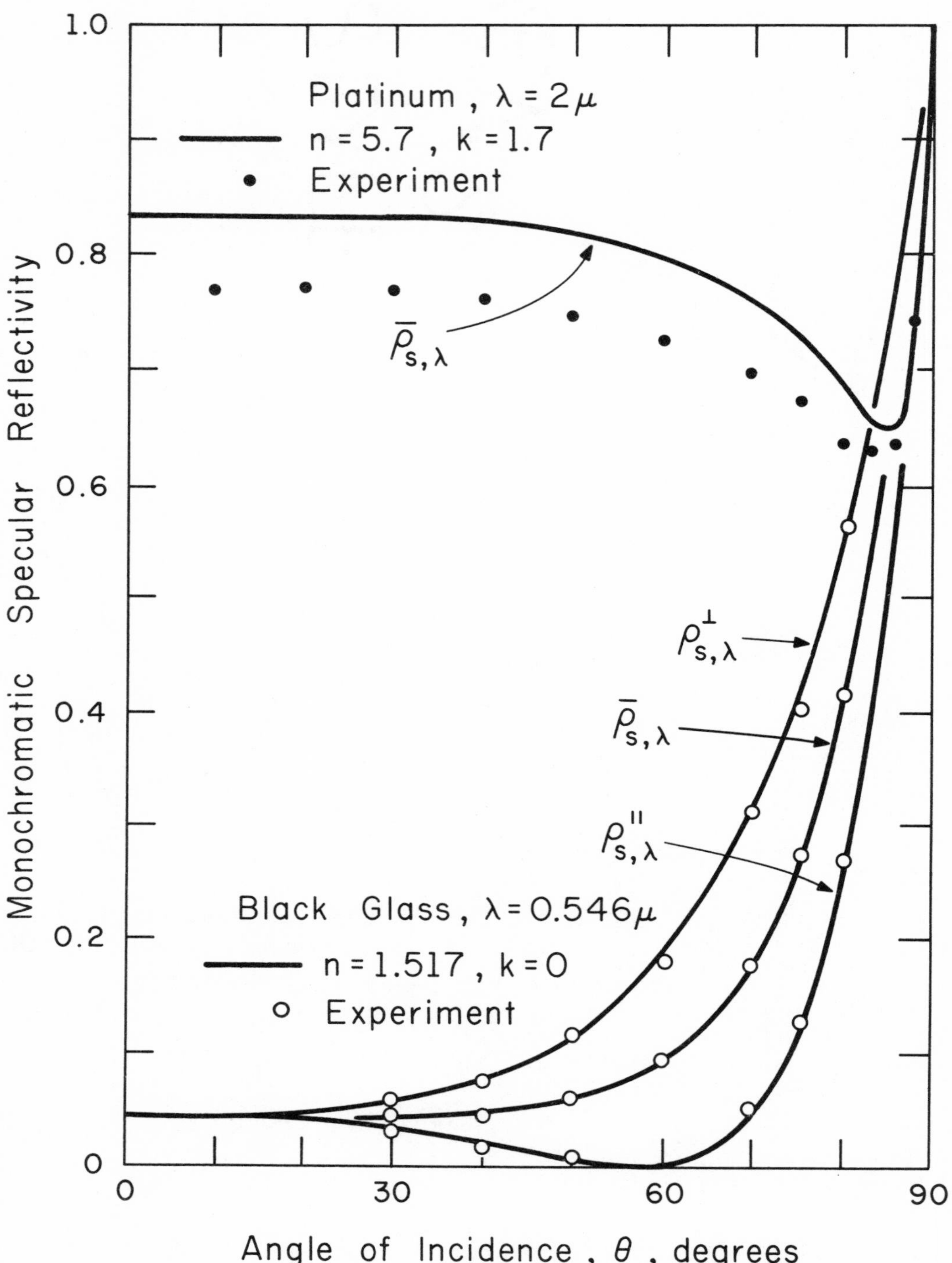

FIGURE 2.9 COMPARISON OF THEORY AND EXPERIMENT FOR SPECULAR REFLECTIVITY (REFERENCE 4).

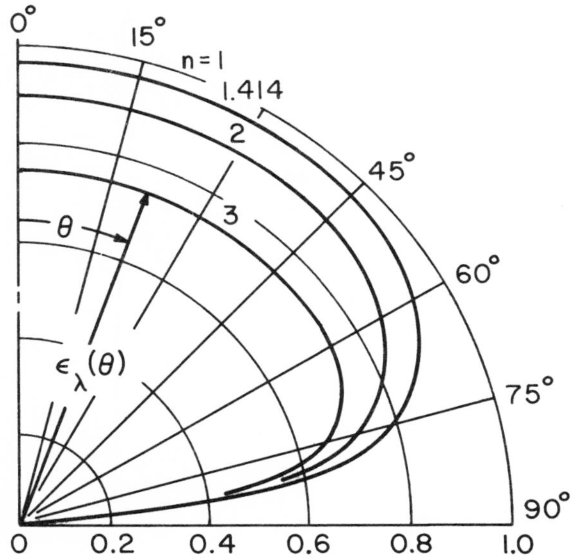

FIGURE 2.10 DIRECTIONAL EMISSIVITY OF ELECTRICAL NONCONDUCTORS.

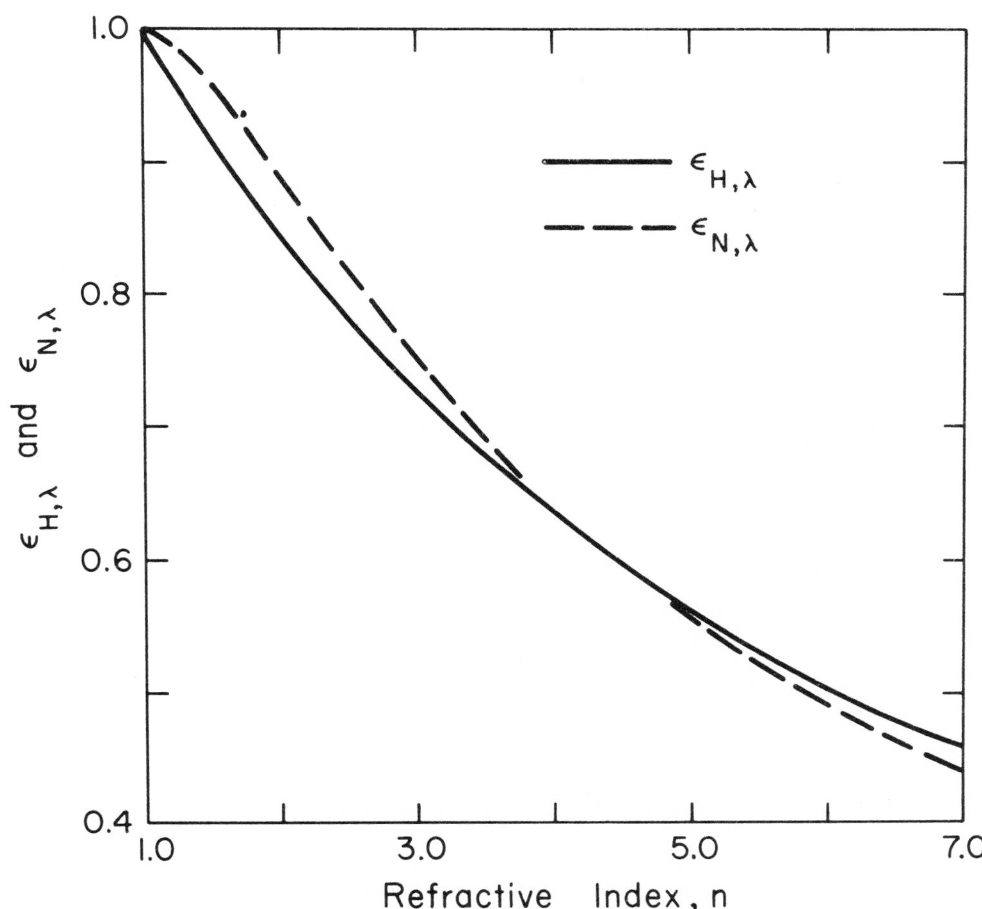

FIGURE 2.11 NORMAL AND HEMISPHERICAL EMISSIVITY OF ELECTRICAL NONCONDUCTORS.

1.1
1.0
0.9
0.4
0.6
0.8
1.0
$\epsilon_{H,\lambda}/\epsilon_{N,\lambda}$
$\epsilon_{N,\lambda}$

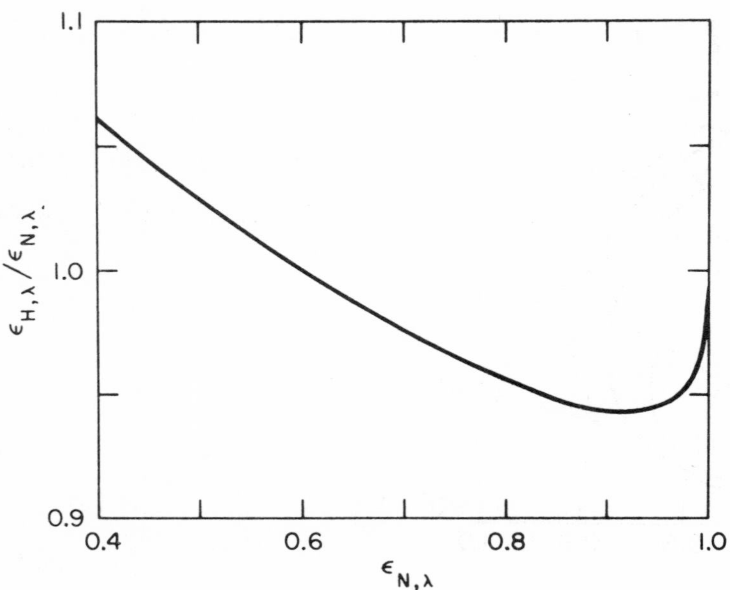

FIGURE 2.12 RATIO $\varepsilon_{H,\lambda}/\varepsilon_{N,\lambda}$ FOR ELECTRICAL NONCONDUCTORS.

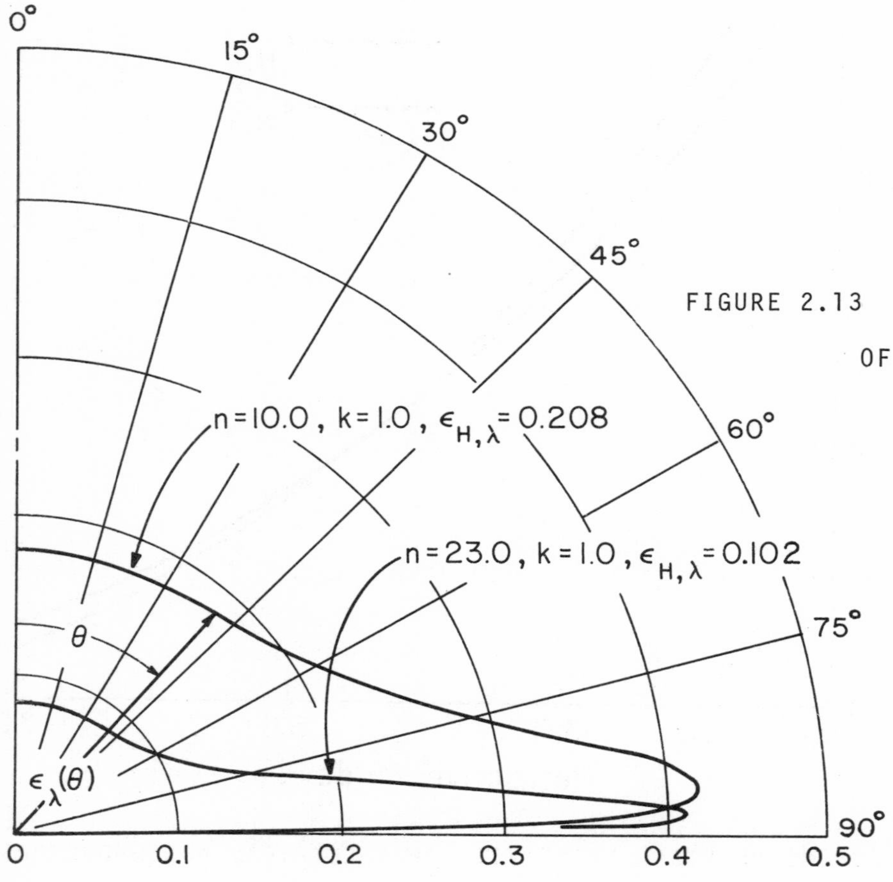

0°
15°
30°
45°
60°
75°
90°
n = 10.0, k = 1.0, $\epsilon_{H,\lambda}$ = 0.208
n = 23.0, k = 1.0, $\epsilon_{H,\lambda}$ = 0.102
θ
$\epsilon_\lambda(\theta)$
0
0.1
0.2
0.3
0.4
0.5

FIGURE 2.13 DIRECTIONAL EMISSIVITY OF ELECTRICAL CONDUCTORS.

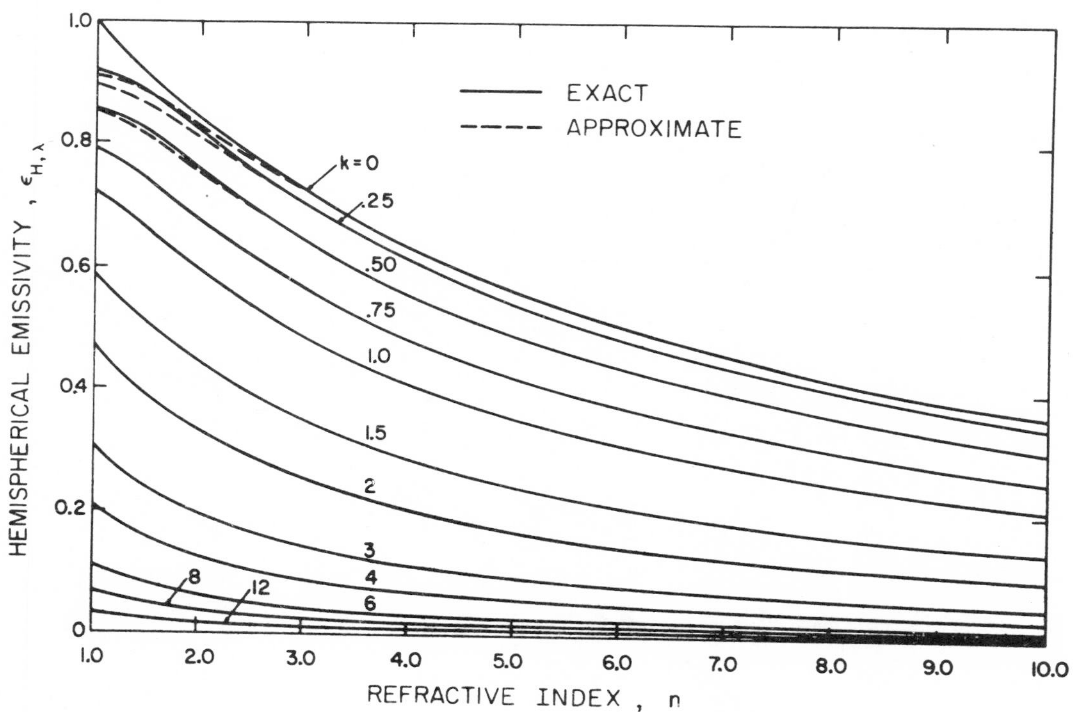

FIGURE 2.14 HEMISPHERICAL EMISSIVITY OF ELECTRICAL CONDUCTORS ($n \geq 1$).

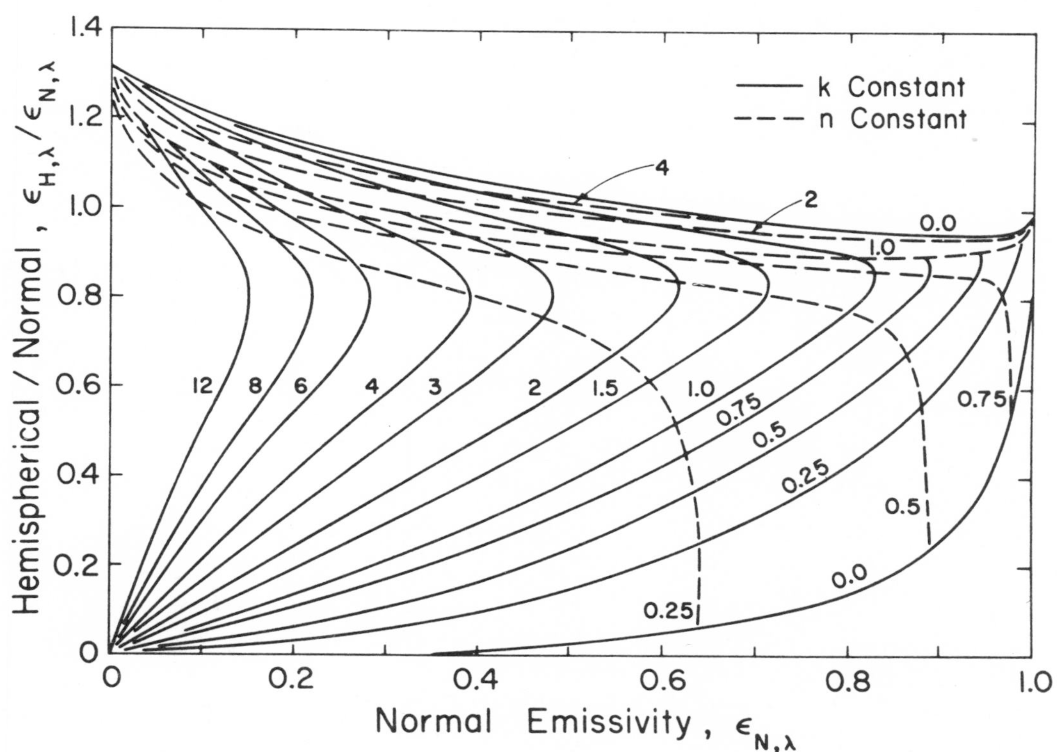

FIGURE 2.15 RATIO $\varepsilon_{H,\lambda}/\varepsilon_{N,\lambda}$ FOR ELECTRICAL CONDUCTORS.

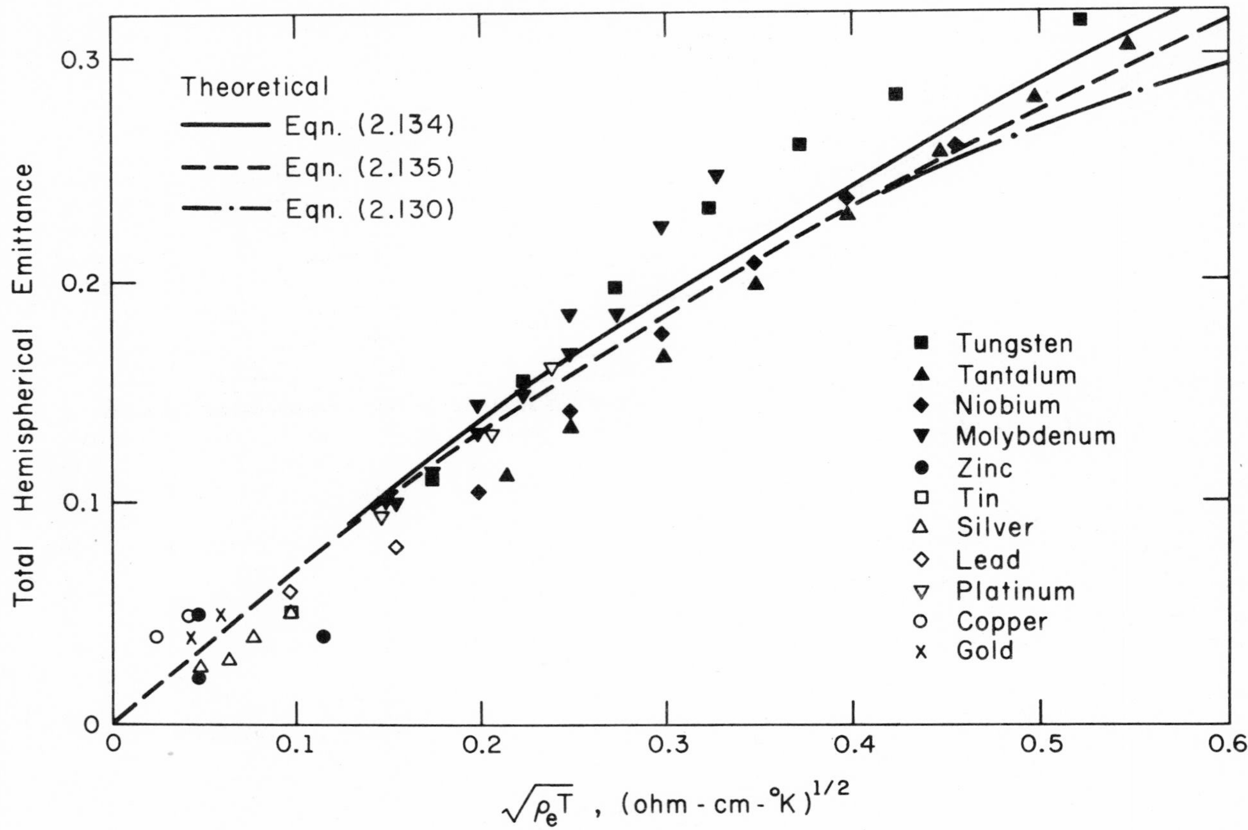

FIGURE 2.16 TOTAL HEMISPHERICAL EMITTANCE OF VARIOUS METALS (REFERENCE 8).

3. RADIATIVE TRANSFER BETWEEN SURFACES

3.1 INTRODUCTION

Recent technological applications, particularly those connected with space flight, have increased the need for precise evaluation of radiant heat transfer between surfaces. The achievement of greater accuracy can be foreseen by employing methods of analysis which account for the directional and spectral property dependence exhibited by engineering surfaces. To include such effects in the formulation of the radiant transfer problem requires a consideration of the radiant exchange process on a more fundamental level than is usually undertaken. Surface properties must be introduced which can characterize the spatial and spectral nonuniformity of surface emission, absorption, and reflection processes. It is the intent here to present a formulation for radiant transfer between surfaces which takes into account the aforementioned real surface characteristics. To facilitate the discussion, a brief review of surface radiation properties and their interrelationships is presented. This is followed by the formulation of the radiant transfer problem. The simplifications introduced by some of the more commonly employed methods of radiant transfer analysis are illustrated by considering a recently proposed surface property model. Consideration is limited throughout to the steady transfer between opaque surfaces in the presence of a medium transparent to thermal radiation with constant refractive index. The influence of polarization on radiant transfer is neglected.

3.2 RADIATION PROPERTIES OF SURFACES

It is customary to introduce certain quantities which characterize the ability of a material to reflect, absorb, and emit thermal radiation. To facilitate the subsequent formulation, these surface properties are introduced with sufficient generality to account for direction and wavelength dependence. Although the properties are not defined on a spectral basis, their spectral counterparts follow directly from those defined by limiting the stated quantities to a sufficiently small spectral interval about a selected wavelength. Consideration is given first to reflection properties, then to absorption properties, and finally to emission properties. All property definitions are based on the coordinate system illustrated in Figure 3.1.

The intensity of radiation incident from a direction $\vec{s}'$ is denoted by $I^-(\vec{s}')$, while that of radiation leaving the element in the $\vec{s}$-direction is $I^+(\vec{s})$. In general, primed variables denote quantities associated with incident energy and unprimed, those associated with leaving energy.

3.2.1 Reflectance

The fundamental surface reflection property required for the analysis to follow is the bidirectional (or biangular) reflectance. This reflectance essentially gives the contribution of the radiation incident on a surface within a small solid angle about an arbitrary direction to the reflected energy leaving the surface within a small solid angle about another prescribed direction. To arrive at a consistent set of definitions, the symbol $d\mathcal{E}$ is used to denote the flow of radiant energy per unit time and unit surface area contained within an infinitesimal solid angle. Incident, reflected, and emitted energies, as well as the corresponding intensities when necessary, are distinguished by the subscripts i, r, and e, respectively. Thus, for example, the intensity of the incident energy within $d\omega'$ about the $\vec{s}'$-direction is

$$I^-(\vec{s}') = \frac{d\mathcal{E}_i(\vec{s}')}{\cos\theta' \, d\omega'} , \tag{3.1}$$

while that for reflected energy within solid angle $d\omega$ about the $\vec{s}$-direction is

$$I_r^+(\vec{s}) = \frac{d\mathcal{E}_r(\vec{s})}{\cos\theta \, d\omega} . \tag{3.2}$$

Except for notational changes, the surface property definitions suggested by Torrance and Sparrow in the discussion appended to Reference 1 are used.

The <u>bidirectional</u> reflectance, $\rho_{bd}(\vec{s}';\vec{s})$, corresponding to energy incident from the $\vec{s}'$-direction and reflection into the $\vec{s}$-direction is defined as

$$\rho_{bd}(\vec{s}';\vec{s}) = \frac{dI_r^+(\vec{s}';\vec{s})}{d\mathcal{E}_i(\vec{s}')}$$

$$= \frac{dI_r^+(\vec{s}';\vec{s})}{I^-(\vec{s}')\cos\theta' \, d\omega'} . \tag{3.3}$$

The intensity of reflected energy is taken as a differential quantity to indicate that the radiation incident from a prescribed direction is generally scattered throughout hemispherical space after reflection and, therefore, only a small portion of this incident energy is found in the solid angle $d\omega$. In view of the foregoing definition, the contribution to the reflected energy in $d\omega$ by the energy incident within $d\omega'$ is

$$dI_r^+(\vec{s}'\,;\,\vec{s})\cos\theta\, d\omega = \rho_{bd}(\vec{s}'\,;\,\vec{s})\, I^-(\vec{s}')\cos\theta\cos\theta'\, d\omega\, d\omega' \,. \tag{3.4}$$

The product $\rho_{bd}(\vec{s}'\,;\,\vec{s})\cos\theta\, d\omega$ obviously possesses a physical interpretation as the fraction of the energy incident within $d\omega'$ about the $\vec{s}'$-direction which is reflected by the surface into solid angle $d\omega$ about the $\vec{s}$-direction. On the basis of thermodynamic reasoning, it can be shown that the bidirectional reflectance must satisfy the Helmholtz reciprocity principle. That is,

$$\rho_{bd}(\vec{s}'\,;\,\vec{s}) = \rho_{bd}(-\vec{s};\,-\vec{s}') \,. \tag{3.5}$$

The above relation expresses the fact that the decrease in intensity which a ray experiences in reflection from the $\vec{s}'$-direction to the $\vec{s}$-direction is equal to that experienced by a ray traveling the oppositely directed path.

The directional-hemispherical (or angular-hemispherical) reflectance, $\rho_{dh}(\vec{s}')$, is defined as the fraction of the incident radiant energy contained in $d\omega'$ about the $\vec{s}'$-direction which is reflected into hemispherical space. Then

$$\rho_{dh}(\vec{s}') = \frac{d\mathcal{E}_{r,h}}{d\mathcal{E}_i(\vec{s}')} \,, \tag{3.6}$$

where $d\mathcal{E}_{r,h}$ is the reflected energy that is collected over the entire hemisphere. Employing Equations (3.1) and (3.4), one obtains

$$\rho_{dh}(\vec{s}') = \frac{\int_{\triangle_r} \rho_{bd}(\vec{s}'\,;\,\vec{s})\, I^-(\vec{s}')\cos\theta'\, d\omega'\cos\theta\, d\omega}{I^-(\vec{s}')\cos\theta'\, d\omega'}$$

$$= \int_{\triangle_r} \rho_{bd}(\vec{s}'\,;\,\vec{s})\cos\theta\, d\omega \,. \tag{3.7}$$

The subscript $\triangle_r$ on the integral denotes integration over the solid angles of reflection.

The hemispherical-directional (or hemispherical-angular) reflectance, $\rho_{hd}(\vec{s})$, corresponding to hemispherical illumination and collection over a specific angular direction, may be defined as

$$\rho_{hd}(\vec{s}) = \frac{I_r^+(\vec{s})}{\mathcal{E}_{i,h}/\pi} \,, \tag{3.8}$$

where $\mathcal{E}_{i,h}$ is the rate at which energy is incident on the surface per unit area from hemispherical space. Introducing the bidirectional reflectance and intensity of incident energy, $\rho_{hd}(\vec{s})$ may be expressed as follows:

$$\rho_{hd}(\vec{s}) = \frac{\int_{\triangle_i} \rho_{bd}(\vec{s}'\,;\,\vec{s})\, I^-(\vec{s}')\cos\theta'\, d\omega'}{\frac{1}{\pi}\int_{\triangle_i} I^-(\vec{s}')\cos\theta'\, d\omega'} \,. \tag{3.9}$$

The subscript $\triangle_i$ on the integrals denotes integration over all directions for incident energy. When the intensity of incident energy is independent of direction, that is, the illumination is diffuse, the above expression simplifies to

$$\rho_{hd}(\vec{s}) = \int_{\triangle_i} \rho_{bd}(\vec{s}'\,;\vec{s})\cos\theta'\, d\omega' \,. \tag{3.10}$$

Employing the reciprocity principle for the bidirectional reflectance in Equation (3.10) demonstrates the equivalence of the directional-hemispherical reflectance to the hemispherical-directional reflectance for diffuse illumination. Thus

$$\rho_{hd}(\vec{s}) = \rho_{dh}(\vec{s}' = -\vec{s})$$

or in a more often quoted form

$$\rho_{hd}(\theta,\varphi) = \rho_{dh}(\theta' = \theta,\ \varphi' = \varphi) \ . \qquad (3.11)$$

The hemispherical reflectance, ρ_H, is the fraction of the energy incident from throughout hemispherical space which is reflected from the surface.

$$\rho_H = \frac{\mathcal{E}_{r,h}}{\mathcal{E}_{i,h}} \ . \qquad (3.12)$$

In view of the previously defined reflectances, this reflectance may be expressed as

$$\rho_H = \frac{\int_{\Omega_r} I_r^+(\vec{s}) \cos\theta \, d\omega}{\int_{\Omega_i} I^-(\vec{s}')\cos\theta' \, d\omega'} = \frac{\int_{\Omega_r} \int_{\Omega_i} \rho_{bd}(\vec{s}' ; \vec{s}) I^-(\vec{s}')\cos\theta' \cos\theta \, d\omega \, d\omega'}{\int_{\Omega_i} I^-(\vec{s}') \cos\theta' \, d\omega'}$$

$$= \frac{\int_{\Omega_i} \rho_{dh}(\vec{s}') I^-(\vec{s}')\cos\theta' \, d\omega'}{\int_{\Omega_i} I^-(\vec{s}') \cos\theta' \, d\omega'} \ . \qquad (3.13)$$

For diffuse illumination, $I^-(\vec{s}')$ is independent of direction, and the expression for ρ_H simplifies to

$$\rho_H = \frac{1}{\pi} \int_{\Omega_i} \rho_{dh}(\vec{s}')\cos\theta' \, d\omega' \ . \qquad (3.14)$$

Finally, the specular reflectance, ρ_s, is defined as the ratio of the intensity of reflected radiant energy to the intensity of incident energy in a specular reflection.

$$\rho_s(\vec{s}') = \frac{I_r^+(\vec{s}_{sp})}{I^-(\vec{s}')} = \frac{I_r^+(\theta = \theta', \varphi = \varphi' \pm \pi)}{I^-(\theta', \varphi')} \ .$$

In a specular reflection, the reflected beam is contained within a solid angle, $d\omega$, equal to the solid angle of the incident beam, $d\omega'$, and located in the plane of incidence with polar angle of reflection equal to the polar angle of incidence.

3.2.2 Absorptance

The directional absorptance, $\alpha_d(\vec{s}')$, is the fraction of the radiant energy contained in a small solid angle, $d\omega'$, about the $\vec{s}'$-direction which is absorbed.

$$\alpha_d(\vec{s}') = \frac{d\,\mathcal{E}_a}{d\,\mathcal{E}_i(\vec{s}')} \ . \qquad (3.15)$$

In the above, $d\mathcal{E}_a$ represents the rate of absorption of incident energy. Since for an opaque surface the incident energy is either reflected or absorbed, the directional absorptance follows from the directional-hemispherical reflectance as

$$\alpha_d(\vec{s}') = 1 - \rho_{dh}(\vec{s}') \ , \qquad (3.16)$$

or upon introducing the bidirectional reflectance as

$$\alpha_d(\vec{s}') = 1 - \int_{\Omega_r} \rho_{bd}(\vec{s}' ; \vec{s}) \cos\theta \, d\omega \ . \qquad (3.17)$$

The hemispherical absorptance, α_H, is the fraction of the radiation incident from hemispherical space which is absorbed.

$$\alpha_H = \frac{\mathcal{E}_{a,h}}{\mathcal{E}_{i,h}} . \quad (3.18)$$

Expressing the absorbed incident energy in terms of the directional absorptance yields

$$\alpha_H = \frac{\int_{\Omega_i} \alpha_d(\vec{s}') I^-(\vec{s}') \cos\theta' \, d\omega'}{\int_{\Omega_i} I^-(\vec{s}') \cos\theta' \, d\omega'} . \quad (3.19)$$

Employing Equation (3.16) gives the result

$$\alpha_H = 1 - \frac{\int_{\Omega_i} \rho_{dh}(\vec{s}') I^-(\vec{s}') \cos\theta' \, d\omega'}{\int_{\Omega_i} I^-(\vec{s}') \cos\theta' \, d\omega'} , \quad (3.20)$$

from which it follows that

$$\alpha_H = 1 - \rho_H . \quad (3.21)$$

3.2.3 Emittance

The directional emittance, $\epsilon_d(\vec{s})$, is the ratio of the intensity of emission of a surface in a particular direction to the intensity of a black surface at the same temperature in the prescribed direction. Then

$$\epsilon_d(\vec{s}) = \frac{I_e^+(\vec{s})}{I_b} , \quad (3.22)$$

where I_b denotes the direction independent black body intensity.

The hemispherical emittance, ϵ_H, is defined as the ratio of the energy emitted per unit time and unit area throughout hemispherical space, $\mathcal{E}_{e,h}$, to the emissive power of a black surface at the same temperature, $\mathcal{E}_{e,h,b}$. Consequently,

$$\epsilon_H = \frac{\mathcal{E}_{e,h}}{\mathcal{E}_{e,h,b}} . \quad (3.23)$$

In terms of the directional emittance, ϵ_H may be expressed as

$$\epsilon_H = \frac{1}{\pi} \int_{\Omega} \epsilon_d(\vec{s}) \cos\theta \, d\omega . \quad (3.24)$$

For a diffusely emitting surface, the directional emittance is independent of direction and therefore equal to the hemispherical emittance.

3.3 RADIANT TRANSFER IN AN ENCLOSURE

Consider an enclosure consisting of opaque gray surfaces of specified local temperature as illustrated in Figure 3.2. The restriction to gray surfaces of prescribed temperature is not essential to the development and is later relaxed. Let dA_i be an infinitesimal surface area element of the enclosure wall whose position is designated by the position vector, $\vec{r}_i$, from the origin of a convenient reference system.* At dA_i construct a local reference system with one axis aligned with the unit surface normal, $\vec{n}_i$. The net rate of radiant energy transfer from this element per unit area may be expressed as

$$q^r(\vec{r}_i) = \int_{\omega=4\pi} I(\vec{r}_i, \vec{s}_i) \cos\theta \, d\omega , \quad (3.25)$$

or more conveniently, in terms of the one-sided fluxes as

$$q^r(\vec{r}_i) = \int_{\omega=2\pi} I^+(\vec{r}_i, \vec{s}_i) \cos\theta \, d\omega - \int_{\omega'=2\pi} I^-(\vec{r}_i, \vec{s}_i') \cos\theta' \, d\omega' . \quad (3.26)$$

*Throughout the remainder of Section 3, subscript i refers to the area location.

It is customary in radiant transfer analysis between surfaces to let

$$J(\vec{r}_i) = \int_{\omega=2\pi} I^+(\vec{r}_i,\vec{s}_i)\cos\theta \, d\omega, \quad (3.27)$$

$$G(\vec{r}_i) = \int_{\omega'=2\pi} I^-(\vec{r}_i,\vec{s}_i')\cos\theta' \, d\omega', \quad (3.28)$$

where $J(\vec{r}_i)$ and $G(\vec{r}_i)$ are called the radiosity* and irradiation, respectively. The radiosity obviously is the radiant energy emergent from the element per unit time and unit area, whereas the irradiation is the radiant energy incident on the element per unit time and unit area. With the introduction of these, the net radiant flux follows as

$$q^r(\vec{r}_i) = J(\vec{r}_i) - G(\vec{r}_i) \, . \quad (3.29)$$

Clearly, the computation of the local radiant flux depends entirely on the evaluation of the intensity. With this in mind, attention is directed to the calculation of radiant intensity.

The radiant energy, $d\mathcal{E}_\ell$, leaving the surface element dA_i in the arbitrary direction $\vec{s}_i$ within the solid angle $d\omega$ per unit time and unit area may be expressed in terms of the intensity, I^+, as

$$d\mathcal{E}_\ell = I^+(\vec{r}_i, \vec{s}_i) \cos\theta \, d\omega \, . \quad (3.30)$$

This energy is the sum of emitted energy and incident radiant energy which is reflected by dA_i into the considered solid angle. With the definition of directional emissivity, the energy emitted per unit time and per unit area into $d\omega$ is

$$d\mathcal{E}_e = \epsilon_d(\vec{r}_i,\vec{s}_i) I_b(\vec{r}_i)\cos\theta \, d\omega \, . \quad (3.31)$$

To evaluate the contribution of reflected energy, consider first the energy incident on the surface which is contained within the solid angle $d\omega'$ about the arbitrary direction of incidence, $\vec{s}_i'$. This incident energy contributes the amount

$$d^2\mathcal{E}_r = dI_r^+(\vec{r}_i,\vec{s}_i';\vec{s}_i)\cos\theta \, d\omega \, , \quad (3.32)$$

to the reflected energy in $d\omega$. Introducing the bidirectional reflectance of the surface element dA_i, this expression may be written as

$$d^2\mathcal{E}_r = \rho_{bd}(\vec{r}_i, \vec{s}_i'; \vec{s}_i) I^-(\vec{r}_i, \vec{s}_i') \cos\theta' \, d\omega' \cos\theta \, d\omega \, . \quad (3.33)$$

Similar contributions to the reflected energy in $d\omega$ are made by energy incident from all possible directions. Then, the total energy incident on the surface per unit time and per unit area, which after reflection from the surface is leaving within the solid angle $d\omega$ about the $\vec{s}_i$-direction, is simply the integral of $d^2\mathcal{E}_r$ over all directions of incidence.

$$d\mathcal{E}_r = \left[\int_{\omega'=2\pi} \rho_{bd}(\vec{r}_i, \vec{s}_i'; \vec{s}_i) I^-(\vec{r}_i, \vec{s}_i') \cos\theta' \, d\omega' \right] \cos\theta \, d\omega \, . \quad (3.34)$$

Since

$$d\mathcal{E}_\ell = d\mathcal{E}_e + d\mathcal{E}_r \, , \quad (3.35)$$

the substitution of the various expressions for the energy flows in terms of intensity yields

$$I^+(\vec{r}_i, \vec{s}_i) = \epsilon_d(\vec{r}_i, \vec{s}_i) I_b(\vec{r}_i) + \int_{\omega'=2\pi} \rho_{bd}(\vec{r}_i, \vec{s}_i'; \vec{s}_i) I^-(\vec{r}_i, \vec{s}_i') \cos\theta' \, d\omega' \, . \quad (3.36)$$

*The term radiosity is here used in a general sense rather than in the restricted usage (diffusely emitting and diffusely reflecting surfaces) common in the literature.[2]

This fundamental equation expresses the intensity of energy leaving the considered surface area element of the enclosure wall within the solid angle $d\omega$ about the $\vec{s}_i$-direction as the sum of the directional emission and the energy incident from the surroundings which is reflected by the surface into the considered direction. The intensity $I^-(\vec{r}_i, \vec{s}_i')$ must equal the intensity $I^+(\vec{r}_j, \vec{s}_j)$ of a surface area element dA_j in the direction $\vec{s}_j$, since the intensity of radiation is constant along the path in a nonradiating medium with constant index of refraction. Then Equation (3.36) is

$$I^+(\vec{r}_i, \vec{s}_i) = \epsilon_d(\vec{r}_i, \vec{s}_i) I_b(\vec{r}_i) + \int_{\omega'=2\pi} \rho_{bd}(\vec{r}_i, \vec{s}_i'; \vec{s}_i) I^+(\vec{r}_j, \vec{s}_j) \cos\theta' \, d\omega' . \quad (3.37)$$

Since the temperature and enclosure wall properties are specified, the first term to the right in Equation (3.37) representing the local directional emission is known. However, the unknown intensity, I^+, appears not only to the left but also within the integral operator. Thus Equation (3.37) is an integral equation which governs the radiation intensity.

Employing Equation (3.37) in Equation (3.27), and utilizing Equation (3.13), the radiosity may be expressed as

$$J(\vec{r}_i) = \epsilon_H(\vec{r}_i) e_b(\vec{r}_i) + \rho_H(\vec{r}_i) G(\vec{r}_i), \quad (3.38)$$

where $e_b(\vec{r}_i)$ is the black body emissive power given by

$$e_b(\vec{r}_i) = \sigma T^4(\vec{r}_i) , \quad (3.39)$$

with σ the Stefan-Boltzmann radiation constant. Equation (3.38) expresses the intuitive result that the radiant energy emergent from an opaque surface per unit time and unit area is the sum of the emission and reflected incident energy.

With Equation (3.38) either the radiosity or the irradiation may be eliminated from the heat flux expression of Equation (3.29). Proceeding with this replacement, the local radiant heat flux relation transforms to

$$q^r(\vec{r}_i) = \frac{1}{\rho_H(\vec{r}_i)} \left[\epsilon_H(\vec{r}_i) e_b(\vec{r}_i) - \alpha_H(\vec{r}_i) J(\vec{r}_i) \right] = \left[\epsilon_H(\vec{r}_i) e_b(\vec{r}_i) - \alpha_H(\vec{r}_i) G(\vec{r}_i) \right] . \quad (3.40)$$

Once the radiosity or irradiation is determined, the local radiant flux follows from Equation (3.40). Since the radiosity may be evaluated by Equation (3.27) once the solution for the intensity is available, [hence the irradiation through Equation (3.38)] the fundamental problem of the radiant transfer analysis is the determination of the solution to Equation (3.37).

It is often convenient to express the intensity and heat flux relations completely in terms of the surfaces participating in the radiant exchange. For this purpose, consider dA_j to be an element of the finite surface A_j, and let the enclosure consist of the finite surface areas $A_1, A_2, \ldots A_N$. The solid angle subtended by dA_j at dA_i is

$$d\omega' = \frac{\cos\theta_j \, dA_j}{|\vec{r}_i - \vec{r}_j|^2} . \quad (3.41)$$

Introducing the geometry function

$$K(\vec{r}_i, \vec{r}_j) = \frac{\cos\theta' \cos\theta_j}{|\vec{r}_i - \vec{r}_j|^2} = \frac{\cos\theta_i \cos\theta_j}{|\vec{r}_i - \vec{r}_j|^2} \quad (3.42)$$

the integral equation for the local intensity on the i-th surface may be expressed as follows:

$$I^+(\vec{r}_i,\vec{s}_i) = \epsilon_d(\vec{r}_i,\vec{s}_i)I_b(\vec{r}_i) + \sum_{j=1}^{N}\int_{A_j}\rho_{bd}(\vec{r}_i,\vec{s}_i';\vec{s}_i)I^+(\vec{r}_j,\vec{s}_j)K(\vec{r}_i,\vec{r}_j)\,dA_j\ , \quad (1\le i\le N)\ . \tag{3.43}$$

The corresponding expressions for irradiation, radiosity, and heat flux are*

$$G(\vec{r}_i) = \sum_{j=1}^{N}\int_{A_j} I^+(\vec{r}_j,\ \vec{s}_j)K(\vec{r}_i,\ \vec{r}_j)\,dA_j\ , \tag{3.44}$$

$$J(\vec{r}_i) = \sum_{j=1}^{N}\int_{A_j} I^+(\vec{r}_i,\ \vec{s}_i)K(\vec{r}_i,\ \vec{r}_j)\,dA_j\ , \tag{3.45}$$

$$\begin{aligned} q^r(\vec{r}_i) &= \frac{1}{\rho_H(\vec{r}_i)}\left[\epsilon_H(\vec{r}_i)e_b(\vec{r}_i) - \alpha_H(\vec{r}_i)J(\vec{r}_i)\right] \\ &= \left[\epsilon_H(\vec{r}_i)e_b(\vec{r}_i) - \alpha_H(\vec{r}_i)G(\vec{r}_i)\right]\ . \end{aligned} \tag{3.46}$$

The simultaneous integral equations for $I^+(\vec{r}_i,\ \vec{s}_i)$ may be used to evaluate the intensity provided all surfaces of the enclosure have prescribed surface temperature distribution. Should a number of the enclosure surfaces have specified local radiant heat flux and not temperature, the intensity relation of Equation (3.43) for these surfaces is not in a very useful form, since it contains both the unknown black body intensity and I^+. Let surface A_k have prescribed local radiant flux. The black body intensity may be eliminated from the appropriate integral equation for the intensity by using Equation (3.46) to replace $I_b(\vec{r}_k)$. The black body intensity from the local heat flux expression in terms of irradiation is

$$I_b(\vec{r}_k) = \frac{q^r(\vec{r}_k)}{\pi\,\epsilon_H(\vec{r}_k)} + \sum_{j=1}^{N}\int_{A_j}\frac{\alpha_d(\vec{r}_k,\ \vec{s}_k')}{\pi\,\epsilon_H(\vec{r}_k)}\,I^+(\vec{r}_j,\ \vec{s}_j)K(\vec{r}_k,\ \vec{r}_j)\,dA_j\ . \tag{3.47}$$

Employing the above to eliminate $I_b(\vec{r}_k)$ from the applicable intensity equation yields

$$\begin{aligned} I^+(\vec{r}_k,\ \vec{s}_k) = \frac{1}{\pi}\frac{\epsilon_d(\vec{r}_k,\ \vec{s}_k)}{\epsilon_H(\vec{r}_k)}\,q^r(\vec{r}_k) + \sum_{j=1}^{N}\int_{A_j}\Bigg[\frac{1}{\pi}\frac{\epsilon_d(\vec{r}_k,\ \vec{s}_k)\alpha_d(\vec{r}_k,\ \vec{s}_k')}{\epsilon_H(\vec{r}_k)} \\ + \rho_{bd}(\vec{r}_k,\vec{s}_k';\vec{s}_k)\Bigg]\,I^+(\vec{r}_j,\ \vec{s}_j)K(\vec{r}_k,\ \vec{r}_j)\,dA_j\ . \end{aligned} \tag{3.48}$$

*The summation over j includes j = i.

With Equations (3.47) and (3.48) available, the analysis may be extended to include enclosures with both thermal conditions of interest. Let the surfaces $A_1, A_2, \ldots, A_\eta$ of an enclosure of N finite surfaces have prescribed temperature distribution, while the remaining surfaces, $A_{\eta+1}, A_{\eta+2}, \ldots, A_N$, have prescribed radiant heat flux. The relations of Equation (3.43) are appropriate for the surfaces of prescribed temperature, and Equation (3.48) for surfaces of prescribed flux. Both integral equations are of identical form so that the intensity integral equations may be written as

$$I^+(\vec{r}_i,\vec{s}_i) = F(\vec{r}_i,\vec{s}_i) + \sum_{j=1}^{N} \int_{A_j} C(\vec{r}_i,\vec{s}_i\,';\vec{s}_i)\, I^+(\vec{r}_j,\vec{s}_j)\, K(\vec{r}_i,\vec{r}_j)\, dA_j\,, \qquad 1 \le i \le N, \tag{3.49}$$

where

$$F(\vec{r}_i,\vec{s}_i) = \begin{cases} \epsilon_d(\vec{r}_i,\vec{s}_i)\, I_b(\vec{r}_i)\,, & 1 \le i \le \eta \\[2ex] \dfrac{\epsilon_d(\vec{r}_i,\vec{s}_i)}{\epsilon_H(\vec{r}_i)}\, \dfrac{q^r(\vec{r}_i)}{\pi}\,, & \eta < i \le N \end{cases} \tag{3.50}$$

$$C(\vec{r}_i,\vec{s}_i\,';\vec{s}_i) = \begin{cases} \rho_{bd}(\vec{r}_i,\vec{s}_i\,';\vec{s}_i)\,, & 1 \le i \le \eta \\[2ex] \rho_{bd}(\vec{r}_i,\vec{s}_i\,';\vec{s}_i) + \dfrac{\epsilon_d(\vec{r}_i,\vec{s}_i)\,\alpha_d(\vec{r}_i,\vec{s}_i\,')}{\pi\, \epsilon_H(\vec{r}_i)}\,, & \eta < i \le N\,. \end{cases} \tag{3.51}$$

Once the intensities have been determined from the integral equations of (3.49), the local radiant flux for the surfaces of prescribed temperature follows from

$$q^r(\vec{r}_i) = \frac{1}{\rho_H(\vec{r}_i)} \left[\epsilon_H(\vec{r}_i)\, e_b(\vec{r}_i) - \alpha_H(\vec{r}_i)\, J(\vec{r}_i) \right], \qquad 1 \le i \le \eta\,, \tag{3.52}$$

while the local temperature for the prescribed flux surfaces may be evaluated from

$$\sigma T^4(\vec{r}_i) = \frac{\rho_H(\vec{r}_i)\, q^r(\vec{r}_i) + \alpha_H(\vec{r}_i)\, J(\vec{r}_i)}{\epsilon_H(\vec{r}_i)}\,, \qquad \eta < i \le N\,. \tag{3.53}$$

The extension of the previous formulation to account for the wavelength dependence of surface properties can be achieved in principle, by performing the calculations on a monochromatic basis and then integrating over all wavelengths. When all the enclosure surfaces have prescribed temperature, the black body intensity is replaced with the black body spectral intensity available from Planck's law; the solution for the intensity must be carried out for each wavelength. Once the spectral intensity has been so determined, the spectral radiant heat flux follows from the monochromatic counterpart of Equation (3.46). The total (integrated over wavelength) intensity and heat flux are simply obtained from the wavelength integration of the spectral values.

$$I^{+}(\vec{r}_i,\vec{s}_i) = \int_0^\infty I_\lambda^{+}(\vec{r}_i,s_i)\,d\lambda \;, \qquad (3.54)$$

$$q^{r}(\vec{r}_i) = \int_0^\infty q_\lambda^{r}(\vec{r}_i)\,d\lambda \;. \qquad (3.54)$$

On the other hand, certain additional difficulties occur when one or more of the enclosure surfaces have prescribed radiant flux. This complication arises because the total radiant flux and not the monochromatic radiant flux is usually known. Suppose, for example, that the temperatures are specified at all enclosure surfaces except the k-th where the total flux is prescribed. The governing integral equation for the spectral intensity of this surface is simply the monochromatic counterpart to Equation (3.43).

$$I_\lambda^{+}(\vec{r}_k,\vec{s}_k) = \epsilon_{d,\lambda}(\vec{r}_k,\vec{s}_k)\,I_{b,\lambda}(\vec{r}_k,T) + \sum_{j=1}^{N}\int_{A_j}\rho_{bd,\lambda}(\vec{r}_k,\vec{s}_k';s_k)\,I_\lambda^{+}(\vec{r}_j,\vec{s}_j)\,K(\vec{r}_k,\vec{r}_j)\,dA_j \;.$$

The temperature, $T(\vec{r}_k)$, is not known and no simplifications occur by introducing the spectral radiant flux, since it also is not prescribed. Thus it appears that the temperature distribution $T(\vec{r}_k)$ must be assumed and the intensities found therefrom subjected to the condition that

$$q^{r}(\vec{r}_k) = \epsilon_H(\vec{r}_k)\,\sigma T^4(\vec{r}_k) - \int_0^\infty \alpha_{H,\lambda}(\vec{r}_k)\,G_\lambda(\vec{r}_k)\,d\lambda \;,$$

where $G_\lambda(\vec{r}_k)$ is the monochromatic irradiation. Should the prescribed value $q^r(\vec{r}_k)$ not result from the assumed temperature distribution and resulting irradiation, the procedure must be repeated until the prescribed flux is obtained. Further discussion of non-gray analysis may be found in References 2 and 3.

3.4 RADIANT TRANSFER IN AN ENCLOSURE OF SPECULAR-DIFFUSE SURFACES

The formulation of the previous section may be used to evaluate heat transfer between surfaces with direction- and wavelength-dependent surface properties. The complexity of the governing equations, as well as the lack of sufficient data on the surface properties in the analysis, have delayed its application. As a result of these circumstances, various models for the emission and reflection of thermal radiation from opaque surfaces have been suggested and employed for engineering applications. One of the more interesting of these is a specular-diffuse reflectance model.[2] In this model, the directional-hemispherical reflectance is regarded as the sum of a diffuse and a specular component. Both components are taken independent of the direction of incident energy. This model is particularly attractive because it includes both limiting cases of completely diffuse reflection and perfectly specular reflection. Also, it leads to remarkably simple and general expressions for radiant heat transfer. In the development to follow, the emission is taken diffuse and the surfaces are considered gray. The analysis is equally applicable, however, on a spectral basis.

In Section 3.4.1, expressions for evaluating radiant transfer and temperature in an enclosure of specular-diffuse surfaces are developed from the general relations of the previous section. These are simplified to an algebraic or finite difference engineering formulation in Section 3.4.2. Finally, the radiation electrical analog for surfaces of the type here considered is developed and applied to a simple enclosure.

3.4.1 Radiant Transfer

The bidirectional reflectance corresponding to the diffuse-specular reflectance model may be conveniently expressed in terms of a local spherical coordinate system as

$$\rho_{bd}(\vec{s}\,';\vec{s}) = \rho_{bd}(\theta',\varphi';\theta,\varphi) = \frac{\rho_D}{\pi} + 2\rho_s\, \delta[\sin^2\theta' - \sin^2\theta]\, \delta[\varphi' - (\varphi \pm \pi)], \tag{3.55}$$

where δ is the Dirac delta function which has the property that

$$\int_{x_1}^{x_2} f(x)\, \delta(x-a)\, dx = f(a) , \tag{3.56}$$

when $x_1 < a < x_2$ and $f(x)$ is an arbitrary function. In Equation (3.55), ρ_D and ρ_s are the diffuse and the specular component of the reflectance, respectively. It follows from Equation (3.55) and the surface property definitions introduced earlier that the properties are interrelated as follows.

$$\left.\begin{aligned} \rho_{dh} &= \rho_H = \rho_D + \rho_s, \\ \alpha_d &= \alpha_H = 1 - \rho_H = 1 - (\rho_D + \rho_s), \\ \epsilon_d &= \epsilon_H . \end{aligned}\right\} \tag{3.57}$$

For convenience, the diffuse emittance and the hemispherical reflectance are henceforth denoted ϵ and ρ, respectively. For gray surface or monochromatic analysis, the absorptance may be replaced with the emittance through Kirchhoff's law. Then

$$\epsilon = 1 - \rho = 1 - (\rho_D + \rho_s) . \tag{3.58}$$

Attention is now directed to the simplifications introduced into the fundamental intensity equation by this reflectance model. In this approximation, the integral of Equation (3.36) transforms into two parts corresponding to the specular and the diffuse component of the bidirectional reflectance. The specular part transforms as follows:

$$\begin{aligned} Z &= 2\rho_s(\vec{r}_i) \int_0^{2\pi} \int_0^{\pi/2} I^-(\vec{r}_i,\theta',\varphi')\, \delta[\sin^2\theta' - \sin^2\theta]\, \delta[\varphi' - (\varphi \pm \pi)] \cos\theta' \sin\theta'\, d\theta'\, d\varphi' \\ &= 2\rho_s(\vec{r}_i) \int_0^{\pi/2} I^-(\vec{r}_i,\theta',\varphi \pm \pi)\, \delta[\sin^2\theta' - \sin^2\theta] \cos\theta' \sin\theta'\, d\theta' \\ &= \rho_s(\vec{r}_i) I^-(\vec{r}_i,\theta,\varphi \pm \pi) = \rho_s(\vec{r}_i) I^-(\vec{r}_i,\vec{s}\,'_{i,sp}) . \end{aligned} \tag{3.59}$$

The direction $\vec{s}\,'_{i,sp}$ is that which an incident ray must have to result in a specular reflection into the direction $\vec{s}_i$. Note that the direction $\vec{s}\,'_{i,sp}$ is completely specified when the direction $\vec{s}_i$ is selected and, therefore, does not constitute an additional variable.

The integral associated with the diffuse component of the bidirectional reflectance is

$$\frac{\rho_D(\vec{r}_i)}{\pi} \int_{\omega'=2\pi} I^-(\vec{r}_i,\vec{s}\,'_i) \cos\theta'\, d\omega' . \tag{3.60}$$

With the above simplifications, the intensity integral equation expressed in Equation (3.36) transforms to

$$I^+(\vec{r}_i,\vec{s}_i) = \epsilon(\vec{r}_i)I_b(\vec{r}_i) + \frac{\rho_D(\vec{r}_i)}{\pi}\int_{\omega'=2\pi} I^-(\vec{r}_i,\vec{s}_i{}')\cos\theta'\,d\omega' + \rho_s(\vec{r}_i)I^-(\vec{r}_i,\vec{s}_{i,sp}{}') . \quad (3.61)$$

Since the first two terms to the right in Equation (3.61) are independent of direction, they represent the intensity of diffusely distributed energy leaving the surface element. These may be conveniently expressed in terms of a diffuse intensity, $I_D^+(\vec{r}_i)$, defined as follows:

$$I_D^+(\vec{r}_i) = \epsilon(\vec{r}_i)I_b(\vec{r}_i) + \frac{\rho_D(\vec{r}_i)}{\pi}\int_{\omega'=2\pi} I^-(\vec{r}_i,\vec{s}_i{}')\cos\theta'\,d\omega' . \quad (3.62)$$

With the introduction of the diffuse intensity, Equation (3.61) is concisely written as

$$I^+(\vec{r}_i,\vec{s}_i) = I_D^+(\vec{r}_i) + \rho_s(\vec{r}_i)I^-(\vec{r}_i,\vec{s}_{i,sp}{}') . \quad (3.63)$$

This relation with the definition of $I_D^+(\vec{r}_i)$ from Equation (3.62) constitutes the analog to Equation (3.36) for surfaces with the emission and reflection characteristics of the $\rho_D - \rho_s$ model. Later, we give further consideration to these relations, but first attention is directed to the simplifications introduced into the radiosity and local heat flux relations by this model.

In view of the property relations for this model, the radiosity expression of Equation (3.38) becomes

$$J(\vec{r}_i) = J_D(\vec{r}_i) + \rho_s(\vec{r}_i)G(\vec{r}_i) , \quad (3.64)$$

where $J_D(\vec{r}_i)$ is the diffuse radiosity defined as

$$J_D(\vec{r}_i) = \pi I_D^+(\vec{r}_i) = \epsilon(\vec{r}_i)e_b(\vec{r}_i) + \rho_D(\vec{r}_i)G(\vec{r}_i) . \quad (3.65)$$

The irradiation may be eliminated from Equation (3.64) in favor of the diffuse radiosity by using Equation (3.65). The result is

$$J(\vec{r}_i) = \frac{1}{\rho_D(\vec{r}_i)}\left[\rho(\vec{r}_i)J_D(\vec{r}_i) - \rho_s(\vec{r}_i)\epsilon(\vec{r}_i)e_b(\vec{r}_i)\right] . \quad (3.66)$$

The local radiant transfer expression of Equation (3.40) then transforms to

$$q^r(\vec{r}_i) = \frac{\epsilon(\vec{r}_i)}{\rho_D(\vec{r}_i)}\left\{\left[\epsilon(\vec{r}_i) + \rho_D(\vec{r}_i)\right] e_b(\vec{r}_i) - J_D(\vec{r}_i)\right\} . \quad (3.67)$$

It is evident from this expression that if either the flux or temperature is specified on the surfaces of a prescribed enclosure with known surface properties, the local temperature or flux, whichever is unknown, may be determined once the diffuse radiosity has been evaluated. With this in mind attention is directed to the evaluation of the diffuse radiosity. Consideration is initially confined to an enclosure in which the temperature is everywhere specified. Later this restriction is removed.

According to Equation (3.65), diffuse radiosity is simply π times the diffuse intensity. Hence, from Equation (3.62), we have

$$J_D(\vec{r}_i) = \epsilon(\vec{r}_i)e_b(\vec{r}_i) + \pi\,\rho_D(\vec{r}_i) \int_{\omega'=2\pi} I^-(\vec{r}_i,\vec{s}_i')\,\frac{\cos\theta'\,d\omega'}{\pi} . \qquad (3.68)$$

Now the intensity $I^-(\vec{r}_i,\vec{s}_i')$ is that leaving another element of the enclosure wall. For fixed dA_i and $\vec{s}_i$, the location of this element is completely determined by the intersection of the extension of the ray with direction $-\vec{s}_i'$ and the enclosure wall (Figure 3.3). Let the position vector of this element be denoted by $\vec{r}_1$ and denote by $\vec{s}_1$ the direction relative to this surface area normal, $\vec{n}_1$, which corresponds to the direction of incidence, $\vec{s}_i'$, at dA_i. Then

$$I^-(\vec{r}_i,\vec{s}_i') = I^+(\vec{r}_1,\vec{s}_1) . \qquad (3.69)$$

But according to Equation (3.63),

$$I^+(\vec{r}_1,\vec{s}_1)=I_D^+(\vec{r}_1)+\rho_s(\vec{r}_1)I^-(\vec{r}_1,\vec{s}_{1,sp}'). \qquad (3.70)$$

The intensity, $I^-(\vec{r}_1,\vec{s}_{1,sp}')$, is identical to that leaving another element of the enclosure surface. The location of this element is completely determined by the requirement that it lies on the enclosure wall and that energy leaving this element after undergoing a specular reflection from dA_1 arrives at dA_i. Let the position vector of this element be denoted by $\vec{r}_2$. Then

$$I^-(\vec{r}_1,\vec{s}_{1,sp}') = I^+(\vec{r}_2,\vec{s}_2) , \qquad (3.71)$$

where $\vec{s}_2$ is the direction relative to $\vec{n}_2$ that corresponds to $\vec{s}_{1,sp}'$ for the element at $\vec{r}_1$. Again, employing Equation (3.63)

$$I^+(\vec{r}_2,\vec{s}_2)=I_D^+(\vec{r}_2)+\rho_s(\vec{r}_2)I^-(\vec{r}_2,\vec{s}_{2,sp}'). \qquad (3.72)$$

Evidently, the foregoing procedure may be continued and the expression of Equation (3.69) ultimately written as

$$I^-(\vec{r}_i,\vec{s}_i') = I_D^+(\vec{r}_1) + \rho_s(\vec{r}_1)\Big[I_D^+(\vec{r}_2)+\rho_s(\vec{r}_2)\Big[I_D^+(\vec{r}_3)+\rho_s(\vec{r}_3)\Big[I_D^+(\vec{r}_4)+\rho_s(\vec{r}_4)\Big[I_D^+(\vec{r}_5)+\ldots ,$$

or

$$I^-(\vec{r}_i,\vec{s}_i')=I_D^+(\vec{r}_1)+I_D^+(\vec{r}_2)\rho_s(\vec{r}_1)+I_D^+(\vec{r}_3)\rho_s(\vec{r}_2)\rho_s(\vec{r}_1)+I_D^+(\vec{r}_4)\rho_s(\vec{r}_3)\rho_s(\vec{r}_2)\rho_s(\vec{r}_1) + \ldots . \qquad (3.73)$$

Although not explicitly indicated, the right side of the above relation depends only on $\vec{r}_i$ and $\vec{s}_i'$, since the various position vectors are completely determined by these variables and the geometry of the enclosure. Introducing Equation (3.73) into (3.68) yields

$$J_D(\vec{r}_i) = \epsilon(\vec{r}_i)e_b(\vec{r}_i) + \rho_D(\vec{r}_i)\Big[\int_{2\pi} J_D(\vec{r}_1)\,\frac{\cos\theta'\,d\omega'}{\pi} + \int_{2\pi} J_D(\vec{r}_2)\rho_s(\vec{r}_1)\,\frac{\cos\theta'\,d\omega'}{\pi}$$

$$+ \int_{2\pi} J_D(\vec{r}_3)\rho_s(\vec{r}_2)\rho_s(\vec{r}_1)\,\frac{\cos\theta'\,d\omega'}{\pi} + \ldots\Big] . \qquad (3.74)$$

In the above the relation $J_D = \pi\, I_D^+$ has been used to replace the diffuse intensities with diffuse radiosities. Equation (3.74) is the integral equation which governs the diffuse radiosity of the enclosure wall. The specular component of the wall reflectance enters as a multiplicative factor in the kernels of the integrals. Each of the integrals possesses a physical interpretation. The first represents diffuse energy leaving the enclosure walls visible to an observer stationed at dA_i which is directly transported to dA_i per unit time and per unit area of the receiving element. The second represents diffuse energy leaving the enclosure wall which after one specular reflection from that part of the enclosure surface visible from dA_i arrives at dA_i per unit time and area. The third has a similar interpretation as the second but now the energy arrives at dA_i after two specular reflections. The factors $\rho_s(\vec{r}_j)$ account for the reduction in the intensity which occurs at each specular reflection.

The integral expressions of Equation (3.74) may be transformed into a more convenient form. For this purpose recall that the factor $\cos\theta' d\omega'/\pi$ may be expressed as follows:

$$\frac{\cos\theta' d\omega'}{\pi} = \frac{\cos\theta' \cos\theta_1 dA_1}{\pi |\vec{r}_i - \vec{r}_1|^2} = d^2F_{dA_i - dA_1} , \qquad (3.75)$$

where

$$d^2F_{dA_i - dA_1}$$

is the diffuse view factor for transfer from dA_i to dA_1. With this, a typical integral of Equation (3.74), representing energy arriving at dA_i after one or more specular reflections may be written as

$$\frac{1}{dA_i} \int_{A_1} J_D(\vec{r}_{n+1}) \left[\rho_s(\vec{r}_n)\rho_s(\vec{r}_{n-1}) \ldots \rho_s(\vec{r}_1) dA_i d^2F_{dA_i - dA_1} \right] , \qquad (3.76)$$

where A_1 is the portion of the enclosure wall visible to an observer at dA_i. The coefficient of $J_D(\vec{r}_{n+1})$ in the brackets represents the fraction of the diffuse energy leaving the area element, dA_{n+1}, per unit time and area which after (n-1) specular reflections arrives at dA_1 and is specularly reflected to dA_i. By the principle of reciprocity, this coefficient may also be written as

$$\rho_s(\vec{r}_1)\rho_s(\vec{r}_2) \ldots \rho_s(\vec{r}_n) d^2F_{dA_{n+1} - dA_n} \cdot dA_{n+1} , \qquad (3.77)$$

where

$$d^2F_{dA_{n+1} - dA_n}$$

is the diffuse view factor from the element dA_{n+1} to the element dA_n at which the first of the n specular reflections occurs. The location and geometrical disposition of this element is defined by the requirement that after (n-1) specular reflections, the diffuse energy leaving dA_{n+1} illuminates dA_1 and is specularly reflected to dA_i. Other possible paths from the element dA_{n+1} to dA_i with the same number of intervening specular reflections are accounted for in Equation (3.76) by the integration over all possible elements at which the n-th specular reflection could occur. With these ideas in mind, the integral of Equation (3.76) may be rephrased as follows. Select a typical element of the enclosure, say dA_j, as the "emitter" and dA_i as the "receiver." The element dA_j need not be directly visible to an observer at dA_i. There are, in general, a number of different paths by which the diffuse energy leaving this element may arrive at dA_i after n intervening specular reflections on the enclosure wall. Corresponding to the k-th path of n specular reflections, there is a contribution to the energy incident on dA_i per unit time and unit area of the form

$$\frac{J_D(\vec{r}_j) dA_j \left[\rho_s(\vec{r}_{1k})\rho_s(\vec{r}_{2k}) \ldots \rho_s(\vec{r}_{nk}) d^2F_{dA_j - dA^*_{i,nk}} \right]}{dA_i} , \qquad (3.78)$$

where

$$d^2F_{dA_j - dA^*_{i,nk}}$$

denotes the diffuse view factor from dA_j to $dA^*_{i,nk}$. The latter element is that on which the first of n specular reflections occur in this, the k-th path for energy transfer to dA_i. Factors of the form within the brackets must be included for each possible path from dA_j to dA_i with n specular reflections. Accounting for all such paths and denoting the result

$$d^2E_{dA_j - dA_{i,n}},$$

one obtains

$$d^2E_{dA_j - dA_{i,n}} = \sum_{k=1}^{\infty} \rho_s(\vec{r}_{1k})\rho_s(\vec{r}_{2k}) \ldots \rho_s(\vec{r}_{nk}) d^2F_{dA_j - dA^*_{i,nk}} . \tag{3.79}$$

The factor

$$d^2E_{dA_j - dA_{i,n}}$$

may be called the exchange factor for the transport of diffuse energy from dA_j to dA_i via n intervening specular reflections. The exchange factor concept appears to have been first introduced by Sparrow and Lin.[4] Physically, this factor represents the fraction of the diffuse energy leaving dA_j per unit time which after n specular reflections arrives at dA_i. It is seen to depend on purely geometrical quantities and the specular reflectance component of the surfaces. The exchange factors satisfy reciprocity relations of the form

$$dA_j d^2E_{dA_j - dA_{i,n}} = dA_i d^2E_{dA_i - dA_{j,n}} . \tag{3.80}$$

With the exchange factor concept, Equation (3.78) may be written for all possible specular paths of transfer from dA_j to dA_i with n intervening reflections as

$$J_D(\vec{r}_j) \frac{dA_j}{dA_i} d^2E_{dA_j - dA_{i,n}} = J_D(\vec{r}_j) d^2E_{dA_i - dA_{j,n}} . \tag{3.81}$$

Hence the typical integral of Equation (3.74) may be written as

$$\int_A J_D(\vec{r}_j) d^2E_{dA_i - dA_{j,n}} . \tag{3.82}$$

The integration extends over the entire enclosure, including those portions which are not directly visible from dA_i. Since each of the integrals in Equation (3.74) may be expressed in the above form, the radiosity integral equation transforms to

$$J_D(\vec{r}_i) = \epsilon(\vec{r}_i)e_b(\vec{r}_i) + \rho_D(\vec{r}_i) \int_A J_D(\vec{r}_j) d^2E_{dA_i - dA_j} , \tag{3.83}$$

where the exchange factor,

$$d^2E_{dA_i - dA_j} ,$$

is defined as follows:

$$d^2E_{dA_i - dA_j} = \sum_{n=0}^{\infty} d^2E_{dA_i - dA_{j,n}} . \tag{3.84}$$

Evidently this exchange factor gives the fraction of the diffuse energy leaving dA_i

which is incident on dA_i directly (n=0) and by all possible specular reflection paths. The form of the radiosity integral equation is now identical to that applicable for diffusely reflecting surfaces.[2] For diffuse surfaces, however, the exchange factors are replaced with geometrical view factors and the integration extends only over the portion of the enclosure walls visible at dA_i. The dependence of the exchange factors on surface properties is a major distinction between analysis for diffuse surfaces and for surfaces with a specular reflectance component. Also, exchange factors are, in general, difficult to evaluate. The image methods introduced by Eckert and Sparrow[5] for plane surfaces are very useful. However, the evaluation of these factors for curved surfaces can present a formidable undertaking. Further discussion of the exchange factor concept may be found in Reference 2. Once the solution to the integral equation for the diffuse radiosity is available, the local heat transfer for surfaces of specified temperature follows from Equation (3.67).

Again, it is useful to extend the preceding analysis to an enclosure consisting of N finite surfaces $A_1, A_2, \ldots, A_N$. Let the surface $A_1, A_2, \ldots, A_\eta$ have prescribed temperature distribution, while those designated $A_{\eta+1}, A_{\eta+2}, \ldots, A_N$ possess prescribed local radiant heat transfer. The diffuse radiosity functions for the prescribed temperature surfaces are then determined by the following system of integral equations.

$$J_D(\vec{r}_i) = \varepsilon(\vec{r}_i)e_b(\vec{r}_i) + \rho_D(\vec{r}_i)\sum_{j=1}^{N}\int_{A_j} J_D(\vec{r}_j)\,d^2E_{dA_i - dA_j}, \qquad 1 \leq i \leq \eta. \tag{3.85}$$

For surfaces of prescribed local heat transfer, Equation (3.67) may be used to eliminate the local black body emissive power in favor of the known local flux. Then the diffuse radiosity functions of these surfaces are evaluated from the following system:

$$J_D(\vec{r}_i) = q(\vec{r}_i) + \left[\rho_D(\vec{r}_i) + \varepsilon(\vec{r}_i)\right]\sum_{j=1}^{N}\int_{A_j} J_D(\vec{r}_j)\,d^2E_{dA_i - dA_j}, \qquad \eta < i \leq N. \tag{3.86}$$

Once the diffuse radiosity functions have been determined from the integral equations, the local temperature or heat flux, whichever is unknown, follows from the relations

$$q(\vec{r}_i) = \begin{cases} \dfrac{\varepsilon(\vec{r}_i)}{\rho_D(\vec{r}_i)}\left\{\left[\varepsilon(\vec{r}_i) + \rho_D(\vec{r}_i)\right]\sigma T^4(\vec{r}_i) - J_D(\vec{r}_i)\right\}, & \left[\rho_D(\vec{r}_i) \neq 0\right] \\[2ex] \varepsilon(\vec{r}_i)\left[\sigma T^4(\vec{r}_i) - \displaystyle\sum_{j=1}^{N}\int_{A_j} J_D(\vec{r}_j)\,d^2E_{dA_i - dA_j}\right], & \left[\rho_D(\vec{r}_i) = 0\right] \end{cases} \tag{3.87}$$

$$\sigma T^4(\vec{r}_i) = \frac{\varepsilon(\vec{r}_i)\,J_D(\vec{r}_i) + \rho_D(\vec{r}_i)q(\vec{r}_i)}{\varepsilon(\vec{r}_i)\left[\varepsilon(\vec{r}_i) + \rho_D(\vec{r}_i)\right]}. \tag{3.88}$$

3.4.2 Finite Difference Formulation

The integral equation formulation of the preceding section is often difficult to apply in routine engineering calculations. An algebraic or finite difference formulation is in many instances sufficiently accurate and also reasonably simple to use. To obtain such a formulation from the integral equations (3.85) and (3.86), the surface properties on each of the enclosure surfaces are taken uniform. Then multiplying the radiosity integral equation for the i-th surface by (dA_i/A_i) and integrating over the extent of surface A_i yields

$$J_{D,i} = F_i + C_i \sum_{j=1}^{N} \int_{A_j} J_D(r_j) \left[\frac{1}{A_i} \int_{A_i} d^2E_{dA_i - dA_j}\, dA_i \right] dA_j \quad , \qquad (3.89)$$

where

$$F_i = \begin{Bmatrix} \epsilon_i \sigma T_i^4 \ , & 1 \le i \le \eta \\ \dfrac{Q_i}{A_i} \ , & \eta < i \le N \end{Bmatrix} \ , \quad C_i = \begin{Bmatrix} \rho_{D,i} \ , & 1 \le i \le \eta \\ (\rho_{D,i} + \epsilon_i) \ , & \eta < i \le N \end{Bmatrix} \qquad (3.90)$$

Area mean values of functions are denoted by identical symbols with subscripts. For example, the area mean value for $J_D(\vec{r}_i)$ is denoted $J_{D,i}$. The quantity Q_i/A_i represents the mean value of the radiant transfer over the extent of the surface A_i. Further simplifications of Equation (3.89) are possible by first noting that

$$\int_{A_i} d^2E_{dA_i - dA_j} dA_i = A_i\, dE_{A_i - dA_j} \ , \qquad (3.91)$$

where

$$dE_{A_i - dA_j}$$

is the exchange factor for diffuse energy leaving the finite surface A_i and incident on dA_j of A_j. Then Equation (3.89) becomes

$$J_{D,i} = F_i + C_i \sum_{j=1}^{N} \int_{A_j} J_D(\vec{r}_j)\, dE_{A_i - dA_j}\, dA_j \quad . \qquad (3.92)$$

If the enclosure surface elements are taken so that $J_D(\vec{r}_j)$ is sufficiently uniform over the extent of each, the radiosity function may be approximated by its area mean value. Then since

$$\int_{A_j} dE_{A_i - dA_j}\, dA_j = E_{A_i - A_j} \equiv E_{ij} \ , \qquad (3.93)$$

the relation of Equation (3.92) transforms to the following:

$$J_{D,i} = F_i + C_i \sum_{j=1}^{N} J_{D,j} E_{ij} , \quad 1 \le i \le N \ . \qquad (3.94)$$

This system of N simultaneous linear algebraic equations may be written as

$$\sum_{j=1}^{N} a_{ij} J_{D,j} = F_i \ , \qquad (3.95)$$

where

$$a_{ij} = \begin{cases} \left[\delta_{ij} - \rho_{D,i} E_{ij}\right] , & 1 \le i \le \eta , \\ \left[\delta_{ij} - (\epsilon_i + \rho_{D,i}) E_{ij}\right] , & \eta < i \le N , \end{cases} \tag{3.96}$$

with δ_{ij} the Kronecher delta which has a value unity when $i = j$ and zero otherwise. The solution to Equation (3.95) is

$$J_{D,i} = \sum_{j=1}^{\eta} d_{ij}\epsilon_j \sigma T_j^4 + \sum_{j=\eta+1}^{N} d_{ij} \frac{Q_j}{A_j} , \tag{3.97}$$

with d_{ij} the inverse of the coefficient matrix a_{ij}. Note that d_{ij} depends only on the surface properties and the exchange factors for energy transport between the surfaces of the enclosure (E_{ij}).

After the diffuse radiosities have been evaluated, the mean heat transfer and temperature of the surfaces follow from Equations (3.87) and (3.88) as

$$\frac{Q_i}{A_i} = \begin{cases} \dfrac{\epsilon_i}{\rho_{D,i}} \left[(\epsilon_i + \rho_{D,i}) \sigma T_i^4 - J_{D,i} \right] , & (\rho_{D,i} \ne 0) \\ \epsilon_i \left[\sigma T_i^4 - \sum_{j=1}^{N} J_{D,j} E_{ij} \right] , & (\rho_{D,i} = 0) \end{cases} , \quad 1 \le i \le \eta \tag{3.98}$$

$$\sigma T_i^4 = \frac{\epsilon_i J_{D,i} + \rho_{D,i} (Q_i/A_i)}{\epsilon_i(\epsilon_i + \rho_{D,i})} , \qquad \eta < i \le N . \tag{3.99}$$

3.4.3 Electrical Analog

The radiant transfer relations of the finite difference formulation can be recast in a form which suggests an analog with an electrical circuit.(6) According to Equation (3.98), the radiant transfer per unit time from a surface of known temperature may be expressed as

$$Q_i = \frac{\epsilon_i(\epsilon_i + \rho_{D,i})}{\rho_{D,i}} A_i \left[e_{b,i} - \frac{J_{D,i}}{(\epsilon_i + \rho_{D,i})} \right] . \tag{3.100}$$

This transfer Q_i may be interpreted as the current flowing through a resistance with the value

$$\frac{\rho_{D,i}}{\epsilon_i A_i (\epsilon_i + \rho_{D,i})} \tag{3.101}$$

connected between the potentials $e_{b,i}$ and

$$\frac{J_{D,i}}{(\epsilon_i + \rho_{D,i})} .$$

The electrical circuit corresponding to this interpretation is illustrated in Figure 3.4a. For a surface with specified radiant heat transfer, the battery with potential $e_{b,i}$ is replaced with a current source of Q_i. The problem remains to interconnect the potentials

$$\frac{J_{D,i}}{(\epsilon_i + \rho_{D,i})}$$

of the interacting surfaces. To arrive at the radiant transfer relations which describe this interaction, recall that the irradiation on the i-th surface may be expressed in terms of the diffuse radiosities and the exchange factors as follows:

$$G_i = \sum_{j=1}^{N} E_{ij} J_{D,j} \quad . \qquad (3.102)$$

Inserting the value of $J_{D,j}$ in terms of emission and diffusely reflected irradiation, this expression becomes

$$G_i = \sum_{j=1}^{N} E_{ij}(\epsilon_j e_{b,j} + \rho_{D,j} G_j) \quad . \qquad (3.103)$$

In thermodynamic equilibrium $G_i = e_{b,j} = G_j$ for all j. Hence the general result follows

$$\sum_{j=1}^{N} (\epsilon_j + \rho_{D,j}) \; E_{ij} = 1 \quad . \qquad (3.104)$$

Now the finite difference form of Equation (3.65) may be used to eliminate the emissive power from Equation (3.100). The result is

$$Q_i = A_i \left[J_{D,i} - (\epsilon_i + \rho_{D,i}) \; G_i \right] \quad . \qquad (3.105)$$

Expressing G_i in terms of the diffuse surface radiosities through Equation (3.102) and employing the expression of Equation (3.104), transforms this Q_i relation to the following.

$$Q_i = \sum_{j=1}^{N} A_i E_{ij} \; (\epsilon_i + \rho_{D,i})(\epsilon_j + \rho_{D,j}) \left[\frac{J_{D,i}}{(\epsilon_i + \rho_{D,i})} - \frac{J_{D,j}}{(\epsilon_j + \rho_{D,j})} \right] \quad . \qquad (3.106)$$

This result states that the radiant transfer from A_i per unit time may be interpreted as the sum of the currents flowing between the potential

$$\frac{J_{D,i}}{(\epsilon_i + \rho_{D,i})}$$

and the similar potential for all other surfaces of the enclosure. Each pair of such potentials are connected through an electrical resistance equal to the reciprocal of the coefficient in Equation (3.106). This portion of the circuit is illustrated in Figure 3.4b.

The electrical circuit corresponding to an enclosure of three surfaces of prescribed temperature is shown in Figure 3.5. Surface 1 is a $\rho_D - \rho_s$ surface, while surface 2 is diffuse and surface 3 specular.

REFERENCES

1. Birkebak, R. C. and E. R. G. Eckert, "Effects of Roughness of Metal Surfaces on Angular Distribution of Monochromatic Reflected Radiation," Journal of Heat Transfer, C, 87 (1965), p. 85.

2. Sparrow, E. M. and R. D. Cess. Radiation Heat Transfer. Belmont, Calif.: Brooks/Cole Publishing Co., 1966.

3. Wiebelt, J. Engineering Radiation Heat Transfer. New York: Holt, Rinehart, and Winston, Inc., 1966.

4. Sparrow, E. M. and S. L. Lin, "Radiation Heat Transfer at a Surface Having Both Specular and Diffuse Reflectance Components," International Journal of Heat and Mass Transfer, 8 (1965), p. 769.

5. Eckert, E. R. G. and E. M. Sparrow, "Radiative Heat Exchange Between Surfaces with Specular Reflection," International Journal of Heat and Mass Transfer, 2 (1961), p. 42.

6. Zering, M. B. and A. F. Sarofim, "The Electrical Network Analog to Radiative Transfer: Allowance for Specular Reflection," Journal of Heat Transfer, C, 88 (1966), p. 341.

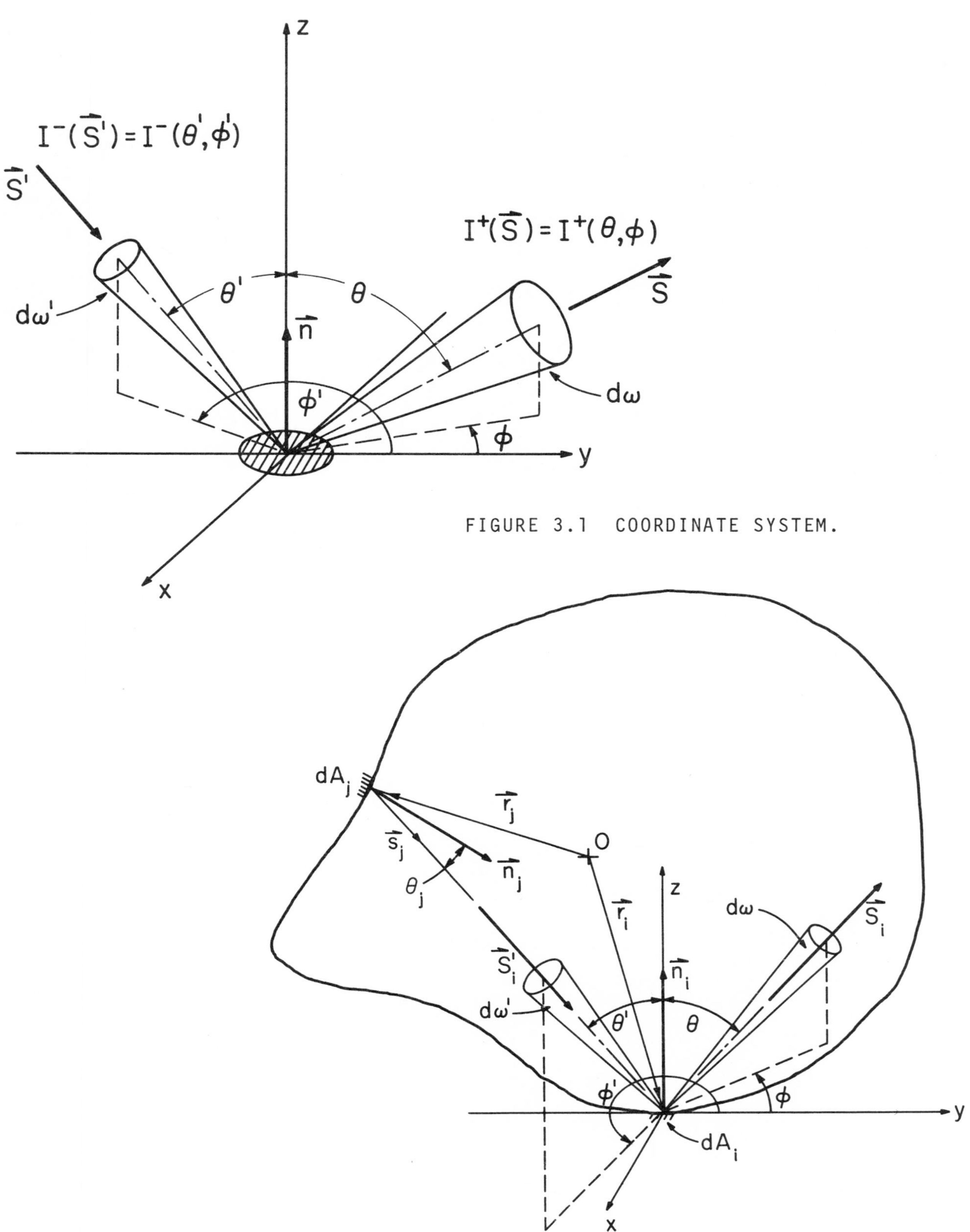

FIGURE 3.1 COORDINATE SYSTEM.

FIGURE 3.2 SCHEMATIC ENCLOSURE FOR RADIANT TRANSFER ANALYSIS.

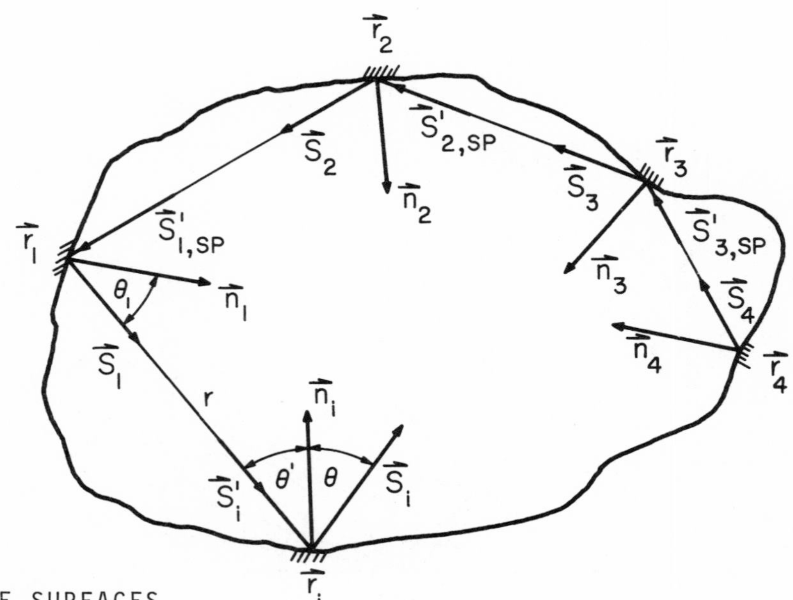

FIGURE 3.3 SCHEMATIC FOR ENCLOSURE OF SPECULAR-DIFFUSE SURFACES.

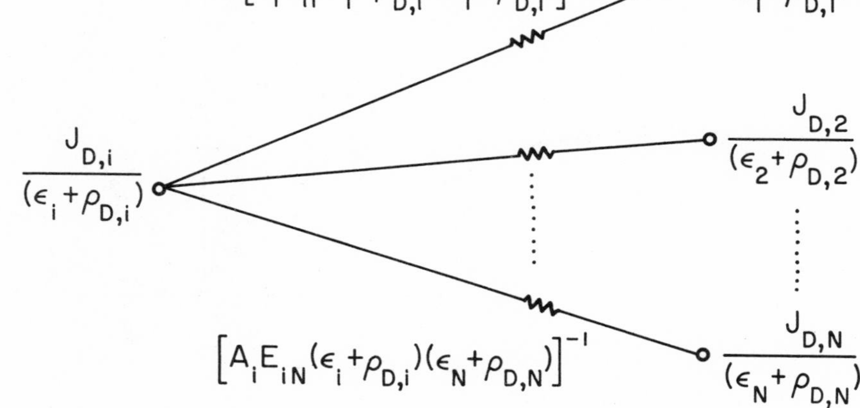

FIGURE 3.4 ELEMENTS OF ELECTRICAL ANALOG FOR RADIANT TRANSFER BETWEEN SPECULAR-DIFFUSE SURFACES.

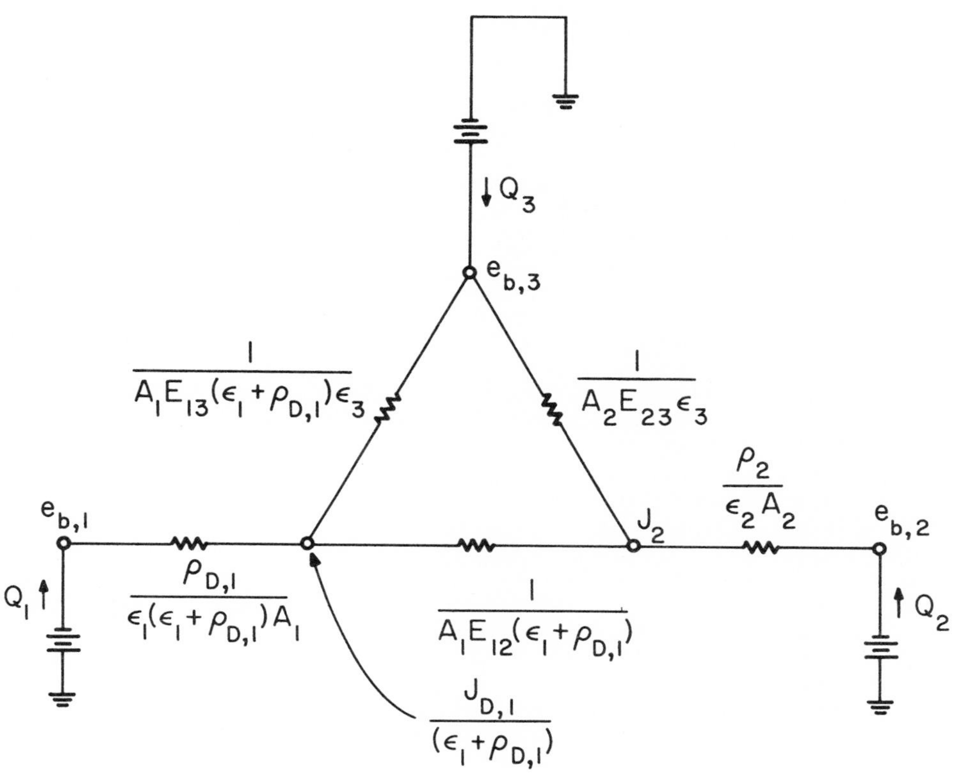

FIGURE 3.5 ELECTRICAL ANALOG FOR THREE SURFACE ENCLOSURE CONSISTING OF A SPECULAR-DIFFUSE SURFACE (1), DIFFUSE SURFACE (2), SPECULAR SURFACE (3).

4. SIMULTANEOUS CONDUCTION, CONVECTION, AND RADIATION

4.1 INTRODUCTION

Current technological problems in high-speed flight, gas-cooled nuclear reactors, and re-entry have resulted in considerable interest in the interaction of radiation with other modes of energy transport. Although astrophysicists have dealt with radiation transfer problems for a number of years, their considerations were usually limited to radiation dominated processes in large masses of radiating materials. These results are not generally applicable to engineering systems and, as a result, an enormous amount of effort has been expended in recent years to obtain information concerning the interaction of radiation with conduction and convection. With the limited space available here, our purpose is to illustrate (a) how radiation enters into the conventional conservation equations, (b) the complexities introduced by the radiation terms, and (c) some typical results. In light of these goals, consideration is limited to a simple geometry consisting of a radiating gray medium confined between infinite parallel black plates. Interested readers are referred to the reviews of Viskanta[1] and Cess[2] for details, other geometries, and more realistic assumptions concerning the radiating character of the radiating materials.

Thermal radiation influences the classical conservation equations in three ways. First, the radiant flux constitutes an additional mechanism by which heat transfer may occur. Second, the molecular energy of the material is augmented by the radiation energy. Finally, the radiation stress tensor supplements the usual mechanical stress tensor. Except at very high temperatures, the latter two effects may be neglected; only the energy equation is altered through the addition of a term representing the energy transfer by radiation. Since the radiative flux enters in a manner identical to that of molecular conduction, one need only interpret the heat flux vector, $\vec{q}$, in Equation (1.26) in B. T. Chao's paper,* as the sum of the conductive flux vector, $\vec{q}^c$, and the radiative flux vector, $\vec{q}^r$, thereby effecting the replacement of $\nabla\cdot\vec{q}$ with $(\nabla\cdot\vec{q}^c + \nabla\cdot\vec{q}^r)$ in the energy equation.

For the purposes of the present discussion, consideration is limited to an incompressible fluid of constant properties in the absence of significant viscous dissipation or energy generation. Under these conditions, the energy equation is

$$\rho c_p \frac{DT}{Dt} = -\nabla\cdot\vec{q}^c - \nabla\cdot\vec{q}^r .$$

The conductive heat flux vector for an isotropic medium is given by Fourier's law as

$$\vec{q}^c = -k\nabla T .$$

In order to investigate simultaneous energy transfer processes, expressions for radiant flux in terms of temperature are required. These are developed in Section 4.2 for the plane layer. Certain simpler approximate expressions are presented next. These preliminary aspects are followed by a consideration of three selected steady heat transfer problems. First, we consider radiation as the only energy transfer mechanism in the plane layer, then combined radiation and conduction, and finally simultaneous conduction, convection, and radiation in a plane parallel wall duct.

4.2 ONE-DIMENSIONAL RADIATION TRANSFER

Consider a system consisting of two uniform temperature infinite black planes separated by distance, L, and enclosing an absorbing and isotropically scattering gray medium (Figure 4.1). In the gray medium approximation it is assumed that the interaction coefficients (Section 1.4) are independent of wavelength. Let the lower and upper surfaces have temperatures T_1 and T_2, respectively. The enclosed medium is homogeneous with unit refractive index and its properties are taken independent of temperature. The distance from the lower surface is measured by the coordinate y. At an arbitrary distance from the lower surface, direction $\vec{s}$ is defined by the polar angle θ measured relative to the positive y-axis and the azimuthal angle φ measured relative to any fixed line in the xz-plane, the z-axis being normal to the plane of the paper. As a consequence of the simplifications introduced, the radiation transfer depends only on the distance measured normal to the surfaces and azimuthal symmetry exists.

The steady state equation of transfer for the $\vec{s}$-direction under the stated conditions is

---

*"Selected Topics on Convective Heat Transfer," page 1 of this book.

$$\frac{dI}{ds} + \beta I = \kappa I_b(T) + \frac{\sigma'}{4\pi} \int_{4\pi} I(s')\,d\omega' \,, \quad (4.1)$$

where β, κ, and σ' are the extinction, absorption, and scattering coefficients, respectively. The symbol $I_b(T)$ denotes the black body intensity evaluated at the local absolute temperature, T; that is, $\sigma T^4(y)/\pi$ where σ is the Stefan-Boltzmann radiation constant. Introducing

$$\left.\begin{aligned} s &= \frac{y}{\mu}\,, \\ \mu &= \cos\theta\,, \\ d\omega' &= \sin\theta'\,d\theta'\,d\varphi' = -\,d\varphi'\,d\mu'\,, \end{aligned}\right\} \quad (4.2)$$

and carrying out the integration over azimuthal angle, transforms Equation (4.1) to

$$\mu\,\frac{dI(y,\mu)}{dy} + \beta I(y,\mu) = \kappa I_b(y) + \frac{\sigma'}{2} \int_{-1}^{+1} I(y,\mu')\,d\mu' \,. \quad (4.3)$$

It is convenient to introduce a new independent variable called the optical depth and defined by

$$\tau = \int_0^y \beta\,dy = \beta y \,. \quad (4.4)$$

With this, the equation of transfer transforms to

$$\mu\,\frac{dI(\tau,\mu)}{d\tau} + I(\tau,\mu) = \frac{\kappa}{\beta} I_b(\tau) + \frac{1}{2}\frac{\sigma'}{\beta} \int_{-1}^{+1} I(\tau,\mu')\,d\mu' \,. \quad (4.5)$$

It is customary to introduce the albedo of scattering, ω_o, defined as

$$\omega_o = \frac{\sigma'}{\beta} = \frac{\sigma'}{\kappa + \sigma'} \,. \quad (4.6)$$

The albedo is a measure of the relative importance of scattering in the attenuation of the radiation beam. When ω_o is zero, no scattering is present and the medium only emits and absorbs radiation, while for the other extreme, $\omega_o = 1$, the medium only scatters and neither emits nor absorbs radiation. The coefficient of the black body intensity may also be expressed in terms of the albedo by the relation

$$1 - \omega_o = \frac{\kappa}{\beta} = \frac{\kappa}{\kappa + \sigma'} \,. \quad (4.7)$$

The right side of Equation (4.5) is often called the source function and is denoted by $S(\tau)$.

$$S(\tau) = (1-\omega_o) I_b(\tau) + \omega_o H(\tau) \,. \quad (4.8)$$

The function $H(\tau)$ is given as

$$H(\tau) = \frac{1}{2} \int_{-1}^{+1} I(\tau,\mu')\,d\mu' \,. \quad (4.9)$$

With the introduction of the above, the one-dimensional equation of transfer may be written as

$$\frac{dI(\tau,\mu)}{d\tau} + \frac{I}{\mu} = \frac{S(\tau)}{\mu} \,. \quad (4.10)$$

This differential equation may be treated as an inhomogeneous equation with parameter $\mu = \cos\theta$. The formal solution satisfying the boundary conditions

$$\left.\begin{aligned} I^+(0,\mu) &= I_b(T_1) = \frac{\sigma T_1^4}{\pi}\,, \\ I^-(\tau_o,\mu) &= I_b(T_2) = \frac{\sigma T_2^4}{\pi}\,, \end{aligned}\right\} \quad (4.11)$$

where $\tau_o(=\beta L)$ is the optical thickness of the medium is

$$\left.\begin{aligned} I^+(\tau,\mu) &= \frac{\sigma T_1^4}{\pi} \exp(-\tau/\mu) + \int_0^{\tau} S(t)\exp[-(\tau-t)/\mu]\,\frac{dt}{\mu}\,, & 0 < \mu \le 1, \\ I^-(\tau,\mu) &= \frac{\sigma T_2^4}{\pi} \exp[(\tau_o-\tau)/\mu] - \int_{\tau}^{\tau_o} S(t)\exp[-(\tau-t)/\mu]\,\frac{dt}{\mu}\,, & -1 \le \mu < 0. \end{aligned}\right\} \quad (4.12)$$

The above results were derived as if the source function were known. In reality it is not. If, in fact, the solution for the intensity is utilized in the expression defining $S(\tau)$, the following integral equation results:

$$\pi S(\tau) = (1-\omega_o)\sigma T^4(\tau) + \frac{\omega_o}{2}\left[\sigma T_1^4 E_2(\tau) + \sigma T_2^4 E_2(\tau_o - \tau) + \int_0^{\tau_o} \pi S(t) E_1\left(|\tau - t|\right) dt\right], \quad (4.13)$$

where

$$E_n(x) = \int_0^1 \mu^{n-2} \exp\left(-\frac{x}{\mu}\right) d\mu \quad (4.14)$$

is the well-known exponential integral function.

With the solution of Equation (4.12) available for the intensity, the radiant flux vector $\vec{q}^r$ may be evaluated in terms of the boundary temperatures and the source function. Recall that the radiant flux vector is given as

$$\vec{q}^r = \int_{\omega=4\pi} I\vec{s}\, d\omega . \quad (4.15)$$

For the situation under study, the flux vector is directed normal to planes parallel to the walls and is considered positive in the positive y-coordinate direction. Its magnitude may be expressed in terms of $I^+(\tau,\mu)$ and $I^-(\tau,\mu)$ as

$$q^r(\tau) = 2\pi\left[\int_0^1 I^+(\tau,\mu)\mu d\mu - \int_0^{-1} I^-(\tau,\mu)\mu d\mu\right] . \quad (4.16)$$

Upon substituting Equations (4.12) into (4.16), the expression for the one-dimensional radiation flux is determined.

$$q^r(\tau) = 2\left[\sigma T_1^4 E_3(\tau) - \sigma T_2^4 E_3(\tau_o - \tau) + \int_0^{\tau} \pi S(t) E_2(\tau - t) dt - \int_{\tau}^{\tau_o} \pi S(t) E_2(t - \tau) dt\right] . \quad (4.17)$$

Operating on the expression of Equation (4.17) yields the following result for the divergence of the radiation flux:

$$-\nabla\cdot\vec{q}^r = -\beta\frac{dq^r}{d\tau} = 2(1-\omega_o)\left[\sigma T_1^4 E_2(\tau) + \sigma T_2^4 E_2(\tau_o - \tau) + \int_0^{\tau_o} \pi S(t) E_1\left(|\tau - t|\right) dt - 2\sigma T^4(\tau)\right] . \quad (4.18)$$

Equations (4.17) and (4.18) are the expressions required for the study of simultaneous energy transfer for one-dimensional radiation.

For convenience, we collect here the basic governing equations.

Energy Equation:

$$\rho\, c_p \frac{DT}{Dt} = \nabla\cdot(k\nabla T) - \nabla\cdot\vec{q}^{\,r} \quad . \tag{4.19}$$

Divergence of Radiation Flux:

$$-\nabla\cdot\vec{q}^{\,r} = 2(1-\omega_o)\left[\sigma T_1^4 E_2(\tau) + \sigma T_2^4 E_2(\tau_o - \tau) + \int_0^{\tau_o} \pi S(t) E_1\left(|\tau - t|\right)dt - 2\sigma T^4(\tau)\right] . \tag{4.20}$$

Integral Equation for Source Function:

$$\pi S(\tau) = (1-\omega_o)\sigma T^4(\tau) + \frac{\omega_o}{2}\left[\sigma T_1^4 E_2(\tau) + \sigma T_2^4 E_2(\tau_o - \tau) + \int_0^{\tau_o} \pi S(t) E_1\left(|\tau - t|\right) dt\right] . \tag{4.21}$$

The following observations may be made concerning the above system. First, since the temperature appears in Equation (4.20) to the fourth power and in Equation (4.19) to the first and fourth powers, the system is in general nonlinear in character. Also, because the source function appears within the integral operator of Equations (4.20) and (4.21), the energy equation is an integro-differential equation with the temperature as the dependent variable. Thus this system of equations governing combined conduction and convection in a radiating medium consists of two nonlinear, simultaneous integro-differential equations. As a consequence of the difficulty in solving such systems of equations, approximate forms for the radiation flux are often considered. Two of these are presented next.

4.3 APPROXIMATE RADIATION FLUX EXPRESSIONS

Numerous approximate radiation flux expressions have been suggested in attempts to circumvent the difficulties inherent in the system of equations governing simultaneous conduction and/or convection in a radiating medium. Two of the more important of these are those corresponding to an optically thin ($\tau_o \ll 1$) and an optically thick ($\tau_o \gg 1$) medium.

Consider first the optically thin situation. The exponential integral functions may be approximated for small values of their arguments by the relations

$$E_2(\tau) \simeq 1 + O(\tau) \; ,$$
$$E_3(\tau) \simeq \frac{1}{2} - \tau + O(\tau^2) \; . \tag{4.22}$$

Employing these approximations in the expression for q^r and neglecting terms of the order τ_o gives[3]

$$q^r = \sigma T_1^4 - \sigma T_2^4 \; . \tag{4.23}$$

The radiation flux under optically thin conditions is identical to that through a nonparticipating medium. A similar procedure applied to the derivative of q^r with respect to τ yields

$$-\frac{dq^r}{d\tau} = 2(1-\omega_o)\left[\sigma T_1^4 + \sigma T_2^4 - 2\sigma T^4(\tau)\right] . \tag{4.24}$$

The first two terms on the right side of Equation (4.24) represent radiation emitted by the walls which is absorbed by an elemental volume; whereas the last term is energy emitted by the volume. Thus, under optically thin conditions, each element of the medium exchanges radiation directly with the boundary surfaces. No attenuation of this radiation occurs. An optically thin medium is often referred to as a medium with negligible self-absorption. The simplifications introduced into the system of equations governing simultaneous energy transport in a radiating medium for optically thin conditions are immediately apparent. Since no integral now appears in the divergence of the radiation flux, the energy equation is a differential equation and although nonlinear, it with appropriate boundary conditions is sufficient to evaluate the temperature distribution.

The radiation flux expression for an optically thick medium may be determined by expanding the source function in a Taylor series about $t = \tau$.

$$S(t) = S(\tau) + \frac{dS}{d\tau}(t-\tau) + \frac{1}{2}\frac{d^2S}{d\tau^2}(t-\tau)^2 + \ldots \; . \tag{4.25}$$

Substituting this expansion into Equation (4.17) for the radiation flux and taking the limit as τ and $(\tau_o - \tau)$ tend to infinity,[3] yields

$$q^r(\tau) = -\frac{4}{3}\frac{d(\sigma T^4)}{d\tau} . \tag{4.26}$$

A similar procedure with the flux gradient expression of Equation (4.18) gives[3]

$$-\frac{dq^r}{d\tau} = \frac{4(1-\omega_o)}{3}\frac{d^2(\sigma T^4)}{d\tau^2} , \tag{4.27}$$

since for an optically thick medium

$$S(\tau) = \frac{\sigma T^4(\tau)}{\pi} .$$

Equation (4.26) is known as the Rosseland, optically thick, or diffusion approximation for the radiation flux. Again, the use of Equation (4.27) for the divergence of the radiation flux in the energy equation reduces the system of governing relations to a nonlinear differential equation. In this limit, radiation undergoes a large attenuation and the radiation flux is dependent on local conditions only.

4.4 RADIATIVE EQUILIBRIUM

One of the simpler problems of radiative transfer for the system under study is that prevalent when both convection and conduction are negligible in comparison to radiant energy transfer. In this situation the steady state energy equation reduces to

$$\nabla \cdot \vec{q}^r = 0 . \tag{4.28}$$

Equation (4.28) defines the state of radiative equilibrium. It follows from Equation (4.20) that

$$2\sigma T^4(\tau) = \sigma T_1^4 E_2(\tau) + \sigma T_2^4 E_2(\tau_o - \tau) + \int_0^{\tau_o} \pi S(t) E_1\left(|\tau - t|\right) dt , \tag{4.29}$$

where the source function is determined by the integral equation

$$\pi S(\tau) = (1-\omega_o)\sigma T^4(\tau) + \frac{\omega_o}{2}\left[\sigma T_1^4 E_2(\tau) + \sigma T_2^4 E_2(\tau_o - \tau) + \int_0^{\tau_o} \pi S(t) E_1\left(|\tau - t|\right) dt\right] . \tag{4.30}$$

If Equation (4.29) is used to replace the integral of the source function in Equation (4.30), the following result obtains

$$\pi S(\tau) = \sigma T^4(\tau) . \tag{4.31}$$

Subsequent use of this result in Equation (4.29) yields the following linear integral equation for $T^4(\tau)$:

$$2T^4(\tau) = T_1^4 E_2(\tau) + T_2^4 E_2(\tau_o - \tau) + \int_0^{\tau_o} T^4(t) E_1\left(|\tau - t|\right) dt . \tag{4.32}$$

It is interesting to note that the only influence of scattering on the temperature distribution is through the optical thickness τ_o. Once the local black body emissive power, $\sigma T^4(\tau)$, has been determined the radiation flux follows from Equation (4.17) as

$$q^r = 2\sigma T_1^4 E_3(\tau) - 2\sigma T_2^4 E_3(\tau_o - \tau) + 2\int_0^{\tau} \sigma T^4(t) E_2(\tau - t)dt - 2\int_{\tau}^{\tau_o} \sigma T^4(t) E_2(t - \tau)dt \ . \quad (4.33)$$

Of course, since

$$\frac{dq^r}{d\tau} = 0 \ ,$$

the radiation flux is constant across the layer and, therefore, is a function only of the optical thickness τ_o and the values for the wall temperatures. It is convenient to put Equations (4.32) and (4.33) in dimensionless form. For this purpose, the following dimensionless quantities are introduced:

$$\varphi(\tau) = \frac{T^4(\tau) - T_2^4}{T_1^4 - T_2^4} \ ,$$

$$Q = \frac{q^r}{\sigma\left(T_1^4 - T_2^4\right)} \ . \quad (4.34)$$

With these, Equations (4.32) and (4.33) reduce, respectively, to

$$2\varphi(\tau) = E_2(\tau) + \int_0^{\tau_o} \varphi(t) E_1\left(|\tau - t|\right) dt \ , \quad (4.35)$$

$$Q = 2E_3(\tau) + 2\int_0^{\tau} \varphi(t) E_2(\tau - t)dt - 2\int_{\tau}^{\tau_o} \varphi(t) E_2(t - \tau)dt \ . \quad (4.36)$$

Although the integral equation for $\varphi(\tau)$ is linear, it is difficult to solve, because the kernel $E_1(|\tau - t|)$ is singular when the argument is zero. Various methods have been employed to obtain solutions.[3] Here we only present solutions for the limiting cases of an optically thin and an optically thick medium. According to Equation (4.24), the local value of T^4 for optically thin conditions is a constant equal to the arithmetic mean of the wall values.

$$T^4 = \frac{T_1^4 + T_2^4}{2} \ . \quad (4.37)$$

In dimensionless form,

$$\varphi(\tau) = \frac{1}{2} \ . \quad (4.38)$$

The dimensionless flux corresponding to an optically thin medium is found by inserting Equation (4.38) in Equation (4.36).

$$Q = 1 - \tau_o \quad (4.39)$$

For an optically thick medium in radiative equilibrium, it follows from Equation (4.27) that

$$T^4(\tau) = \left(T_2^4 - T_1^4\right)\frac{\tau}{\tau_o} + T_1^4 \ , \quad (4.40)$$

or

$$\varphi(\tau) = 1 - \frac{\tau}{\tau_o} \ . \quad (4.41)$$

The dimensionless flux is

$$Q = \frac{4}{3\tau_o} \ . \quad (4.42)$$

Typical results for the dimensionless temperature function, φ, are shown in Figure 4.2 for selected values of optical thickness.[4] The distributions are almost linear across the layer. Of particular interest is the discontinuity in temperature at the gas-surface interface. This temperature discontinuity at the wall is due to the neglect of heat conduction within the radiating medium. In the absence of conduction there is no requirement that the surface and fluid have identical temperature at their interface.

Figure 4.3 illustrates the dimensionless heat transfer as a function of the optical thickness of the layer. As expected, the heat transfer diminishes with increasing optical thickness. This figure also serves to illustrate the limitations of the optically thin and optically thick approximations. It appears that the optically thin results are acceptable for optical thickness values less than 0.1, while the optically thick approximation yields reasonably accurate results for $\tau_o > 10$. The results for intermediate values of optical thickness shown on the figure were obtained by Viskanta and Grosh[4] using an approximate method of solution to the integral equation for $\varphi(\tau)$.

4.5 COMBINED CONDUCTION AND RADIATION

The interaction of conduction and radiation transfer may be demonstrated by considering the steady energy transfer in the medium between the parallel black plates. If convection may be neglected, the energy equation of (4.19) reduces to

$$\frac{d}{dy}\left(q^c + q^r\right) = 0 . \tag{4.43}$$

Thus the sum of the conductive and radiative fluxes is constant across the layer. Taking the conductivity constant in the Fourier law of heat conduction transforms the above to

$$k\frac{d^2T}{dy^2} - \frac{dq^r}{dy} = 0 . \tag{4.44}$$

Introducing the optical depth as the independent variable and the flux derivative expression from Equation (4.20) yields

$$\frac{d^2T}{d\tau^2} = \frac{(1-\omega_o)}{k\beta}\left[4\sigma T^4(\tau) - G(\tau)\right] , \tag{4.45}$$

where $G(\tau)$ is

$$G(\tau) = 2\sigma T_1^4 E_2(\tau) + 2\sigma T_2^4 E_2(\tau_o - \tau) + 2\int_0^{\tau_o}\left[(1-\omega_o)\sigma T^4(t) + \frac{\omega_o}{4}G(t)\right] E_1\left(|\tau - t|\right) dt . \tag{4.46}$$

It is convenient to transform the governing equations to dimensionless form. For this purpose the following dimensionless quantities are introduced:

$$\theta(\tau) = \frac{T(\tau)}{T_1} , \quad \theta_2 = \frac{T_2}{T_1} , \tag{4.47}$$

$$n(\tau) = \frac{G(\tau)}{\sigma T_1^4} , \quad N = \frac{k\beta}{4\sigma T_1^3} . \tag{4.48}$$

With the above definitions, Equations (4.45) and (4.46) reduce, respectively, to

$$N\frac{d^2\theta}{d\tau^2} = (1-\omega_o)\left[\theta^4(\tau) - \frac{1}{4}\eta(\tau)\right] , \tag{4.49}$$

$$\eta(\tau) = 2E_2(\tau) + 2\theta_2^4 E_2(\tau_o - \tau) + 2\int_0^{\tau_o}\left[(1-\omega_o)\theta^4(t) + \frac{\omega_o}{4}\eta(t)\right] E_1\left(|\tau - t|\right) dt . \tag{4.50}$$

The boundary conditions required are

$$\theta(0) = 1, \quad \theta(\tau_o) = \theta_2 . \tag{4.51}$$

The total flux, conductive plus radiative, is given in dimensionless form as

$$\frac{q}{\sigma T_1^4} = -4N\frac{d\theta}{d\tau} + 2E_3(\tau) - 2\theta_2^4 E_3(\tau_o - \tau) + 2\int_0^{\tau}\left[(1-\omega_o)\theta^4(t) + \frac{\omega_o}{4}\eta(t)\right] E_2(\tau - t)dt$$
$$- 2\int_{\tau}^{\tau_o}\left[(1-\omega_o)\theta^4(t) + \frac{\omega_o}{4}\eta(t)\right] E_2(t-\tau)dt . \tag{4.52}$$

A number of parameters influence combined conduction and radiation. These are the optical thickness of the medium, τ_o, the albedo of scattering, ω_o, the temperature ratio of the walls, T_2/T_1 and N. Of these only the parameter N requires further elucidation. The factor N is a measure of the relative role of heat transfer by conduction to that by radiation. When $N \rightarrow \infty$, the heat transfer is purely conductive, while for $N \rightarrow 0$, conduction is unimportant and the previously studied state of radiative equilibrium is present.

Again, the results for the limiting cases of an optically thin and an optically thick medium are of particular interest. When the medium is optically thin, the radiation heat transfer is identical to that for a nonparticipating medium.

$$q^r = \sigma\left(T_1^4 - T_2^4\right) . \tag{4.53}$$

Since both the conductive and radiative flux are independent of position, the energy equation (4.43) may be integrated directly. The result is

$$q = \frac{k(T_1 - T_2)}{L} + \sigma\left(T_1^4 - T_2^4\right) . \tag{4.54}$$

Thus, in the optically thin limit, the total flux is the sum of the independently evaluated conduction and radiation components. Note also that the total flux is independent of the radiative properties of the medium. Under optically thick conditions, the total flux is

$$q = q^c + q^r = -k\frac{dT}{dy} - \frac{4}{3\beta}\frac{d(\sigma T^4)}{dy}$$

$$= -\left(k + \frac{16\sigma T^3}{3\beta}\right)\frac{dT}{dy} . \tag{4.55}$$

Direct integration across the layer yields

$$q = \frac{k(T_1 - T_2)}{L} + \frac{4\sigma}{3\tau_o}\left(T_1^4 - T_2^4\right) . \tag{4.56}$$

This result is of the same form as that for the optically thin limit. Hence, in both limits for optical thickness, the total flux may be determined by superposition of the conduction and radiation fluxes with each evaluated as if the other were not present. Also, it is evident from Equation (4.56) that the effect of scattering enters into the optically thick heat flux result only through the optical thickness $\tau_o(=\beta L)$.

Typical results illustrating the transition of the dimensionless heat transfer from radiative to conductive in a nonscattering medium ($\omega_o = 0$) are illustrated in Figure 4.4 for $\theta_2 = 0.1$. The dashed lines represent the limits of pure radiation (N = 0) and the dash-dot lines represent those for pure conduction ($N = \infty$). The other results were obtained by various approximate or numerical methods of solution to the governing system of equations.[3] It has been found that for high emittance surfaces, the superposition principle gives accurate results even for intermediate values of optical thickness, although there is no theoretical justification for this procedure.[3] In Figure 4.5, the influence of albedo on the heat transfer for $\tau_o = 1$ is shown. In general, the heat transfer is a weak function of ω_o.

4.6 COMBINED CONDUCTION, CONVECTION, AND RADIATION

As an example of the interaction of conduction, convection, and radiation, consider the flow of a radiating fluid in a plane duct consisting of black parallel planes of infinite extent.[5] Neglecting conduction and radiation in the flow direction, which is taken as the x-coordinate direction, the energy equation for steady laminar flow simplifies to

$$\rho c_p u(y)\frac{\partial T}{\partial x} = k\frac{\partial^2 T}{\partial y^2} - \frac{\partial q^r}{\partial y} . \tag{4.57}$$

In writing Equation (4.57), the fluid properties have been considered constant and viscous dissipation and energy generation processes ignored. For fully-developed flow at distances remote from the duct entrance, the axial temperature gradient may be replaced by

$$\frac{\partial T}{\partial x} = \left(\frac{T_w - T}{T_w - T_b}\right)\frac{dT_b}{dx}$$

$$= 2\left(\frac{T_w - T}{T_w - T_b}\right)\frac{q_w}{\rho c_p \bar{u} L} , \tag{4.58}$$

where T_w, T_b, q_w, and $\bar{u}$ are the duct wall temperature, fluid mixing cup temperature, wall heat flux, and fluid mass mean velocity, respectively. Introducing Equation (4.58) into (4.57) yields

$$2\left(\frac{T_w - T}{T_w - T_b}\right)\frac{u(y)}{\bar{u}}\frac{q_w}{L} = k\frac{d^2T}{dy^2} - \frac{dq^r}{dy}. \quad (4.59)$$

The velocity ratio for Poiseuille flow is

$$\frac{u(y)}{\bar{u}} = 6\left[\left(\frac{y}{L}\right) - \left(\frac{y}{L}\right)^2\right]. \quad (4.60)$$

For a non-scattering fluid and equal temperature black walls, the derivative of the radiation flux for one-dimensional transfer is

$$-\frac{dq^r}{d\tau} = 2\sigma T_w^4\left[E_2(\tau) + E_2(\tau_o - \tau)\right] + 2\sigma\int_0^{\tau_o} T^4(t)E_1\left(|\tau - t|\right)dt - 4\sigma T^4(\tau), \quad (4.61)$$

and the total wall heat flux is

$$q_w = -k\left.\frac{\partial T}{\partial y}\right|_{y=0} + 2\sigma T_w^4\left[1 - E_3(\tau_o)\right] - 2\sigma\int_0^{\tau_o} T^4(t)E_2(t)dt. \quad (4.62)$$

To put the equations in dimensionless form, the following dimensionless quantities are defined:

$$\left.\begin{aligned} N &= \frac{k\kappa}{4\sigma T_w^3}, \\ \psi &= \frac{q_w}{\sigma T_w^4}, \\ \theta &= \frac{T}{T_w}, \\ \theta_b &= \frac{T_b}{T_w}. \end{aligned}\right\} \quad (4.63)$$

Combining Equations (4.59), (4.60), and (4.61), and introducing the dimensionless variables yields the energy equation as(3)

$$2N\frac{d^2\theta}{d\tau^2} - \frac{6\psi}{\tau_o}\left(\frac{1-\theta}{1-\theta_b}\right)\left(\frac{\tau}{\tau_o} - \frac{\tau^2}{\tau_o^2}\right) = 2\theta^4(\tau) - E_2(\tau) - E_2(\tau_o - \tau) - \int_0^{\tau_o}\theta^4(t)E_1\left(|\tau - t|\right)dt. \quad (4.64)$$

The boundary conditions are

$$\theta(0) = \theta(\tau_o) = 1. \quad (4.65)$$

The wall heat flux in dimensionless form is

$$\psi = -4N\left(\frac{d\theta}{d\tau}\right)_{\tau=0} + 2 - 2E_3(\tau_o) - 2\int_0^{\tau_o}\theta^4(t)E_2(t)dt. \quad (4.66)$$

It is customary to express the wall heat flux in terms of a Nusselt number, Nu. For the parallel plate channel, the conventional definition is

$$Nu = \frac{2q_wL}{k(T_w - T_b)}. \quad (4.67)$$

In terms of the previously introduced dimensionless quantities, the Nusselt number is

$$Nu = \frac{\psi\,\tau_o}{2N(1 - \theta_b)}. \quad (4.68)$$

The parameters governing the heat transfer are the optical thickness τ_o, the factor N which measures the relative importance of conduction to radiation heat transfer and the ratio of bulk fluid temperature to the wall temperature, T_b/T_w.

For the limiting cases of an optically thin and an optically thick medium, the energy equation transforms from a nonlinear integro-differential equation to a nonlinear differential equation. For optically thin conditions, the energy equation is

$$2N \frac{d^2\theta}{d\tau^2} - \frac{6\psi}{\tau_o}\left(\frac{1-\theta}{1-\theta_b}\right)\left(\frac{\tau}{\tau_o} - \frac{\tau^2}{\tau_o^2}\right) = 2\left[\theta^4(\tau) - 1\right] \; ; \tag{4.69}$$

while for optically thick conditions, the energy equation becomes

$$\frac{d}{d\tau}\left[\left(2N + \frac{8\theta^3}{3}\right)\frac{d\theta}{d\tau}\right] - \frac{6\psi}{\tau_o}\left(\frac{1-\theta}{1-\theta_b}\right)\left(\frac{\tau}{\tau_o} - \frac{\tau^2}{\tau_o^2}\right) = 0 \; . \tag{4.70}$$

Figure 4.6 illustrates results for Nusselt number obtained from numerical solutions to the energy equation.(5) The results for $N = \infty$ correspond to those of a nonradiating fluid. The results obtained from numerical solutions to the energy equations for the approximate radiation flux expressions are also shown. Viskanta chose to express his results in terms of θ_c, the ratio of the center line temperature to the wall temperature rather than θ_b. It may be noted that both the optically thin and optically thick results are greater than those predicted by the exact expression. The diffusion approximation yields results which show large discrepancies even for optical thickness values as large as ten.

REFERENCES

1. Viskanta, R., "Radiation Transfer and Interaction of Convection with Radiation Heat Transfer," in Advances in Heat Transfer, Vol. III. New York: Academic Press, 1966.

2. Cess, R. D., "The Interaction of Thermal Radiation with Conduction and Convection Heat Transfer," in Advances in Heat Transfer, Vol. I. New York: Academic Press, 1964.

3. Sparrow, E. M. and R. D. Cess. Radiation Heat Transfer. Belmont, Calif.: Brooks/Cole Publishing Co., 1966.

4. Viskanta, R. and R. J. Grosh, "Heat Transfer in a Thermal Radiation Absorbing and Scattering Medium," Proceedings, International Heat Transfer Conference, Boulder, Colorado, 1961.

5. Viskanta, R., "Interaction of Heat Transfer by Conduction, Convection, and Radiation in a Radiating Fluid," Journal of Heat Transfer, C, 85 (1963), pp. 318-328.

6. Lick, W., "Energy Transfer by Radiation and Conduction," Proceedings, 1963 Heat Transfer and Fluid Mechanics Institute. Palo Alto, Calif.: Stanford University Press, 1963, pp. 14-26.

7. Viskanta, R., "Heat Transfer by Conduction and Radiation in Absorbing and Scattering Materials," Journal of Heat Transfer, C, 87 (1965), pp. 143-150.

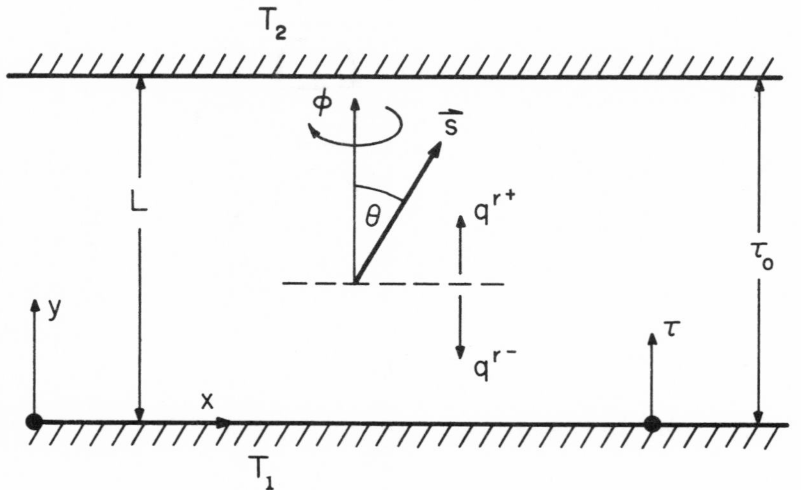

FIGURE 4.1 PARALLEL PLATE SYSTEM.

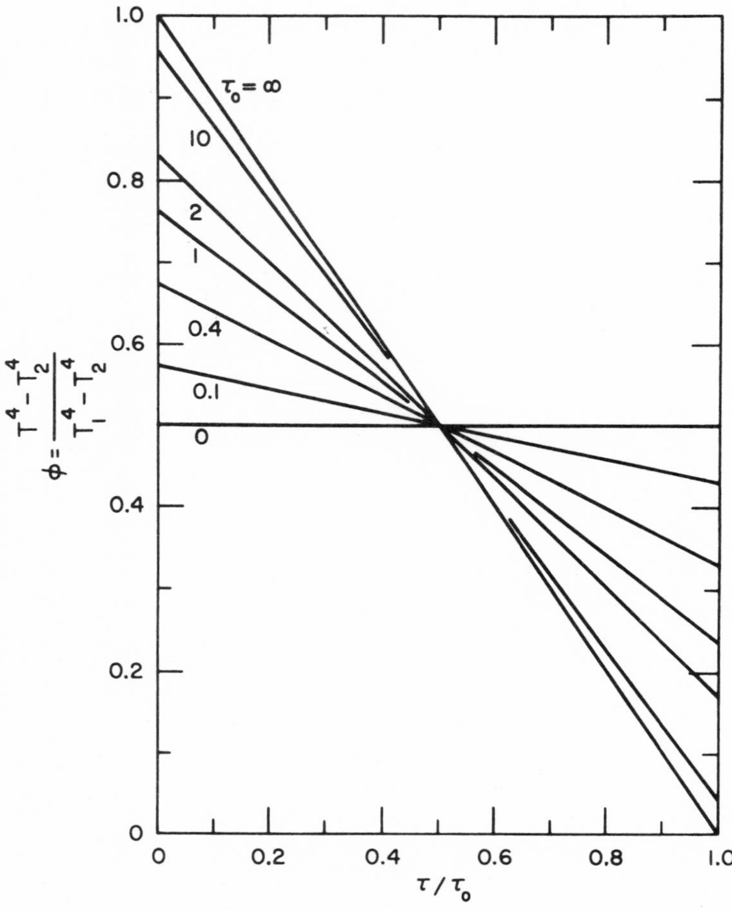

FIGURE 4.2 DIMENSIONLESS BLACK BODY EMISSIVE POWER DISTRIBUTION BETWEEN BLACK PARALLEL PLATES (REFERENCE 4).

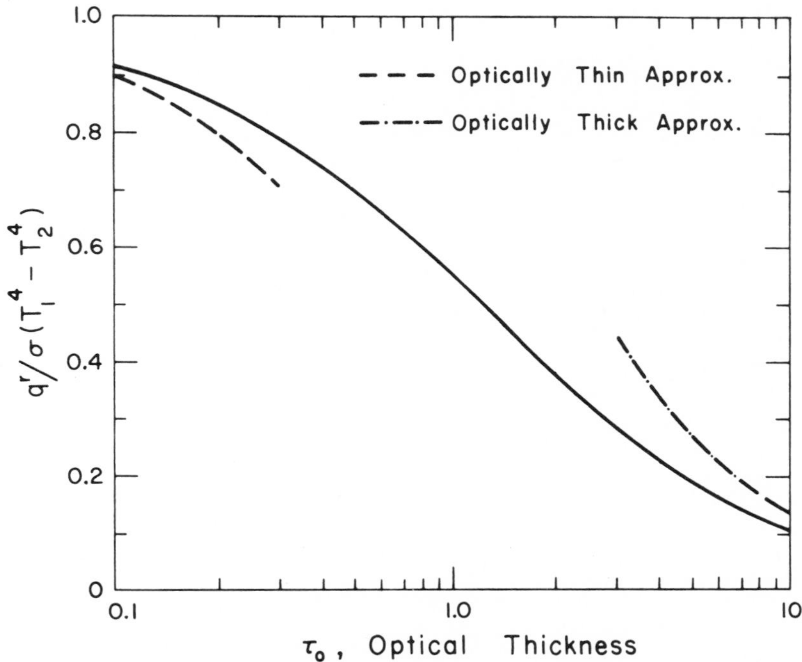

FIGURE 4.3 HEAT TRANSFER BETWEEN PARALLEL BLACK PLATES -- RADIATIVE EQUILIBRIUM (REFERENCE 4).

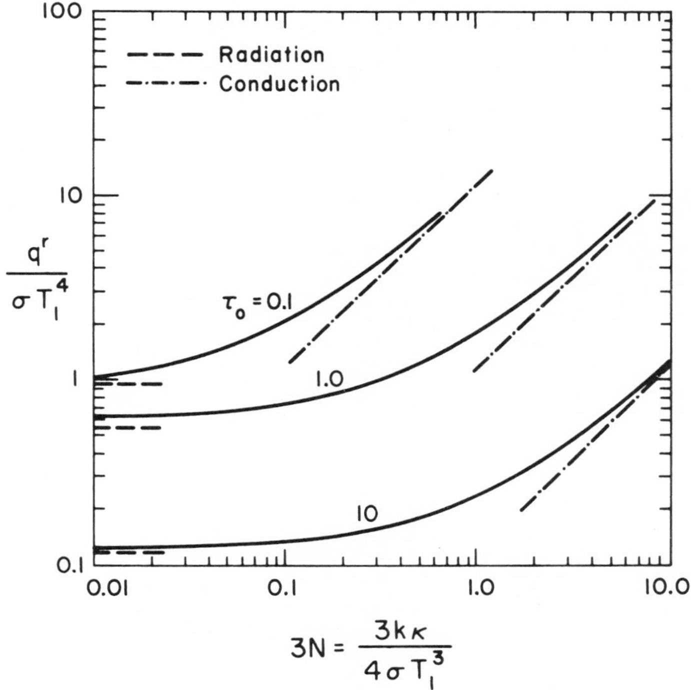

FIGURE 4.4 HEAT TRANSFER BETWEEN PARALLEL BLACK PLATES FOR $\Theta_2 = 0.1$ -- CONDUCTION AND RADIATION (REFERENCE 6).

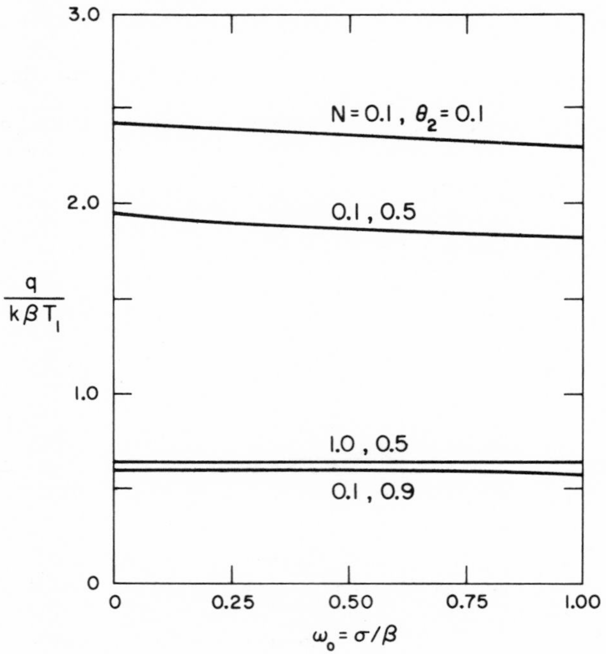

FIGURE 4.5 EFFECT OF ALBEDO ON HEAT TRANSFER BETWEEN PARALLEL BLACK PLATES FOR τ_0 = 1 (REFERENCE 7).

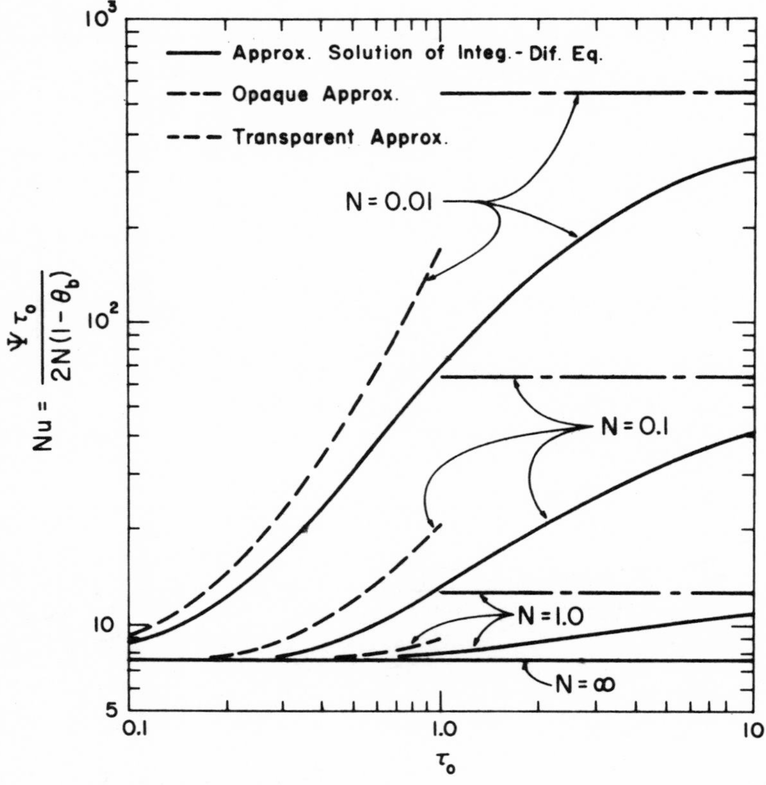

FIGURE 4.6 NUSSELT NUMBER FOR Θ_c = 0.5 AND BLACK WALLS (REFERENCE 5).

CONTENTS

NUMERICAL METHODS IN HEAT TRANSFER
by A. M. Clausing

1. INTRODUCTION 157

2. DERIVATION OF APPLICABLE EQUATIONS AND BOUNDARY CONDITIONS IN DIFFERENCE FORM. 158

2.1 Notation and Mathematical Preliminaries 158
2.2 Governing Difference Relationships. 164
2.3 Boundary Conditions in Difference Form. 169
2.4 An Example. 173

3. STEADY STATE -- NUMERICAL PROCEDURES 182

3.1 Direct Methods. 183
3.2 Iterative Methods 184
3.3 Global Accuracy and Convergence Tests for Iterative Methods 192

4. TRANSIENTS -- NUMERICAL PROCEDURES 196

4.1 Standard Explicit Method. 197
4.2 Standard Implicit and Crank-Nicolson Methods. 200
4.3 DuFort-Frankel Method 202
4.4 Alternating Direction Methods 203
4.5 Exponential Approximation 208
4.6 Comparisons and Conclusions 209

NUMERICAL METHODS IN HEAT TRANSFER*

A. M. Clausing**

1. INTRODUCTION

The exploration of space, utilization of nuclear power, supersonic flight, high-temperature processing plants, and other areas of recent interest have resulted in a vast number of complex heat transfer problems. The study of heat transfer processes experimentally, which has played a major role in the past, is becoming more difficult and expensive. Thus considerable impetus is present for the development of our analytical, numerical, and analog capabilities.

Today's heat transfer problems are formidable ones, not only because of the complex nature of the processes under study, but also because the accuracy required often excludes the assumption of constant properties or prohibits the employment of a simplified model or geometry. Previously employed procedures are for these reasons not always applicable.

Three general methods of solving heat transfer problems are available: analytical, numerical, and analog. Let us consider briefly the advantages and disadvantages of these methods in the solution of heat conduction problems.

*This paper was delivered as three lectures at the conference.

**Assistant Professor of Mechanical Engineering, University of Illinois, Urbana, Illinois.

The exact solution of heat conduction problems dates back many years, and excellent collections of solutions are available in the literature (e.g., Reference 1). These solutions are almost without exception confined to linear problems and regions of simple geometry -- a devastating weakness of exact procedures. It is often argued that exact solutions show more vividly the effects of various parameters; however, the significant dimensionless groups which yield much valuable information about a process can be obtained from the governing <u>difference</u> equations. Thus one need not even obtain the governing differential equations to obtain these groups. Due to the complexity of the infinite series solutions which normally arise in exact procedures, an inspection of the solutions themselves reveal, in general, very little additional information. Also, this complexity often results in the necessity of employing a digital computer to evaluate the solution. However, such computer programs are difficult to write; they are only valid or of use for the specific solution programmed, and errors in these programs are difficult to detect. If the exact solution is to be evaluated at many points, an all-numerical method would probably result in a reduction of computational effort.

Of course, the engineer must certainly study analytical techniques. These techniques contribute greatly to our overall understanding, provide many transformations and procedures of value in numerical work, and supply the tools required in the analysis of the truncation error and stability of difference approximations. In addition, exact solutions provide an excellent means of checking the accuracy of various approximate techniques.

At one time analog techniques for the solution of heat transfer problems received considerable attention. Even today it is difficult to compete with the simplicity of the conductive sheet electrical analog in the solution of two-dimensional steady state conduction problems in complex geometries if the boundary conditions are relatively simple. The active analog computer is convenient for the solution of transient problems which give rise to either linear or nonlinear ordinary differential equations. On the other hand, active analog computers are not well suited for the solution of the partial differential equations which arise in heat transfer. Special R-C analogs have sometimes been built for the solution of transient conduction problems; however, such facilities are not readily available nor are they very versatile or powerful. It is difficult on such devices, for example, to treat nonlinear problems such as radiative boundary conditions or temperature dependent properties. It is even difficult to include internal generation within the region. Since the advent of the digital computer, the pressure which forced the development of passive analog facilities has been removed. To the author's knowledge, the digital computer is capable of solving every heat transfer problem which could be solved on these passive analog devices and many more.

The biggest advantage of numerical methods is their flexibility. For example, problems involving complex geometries, mixed, nonlinear, or other complex boundary conditions, temperature dependent properties, changes of phase, combined convection, radiation and conduction, etc., have been successfully treated with numerical procedures. This does not mean to say, however, that techniques are available today to solve any and all such problems. Many difficulties are also encountered in numerical calculations and much work remains to be done.

In the past, numerical methods were criticized because they generally necessitated a digital computer and because computer solutions were very expensive. However, significant changes are taking place which are removing the basis for these criticisms and are beginning to question our present emphasis on analytical methods. In 1950 there were only 10 or 15 computers at work in the United States. Today there are over 35,000.[(2)*] In 1950, a fast computer was approximately 10,000 times slower than the fastest contemporary computers.[(2)] For example, an addition takes 0.18 μ sec on the new IBM 360. Also, this machine is capable of performing many operations simultaneously. Since the hourly cost of these computers has not changed significantly, the cost of an answer has been greatly reduced.

Real impetus to numerical solutions of practical problems in engineering and physics was given by Southwell[(3)] and later by Allen,[(4)] the pioneers of relaxation methods. Dusinberre[(5)] contributed greatly to the application of relaxation methods and other techniques in the numerical solution of heat transfer problems. These techniques, which are also discussed in almost every heat conduction textbook, were used for many years by those who used hand calculation procedures. However, relaxation methods require human insight which is generally difficult to incorporate efficiently into computer programs.

The purpose of this work is to present a unified and modern discussion of numerical methods in heat transfer. The emphasis will be placed on steady state and transient heat conduction problems. Only those methods which are suitable for use on a high-speed digital computer will be stressed. The notation which will be employed is designed to simplify transcription to a computer language such as FORTRAN. A discussion of the physical and mathematical derivation of the appropriate difference equations will be given. Algorithms for executing these numerical solutions will be discussed. Consideration will be given to the determination of convergence rates, stability, and accuracy of the various difference analogs. Some numerical examples will be given.

2. DERIVATION OF APPLICABLE EQUATIONS AND BOUNDARY CONDITIONS IN DIFFERENCE FORM

2.1 NOTATION AND MATHEMATICAL PRELIMINARIES

In finite difference methods, the continuous field and the continuous independent variables are replaced by a discretized system. The resulting solution no longer yields the continuous variation of the dependent variable, φ, but it gives its value at a finite number of discrete grid points

*Superscript numbers in parentheses refer to References at the end of this paper.

which are distributed throughout the region. Considering a cartesian coordinate system, the continuous independent variables will be replaced by:

$$\left.\begin{aligned} x &= i(\Delta x), \\ y &= j(\Delta y), \\ z &= k(\Delta z), \\ \text{and} \\ t &= n(\Delta t), \end{aligned}\right\} \qquad (2.1)^*$$

where t is the independent variable, time. The numbers i, j, k, and n are integers restricted as follows:

$$\begin{aligned} 0 &\le i \le \frac{x_1}{\Delta x} = p, \\ 0 &\le j \le \frac{y_1}{\Delta y} = q, \\ 0 &\le k \le \frac{z_1}{\Delta z} = s, \\ 0 &\le n. \end{aligned} \qquad (2.2)$$

It is assumed that Δx, Δy, and Δz do not vary throughout the discretized region. The dependent variable in the discretized system is written as:

$$\begin{aligned} \varphi_{i,j,k,n} &\equiv \varphi\,(i\Delta x,\ j\Delta y,\ k\Delta z,\ n\Delta t)\ , \\ \varphi_{i+1,\ j+1,\ k,\ n} &\equiv \varphi\left[(i+1)\Delta x,\ (j+1)\Delta y,\ k\Delta z,\ n\Delta t\right], \end{aligned} \qquad (2.3)$$

etc. If the space indices are suppressed, it is understood that the point i, j, k is under consideration.

Difference operators will now be introduced in order to simplify the writing and manipulation of the difference relationships which will arise. Consider for brevity a function of two independent variables, $\varphi = \varphi(x,y)$.

The first forward difference with respect to x is defined as:

$$\Delta_x \varphi_{i,j} = \varphi_{i+1,j} - \varphi_{i,j} = \varphi(x+\Delta x, y) - \varphi(x,y)\ , \qquad (2.4)$$

where Δ is the forward difference operator. It can be distinguished from the spatial increments, Δx, Δy, etc., by its subscript. The first forward difference with respect to y is:

$$\Delta_y \varphi_{i,j} = \varphi_{i,j+1} - \varphi_{i,j};$$

the second forward difference with respect to x is defined as:

$$\begin{aligned} \Delta_x^2 \varphi_{i,j} &= \Delta_x \varphi_{i+1,j} - \Delta_x \varphi_{i,j} \\ &= \varphi_{i+2,j} - 2\varphi_{i+1,j} + \varphi_{i,j}\ . \end{aligned} \qquad (2.5)$$

Higher order differences are defined by iteration. It is to be noted that a forward difference with respect to x could not be employed at the right boundary, i = p, without introducing fictitious grid points.

The backward difference operator is del, ∇, and the first backward difference with respect to x is defined as:

$$\nabla_x \varphi_{i,j} = \varphi_{i,j} - \varphi_{i-1,j}\ . \qquad (2.6)$$

The r-th backward difference with respect to x is:

*Some computer languages such as FORTRAN do not permit a subscript of zero. In this case, it is usually more convenient to define the space variables as: $x = (i-1)\Delta x$, etc.

$$\nabla_x^r \varphi_{i,j} = \nabla_x^{r-1} \varphi_{i,j} - \nabla_x^{r-1} \varphi_{i-1,j} . \quad (2.7)$$

A backward difference with respect to x cannot be employed at the left boundary, i = 0.

The third difference operator which will be introduced will be the central difference operator, δ. The first central difference with respect to x is defined as:

$$\delta_x \varphi_{i,j} = \varphi_{i+1/2,j} - \varphi_{i-1/2,j} , \quad (2.8)$$

and the second central difference is:

$$\delta_x^2 \varphi_{i,j} = \delta_x \varphi_{i+1/2,j} - \delta_x \varphi_{i-1/2,j}$$
$$= \varphi_{i+1,j} - 2\varphi_{i,j} + \varphi_{i-1,j} . \quad (2.9)$$

It is seen that only the even central differences involve the dependent variable at defined grid points. Occasionally, the shifting operator, E, will also be of use. It is defined as:

$$E_x \varphi_{i,j} = \varphi_{i+1,j} .$$

Also,

$$E_x^m \varphi_{i,j} = \varphi_{i+m,j} . \quad (2.10)$$

The differential operator, D, will be frequently employed where

$$D_x \varphi_{i,j} = \left.\frac{\partial\varphi}{\partial x}\right|_{i,j} . \quad (2.11)$$

It is understood that the derivative is evaluated at the point i, j if the indices are suppressed.

The five operators defined here possess the distributive, commutative, and associative properties of real numbers. In order to relate the difference operators to the differential operator, consider Taylor's expansion:

$$\varphi(x+\Delta x,y) = \varphi(x,y) + \frac{\Delta x}{1!}\varphi_x(x,y) + \frac{(\Delta x)^2}{2!}\varphi_{xx}(x,y) + \ldots , \quad (2.12)$$

or

$$\varphi_{i+1,j} = \left(1 + \frac{\Delta x}{1!} D_x + \frac{(\Delta x)^2}{2!} D_x^2 + \frac{(\Delta x)^3}{3!} D_x^3 + \ldots\right)\varphi_{i,j} .$$

It is seen that Taylor's expansion can be written in the compact form:

$$\varphi_{i+1,j} = e^{\Delta x D_x} \varphi_{i,j} , \quad (2.13)$$

also

$$\varphi_{i-1,j} = e^{-\Delta x D_x} \varphi_{i,j} . \quad (2.14)$$

From Equation (2.13) one can also obtain by iteration the useful result:

$$\varphi_{i+1,j+1} = e^{(\Delta x D_x + \Delta y D_y)} \varphi_{i,j} , \quad (2.15)$$

which can easily be extended to a function of n independent variables. Considerable use will be given to Equations (2.13) to (2.15) in the analysis of truncation errors. Let us now again consider relating D to Δ, ∇, and δ.

Following the procedure employed by Hildebrand,[6] Equations (2.4) and (2.13) can be combined to give:

$$(\Delta_x + 1)\varphi_{i,j} = e^{\Delta x D_x} \varphi_{i,j} ,$$

or

$$D_x \varphi_{i,j} = \frac{\partial\varphi}{\partial x}\bigg|_{i,j} = \frac{1}{\Delta x} \ln(1+\Delta_x)\, \varphi_{i,j}$$

$$= \frac{1}{\Delta x}\left(\Delta_x - \frac{\Delta_x^2}{2} + \frac{\Delta_x^3}{3} - \ldots\right) \varphi_{i,j} \; . \qquad (2.16)$$

In a similar manner, one can obtain:

$$D_x \varphi_{i,j} = \frac{1}{\Delta x}\left(\nabla_x + \frac{\nabla_x^2}{2} + \frac{\nabla_x^3}{3} + \ldots\right) \varphi_{i,j} \; . \qquad (2.17)$$

Higher order partial derivatives can be obtained from Equations (2.16) and (2.17) by iteration. Approximations for derivatives at boundaries, which are obtained by employing one or more terms of these expressions, are often found in the literature. Probably the most common formulas are obtained by employing only the first term. In this case, Equation (2.16) simplifies to:

$$\frac{\partial\varphi}{\partial x} \simeq \frac{\varphi_{i+1,j} - \varphi_{i,j}}{\Delta x} \; , \qquad (2.18)$$

and Equation (2.17) becomes:

$$\frac{\partial\varphi}{\partial x} \simeq \frac{\varphi_{i,j} - \varphi_{i-1,j}}{\Delta x} \; . \qquad (2.19)$$

Equation (2.18) is known as the first forward difference quotient, and Equation (2.19) is known as the first backward difference quotient. It is seen that if one takes the limit of Equation (2.18) as Δx approaches zero, the definition of the partial derivative is obtained.

Expressions can also be obtained in a similar fashion in terms of central differences, which are useful in the interior of the region. Hildebrand[6] gave the following relationship for the second derivative:

$$\frac{\partial^2\varphi}{\partial x^2} = \frac{1}{(\Delta x)^2}\left(\delta_x^2 - \frac{\delta_x^4}{12} + \frac{\delta_x^6}{90} - \ldots\right) \varphi_{i,j} \; . \qquad (2.20)$$

If only the first term of Equation (2.20) is employed, the following approximation, known as the second central difference quotient, results.

$$\frac{\partial^2\varphi}{\partial x^2} \simeq \frac{\varphi_{i+1,j} - 2\varphi_{i,j} + \varphi_{i-1,j}}{(\Delta x)^2} \; . \qquad (2.21)$$

This approximation is frequently employed in the numerical solution of partial differential equations. The magnitude of the error committed in the approximation of Equation (2.21) can be estimated from the relative magnitude of the first neglected term,

$$\frac{\delta_x^4 \varphi_{i,j}}{12(\Delta x)^2} \; .$$

It is often more useful if the error, which is committed when a derivative is approximated by a difference quotient, is obtained in terms of derivatives instead of differences. This is possible by considering Taylor's expansion directly. For example, consider the following two expansions:

$$\varphi_{i+1,j} = \varphi_{i,j} + \Delta x \frac{\partial \varphi}{\partial x} + \frac{(\Delta x)^2}{2!} \frac{\partial^2 \varphi}{\partial x^2} + \frac{(\Delta x)^3}{3!} \frac{\partial^3 \varphi}{\partial x^3} + \frac{(\Delta x)^4}{4!} \frac{\partial^4 \varphi}{\partial x^4} + \ldots , \tag{2.22a}$$

and

$$\varphi_{i-1,j} = \varphi_{i,j} - \Delta x \frac{\partial \varphi}{\partial x} + \frac{(\Delta x)^2}{2!} \frac{\partial^2 \varphi}{\partial x^2} - \frac{(\Delta x)^3}{3!} \frac{\partial^3 \varphi}{\partial x^3} + \frac{(\Delta x)^4}{4!} \frac{\partial^4 \varphi}{\partial x^4} - \ldots . \tag{2.22b}$$

Equation (2.22a) can be written as:

$$\frac{\partial \varphi}{\partial x} = \frac{\varphi_{i+1,j} - \varphi_{i,j}}{\Delta x} - \frac{\Delta x}{2} \left[D_x^2 + \frac{(\Delta x)}{3} D_x^3 + \ldots \right] \varphi_{i,j} . \tag{2.23}$$

In approximating a partial derivative or a partial differential equation by a difference expression, the truncation error will be defined as:

$$E_T \equiv \left\{ f_1 \left[\begin{matrix} \text{Differential} \\ \text{operators} \end{matrix} \right] - f_2 \left[\begin{matrix} \text{Difference} \\ \text{operators} \end{matrix} \right] \right\} \varphi_{i,j,k,n} . \tag{2.24}$$

For example, if the first derivative is approximated by the forward difference quotient, the truncation error would be:

$$E_T = \left(D_x - \frac{\Delta_x}{\Delta x} \right) \varphi_{i,j} .$$

The truncation error of an acceptable difference approximation of a partial differential equation must approach zero as the spatial and time increments approach zero. This property is called consistency. Consistency may require, for example, that $\Delta t/\Delta x$ and $\Delta t/\Delta y$ vanish as Δt, Δx, and Δy approach zero; and a certain consistency condition (for instance, a requirement that $\Delta t/\Delta y$ and $\Delta t/\Delta x$ are small) must be satisfied in the numerical calculations to insure that the difference relationships approximate the differential equation under study.*

If the first derivative is approximated by the first forward difference quotient, Equations (2.23) and (2.24) show that the truncation error is:

$$E_T = - \frac{\Delta x}{2} \left[D_x^2 + \frac{(\Delta x)}{3} D_x^3 + \ldots \right] \varphi_{i,j} ; \tag{2.25}$$

or employing the Lagrangian form of the remainder,

$$E_T = - \frac{\Delta x}{2} D_x^2 \varphi_{i+\xi_1,j} , \qquad 0 < \xi_1 < 1 . \tag{2.26}$$

Likewise, Equation (2.22b) gives:

$$\frac{\partial \varphi}{\partial x} = \frac{\varphi_{i,j} - \varphi_{i-1,j}}{\Delta x} + \frac{\Delta x}{2} \left[D_x^2 + \frac{(\Delta x)}{3} D_x^3 + \ldots \right] \varphi_{i,j} . \tag{2.27}$$

---

*For examples of such requirements see Sections 4.3, 4.4, and 4.5.

Thus approximating the first derivative by the first backward difference quotient results in the following truncation error:

$$E_T = + \frac{\Delta x}{2} D_x^2 \varphi_{i-\xi_2,j} , \qquad 0 < \xi_2 < 1 . \tag{2.28}$$

If Equation (2.22b) is subtracted from Equation (2.22a) and rearranged, the following expression is obtained:

$$\frac{\partial \varphi}{\partial x} = \frac{\varphi_{i+1,j} - \varphi_{i-1,j}}{2\Delta x} - \frac{(\Delta x)^2}{6} D_x^3 \varphi_{i+\xi_3,j} , \qquad -1 < \xi_3 < 1 . \tag{2.29}$$

The first term on the right side of Equation (2.29) is the average of the first forward and first backward difference quotients. The average difference quotient results in a truncation error of order $(\Delta x)^2$ instead of order (Δx). Finally, if one adds Equations (2.22a) and (2.22b), the following expression for the second partial derivative in terms of the second central difference quotient is obtained:

$$\frac{\partial^2 \varphi}{\partial x^2} = \frac{\varphi_{i+1,j} - 2\varphi_{i,j} + \varphi_{i-1,j}}{(\Delta x)^2} - \frac{\Delta x^2}{12} \left[D_x^4 + \frac{(\Delta x)^2}{30} D_x^6 + \dots \right] \varphi_{i,j} , \tag{2.30}$$

the truncation error follows as:

$$E_T = - \frac{(\Delta x)^2}{12} \left[D_x^4 + \frac{(\Delta x)^2}{30} D_x^6 + \dots \right] \varphi_{i,j}$$

$$= - \frac{(\Delta x)^2}{12} D_x^4 \varphi_{i+\xi_4,j} , \qquad -1 < \xi_4 < 1 . \tag{2.31}$$

Thus the error involved in approximating the second derivative by a second central difference quotient is of order $(\Delta x)^2$.

As an example of the use of these results, consider the Laplace equation,

$$\frac{\partial^2 \varphi}{\partial x^2} + \frac{\partial^2 \varphi}{\partial y^2} = 0 . \tag{2.32}$$

If second central difference quotients are employed to approximate these second derivatives with $\Delta x = \Delta y = h$, the difference analog becomes:

$$\frac{1}{h^2} \left(\delta_x^2 + \delta_y^2 \right) \varphi_{i,j} = 0 . \tag{2.33}$$

The truncation error is:

$$E_T = \left[\left(D_x^2 + D_y^2 \right) - \frac{1}{h^2} \left(\delta_x^2 + \delta_y^2 \right) \right] \varphi_{i,j}$$

$$= - \frac{h^2}{12} \left(D_x^4 + D_y^4 + \dots \right) \varphi_{i,j} . \tag{2.34}$$

It is easily seen that as Δx and Δy approach zero, the truncation error vanishes; therefore the approximation is consistent.

2.2 GOVERNING DIFFERENCE RELATIONSHIPS

The energy equation is sometimes derived by performing an energy balance on a small element of volume fixed in space. The shape of this element depends on the coordinate system being employed; for example, it is a rectangular parallelepiped for a cartesian coordinate system. Let us consider the application of the first law of thermodynamics for a solid element.

Verbally, this statement of the conservation of energy is:

$$\begin{bmatrix}\text{the change in the}\\ \text{internal energy}\\ \text{of the element}\\ \text{during } \Delta t\end{bmatrix} = \begin{bmatrix}\text{the net heat}\\ \text{conducted into}\\ \text{the element}\\ \text{during } \Delta t\end{bmatrix} + \begin{bmatrix}\text{the energy}\\ \text{generated within}\\ \text{the element}\\ \text{during } \Delta t\end{bmatrix}, \quad (2.35)$$

where the left side of Equation (2.35) is zero for a steady state problem.

If one lets the size of the element approach zero after the energy balance is written, the partial differential equation which describes the continuous variation of the dependent variable, temperature, throughout the continuous field is obtained. All the parameters are distributed continuously throughout the space, and the space and time variables are continuous independent variables.

The method of deriving the appropriate finite difference expression is the same except that the finite size of the element is retained; that is, the entire region is discretized. The resulting solution will no longer yield the continuous variation of the dependent variable, but will give its value at a finite number of discrete grid points or nodes throughout the region.

Consider an energy balance on any interior element in a region for a cartesian coordinate system. The center of the element is at (x,y,z) or $(i\Delta x, j\Delta y, k\Delta z)$ since uniform spatial increments are assumed. For the parallelepiped shown in Figure 2.1, Equation (2.35) becomes:

$$\rho\, c_p (T_{t+\Delta t} - T_t)\, \Delta V = \Delta t \left[q_x\Big|_{x-\frac{\Delta x}{2}} - q_x\Big|_{x+\frac{\Delta x}{2}} + q_y\Big|_{y-\frac{\Delta y}{2}} - q_y\Big|_{y+\frac{\Delta y}{2}} + q_z\Big|_{z-\frac{\Delta z}{2}} - q_z\Big|_{z+\frac{\Delta z}{2}} + q'''\, \Delta V \right], \quad (2.36)$$

where q is the rate of heat flow, q''' is the rate of heat generated per unit volume, ΔV is the volume of the element, ρ is the density, c_p is the specific heat, and T is the temperature. If it is assumed that the medium is isotropic the rate of heat flow through any element of area ΔA is given by Fourier's law as:

$$q_n = -\, k\, \Delta A \frac{\partial T}{\partial n}, \quad (2.37)$$

where n is the direction normal to the surface, q_n is positive if it is flowing in the positive n-direction, and k is the thermal conductivity. Fourier's law in the discretized system will be written as:

$$q_n = (\bar{k})(\text{area of face}) \frac{[T\ (\text{adjacent element}) - T\ (\text{element})]}{(\text{distance between centers of elements})}, \quad (2.38)$$

where $\bar{k}$ is based on the average of the two relevant nodal temperatures. In this case, q_n is assumed positive if heat is flowing to the element. Equation (2.38) is applicable to any orthogonal coordinate system and can also be employed at boundary nodes. It is also useful to interpret Equation (2.38) as a potential difference divided by a resistance.

The rate of heat flow into the element through the face at $x-(\Delta x)/2$ follows from Equation (2.38) as:

$$q_x \Big|_{x-\frac{\Delta x}{2}} = k_{i-1/2} \, \Delta y \Delta z \, \frac{T_{i-1,j,k} - T_{i,j,k}}{\Delta x} , \tag{2.39}$$

or

$$q_x \Big|_{x-\frac{\Delta x}{2}} = q_{(i-1)\to i} = \frac{T_{i-1,j,k} - T_{i,j,k}}{R_{i-1/2}} , \tag{2.40}$$

where

$$R_{i-1/2} = \frac{\Delta x}{k_{i-1/2} \, \Delta y \Delta z} ,$$

and $k_{i-1/2}$ is the thermal conductivity evaluated at

$$\frac{T_{i-1,j,k} + T_{i,j,k}}{2} .$$

A single index is used on the resistances and conductivities for brevity. The heat flow rate

$$q_{(i-1)\to i} = - q_{i\to(i-1)}$$

is defined as the rate of heat flow from the element (i-1, j, k) to the element (i, j, k). Similarly,

$$q_{(i+1)\to i} = - q_x \Big|_{x+\frac{\Delta x}{2}} = k_{i+1/2} \, \Delta y \Delta z \, \frac{T_{i+1,j,k} - T_{i,j,k}}{\Delta x} . \tag{2.41}$$

If one proceeds in this manner in the y- and z-directions, substitutes these results into Equation (2.36), divides by Δt, and takes the limit as Δt approaches zero, one obtains:

$$\rho \, c_p \, \Delta V \, \frac{dT_{i,j,k}}{dt} = \frac{T_{i-1,j,k} - T_{i,j,k}}{R_{i-1/2}} + \frac{T_{i+1,j,k} - T_{i,j,k}}{R_{i+1/2}} + \frac{T_{i,j-1,k} - T_{i,j,k}}{R_{j-1/2}} + \frac{T_{i,j+1,k} - T_{i,j,k}}{R_{j+1/2}} + \frac{T_{i,j,k-1} - T_{i,j,k}}{R_{k-1/2}} + \frac{T_{i,j,k+1} - T_{i,j,k}}{R_{k+1/2}} + q''' \, \Delta V , \tag{2.42}$$

where the values of the resistances $R_{i+1/2}$, $R_{i-1/2}$, ... are given in Table 2.1. Since continuous time is being employed, Equation (2.42) represents the form of most of the equations in the set of simultaneous ordinary differential equations which must be solved. This set would be employed, for example, in an analog computer solution. The time variable must also be discretized in order to employ digital computational methods. Discretized time will be considered in Section 4.

Equation (2.42) can also be written in the following form:

TABLE 2.1

COMPONENTS OF EQUATION (2.42) FOR VARIOUS COORDINATE SYSTEMS

	Cartesian	Cylindrical	Spherical
Coordinate System	$x = \Delta x(i)$ $y = \Delta y(j)$ $z = \Delta z(k)$	$r = \Delta r(i)$ $\theta = \Delta\theta(j)$ $z = \Delta z(k)$	$r = r(i)$ $\theta = \Delta\theta(j)$, azimithal angle $\varphi = \Delta\varphi(k)$, polar angle
ΔV	$\Delta x\ \Delta y\ \Delta z$	$r_i\ \Delta\theta\Delta r\ \Delta z$	$r_i^2(\sin\varphi_k)\ \Delta r\ \Delta\varphi\ \Delta\theta$
$R_{i-1/2}$	$\frac{\Delta x}{\Delta y\Delta z\ k_{i-1/2}}$	$\frac{\Delta r}{(r_i - \Delta r/2)\Delta\theta\Delta z\ k_{i-1/2}}$	$\frac{\Delta r}{(r_i - \Delta r/2)^2(\sin\varphi_k)\Delta\varphi\Delta\theta\ k_{i-1/2}}$
$R_{i+1/2}$	$\frac{\Delta x}{\Delta y\Delta z\ k_{i+1/2}}$	$\frac{\Delta r}{(r_i + \Delta r/2)\Delta\theta\Delta z\ k_{i+1/2}}$	$\frac{\Delta r}{(r_i + \Delta r/2)^2(\sin\varphi_k)\Delta\varphi\Delta\theta\ k_{i+1/2}}$
$R_{j-1/2}$	$\frac{\Delta y}{\Delta x\Delta z\ k_{j-1/2}}$	$\frac{r_i\Delta\theta}{\Delta r\ \Delta z\ k_{j-1/2}}$	$\frac{\Delta\theta\ \sin\varphi_k}{\Delta r\ \Delta\varphi\ k_{j-1/2}}$
$R_{j+1/2}$	$\frac{\Delta y}{\Delta x\Delta z\ k_{j+1/2}}$	$\frac{r_i\Delta\theta}{\Delta r\ \Delta z\ k_{j+1/2}}$	$\frac{\Delta\theta\ \sin\varphi_k}{\Delta r\ \Delta\varphi\ k_{j+1/2}}$
$R_{k-1/2}$	$\frac{\Delta z}{\Delta x\Delta y\ k_{k-1/2}}$	$\frac{\Delta z}{\Delta r\ \Delta\theta\ r_i k_{k-1/2}}$	$\frac{\Delta\varphi}{\sin(\varphi_k - \Delta\varphi/2)\Delta r\Delta\theta\ k_{k-1/2}}$
$R_{k+1/2}$	$\frac{\Delta z}{\Delta x\Delta y\ k_{k+1/2}}$	$\frac{\Delta z}{\Delta r\ \Delta\theta\ r_i k_{k+1/2}}$	$\frac{\Delta\varphi}{\sin(\varphi_k + \Delta\varphi/2)\Delta r\Delta\theta\ k_{k+1/2}}$

$$C\frac{dT_{i,j,k}}{dt} = q_{i+1\to i} + q_{i-1\to i} + q_{j-1\to j} + q_{j+1\to j} + q_{k-1\to k} + q_{k+1\to k} + q''' \, \Delta V \,, \tag{2.43}$$

where C is the thermal capacitance of the element, $\rho c_p \Delta V$. In the derivation of the difference equations for interior as well as boundary nodes, it is often convenient to begin with an energy balance written in the form of Equation (2.43). Equations (2.42) and (2.43) are valid for all coordinate systems;* however, the values of ΔV and the resistances, $R_{i+1/2}$, $R_{i-1/2}$, etc., will vary with the coordinate system. The expression for the resistances for other orthogonal coordinate systems can be obtained from Equation (2.38) and the definition of R. Figures 2.2 and 2.3 are provided to aid in obtaining these values for cylindrical and spherical coordinates. Table 2.1 gives the corresponding values for cartesian, cylindrical, and spherical coordinates. In the cylindrical and spherical coordinate systems, the volume of the element and the corresponding resistances are, in general, dependent upon the location of the element, i.e., on the values of i, j, and k.

If the thermal conductivity is independent of the temperature and location of the element, Equation (2.42) can be divided by $k\Delta V$. The corresponding forms in this case for the three coordinate systems are:

Cartesian:

$$\frac{1}{(\Delta x)^2}\left(T_{i-1,j,k} + T_{i+1,j,k} - 2T_{i,j,k}\right) + \frac{1}{(\Delta y)^2}\left(T_{i,j-1,k} + T_{i,j+1,k} - 2T_{i,j,k}\right) + \frac{1}{(\Delta z)^2}\left(T_{i,j,k-1} + T_{i,j,k+1} - 2T_{i,j,k}\right) + \frac{q'''}{k} = \frac{1}{\alpha}\frac{dT_{i,j,k}}{dT} \;; \tag{2.44}$$

Cylindrical:

$$\frac{1}{(\Delta r)^2}\left[T_{i-1,j,k}\left(1 - \frac{\Delta r}{2r_i}\right) + T_{i+1,j,k}\left(1 + \frac{\Delta r}{2r_i}\right) - 2T_{i,j,k}\right] + \frac{1}{(\Delta\theta)^2 r_i^2}\left[T_{i,j-1,k} + T_{i,j+1,k} - 2T_{i,j,k}\right] + \frac{1}{(\Delta z)^2}\left[T_{i,j,k-1} + T_{i,j,k+1} - 2T_{i,j,k}\right] + \frac{q'''}{k} = \frac{1}{\alpha}\frac{dT_{i,j,k}}{dt} \;; \tag{2.45}$$

*For nonorthogonal coordinate systems, the computation of the respective resistances is not trivial (see, e.g., Reference 5, p. 85).

Spherical:

$$\frac{1}{(\Delta r)^2}\left[\left(T_{i-1,j,k} - T_{i,j,k}\right)\left(1 - \frac{\Delta r}{2r_i}\right)^2 + \left(T_{i+1,j,k} - T_{i,j,k}\right)\left(1 + \frac{\Delta r}{2r_i}\right)^2\right] + \frac{1}{r_i^2 \sin^2\varphi_k (\Delta\theta)^2}\left[T_{i,j-1,k} + T_{i,j+1,k} - 2T_{i,j,k}\right]$$

$$+ \frac{1}{r_i^2 \sin\varphi_k (\Delta\varphi)^2}\left[\left(T_{i,j,k-1} - T_{i,j,k}\right)\sin\left(\varphi_k - \frac{\Delta\varphi}{2}\right) + \left(T_{i,j,k+1} - T_{i,j,k}\right)\sin\left(\varphi_k + \frac{\Delta\varphi}{2}\right)\right] + \frac{q'''}{k} = \frac{1}{\alpha}\frac{dT_{i,j,k}}{dt} \quad ; \tag{2.46}$$

where α is the thermal diffusivity.

In order to get a more complete picture of the problem at hand, consider the determination of the two-dimensional temperature distribution throughout a unit square whose surface temperature is prescribed everywhere. The nodal points are indicated by the solid dots in Figure 2.4 and the boundaries of the elements are indicated by dotted lines. (An alternate network for this problem is discussed in Section 2.3.) The maximum values of i and j were chosen equal to 4; thus $\Delta x = \Delta y = 1/4$. It is seen that the heat capacity is no longer distributed uniformly throughout the region but is lumped at the nodal points. Heat can no longer flow everywhere and in any direction but must now flow only along the resistors where the distributed resistance has been lumped. The temperature at any node represents the temperature at that point in the continuous region. Considering this temperature to be the average temperature of the element is acceptable in the interior of the region but can lead to appreciable errors when one is considering boundary elements. If there is no internal generation in this region and its thermal conductivity is constant, the governing difference/differential equation is:

$$\left(\delta_x^2 + \delta_y^2\right) T_{i,j} = \frac{(\Delta x)^2}{\alpha}\frac{dT_{i,j}}{dt} , \tag{2.47}$$

where i = 1, 2, or 3

j = 1, 2, or 3.

This equation represents the set of nine differential equations which must be solved simultaneously. For a steady state problem, nine algebraic equations are obtained.

We have seen how to obtain the governing difference equations from an energy balance and have indicated the physical approximations which these equations represent. Now consider starting from the governing partial differential equations. This approach might seem illogical since a possible method of deriving the governing equation is to take the limit of Equation (2.36) as Δx, Δy, Δz, and Δt approach zero. We now want to reverse this procedure in order to go from the resulting partial differential equation back to a difference equation and a discretized system. It will be seen that, in general, if two or three point central difference quotients are employed to approximate the derivatives, the same results will be obtained.

Since discretized time will be considered in Section 4, steady heat flow in two dimensions will be considered at this time. Internal generation will not be considered since the treatment of this term is straightforward. The governing differential equation in cartesian coordinates is:

$$\frac{\partial}{\partial x}\left(k\frac{\partial T}{\partial x}\right) + \frac{\partial}{\partial y}\left(k\frac{\partial T}{\partial y}\right) = 0 .$$

If first central difference quotients are employed to replace the first differentiation, one obtains:

$$\frac{k_{i+1/2}\left[\partial T/\partial x\right]_{i+1/2,j} - k_{i-1/2}\left[\partial T/\partial x\right]_{i-1/2,j}}{\Delta x}$$

$$+ \frac{k_{j+1/2}\left[\partial T/\partial y\right]_{i,j+1/2} - k_{j-1/2}\left[\partial T/\partial y\right]_{i,j-1/2}}{\Delta y} = 0 \,. \qquad (2.48)$$

If the first derivatives are again replaced by first central difference quotients, Equation (2.48) becomes:

$$\frac{k_{i+1/2}}{(\Delta x)^2}\left(T_{i+1,j} - T_{i,j}\right) + \frac{k_{i-1/2}}{(\Delta x)^2}\left(T_{i-1,j} - T_{i,j}\right)$$

$$+ \frac{k_{j+1/2}}{(\Delta y)^2}\left(T_{i,j+1} - T_{i,j}\right) + \frac{k_{j-1/2}}{(\Delta y)^2}\left(T_{i,j-1} - T_{i,j}\right) = 0 \,. \qquad (2.49)$$

Equation (2.49) is identical to Equation (2.42) if the latter is simplified to the problem being considered. If the thermal conductivity is a constant, Equation (2.49) becomes:

$$\left(\frac{\delta_x^2}{(\Delta x)^2} + \frac{\delta_y^2}{(\Delta y)^2}\right) T_{i,j} = 0 \,. \qquad (2.50)$$

This approximation for the Laplace equation is identical to the result obtained employing an energy balance as is seen by simplifying Equation (2.44) to the case being considered.

In a similar manner, results could also be obtained for cylindrical and spherical coordinates which are identical to those given by Equation (2.42) and Table 2.1 for variable conductivity, or Equations (2.45) and (2.46) for constant conductivity. The next section will consider expressing the boundary conditions in finite difference form. It will be seen that the straightforward substitution of difference quotients will not always result in the same relationship as that which is obtained employing an energy balance. The advantage of the latter method should become obvious.

2.3 BOUNDARY CONDITIONS IN DIFFERENCE FORM

The derivation of a difference form of the boundary conditions using either an energy balance or a Taylor's expansion is not difficult. However, only one of the possible representations of a given boundary condition is physically consistent with the difference approximation employed in the interior of the region. In addition, the most obvious set of physically consistent equations will not always result in the best approximation. In these cases different networks will be introduced to improve the approximation. The definition of physical consistency will now be given.

Most of the individual network resistors (e.g., consider the networks of Figure 2.4) occur in two equations of the resulting set of simultaneous equations including, to be sure, those equations which result from the boundary conditions. However, the difference equations can be derived without any reference to such a network. If the resulting set of equations so derived are then written in the form of Equation (2.42) and compared with a resistance network, the set of equations is physically consistent if all the resulting resistors are uniquely defined. If the set is not physically consistent, the network representation of Figure 2.4 is not possible. In addition, it is impossible to write an overall energy balance for the system. This is a severe disadvantage, since such an energy balance provides an excellent check of the correctness of the derived difference equations and the computer program. It also provides a meaningful criterion for the termination of an iterative process as is discussed further in Section 3.3.

If one employs an energy balance on finite elements in conjunction with a pictorial R-C network, it is a simple matter to obtain a physically consistent set of difference equations. This method will be referred to as the physical approach. On the other hand, if one blindly employs Taylor's expansions or various difference quotients to obtain the required difference representations, a physically consistent set will probably not result. Even if the order of the truncation error is the same in all the difference approximations employed, this will not guarantee that a physically consistent set of difference equations will be obtained. Although it may be possible to obtain a physically consistent set employing Taylor's expansions, it is not always clear how this is done nor how one would be assured of this consistency without making a comparison with the physically derived relationships.

Consider first a boundary condition of prescribed surface temperature. This is the easiest boundary condition to work with in numerical methods as was also the case in analytical methods. The values of the temperatures for the boundary nodes are determined by simply substituting the corresponding values of the discretized independent variables in the prescribed function. The initial nodal temperatures in transient problems would be applied in an identical manner. The resulting network, for example, that shown in Figure 2.4, would be physically consistent. Although this network is excellent for the case of steady state problems, an examination of the physical approximation being made will show that the network can result in serious errors in transient problems at small values of time.

For instance, consider again the problem of Section 2.2 and the network of Figure 2.4. The region will be assumed to be initially at temperature $T_i = 0$, and at time $t = 0$ all the boundaries are suddenly changed to T_s, a constant. It is seen that this boundary condition results in all the surface capacitors being charged instantaneously. In effect, this capacitance is being neglected which in the present case is 41 per cent of the total heat capacity of the region. If the heat flow through all the resistors leading into the interior of the region were integrated to give the change in the internal energy of the region, this value would be in error by approximately 40 per cent at steady state and by a considerably greater amount at small values of time. An unnecessarily fine network would be required to avoid these errors, i.e., with the present grid arrangement.

The problems arising from the network of Figure 2.4 with the prescribed temperature boundary condition would be eliminated if the network shown in Figure 2.5 is employed. No surface capacitors are present in this network, and for this reason it would represent a better approximation. In this case

$$x = (i - \tfrac{1}{2})\ \Delta x\ ,$$

$$y = (j - \tfrac{1}{2})\ \Delta y\ ,$$

and Equation (2.47) would only be valid for the four innermost nodal points. The elements adjacent to the boundary would result in slightly different equations. For example, an energy balance on element (1,2) gives:

$$\left(T_{2,2} + T_{1,1} + T_{1,3} + 2T_{1/2,2} - 5T_{1,2}\right) = \frac{(\Delta x)^2}{\alpha}\ \frac{dT_{1,2}}{dt}\ . \qquad (2.51)$$

Consider next the same problem and the network of Figure 2.5 with the following boundary conditions along the boundary $x = 0,\ 0 \leq y \leq 1$.

(a) Prescribed Flux:

$$-\ k\left.\frac{\partial T}{\partial x}\right|_{x=0} = f(y,t)\ , \qquad (2.52a)$$

where f is the surface heat flux which is assumed to be a function of y and t. In this case an energy balance for node (1,2) gives:

$$\Delta A f + q_{(1,1)\to(1,2)} + q_{(2,2)\to(1,2)} + q_{(1,3)\to(1,2)} = \rho\ c_p\ \Delta V \frac{dT_{1,2}}{dt}\ ,$$

or

$$\frac{\Delta x}{k} f\left(\frac{3\Delta y}{2}, t\right) + \left(T_{2,2} + T_{1,3} + T_{1,1} - 3T_{1,2}\right) = \frac{(\Delta x)^2}{\alpha} \frac{dT_{1,2}}{dt} . \tag{2.52b}$$

(b) Convective Boundary Condition:

$$- k \left.\frac{\partial T}{\partial x}\right|_{x=0} = \tilde{h} (T - T_a) , \tag{2.53a}$$

where T_a is the temperature of the ambient fluid and $\tilde{h}$ is the convective heat transfer coefficient. Both of these quantities could also be a function of the location on the boundary and time. In this case an energy balance gives

$$\frac{2Bi}{2+Bi} T_a + T_{2,2} + T_{1,3} + T_{1,2} - \left(3 + \frac{2Bi}{2+Bi}\right) T_{1,2} = \frac{(\Delta x)^2}{\alpha} \frac{dT_{1,2}}{dt} , \tag{2.53b}$$

where Bi is the Biot number $\tilde{h}\Delta x/k$. If T_a is a constant and the thermal properties are constant, a simplification could be effected by shifting the temperature scale to reduce T_a to zero.

(c) Radiative Boundary Condition:

$$- k \Delta A \left.\frac{\partial T}{\partial x}\right|_{x=0} = \sigma \mathcal{F}_{\Delta A-S} \left(T^4 - T_s^4\right) \Delta A, \tag{2.54a}$$

where $\mathcal{F}_{\Delta A-S}$ is the gray body shape factor between the element of area of interest and its surroundings and T_s is an equivalent temperature of the surroundings.

It is seen that the surface temperature is required. The surface temperature for this or the other boundary conditions can be obtained from a simple energy balance on the <u>zero capacitance</u> surface node; that is,

$$q_{\text{surroundings} \to (1/2,2)} + q_{(1,2) \to (1/2,2)} = 0 .$$

Once the surface temperature has been determined, an energy balance on (1, 2) gives

$$\frac{\Delta x}{k} \sigma \mathcal{F}_{\Delta A-s} \left(T_s^4 - T_{1/2,2}^4\right) + T_{2,2} + T_{1,3} + T_{1,1} - 3T_{1,2} = \frac{(\Delta x)^2}{\alpha} \frac{dT_{1,2}}{dt} . \tag{2.54b}$$

If

$$T_s^4 \gg T_{1/2,2}^4 ,$$

Equation (2.54b) would reduce to Equation (2.52b); that is, the boundary condition would reduce to a specified flux boundary condition. In some cases it is advantageous to write the radiative boundary condition in terms of the convective boundary condition. In this case the convective heat transfer coefficient becomes:

$$\tilde{h} = \frac{\sigma \mathcal{F}_{\Delta A-s} \left(T_{1/2,2}^4 - T_s^4\right)}{\left(T_{1/2,2} - T_s\right)} \tag{2.55}$$

or if $T_{1/2,2} - T_s$ is small:

$$\tilde{h} \simeq 4 \sigma \mathcal{F}_{\Delta A-s} T_m^3 , \tag{2.56}$$

where

$$T_m = \frac{T_{1/2,2} + T_s}{2} .$$

The radiative boundary condition, or a convective boundary condition with a film coefficient which is temperature dependent, results in a nonlinear difference equation. Some solution methods can treat nonlinear terms without much added difficulty; however, other methods will not be directly applicable to a set of nonlinear difference equations. Equations (2.52b), (2.53b), and (2.54b) could have been obtained by employing Taylor's expansions; however, the method might not appear straightforward.

The network of Figure 2.4 results in greater simplicity, but it would appear to be less accurate for most transient problems. Since it is an equally acceptable network for steady state problems and since it results in greater simplicity, the difference relations for a left boundary node (0, j) will be given. For prescribed flux, convection heat transfer, and radiative heat transfer, the equations corresponding to Equations (2.52b), (2.53b) and (2.54b) are:

$$2\frac{\Delta x}{k} f(j\Delta y, t) + T_{0,j+1} + T_{0,j-1} + 2T_{1,j} - 4T_{0,j} = \frac{(\Delta x)^2}{\alpha} \frac{dT_{0,j}}{dt} , \tag{2.52c}$$

$$2 \text{ Bi } T_a + T_{0,j+1} + T_{0,j-1} + 2T_{1,j} - (4+2\text{Bi})T_{0,j} = \frac{(\Delta x)^2}{\alpha} \frac{dT_{0,j}}{dt} , \tag{2.53c}$$

and

$$\frac{2\Delta x \sigma \mathcal{F}_{\Delta A-s}}{k} \left(T_s^4 - T_{0,j}^4\right) + T_{0,j+1} + T_{0,j-1} + 2T_{1,j} - 4T_{0,j} = \frac{\Delta x^2}{\alpha} \frac{dT_{0,j}}{dt} . \tag{2.54c}$$

Again, these equations could have been obtained by employing either an energy balance or Taylor's expansions. It is noted that the treatment of the boundary conditions is somewhat simpler if the network of Figure 2.4 is employed. In addition, the surface capacitors are no longer charged instantaneously for the above boundary conditions as they were for the prescribed temperature case.

Cylindrical and spherical regions give rise to difficulties not experienced in cartesian coordinates. For example, neither Equation (2.45) nor Equation (2.46) are defined at the centerline, $r_i = 0$. Consider as an example an axially symmetrical problem -- a right solid cylinder. Since it is known that $\partial T/\partial r = 0$ at the centerline, one might be tempted to employ a forward difference quotient to satisfy this condition. However, the resulting set of difference equations would not be physically consistent. To be consistent, the resistor between the nodes (0, k) and (1, k) must be the same as the value used for the node (1, k) which was:

$$\frac{\Delta r}{k\, 2\pi\, (r_1 - \Delta r/2)\, \Delta z},$$

or if $\Delta r = \Delta z$ this becomes $2/(k\pi\Delta r)$. A logical resistance to employ in the z-direction is:

$$\frac{L}{kA} = \frac{4}{k\pi\Delta r} .$$

Employing the element's actual volume, the resulting difference equation is:

$$T_{0,k+1} + T_{0,k-1} + 2T_{1,k} - 4T_{0,k} = \frac{(\Delta V_0)^2}{\alpha} \frac{dT_{0,k}}{dt} . \tag{2.57}$$

Consider next the cylindrical outer surface which is assumed to be perfectly insulated. Equation (2.45) for this case contains the fictitious temperature $T_{p+1,k}$, which must be eliminated with the aid of the boundary condition

$$\left.\frac{\partial T}{\partial r}\right|_{r=r_p} = 0 .$$

If an average of the forward and backward difference quotients are employed, the result is:

$$T_{p+1,k} = T_{p-1,k} .$$

The resulting difference equation would be:

$$2T_{p-1,k} + T_{p,k-1} + T_{p,k+1} - 4T_{p,k} = \frac{(\Delta v_p)^2}{\alpha}\frac{dT_{p,k}}{dt} . \quad (2.58)$$

If an energy balance is employed, the resulting difference equation would be:

$$2\left(1-\frac{\Delta r}{2r_p}\right)\left(T_{p-1,k}-T_{p,k}\right) + \left(1-\frac{\Delta r}{4r_p}\right)\left(T_{p,k+1}+T_{p,k-1}-2T_{p,k}\right) = \frac{(\Delta r)^2\left[1-(\Delta r/4r_p)\right]}{\alpha}\frac{dT_{p,k}}{dt}. \quad (2.59)$$

The latter equation, Equation (2.59), is consistent with the approximation employed in the interior of the region. On the other hand, the use of Equation (2.58) would result in a physically inconsistent set of equations. If Equation (2.58) were employed, an overall energy balance for the system would not be satisfied, since some energy would be passing through the outer boundary. This inconsistency could require the employment of an appreciably finer network to obtain the same accuracy. Perhaps a more thorough study would reveal that Equation (2.59) could be obtained from a Taylor's expansion; however, the fact remains that many approximations which suggest themselves are physically inconsistent with those employed in the interior, Equation (2.45).

2.4 AN EXAMPLE

The following simple example is given to demonstrate the degree of accuracy which can be obtained with an extremely course network if one is physically consistent, but a much finer network is required to obtain similar accuracy with inconsistent approximations. The problem of the steady state heat exchange between a pin or rectangular fin and a fluid medium will be considered.

It is assumed that the fin has a constant cross-sectional area A, perimeter P, length L, and known base temperature T_o. The heat flow in the fin is assumed to be one-dimensional; the film coefficient, $\tilde{h}$, and the thermal conductivity, k, are also assumed to be constants. The ambient fluid temperature is T_a. The region is shown in Figure 2.6.

The differential equation which governs the temperature distribution in this fin is

$$\frac{d^2T^*}{dx^{*2}} - m^{*2}T^* = 0 ,$$

where

$$T^* = \frac{T - T_a}{T_o - T_a} ,$$

$$x^* = \frac{x}{L} ,$$

and

$$m^{*2} = \frac{\tilde{h}PL^2}{kA} .$$

The asterisk is employed to denote a dimensionless quantity.

If one employs Equation (2.30) to eliminate the second derivative, one obtains:

$$\delta_x^2 T_i^* - (\Delta x^*)^2\, m^{*2} T_i^* = 0 , \quad (2.60)$$

$$i = 1, 2, \ldots p-1,$$

where the truncation error is

$$E_T = -\frac{(\Delta x^*)^2}{12}\left.\frac{d^4T^*}{dx^{*4}}\right|_{i,j} + \ldots .$$

If an energy balance on an interior element is performed, the same difference equation would result.

Equation (2.60) represents (p-1) equations, but it contains p unknowns. The additional equation which is required can be obtained from the boundary condition at the end of the fin. To avoid the introduction of an additional dimensionless group, the heat loss from the end of the fin will be neglected. This boundary condition results in significantly less algebra in the determination of the exact solution. In a finite difference method it causes little difficulty.

Introducing a fictitious node (p+1) and employing the following approximation:

$$\left.\frac{dT^*}{dx^*}\right|_p = 0 = \frac{T^*_{p+1} - T^*_{p-1}}{2\Delta x^*}$$

gives

$$T^*_{p+1} = T^*_{p-1} .$$

If this is substituted into Equation (2.60), one obtains the additional required equation.

$$T^*_{p-1} - \left[1 + (\Delta x^*)^2 \frac{m^{*2}}{2}\right] T^*_p = 0. \qquad (2.61)$$

An energy balance on this element would give the same result if the convective exchange from the element were based on the temperature at the end of the element. Alternately one could employ the interpolated average temperature of the element which would result in the following difference equation:

$$T^*_{p-1} - \frac{1 + (3/8)m^{*2}(\Delta x^*)^2}{1 - \left[m^{*2}(\Delta x^*)^2\right]/8} T^*_p = 0. \qquad (2.62)$$

Due to the nature of the boundary condition, the difference between the results employing Equations (2.61) and (2.62) are small. However, it will be seen that these considerations are of greater importance at the base of the fin and would be of importance at the tip if it were not assumed to be insulated.

The set of simultaneous linear algebraic equations represented by Equations (2.60) and (2.62) were solved by employing the algorithm of Section 3.1 for three different values of m^*, 0.5, 1.42, and 4.0. The optimum m^* for maximum heat exchange per unit mass of fin is 1.42. The choice of $m^* = 4.0$ represents a very poor thermal design but was employed, along with a $\Delta x^* = 1/3$, to exaggerate the truncation error. The results of the calculations are given in Table 2.2. The errors in the calculated temperatures are seen to be small even for this coarse network. In one

TABLE 2.2

COMPARISON OF EXACT AND NUMERICAL TEMPERATURE DISTRIBUTIONS IN A RECTANGULAR FIN

m^*		Dimensionless Temperatures				
		$x^* = 0$	$x^* = 1/3$	$x^* = 2/3$	$x^* = 1$	maximum per cent error*
0.5	Exact	1.0	.937	.899	.887	0.01
	Numerical	1.0	.937	.899	.887	
1.42	Exact	1.0	.677	.509	.457	0.34
	Numerical	1.0	.680	.512	.459	
4.0	Exact	1.0	.265	.074	.037	2.30
	Numerical	1.0	.288	.087	.041	

*Per cent error is based on the root temperature.

case, 73 per cent of the dimensionless temperature drop takes place before the first internal node, and the maximum error in the calculated temperatures is still only 0.023.

Consider next the calculation of the heat exchanged by the fin. A dimensionless value will be employed and is defined as:

$$q^* = \frac{qL}{kA\,(T_o - T_a)} = -\left.\frac{dT^*}{dx^*}\right|_{x^*=0} .$$

In order to approximate this derivative at a left boundary, one might naturally consider employing forward differences. If one uses the first forward difference quotient, one obtains

$$q^* = \frac{1 - T_1^*}{\Delta x^*} . \tag{2.63}$$

A three-point formula, that is, the first two terms of Equation (2.16), gives

$$q^* = \frac{3 - 4T_1^* + T_2^*}{2\Delta x^*} . \tag{2.64}$$

If, on the other hand, one would employ a physical approach using an approximation consistent with that employed to determine the temperature distribution, i.e., the heat flow $q_{0\to 1}$ plus the convective loss of the first half-element based on its interpolated average temperature, one obtains:

$$q^* = \frac{1 - T_1^*}{\Delta x^*} + \frac{m^{*2}\,\Delta x^*}{8}\,(3 + T_1^*) . \tag{2.65}$$

A comparison of the results employing Equations (2.63) through (2.65) is given in Table 2.3. These results show that the physically consistent equation, Equation (2.65), gives values with an x^* of 1/3 which are generally more accurate than those obtained with the inconsistent equation, Equation (2.63), with a Δx^* of 1/30.

Although inconsistencies in other cases might be less important, the advantage of what might seem a minor change to improve the consistency of the physical model is obvious. Equation (2.65) could have been obtained by employing a Taylor series expansion with the Lagrangian form of the remainder to determine an approximation for

$$\left.\frac{dT^*}{dx^*}\right|_{x^*=0}$$

by the following procedure:

$$T_1^* = T_o^* + (\Delta x^*)\left.\frac{dT^*}{dx^*}\right|_o + \frac{(\Delta x^*)^2}{2}\left.\frac{d^2T^*}{dx^{*2}}\right|_\xi ,$$

$$0 < \xi < \Delta x .$$

If the governing differential equation is employed to eliminate the second derivative, and if $\xi = \Delta x^*/2$ is employed, Equation (2.65) will result.

TABLE 2.3

COMPARISON OF THE ACCURACY OF CONSISTENT AND INCONSISTENT APPROXIMATIONS

	Per Cent Error in q^*				
m^*	Equation (2.63) ($\Delta x^* = 1/3$)	Equation (2.64) ($\Delta x^* = 1/3$)	Equation (2.65) ($\Delta x^* = 1/3$)	Equation (2.63) ($\Delta x^* = 1/30$)	q^* Exact
0.5	17.74	0.85	0.01	1.81	.2311
1.42	23.99	5.91	-0.48	2.64	1.2632
4.0	46.54	27.35	-8.29	6.45	3.997

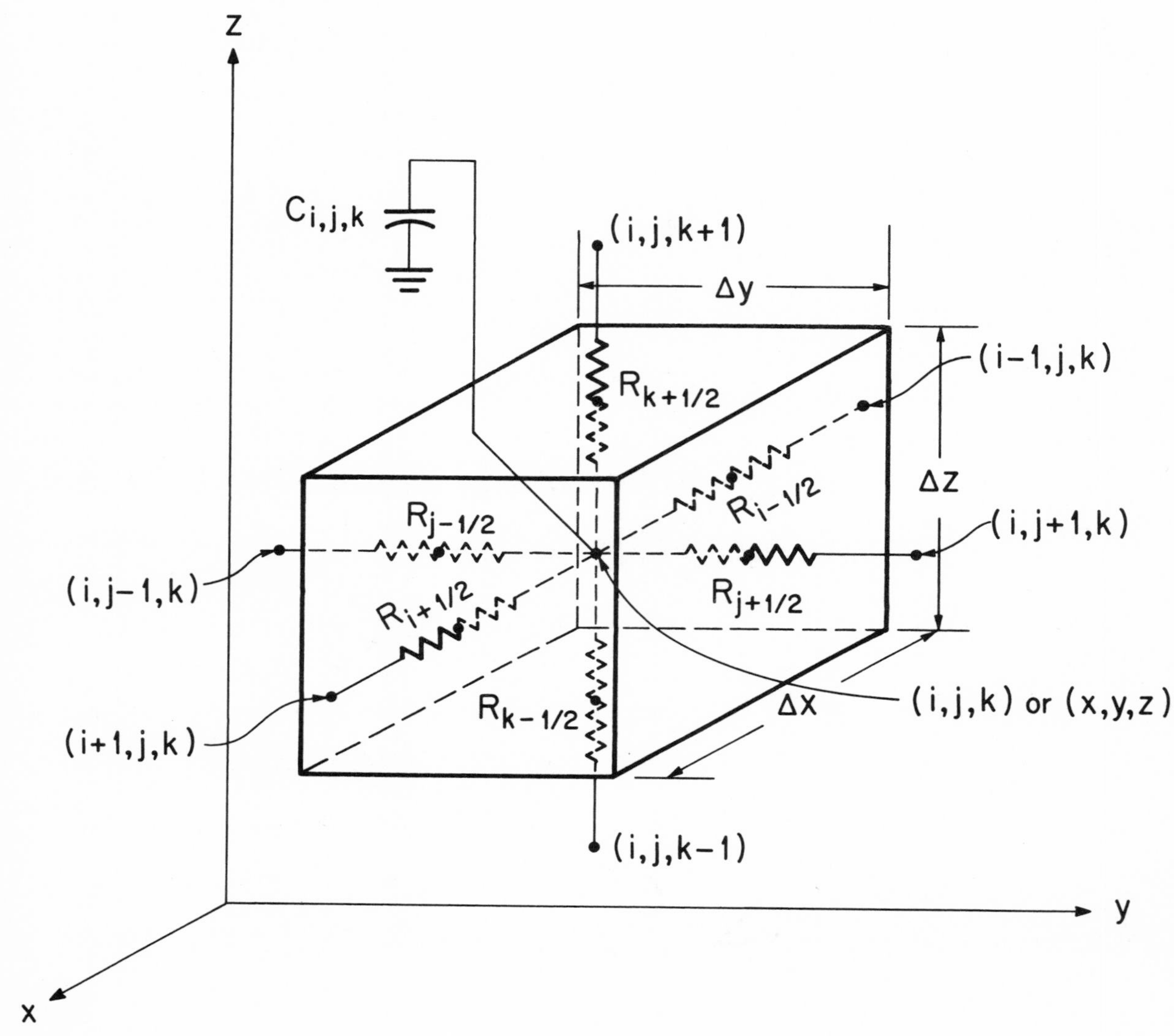

FIGURE 2.1 VOLUME ELEMENT -- CARTESIAN COORDINATES.

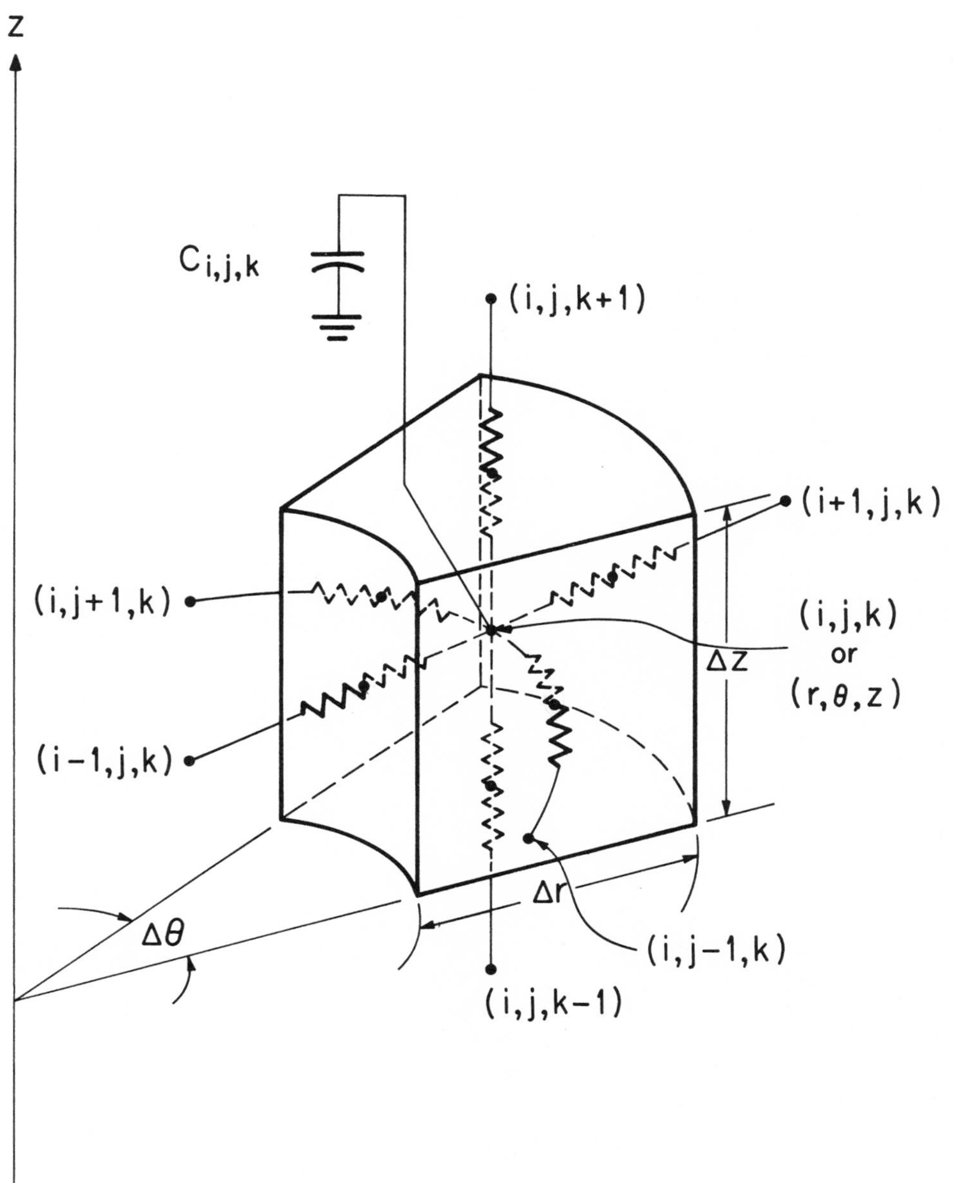

FIGURE 2.2 VOLUME ELEMENT -- CYLINDRICAL COORDINATES.

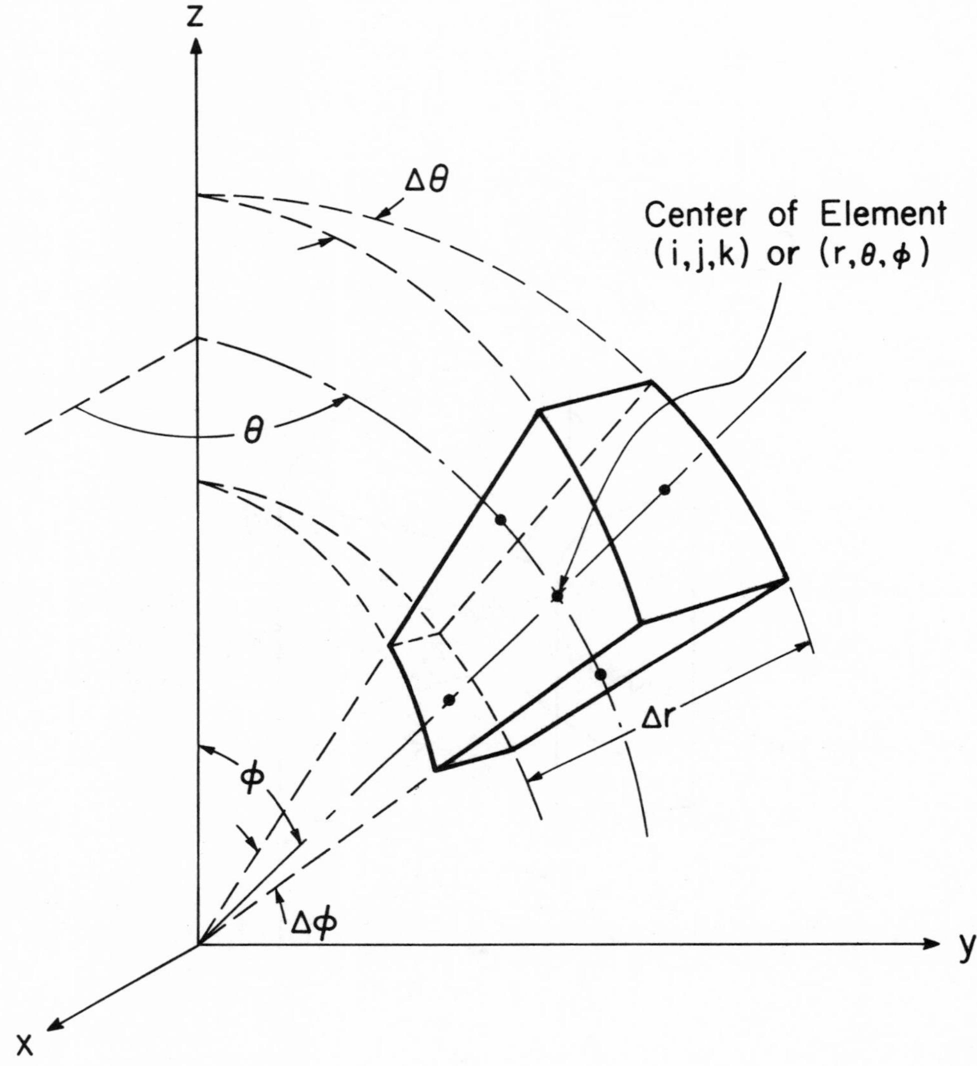

FIGURE 2.3 SPHERICAL COORDINATE SYSTEM.

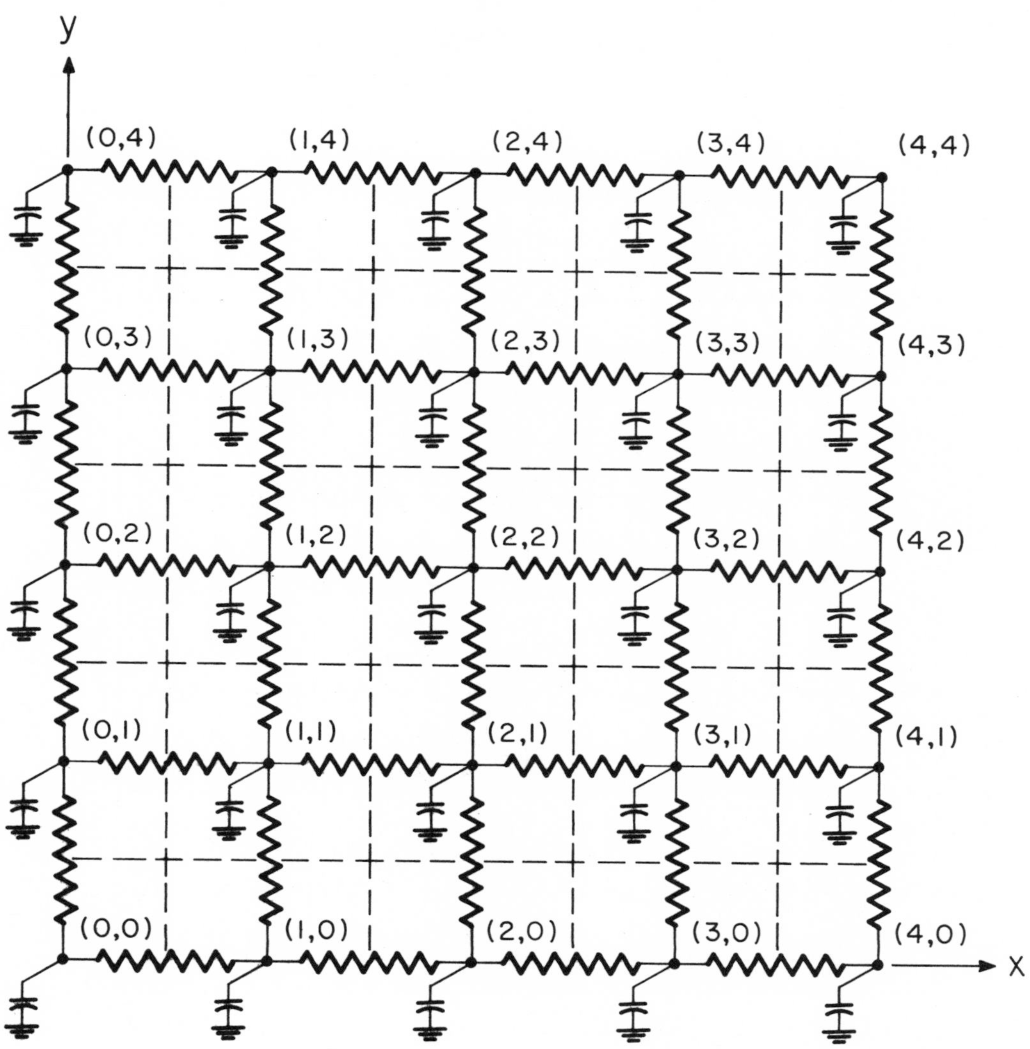

FIGURE 2.4 R-C NETWORK WITH FINITE CAPACITANCE SURFACE NODES.

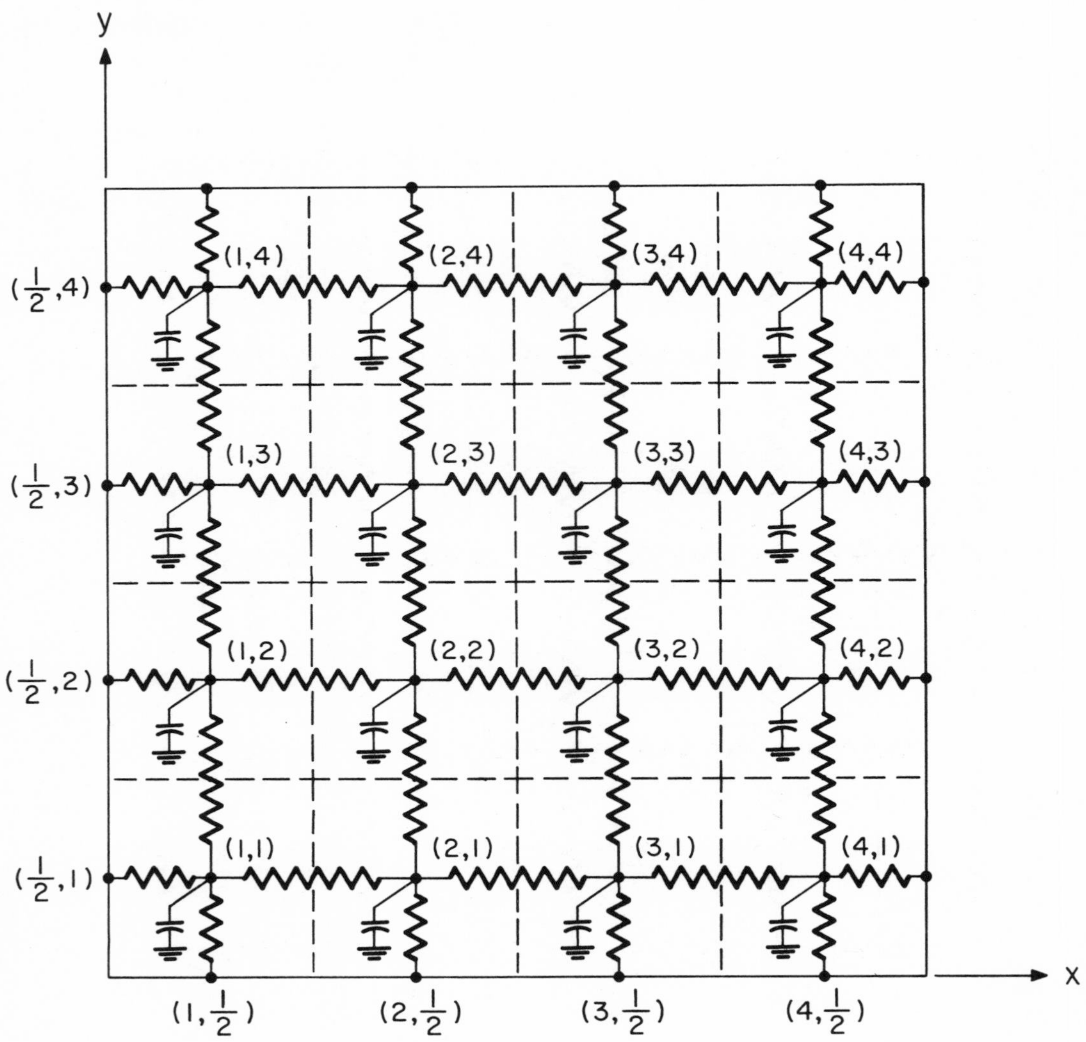

FIGURE 2.5 R-C NETWORK WITH ZERO CAPACITANCE SURFACE NODES.

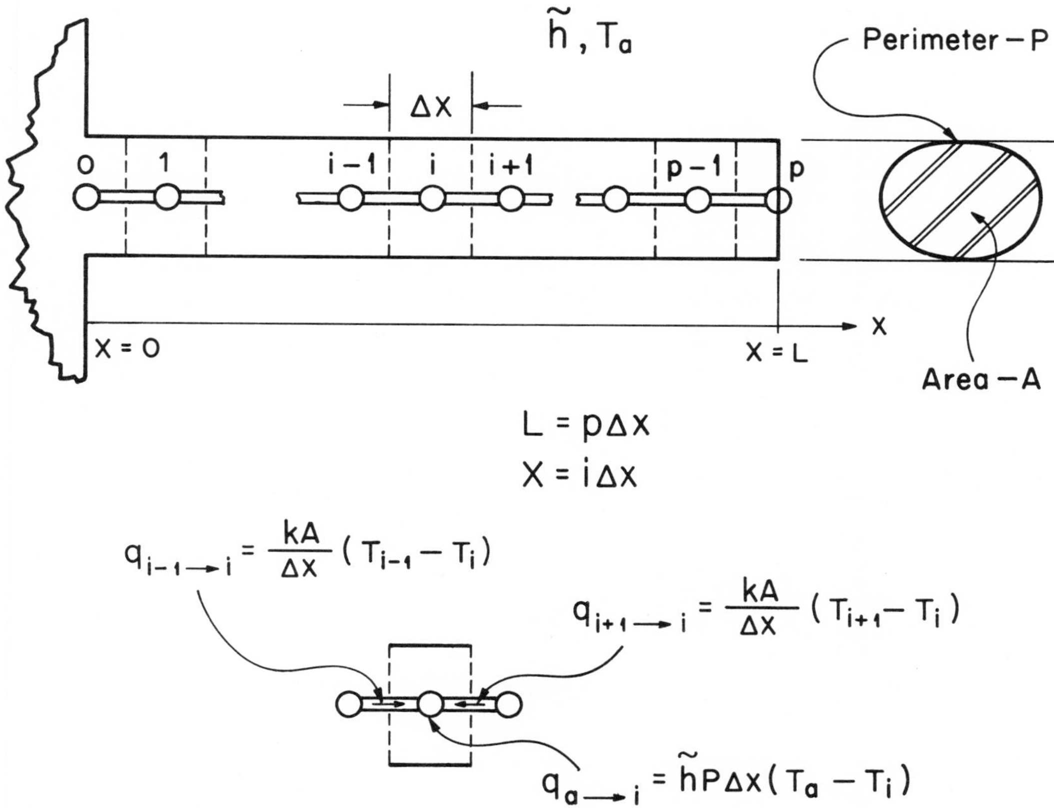

FIGURE 2.6 HEAT TRANSFER FROM AN EXTENDED SURFACE.

3. STEADY STATE -- NUMERICAL PROCEDURES

The general equations derived in Section 2 were, in general, a set of simultaneous ordinary differential equations. In steady state heat transfer problems, the derivatives of the nodal temperatures with respect to time are identically zero, and simultaneous algebraic equations are obtained. Initially only linear boundary conditions and constant thermal conductivity will be considered. Thus the resulting difference equations are linear. The solution of the partial differential equation has been reduced to the solution of a set of simultaneous linear algebraic equations. These equations are of the form:

$$[A]\,[X] = [B] \,, \tag{3.1}$$

where $[X]$ is the unknown column vector or matrix, $[B]$ is a constant column matrix, and $[A]$ is the coefficient matrix. Alternately, Equation (3.1) can be written as:

$$\left.\begin{array}{l} a_{11}\,x_1 + a_{12}\,x_2 + \ldots + a_{1N}\,x_N = b_1 \\ a_{21}\,x_1 + a_{22}\,x_2 + \ldots + a_{2N}\,x_N = b_2 \\ \cdot\quad\cdot\quad\cdot\quad\cdot\quad\cdot\quad\cdot\quad\cdot\quad\cdot\quad\cdot\quad\cdot\quad\cdot \\ a_{N1}\,x_1 + a_{N2}\,x_2 + \ldots + a_{NN}\,x_N = b_N \end{array}\right\} \tag{3.2}$$

The problem of obtaining the solution of a set of simultaneous linear algebraic equations is an old and common one. Many different procedures are available, anyone of which could be quite advantageous for some special system of equations. Before we consider what methods are most suitable for the resulting systems of equations which arise in the numerical solution of partial differential equations, the special characteristics of such sets will be given. They are:

(a) The number of simultaneous algebraic equations which arise is often large. For example, the two-dimensional mixed boundary value problem of Reference 7 required the solution of over 2000 simultaneous equations. If a solution procedure required the storage of the complete coefficient matrix including the zero elements, 4×10^6 storage locations would be required. This would far exceed the high-speed storage space of most computers.

(b) The resulting coefficient matrix is sparse, i.e., a large proportion of its elements is zero. In general, a one-dimensional problem would result in three nonzero elements in each row (a tridiagonal matrix), a two-dimensional problem would contain five, and a three-dimensional case would contain seven.

(c) The coefficient matrix, $[A]$, is strongly diagonal, i.e., the diagonal elements are large relative to the off-diagonal elements. This property permits the employment of iterative methods which are, in general, limited to such systems. It also is a desirable property if an elimination procedure is employed. In many cases, the following inequality is satisfied for all rows:

$$|a_{ii}| \geq |a_{i1}| + \ldots + |a_{i,i-1}| + |a_{i,i+1}| + \ldots + |a_{iN}| \,.$$

It is desirable to choose methods which are suitable for these properties and which will take advantage of their peculiarities.

The many procedures available for solving simultaneous linear algebraic equations can be broken down into two general types: direct methods which require a finite number of steps and indirect methods which require, in theory, an infinite number of steps to effect the solution. Generally, it is dangerous to state that direct methods are better than indirect methods or vice versa. However, it is generally true that: (a) direct methods are more suitable to general matrices, in theory, for any nonsingular set, whereas iterative methods are successful with sparse matrices, (b) direct methods are more susceptible to inaccuracies due to roundoff errors than indirect methods, (c) indirect methods may not converge or may require a large number of operations to converge, (d) indirect methods are generally easier to program and require less computer storage, and (e) indirect methods can often be extended to nonlinear algebraic equations, whereas direct methods are limited to linear problems. Indirect methods are probably more widely used in heat transfer studies; however, several cases arise where direct methods are quite useful. These will be considered next.

3.1 DIRECT METHODS

In the solution of one-dimensional steady state problems and in some of the implicit schemes in transient analyses, the coefficient matrix of the set of simultaneous linear algebraic equations which arises is tridiagonal. A tridiagonal matrix is one which has no nonzero elements other than those on the main diagonal and the two diagonals which are adjacent to the main diagonal. For such systems of equations a very efficient algorithm is available for their solution. This algorithm, which is an elimination procedure, will now be presented.

Consider the set of linear equations:

$$[A]\,[X] = [B]\ ,$$

where

$$[A] = \begin{bmatrix} B_1 & C_1 & 0 & \cdots & & & 0 \\ A_2 & B_2 & C_2 & 0 & \cdots & & 0 \\ 0 & A_3 & B_3 & C_3 & 0 & \cdots & 0 \\ \cdots & & & & & & \\ 0 & \cdots & & 0 & A_{N-1} & B_{N-1} & C_{N-1} \\ 0 & \cdots & & & 0 & A_N & B_N \end{bmatrix} \quad \text{and } [B] = \begin{bmatrix} b_1 \\ b_2 \\ \cdot \\ \cdot \\ \cdot \\ b_{N-1} \\ b_N \end{bmatrix}$$

The procedure is as follows. The first equation is divided by B_1 to reduce the coefficient of x_1 to one. This equation is then employed to eliminate x_1 from the second equation. Next, one divides by the resulting coefficient of x_2. By continuing this procedure, which is known as Gauss elimination, one obtains:

$$\left.\begin{aligned} x_1 + \frac{C_1}{G_1}\,x_2 &= F_1 \\ x_2 + \frac{C_2}{G_2}\,x_3 &= F_2 \\ \cdots\ &\cdots \\ x_{N-1} + \frac{C_{N-1}}{G_{N-1}}\,x_N &= F_{N-1} \\ x_N &= F_N \end{aligned}\right\} \qquad (3.3)$$

where

$$\left.\begin{aligned} & G_1 = B_1 \quad ; \quad F_1 = \frac{b_1}{G_1} \\ & \text{and} \\ & \left.\begin{aligned} D_r &= \frac{C_{r-1}}{G_{r-1}} \\ G_r &= B_r - A_r D_r \\ F_r &= \frac{b_r - A_r F_{r-1}}{G_r} \end{aligned}\right\} \quad 1 < r \le N \end{aligned}\right\} \qquad (3.4)$$

Proceeding in reverse order, one obtains by back-substitution:

$$\begin{aligned} x_N &= F_N \\ x_{r-1} &= F_{r-1} - D_r x_r\ , \quad 1 < r \le N\ . \end{aligned} \qquad (3.5)$$

Thus the D's, G's, and F's are first calculated by employing Equation (3.4); the unknowns are then calculated from Equation

(3.5) by beginning with x_N and proceeding in reverse order.

It is seen that this algorithm requires approximately 8N operations to effect the solution; whereas a general matrix requires $N^3/3$ operations. The algorithm compared with other known direct and indirect methods is extremely efficient; therefore its use is recommended whenever it is applicable.

The coefficient matrix which results in the solution of multidimensional problems will also be banded. If the bandwidth of this matrix can be kept small, Gauss elimination would require considerably fewer than $N^3/3$ operations and could be an efficient process. Care must be exercised in numbering the equations in order to obtain the smallest possible bandwidth. If the bandwidth is not a constant, one should attempt to arrange the equations such that the larger bandwidths occur in the latter stages of the elimination. This would minimize the number of operations required.

As an example, consider the problem of determining the two-dimensional steady state temperature distribution in a rectangular region with the potential specified along the boundary. The governing differential equation is the Laplace equation:

$$\frac{\partial^2 T}{\partial x^2} + \frac{\partial^2 T}{\partial y^2} = 0 .$$

A difference analog was found to be

$$(\delta_x^2 + \delta_y^2)\, T_{i,j} = 0 . \tag{3.6}$$

Equation (3.6) is valid for every internal node. Now consider the renumbering of the internal nodal points in terms of a single subscript. That is, Equation (3.6) must be rewritten in the form of Equation (3.2). Assume the rectangle has m rows and n columns in the interior of the region, and that it is oriented such that $n \leq m$. The total number of unknowns is $N = mn$. The minimum bandwidth will be $(2n+1)$, and the numbering system to obtain this bandwidth is shown in Table 3.1. The resulting coefficient matrix is quindiagonal. It has diagonal elements, elements on either side of the diagonal elements, and elements n columns on either side of the diagonal elements. All the other elements are zero.

As the elimination proceeds, the entire band will eventually contain nonzero elements. The number of operations required for the solution was found to be approximately $(2n^2+5n+1)N$. Thus, for a rectangle, the approximate number of operations will vary from 8N for $n = 1$ to $2N^2+5N^{3/2}$ for $m = n$.

The advantages of elimination for narrow bandwidths are obvious. It must be remembered, however, that elimination procedures will generally be more difficult to program, especially for nonrectangular regions, than some of the iterative techniques presented in Section 3.2. In addition, only linear equations can be solved by this procedure, i.e., without iteration. In elimination procedures a long succession of calculations are made and the final answers are dependent on the accuracy of all the previous calculations; thus round-off errors could become important for large values of N. The importance of round-off error can be easily assessed by the methods of Section 3.3.

3.2 ITERATIVE METHODS

Iterative methods are indirect methods of solution which are popular in many fields and can be employed for the solution of both

TABLE 3.1

NUMBERING OF NODAL POINTS FOR MINIMUM BANDWIDTH

1	2	3	n
(n+1)	(n+2)	(n+3)	2n
. . . . .	. . . . .	. . . .	.
. . . . .	. . . . .	. . . .	.
n(m-1)+1	n(m-1)+2	n(m-1)+3	mn

linear and nonlinear equations. The relaxation method, a noncyclic iterative scheme, will be considered first. The other methods presented are all cyclic iterative techniques.

Since iterative and other indirect methods require, in theory, an infinite number of operations to obtain the exact solution of the set of algebraic equations, a means of determining the error which results from the termination of this process is required. A meaningful estimate of this truncation error is not difficult in our application and is considered in Section 3.3.

3.2.1 Relaxation Method

Relaxation methods were already popular numerical procedures in engineering 25 years ago. These methods are excellent for hand calculation or for use in conjunction with a desk calculator; however, their use requires human insight which is difficult to incorporate efficiently into computer programs. Relaxation methods are discussed in almost every heat conduction textbook, and the reader is referred to these sources for applications of the relaxation technique. Emphasis in this discussion will be placed on methods which are suitable for use on digital computers; however, the relaxation method will be outlined: (a) to show why it cannot be efficiently utilized in computations on digital computers, and (b) to enable later comparisons with several of the cyclic iterative methods which will be found to be quite analogous to the relaxation process.

Consider a set of linear algebraic equations in matrix form:

$$[A]\,[X] = [B]\ .$$

Residuals will be defined by the equation

$$[A]\,[X] - [B] = [R]\ , \tag{3.7}$$

or

$$a_{11}\,x_1 + a_{12}\,x_2 + \ .\ .\ .\ + a_{1N}\,x_N - b_1 = R_1$$

$$.\quad .\quad .\quad .\quad .\quad .\quad .\quad .\quad .$$

$$a_{N1}\,x_1 + a_{N2}\,x_2 + \ .\ .\ .\ + a_{NN}\,x_N - b_N = R_N\ .$$

The relaxation method consists of the following procedure:

(1) Determine an initial approximation for all the unknowns $x_1,\ x_2\ \ldots\ x_N$. This approximation could be determined from the physical problem or from a previous solution with a coarse network.

(2) Calculate the residuals as defined by Equation (3.7).

(3) Find the maximum residual, $(R_i)_{max}$.

(4) Reduce the absolute magnitude of this residual $(R_i)_{max}$ by changing the corresponding value of $(x_i)_{max}$. If the residual is not reduced to zero, the process is known as underrelaxation or overrelaxation depending upon whether its sign remains the same or is changed, respectively.

(5) Determine which residuals have changed due to the change in $(x_i)_{max}$ and recalculate these residuals.

(6) Repeat operations 3, 4, and 5 until the process has converged.

The determination of the maximum residual, operation 3, is a simple task if the calculations are being performed by hand. One need only inspect the tabulated residuals. On the other hand, this determination on a digital computer is alone sufficient to make the process unattractive. A set of N simultaneous equations would require (N-1) subtractions to determine a single maximum residual. If each residual were reduced only once, approximately N^2 operations would be required simply to determine the maximum residuals. Obviously, each residual would have to be reduced more than once.

In hand calculations a relaxation table is employed to aid in the determination of the changes in the successive residuals corresponding to a given change in $(x_i)_{max}$. The relaxation table is the transpose of the coefficient matrix; that is,

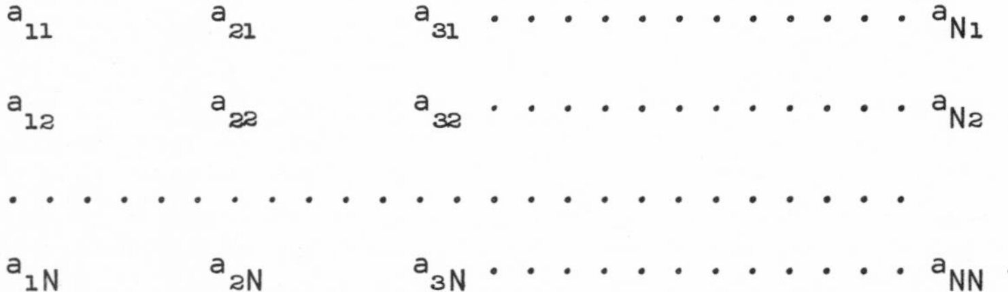

$$
\begin{matrix}
a_{11} & a_{21} & a_{31} & \cdots & a_{N1} \\
a_{12} & a_{22} & a_{32} & \cdots & a_{N2} \\
\cdots & \cdots & \cdots & \cdots & \\
a_{1N} & a_{2N} & a_{3N} & \cdots & a_{NN} .
\end{matrix}
$$

The entries in the r-th row of this table represent the change in the corresponding residuals due to a unit increase in x_r. If a hand calculation procedure is employed, the particular residuals involved and their respective changes for a unit change in $(x_i)_{max}$ are easily determined by inspection. Again this process is difficult to incorporate efficiently into a digital computer program. It would be a much easier programming task if one simply recalculated all the residuals; however, such wasteful computation could not normally be tolerated.

Important variations of the relaxation process described are the so-called "block" and "group" relaxations. In the block relaxation the sum of all the residuals are reduced to zero. In a group relaxation, a group of residuals are reduced to zero in each step. Cyclic iterative methods which are more readily programmed for digital computation will now be considered.

3.2.2 Richardson Iteration

The Richardson method is one of the most elementary iterative procedures. It is also known as the method of simultaneous displacements and bears a close resemblance to first-order methods for solving nonlinear equations. The Richardson method is seldom used in practice since more efficient methods are available. However, a brief discussion of this method is warranted, since it will reveal many of the characteristics of iterative procedures which will aid in the general understanding of the application of such methods to the solution of difference analogs of partial differential equations.

Although the equations to be solved can be written in the form $[A]\,[X] = [B]$, it is more convenient to employ multiple indices and the notation introduced in Section 2.1. Thus $T_{i,j,k}$ represents the unknown nodal temperatures; $T^0_{i,j,k}$ is the initial approximation; and $T^n_{i,j,k}$ is the n-th approximation which is calculated by some process which guarantees the convergence of $T^n_{i,j,k}$ to $T_{i,j,k}$ as n is increased.

Consider the problem of determining the steady state temperature distribution in a two-dimensional region with constant thermal conductivity and no internal heat generation. The temperature is specified everywhere on the boundaries of the region; i.e., the Dirichlet problem will be considered. The potential distribution in this case satisfies the Laplace equation and the difference analog derived in Section 2.2, Equation (2.44) reduces to (assuming $\Delta x = \Delta y = h$):

$$(\delta_x^2 + \delta_y^2)\, T_{i,j} = 0 , \qquad (3.8)$$

$$i = 1, 2, \ldots p-1$$

$$j = 1, 2, \ldots q-1 .$$

The boundary values are known and will be represented by

$$T_{i,j} = b_{i,j} , \qquad (3.9)$$

$$i = 0 \text{ or } p$$

$$j = 0 \text{ or } q .$$

The correction process in the Richardson iteration is:

$$T^{n+1}_{i,j} = T^n_{i,j} + \beta \left(\delta_x^2 + \delta_y^2\right) T^n_{i,j} \qquad \text{(interior points)}$$

$$T^{n+1}_{i,j} = T^n_{i,j} = b_{i,j} \qquad \text{(boundary points)} , \qquad (3.10)$$

where β is a positive constant. The Richardson process is a total step iteration which means that only the unknowns of the previous approximation $T^n_{i,j}$ are employed in the calculation of the new approximation $T^{n+1}_{i,j}$. A process which employs both the old and new values is known as a single step iteration.

Frankel[8] showed that this iterative scheme converges most rapidly for $\beta = 1/4$ which is the most familiar form of the Richardson method and is known as the Jacobi method. Equation (3.10) with $\beta = 1/4$ is the same equation as one would have obtained if Equation (3.8) were solved for $T_{i,j}$. This equation is

$$T^{n+1}_{i,j} = 1/4 \left(T^n_{i+1,j} + T^n_{i-1,j} + T^n_{i,j+1} + T^n_{i,j-1}\right) \quad \text{(interior points)}$$

$$T^{n+1}_{i,j} = b_{i,j} \quad \text{(boundary points)} . \tag{3.11}$$

If one were to determine the transient temperature distribution in this same region with the same boundary conditions, the resulting difference/differential equation previously derived, Equation (2.44), simplifies to

$$\left(\delta^2_x + \delta^2_y\right) T_{i,j} = \frac{h^2}{\alpha} \frac{dT_{i,j}}{dt} . \tag{3.12}$$

If a first forward difference quotient is employed for an approximation of $dT_{i,j}/dt$ and if the right side is evaluated at time n, one obtains:

$$T_{i,j,n+1} = T_{i,j,n} + \frac{1}{M}\left(\delta^2_x + \delta^2_y\right) T_{i,j,n},$$

where $M = h^2/\alpha\Delta t$. If $M = 4$, one obtains:

$$T_{i,j,n+1} = \frac{1}{4}\left(T_{i+1,j,n} + T_{i-1,j,n} + T_{i,j+1,n} + T_{i,j-1,n}\right) \quad \text{(interior points)}$$

$$T_{i,j,n+1} = b_{i,j} \quad \text{(boundary points)} . \tag{3.13}$$

Equation (3.13) is known as the standard explicit approximation and is discussed in detail in Section 4.1, where it is shown that this approximation is stable. The time truncation error follows from Equation (2.23) as:

$$E_T = -\frac{\Delta t}{2}\left[\left.\frac{\partial^2 T}{\partial t^2}\right|_{i,j,n} + \ldots\right] .$$

It is seen that this error will vanish as steady state is attained.

Equation (3.11), the Jacobi iteration, is identical to Equation (3.13), the equation for the stepwise, line by line, calculation of the transient temperature variation starting from the initial condition at $n = 0$. Thus successive iterations can be regarded as time steps. The error at any point in the n-th approximation can be regarded as a transient which has not damped out. The initial condition, $T_{i,j,0}$, is seen to be analogous to the initial approximation $T^0_{i,j}$.

Most methods of determining rates of convergence are complex, abstract mathematical procedures (see Reference 8). Instead of performing such an analysis, the analogy between the Jacobi method and the standard explicit scheme, which was just demonstrated, will be exploited to show the influence of various parameters on the rate of convergence. This analogy can be applied to any geometry and a variety of boundary conditions. Internal generation could also be included. Thus, in some ways, this analogy will be more fruitful. It will require only a knowledge of the transient behavior of these problems.

First, consider the effect of the initial approximation, $T^0_{i,j}$, which is analogous to the initial condition. An initial condition is desired that will result in a rapid decay of the transients. Of course, this is easier said than done. An obvious thing to avoid is either overestimating or underestimating all the unknowns. This would result in all the residuals being of the same sign, and a large number of iterations would be required since a large

TABLE 3.2

THE INFLUENCE OF BOUNDARY CONDITIONS, SPATIAL INCREMENT, AND REGION SIZE ON THE CONVERGENCE RATE

Problem Type	I	II	II	III
Region Size	1 x 1	1/2 x 1/2	1/2 x 1/2	1 x 1
Spatial Increment	h	h	h/2	h
Number of Iterations Required	n	n	4n	4n
Number of Unknowns	N	N/4	N	N

amount of time would be necessary to add or remove this energy from the entire region. On the other hand, if some potentials were too large and others too small, the energy would only have to be redistributed, which would take place in a relatively short time. A solution from a coarse network would provide an excellent method of generating a good initial approximation.

Next consider the effect of the type of boundary conditions on the rate of convergence. Assume the two-dimensional Dirichlet problem being considered has boundary conditions which are symmetrical with respect to the centroidal axes and an initial approximation with similar symmetry is being employed. The solution of this problem, which will be referred to as problem I, is assumed to require n iterations to solve the N equations which result for a spatial increment, h.

Because of symmetry, only a quarter of the region of problem I need be considered. Problem II will be this quarter region which has two adiabatic boundaries; that is, the region, $0 \leq x \leq 1/2$ and $0 \leq y \leq 1/2$. The solution of problem II would also require n iterations; however, N/4 equations would be involved for the same spatial increment, h. If this spatial increment were halved, 4n iterations would be required to solve approximately N equations.

Problem III will be a region 1 by 1 with the same boundary conditions at homologous points as problem II. For a spatial increment of h, the same set of equations will result as those for problem II with a spatial increment of h/2. Thus the same number of iterations, 4n, would be required to solve this set of N equations. These results are summarized in Table 3.2; the respective regions are shown in Figure 3.1.

The conclusions drawn from these examples are:

(1) The number of iterations required for the square region with four prescribed temperature boundaries was only one-fourth the number required for the similar problem with two of the boundaries replaced by adiabatic surfaces. Furthermore, if the boundary at y = 0 in problem III were replaced by an adiabatic surface over the region $0 \leq x \leq a < 1$, and likewise the boundary at x = 0 were adiabatic over the region $0 \leq y \leq b < 1$, the number of iterations required would approach infinity as a and b both approached one (see problem IV of Figure 3.1). This leads to the conclusion that the rate of convergence depends strongly on the degree of ill-conditionedness of the equations.*

(2) If the same initial approximation and boundary conditions are employed, the number of iterations required for convergence increases as h^{-2}. Thus the number of iterations is proportional to N^2 for a one-dimensional problem, to N for a two-dimensional problem, and to $N^{2/3}$ for a three-dimensional problem.

(3) The number of iterations required for convergence for a fixed number of equations is the same for all geometrically similar regions with similar boundary conditions and initial approximations.

The comparison of the Jacobi iteration with the standard explicit approximation

---

*A set of equations is ill-conditioned if its coefficient matrix is almost singular. Physically, an ill-conditioned set will result if the number of boundary nodes which determine the temperature level is small relative to the total number of unknowns. A finer nodal spacing will always increase the degree of ill-conditionedness.

for the diffusion equation has indicated the quantities which influence the rate of convergence. However, the prediction of the exact number of iterations required to obtain a specified accuracy is not possible unless the transient solution is available, in which case there would be no need for a numerical solution. The absence of this information is no serious handicap, since meaningful tests for convergence are available (see Section 3.3). Some additional properties of the Jacobi iteration are given in the comparison with the extrapolated Liebmann method in the next section.

3.2.3 Extrapolated Liebmann Method

The Liebmann method is a correction process similar to the Richardson method with the exception that the corrected values are employed in subsequent operations as soon as they are available. Hence the Liebmann method is a single step iteration. It is also known as the method of successive displacements. The Gauss-Seidel iteration will be seen to be a special case of the extrapolated Liebmann method.

If the Liebmann method is applied to the two-dimensional Dirichlet problem considered in Section 3.2.2, the process is:

$$T^{n+1}_{i,j} = T^{n}_{i,j} + \beta \left[T^{n+1}_{i-1,j} + T^{n+1}_{i,j-1} + T^{n}_{i+1,j} + T^{n}_{i,j+1} - 4T^{n}_{i,j} \right] \quad \text{(interior points)}, \tag{3.14a}$$

$$T^{n+1}_{i,j} = b_{i,j} \quad \text{(boundary points)}, \tag{3.15}$$

where it is assumed that the calculation is proceeding in the direction of increasing i and j, i.e., the rows (or columns) are scanned in the same direction and taken successively. Equation (3.14a) can be alternately written as:

$$T^{n+1}_{i,j} = (1-\omega)\, T^{n}_{i,j} + \frac{\omega}{4} \left(T^{n+1}_{i-1,j} + T^{n+1}_{i,j-1} + T^{n}_{i+1,j} + T^{n}_{i,j+1} \right), \tag{3.14b}$$

where ω (= 4β) is called the <u>relaxation parameter</u>. If $\omega = 1$, the extrapolated Liebmann process reduces to the Gauss-Seidel iteration:

$$T^{n+1}_{i,j} = \frac{1}{4} \left[T^{n+1}_{i-1,j} + T^{n+1}_{i,j-1} + T^{n}_{i+1,j} + T^{n}_{i,j+1} \right] . \tag{3.16}$$

A comparison of the extrapolated Liebmann method with the relaxation process, Section 3.2.1, shows that it is a mechanical, cyclic application of the relaxation process. Successive reduction of the residuals to zero corresponds to the Gauss-Seidel method, $\omega = 1$. Values of ω greater than one correspond to overrelaxation, and values less than one correspond to underrelaxation.

The problem of determining the value of ω which results in the maximum rate of convergence (ω optimum) is one of great importance and interest. This problem, along with an examination of the convergence rates for this method, the Richardson method, and the Gauss-Seidel iteration was investigated by Frankel[8] (see also Garabedian[9]) for the two-dimensional Dirichlet problem being considered in this section. He determined the rate of decay of the error-eigenfunction components which are multiplied by the eigenfunction of the iteration operation at each iteration. The ultimate convergence rate is determined by the maximum magnitude of the eigenfunction which he called K^*. He found: (a) the greatest rate of decay of the error occurred for the value of ω which is the smaller root of:

$$U^2_{max}\, \omega^2 - \omega + 1 = 0 , \tag{3.17}$$

where

$$U^2_{max} = \frac{1}{16} \left(\cos \frac{\pi}{p} + \cos \frac{\pi}{q} \right)^2 .$$

If π/p and π/q are much less than one, ω_{opt} becomes:

$$\omega_{opt} \simeq 2 - \sqrt{2} \quad \pi \left(\frac{1}{p^2} + \frac{1}{q^2} \right)^{1/2} , \quad (3.18)$$

and (b) the maximum magnitude of the eigenfunction of the iteration operation for large values of p and q were found to be:

(1) Jacobi Method

$$K^* \simeq 1 - \frac{\pi^2}{4} \left(\frac{1}{p^2} + \frac{1}{q^2} \right) , \quad (3.19a)$$

(2) Gauss-Seidel Method, $\omega = 1$

$$K^* \simeq 1 - \frac{\pi^2}{2} \left(\frac{1}{p^2} + \frac{1}{q^2} \right) , \quad (3.19b)$$

(3) Extrapolated Liebmann Method, ω_{opt}

$$K^* \simeq 1 - \sqrt{2} \quad \pi \left(\frac{1}{p^2} + \frac{1}{q^2} \right)^{1/2} . \quad (3.19c)$$

These relationships are all of the form $K^* = 1 - \epsilon$, where ϵ is much less than one. If the iteration is continued until all errors are reduced by a factor of e^{-m}, the number of iterations, n, required is

$$(K^*)^n = e^{-m}$$

or

$$n \simeq \frac{m}{\epsilon} . \quad (3.20)$$

Thus the number of iterations required is

$$n \text{ (Jacobi)} \simeq \frac{4m}{\pi^2 \ (1/p^2 + 1/q^2)} , \quad (3.21a)$$

$$n \text{ (Gauss-Seidel)} \simeq \frac{2m}{\pi^2 \ (1/p^2 + 1/q^2)} , \quad (3.21b)$$

$$n \text{ (Extrapolated Liebmann)} \simeq \frac{m}{\sqrt{2} \ \pi \left(1/p^2 + 1/q^2\right)^{1/2}} . \quad (3.21c)$$

It is concluded that the error is reduced by as much in one cycle of the Gauss-Seidel iteration as in two cycles of the Jacobi method. The number of iterations required in both of these procedures increases quadratically with p and q; however, the number required for the extrapolated Liebmann method increases linearly with p and q. These conclusions are based on p and q much larger than one.

Since the error in the initial approximation is not known, Equations (3.21) are of limited value and would seldom be employed, especially since meaningful tests for convergence are available. (However, they do show the relative rates of convergence of the various iterative schemes.) On the other hand, the ability to calculate the optimum value of ω is highly desirable. Equations (3.21) clearly show the tremendous savings in computational effort possible, especially for large values of p and q. Unfortunately, the optimum value of ω given by Equation (3.17) or (3.18) is of limited value, since it is only applicable to the simple problem being considered. For example, if the x = 0 and y = 0 boundaries were perfectly insulated, ω_{opt} for large p and q becomes:

$$\omega_{opt} \simeq 2 - \frac{\sqrt{2}\,\pi}{2} \left(\frac{1}{p^2} + \frac{1}{q^2} \right)^{1/2} . \quad (3.22)$$

The analytical determination of ω_{opt} for complex problems is difficult.

The influence of ω on the number of iterations can also be investigated by trial and error on the computer. Such a procedure, of course, would not be profitable for the solution of a single problem. The optimum value for the problem described by the following differential equation and boundary conditions was investigated in this manner.

$$\frac{\partial^2 T}{\partial x^2} + \frac{\partial^2 T}{\partial y^2} = 0$$

$$\text{(a)} \begin{cases} T(x,0) = T_o, & 0 \le x \le 1/2 \\ \dfrac{\partial T}{\partial y}(x,0) = 0, & 1/2 < x \le L \end{cases}$$

$$\text{(b)} \quad T(x,L) = T_L, \quad 0 \le x \le L$$

$$\text{(c)} \quad \frac{\partial T}{\partial x}(o,y) = 0, \quad 0 < y < L$$

$$\text{(d)} \quad \frac{\partial T}{\partial x}(L,y) = 0, \quad 0 < y < L$$

An overall energy balance was performed every six iterations and the results were printed when the rate of heat flow through the upper boundary was within 2 per cent and 1/3 per cent of that through the lower boundary. Equal spatial increments were employed, $\Delta x = \Delta y = h$ and $p = q = 19$. The total number of unknowns was 370. The results are given in Table 3.3. The results showed that (a) $\omega_{opt} \simeq 1.9$, (b) the number of iterations required for the Gauss-Seidel method for this example is about 14 times the number required for the extrapolated Liebmann method with ω_{opt}, and (c) the computational time required to obtain an energy balance of 1/3 per cent on the IBM 7094 digital computer with an optimum value of ω, i.e., the time needed to solve the 370 simultaneous equations was about six seconds.

It is noted that neither Equation (3.18) nor Equation (3.22) are applicable to this problem. Consider the empirical modification of Equation (3.18) to account for adiabatic boundaries. If $\Delta x = \Delta y = h$, Equation (3.18) becomes for a square region:

$$\omega_{opt} \simeq 2\,(1 - \pi h) .$$

If one defines an effective increment, h_e, as

$$h_e = (h)\,(\text{Fraction of the total boundary which is not adiabatic}),$$

with

$$\omega_{opt} \simeq 2\,(1 - \pi h_e) , \tag{3.23}$$

Equation (3.23) gives the correct results for the cases represented by Equations (3.18) and (3.22). It gives a value of ω optimum of 1.88 for the computer optimized problem, which is in agreement with the results of Table 3.3. A more detailed study of the problem of determining the optimum value of ω remains to be performed. The reader is also referred to Reference 11.

The extrapolated Liebmann method is not only an efficient method, but it is also an easy method to program, especially compared to elimination procedures, or other iterative methods. The Liebmann method requires approximately N storage locations, excluding those required for the storage of the relatively simple program. The Richardson method would require 2N, and elimination procedures generally require

TABLE 3.3

THE INFLUENCE OF THE RELAXATION PARAMETER ON THE RATE OF CONVERGENCE

Energy Balance to:	ω 1.0	1.3	1.5	1.8	1.9	1.95
2 per cent	600	324	216	60	42	84
1/3 per cent	858	462	288	84	66	132

considerably more than 2N locations. For a two-dimensional square region with equal spatial increments, the Gauss-Seidel method requires 4N operations per cycle, whereas the extrapolated Liebmann method requires 6N. However, the total number of operations is of the order of N^2 for Gauss-Seidel or elimination, and of the order $N^{3/2}$ for the extrapolated Liebmann method with optimum ω.

It will be seen in Section 4.4 that the Gauss-Seidel and the extrapolated Liebmann methods are also analogous to difference analogs of the diffusion equation. As in the analysis of the Richardson iteration, this fact can be employed with great reward in the study of the convergence rates of these iterations.

The Peaceman-Rachford alternating direction method will be considered next. The number of iterations required for this method for the two-dimensional Dirichlet problem is of the order of N ln N; thus it is more efficient than the other procedures considered here for sufficiently large N. However, it is a difficult method to program and it is limited to a rather specialized class of problems. Its employment, if applicable, would appear practical only for extremely large values of N.

3.2.4 Peaceman-Rachford Method

Peaceman and Rachford[10] suggested a method for the solution of transient heat conduction problems. This method is implicit in the one direction and explicit in the other direction for all even time steps. The implicit and explicit directions are reversed for odd time steps (odd n). Of course, this method like any method for the solution of the diffusion equation, can also be employed to obtain steady state solutions.

The similarity of the Peaceman-Rachford method to the relaxation process, or the extrapolated Liebmann method, can be seen by writing the two difference equations which are alternately employed in their scheme in the following form:

$$T_{i,j}^{n+1/2} = (1-\omega)\, T_{i,j}^{n} + \frac{\omega}{4}\left(T_{i-1,j}^{n+1/2} + T_{i+1,j}^{n+1/2} + T_{i,j-1}^{n} + T_{i,j+1}^{n}\right), \tag{3.24a}$$

and

$$T_{i,j}^{n+1} = (1-\omega)\, T_{i,j}^{n+1/2} + \frac{\omega}{4}\left(T_{i-1,j}^{n+1/2} + T_{i+1,j}^{n+1/2} + T_{i,j-1}^{n+1} + T_{i,j+1}^{n+1}\right). \tag{3.24b}$$

If the above equations are compared with the form they presented, one sees that $\omega = 4/(M+2)$, where

$$M = \frac{\Delta x^2}{\alpha\Delta t} = \frac{\Delta y^2}{\alpha\Delta t} = \frac{h^2}{\alpha\Delta t}.$$

There are two main differences between the above procedure and the extrapolated Liebmann method. First, the procedure alternates the direction of calculation every half-step; second, the previous half-cycle values are employed in one direction, i.e., at $T_{i,j+1}$ and $T_{i,j-1}$, instead of at $T_{i+1,j}$ and $T_{i,j+1}$. This necessitates considering a complete row or column as a group. For example, if $\omega = 1$, it would correspond to reducing the residuals to zero for all nodal points in a given row or column simultaneously, which necessitates solving simultaneous equations. The simultaneous algebraic equations for linear problems are easily solved employing the algorithm of Section 3.1, since the coefficient matrix is tridiagonal. Thus one cycles through the region considering, during the first half-cycle, a row at a time and, during the next half-cycle, a column at a time. The same value of ω must be employed for both half-cycles for stability.

The major problem in the Peachman-Rachford method is to determine a set of time steps which will result in a maximum rate of convergence. The authors outlined a procedure for the determination of such a set for a rather simple problem, and the reader is referred to Reference 10 for the details of this analysis. An extension of their procedure to more complex boundary conditions and regions is by no means obvious.

3.3 GLOBAL ACCURACY AND CONVERGENCE TESTS FOR ITERATIVE METHODS

Consider the Laplace equation in two dimensions:

$$\frac{\partial^2 T}{\partial x^2} + \frac{\partial^2 T}{\partial y^2} = 0 \,. \qquad (3.25)$$

A difference analog for this equation was found to be:

$$\frac{1}{h^2}\left(\delta_x^2 + \delta_y^2\right) T_{i,j} = 0 \,, \qquad (3.26)$$

where $\Delta x = \Delta y = h$. The truncation error is given by Equation (2.34) and is:

$$E_T = -\frac{h^2}{12}\left\{D_x^4 + D_y^4 + \ldots.\right\} T_{i,j}. \qquad (3.27)$$

This expression, however, is not very useful since the derivatives are unknown quantities. Even if these derivatives were estimated by difference quotients, Equation (3.27) still represents the error committed in approximating Equation (3.25) by Equation (3.26), the local error, and not the desired error in the nodal temperatures, the global error.

A technique used in estimating the global error is to assume this error is proportional to the square of the spatial increment, which is suggested by the local error. For example, assume that a heat flow rate, q, were desired. If q represents the exact value, and q_1 represents the value obtained from the numerical calculation with a spatial increment of h_1, one obtains:

$$q - q_1 = C\,h_1^2 \,,$$

where C is an unknown constant. If the problem is resolved with a spatial increment of h_2, one obtains:

$$q = q_1 + \frac{q_2 - q_1}{1 - \left(h_2/h_1\right)^2} \,.$$

This equation represents a parabolic extrapolation to a zero grid size. Although such an extrapolation is somewhat dangerous, it often gives a significant improvement in the accuracy, and at least an indication of the magnitude of the error.

A second computation, with a spatial increment of 1/2 or 1/3 of the initial value in order to effect a meaningful extrapolation, might appear to be a tremendous waste of effort. However, if an iterative method is employed, the first solution can be used as an initial approximation for the second solution. One would probably employ

$$h_2 = \frac{h_1}{2} \quad \text{or} \quad \frac{h_1}{3} \,,$$

in which case the potential at additional nodal points in the finer network could be easily determined by interpolation. The solution of the finer network, which would normally dominate the computational effort required, would converge relatively quickly, since the initial approximation would contain only a small truncation error. Thus this procedure will not only supply a rational error estimate, but it will also reduce the computational effort required over that which would be necessary to solve only the finer network.

Another obvious method of determining the global accuracy of a numerical solution is to compare it with an exact solution. This procedure is often used, but it cannot be used in practical cases. If the exact solution were available, there would be no need for the numerical procedure. In general, nothing seems to be available for error estimation which comes close to satisfying the needs of the practical computer user.(11)

Next consider the determination of the importance of round off error if a direct method is employed, or the importance of the error caused by truncation of the infinite process in indirect methods. The termination of iterative processes is often based on one of the following: (a) the magnitude of the residuals or of a particular residual such as R_{max}, (b) the magnitude of the difference (or the sum of the absolute value of all the differences) between the values of the dependent variable obtained from successive cycles or between a fixed number of cycles, m, or (c) a visual comparison of the entire potential distribution after every m cycles to determine whether any "appreciable change" has occurred. In essence, these tests are only minor variations of each other, and they all suffer from the same basic weakness. Their meaning depends on a rate of convergence; i.e., it is difficult to define a meaningful ϵ such that:

$$|R_{max}| < \epsilon, \text{ or } \sum_{i,j,k} |R_{i,j,k}| < \epsilon \,,$$

or

$$\sum_{i,j,k} |T_{i,j,k}^{n+m} - T_{i,j,k}^{n}| < \epsilon$$

will give the accuracy desired with minimum effort. The rate of convergence was seen to depend upon: (a) the iterative method being employed, (b) the spatial increment, (c) the boundary conditions, (d) the initial approximation, and (e) the particular differential equation under consideration. If a meaningful criterion were developed, a change in any of these quantities would necessitate a change in the criterion.

The tests described above are employed out of necessity by mathematicians solving general systems without knowledge of the physical problem, if any, represented by these equations. In the present case, a meaningful test can be derived from the physical problem, mainly, an energy balance for the overall system. An overall energy balance is a more difficult test to apply; however, its significance is easily worth the added effort required. In addition, the quantities required in this energy balance are often part of the solution being sought.

An energy balance will not only provide a criterion for the termination of an iterative method, it can also be used:

(a) to deduce the magnitude of round off errors if direct procedures are employed,

(b) to provide a check on the computer program and the derived set of difference equations, and

(c) to indicate the presence of physical inconsistencies in the approximations being employed.

The method is applicable to both steady state problems and to transient problems where it might be employed if an iterative method were used in conjunction with an implicit scheme.

An energy balance on a fixed region or control volume gives:

$$\iint_{c.s} k \frac{\partial T}{\partial n} dA + \iiint_{c.v} q''' \, dV = \frac{\partial}{\partial t} \iiint_{c.v} \rho c T \, dV , \qquad (3.28)$$

or

$$\left\{ \begin{array}{l} \text{net rate of heat} \\ \text{conducted into} \\ \text{the control volume} \end{array} \right\} + \left\{ \begin{array}{l} \text{rate of heat} \\ \text{generated within} \\ \text{the control volume} \end{array} \right\} = \left\{ \begin{array}{l} \text{rate of change of the} \\ \text{internal energy within} \\ \text{the control volume} \end{array} \right\} .$$

In a discretized system, the integrals in Equation (3.28) become summations and the partial derivatives become difference quotients. The use of Equation (3.28) is straightforward if a physical approach is employed; however, one must be careful, otherwise inconsistencies will result and the equation will not be satisfied. A straightforward substitution of difference quotients followed by appropriate summations would probably give an incorrect result. In difference analogs of the diffusion equation, Equation (3.28) would be applied each time step. The particular form these integrals would take depends upon the difference scheme utilized.

An energy balance was employed to terminate the iteration process in the solution of a steady state problem in Reference 7. In this case, heat was flowing into the region across one isothermal boundary, q_{in}, and out across another boundary, q_{out}. The iteration process was continued until:

$$100 \left| \frac{q_{in} - q_{out}}{(q_{in} + q_{out})/2} \right| < \epsilon,$$

where ϵ is the per cent deviation. The results were printed for ϵ, $\epsilon/3$, and $\epsilon/9$. The test was applied approximately every N/4 cycles to reduce the computational effort required to apply the convergence test.*

Since all the iterative methods considered in this section are possible difference analogs of the diffusion equation, $(q_{in} - q_{out})$ is an indication of the rate of change of the internal energy of the system. Convergence of the iterative technique is analogous to reaching steady state. The rare possibility that the right side of Equation (3.28) and hence ϵ might be

*The optimum frequency of application of the test would be quite difficult to determine; obviously, it would be illogical to apply it every cycle.

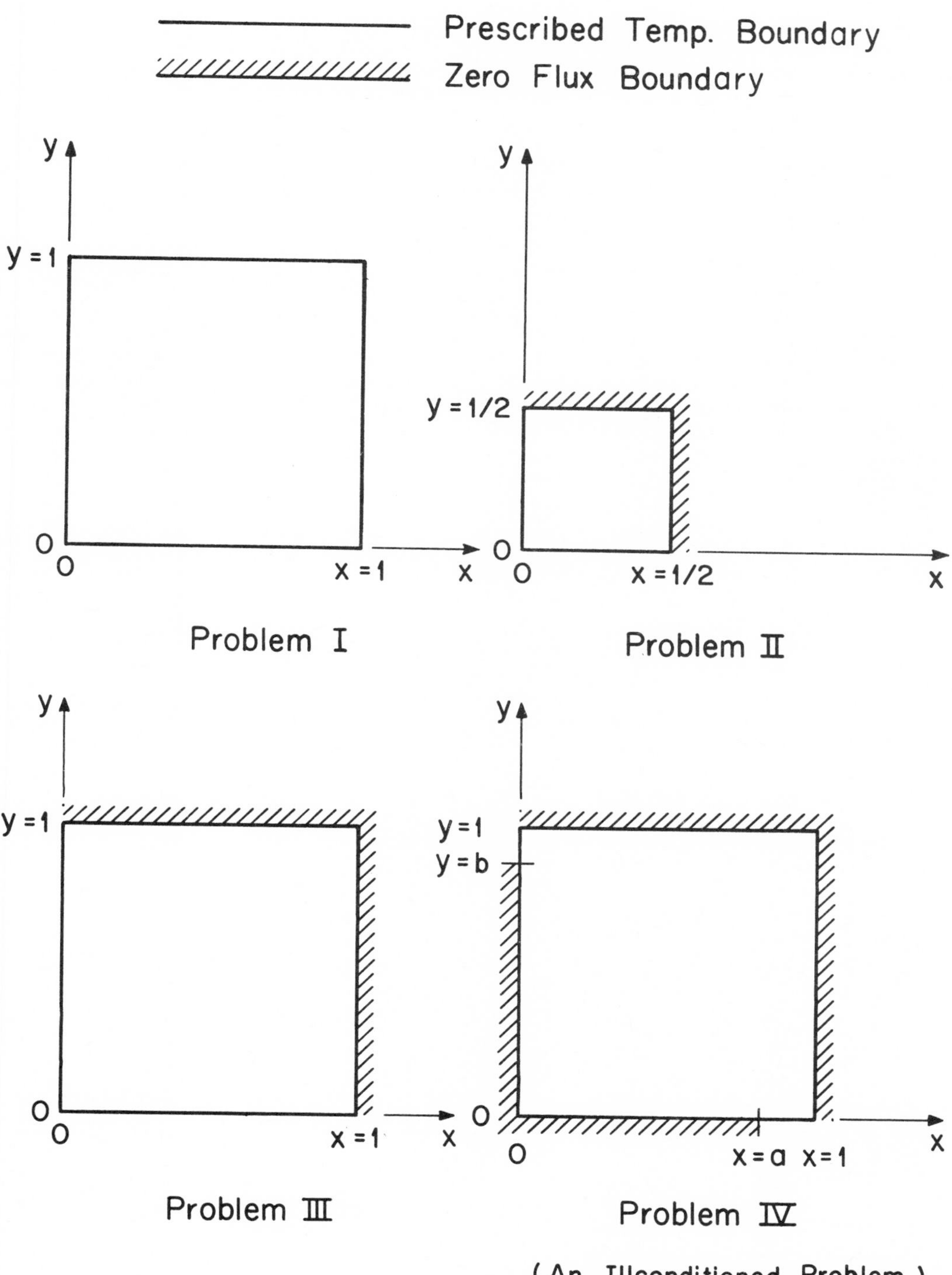

FIGURE 3.1 REGIONS EMPLOYED TO SHOW THE INFLUENCE OF BOUNDARY CONDITIONS ON CONVERGENCE RATES.

momentarily zero before the solution has converged will not cause any difficulty if several different values of ϵ are employed as was done in Reference 7. The test is seen to have considerable physical meaning, which enables one to save computation cost by requiring only that degree of accuracy which is consistent with the physical parameters.

4. TRANSIENTS -- NUMERICAL PROCEDURES

The governing partial differential equations in the analysis of steady state heat conduction problems are <u>elliptic</u> equations. The conditions along the boundary of the regions were specified everywhere and independent of time. In contrast, the governing partial differential equations in transient heat conduction problems, or similar diffusion processes, are <u>parabolic</u> equations. The conditions along the boundaries are again specified everywhere but may be functions of time. The initial potential distribution throughout the region, the initial condition, must also be specified. The problem is to determine the future potential distributions for the given set of initial and boundary conditions.

In the numerical solution of elliptic equations, the solution must be determined everywhere simultaneously. The solution to parabolic equations, in contrast, is constructed step-by-step, marching forward with respect to time. The step-by-step construction of the solution for parabolic equations introduces a difficulty not experienced in elliptic problems -- instabilities. If the round off error, or the so-called numerical error, grows as the solution progresses, the solution is unstable; conversely, the procedure is stable if these errors decay.

If the physical process of diffusion is considered, instabilities might seem unnatural. If one sequentially extrapolates ahead in time, it is not difficult to see why the time truncation errors might accumulate. This would cause a smooth, gradual divergence from the exact solution such as is also experienced in the solution of ordinary differential equations by Euler's method. Many transient problems eventually attain a steady state condition. Since the time truncation errors vanish as steady state is reached, the solution would in this case again converge to the lumped steady state solution. Such error accumulation is caused by <u>truncation errors</u> and is to be expected. On the other hand, instability was exhibited by a growth of numerical or round off errors; however, one would think that the diffusion process should tend to smooth such errors if they were introduced. The energy level of the system would not be expected to change due to the randomness of numerical errors. Why then do some difference schemes become unstable?

It is the aim of this treatment to show that the presence of instabilities appears to stem, in the cases which were examined, from violations of the first or second law of thermodynamics. If these laws are violated when a step forward in time is made, the answers obtained will lose their physical significance and the process may become unstable. On the other hand, if these laws are not violated, the scheme will be stable, and any round off or other error which is introduced will be damped out with time as is natural in a diffusion process. If an unstable scheme is employed, the solution will generally oscillate, and the amplitude of the oscillations will grow continuously as the calculation proceeds. These large meaningless fluctuations generally do not stem from the growth of round off errors, but from the absurd results obtained from calculations in violation of the first or second law. This fact has not been fully appreciated.

Stability analyses of difference analogs of the diffusion equation in the literature are usually performed by a method attributed to von Neumann which is described, for example, by O'Brien, <u>et al</u>.,(12) and Richtmyer.(13) This method was usually employed in the stability analyses of the new difference analogs of the diffusion equation which were introduced in recent years (see, e.g., Peaceman and Rachford(10) and Larkin(14)). The method assumes that the error, $\epsilon_{i,j,k,n}$, at a particular time, e.g., $n = 0$, can be expanded in a harmonic series. This series is then analyzed to determine if these initial errors grow or decay with time. The reader is referred to the above references for the details of this valuable technique for the analysis of stability.

The analysis of stability from physical considerations was employed by Dusinberre(5) in conjunction with the standard explicit approximation. He realized that the presence of negative coefficients in the scheme "would be absurd, physically," and laid down the simple rule: "avoid negative coefficients." This rule is often employed in textbooks to determine if the standard explicit method is stable. A detailed discussion of stability from considerations of the first and second laws of thermodynamics does not appear to have been given, nor applied to other difference analogs of the diffusion equation. Many of these difference relations have not even been given a physical interpretation. The arguments which will be given here are not intended as proofs. They are given with the aim of clarifying the source of instabilities and to give a physical basis for what might have appeared as a rather bewildering phenomenon.

A large number of difference analogs for the diffusion equation are available. Many of these schemes have recently been proposed and are not well known. To familiarize the reader with available methods both the new and old difference representations will be described and briefly compared. In general, it cannot be said that one method is much better or worse than the next. Much depends on the problem's complexity and size. All transient schemes presented can be employed to solve steady state problems, and it will be seen that the schemes which are presented encompass all the cyclic iterative methods of Section 3.

Although the methods which will be presented are applicable to more general problems, the comparisons will be made employing for simplicity the problem of determining the transient temperature distribution in a two-dimensional square solid region with constant thermal properties and no internal heat generation. The region will be assumed to be isotropic and homogeneous. The governing partial differential equation is:

$$\frac{1}{\alpha}\frac{\partial T}{\partial t} = \frac{\partial^2 T}{\partial x^2} + \frac{\partial^2 T}{\partial y^2} \quad . \tag{4.1}$$

Prescribed temperature boundary conditions will be employed unless otherwise stated. These prescribed functions could be time dependent. Equal spatial increments will be employed for simplicity, i.e., $\Delta x = \Delta y = h$. A new time variable will be introduced, $\tau = \alpha t$; thus Equation (4.1) becomes

$$\frac{\partial T}{\partial \tau} = \frac{\partial^2 T}{\partial x^2} + \frac{\partial^2 T}{\partial y^2} \quad . \tag{4.2}$$

One can either proceed from Equation (4.2) or from the relationship (2.44) derived in Section 2.2 which for this case becomes (employing the network of Figure 2.4):

$$\frac{dT_{i,j}}{d\tau} = \frac{1}{h^2}\left(\delta_x^2 + \delta_y^2\right) T_{i,j} \ , \tag{4.3}$$

where $i = 1, 2, \ldots p-1$

$j = 1, 2, \ldots q-1$.

The boundary and initial conditions in the discretized system are to be evaluated at the respective nodal points of this system, $(i\Delta x, j\Delta y, k\Delta z, \tau)$. Discretized time will now be introduced.

4.1 STANDARD EXPLICIT METHOD

One of the most popular and frequently described methods is the standard explicit method which is also referred to as the forward difference method. In this method, the rate of change in the internal energy of the element during the time interval $n\Delta t$ to $(n+1)\Delta t$ is based on the net rate of heat flowing to the element at time $n\Delta t$. In equation form, this statement becomes:

$$\left(\Delta U_{n\to n+1}\right) = \Delta t \left(q_{net,n}\right) + \Delta t \left(q_n'''\right) \Delta V \ , \tag{4.4}$$

where $\Delta U_{n\to n+1}$ represents the change in the internal energy during the time increment; $q_{net,n}$ is the net rate of heat flow to the element from adjacent elements at time $n\Delta t$; and q_n''' represents any constant, time-dependent, or temperature-dependent term such as (a) an internal generation term, (b) the convective loss term in a fin problem, or (c) the respective quantity required to satisfy the boundary condition when considering elements which are part of the region's boundary.

For the problem being considerea, Equation (4.4) or (4.3) becomes:

$$\frac{T_{i,j,n+1} - T_{i,j,n}}{\Delta \tau} = \frac{1}{h^2}\left(\delta_x^2 + \delta_y^2\right) T_{i,j,n} \ . \tag{4.5}$$

It is to be noted that the standard explicit method results in the same expressions as those which result if Euler's method were employed to solve the set of simultaneous ordinary differential equations represented by Equation (4.3). If Equation (4.5) is re-written as:

$$T_{i,j,n+1} = \frac{1}{M}\left(T_{i+1,j,n} + T_{i-1,j,n} + T_{i,j+1,n} + T_{i,j-1,n} + (M-4)\,T_{i,j,n}\right), \qquad (4.6)$$

where $M = h^2/\Delta\tau$, it is easily seen that $T_{i,j,n+1}$ is an explicit function of the known temperatures at time $n\Delta\tau$. The future temperature at (i,j) is seen to be a weighted average of the present temperatures at (i,j) and the adjacent elements. The temperatures at time (n+1) can be calculated in any desirable order or simultaneously, which could be a desirable feature if one were to utilize the parallel capabilities being developed into modern digital computers.

The assumption that the change in the internal energy over the time interval $n\Delta t$ to $(n+1)\Delta t$ can be based on the heat flow rates at time $n\Delta t$ will only be valid in the limit as Δt approaches zero. Consider what might happen if large values of Δt are employed. The second term on the right side of Equation (4.4) is zero for the present case. Therefore Equation (4.4) shows that if $q_{net,n}$ is positive, the internal energy will increase; if $q_{net,n}$ is zero, the internal energy will remain constant; and if $q_{net,n}$ is negative, the internal energy will decrease. Thus the scheme is in agreement with the first law of thermodynamics. Consider again the case of a negative value of $q_{net,n}$, and assume for simplicity that $T_{i,j,n} = 1$ and that the temperature of all the surrounding elements is zero for all values of n, i.e., let them be the boundary nodes. If Δt is sufficiently large, the decrease in internal energy of the element will be so large that $T_{i,j,n+1}$ will be less than zero. This would be a violation of the second law of thermodynamics.

More generally, consider any interior or boundary element, o, which is exchanging heat with r surrounding solid elements or other media whose temperatures will be denoted by $T_{s,n}$ where the index s stands for the surroundings and takes on all values from 1 through r. The "resistances" connecting the element, o, with the surrounding elements will be denoted by R_{s-o}. For this case, Equation (4.4) gives:

$$\frac{C_o\left(T_{o,n+1} - T_{o,n}\right)}{\Delta t} = \sum_{s=1}^{r} \frac{T_{s,n} - T_{o,n}}{R_{s-o}},$$

where $C_o(=\rho c_p \Delta V)$ is the thermal capacitance of element o. If a potential violation of the second law of thermodynamics is to be avoided, the time step must be sufficiently small so that the right side of Equation (4.4) or of the above equation remains of the same sign when $T_{o,n}$ is replaced by $T_{o,n+1}$. Thus, if

$$\sum_{s=1}^{r} \frac{T_{s,n} - T_{o,n}}{R_{s-o}} > 0,$$

then

$$\sum_{s=1}^{r} \frac{T_{s,n} - T_{o,n+1}}{R_{s-o}} \geq 0$$

If the second inequality is substituted into Equation (4.1), the following inequality results:

$$\frac{C_o}{\Delta t} \geq \sum_{s=1}^{r} \frac{1}{R_{s-o}}. \qquad (4.7a)$$

Equation (4.7a) represents a general criterion which must be satisfied if a potential violation of the second law is to be avoided. Equation (4.7a) results in the following criterion for an internal node of the two-dimensional problem under consideration:

$$\frac{CR}{\Delta t} = \frac{h^2}{\alpha\Delta t} = M \geq 4,$$

where

$$R = \frac{h}{k\Delta A} \text{ and } C = \rho c_p\,(h\Delta A).$$

It is to be noted that M is a dimensionless time constant of the system. Equation (4.7a) clearly shows that the boundary elements may result in different criteria than the interior elements. For example, consider a convective boundary condition and an element on the left boundary of the network of Figure 2.4. Equation (4.7) gives:

$$M \geq 4 + 2Bi \ ,$$

where Bi is the Biot modulus, $\tilde{h}\, \Delta y/k$. It is obvious that a large Biot modulus will require a small time step, since in this case the boundary resistor, $1/\tilde{h}\Delta A$ is very small. On the other hand, if the network of Figure 2.5 is employed, the film resistance adds to the boundary resistor and makes the criterion less stringent than with an isothermal boundary condition. The criterion follows from Equation (4.7a) as:

$$M \geq 3 + \frac{2Bi}{2 + Bi} \ .$$

The advantage of the network of Figure 2.5 for large Biot numbers is obvious.

If one were to employ von Neumann's method, stability criteria slightly less stringent than the conditions given by Equation (4.7a) may result. This indicates that the nonsense introduced by round off error or generated by a violation of the second law would not grow but would be damped out. For instance, a trial with a single internal element (the case previously considered) would show that these errors would grow if $M < 2$ and would be damped out for $M > 2$. For large networks, where the stabilizing influence of the boundaries is farther removed, M could be not much less than 4. The second law, however, would be violated in this case for all values of M less than 4.

Consider next the truncation error which results if Equation (4.2) is approximated by Equation (4.5), the standard explicit approximation. The truncation error in this case is:

$$E_T = \left[\left(D_\tau - D_x^2 - D_y^2\right) - \left(\frac{\Delta_\tau}{\Delta\tau} - \frac{\delta_x^2 + \delta_y^2}{h^2}\right)\right] T_{i,j,n} \ .$$

Employing Taylor expansions (see Section 2.1), one obtains

$$E_T = -\frac{\Delta\tau}{2}\frac{\partial^2 T}{\partial\tau^2}\bigg|_{i,j,n} + \frac{h^2}{12}\left(\frac{\partial^4 T}{\partial x^4}\bigg|_{i,j,n} + \frac{\partial^4 T}{\partial y^4}\bigg|_{i,j,n}\right) + \dots \ , \qquad (4.8a)$$

or simply,

$$E_T = 0\ (\Delta\tau + h^2) \ . \qquad (4.8b)$$

Equation (4.8a) shows that: (a) as steady state is approached, the time truncation errors will vanish, (b) the consistency requirement is satisfied, (c) the time truncation error is of the order $\Delta\tau$, and (d) the spatial truncation error is of the order h^2.

If the error is denoted by $\epsilon_{i,j,n}$, i.e.,

$$\epsilon_{i,j,n} = T_{i,j,n}(\text{exact}) - T_{i,j,n}(\text{numerical}),$$

it follows from Equations (4.5) and (4.8) that:

$$\Delta_\tau\, \epsilon_{i,j,n} = \frac{1}{M}\left(\delta_x^2 + \delta_y^2\right)\epsilon_{i,j,n} + 0\left[(\Delta\tau)^2 + h^2\ (\Delta\tau)\right] \ , \qquad (4.8c)$$

where $\epsilon_{i,j,n} = 0$ along the boundaries. Following the maximum principle employed by Douglas,[15] one sees that if the partial derivatives in E_T are bounded and if $M \geq 4$, Equation (4.8c) leads to the inequality:

$$|\epsilon_{i,j,n+1}| \leq \left(1 - \frac{4}{M}\right) |\epsilon_{i,j,n}| + \frac{1}{M}|\epsilon_{i+1,j,n} + \epsilon_{i-1,j,n} + \epsilon_{i,j+1,n}$$

$$+ \epsilon_{i,j-1,n}| + A\left[(\Delta\tau)^2 + h^2\,\Delta\tau\right],$$

where A is a positive constant. Letting $\epsilon_{max,n}$ be the maximum value of $|\epsilon_{i,j,n}|$ for any i or j, one obtains:

$$|\epsilon_{i,j,n+1}| \leq \epsilon_{max,n} + A\left[(\Delta\tau)^2 + h^2\,\Delta\tau\right].$$

Since $\epsilon_{max,0} = 0$, one obtains by iteration:

$$\epsilon_{max,n} \leq n\,(\Delta\tau)\,A\,(\Delta\tau + h^2)\,.$$

If τ is less than τ_1, this error becomes:

$$\epsilon_{max,n} \leq \tau_1\,A(\Delta\tau + h^2)\,.$$

This equation shows that the solution of the difference analog converges to the solution of the partial differential equation as Δx and $\Delta\tau$ approach zero with $M \geq 4$. The process must, therefore, also be stable for $M \geq 4$.

Although the standard explicit method is probably the simplest to use in practice, it is sometimes found that the small time steps required for stability result in excessive computational time. This is especially true if the steady state solution is also being sought. Several implicit schemes will be considered next which remove these stringent stability requirements.

4.2 STANDARD IMPLICIT AND CRANK-NICOLSON METHODS

Consider first the standard implicit method which is also referred to as the backward difference method. It differs from the standard explicit scheme in that the right side of Equation (4.4) is evaluated at time $(n+1)\Delta t$ instead of $n\Delta t$, i.e.,

$$\left(\Delta U_{n\to n+1}\right) = \Delta t\left(q_{net,n+1}\right) + \Delta t\left(q'''_{n+1}\right)\Delta V\,. \tag{4.9}$$

The resulting difference equation for the two-dimensional problem being considered is:

$$\frac{T_{i,j,n+1} - T_{i,j,n}}{\Delta\tau} = \frac{1}{h^2}\left(\delta_x^2 + \delta_y^2\right) T_{i,j,n+1}\,. \tag{4.10}$$

This might seem like a minor variation of the standard explicit method; however, there are two major differences between these methods.

The first is that Equation (4.10) is an implicit equation; i.e., it is not possible to calculate the temperature distributions without solving a set of (p-1)(q-1) simultaneous linear algebraic equations to effect each time step. This would cause a tremendous increase in the number of calculations required per time step for two- and three-dimensional problems. For one-dimensional problems, the algorithm given in Section 3.1 can be employed to solve the resulting simultaneous equations. For this case, the implicit and explicit schemes require a comparable amount of effort per time step.

The second major difference lies in the stability of this scheme. In contrast to the standard explicit method, the change in the internal energy is no longer directly proportional to Δt. As Δt becomes large $q_{net,n+1}$ becomes small. This, in contrast to the standard explicit method, places a bound on ΔU and eliminates the possibility of a violation of the second law of thermodynamics. Consider a case where the boundary conditions are independent of time and let Δt approach infinity. In this case $q_{net,n+1}$ would approach zero for all values of i and j and the solution of the set of equations represented by Equation (4.10) would approach the steady state solution. Since the physical laws cannot be violated with this scheme for any size time step, it would appear unconditionally stable. Formal stability analyses have shown this to be the case.

The truncation error for the standard implicit scheme is:

$$E_T = + \frac{\Delta\tau}{2} \frac{\partial^2 T}{\partial \tau^2}\bigg|_{i,j,n+1} + \frac{h^2}{12} \left(\frac{\partial^4 T}{\partial x^4}\bigg|_{i,j,n+1} + \frac{\partial^4 T}{\partial y^4}\bigg|_{i,j,n+1} + \dots \right) . \quad (4.11)$$

This truncation error is very similar to that obtained for the standard explicit approximation, Equations (4.8); however, the time truncation error is of opposite sign.

The analysis of the convergence of Equation (4.10) follows similarly to that of the explicit method. Employing the fact that the second central spatial differences are negative at $\epsilon_{max,n}$, one again obtains:

$$\epsilon_{max,n} \le \tau_1 \; A \; (\Delta\tau + h^2) \; . \quad (4.12)$$

In this case, Equation (4.12) is valid for all values of M; therefore the solution of Equation (4.10) converges to the solution of the partial differential equation. Equation (4.2), as $\Delta\tau$ and h approach zero independently and is stable for all values of M.

Consider next the method suggested by Crank and Nicolson in 1946.(16) Since the sign of the time truncation error in the standard explicit method is opposite to that in the standard implicit method, an average of the two procedures is suggested. An average of Equations (4.4) and (4.9) for the case of interest gives:

$$\left(\Delta U_{n \to n+1}\right) = \frac{\Delta t}{2} \left(q_{net,n+1}\right) + \frac{\Delta t}{2} \left(q_{net,n}\right) . \quad (4.13)$$

The resulting difference approximation which is known as the Crank-Nicolson method is:

$$\frac{T_{i,j,n+1} - T_{i,j,n}}{\Delta\tau} = \frac{1}{2h^2} \left[\left(\delta_x^2 + \delta_y^2\right) T_{i,j,n+1} + \left(\delta_x^2 + \delta_y^2\right) T_{i,j,n} \right] . \quad (4.14)$$

Employing Taylor's expansions gives:

$$\frac{\partial T}{\partial \tau}\bigg|_{i,j,n+1/2} = \frac{T_{i,j,n+1} - T_{i,j,n}}{\Delta\tau} - \frac{(\Delta\tau)^2}{24} \frac{\partial^3 T}{\partial \tau^3}\bigg|_{i,j,n+1/2} + \dots ,$$

and

$$\frac{\partial^2 T}{\partial x^2}\bigg|_{i,j,n+1/2} = \frac{1}{2} \left(\frac{\partial^2 T}{\partial x^2}\bigg|_{i,j,n+1} + \frac{\partial^2 T}{\partial x^2}\bigg|_{i,j,n} \right) - \frac{(\Delta\tau)^2}{8} \frac{\partial^4 T}{\partial \tau^2 \partial x^2}\bigg|_{i,j,n+1/2} - \dots .$$

It follows that the truncation error in approximating Equation (4.2) by Equation (4.14) is:

$$E_T = 0 \; (\Delta\tau)^2 + 0 \; (h^2) \; .$$

Thus the local time truncation error is now of second order in contrast to the first order error of the methods previously considered.

The Crank-Nicolson difference method is also an implicit scheme; thus a set of simultaneous equations must be solved each time step. This set of equations is essentially the same as those which were obtained in the standard implicit method, and the same elimination or iteration procedures can be employed. Thus the effort required per time step is not significantly greater, but the time truncation error has been reduced.

Consider next the stability of the Crank-Nicolson method. Equation (4.13) shows that if Δt is large, the second term on the right side will also be large in magnitude, which would tend to cause a large change in the internal energy of the element. However, the first term on the right side will, in this case, change sign, which again places a bound on the maximum possible value of $\Delta U_{n \to n+1}$.

Consider as in Section 4.1 any interior or boundary node, o, which is exchanging heat with r surrounding nodal points. Further, consider a sufficiently large time step such that some $q_{net,n+1}$ [see Equation (4.13)] becomes zero. A further change in internal energy of the element might appear as a violation of the first law except for the fact that this change is also being based on $q_{net,n}$. However, if a potential violation of the second law of thermodynamics is to be avoided, this term must remain of the same sign when $T_{o,n}$ is replaced by $T_{o,n+1}$. The combination of these two conditions results in the following criterion which must be satisfied:

$$\frac{C_o}{\Delta t} \geq \frac{1}{2} \sum_{s=1}^{r} \frac{1}{R_{s-o}} \quad . \qquad (4.7b)$$

For the case of a single internal node, Equation (4.7b) shows that the second law will be violated for $M < 2$; i.e., a time step twice as large as that permissible in the standard explicit method can be employed. At the same time, this single node case shows that the error introduced with any finite time increment is bounded and will be damped out. This is in direct contrast to the standard explicit method and an indication that the method is unconditionally stable. Formal stability analyses have shown this to be the case.[12,16]

The question which needs to be answered is: If the method is unconditionally stable, of what significance is the criterion (4.7b)? It would appear to indicate that the Crank-Nicolson method is not as attractive as indicated by formal stability analyses, since physically meaningless fluctuations might still occur if Equation (4.7b) is not satisfied. Gaumer's results (see Reference 25) showed such fluctuations only for cases when Equation (4.7b) was not satisfied. These fluctuations were, of course, damped with increasing time; however, they caused the accuracy of the Crank-Nicolson method to be less than that of the standard implicit technique.

4.3 DUFORT-FRANKEL METHOD

In 1910 Richardson[17] proposed the following difference approximation to (4.2):

$$\frac{T_{i,j,n+1} - T_{i,j,n-1}}{2\Delta\tau} = \frac{1}{h^2}\left(\delta_x^2 + \delta_y^2\right) T_{i,j,n} \quad . \qquad (4.15)$$

Equation (4.15) is a natural approximation to suggest, since the average difference quotient employed for the time derivative is obviously more accurate than the forward difference quotient employed in the standard explicit approximation. The truncation error for this case is:

$$E_T = 0\ (\Delta\tau^2 + h^2) \ .$$

Equation (4.15) represents the following energy balance:

$$\left(\Delta U_{n \to n+1}\right) = 2\Delta t \left(q_{net,n}\right) - \left(\Delta U_{n-1 \to n}\right) . \qquad (4.16)$$

Equation (4.16) clearly shows that a violation of the first law could occur very easily. For example, if $\Delta U_{n-1 \to n}$ were large and positive, a reduction in the internal energy of the element could occur in spite of the fact that heat is flowing to the element. If $q_{net,n}$ ever becomes zero, a violation of the first law would again occur regardless of the size of the time step. An analysis of a single interior element will show that the method is unstable for all values of M. The method has become a classic for this reason.

Consider next what might appear as a trivial modification of Equation (4.15); that is, replace $T_{i,j,n}$ by $(T_{i,j,n+1} + T_{i,j,n-1})/2$ to give:

$$T_{i,j,n+1} = \frac{M-4}{M+4} T_{i,j,n-1} + \frac{2}{M+4} \Big(T_{i+1,j,n} + T_{i-1,j,n} + T_{i,j+1,n} + T_{i,j-1,n}\Big) . \quad (4.17)$$

This form was suggested by DuFort and Frankel in 1953.[18] Equation (4.16) with this modification becomes:

$$\Big(\Delta U_{n\to n+1}\Big) = \Delta t \Big[q_{net,n}\ (T_{i,j,n+1})\Big] + \Delta t \Big[q_{net,n}\ (T_{i,j,n-1})\Big] - \Big(\Delta U_{n-1\to n}\Big) , \quad (4.18)$$

where $q_{net,n}(T_{i,j,n+1})$ means the net rate of heat flow to (i,j) based on the surrounding temperatures evaluated at time $n\Delta t$ and its own temperature at time $(n+1)\Delta t$. In this case if the third term on the right side of Equation (4.18) were large, the second term would also be large and of opposite sign. In the case of Equation (4.18), a violation of the first law no longer appears possible. An examination of a single interior element gives results similar to the Crank-Nicolson method with the exception that the second law could be violated for $M < 4$ instead of 2, but the error would damp out for all finite values of $\Delta\tau$. Thus the method generally appears unconditionally stable as is verified by DuFort and Frankel's analysis.

Consider next the truncation error of Equation (4.17). The truncation error of Equation (4.15) was $0(\Delta\tau^2 + h^2)$, and Taylor's expansions give:

$$- 4T_{i,j,n} = - 2\ (T_{i,j,n+1} + T_{i,j,n-1}) + 2(\Delta\tau)^2 \left.\frac{\partial^2 T}{\partial\tau^2}\right|_{i,j,n} + 0\ (\Delta\tau)^4 .$$

Thus the truncation error in the DuFort-Frankel method becomes:

$$E_T = -2\,\frac{(\Delta\tau)^2}{h^2} \left.\frac{\partial^2 T}{\partial\tau^2}\right|_{i,j,n} + 0\ (\Delta\tau^2 + h^2) . \quad (4.19)$$

Consistency requires that $\Delta\tau/h$ approach zero as $\Delta\tau$ and h vanish. If h and $\Delta\tau$ approached zero with a constant ratio, $c = \Delta\tau/h$, Equation (4.17) would approximate the hyperbolic equation:

$$2c^2 \frac{\partial^2 T}{\partial\tau^2} + \frac{\partial T}{\partial\tau} = \frac{\partial^2 T}{\partial x^2} + \frac{\partial^2 T}{\partial y^2} .$$

If Equation (4.18) is to be an approximation for Equation (4.2), c^2 must be small. If $\Delta\tau$ is of the order of h^2, then Equation (4.19) becomes:

$$E_T = 0\ (h^2) .$$

Equation (4.17) is a three time level formula; thus there is a problem in getting the solution started. One could employ a different approximation to begin the calculations. As an alternate procedure, Larkin[19] suggests using, "the initial condition as the first time step, but with a very small increment. Then this small increment could be increased by factors of 2 in subsequent calculations until the increment becomes large enough for efficient calculation."

It is to be noted that for M = 4 this method is identical to the standard explicit method for the problem under consideration. In addition, $T_{i,j,n}$ may be evaluated only for even values of $(i+j)$ for one cycle, and for odd values of $(i+j)$ during the next cycle. This technique would reduce the number of calculations required by two, but it would be considerably more difficult to program.

4.4 ALTERNATING DIRECTION METHODS

In 1954 Peaceman and Rachford[10] presented an alternating direction implicit approximation for the solution of the diffusion equation in two space variables. In this method two difference equations are employed alternately. The following physical approximations are employed:

$$\left(\Delta U_{n\to n+1/2}\right) = \frac{\Delta t}{2}\left(q_{x-net,n} + q_{y-net,n+1/2}\right) , \tag{4.20a}$$

and

$$\left(\Delta U_{n+1/2\to n+1}\right) = \frac{\Delta t}{2}\left(q_{x-net,n+1} + q_{y-net,n+1/2}\right) , \tag{4.20b}$$

where, e.g., $q_{x-net,n}$ represents the net rate of heat flow at time $n\Delta t$ to the element (i,j) from elements $(i+1,j)$ and $(i-1,j)$, i.e., the net rate in the x-direction. The other quantities follow similarly. Equation (4.20a) is explicit in the x-direction and implicit in the y-direction. Its implicit form would prevent violations of the second law of thermodynamics in the y-direction, but since it is explicit in the x-direction, a violation of the second law could occur for the heat exchanged by the i-th column with the i+1 and i-1 columns. Thus, if Equation (4.20a) were applied alone, one would expect the stability criterion to be the same as that which results in the solution of a one-dimensional problem by the standard explicit method -- $M \geq 2$. This is verified by the stability analysis of Peaceman and Rachford.[10] On the other hand, they showed that if Equations (4.20a) and (4.20b) are applied alternately, the method is stable for all values of M. In difference form, Equations (4.20a) and (4.20b) become:

$$\frac{T_{i,j,n+1/2} - T_{i,j,n}}{\Delta\tau/2} = \frac{1}{h^2}\left(\delta_x^2\, T_{i,j,n} + \delta_y^2\, T_{i,j,n+1/2}\right) , \tag{4.21a}$$

$$\frac{T_{i,j,n+1} - T_{i,j,n+1/2}}{\Delta\tau/2} = \frac{1}{h^2}\left(\delta_x^2\, T_{i,j,n+1} + \delta_y^2\, T_{i,j,n+1/2}\right) . \tag{4.21b}$$

Douglas[20] demonstrated that the Peaceman-Rachford alternating direction method is a perturbation of the Crank-Nicolson method. The advantage of the former method is that the sets of simultaneous equations which result each half-step are tridiagonal; hence they can be readily solved line by line or row by row employing the algorithm presented in Section 3.1. If the Crank-Nicolson method were employed, one set of (p-1) (q-1) equations with a quindiagonal matrix would have to be solved each time step. In the Peaceman-Rachford method the first half-step requires the solution of (p-1) systems of order (q-1) with tri-diagonal matrices and the second half-step (q-1) systems of order (p-1).

Douglas[20] suggested an alternating direction procedure which is equivalent to the Peaceman-Rachford method on a rectangle; however, the intermediate values are different. In this method, one moves forward in time in the x-term employing:

$$\frac{\tilde{T}_{n+1} - T_n}{\Delta\tau} = \frac{\delta_x^2}{2h^2}\left(\tilde{T}_{n+1} + T_n\right) + \frac{\delta_y^2}{h^2}\, T_n , \tag{4.22a}$$

where the i and j indices have been suppressed. $\tilde{T}_{n+1}$ is then corrected by moving forward in time in the y-term employing:

$$\frac{T_{n+1} - T_n}{\Delta\tau} = \frac{\delta_x^2}{2h^2}(\tilde{T}_{n+1} + T_n) + \frac{\delta_y^2}{2h^2}(T_{n+1} + T_n) \; . \qquad (4.22b)$$

Equation (4.22b) is seen to be identical to the Crank-Nicolson difference analog with the exception that T_{n+1} has been replaced by $\tilde{T}_{n+1}$. The advantage of Douglas' procedure over the Peaceman-Rachford procedure is that it can be extended to three dimensions, whereas the obvious extension of the Peaceman-Rachford method to three dimensions is unstable for M less than 2/3.[15]

A slight variation of the Peaceman-Rachford method will now be considered which results in a method which is entirely explicit and remains unconditionally stable. In contrast to the Peaceman-Rachford method, this technique is also applicable to one- and three-dimensional problems. Its extension to these cases is straightforward; thus the two-dimensional problem previously described will again be considered. The method is attributed to Saul'yev[21] and is described in detail by Larkin.[14]

Let us assume that the calculation of the grid temperatures is proceeding in the northeast direction. In this case, $T_{i-1,j,n+1}$ and $T_{i,j-1,n+1}$ are known quantities; hence the heat flow rates in the north and east directions could be based on time $(n+1)\Delta t$. The net flow rate in these directions will be denoted as: $q_{NE\text{-}net,n+1}$. The change of the internal energy of the element is based on the following energy balance:

$$\left[\Delta U_{n\to n+1}\right] = \Delta t\left[q_{NE\text{-}net,n+1}\right] + \Delta t\left[q_{SW\text{-}net,n}\right] \; . \qquad (4.23a)$$

Alternately, if the calculations were proceeding in the south and west directions, the following energy balance could be employed:

$$\left[\Delta U_{n\to n+1}\right] = \Delta t\left[q_{NE\text{-}net,n}\right] + \Delta t\left[q_{SW\text{-}net,n+1}\right] \; . \qquad (4.24a)$$

Again the presence of heat flow rates evaluated at time $(n+1)\Delta t$ bounds the maximum change in the internal energy and prohibits the error growth. In the Peaceman-Rachford method, the stabilizing influence of the implicit terms was limited to a single column (or row); in contrast, this influence is extended throughout the entire region if either Equations (4.23a) or (4.24a) is employed. A transient introduced at a boundary of the system would be felt throughout the entire system in only one time step, if the calculation proceeded in a direction away from this boundary.

In terms of the difference operators, Equations (4.23a) and (4.24a) become, respectively, (suppressing the i and j indices):

$$\frac{\Delta_\tau T_n}{\Delta\tau} = \frac{\Delta_x T_n - \nabla_x T_{n+1}}{(\Delta x)^2} + \frac{\Delta_y T_n - \nabla_y T_{n+1}}{(\Delta y)^2} \; , \qquad (4.23b)$$

and

$$\frac{\Delta_\tau T_n}{\Delta\tau} = \frac{\Delta_x T_{n+1} - \nabla_x T_n}{(\Delta x)^2} + \frac{\Delta_y T_{n+1} - \nabla_y T_n}{(\Delta y)^2} \; . \qquad (4.24b)$$

Expanding the above operators and rearranging gives, respectively:

$$T_{i,j,n+1} = \frac{M-2}{M+2} T_{i,j,n} + \frac{1}{M+2} \Big(T_{i+1,j,n} + T_{i,j+1,n} + T_{i-1,j,n+1} + T_{i,j-1,n+1} \Big) , \quad (4.23c)$$

and

$$T_{i,j,n+1} = \frac{M-2}{M+2} T_{i,j,n} + \frac{1}{M+2} \Big(T_{i+1,j,n+1} + T_{i,j+1,n+1} + T_{i-1,j,n} + T_{i,j-1,n} \Big) . \quad (4.24c)$$

Similar relationships could be obtained in an identical manner for calculating in the northwest or southeast directions. Both Equations (4.23) and (4.24) were shown to be stable for all values of M.[14] Thus, unlike the Peaceman-Rachford difference equations which are only stable if they are applied alternately, either Equation (4.23) or (4.24) can be employed by itself. However, one can see from the physical nature of the approximation that a more accurate result could be obtained by employing Equations (4.23) and (4.24) alternately or by using both in each time step and averaging the results.

Let us consider the truncation error introduced by Equations (4.23b) and (4.24b). Equations (2.13) and (2.15) can be employed to give:

$$\frac{\Delta_\tau T_n}{\Delta\tau} = \left(D_\tau + \frac{(\Delta\tau)}{2} D_\tau^2 + \frac{(\Delta\tau)^2}{6} D_\tau^3 + \dots \right) T_n , \quad (4.25a)$$

$$\frac{\Delta_x T_n}{(\Delta x)^2} = \left(\frac{D_x}{\Delta x} + \frac{D_x^2}{2} + \frac{(\Delta x) D_x^3}{6} + \frac{(\Delta x)^2 D_x^4}{24} + \dots \right) T_n , \quad (4.25b)$$

and

$$\frac{E_\tau \nabla_x T_n}{(\Delta x)^2} = \frac{\nabla_x T_{n+1}}{(\Delta x)^2} = \left(\frac{D_x}{\Delta x} - \frac{D_x^2}{2} + \frac{\Delta\tau}{\Delta x} D_x D_\tau + \frac{\Delta x}{6} D_x^3 - \frac{\Delta\tau}{2} D_x^2 D_\tau + \frac{(\Delta\tau)^2}{2\Delta x} D_x D_\tau^2 - \frac{(\Delta x)^2}{24} D_x^4 + \dots \right) T_n , \quad (4.25c)$$

where for brevity a one-dimensional case is being considered. The y-direction would only introduce terms identical to the x-terms. Employing the one-dimensional form of Equation (4.23b), the truncation error is:

$$E_T = \left[(D_\tau - D_x^2) - \left(\frac{\Delta_\tau}{\Delta\tau} - \frac{\Delta_x - E_\tau \nabla_x}{(\Delta x)^2} \right) \right] T_n ,$$

or upon substituting Equations (4.25a,b,c):

$$E_T = \left[-\frac{\Delta\tau}{2} D_\tau \left(D_\tau - D_x^2 \right) - \frac{\Delta\tau}{\Delta x} D_\tau D_x - \frac{(\Delta\tau)^2}{2\Delta x} D_\tau^2 D_x - \frac{(\Delta\tau)^2}{6} D_\tau^3 - \frac{(\Delta\tau)^3}{6\Delta x} D_\tau^3 D_x + \frac{(\Delta\tau)^2}{4} D_\tau^2 D_x^2 - \frac{(\Delta\tau)(\Delta x)}{6} D_\tau D_x^3 + \frac{(\Delta x)^2}{12} D_x^4 + \ldots \right] T_n . \qquad (4.26a)$$

Equation (4.24b) results in the following truncation error:

$$E_T = \left[-\frac{\Delta\tau}{2} D_\tau \left(D_\tau - D_x^2 \right) + \frac{\Delta\tau}{\Delta x} D_\tau D_x + \frac{(\Delta\tau)^2}{2\Delta x} D_\tau^2 D_x - \frac{(\Delta x)^2}{6} D_\tau^3 + \frac{(\Delta\tau)^3}{6\Delta x} D_\tau^3 D_x + \frac{(\Delta\tau)^2}{4} D_\tau^2 D_x^2 + \frac{(\Delta\tau)(\Delta x)}{6} D_\tau D_x^3 + \frac{(\Delta x)^2}{12} D_x^4 + \ldots \right] T_n , \qquad (4.26b)$$

where the first terms of Equations (4.26a) and (4.26b) are obviously small. Consistency of Equations (4.23) or (4.24) with the diffusion equation is seen to require that $\Delta\tau/\Delta x$ and $\Delta\tau/\Delta y$ vanish as $\Delta\tau$, Δx, and Δy approach zero. If

$$\Delta\tau = 0(h^2) ,$$

the truncation error of either Equation (4.23) or (4.24) would be:

$$E_T = 0\ (h) .$$

Conversely, if the results of Equations (4.23) and (4.24) were averaged each time step, the truncation error follows from Equations (4.26) as:

$$E_T = \left[-\frac{\Delta\tau}{2} D_\tau \left(D_\tau - D_x^2 \right) - \frac{(\Delta\tau)^2}{6} D_\tau^3 + \frac{(\Delta x)^2}{12} D_x^4 + \frac{(\Delta\tau)^2}{4} D_\tau^2 D_x^2 + \ldots \right] T_n ,$$

or

$$E_T = 0 \left[(\Delta\tau)^2 + h^2 \right] .$$

It is seen that many of the error terms cancel and the accuracy and consistency is greatly improved. Of course, if Equations (4.23) and (4.24) were used alternately, the errors would also tend to cancel. The slight loss in accuracy obtained by using the alternating direction procedure is probably well compensated for by the greater than 50 per cent reduction in the required number of operations. In general this procedure looks very attractive; it is stable, efficient, and accurate. This combination is rare.

Although Equation (4.23c) was proposed only recently, it now will be shown that it has been employed for many years. If ω is set equal to $4/(M + 2)$, then

$$1 - \omega = \frac{M - 2}{M + 2}$$

and Equation (4.23c) is seen to be analogous to Equation (3.14b). Thus, at least in the interior of the region, the scheme represented by Equation (4.23c) is analogous to the extrapolated Liebmann iterative method for solving steady state problems. The n-th approximation is again seen to correspond to the n-th time step. An infinite $\Delta\tau$ corresponds to $\omega = 2$; whereas an $\Delta\tau$ approaches zero, ω also approaches zero. The Gauss-Seidel iteration corresponds to $M = 2$. This fixed value of M implies that the number of iterations required for convergence with the Gauss-Seidel iteration varies as h^{-2}. [Compare with Equation (3.21b).] The Jacobi iteration is analogous to the standard explicit method with $M = 4$; hence it takes twice the number of iterations required with the Gauss-Seidel iteration.

The application of Equations (4.23c) and (4.24c) alternately for the solution of transient problems also appears attractive from the standpoint that such a program could be easily modified to the use of only Equation (4.23c) or (4.24c), which would represent the extrapolated Liebmann method -- an efficient proven technique for the solution of steady state problems. Initial results have indicated that the use of these equations alternately for the solution of steady state problems is slightly less efficient. Additional studies in this area are necessary and are being considered.

4.5 EXPONENTIAL APPROXIMATION

No known published description of this method exists other than the brief outline of the technique by Larkin.[19] He also stated[22] that this procedure or some modification of it has been widely employed in the aerospace industry. Additional correspondence further substantiated this statement.[23] This method, along with its stability, truncation error, and consistency requirement, will now be considered.

Again, discretized spatial coordinates will be employed; thus the energy balance for the problem under consideration results in a set of ordinary differential equations of the form:

$$\frac{dT_{i,j}}{d\tau} = \frac{1}{h^2}\left(\delta_x^2 + \delta_y^2\right) T_{i,j} \,. \tag{4.3}$$

This method, however, differs from all the others in that the time and the temperature, $T_{i,j}$, are treated as continuous variables during the time interval τ to $\tau + \Delta\tau$. On the other hand, the temperatures of the surrounding elements are assumed to be constants during this integration. The physical approximation being made is obviously less accurate than, for example, that made in the standard explicit approximation. However, by allowing $T_{i,j}$ to vary during this integration, neither the first nor the second law of thermodynamics can be violated; hence the method is very stable.

If the calculation is proceeding in the direction of increasing i and j, the calculation procedure is simplest and fastest if the following constant values are assumed for the temperatures of the surrounding elements:

$$T_{i+1,j,n} \,,$$

$$T_{i-1,j,n+1} \,,$$

$$T_{i,j+1,n} \,,$$

and

$$T_{i,j-1,n+1} \,.$$

The integration of Equation (4.3) in this case gives:

$$T_{i,j,n+1} = \frac{1}{4}\left(1 - e^{-4/M}\right)\left(T_{i+1,j,n} + T_{i-1,j,n+1} + T_{i,j+1,n} + T_{i,j-1,n+1}\right) + e^{-4/M}\, T_{i,j,n} \,. \tag{4.27}$$

The exponentials on the right side of Equation (4.27) are the basis for the name, exponential method. A calculation of the truncation error of this approximation gives:

$$E_T = \left\{\left(-\frac{\Delta\tau}{2} D_\tau^2 + \frac{2\Delta\tau}{h^2} D_\tau + \ldots\right) + \left(-\frac{4/M}{2} + \frac{(4/M)^2}{6} - \frac{(4/M)^3}{24} + \ldots\right)\left(\frac{2\Delta\tau}{h^2} D_\tau + D_x^2 + D_y^2 + \ldots\right)\right\} T_{i,j,n} \,. \tag{4.28}$$

Therefore, if Equation (4.27) is to be an approximation for Equation (4.2), $\Delta\tau/h^2$ must be small. If $\Delta\tau = 0(h^3)$, then Equation (4.28) becomes:

$$E_T = 0\ (h) \,.$$

This consistency requirement is indeed severe and makes the method appear relatively unattractive.

One might think that the accuracy of the method would be improved if the temperature at time $n\Delta t$ were employed for all the surrounding elements or if the direction of calculation were alternated with each time step. However, an analysis of the truncation error for these cases results in the same consistency criterion, although the expressions for the truncation errors are, of course, different. Computer trial results also exhibited negligible differences in the resulting errors with these variations. One might ask: What is the value of a method with such a severe consistency criterion? This question will be considered in the next section in order to use the comparisons given there.

4.6 COMPARISONS AND CONCLUSIONS

All the techniques available for the solution of the diffusion equation have their advantages and disadvantages. For this reason, available procedures should be carefully studied and weighed in relation to individual requirements. To aid in this evaluation a linear, two-dimensional transient problem will be employed for a comparison of the accuracy and computer execution times of some of the schemes. This simple problem obviously does not represent a thorough test of the methods, but represents a starting point for choosing a scheme suitable for individual requirements.

The region geometry, the boundary conditions, and the initial condition of the example chosen are given in Figure 4.1. Dimensionless time,

$$t^* = \frac{\alpha t}{L^2} ,$$

and dimensionless temperature,

$$T^* = \frac{T - T_i}{T_L - T_i} ,$$

will be employed. A network representation similar to Figure 2.4 was used, except that

$$\Delta x \; (= \Delta y = h) = \frac{L}{9} ;$$

thus the grid contains 100 points. A fixed value of h was used throughout the calculations, since this study was concerned only with the time truncation error, i.e., the error caused by replacing the left side of Equation (4.3) by a difference expression.

Consider first a comparison of the computer execution time requirements. All calculations were performed on an IBM 7094 computer. The programs were written in FORTRAN II. Table 4.1 compares the execution time required to carry the calculations to $t^* = 1.18$. For this value of t^*, the heat flow rates across the isothermal boundaries are within 2 per cent of their steady state values. A value of M = 16 was employed; hence $t^* = 1.18$ represents 1536 time steps. Caution was employed to write all programs as efficiently as possible so that the results of Table 4.1 would be meaningful. At the same time, it is not difficult to analyze the number of operations required for any given scheme and then

TABLE 4.1

COMPUTATIONAL TIME REQUIREMENTS OF VARIOUS SCHEMES

Method	Computational Time, $t^* = 1.18$ (n = 1536)
Standard Explicit	10.33 sec ($\simeq 10^6$ operations)
Exponential	8.93 sec
DuFort-Frankel	4.65 sec
Larkin, alternating	16.47 sec
Larkin, averaging	36.62 sec

estimate the total execution time based on the speed of the computer being employed. The values given in Table 4.1 are in approximate agreement with such calculations. In this table, as well as Figures 4.2 and 4.3, the scheme which employs Equations (4.23c) and (4.24c) alternately is labeled: Larkin, alternating. The scheme which employs both of these equations during each time step and averages the two results is labeled: Larkin, averaging. The reported time in all cases excludes that portion required for the input and the output of data.

The small execution time of the DuFort-Frankel method is due to the fact that $T_{i,j,n}$ was evaluated only for even values of (i + j) on one cycle and for odd values of (i + j) on alternate cycles. This leap-frogging technique reduces the computational time by approximately 50 per cent, but it: (a) increases the program complexity, (b) results in less information, and (c) may not be possible for more complex cases. Hence this computation time must be interpreted accordingly.

Consider next the accuracy of the various schemes. Graphical comparisons are given for M = 0.67 and M = 2.0 in Figures 4.2 and 4.3, respectively. The solid curve represents the Larkin averaging technique with M = 16. The results clearly show that the time truncation error is negligible for this case. The standard explicit method is not included in these comparisons since it is unstable. Values of $M \geq 4$ resulted in such small errors that graphical comparisons for these cases were unsuitable. Due to the step change in the temperature at the boundary at x = L, the greatest errors occur at small values in time. A comparison of the percentage error for M = 4 is given in Table 4.2 for the first printout, $t^* = 0.037$.

The following conclusions were drawn from the results of this example:

(1) The additional accuracy obtained when the results of Equations (4.23c) and (4.24c) were averaged over that obtained when these equations were used alternately does not warrant the more than 100 per cent increase in computational effort. Larkin[14] presented a one-dimensional example in which the results of the averaging scheme were slightly <u>less</u> accurate, which is further evidence against the averaging scheme.

(2) The accuracy of the exponential method was poor. This was expected since the consistency requirement that $\Delta\tau/h^2$ (= 1/M) be small was not satisfied.

(3) As one would expect from the truncation error analyses, the DuFort-Frankel method was less accurate than the alternating direction explicit approximation, but slightly more accurate than the standard explicit method when the latter method was stable and not identical, i.e., for M > 4.

(4) The schemes if listed according to ease of programming fall in the following order:

Exponential.................. easiest
Standard Explicit............
Larkin, alternating..........
Larkin, averaging............
DuFort-Frankel...............
Implicit schemes............. hardest

TABLE 4.2

PER CENT ERROR IN q AND T* = 0.037

METHOD	PER CENT ERROR	
	q (Boundary at x = L)	T^* (2L/3, 0)
Larkin, averaging	0.4	0.5
Larkin, alternating	1.1	1.3
DuFort-Frankel*	2.6	3.1
Standard Explicit*	2.6	3.1
Exponential	3.7	23.9

*These methods are identical for M = 4.

A different order might arise for different problems, especially if a nonlinear problem were considered.

Consider again the exponential approximation. It was seen that the method is very stable, easy to program, and requires relatively little computation time for any given time increment. On the other hand, the severe consistency requirement generally forces one to excessively small time increments; hence, of what value is this method? Lockheed[(23)] modified this technique by approximating the exponential by the first two terms of its Taylor series expansion. This, of course, effects both its stability and accuracy. Although this modification was not analyzed, the accuracy, according to Reference 23, is improved. Alternately, consider such problems as a thin fin, a thin radiating member, or a region with a large internal generation term which is temperature dependent. In these cases, the accuracy is strongly dependent upon the other terms appearing in the equation, and the conductive heat exchange with the surrounding elements is of less importance. This appears to be precisely the case of the comparison made by Larkin.[(19)] He compared the DuFort-Frankel, the exponential, and the explicit alternating direction methods for the case of heat flow in a thin-wall, hollow black box. The box was exchanging heat internally by radiation and externally by radiation and convection. Conduction occurred along the walls of the box. The accuracy of the exponential method was as good as or better than the other methods for this case. His reported accuracy of the exponential method compared to other techniques can, in the author's opinion, only be explained by the minor importance of conduction. At the same time, this points to problems in which the exponential method is competitive without modification.

The implicit methods were not included in these comparisons partly due to the relative difficulty of programming these techniques. In addition, many good comparisons of these methods exist in the literature (see, e.g., References 10, 21, and 25).

The following general conclusions are based on the analyses of the schemes given in Sections 4.1 through 4.5.

(1) A scheme which under all conditions is in agreement with the laws of thermodynamics is also unconditionally stable. However, a scheme which has been shown in the literature to be unconditionally stable may still under certain conditions result in a violation of the second law.

(2) Consider a scheme which results in a difference equation of the form:

$$T_{o,n+1} = A\,T_{o,n} + B\,T_{o,n-1} + \sum_{s=1}^{r} \left[C_{s-o}\,T_{s,n} + D_{s-o}\,T_{s,n-1} \right] ,$$

where o represents the node of interest and s represents one of the r surrounding nodal points. The scheme will be unconditionally stable if A and B are both positive under all conditions for all values of o. However, it may or may not be unconditionally stable if negative values of either A or B are possible. If A and B are both positive, the second law of thermodynamics will not be violated; if A or B are negative, a violation of the second law is possible, and physically meaningless fluctuations may occur.

(3) It appears as if only two of the schemes which were considered will satisfy the second law of thermodynamics under any and all conditions. These are the standard implicit and the exponential methods.

(4) The significance of a procedure which is unconditionally stable but yet may result in a violation of the second law is presently unclear. Gaumer's results showed meaningless fluctuations and consequently poor accuarcy for such cases. The results of the example of this section with M = 0.67 did not exhibit such fluctuations when successive time steps were compared. (This conclusion excludes the DuFort-Frankel method, since values of T^* (2L/3, 0) were only calculated for even time steps.) Additional studies of such procedures are in progress.

The procedures which have been discussed for the solution of the diffusion equation do not, by any means, encompass all known methods. The reader is referred to References 11, 13, 15, 21, and 24 for discussions of additional techniques. Some of these references also contain large bibliographies.

REFERENCES

1. Carslaw, H. S. and J. C. Jaeger. Conduction of Heat in Solids, 2nd ed. Oxford, England: Oxford University Press, 1959.

2. McCarthy, J., "Information," Scientific American (September, 1966), p. 67.

3. Southwell, R. V. Relaxation Methods in Engineering Science. Oxford, England: Oxford University Press, 1940.

4. Allen, D. N. G. Relaxation Methods. New York: McGraw-Hill Book Co., 1954.

5. Dusinberre, G. M. Heat Transfer Calculations by Finite Differences, 2nd ed. Scranton, Penn.: International Textbook Co., 1961.

6. Hildebrand, F. B. Introduction to Numerical Analysis. New York: McGraw-Hill Book Co., 1956, pp. 134-137.

7. Clausing, A. M., "Some Influences of Microscopic Constructions on the Thermal Contact Resistance," Technical Report ME-TN-242-2. Urbana, Ill.: Department of Mechanical Engineering, University of Illinois, 1965.

8. Frankel, S. P., "Convergence Rates of Iterative Treatments of Partial Differential Equations," Mathematical Tables and Other Aids to Computation, 4 (1950), pp. 65-75.

9. Garabedian, P. R., "Estimation of the Relaxation Factor for Small Mesh Size," Mathematical Tables and Other Aids to Computation, 10:56 (October, 1956), pp. 183-185.

10. Peaceman, D. W. and H. H. Rachford, Jr., "The Numerical Solution of Parabolic and Elliptic Differential Equations," Journal of the Society of Industrial and Applied Mathematics, 3 (1955), pp. 28-41.

11. Young, D., "The Numerical Solution of Elliptic and Parabolic Partial Differential Equations," Survey of Numerical Analysis, J. Todd, ed. New York: McGraw-Hill Book Co., 1962, pp. 380-435.

12. O'Brien, G. G., M. A. Hyman, and S. Kaplan. "A Study of the Numerical Solution of Partial Differential Equations," Journal of Mathematical Physics, 29:4 (1951), pp. 223-251.

13. Richtmyer, R. D. Difference Methods for Initial Value Problems. New York: Interscience Publishers, Inc., 1957.

14. Larkin, B. K., "Some Stable Explicit Difference Approximations to the Diffusion Equation," Mathematics of Computation, 18 (1964), pp. 196-202.

15. Douglas, J., Jr., "A Survey of Numerical Methods for Parabolic Differential Equations," Advances in Computers, F. L. Alt, ed. Vol. II. New York: Academic Press, 1961, pp. 1-54.

16. Crank, J. and P. Nicolson, "A Practical Method for Numerical Evaluation of Solutions of Partial Differential Equations of the Heat Conduction Type," Proceedings, Cambridge Philosophical Society, 43 (1947), pp. 50-64.

17. Richardson, L. F., "The Approximate Arithmetical Solution by Finite Differences of Physical Problems Involving Differential Equations, with an Application to the Stresses in a Masonry Dam," Philosophical Transactions, Royal Society of London, A, 210 (1910), pp. 307-357.

18. DuFort, E. C. and S. P. Frankel, "Stability Conditions in the Numerical Treatment of Parabolic Differential Equations," Mathematical Tables and Other Aids to Computation, 7 (1953), pp. 135-152.

19. Larkin, B. K., "Some Finite Difference Methods for Problems in Transient Heat Flow," AIChE Paper No. 16, 1964.

20. Douglas, J., Jr., "Alternating Direction Methods for Three Space Variables," Numerische Mathematik, 4 (1962), pp. 41-63.

21. Saul'yev, V. K. Integration of Equations of Parabolic Type by the Method of Nets (Russian translation) New York: MacMillan Co., 1964.

22. Larkin, B. K., Denver Division of the Martin Company, May, 1965, personal correspondence.

23. Prince, R. K., Lockheed-California Co., May, 1965, personal correspondence.

24. Ames, W. F. Nonlinear Partial Differential Equations in Engineering, New York: Academic Press, 1965.

25. Gaumer, G. R., "The Stability of Three Finite Difference Methods of Solving for Transient Temperatures," Fifth U.S. Navy Symposium on Aeroballistics, 1961.

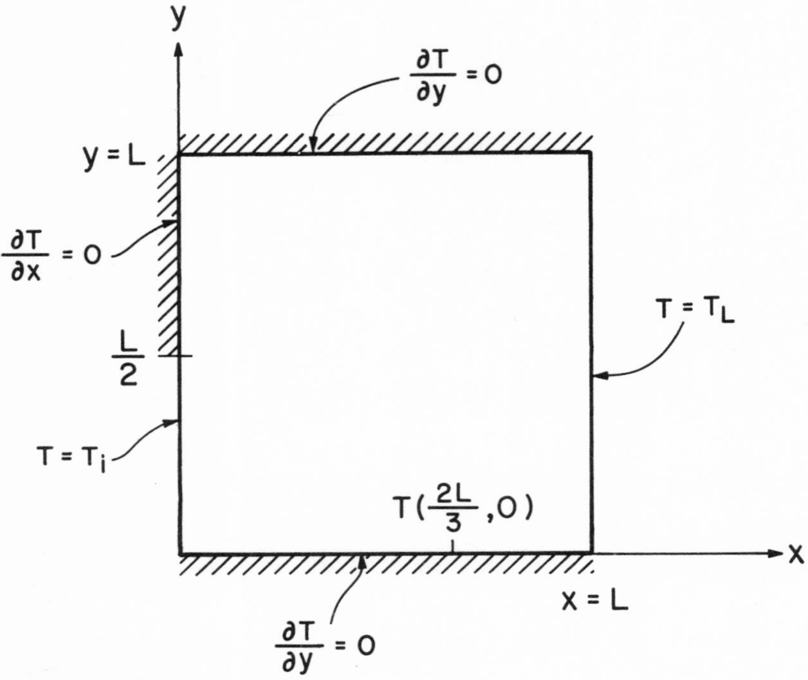

Governing Equation

$$\frac{1}{\alpha}\frac{\partial T}{\partial t} = \frac{\partial^2 T}{\partial x^2} + \frac{\partial^2 T}{\partial x^2}$$

Initial Condition

$$T(x, y, 0) = T_i$$

FIGURE 4.1 EXAMPLE USED IN COMPARISONS.

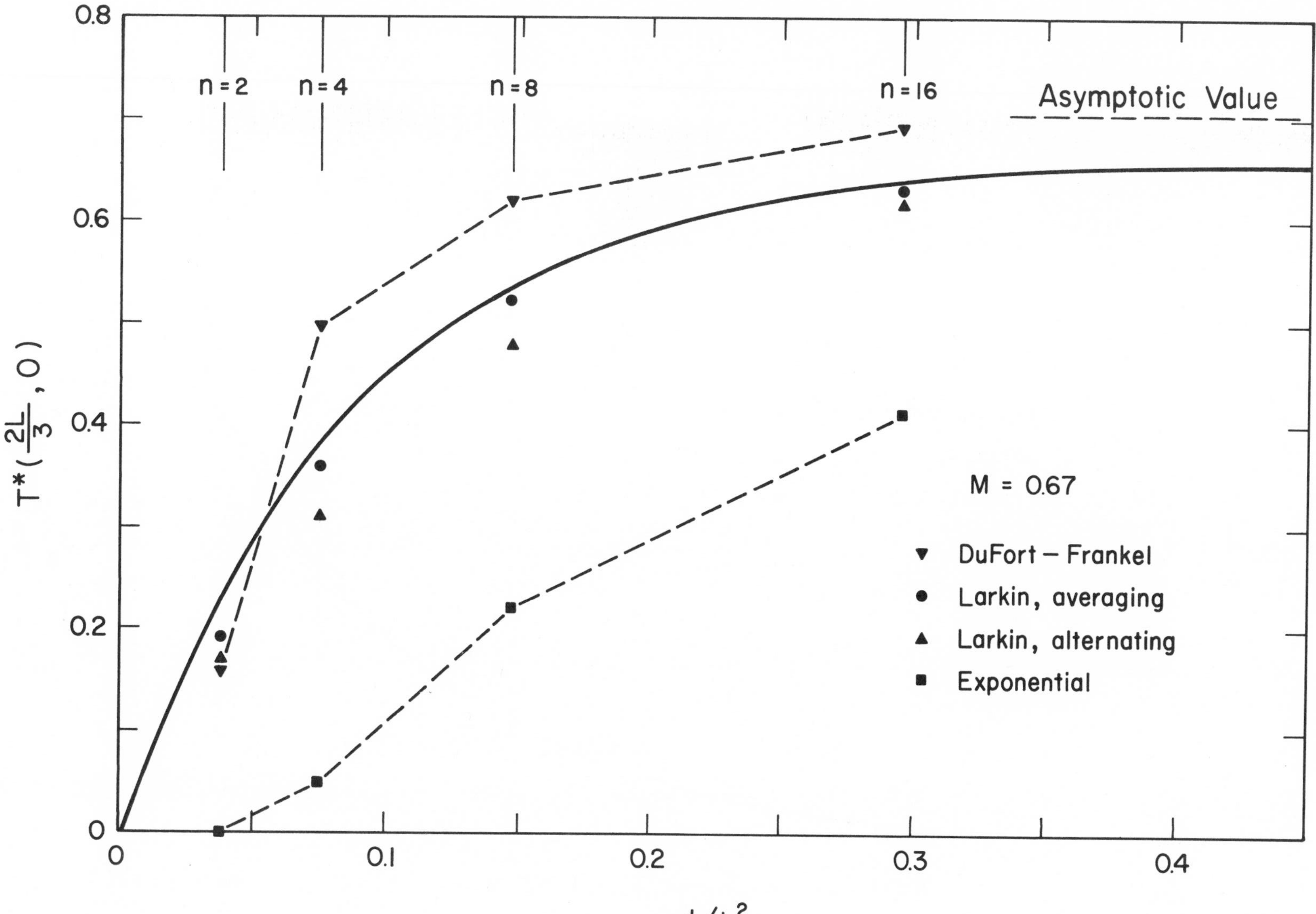

FIGURE 4.2 A COMPARISON OF RESULTS FOR M = 0.67.

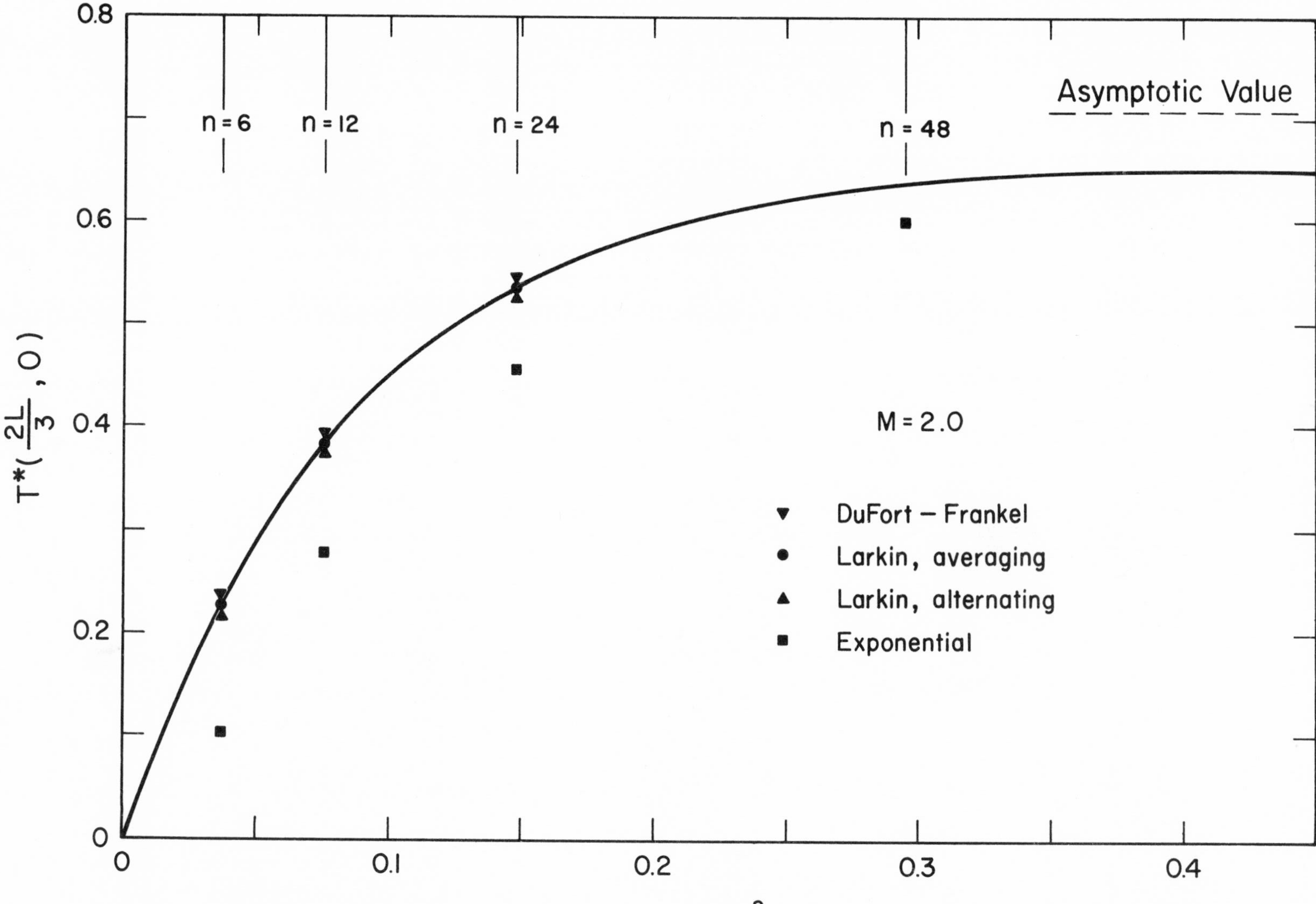

FIGURE 4.3 A COMPARISON OF RESULTS FOR M = 2.0.

NUCLEATE POOL BOILING

J. W. Westwater*

ABSTRACT

Boiling is a process of extreme complexity. Consider, for example, boiling during two-phase flow of a multicomponent fluid in a curved channel using high pressure, high velocity, and high subcooling. At present no one can design a heat exchanger for such a system. In fact, an overdesign by a factor of at least two often is needed to design for the simplest case imaginable: a one-component liquid, at its saturation temperature, in a pool with no forced convection. This article discusses the reason for this unsatisfactory state of affairs for nucleate pool boiling.

INTRODUCTION

Nucleate pool boiling, the simplest kind of boiling, has several distinguishing characteristics, namely: the fluid is a pool of liquid with its bulk at the saturation temperature; no forced convection is employed; the heat source is a hot solid surface; and, for the purposes of this paper, the liquid is considered to be a single pure component. To introduce additional simplifications, let us not be concerned with the peak in the boiling curve. Our problem is to express the functionality between the heat flux and the imposed operating conditions during nucleate boiling. If we cannot master this case, we cannot master the more complicated cases.

The fact that the design of boiling equipment requires empiricism is no cause for special apology. None of the convective heat processes can be handled by straight analytical techniques. Conduction heat transfer and radiation heat transfer are processes which are amenable to mathematics. Even these require empiricism; thermal conductivities and emissivities are coefficients (found experimentally) which must be used in the basic equations.

THE VARIABLES

The number of possible variables is staggering when one considers boiling in the most general sense. Suppose liquid containing some vapor is in relative motion past a hot solid of undefined shape. What will be the rate of heat transfer to the liquid? The shape of the solid is undoubtedly important; it is for all convective processes. The relative velocity between the liquid and the solid and between the vapor and the solid seem important. The

*Professor and Head, Division of Chemical Engineering, University of Illinois, Urbana, Illinois.

fact that the liquid and vapor velocities will be different (the difference is called the slip velocity) is a matter of deep concern. The temperature difference driving force is obviously a prime variable, but even this can be complex. For example, if water at 212°F at atmospheric pressure is in contact with a hot solid at 252°F, the ΔT is of course 40°F. However, the resulting heat flux is not the same as when a ΔT of 40°F is achieved with water at a bulk temperature of 202°F in contact with a solid at 242°F. The latter case is subcooled boiling. In designating the boiling-side ΔT, one must specify how much of this is above the saturation temperature and how much is below.

The type of liquid used is significant. This is but an indirect way of saying that some of the physical properties of the liquid and vapor must be important. In a recent article Hughmark(1)* lists 32 variables that occur to him for nucleate boiling, including for the liquid: the density at the bulk temperature and the superheated temperature, viscosity, surface tension, specific heat, thermal conductivity, and latent heat of vaporization at each temperature; and for the vapor: the density, viscosity, specific heat, and thermal conductivity at each of the two temperatures. Other parameters include the difference in the liquid and vapor density, the wall temperature, saturation temperature, the two reduced temperatures, the vapor pressures at the two temperatures, the two reduced pressures, and also the ΔT and the Δp. Hughmark did not include a number of possible factors such as the contact angle, the coefficient of expansion, the size of the heating surface, gravity, the size of the surface pits, the size of the break-off bubbles, the amount of subcooling, velocity of the liquid, velocity vapor, or the slip velocity. If all of these are significant we are faced with a desperate situation. Dimensional analysis, with mass-length-time-temperature, will give 38 dimensionless groups. Relating these properly would be a monumental, discouraging task.

May thermodynamics be used to relate variables to one another and thus reduce the number of parameters to be studied? One of the vexing features of boiling is the inapplicability of thermodynamics. All transport processes, including heat transfer, are nonequilibrium processes. Thus thermodynamics, the science of equilibrium processes, is of very limited use for these.

Although a liquid may be boiling, we cannot believe that all of the liquid is at the saturation temperature. The liquid close to the heat source is always superheated; the liquid far away may be subcooled. Vapor bubbles in the liquid need not be in equilibrium with the liquid. The vapor temperature can be any value within the maximum and minimum temperature limits of the liquid field. The pressure inside a bubble growing or collapsing in a liquid is not predictable from thermodynamics. This pressure does not have to be the vapor pressure of the liquid nor the ambient pressure. In fact, a bubble could never grow or collapse at finite rates if the bubble pressure were exactly the ambient pressure.

One approach for eliminating spurious variables is laboratory experimentation. Usually the experimenter controls as many variables as possible and then examines the test results to decide the effects of specific variables. This sounds straightforward, but in practice one may not be able to eliminate confounding. For instance, what is the effect of pressure (just pressure, no other variable) on boiling? Tests with water at pressures up to 2000 psi are available. However, the physical properties of saturated water and steam change drastically over the accompanying temperature range. Are the changes in the heat transfer rates caused by the pressure *per se*, or are they caused by the changes in viscosity, density, surface tension, latent heat of vaporization, and so forth?

This issue is not resolved. Some workers believe pressure is a parameter. For example, it is included in Equations (1), (4), (8), and (10) given later. Other persons believe that the only role of pressure lies in its influence on the boiling temperature and physical properties of the fluid. This writer favors the latter view.

To prove experimentally, in an unequivocal manner, whether pressure *per se* is a parameter, one may consider either of two imaginary programs. One would require tests made at two or more pressure levels with a liquid whose physical properties were independent of pressure. No such liquid exists. The alternate scheme is to select two liquids whose physical properties are identical when one liquid is boiling at a specific pressure and the second liquid is boiling at a different pressure. The only difference in the two test conditions would be the pressure. The chances of finding two such liquids are quite remote. No

*Superscript numbers in parentheses refer to References at the end of this paper.

practical, direct experimental method is available for separating the effect of pressure from the effects of physical properties.

The usual way for testing the effect of physical properties is by changing the pressure or by changing liquids. In either case not one but rather a whole set of physical properties are changed simultaneously. Thus the effects are confounded invariably. In spite of the difficulties, tests do indicate that some variables are of prime importance (the surface texture, the latent heat of evaporation, and the acceleration or gravity are examples), while some variables are of much less importance (the vapor viscosity seems to be in this class).

Early experiments were not sensitive enough to detect the effect of weak variables. Some early writers state that during subcooled nucleate boiling, the liquid flow rate, the amount of subcooling, and the geometry of the test surfaces are of no importance. Now that millions of dollars worth of tests are behind us, ample evidence exists to show that all these are real variables, although at a low level. Many nuclear reactors have boiling-water cooling designed to take advantage of the helpful aspects of velocity and subcooling.

Although the effects of geometry are weak, they are real. For example, one set of tests(2) involved the use of a steam-heated bayonet tube immersed in boiling methanol. The apparatus could be rotated by 90 degrees. Tests were made with the same tube, horizontal and vertical, but with no other changes. The resulting boiling curves were slightly different. Tests have been made also(3) with liquid boiling on one face of a heated ribbon. The heater could be pivoted to present the face in an up, down, or vertical position. The results were a function of the geometry. The face-up position gave the greatest heat flux, and the face-down the smallest, both at the same ΔT. Other geometry effects one must include are the equivalent diameter of a flow channel and the length to diameter ratio. Data seem to show that these too are real but weak variables.

EQUATIONS FOR NUCLEATE BOILING

Many persons have responded to the challenge of relating the variables for nucleate boiling heat transfer. Equations were published more than 15 years ago by Jakob and Linke,(4) Cryder and Gilliland,(5) Cryder and Finalborgo,(6) Insinger and Bliss,(7) Bonilla and Perry,(8) Jakob,(9) and others. The present article will be concerned with more recent progress. The equations which follow do not constitute a complete record. They were selected to represent work at scattered locations throughout the world, and are intended to apply to nucleate boiling with no forced convection and no subcooling. For discussion purposes they will be grouped into two classes, depending on whether surface characteristics of the heat source are considered significant.

EXCLUSION OF A SURFACE PARAMETER

One point of view is that the character of the heating surface is of negligible importance (or that there is no way to account for it). Equations (1) through (8) are of this type.

The expression of McNelly(10) contains five dimensionless groups. The exponents and the coefficient were selected to give a good fit with data for a number of liquids. Note that D and μ_L can be cancelled.

$$\frac{hD}{k_L} = 0.225 \left(\frac{Dq}{\lambda\mu_L}\right)^{0.69} \left(\frac{PD}{\sigma}\right)^{0.31} \left(\frac{\rho_L}{\rho_V} - 1\right)^{0.33} \left(\frac{C_L\mu_L}{k_L}\right)^{0.69} . \tag{1}$$

(McNelly, 1963)

Sterman(11) presents an argument that the correct dimensionless groups are the ones of Equation (2).

$$\frac{hD}{k_L} = f\left[\left(\frac{D^3\ \rho_L^2\ (\rho_L - \rho_V)\ g}{\mu_L^2\ \rho_V}\right), \left(\frac{C_L\ \mu_L}{k_L}\right), \left(\frac{qD\ \rho_L\ C_L}{k_L \lambda \rho_V}\right), \left(\frac{\lambda}{C_L\ T_s}\right)\right] . \tag{2}$$

(Sterman, 1953)

Sterman used some of the Cichelli-Bonilla data[12] to discover a functionality satisfying Equation (1) at the peak heat flux. He obtained the same equation as the well-known peak-flux equation of Kutateladze.

The Forster-Zuber equation[13] was developed to describe the peak heat flux and its ΔT. Subsequently, it was extended to include q versus ΔT for the rest of the nucleate boiling region. For some systems the equation gives a good prediction; however, it has not received widespread testing.

$$\left(\frac{C_L \rho_L \sqrt{\pi \alpha_L}\, q}{k_L \lambda \rho_V}\right)\left(\frac{2\sigma}{\Delta p}\right)^{1/2}\left(\frac{\rho_L}{\Delta p}\right)^{1/4} = 0.0015\left[\frac{\rho_L}{\mu_L}\left(\frac{C_L \rho_L \Delta T \sqrt{\pi \alpha_L}}{\lambda \rho_V}\right)^2\right]^{0.62}\left(\frac{C_L \mu_L}{k_L}\right)^{1/3} . \quad (3)$$

(Forster-Zuber)

Gilmour[14] obtained an empirical equation by arranging the suspected variables to give familiar groups.

$$\left(\frac{h}{C_L G}\right)\left(\frac{C_L \mu_L}{k_L}\right)^{0.6}\left(\frac{\rho_L \sigma}{P^2}\right)^{0.425}$$

$$= 0.001\left(\frac{DG}{\mu_L}\right)^{-0.3} . \quad (4)$$

Here G is a carefully defined value, representing the supposed velocity of the liquid as it moves toward the heat transfer surface to replace the escaping vapor. Thus

$$G = \frac{q\rho_L}{\lambda \rho_V} .$$

If we substitute for G, and also use $h = q/\Delta T$, Equation (4a), which relates q to ΔT, is obtained.

$$\left(\frac{\lambda \rho_V}{C_L \rho_L \Delta T}\right)\left(\frac{C_L \mu_L}{k_L}\right)^{0.6}\left(\frac{\rho_L \sigma}{P^2}\right)^{0.425}$$

$$= 0.001\left[\frac{D q \rho_L}{\mu_L \lambda \rho_V}\right]^{-0.3} . \quad (4a)$$

(Gilmour, 1958)

Figure 1 shows Equation (4a) tested with data for 15 liquids from eight papers. No disagreement in the graph is worse than 100 per cent. However, the restrictions stated by Gilmour must be kept in mind. The disagreement increases if data for polished surfaces are included. Also, when discs are used one must include a multiplier with the diameter to obtain the correct significant length, D. Thus, for a 7.875-in. disc, D is selected as 4.0 in.

Three of the groups in Equations (4) and (4a) are dimensionless; the pressure group, however, is not. It can be made dimensionless by including the factor g/g_c so as to obtain

$$\frac{\rho_L \sigma g}{P^2 g_c} .$$

If this is done the coefficient is dimensionless. Otherwise, the coefficient 0.001 is dimensional and is correct only if the units used are the exact ones which were adopted by Gilmour.

Forster and Greif[15] formulated a model and used dimensional analysis. Published data were utilized to fix the constants. Forster and Greif state that their expression is intended for clean surfaces only. They show that Equation (5) is adequate for H_2O from 1 to 50 atm, and for butanol, aniline, and mercury. Hughmark[1] tested the equation with five additional liquids. Some sizable deviations from Equation (5) were discovered: 163 per cent for ethanol at 1 atm, 190 per cent for nitrogen at 1 atm, 210 per cent for propane at 170 psia, and 450 per cent for propane at 475 psia.

$$q = 1.2 \times 10^{-3}\left(\frac{\alpha_L C_L T_s}{J\lambda \rho_V \sigma^{1/2}}\right)\left(\frac{C_L T_s \alpha_L^{1/2}}{J \lambda^2 \rho_V^2}\right)^{1/4}\left(\frac{\rho_L}{\mu_L}\right)^{5/8}\left(\frac{C_L \mu_L}{k_L}\right)^{1/3}(\Delta p)^2 . \quad (5)$$

(Forster-Greif, 1959)

Levy[16] assumed a model whereby bubbles grow in a superheated film of constant temperature. For a given set of operating conditions, the ratio of the final size to initial size for all bubbles is assumed to be constant.

$$q = \frac{k_L\, C_L\, \rho_L^2\, (\Delta T)^3}{B_L\, \sigma\, T_s\, (\rho_L - \rho_V)} \quad . \tag{6}$$

(Levy, 1959)

The dimensionless coefficient, B_L, is not constant. It is a function of the product $\lambda\rho_V$. Levy presents a graph giving value of $1/B_L$ for various values of $\lambda\rho_V$ and includes a recommended line for subsequent users. In several cases the deviation between this line and actual data is as great as 100 per cent. Levy tested his equation with water, benzene, heptane, ethanol, isopropanol, butanol, carbon tetrachloride, and a potassium carbonate solution. Hughmark tested Levy's equation with additional data. He shows that Equation (6) is reliable within 90 per cent for all cases except propane at 475 psi. For this one case the error is 390 per cent.

Chang and Snyder[17] used dimensional analysis to obtain Equation (7). An adjustable coefficient is C, but it is not stated to be a function of the heating surface.

$$h = 0.146 k_L \left\{ 1 + Ch \left[\frac{\Delta p}{C_L\, \mu_L\, k_L\, \sigma} \right]^{1/2} \left[\frac{C_L\, T_s\, (\rho_L - \rho_V)\, \Delta p}{(\lambda\, \rho_V)^2} \right]^{1/5} - \left[\frac{\mu_L^2}{\rho_L^2\, g\, \beta_L\, \Delta T} \right]^{1/2} \right\}^{2/3} \left\{ \frac{g\, \beta_L\, \Delta T\, \rho_L}{\mu_L\, \alpha_L} \right\}^{1/3} \quad . \tag{7}$$

(Chang and Snyder, 1960)

They show that Equation (7) gives a good fit for water at atmospheric pressure. The reliability for other conditions is unknown. At high heat fluxes, Equation (7) reduces to Equation (7a).

$$h = \text{const}\ \frac{k_L\, \Delta p^{1.4} \left[C_L\, T_s\, (\rho_L - \rho_V) \right]^{0.4}}{\sigma\, (\rho_V\, \lambda)^{0.8}} \quad . \tag{7a}$$

Hughmark[1] made no attempt to set up a model for the mechanism of nucleate boiling heat transfer. He started with a list of 32 variables, mentioned earlier, which affect the heat flux, and then eliminated all but eight of these in order to simplify the job. Then an assumption was made that the remaining variables were related by simple power functions, and a digital computer was used to obtain the best exponents to fit the 23 liquids listed in Figure 2. Figure 2 indicates that Hughmark's result for saturated pool boiling, Equation (8), is a reasonable expression, but that deviations of more than 100 per cent are common.

$$q = 2.67 \times 10^{-7} \left[\frac{(\Delta p)^{1.867}\, (\rho_L - \rho_V)_W^{2.27}\, C_{L,W}^{0.945}\, T_s^{1.618}}{\rho_{V,W}^{1.385}\, \lambda_W^{1.15}\, \mu_{L,W}^{1.630}\, (P/P_C)^{0.202}} \right] \quad . \tag{8}$$

(Hughmark, 1961)

The subscript, W, for the variables indicates that they are to be evaluated at the wall surface temperature, T_S, of the heat source.

INCLUSION OF A SURFACE PARAMETER

The point of view that the surface characteristics of the heat source are significant was advanced by Rohsenow and others. Rohsenow's expression is Equation (9). It contains two parameters of interest. The contact angle, β, depends on the liquid and the solid; so does the arbitrary coefficient C_{SF}. The equation has been used widely. Usually the two parameters are combined into one which is determined empirically. The graphs published by Rohsenow[18] show that data may deviate from the equation by as much as 100 per cent in some cases.

$$\left(\frac{h\beta}{k_L}\right)\left(\frac{g_c\,\sigma}{g\,(\rho_L - \rho_V)}\right)^{1/2} = \frac{1}{C_{SF}}\left[\left(\frac{\beta q}{\mu_L\,\lambda}\right)\left(\frac{g_c\,\sigma}{g\,(\rho_L - \rho_V)}\right)^{1/2}\right]^{2/3}\left(\frac{C_L\mu_L}{k_L}\right)^{-0.7} . \qquad (9)$$

(Rohsenow, 1952)

The exponents 2/3 and -0.7 were chosen to give a good fit to data for water, benzene, ethanol, and pentane.

Nishikawa and co-workers have given extensive attention to the development of an equation for nucleate boiling. Their first equation contained the number of bubble-producing sites and also the stirring length for a bubble.[19] Later papers showed how to express these two items in terms of other variables.[20,21,22,23] The final result is Equation (10).

$$\frac{hD}{k_L} = 8.0\left[(F_S)^{1/2}\left(\frac{P}{P_o}\right)\left(\frac{C_L\,\rho_L^2}{M^2\;P\;k_L\,\sigma\,\lambda\,\rho_V}\right)^{1/2} D^{3/2}\;q\right]^{2/3} . \qquad (10)$$

(Nishikawa, <u>et al.</u>, 1956)

The factor F_S is called the foamability. It is meant to be descriptive of the nucleation characteristics of the surface. The value of F_S is 1.0 for clean smooth surfaces; it can be as high as 3.6 for rough surfaces. Apparently it was hoped that F_S would depend on the solid only and not on the liquid, and one paper[20] makes such a claim. However, the published values indicate that the physical properties of the liquid do affect F_S, and thus it is similar to Rohsenow's arbitrary constant. Figure 3 reproduces Nishikawa's correlation with data for water and seven organic liquids. All these data are for systems having assumed values of $F_S = 1.0$. The heat transfer surfaces include wires, tubes, and plates. The pressure covers a wide range; water, for example, is represented from 15 to 2500 psia.

The correlation appears to be very good in Figure 3. In only two cases do the data deviate from the equation by as much as 100 per cent. But appearances may be deceiving when q versus h is graphed. Suppose we choose a ΔT. Then an error in h by a factor of two increases the ordinate by a factor of two. At the same time the abscissa will be increased by a factor of $(2)^{2/3}$, because it contains q and this is proportional to h. If the erroneous point were plotted in the graph it would be in error along both axes, and its final deviation from the graphical line would show the ordinates as being off by the factor 1.59. The real error would be 100 per cent, whereas Figure 3 would give an impression that the error was only 59 per cent. If Figure 3 were replotted to show q versus ΔT, the scatter of the data would increase.

Nishikawa's equation is seriously wrong for hydrogen, by about an order of magnitude.

Equation (10) may be misleading from another standpoint. It contains a significant length, D, which seems to indicate that nucleate boiling is sensitive to the size of the heating surface. But a closer inspection shows that D can be cancelled. In fact, Equation (10) may be reduced to Equation (10a), containing no significant length.

$$q = \text{const}\left(\frac{F_S\;P\;C_L\;\rho_L^2\;k_L^2\;(\Delta T)^3}{\sigma\,\lambda\,\rho_V}\right) . \qquad (10a)$$

(Nishikawa)

Hughmark tested Equation (10) with additional liquids not used by Nishikawa. The equation gave predictions for q which were correct within 100 per cent except for two cases. Propane at 475 psia gave an error of 178 per cent, and mercury at atmospheric pressure gave an error of 13,000 per cent. Hughmark's Equation (8) is off by 13 and 23 per cent for these two cases.

Ruckenstein[24] considers that bubbles form at cavities in the heating surface. These are spaced at random. The final expression is a statement of the necessary dimensionless groups and contains two arbitrary constants which are intended to represent the surface structure. Let L be a significant length representing the mean distance between active sites, and let ℓ_o be a significant length representing the mean diameter of sites.

$$\frac{hD_b}{k_L} = f\left[\left(\frac{q\,\rho_L\,D_b}{\lambda\,\rho_V\,\mu_L}\right), \left(\frac{C_L\mu_L}{k_L}\right), \left(\frac{2\,\sigma\,T_L}{\ell_o\,\lambda\,\rho_V\,\Delta T_C}\right), \left(\frac{\Delta T}{\Delta T_C}\right)\right] \quad . \tag{11}$$

(Ruckenstein, 1959)

It is assumed that the ΔT_C value is determined by the value of L. Ruckenstein gives no comparison with data, and he gives no numerical values for L or ℓ_o. The significant aspect of the paper is the claim that at least two constants are needed to define the surface.

In 1961 Miyauchi and Yagi[25] published a study which was presented before the Society of Chemical Engineers, Japan, in 1951. Artificial nucleation sites were drilled in the heating surface, and the functionality between h and the number of active sites was determined. For a population of active sites greater than 32 per sq cm (206 per sq in.), the boiling curves, h versus ΔT, were "normal." The following correlation was obtained for normal curves, at atmospheric pressure. Two of the groups are dimensional.

$$\frac{h}{k} = C_1 \left(\frac{q\,\rho_L}{\lambda\,\rho_V\,\mu_L}\right)^{0.74} \left(\frac{\rho_V}{\rho_o}\right)^{0.69} \left(\frac{C_L\mu_L}{k_L}\right)^{0.63} \tag{12}$$

(Miyauchi-Yagi, 1961)

The parameter C_1 also is dimensional. It depends on the liquid used and on the smoothness of the heating surface. Organic liquids on the "idealized surfaces" have $C_1 = 1.3$ meter$^{-0.26}$. For other surfaces and for water, C_1 varies from 0.7 to 1.9. Miyauchi and Yagi present a graph to demonstrate the good fit of Equation (12). The data used come from nine papers and include ten liquids. The right side of Equation (12) ranges through four orders of magnitude. The fit is about as good as that of Nishikawa. The reliability at pressures other than atmospheric is unknown.

Chang[26] stated that the phenomenon of nucleation must be accounted for in any expression for nucleate boiling heat transfer. He uses an expression from Volmer as modified by Eyring. The result is Equation (13).

$$q = \frac{C_1\,\rho_L\,C_L\,\Delta T N_A\,(KT_S)^{3/2}}{H\,\sigma^{1/2}} \cdot e^{-nxy^{-m}} \quad , \tag{13}$$

(Chang, vigorous boiling, 1961)

where

$$x = \frac{16\,\pi\,\sigma^3}{3\,K\,T_S\,(\Delta p)^2}$$

$$y = \frac{\rho_L\,C_L\,\Delta T}{\lambda\,\rho_V} \quad .$$

Equation (13) is for vigorous boiling only. The parameters C_1, n, and m are arbitrary constants for nucleation. They depend on the liquid and the heating surface and must be computed from experimental data. The range of values for m is small; organic liquids seem to give m = 1, while water and mercury give m = 2. The values of C_1 and n vary considerably. Figure 4 shows the fit for benzene.

If the boiling is feeble, Chang states that free convection is dominant and Equation (14) should be used (except for liquid metals). The ratio ϵ/α_L is selected so as to match Equation (14) to Equation (13) at an appropriate value of Δ T.

$$q = 0.145 \frac{k_L}{D} \left(1 + \frac{\epsilon}{\alpha_L}\right)^{2/3} \Delta T \, (Gr \cdot Pr)^{1/3} \quad . \tag{14}$$

(Chang, feeble boiling, 1961)

PROBLEM OF THE SURFACE

Which of the equations is correct? Some of the authors (Hughmark is a good example) used brute-force correlation of variables, ignoring all models and theories. The results are sometimes reliable, but deviations in q by factors of two or more do occur with an uncomfortable frequency.

Some of the equations are derived, at least in part, for an assumed model for the mechanism of boiling heat transfer. At present there is serious doubt that any of the models can be correct. A few are admittedly incorrect. Rohsenow's derivation, for example, uses a direct proportionality between q and the number of active sites, whereas recent data[27] show that a better relation employs the square root of the number of active sites, at least for well-polished surfaces.

Table 1 shows an interesting comparison of the equations which are simple power functions. Obviously, the theoretical models used by Forster-Zuber, Forster-Greif, Levy, Rohsenow, Nishikawa, Gilmour, and Chang-Snyder are in substantial disagreement.

Possibly several stumbling blocks are preventing the development of a successful equation for nucleate boiling. The chief obstacle is the problem of the solid surface. Bubbles form only at nucleation sites during nucleate boiling. When the bulk liquid is at its saturation temperature, the rate of heat transfer is equal to the number of active sites multiplied by the rate of phase change per site. The latter value can be computed if one knows the size of the bubbles at break-off and the frequency of bubble generation per site.

Active sites are known to be imperfections in the heating surface,[28] primarily microscopic pits and scratches. The number of these is a function of the surface texture; rough surfaces give higher heat fluxes than smooth ones at the same Δ T.

An alternate route for predicting the rate of heat transfer at a nucleation site is to consider the average rate of growth of bubbles there. This approach has received encouraging success. The basic differential equations to be solved simultaneously consist of a force balance, a heat balance, and (if mixtures are considered) a mass balance. The principal theoretical paper is by Scriven.[29] Experimental verification of the predictions are included in a number of recent papers such as References 30 to 36.

A comparison of theory with experiment for the growth of 86 successive bubbles of pentane at one nucleation site[30] is given in Figure 5. The scatter of the experimental data indicates the natural variability of the phenomenon. The best theoretical prediction here is labelled Plesset-Zweck, and happens to be identical with that of Scriven. Several other modifications of the theory are shown by the lines labelled Forster-Zuber and Griffith. Reference 30 gives details concerning the modifications. The Scriven expression is Equation (15).

$$R = \frac{2 \Delta T}{\lambda \rho_V} \sqrt{\frac{3 k_L \rho_L C_L \theta}{\pi}} \quad . \tag{15}$$

The agreement between theory and experiment is within about 30 per cent for average bubbles, which is better than the agreement between data and empirical Equations (1) to (14).

Now the main problem with the bubble-mechanics approach is not the prediction of the rate of growth. The crucial problem now is to predict exactly how many nucleation sites will actually produce bubbles under a given set of conditions.

At a given Δ T all pits and scratches are not active. For a site to produce bubbles it must contain some trapped gas and it must have a size which is larger than the critical size for that Δ T. The critical size is determined by the surface tension and by the vapor pressure relationship for the liquid in question. A first-order estimate of the critical size is furnished by the Gibbs equation,

TABLE 1

COMPARISON OF NUCLEATE BOILING EQUATIONS

Exponent on:

Ref.	ΔT	Δp	λ	ρ_V	σ	k_L	ρ_L	C_L	μ_L	P	P_C	T_S	g	D	$\rho_L-\rho_V$
1		1.867	-1.15	-1.385				0.945	-1.630	-0.202	0.202	1.618			2.27
10	3.22		-2.22	-1.06	-1	1		2.22		1					1.06
13	0.24	0.75	-0.24	-0.24	-0.5	0.79	0.49	0.45	-0.29						
14	3.33		-2.33	-2.33	-1.417	2	0.917	1.333	-1	2.83				-3.33	
15		2	-1.5	-1.5	-0.5	0.79	0.5	0.46	-0.29			1.25			
16	3				-1	1	2	1				-1			-1
18	3		-2		-0.5	5.1		-2.1	-4.1				0.5		0.5
19-22	3		-1	-1	-1	2	2	1							
25	3.85		-2.85	-0.19		-2.43	2.85	2.43	-0.42						
17	1	1.4	-0.8	-0.8	-1	1		0.4				0.4			0.4

$$\Delta p = \frac{2\sigma}{R_o} ,$$

where R_o is the pit radius, σ is surface tension, and Δp is the vapor pressure difference corresponding to the given ΔT. A great deal of work is under way at present to improve the relationship for nucleation sites in boiling liquids.

Suppose a heat transfer test were made with a surface having many pits, for instance, 1000 per sq in., all containing gas and all exactly the same diameter, 0.001 in. At low superheats all of the heat transfer would be free convection only. But once the specific ΔT for nucleation were exceeded, all the sites would become active, vigorous boiling would occur, and suddenly the heat flux would become very great. The boiling curve, q versus ΔT, would be almost vertical. None of the present boiling equations permit such a curve.

Of course no one has made such an idealized heat transfer surface. But important headway is being made in the deliberate production of nucleation sites. (See References 31, 35, 36, 37, 38, 39, and 40.)

Every solid surface that is manufactured does have imperfections, and no two surfaces are alike in respect to the sizes, numbers, and spacing of these flaws. It is not reasonable to expect two surfaces to give identical boiling curves. Each equation for nucleate boiling employs some functionality between q and ΔT. The Rohsenow equation, for example, gives q proportional to $(\Delta T)^3$; Gilmour uses $(\Delta T)^{3.33}$; Miyauchi and Yagi use $(\Delta T)^{3.85}$; Forster and Greif use $(\Delta P)^2$; and so on. Each of these functionalities can be precise for some one special distribution of active sites. But a distribution which will make one equation correct will invalidate the others.

Is there any way out of this dilemma? An analogous situation arose in the field of radiation heat transfer many years ago. After Boltzmann and Planck had derived the fundamental equations, the matter of theory versus real data was considered. Almost no solids obey the theoretical expressions for radiation. A result was the invention of the terms non-black body, gray body, and emissivity. The emissivity is of particular interest; it is an arbitrary constant and must be found by experimentation. Its value depends on the chemical composition of the solid, the surface roughness, and the wave length of the radiation. Designers use tables of emissivities. The values are typical values only, for it is almost impossible to construct two surfaces which will have identical emissivities, and hence emissivity prevents high accuracy in design.

We may anticipate a similar outcome for boiling heat transfer. Perhaps future handbooks will have tables of "nucleation factors" for various solids. These could be subdivided according to the degree of polish. Two or three numbers may be needed to characterize a surface. For example, a mean pit size, standard deviation of the pit size, and the total number of sites per unit area are three factors that could describe a "normal" pitted surface. Of the empirical nucleate boiling expressions in this paper, two use more than one constant. Ruckenstein uses two constants, but his expression is unfinished. Chang[17] uses three constants. The equations using just one constant cannot be correct.

By adjusting his constants, Chang was able to fit his equations to data for water, ethanol, isopropanol, butanol, acetone, benzene, carbon tetrachloride, three terphenyls, and mercury. A wide variety of pressures was used. The metals were chromium, copper, and platinum, and the shapes included wires, tubes, and plates. Engineers who do curve-fitting know that three arbitrary constants permit a better fit than can be obtained with fewer constants. Therefore, Chang's equation should give a better fit than the others. Most of his graphs show data scatter well below the 100 per cent encountered by others. Typical are Chang's curves for benzene in Figure 4.

CONCLUSION

Studies of nucleate pool boiling still have far to go. Pool boiling design can be carried out with correlations such as those of Gilmour, Hughmark, or others, but a safety factor of 100 per cent or more must be included. Some unreliability is a result of uncertainty as to what the true variables are. But to a much greater degree, uncertainty is caused by a lack of knowledge of how to characterize the heating surface. Strong indications exist that two or possibly three constants are needed for this. The equation of Chang and the proposal of Ruckenstein propose two ways to characterize the surface.

ACKNOWLEDGMENT

The experimental work reported in ten of the references was carried out at the University of Illinois under grants from the National Science Foundation.

NOTATION

B_L = arbitrary constant in Levy's equation

C = arbitrary constant in Chang-Snyder equation

C_L = heat capacity of liquid, Btu/lb-F

C_{SF} = arbitrary constant in Rohsenow's equation

C_1 = arbitrary constant in Chang's equation and in Miyauchi-Yagi's equation

D = significant length for heating surface, ft

D_b = bubble diameter at break-off, ft

F_S = foamability, arbitrary constant in Nishikawa's equation

g = acceleration of gravity, ft/hr^2

g_C = conversion factor, 4.18×10^8 lb mass-ft/lb force-hr^2

G = Gilmour's choice of velocity, $(q\rho_L)/(\lambda\rho_V)$

Gr = Chang's Grashof number for boiling, dimensionless

h = boiling individual heat transfer coefficient, Btu/hr sq ft-F

H = Planck's universal constant, Btu hr

J = conversion factor, 778 ft lb/Btu

k_L = liquid thermal conductivity, Btu/hr-ft-F

K = Boltzmann's universal constant, Btu/molecule-F

L = Ruckenstein's constant for cavity spacing, ft

ℓ_o = Ruckenstein's constant for cavity size, ft

M = constant in Nishikawa's equation, 900 m^{-1}, or 274 ft^{-1}

m = arbitrary constant in Chang's equation

n = arbitrary constant in Chang's equation

N_A = Avagadro's number, molecules/mole

Pr = Prandtl number for liquid, dimensionless

P = ambient pressure on the liquid, lb/sq ft; also Nishikawa's parameter (pressure-dependent), Btu/hr

P_C = thermodynamic critical pressure, lb/sq ft

P_o = Nishikawa's parameter at 1 atmosphere

Δp = vapor pressure difference corresponding to ΔT, lb/sq ft

q = heat flux, Btu/hr-sq ft

R = bubble radius, ft

R_o = pit radius, ft

T_L = temperature of bulk liquid, F

T_S = temperature of heating surface, F

ΔT = temperature difference driving force, F

ΔT_C = value of ΔT at peak heat flux, F

x = defined below Equation (13), dimensionless

X = general expression defined in captions for graphs

y = defined below Equation (13), dimensionless

Y = general expression defined in captions for graphs

Greek Symbols

α_L = thermal diffusivity, $k_L/(\rho_L C_L)$, ft^2/hr

β = contact angle

β_L = liquid coefficient of expansion, 1/F

ϵ = Chang's eddy diffusivity for boiling heat transfer, ft^2/hr

λ = latent heat of vaporization, Btu/lb

μ_L = liquid viscosity, lb/ft-hr

ρ_L = liquid density, lb/cu ft

ρ_o = vapor density at atmospheric pressure, lb/cu ft

ρ_V = vapor density, lb/cu ft

σ = surface tension, lb/ft

θ = time, hr

REFERENCES

1. Hughmark, G. A., "A Statistical Analysis of Nucleate Pool Boiling Data," Preprint No. 41, AIChE National Meeting, Cleveland, Ohio, May, 1961.

2. Nelson, C. D., "Boiling from a Vertical Tube," M.S. thesis, Division of Chemical Engineering, University of Illinois, Urbana, Illinois, 1955.

3. Styrikovich, M. A. and G. M. Polyakov, "The Critical Thermal Load in the Boiling of a Large Volume of Liquids," Izvestiia Akademiia Nauk SSSR, Otdel, Tekh. Nauk (1951), pp. 652-656.

4. Jakob, M. and W. Linke, "Der Wärmeübergang beim Verdampfen von Flüssigkeiten an senkrechten und waagerechten Flächen," Zeitschrift für Physik, 36 (1935), pp. 267-280.

5. Cryder, D. S. and E. R. Gilliland, "Heat Transmission from Metal Surfaces to Boiling Liquids," Industrial and Engineering Chemistry, 24 (1932), pp. 1382-1387.

6. Cryder, D. S. and A. C. Finalborgo, "Heat Transfer from Metal Surfaces to Boiling Liquids," Transactions, AIChE, 33 (1937), pp. 346-391.

7. Insinger, T. H. and H. Bliss, "Transmission of Heat to Boiling Liquids," Transactions, AIChE, 36 (1940), pp. 491-516.

8. Bonilla, C. F. and C. W. Perry, "Heat Transmission to Boiling Binary Liquid Mixtures," Transactions, AIChE, 37 (1941), pp. 685-705.

9. Jakob, M. Heat Transfer. New York: John Wiley & Sons, 1949, p. 645.

10. McNelly, M. J., "A Correlation of the Rates of Heat Transfer to Nucleate Boiling Liquids," Journal of the Imperial College Chemical Engineering Society, 7 (1953), p. 18-34.

11. Sterman, L. S., "On the Theory of Heat Transfer from a Boiling Liquid," Zhurnal Tekhnicheskoi Fiziki, 23 (1953), pp. 341-352.

12. Cichelli, M. T. and C. F. Bonilla, "Heat Transfer to Liquids Boiling Under Pressure," Transactions, AIChE, 41 (1945), pp. 755-787.

13. Forster, H. K. and N. Zuber, "Dynamics of Vapor Bubbles and Boiling Heat Transfer," AIChE Journal, 1 (1955), pp. 531-535.

14. Gilmour, C. H., "Nucleate Boiling -- A Correlation," Chemical Engineering Progress, 54 (1958), pp. 77-79.

15. Forster, H. K. and R. Greif, "Heat Transfer to a Boiling Liquid -- Mechanism and Correlations," Journal of Heat Transfer, C, 81 (1959), pp. 43-53.

16. Levy, S., "Generalized Correlation of Boiling Heat Transfer," Journal of Heat Transfer, C, 81 (1959), pp. 37-42.

17. Chang, Y. P. and N. W. Snyder, "Heat Transfer in Saturated Boiling," Chemical Engineering Progress, Symposium Series, 56 (1960), pp. 25-38.

18. Rohsenow, W. M., "A Méthod of Correlating Heat Transfer Data for Surface Boiling of Liquids," Transactions, ASME, 74 (1952), pp. 969-967.

19. Yamagata, K., F. Hirano, K. Nishikawa, and H. Matsuoka, "Nucleate Boiling of Water on the Horizontal Heating Surface," Memoirs of Faculty of Engineering, Kyushu University, 15 (1955), pp. 97-163.

20. Nishikawa, K., "Studies on Heat Transfer in Nucleate Boiling," Memoirs of Faculty of Engineering, Kyushu University, 16 (1956), pp. 1-28.

21. ————, "Nucleate Boiling Heat Transfer of Water on the Horizontal Roughened Surface," Memoirs of Faculty of Engineering, Kyushu University, 17 (1958), pp. 85-103.

22. ————, "An Experiment of Nucleate Boiling Under Reduced Pressure," Memoirs of Faculty of Engineering, Kyushu University, 19 (1960), pp. 63-71.

23. Nishikawa, K. and K. Yamagata, "On the Correlation of Nucleate Boiling Heat Transfer," International Journal of Heat and Mass Transfer, 1 (1960), pp. 219-235.

24. Ruckenstein, E., "On the Correlation of the Experimental Results in the Case of Nucleate Boiling," Buletinul Institutului Politehnic Bucuresti, 21 (1959), pp. 113-124.

25. Miyauchi, T. and S. Yagi, "Nucleate Boiling Heat Transfer on Horizontal Flat Surfaces," Society of Chemical Engineers Japan, 25 (1961) pp. 18-30.

26. Chang, Y. P., "An Empirical Modification of Nucleation Theory and Its Application to Boiling Heat Transfer," Argonne National Laboratory Report, February, 1961, 34 pages.

27. Gaertner, R. F. and J. W. Westwater, "Population of Active Sites in Nucleate Boiling Heat Transfer," Chemical Engineering Progress, Symposium Series, 56 (1960), pp. 39-48.

28. Clark, H. B., P. H. Strenge, and J. W. Westwater, "Active Sites for Nucleate Boiling," Chemical Engineering Progress, Symposium Series, 55 (1959), pp. 103-110.

29. Scriven, L. E., "On the Dynamics of Phase Growth," Chemical Engineering Science, 10 (1959), p. 1.

30. Strenge, P. H., A. Orell, and J. W. Westwater, "Microscopic Study of Bubble Growth During Nucleate Boiling," AIChE Journal, 7 (1961), p. 578.

31. Benjamin, J. E. and J. W. Westwater, "Bubble Growth in Nucleate Boiling of a Binary Mixture," in International Developments in Heat Transfer, Proceedings, 1961-1962 International Heat Transfer Conference, New York: ASME, 1963, p. 212.

32. Westerheide, D. E. and J. W. Westwater, "Isothermal Growth of Hydrogen Bubbles During Electrolysis," AIChE Journal, 7 (1961), p. 357.

33. Westwater, J. W., "Measurements of Bubble Growth During Mass Transfer," in Cavitation in Real Liquids, R. Davies, ed. Amsterdam: Elsevier Publishing Co., 1964, p. 34.

34. Glas, J. P. and J. W. Westwater, "Measurements of the Growth of Electrolytic Bubbles," International Journal of Heat and Mass Transfer, 7 (1964), p. 1427.

35. Yatabe, J. M. and J. W. Westwater, "Bubble Growth Rates for Ethanol-Water and Ethanol-Propanol Mixtures," Chemical Engineering Progress, Symposium Series, 62 (1966), p. 17.

36. Buehl, W. M. and J. W. Westwater, "Bubble Growth by Dissolution: Influence of Contact Angle," AIChE Journal, 12 (1966), p. 571.

37. Griffith, P. and J. D. Wallis, "The Role of Surface Conditions in Nucleate Boiling," Chemical Engineering Progress, Symposium Series, 56 (1960), p. 49.

38. Young, R. K. and R. L. Hummel, "Improved Nucleate Boiling Heat Transfer," Chemical Engineering Progress, 60 (1964), p. 53.

39. Howell, J. R. and R. Siegel, "Incipience, Growth, and Detachment of Boiling Bubbles in Saturated Water from Artificial Nucleation Sites of Known Geometry and Size," Proceedings, Third International Heat Transfer Conference, AIChE, IV, New York, 1966, p.24.

40. Hatton, A. P. and I. S. Hall, "Photographic Study of Boiling on Prepared Surfaces," Proceedings, Third International Heat Transfer Conference, AIChE, IV, New York, 1966, p. 24.

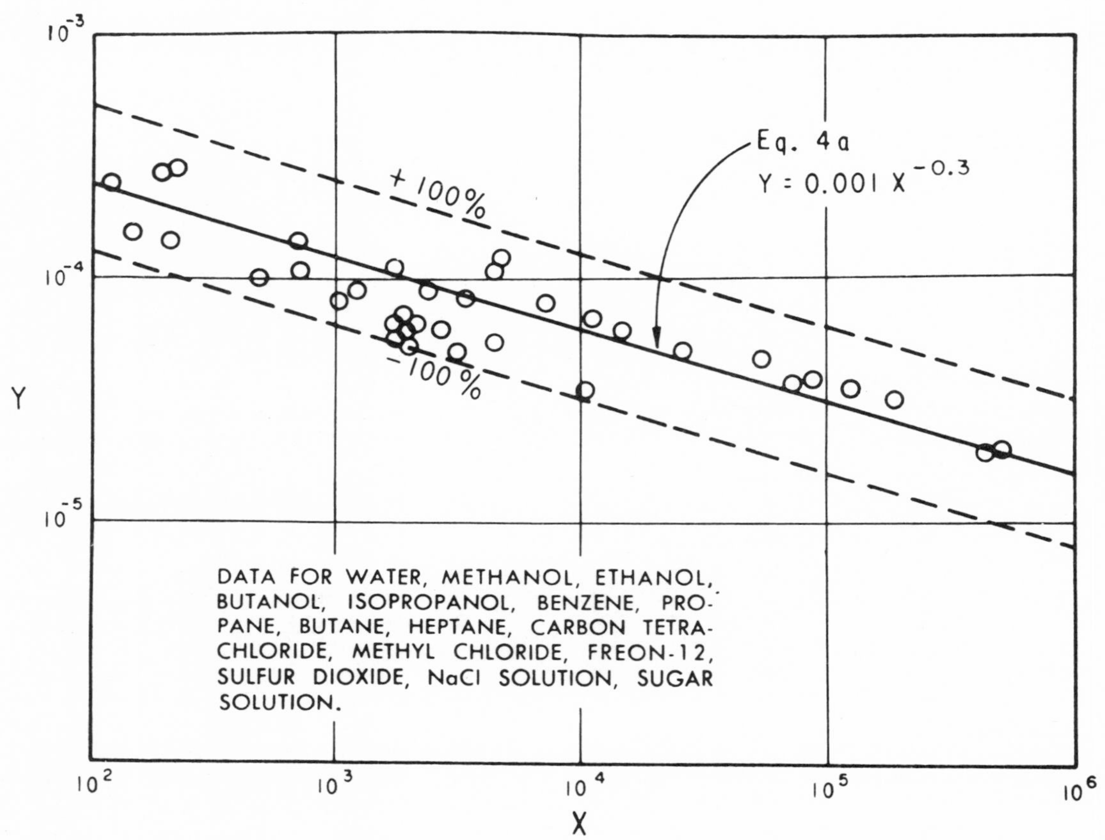

FIGURE 1. CORRELATION OF GILMOUR. HERE X REPRESENTS THE PORTION ENCLOSED IN BRACKETS ON THE RIGHT SIDE OF EQUATION (4a): Y REPRESENTS THE ENTIRE LEFT SIDE.

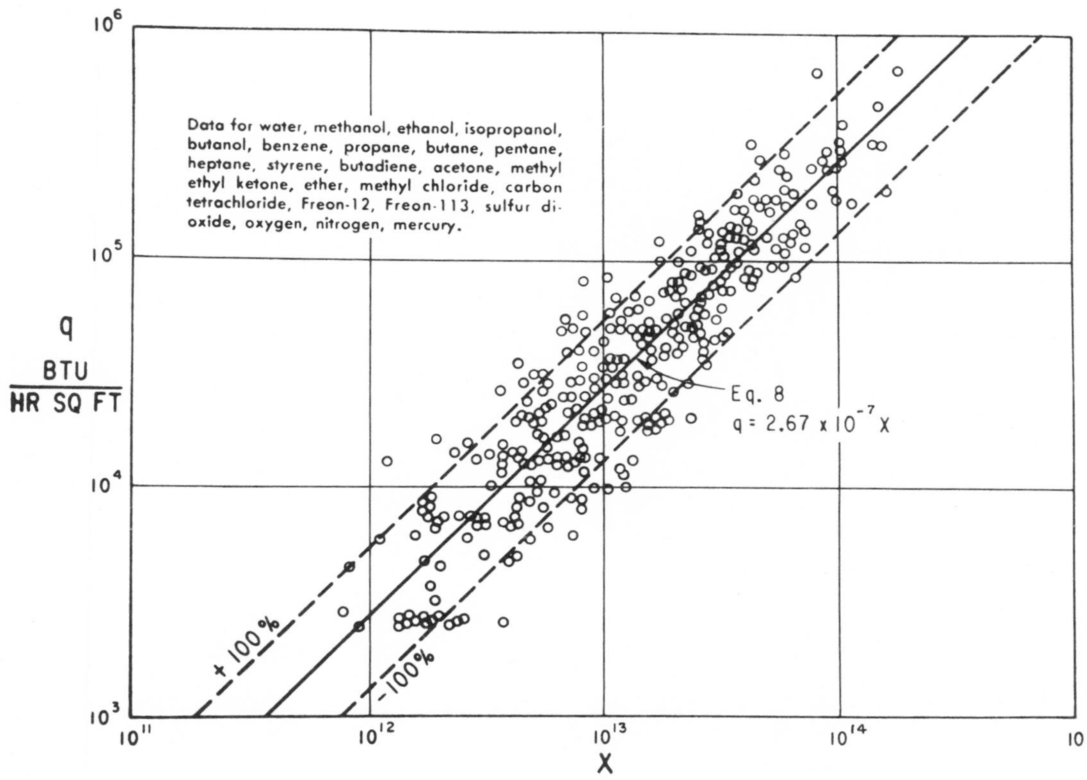

FIGURE 2. CORRELATION OF HUGHMARK. THE SYMBOL X IS THE RIGHT SIDE OF EQUATION (8).

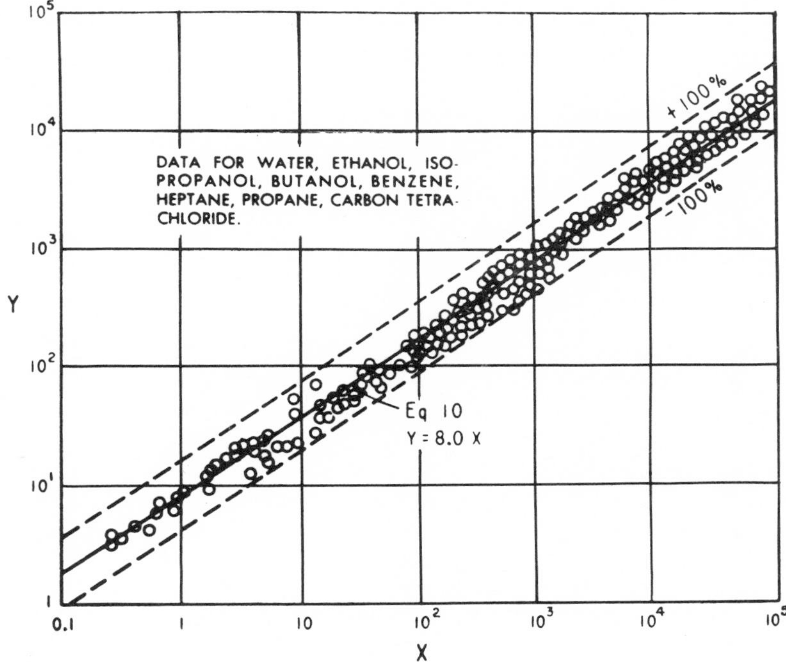

FIGURE 3. CORRELATION OF NISHIKAWA. THE SYMBOL X IS THE PORTION ENCLOSED IN BRACKETS ON THE RIGHT SIDE OF EQUATION (10); Y IS THE LEFT SIDE.

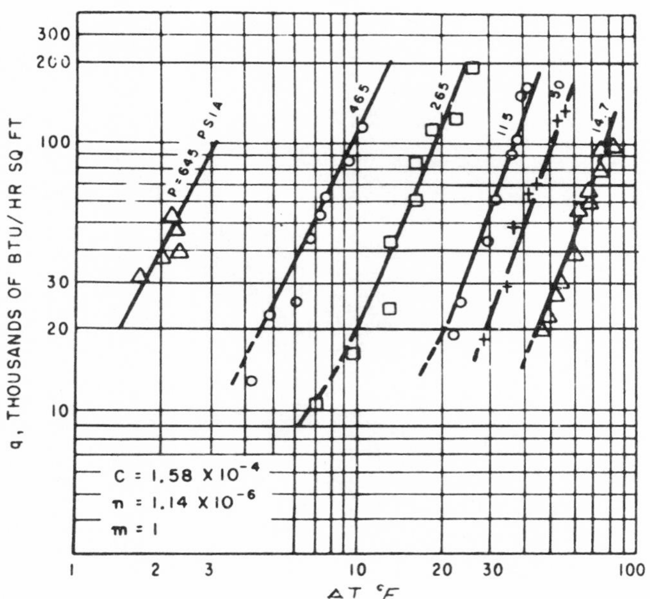

FIGURE 4. CHANG'S CORRELATION, EQUATION (13), USING THE BENZENE DATA OF CICHELLI AND BONILLA. THE DASHED LINES ARE EQUATION (14).

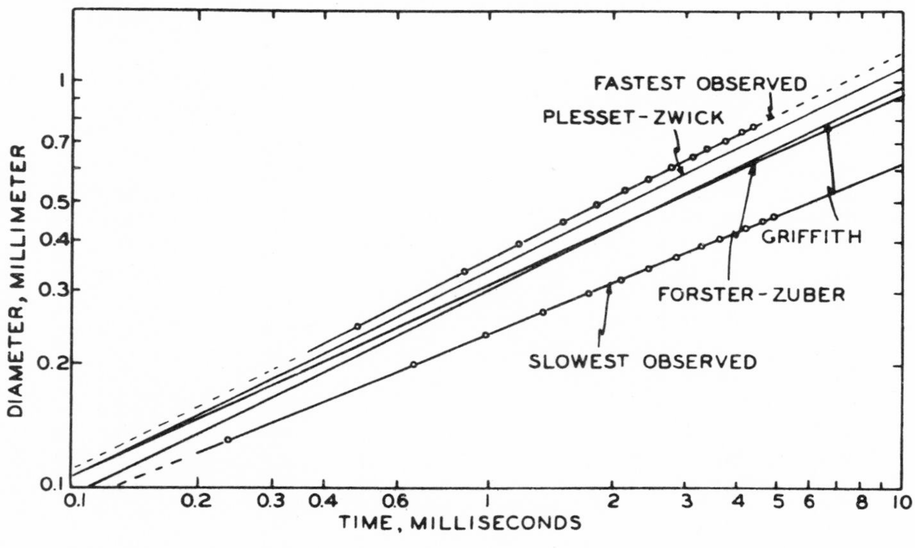

FIGURE 5. ENVELOPE OF GROWTH CURVES FOR 86 CONSECUTIVE BUBBLES OF PENTANE FORMING AT 0.0012 INCH PIT IN ZINC AT A ΔT OF 12°F, ATMOSPHERIC PRESSURE.

DROPWISE CONDENSATION

J. W. Westwater*

ABSTRACT

Dropwise condensation is a nucleation phenomenon. Active sites for drop formation are microscopic pits, scratches, and wettable solid particles. The sizes of the active sites are between upper and lower limits which are determined by the Δ T and the physical properties of the fluid. Drops grow by a dual process. They capture vapor molecules directly, and vapor molecules are also adsorbed onto the bare area between drops and then diffuse over the surface to the drops. The heat transfer coefficient for dropwise condensation is a strong function of the surface texture of the condenser surface; that is, the distribution of nucleation sites. The role of the promoter is to reduce the number of nucleation sites sufficiently to prevent perpetual flooding. The usable population of active sites is roughly a few millions per square inch. The heat flux for dropwise condensation is usually greater than for filmwise condensation, but in principle it can be smaller.

INTRODUCTION

The first scientific study of dropwise condensation was by Ernst Schmidt(1)** in 1930. He presented a photograph and a qualitative discussion of the phenomenon. The appearance of dropwise condensation is contrasted with filmwise condensation clearly in Figure 1 from a more recent study.(2) Here steam is condensing on a vertical copper surface, the right side of which is clean, but the left side of which has a very thin coating of a promoter, cupric oleate. Filmwise condensation occurs on the clean copper; the heat transfer rate there is computed accurately by Nusselt's equation given in numerous texts. Dropwise condensation occurs on the contaminated surface; its heat transfer is more difficult to predict.

A study of Figure 1 gives rise to the pertinent questions for dropwise condensation. How do the drops originate? How do the drops grow? What happens between the drops? How can we predict the heat transfer rate? The object of this writing is to summarize the present knowledge concerning these topics.

BACKGROUND

Figure 2 shows data(3) for dropwise condensation and filmwise condensation for

*Professor and Head, Division of Chemical Engineering, University of Illinois, Urbana, Illinois.

**Superscript numbers in parentheses refer to References at the end of this paper.

the copper surface of Figure 1. In this case dropwise condensation gives higher heat fluxes than the film type by approximately 100 per cent. Some writers have reported far greater differences. The commercial interest in dropwise condensation is a result of the very large heat transfer coefficients which have been reported on occasion.

Textbooks state that only steam and liquid metals can condense in the dropwise manner, but this is incorrect. A number of organic vapors have been condensed successfully in the dropwise manner. A detailed quantitative study of dropwise condensation has been carried out[4] for ethylene glycol on copper with oleic acid as the promoter. Photographs[2] of ethylene glycol and also of glycerol during dropwise condensation on 1/2-in.-diameter, copper cold fingers are shown in Figure 3a. Tests with quinoline are shown also in the illustration. With no promoter, filmwise condensation occurred. With promoter, a strange result occurred, which has been called mixed condensation, for want of a more certain name. Mixed condensation has been reported also for bromoform, aniline, and nitrobenzene, and these are illustrated in Figure 3b. It is neither perfect filmwise condensation nor perfect dropwise condensation.

Certain organic vapors condense in the filmwise manner on copper even in the presence of cupric oleate promoter. Figure 3c shows no difference in the behavior of benzene or ethanol with and without this promoter. Mercury, on the other hand, condenses on copper or on stainless steel in the dropwise fashion, and no promoter is required. This is illustrated in Figure 3c. Erb[5] reports that no promoter is required to achieve dropwise condensation of steam on the noble metals gold, palladium, rhodium, and silver.

Very little is known about the condensation of mixed vapors. Mirkovich and Missen[6] found that some mixtures give an irregular, streaky appearance during condensation. An example is shown in Figure 4, which illustrates pentane plus ethylene dichloride. No promoter was used. Mixed condensation is poorly understood, and much new work is needed on the subject.

For well-developed dropwise condensation, what are the variables? Figure 5 from Tanner, Pope, Potter, and West[7] shows that the promoter can wash off a surface and that different promoters have different rates of wash-off. In Figure 5, stearic acid is removed more rapidly than montanic acid by condensing steam, after a single initial application of each promoter. We conclude that the heat flux during dropwise condensation is a function of the promoter used, how the promoter is applied, and whether or not the promoter is renewed during its usage.

The metal used also has an effect. Figure 6 shows the results obtained[7] with steam on stainless steel and on copper. In both cases two equivalent monolayers of montanic acid had been applied to the metal at the start. Griffith and Lee[8] and Fatica and Katz[9] have also shown that the metal has an important effect.

Pressure is a strong variable during dropwise condensation as shown[4] in Figure 7. The data presented here are the first to demonstrate that pressure must be considered. Clearly, increases in pressure cause increases in the heat flux at any ΔT.

The data in Figure 7 were taken at vapor velocities below 17 ft per sec. Other workers who have used higher velocities have shown that vapor velocity is also a real variable for dropwise condensation.

Intuitively one would expect the presence of a noncondensible gas to be important. Figure 8 shows data from McCormick and Westwater[10] and from Hampson.[11] The two laboratories used different pressures and different geometries, but the trends of the results are unmistakable; an increase in the amount of noncondensible gas (air) present in steam causes a decrease in the heat transfer coefficient. Here h is the coefficient when the air concentration is 40 ppm. The presence of air has been a troublesome factor in a number of studies on dropwise condensation. Unfortunately, some workers have been unaware of its importance, have neither controlled it nor measured it, and have thereby published many useless data. Even today a minority of the research laboratories have provisions for analyzing the air content of their vapors. An argument that the vapor is "free of air" is rather unconvincing in the absence of analysis.

Figure 9 shows the difficulty in comparing data from different experimenters. This graph from Ruckenstein and Metiu[12] is for steam in dropwise condensation. It is unreasonable to expect these data to agree. They were obtained using different metals, promoters, vapor velocities, geometries, pressures, and amounts of inert gas. It is interesting to note that the range of the accommodation coefficient (also called the condensation coefficient, but not to be confused with the engineering term, heat transfer coefficient) for most of this graph is from about 0.02 to 0.15. Thus, in no case did more than 15 per cent of the vapor molecules condense when they struck the condensing surface. In principle, therefore, the maximum heat transfer coefficient seen here, 45,000 Btu per hr sq ft deg F is but 15 per cent of the possible limit of 300,000.

PRIOR THEORIES

Probably the most simplified theory for predicting the heat transfer during dropwise condensation is the one first proposed by Fatica and Katz(9) in 1949. According to their model, the area between drops is completely bare, as shown in Figure 10. This bare area is assumed to have a heat transfer coefficient, h, of infinity. The drops accumulate molecules of vapor and therefore grow. The rate of heat transfer through a drop is found by a pure conduction calculation. Figure 11 shows the plan for the conduction calculation. A network of isotherms and equiflux lines are established. The base of the drop is taken to be at a constant temperature; the liquid-vapor interface is taken to be at a different constant temperature. The difference between these two temperatures is the driving force of the condensation. The rate of conduction between these two surfaces is computed; the rate depends on the contact angle and diameter of the drop. Fatica and Katz assumed that at steady state one may consider all drops as having one representative diameter and one contact angle. Their final equation for the heat flux is not reproduced here. Admittedly it is oversimplified and is rarely used today.

Emmons(13) in 1938 gave a brief discussion of dropwise condensation. Figure 12 is a reproduction of his model. A blanket of subcooled vapor lies on the solid surface between the drops. This cold vapor condenses on the drops and creates violent eddy currents. A turbulent field exists. These interesting ideas were never put into a quantitative form.

Deryagin and Zorin(14) in 1955 published some studies on adsorption that are probably pertinent to dropwise condensation. They used an optical technique to measure the thickness of adsorbed layers on a solid. Elliptically polarized light is directed onto the surface in question. The intensity of the reflected beam is affected by the presence of a transparent layer on the solid. The thickness of the transparent layer can be computed from the optical measurements. One of the results is illustrated in Figure 13. Propyl alcohol vapor in a hydrogen atmosphere was in contact with glass, the entire system being at 25°C. At saturation, a layer of alcohol about 60 Å thick was detected. At alcohol vapor pressure below saturation, a thinner layer was adsorbed. These tests raise the question as to whether an adsorbed layer exists between the drops during dropwise condensation. *A priori*, one cannot be sure, inasmuch as Deryagin and Zorin had no promoter in their system and took measurements during thermodynamic equilibrium. During dropwise condensation a promoter is usually present, and of course, nonequilibrium conditions are maintained.

Jakob(15) described a model for dropwise condensation that is consistent with the idea of a liquid layer resting on the solid between the drops. The sketches given in Figure 14 illustrate the idea which Jakob expressed in words. The upper sketch shows a very thin, submicroscopic, liquid layer on the solid surface. This layer contacts the drop. The combination of drops and a thin layer may be impossible for an isothermal system at equilibrium. However, Sketch 1 does not assume isothermal or equilibrium conditions. Vapor molecules strike the drops, causing them to grow. Vapor molecules also strike the liquid layer, causing it to increase in thickness. Ultimately the layer becomes so thick that it becomes unstable hydrodynamically. The layer starts to wrinkle as indicated in Sketch 2. Finally it ruptures into discrete drops as shown in Sketch 3, and the whole process is repeated. Heat transfer by conduction occurs through the drops and through the liquid film. But most of the drops are much, much bigger than the liquid film thickness, so the conduction through the drops is unimportant compared to the conduction through the liquid film.

Welch and Westwater(3) considered the Jakob model and showed that if it is correct, the liquid film thickness is less than the wavelength of light. These authors studied the dropwise condensation of water on copper by means of high-speed motion picture photography through a microscope. (See Figure 15.(16)) This was taken at 4080 frames per second, and the magnification is such that the height of each frame represents 0.02 in. Note the coalescence of several conspicuous drops in frame 1. The new drop then oscillates, sweeping up the nearby liquid, and leaving a lustrous area around it. The lustre fades in each successive frame. By frame 7, the shiny area has almost disappeared. Welch and Westwater argued that this visual evidence can be interpreted according to the Jakob model. The fading of the lustre can be said to correspond to the growth in thickness of a liquid film. The final disappearance of the lustre can be said to correspond to the final rupture of the film. The Jakob model is attractive because one can hope to formulate it in a quantitative manner. The model has been supported also by Ruckenstein, Baer, and others.

Umur and Griffith[17] published a paper in 1965 which destroys the Jakob model. They condensed steam on gold (with no promoter). The polarized-light, optical system of Deryagin and Zorin was used to examine the surface. The middle graph of Figure 16 shows the record of the resulting optical signal. No increase in intensity above the base-line intensity was observed. Thus no film thicker than one monolayer could have existed. The lower graph of Figure 16 shows what happened when alcohol was injected on the surface and then evaporated off. The alcohol formed a continuous film on the solid. The sinusoidal record corresponds to a decreasing film thickness. The upper graph of Figure 16 shows what happened when steam containing cupric oleate was brought into contact with cold copper. At first a layer built up to a thickness of 25 Å. The authors argue that this is a layer of adsorbed cupric oleate. Then as water drops formed, the intensity of the reflected polarized light began to fall. No further build-up occurred. The final conclusion by these authors is that a layer of promoter may exist between drops, but that no layer (thicker than a monolayer) of condensed vapor exists between the drops.

NUCLEATION PHENOMENON

McCormick and Westwater[10,18] in 1965, and subsequently Erb,[19] and Peterson and Westwater[4] have demonstrated conclusively that dropwise condensation is a nucleation phenomenon. It is demonstrated in a striking fashion by recent motion pictures, taken at 4000 frames per second through a microscope with 20 X enlargement on the negative, which were shown with the presentation of this paper. Repeated formation of one drop after another at nucleation sites clearly may be detected.

Drop population and drop growth rates are determined from motion pictures in a tedious but straightforward manner. It is perhaps surprising that the maximum population of drops on a surface is found to be in the order of millions of drops per square inch.[4] The number of coalescences which a drop undergoes before sliding down a vertical surface is in the order of 400,000.

IDENTIFICATION OF NUCLEATION SITES

Microscopic techniques are required in order to discover the identity of nucleation sites for dropwise condensation. A suitable test system is illustrated in Figure 17. The test surface is a horizontal copper plate cooled on its underside by water and exposed to saturated steam at the upper surface. A microscope views the surface through a window. Auxiliaries permit measurement of the vapor temperature profile near the surface, the metal surface temperature, the heat flux, and the pressure. The temperature difference between the saturated vapor and the metal surface could be varied from 0.02 to 6° C. Both perpendicular (metallographic) illumination (Figure 18) and oblique illumination (from an inclined mirror in Figure 19) were obtained as desired.

Figure 20 shows how natural nucleation sites are detected.[20] The upper scene was photographed with metallographic lighting. It shows some flaws or irregularities in a polished copper surface which had been coated with a monomolecular layer of benzyl mercaptan. A view, by oblique lighting, of the same area serves to distinguish between pits and protuberances and between ridges and scratches. This is the middle photograph. The lower photograph shows the same area after steam was admitted; drops formed on natural, microscopic pits. The area covered in Figure 20 measured 120 x 160 microns.

One way to prove that pits really do serve as nucleation sites is to make the pits artificially. In the upper photograph of Figure 21 is seen a pit 43 microns in diameter made by a spark-erosion technique. After the surface was coated with benzyl mercaptan and steam was admitted to the surface, a drop began to form at the pit as seen in the middle photograph. In the lower photograph, the drop covers the pit completely. Several other nucleation sites (natural ones) are active also in the area of view.

Nucleation can occur at various points in a scratch. In the upper photograph of Figure 22, a scratch, 6 microns wide, made with a scalpel is seen. The middle photograph shows drops of condensate growing along the scratch after steam was admitted. The surface could be dried by heat and retested. The same active sites produced drops again, as shown in the lower photograph.

Certain dust particles can act as nucleation sites. Figure 23 shows three particles of ground glass on a mercaptan-coated copper surface. The view measures 180 x 190 microns. When steam was admitted, drops formed on the glass particles.

On the other hand, certain dust particles cannot act as nucleation sites. Figure 24 shows two particles of carbon on a mercaptan-coated copper surface. When steam was admitted, drops formed at some natural sites but not at the carbon.

Many kinds of particulate matter were studied. The size was found to be important. A particle can be too small to be active. Figure 25 shows the lower limit for water vapor to nucleate on powdered glass and powdered platinum. The triangles show the largest diameters which did not nucleate, whereas the circles show the smallest diameters which did nucleate. The line separates the nucleation region from the inactive region.

An explanation of the minimum size is possible. Equation (1) is the Kelvin equation which relates the subcooling of a stable spherical drop to its radius and surface tension at equilibrium.

$$T_s - T_d = \frac{2\,\sigma\,T_s}{\lambda \rho_L\, r_n} \quad . \tag{1}$$

The drop is imagined to be surrounded by saturated vapor. In reality, during condensation a drop exists on a wall in a non-isothermal field of vapor. If the vapor temperature is imagined to be linear with distance from the wall, Equation (2) can be used. The boundary layer thickness,

$$\frac{T_v - T_d}{T_v - T_w} = 1 - \frac{2r_n}{\delta} \ , \tag{2}$$

is given by δ. If we combine Equations (1) and (2), we obtain Equation (3) which gives two critical radii.

$$r^* = \frac{\delta}{4}\left[(1-\theta_s) \pm \sqrt{(1-\theta_s)^2 - \frac{8A}{\delta \Delta T_w}}\right] ,$$

where (3)

$$\theta_s = \frac{T_v - T_s}{T_v - T_w} \ ,$$

$$A = \frac{2\sigma T_s}{\lambda\,\rho_L} \ ,$$

and

$$\Delta T_w = T_v - T_w \quad .$$

This equation is graphed in Figure 26 for several assumed, reasonable values for the Δ T and the boundary layer thickness. The dashed line labelled "data" shows the upper limit found experimentally. The lower limit is indicated by the triangular and circular data points as before.

If particles are too small, they cannot serve as nucleation sites; the newborn drop cannot increase its size against the crushing force of surface tension. If particles are too large, they cannot serve as nucleation sites; although the drop rests on a cold surface, its top extends beyond the boundary layer out into the warm bulk vapor. In other words, the energy relationships are such that a nucleation site must be larger than a specific minimum and smaller than a specific maximum. For insoluble particles the active ones are in the size range of about 2 to 3 microns. The values are a function of the Δ T, the boundary layer thickness, and the surface tension. Not only is this true for drop nucleation on particles; it can be shown to be true for nucleation on pits.

To determine the relative order of the nucleation abilities of different kinds of particles, two sets of particles of about equal size were placed on a surface and tested simultaneously. The substance which nucleated a drop first was defined as a better nucleation site. About 50 sets of two kinds of materials were investigated. The results are shown in Table 1 from Reference 20. Table 2 shows that the property which fixed the relative order of nucleation sites is the net heat of adsorption (heat of wetting). The correct order cannot be obtained by a consideration of particle density, drop contact angle, or any other parameter yet suggested.

The apparatus depicted in Figures 17, 18, and 19 was used[20] to obtain a large number of counts of the active nucleation site population along with the corresponding Δ T and heat flux. Each such run resulted in one data point in Figure 27. The air concentration was kept below 100 ppm for all runs. This graph is very significant. The heat flux at a particular pressure and Δ T depends on the population of

TABLE 1

RELATIVE ORDER OF NUCLEATION ABILITIES OF PARTICLES

Nucleation ability	Particle	Nucleation frequency	Particle size, microns	Number of particles examined.
Excellent	sodium chloride	always	5-50	20
Very good	glass	always (when size in effective range)	2-40	275
Very good	aluminium oxide	" "	2-50	146
Very good	sodium benzoate	always	10-65	82
Very good	platinum	always (when size in effective range)	1-100	75
Very good	diamond	" "	6-12	223
Very good	boron nitride	" "	7-50	16
Very good	starch	always	9-35	20
Good to very good	surface cavities		1-5	
Good	talc	almost always	2-40	530
Good	bone charcoal	almost always	3-33	56
Fair	polymethyl methacrylate	almost always	7-50	55
Fair	clean copper	sometimes	3-50	876
Poor	anthracene	sometimes	12-120	13
Poor	silver iodide	sometimes	2-24	90
Poor	titanium dioxide	sometimes	10-25	20
Very poor	mercury	very few	4-70	43
Very poor	graphite	very few	7-44	21
Very poor	lampblack	very few	4-44	15
Very poor	Teflon	very few	6-43	45
Rare	copper coated with benzyl mercaptan	hardly ever	5-50	107
None	coconut charcoal	never	11-29	15
None	wood charcoal	never	1-40	443
None	pyrolytic graphite	never	5-50	36
None	stearic acid	never	12-60	25

TABLE 2

ORDER OF NUCLEATION ACTIVITY BY THE HEAT OF ADSORPTION

Material	Net heat of adsorption, $q_n - \lambda$ kcal/mol	n, layers	T, °C	Reference*
Sodium chloride	25.5, soluble	large	25	24
Sodium benzoate	soluble			25
Platinum	5.5	small	20	26
Diamond	large			27
Glass	2.7	5	25	28
Aluminium oxide	1.6	small	25	29
Starch	0.5	small	20	30
Bone charcoal	0.5		25	31, 9
Silver iodide	0.4	50	25	22
Graphite	0.05		25	32, 33
Titanium dioxide	0.03	10	25	34
Mercury	0+	large	20	35
Teflon	0+	large	20	36
Coconut charcoal	-0.85	1	25	37
Graphon (pyrolytic graphite)	-2.0		25	38

*Listed in Reference 20.

active sites; it appears that almost any desired flux less than some maximum can be obtained. If there were only a few sites, say one per sq cm, the flux would be in the order of 1 to 5 Btu per hr sq ft. It depends on the population of active sites; this in turn depends on the surface texture. The Nusselt equation for film condensation on a 3-in. vertical wall at 20 mm pressure predicts a flux of about 14,000 for a Δ T of 3°F.

Most experiments with dropwise condensation are carried out with a series of Δ T values. In Figure 27, the heat flux corresponding to such a test would go from one curve at a low population of active sites, to another curve at a higher population. The heat flux progresses in a stepwise fashion from curve to curve as the Δ T is increased. A consideration of the possibilities shows that the heat transfer coefficient, h, can increase, decrease, or remain constant as Δ T increases. The result is determined by the number and size distribution of the active sites. Figure 28, a graph of h versus heat flux, shows that all these possibilities actually occur.[21] The data of thirteen workers are summarized in this graph. It is fruitless to argue that certain of these lines are correct and others are incorrect. All are possible if we assume a sufficient variation in the texture of the condensing surface.

GROWTH RATES OF DROPS

The correct model for the phenomena occurring during dropwise condensation is believed to be that indicated in Figure 29. Drops exist at active nucleation sites. In the figure, one site is a concial pit and one is a pointed particle on the surface. Vapor molecules strike the surfaces of the drops; some condense and some rebound, but the net result is an increase in the liquid phase. Vapor molecules also strike the solid surface between the drops. Some are adsorbed, and some rebound. The adsorbed molecules release energy, the heat of wetting. The adsorbed molecules diffuse in all directions on the solid surface. When they contact a drop, they are captured and become part of the liquid phase.

Thus the rate of growth of a drop is determined by two mechanisms: (1) the direct capture of vapor molecules by the drop and (2) the adsorption of molecules by the solid followed by surface diffusion and final capture of the molecules by the drop.

The first mechanism is determined by the kinetic theory of a three-dimensional gas. The second mechanism is determined by the kinetic theory of a two-dimensional fluid phase. The coupling of these two mechanisms will be difficult; at present little headway has been made.

PRIOR THEORETICAL EQUATIONS

Scriven[22] has published an excellent analysis of spherical phase growth in an infinite fluid medium. His model assumes the existence of a single drop in a vapor which is initially at a uniform temperature. Equation (4) is the final equation.

$$\frac{D^2}{t} = \frac{8\ k_v}{\rho_L\ \lambda}\ (T_\infty - T_s)\ . \tag{4}$$

It predicts drop growth rates which are too small for the few cases in which it has been tested for dropwise condensation. It does not account for surface diffusion.

McCormick and Baer[23] suggested modifications of the Scriven equation to account for geometrical effects. Equation (5) is their relation.

$$\frac{D^2 - D_o^2}{t - t_o} = \frac{4k_L\ f_1\ (\theta)}{\rho_L\ \lambda\ f_2\ (\theta)}\ (T_1 - T_2)\ . \tag{5}$$

Umur and Griffith[17] developed Equation (6):

$$\frac{dr}{dt} = \frac{k_L\ (T_v - T_s)}{\rho_L\ \lambda\ r} \sum_{n=0}^{\infty} \frac{(2n+1)\ (4n+3)}{1 + [k_L\ (2n+1)]/(h_e r)} \times \left[\int_0^1 P_{2n+1}\ (x)\ dx \right]^2 . \tag{6}$$

Here P_n are Legendre polynomials. The equation is rather complicated to use. It does not predict the diameter proportional to the square root of time.

Drop-growth data[10] are shown in Figure 30. It is clear that D^2 versus t is a good relationship to consider and that the proportionality constant is a function of the Δ T. Unfortunately, the proportionality constant is a function also of the distance to the nearest neighbor. The reason for this is that neighboring drops compete for the same molecules in the vapor phase and in the adsorbed phase. None of the available theoretical equations for drop growth include the important effect of nearby neighbors. Conduction through the metal substrate probably is influenced also by the spacing of drops. Therefore the data of Figure 30, which are for water on benzylmercaptan-promoted copper, may not apply exactly to other metals. It is possible to correlate all the data of Figure 30 into one graphical relationship. This is shown in Figure 31. The parameter, G, is given by Equation (7), where Δ T is ^oF and T_s is ^oR.

$$G = \left(\frac{D^2}{t}\right) \left(\frac{T_s}{0.79 + \Delta\ T}\right) \left(\frac{\rho_L}{\rho_v}\right) \ . \tag{7}$$

Figure 31 includes the effect of pressure, Δ T, and nearest-neighbor distance. It does show D^2 linear with time when the other factors are held constant.

RATE OF HEAT TRANSFER

In principle it is possible to use an iteration technique, taking into account nucleation and drop mechanics, to compute the heat flux. The nucleation and growth of every drop would need to be followed. The population distribution of all the active sites must be known as a function of Δ T. The contact angles of the drops must be known. The idea of the iteration may be understood by considering an instant in time. At that time the location of every drop and its size is known. For each drop, the effective distance to the nearest neighbor is computed. Then the growth rate of each drop is computed by use of Figure 31 (for steam on a horizontal copper plate). The growth rate of a drop is easily converted into an instantaneous heat duty for the drop. Addition of the individual heat duties gives the total heat transfer rate. Interfering (touching) drops coalesce, and new drop sizes and locations are accounted for. New drops nucleate in bare areas that are created by coalescences. The iteration continues.

The iteration process would logically start at time zero, when no liquid was present. The locations of the active sites seem to obey a Poisson distribution. The iteration would proceed until the heat flux did not change with time. The main attraction of this mode of approach is that it employs a physical understanding of the phenomenon. The main drawback is that the job appears immense; the number of drops to be accounted for in a square inch is in the millions; the number of coalescences to be accounted for in a square inch during a few minutes might also be in the millions. A second drawback is that, for an inclined surface, one would need to understand the sliding of drops, their sizes, and the areas swept by a slide.

An alternate method for predicting the heat transfer during dropwise condensation is by equations or graphs based on correlation of data. An attempt is that of Isachenko[24,25] who gives Equations (8) and (9).

$$Nu = (1.6 \times 10^{-4}) Re_x^{-0.84} \pi_k^{1.16} Pr^{0.33} \quad \text{for } 8 \times 10^{-4} < Re_x < 3.3 \times 10^{-3}, \tag{8}$$

and

$$Nu = (2.5 \times 10^{-6}) Re_x^{-1.57} \pi_k^{1.16} Pr^{0.33} \quad \text{for } 3.3 \times 10^{-3} < Re_x < 1.8 \times 10^{-2}. \tag{9}$$

These contain four dimensional groups defined below:

$$Nu = \frac{2\sigma T_v h}{\lambda \rho_L k_L \Delta T_v},$$

$$Pr = \frac{c_L \mu_L}{k_L},$$

$$Re_x = \frac{k_L \Delta T_v}{\mu_L \lambda},$$

and

$$\pi_k = \frac{2 g_c (d\sigma/dT) \sigma T_v}{\lambda \mu_L^2}.$$

Figure 32 shows a test of the equations by means of the results[10] for steam on a vertical copper surface at atmospheric pressure. These data are in fair agreement with the correlation at the higher values of ΔT.

The data for ethylene glycol at a variety of pressures may be well represented by Equation (9). This is demonstrated in Figure 33. An even closer fit may be obtained with Equation (10), which is a slight modification.

$$Nu = 1.46 \times 10^{-6} (Re_x)^{-1.63} \pi_k^{1.16} Pr^{0.5}. \tag{10}$$

Although Equation (10) is good for these data, it may not fit other data. In fact, when a particular correlation does apply, we are lucky. In principle a correlation cannot be correct unless it includes a description of the surface texture. So far no correlation is of this type.

ROLE OF THE PROMOTER

Promoters used to bring about dropwise condensation are all nonwetting materials. For use with water, the promoters typically are waxy or greasy substances. Coatings of Teflon or silicones are good promoters. The behavior of the noble metals is not understood and is excluded from consideration in this section of this writing. The usual practice is to apply a very thin coating of promoter on a metallic, non-noble, wettable surface.

If no promoter is used, filmwise condensation will occur. In such a case, vapor molecules are easily adsorbed every place on the surface, and a liquid film soon builds up to a thickness of tens of thousands of molecules. At the start of filmwise condensation, every place on the surface is a nucleation site. But after steady state operation is established, all the sites are submerged, and none are active.

The role of the promoter is to provide a heterogeneous surface with some areas nonwettable and some areas wettable. The wettable places may be microscopic specks of wettable particles, or they may be tiny pits or scratches. It is very likely that tiny pits and scratches do not become covered with promoter during its application. Thus drops are born on the bits of uncoated metal in the pits or scratches. If this explanation is wrong for pits and scratches, then an alternate description of these sites is that they must have special geometric shapes such that they are excellent mechanical traps for bits of liquid.

For dropwise condensation to be effective, the number of nucleation sites must not be too great. If the population is too great, the surface will be flooded and the mechanism changes to filmwise condensation. If the population is satisfactory, discrete drops will grow on the surface and, after numerous coalescences, finally slide off the surface. The liquid must be removed as sliding drops and not as a flowing sheet. Thus the role of the promoter is to limit the number of nucleation sites, keeping the count below infinity and at a "small" number like a few million per square inch.

We may speculate as to what would happen if one could apply a "perfect" coating of promoter on a metal, resulting in an absolutely smooth homogeneous surface. If no dust particles were present, no active sites would exist. The vapor would tend to remain as vapor, even with a huge amount of subcooling. If the subcooling were made great enough, spontaneous homogeneous nucleation would occur finally in the vapor phase and a mass of liquid would form with great suddenness. Such an experiment seems feasible. The difficulty would consist in getting a truly flat surface and in getting the surface and vapor free of all dust particles.

ACKNOWLEDGMENT

The experimental work reported in seven of the references was carried out at the University of Illinois under grants from the National Science Foundation.

NOTATION

c_L = heat capacity of liquid

D = drop diameter

D_o = drop diameter at zero time

$f_1(\theta)$ = constant in Equation (5), defined by Reference 9, dimensionless

$f_2(\theta)$ = $(2-3\cos\theta + \cos^3\theta)/\sin^3\theta$, function of contact angle

g_c = conversion factor

G = defined by Equation (7)

h = individual heat transfer coefficient

h_A = individual heat transfer coefficient with air present

k_L, k_v = thermal conductivity of liquid, vapor

$\bar{L}$ = average distance to nearest neighbor

N = number of active sites per unit area

Nu = Nusselt number, defined under Equation (9)

P = pressure

Pr = Prandtl number, defined under Equation (9)

Q = heat flux

r = radius of drop

r^* = critical radius of drop

r_n = radius of a nucleus

Re_x = dimensionless group defined under Equation (9)

t = time

t_o = zero time

T = temperature

T_d = temperature of drop at distance, d, from cold surface

T_s = saturation temperature of vapor

T_v = bulk vapor temperature

T_w = wall temperature

T_1 = constant temperature at liquid-vapor interface

T_2 = constant temperature at base of drop

ΔT = temperature difference driving force

ΔT_v = $T_v - T_w$

δ = thermal boundary layer thickness

θ = contact angle, in Equation (5)

λ = latent heat of vaporization

μ_L = viscosity of liquid

π_k = dimensionless group, defined under Equation (9)

ρ_L = density of liquid

σ = liquid-vapor surface tension

REFERENCES

1. Schmidt, E., W. Schurig, and W. Sellschopp, "Experiments About the Condensation of Water Vapor in Film and Drop Form," Technische Mechanik und Thermodynamik, 1 (1930), p. 53.

2. Welch, J. F., "Microscopic Study of Dropwise Condensation," Ph.D. thesis, Division of Chemical Engineering, University of Illinois, Urbana, Illinois, 1960.

3. Welch, J. F. and J. W. Westwater, "Microscopic Study of Dropwise Condensation," in International Developments in Heat Transfer, Proceedings, 1961-1962 International Heat Transfer Conference, ASME, Boulder, Colorado, 1963, p. 302.

4. Peterson, A. C. and J. W. Westwater, "Dropwise Condensation of Ethylene Glycol," Chemical Engineering Progress Symposium Series, 62 (1966), p. 135.

5. Erb, R. A. and E. Thelen, "Promoting Permanent Dropwise Condensation," Industrial and Engineering Chemistry, 57 (1965), p. 49.

6. Mirkovich, V. V. and R. W. Missen, "Non-Filmwise Condensation of Binary Vapors of Miscible Liquids," Canadian Journal of Chemical Engineering, 39 (1961), p. 86.

7. Tanner, D. W., D. Pope, C. J. Potter, and D. West, "Heat Transfer in Dropwise Condensation, Part II," International Journal of Heat and Mass Transfer, 8 (1965), p. 427.

8. Griffith, P. and M. S. Lee, "The Effect of Surface Thermal Properties and Finish on Dropwise Condensation," submitted to International Journal of Heat and Mass Transfer.

9. Fatica, N. and D. L. Katz, "Dropwise Condensation," Chemical Engineering Progress, 45 (1949), p. 661.

10. McCormick, J. L. and J. W. Westwater, "Drop Dynamics and Heat Transfer During Dropwise Condensation of Water Vapor on a Horizontal Surface," Chemical Engineering Progress Symposium Series, 62 (1966), p. 120.

11. Hampson, H., "The Condensation of Steam on a Tube with Filmwise or Dropwise Condensation and in the Presence of a Non-Condensible Gas," International Developments in Heat Transfer, Proceedings, 1961-1962 International Heat Transfer Conference, ASME, 1963, p. 310.

12. Ruckenstein, E. and H. Metiu, "On Dropwise Condensation on a Solid Surface," Chemical Engineering Science, 20(1965), p. 173.

13. Emmons, H., "The Mechanism of Drop Condensation," Transactions, American Institute of Chemical Engineers, 35 (1939), p. 109.

14. Deryagin, B. V. and Z. M. Zorin, "Investigations of the Surface Condensation and Adsorption of Vapors Near Saturation by the Optical Micropolarization Method," Zhurnal Fizicheskoi Khimii SSSR, 29 (1955) p. 1755.

15. Jakob, M., "Heat Transfer in Evaporation and Condensation," Mechanical Engineering, 58 (1936), p. 738.

16. Westwater, J. W. and J. F. Welch, Microscopic Study of Dropwise Condensation, motion picture, University of Illinois, Urbana, Illinois, 1960.

17. Umur, A. and P. Griffith, "Mechanism of Dropwise Condensation," Journal of Heat Transfer, 87 (1965), p. 275.

18. McCormick, J. L., "Drop Condensation on a Horizontal Plate," Ph.D. thesis, Division of Chemical Engineering, University of Illinois, Urbana, Illinois, 1965.

19. Erb, A., "Heterogeneous Nucleation on Single-Crystal Silver and Gold Substrates in Cyclic Condensation of Water Vapor," Ph.D. thesis, Department of Chemistry, Temple University, Philadelphia, Pennsylvania, 1965.

20. McCormick, J. L. and J. W. Westwater, "Nucleation Sites for Dropwise Condensation," Chemical Engineering Science, 20 (1965), p. 1021.

21. Tanner, D. W., C. J. Potter, D. Pope, and D. West, "Heat Transfer in Dropwise Condensation, Part I," International Journal of Heat and Mass Transfer, 8 (1965), p. 419.

22. Scriven, L. E., "On the Dynamics of Phase Growth," Chemical Engineering Science, 10 (1959), p. 1; 17 (1962), p. 55.

23. McCormick, J. L. and E. Baer, "On the Mechanism of Heat Transfer in Dropwise Condensation," Journal of Colloid Science, 18 (1963), p. 208.

24. Isachenko, V. P., "Mechanism and Criterial Equations of Heat Transfer in the Condensation of Steam in Drops," Teploenergetika, 9 (1962), p. 81.

25. ———, "Heat Transfer in the Condensation of Water Vapor in the Form of Drops," Teploenergetika, 12 (1962), p. 54.

FIGURE 1. VISUAL DIFFERENCE BETWEEN DROPWISE CONDENSATION (LEFT SIDE) AND FILMWISE CONDENSATION (RIGHT). A THERMOCUPLE PROBE IS VISIBLE ALSO. (WELCH)

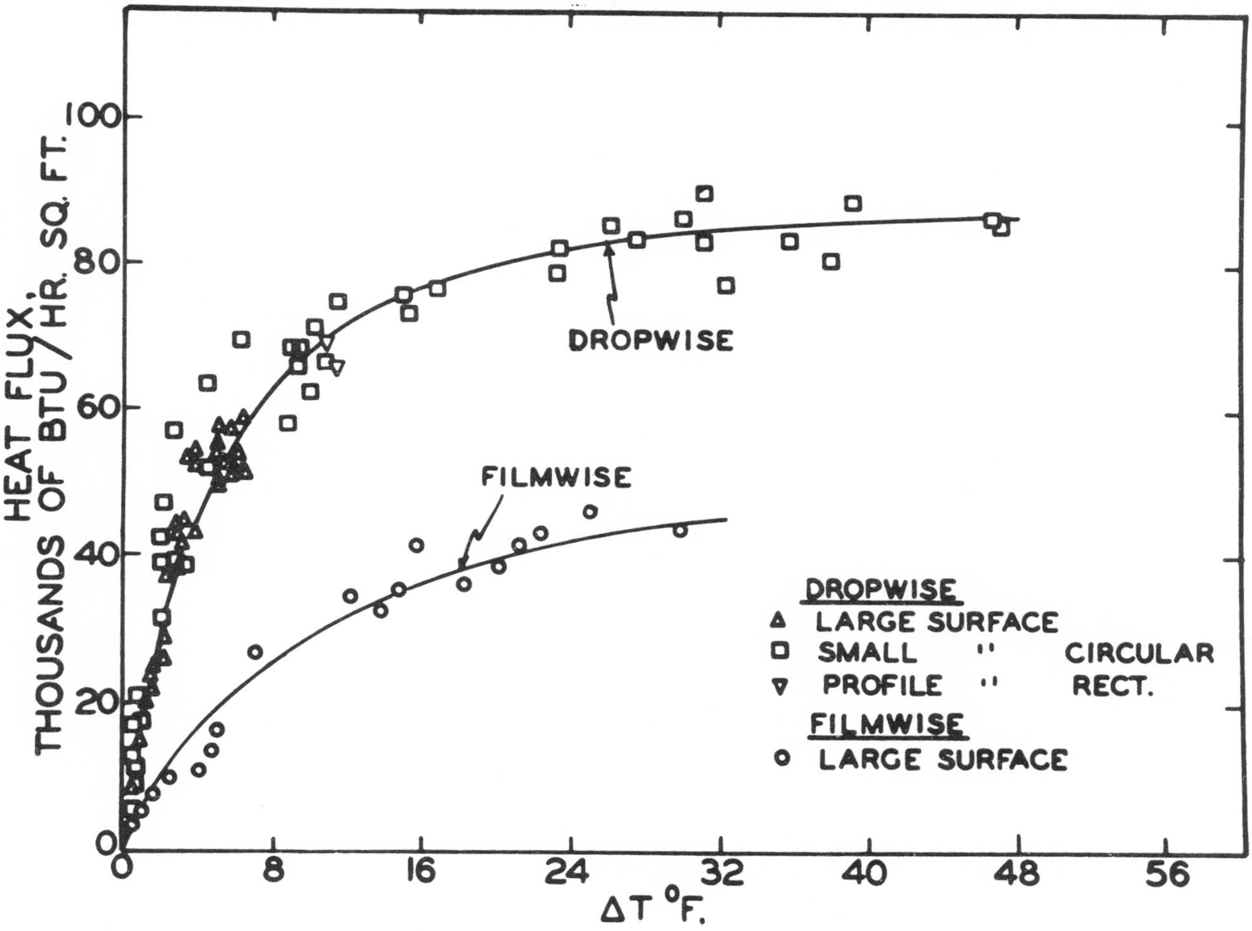

HEAT FLUX, THOUSANDS OF BTU/HR. SQ. FT.
0
20
40
60
80
100
0
8
16
24
32
40
48
56
ΔT °F.
DROPWISE
FILMWISE
DROPWISE
△ LARGE SURFACE
□ SMALL " CIRCULAR
▽ PROFILE " RECT.
FILMWISE
○ LARGE SURFACE

FIGURE 2. COMPARISON OF DROPWISE AND FILMWISE CONDENSATION OF STEAM ON VERTICAL COPPER SURFACES AT ATMOSPHERIC PRESSURE. THE PROMOTER IS CUPRIC OLEATE. (WELCH AND WESTWATER)

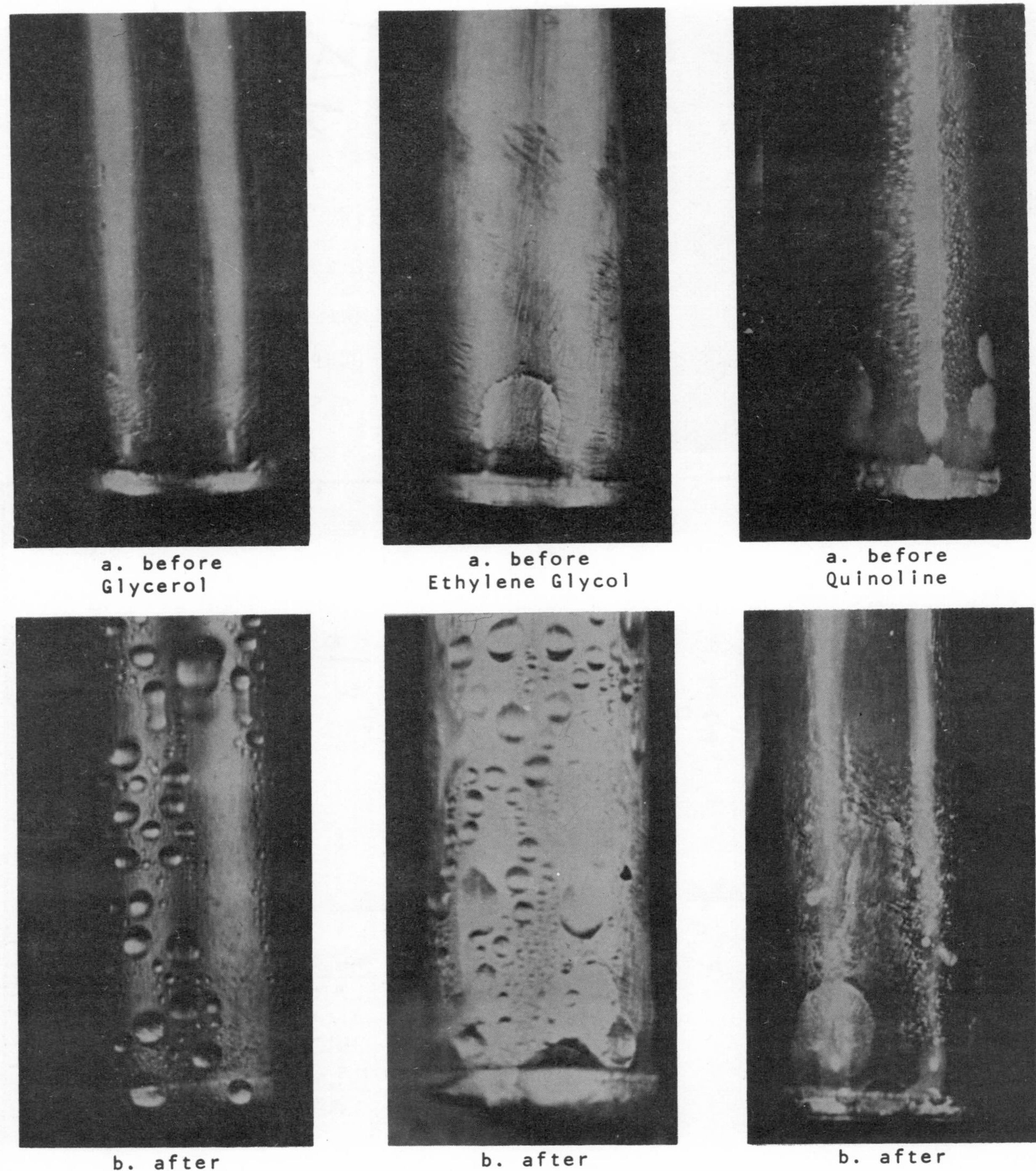

FIGURE 3a. DROPWISE CONDENSATION OF GLYCEROL (LOWER LEFT) AND ETHYLENE GLYCOL (LOWER CENTER), AND STREAKY CONDENSATION OF QUINOLINE (LOWER RIGHT) ON COPPER WITH CUPRIC OLEATE PROMOTER. UPPER THREE PHOTOS SHOW FILMWISE CONDENSATION BEFORE ADDITION OF PROMOTER. (WELCH)

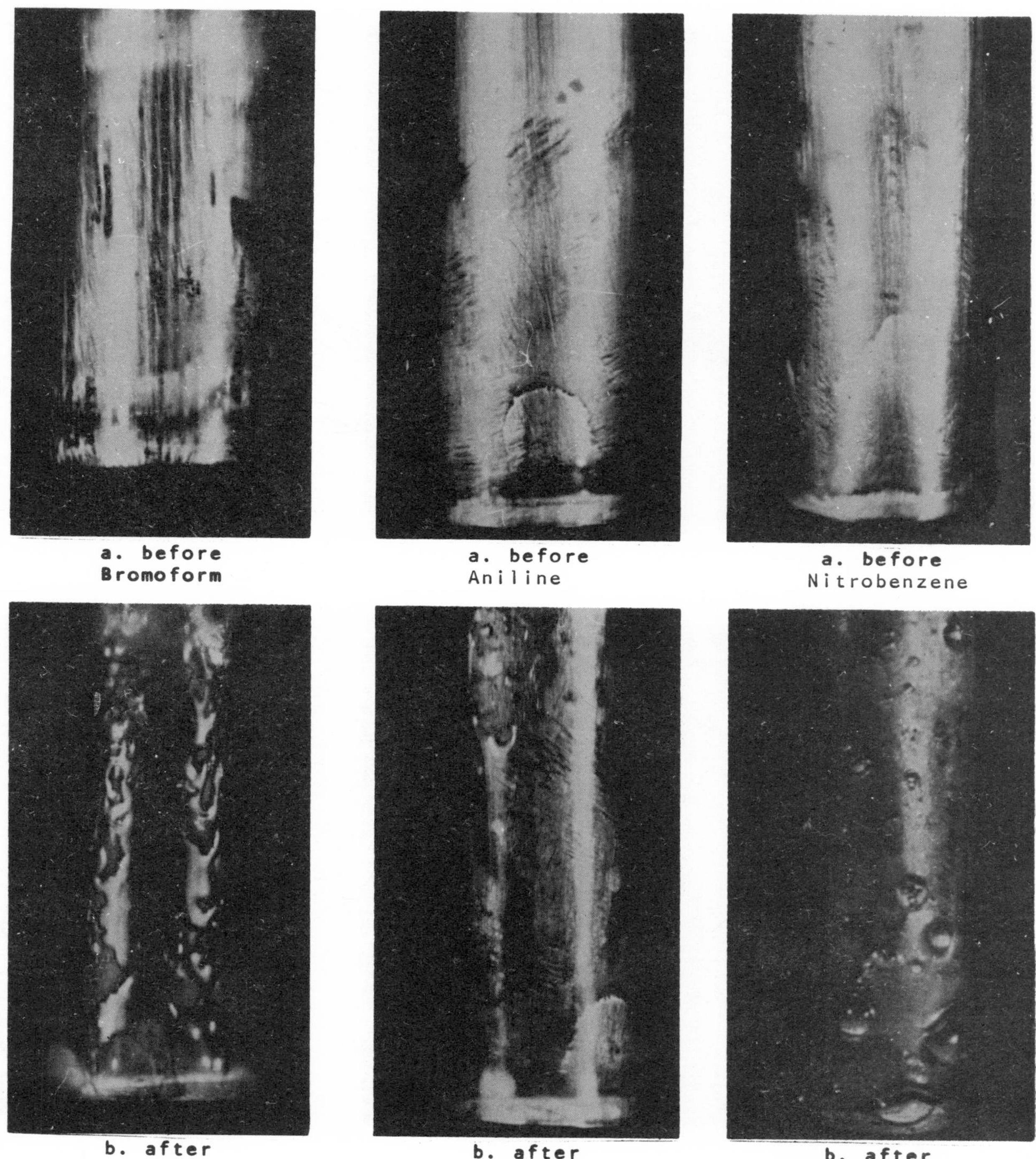

FIGURE 3b. MIXED CONDENSATION IS SEEN IN THE LOWER PHOTOS FOR BROMOFORM, ANILINE, AND NITROBENZENE AFTER ADDITION OF PROMOTER. (WELCH)

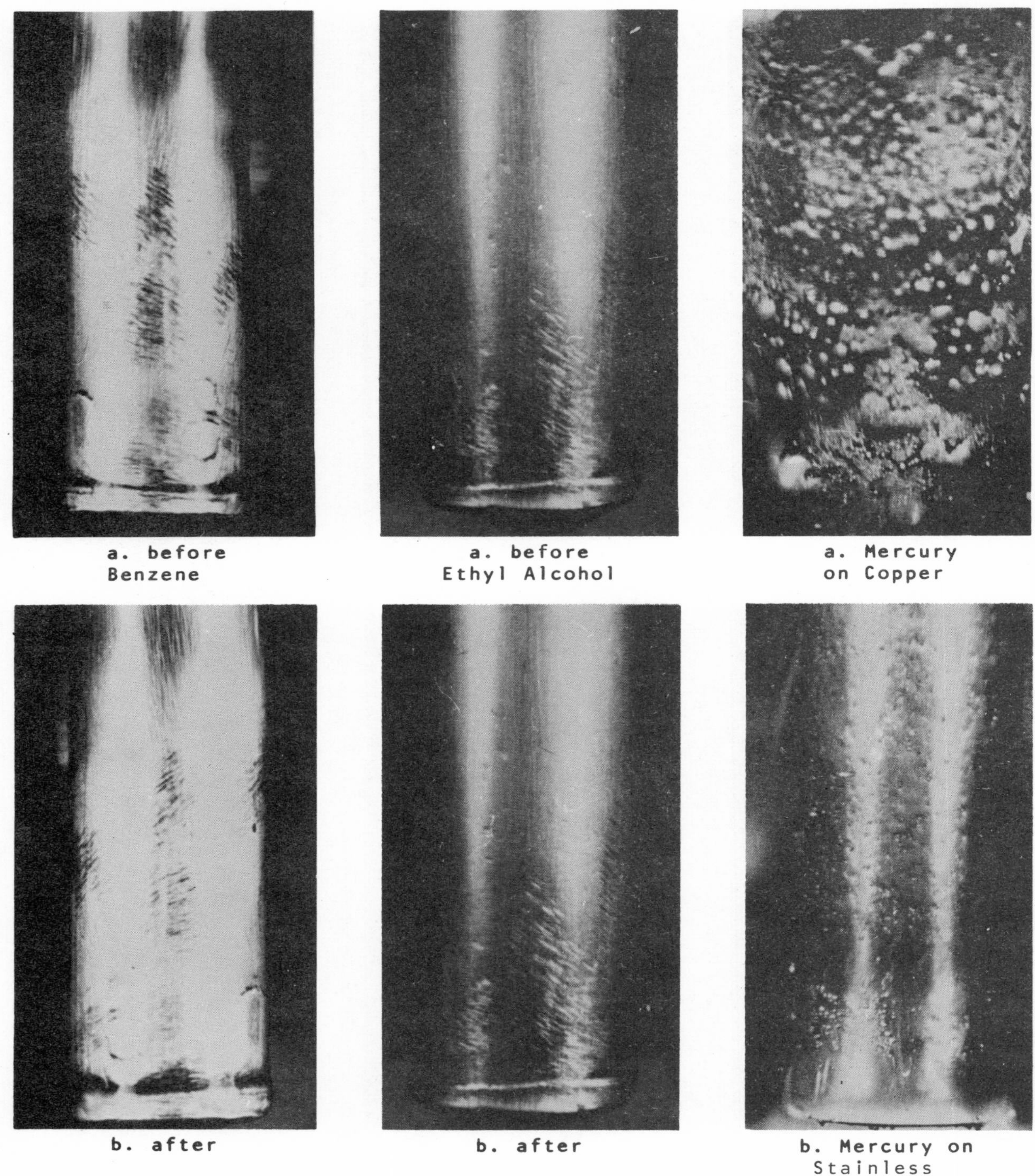

FIGURE 3c. BENZENE (LEFT) AND ETHYL ALCOHOL (CENTER) STILL GIVE FILMWISE CONDENSATION EVEN AFTER ADDITION OF PROMOTER. MERCURY REQUIRES NO PROMOTER FOR DROPWISE CONDENSATION ON COPPER OR STAINLESS STEEL. (WELCH)

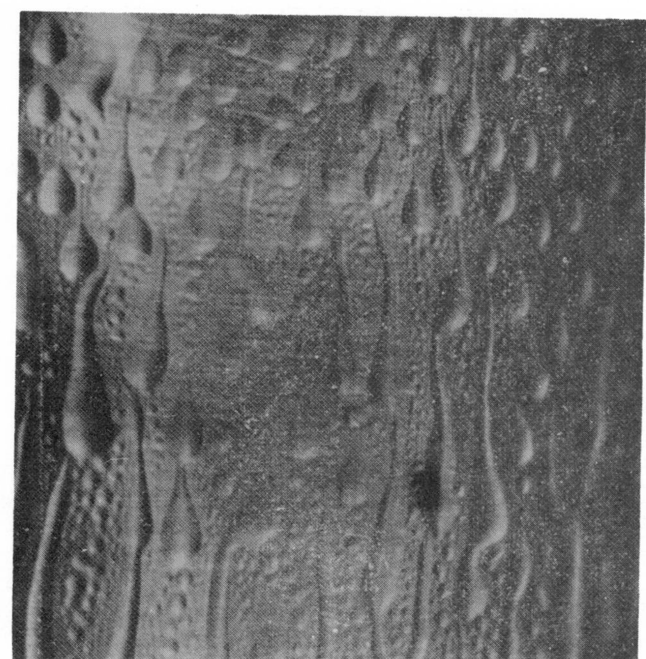

FIGURE 4. NON-FILMWISE, NON-DROPWISE CONDENSATION OF PENTANE-ETHYLENE DICHLORIDE MIXTURE. (MIRKOVICH AND MISSEN)

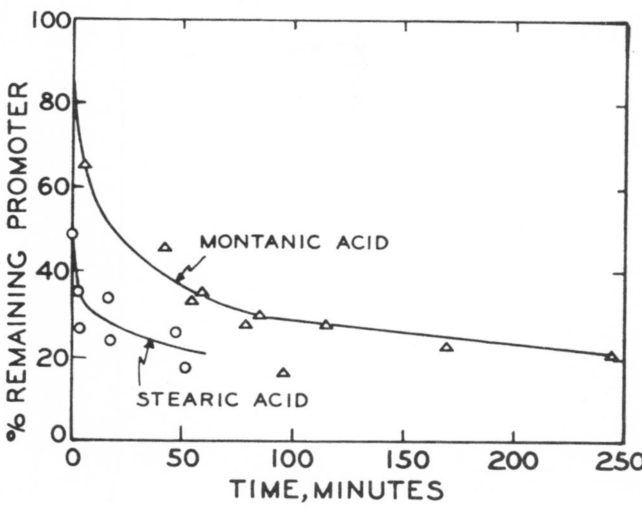

FIGURE 5. COMPARISON OF WASH-OFF OF MONTANIC-ACID PROMOTER AND STEARIC ACID FROM COPPER. (TANNER, POPE, POTTER, AND WEST)

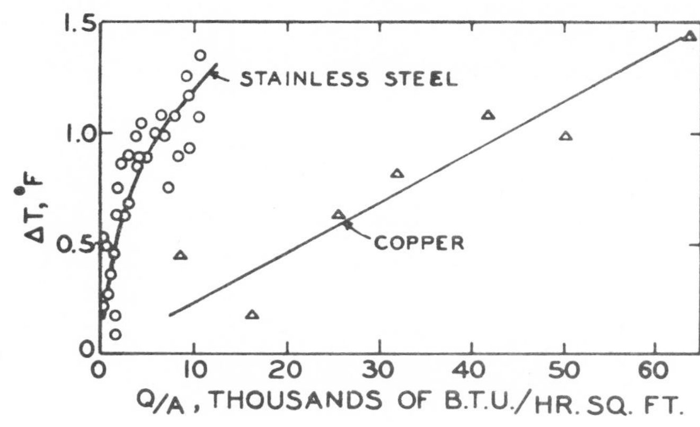

FIGURE 6. COMPARISON OF DIFFERENT METAL SUBSTRATES WITH SAME PROMOTER. (TANNER, POPE, POTTER, AND WEST)

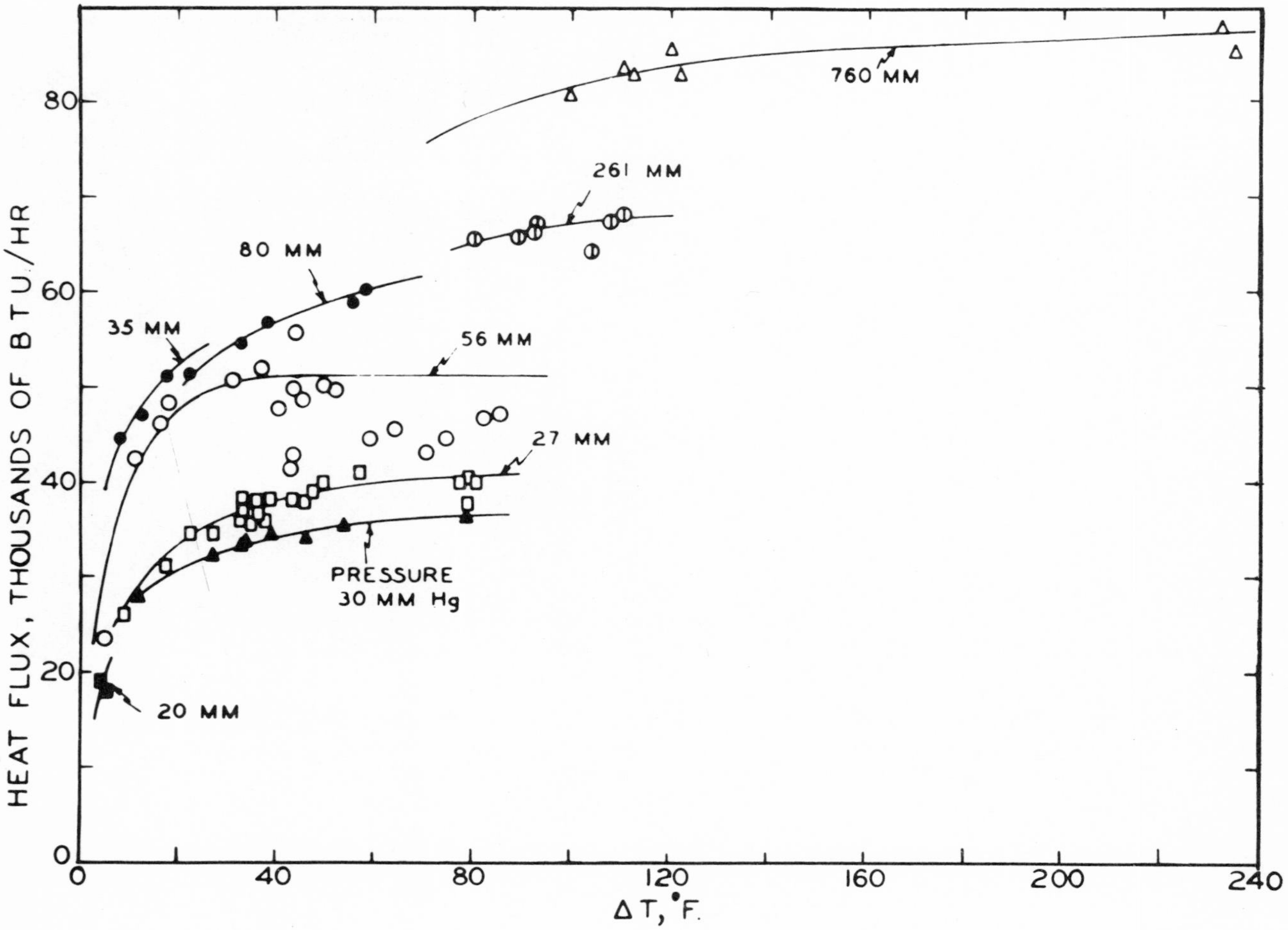

HEAT FLUX, THOUSANDS OF B T U /HR

Δ T, °F.

760 MM
261 MM
80 MM
35 MM
56 MM
27 MM
PRESSURE 30 MM Hg
20 MM

FIGURE 7. EFFECT OF PRESSURE ON HEAT FLUX DURING DROPWISE CONDENSATION OF ETHYLENE GLYCOL ON COPPER WITH OLEIC ACID PROMOTER. (PETERSON AND WESTWATER)

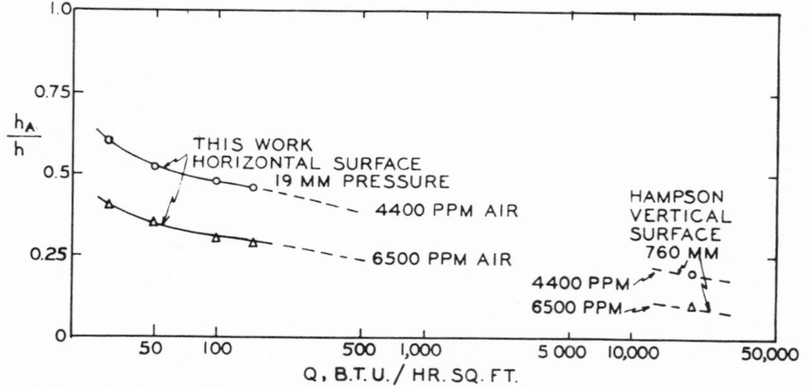

FIGURE 8. EFFECT OF AIR CONCENTRATION ON DROPWISE CONDENSATION OF STEAM. (McCORMICK AND WESTWATER)

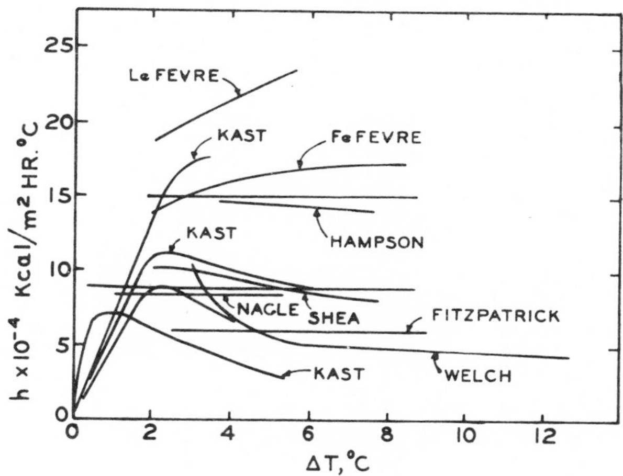

FIGURE 9. SUMMARY OF DATA FOR VARIOUS WORKERS. (RUCKENSTEIN AND METIU)

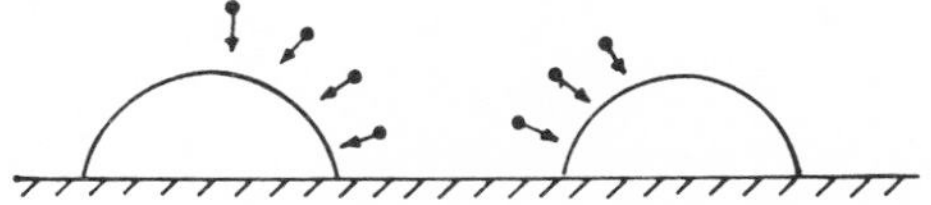

FIGURE 10. SKETCH OF FATICA-KATZ MODEL.

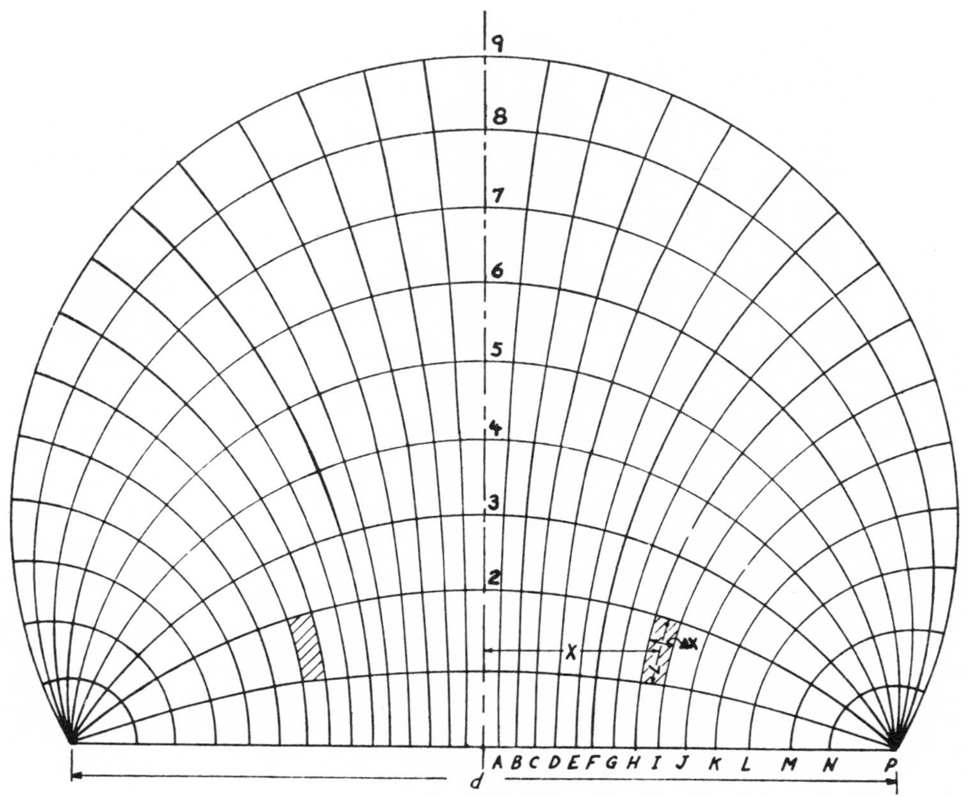

FIGURE 11. ISOTHERMS AND EQUIFLUX LINES FOR CONDUCTION CALCULATION. (FATICA AND KATZ)

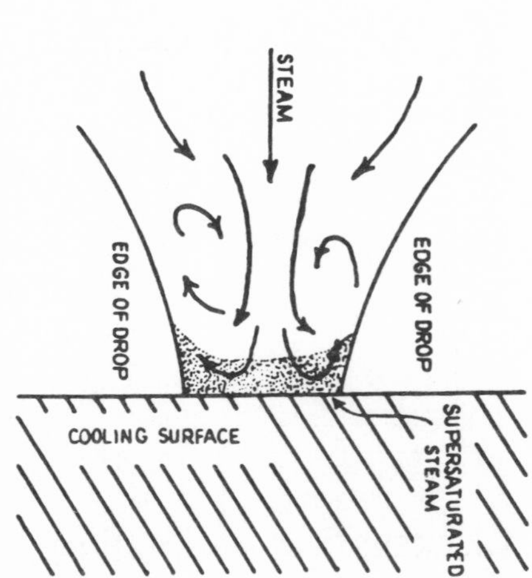

FIGURE 12. MODEL USED BY EMMONS.

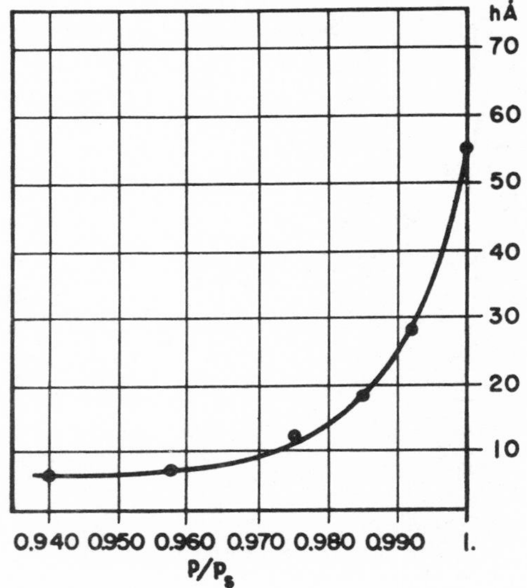

FIGURE 13. ADSORPTION OF PROPYL ALCOHOL ON GLASS. (DERYAGIN AND ZORIN)

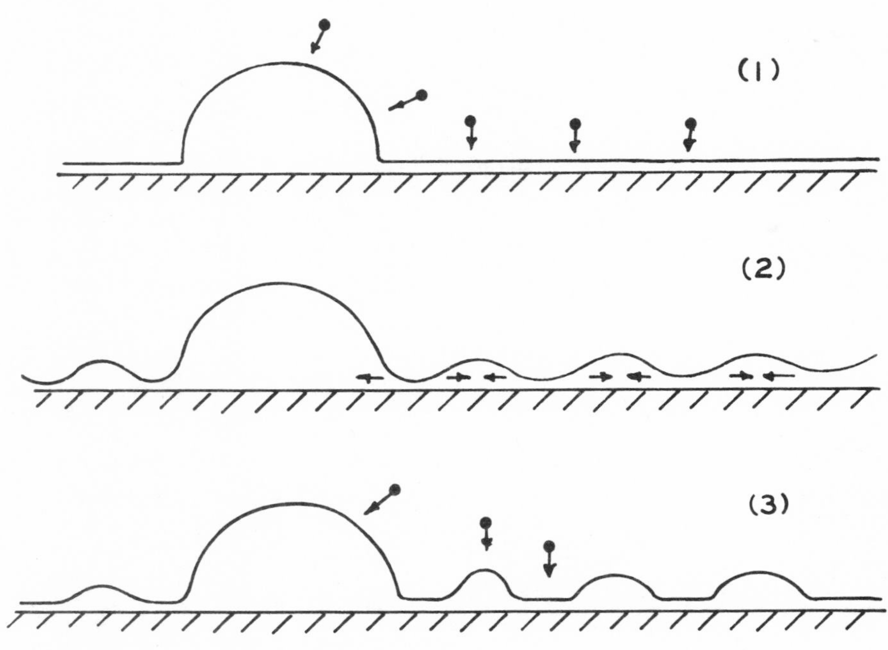

FIGURE 14. MODEL BASED ON LIQUID-FILM RUPTURE. (JAKOB)

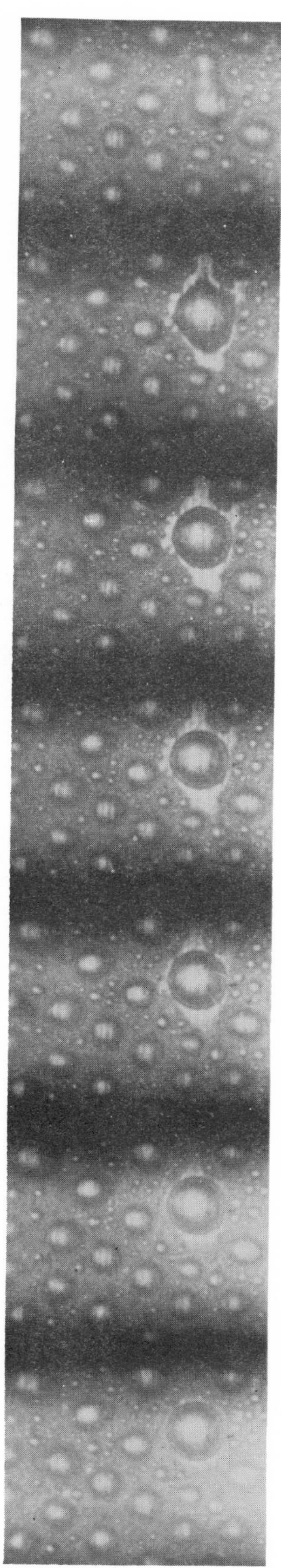

FIGURE 15. MOVIE STRIP TAKEN AT 4080 FRAMES PER SECOND THROUGH A MICROSCOPE. STEAM ON COPPER WITH CUPRIC OLEATE, WITH A ΔT OF 2.9°F AND A HEAT FLUX OF 44,000 BTU/HR SQ FT. (WESTWATER AND WELCH)

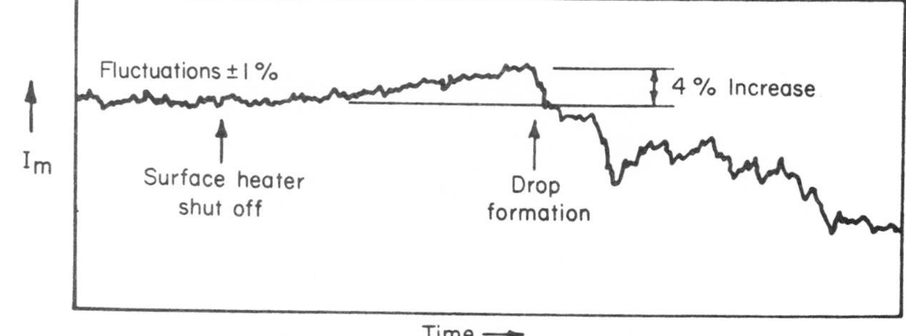

(a) Copper surface with promoted steam

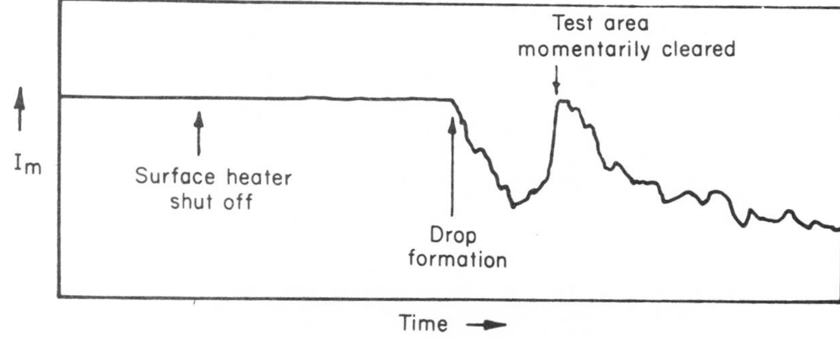

(b) Gold surface with pure steam

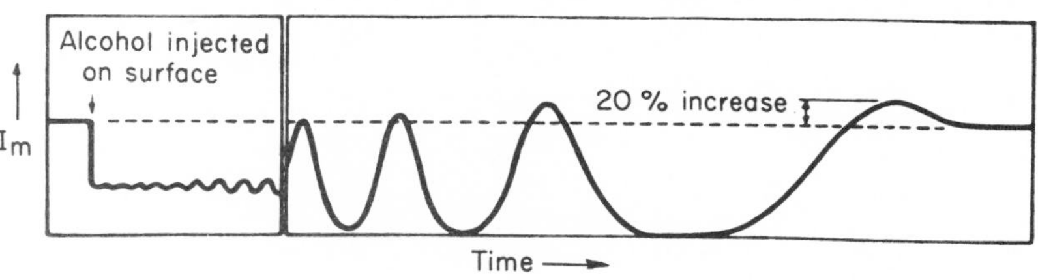

FIGURE 16. MEASUREMENTS OF INTENSITY OF REFLECTED, ELLIPTICALLY POLARIZED LIGHT, USED TO PROVE ABSENCE OF LIQUID LAYER BETWEEN DROPS. (UMUR AND GRIFFITH)

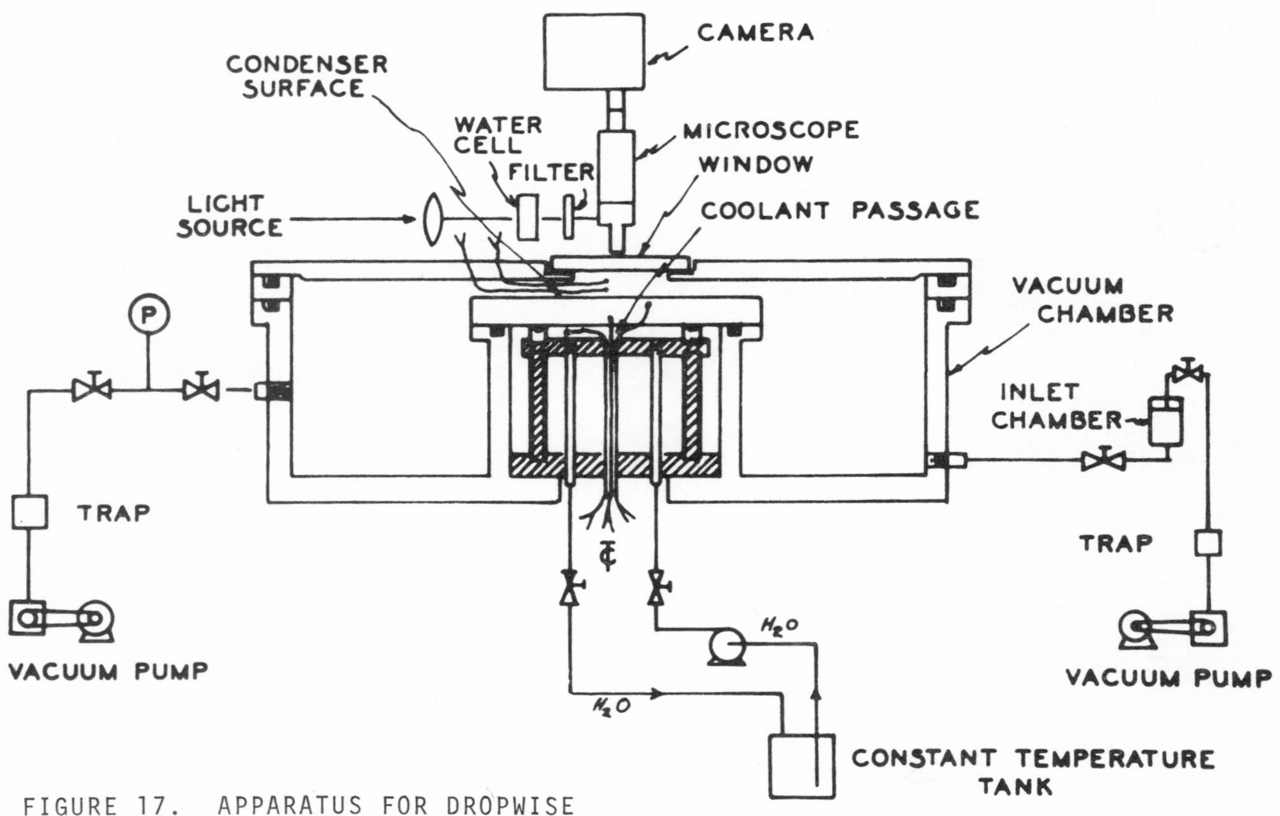

FIGURE 17. APPARATUS FOR DROPWISE CONDENSATION ON A HORIZONTAL PLATE WITH ACCURATE CONTROL OF PRESSURE AND TEMPERATURES. (McCORMICK)

FIGURE 18. USE OF APPARATUS IN FIGURE 17 WITH METALLOGRAPHIC ILLUMINATION AND MOTION PICTURE CAMERA. (McCORMICK)

FIGURE 19. USING OBLIQUE ILLUMINATION AND A STILL CAMERA. (McCORMICK)

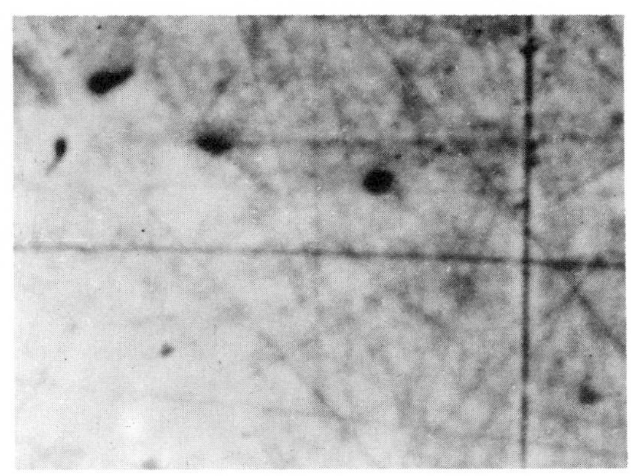

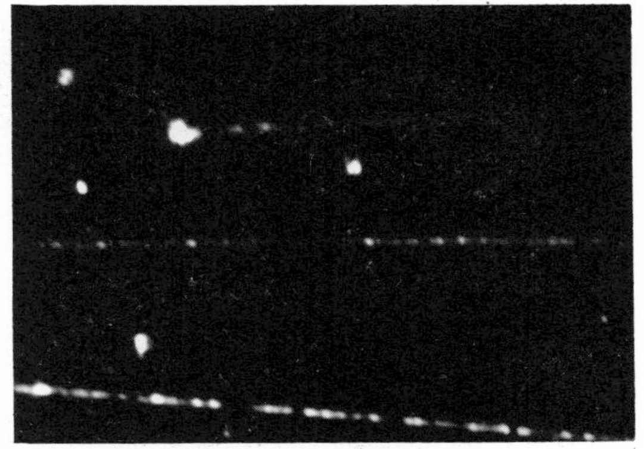

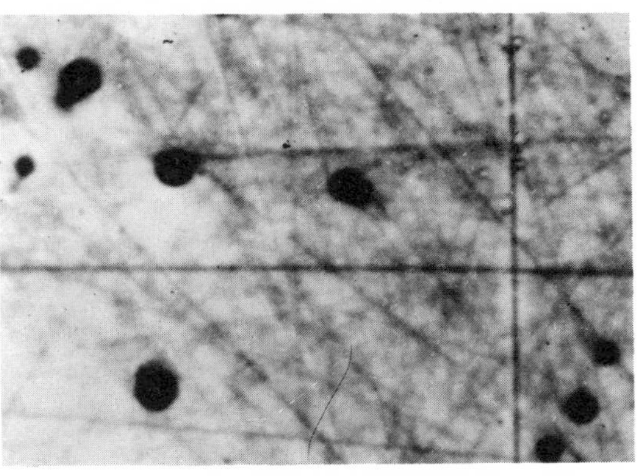

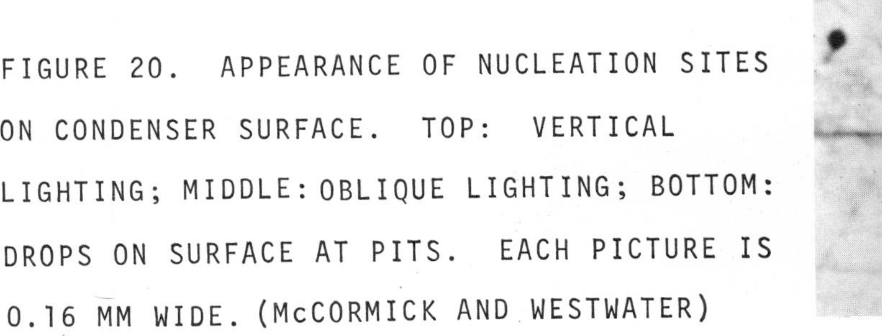

FIGURE 20. APPEARANCE OF NUCLEATION SITES ON CONDENSER SURFACE. TOP: VERTICAL LIGHTING; MIDDLE: OBLIQUE LIGHTING; BOTTOM: DROPS ON SURFACE AT PITS. EACH PICTURE IS 0.16 MM WIDE. (McCORMICK AND WESTWATER)

FIGURE 21. ARTIFICIAL NUCLEATION SITE. AT TOP IS A SPARK EROSION PIT 43 MICRONS WIDE. LOWER PHOTOS SHOW DROP FORMATION AT THE PIT. (McCORMICK)

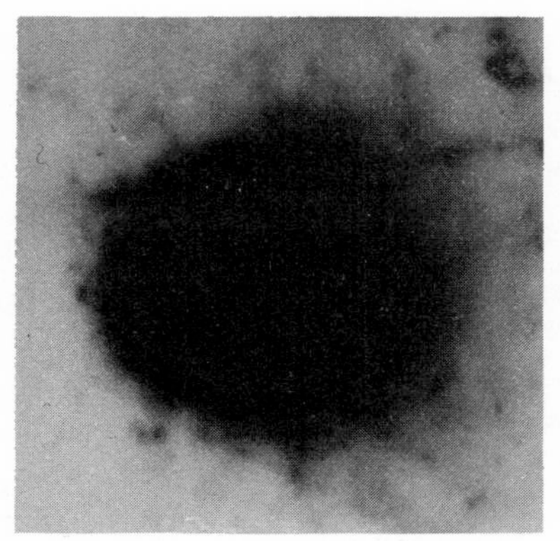

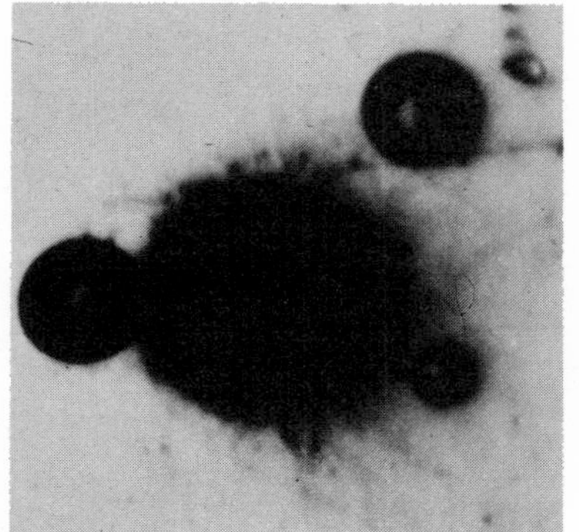

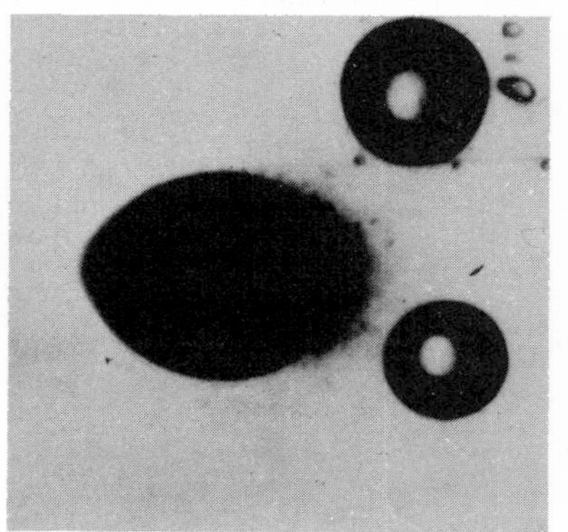

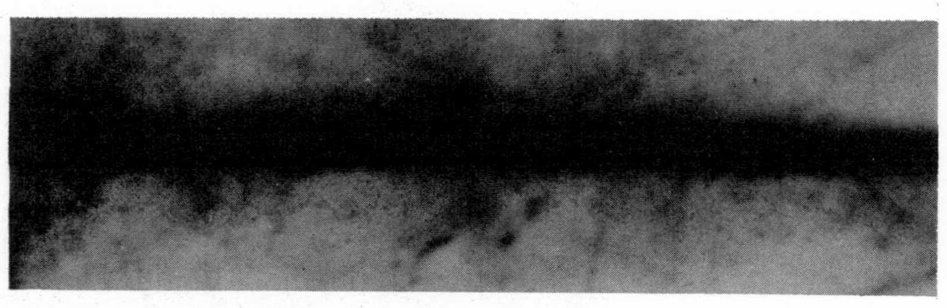

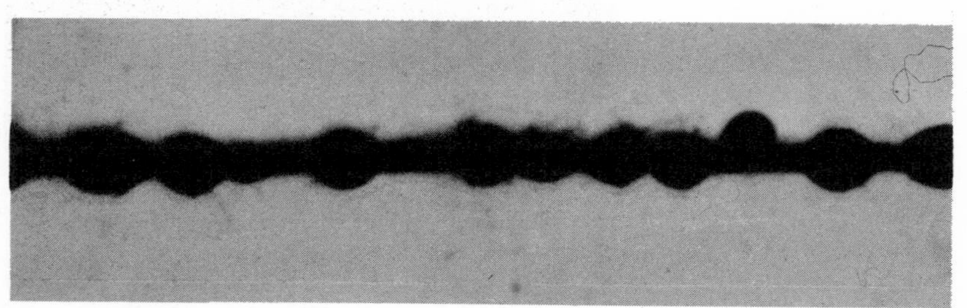

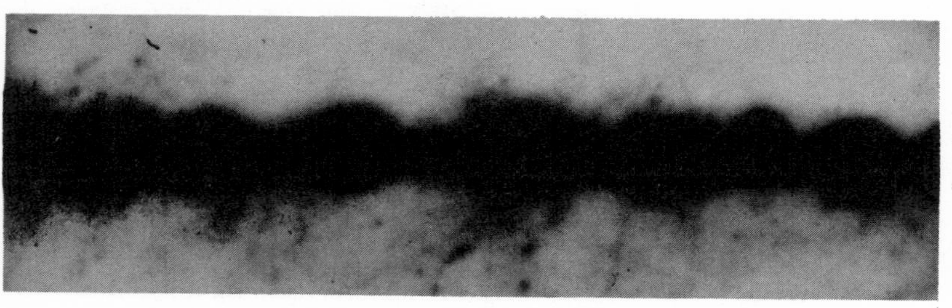

FIGURE 22. NUCLEATION AT SCRATCH SIX MICRONS WIDE. THE SURFACE SUBCOOLING IS 0.3°C. (McCORMICK AND WESTWATER)

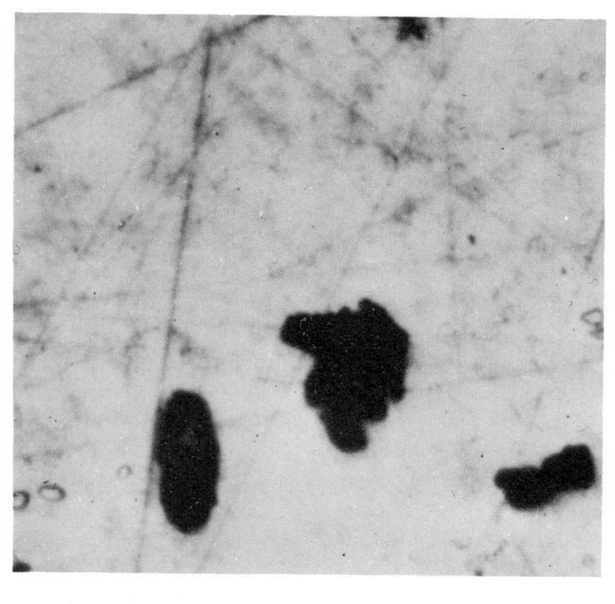

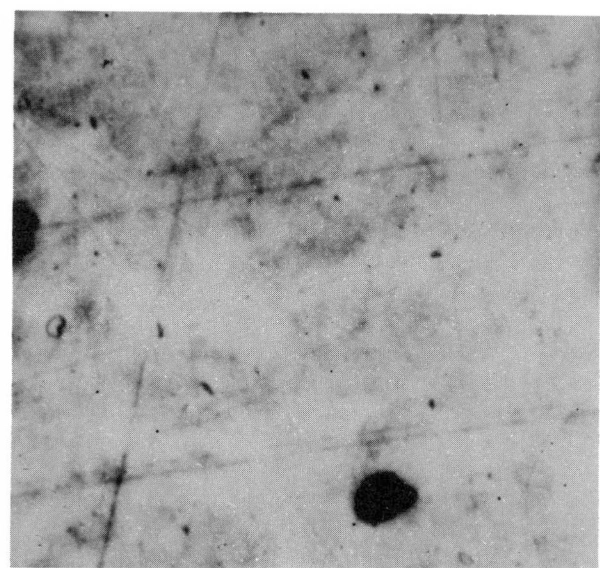

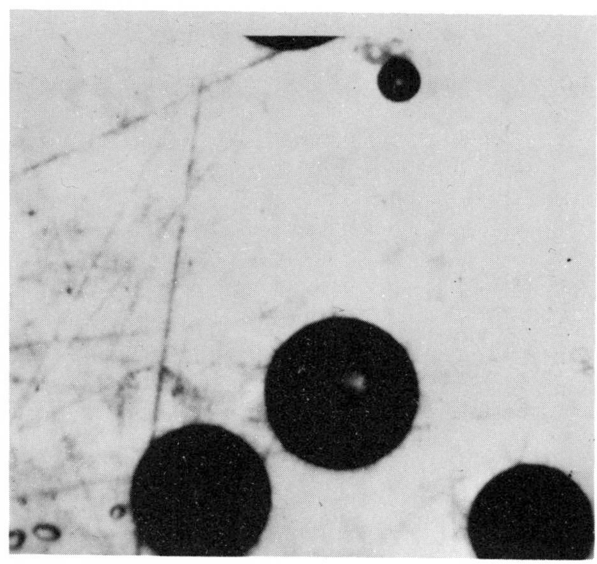

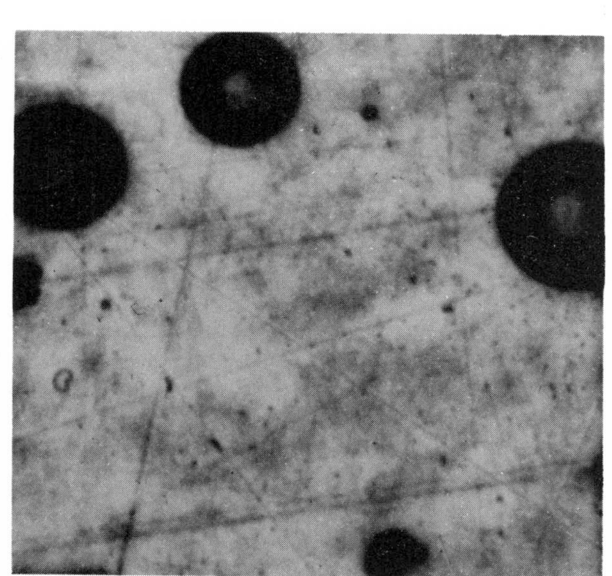

FIGURE 23. NUCLEATION OF GLASS POWDER. TOP PHOTO IS BEFORE WATER VAPOR IS INTRODUCED; LOWER PHOTO IS AFTERWARDS. EACH PICTURE IS 0.19 MM WIDE. (McCORMICK AND WESTWATER)

FIGURE 24. NO NUCLEATION ON CARBON PARTICLES. TOP: PARTICLES; BOTTOM: NUCLEATION AT NATURAL SITES. EACH PICTURE IS 0.20 MM WIDE. (McCORMICK AND WESTWATER)

FIGURE 25. SMALLEST PARTICLES THAT NUCLEATE WATER DROPS. TOTAL PRESSURE IS 19 MM; THE PARTICLES ARE G' ASS AND PLATINUM. UPPER GRAPH IS BASED ON OVERALL TEMPERATURE DIFFERENCE; LOWER BASED ON BOUNDARY LAYER ΔT. (McCORMICK AND WESTWATER)

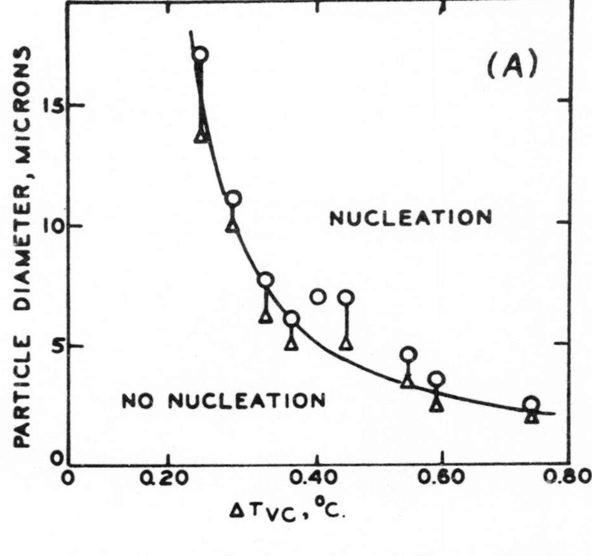

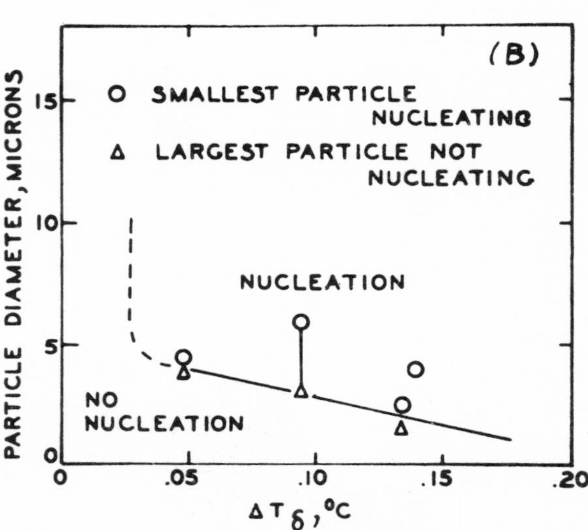

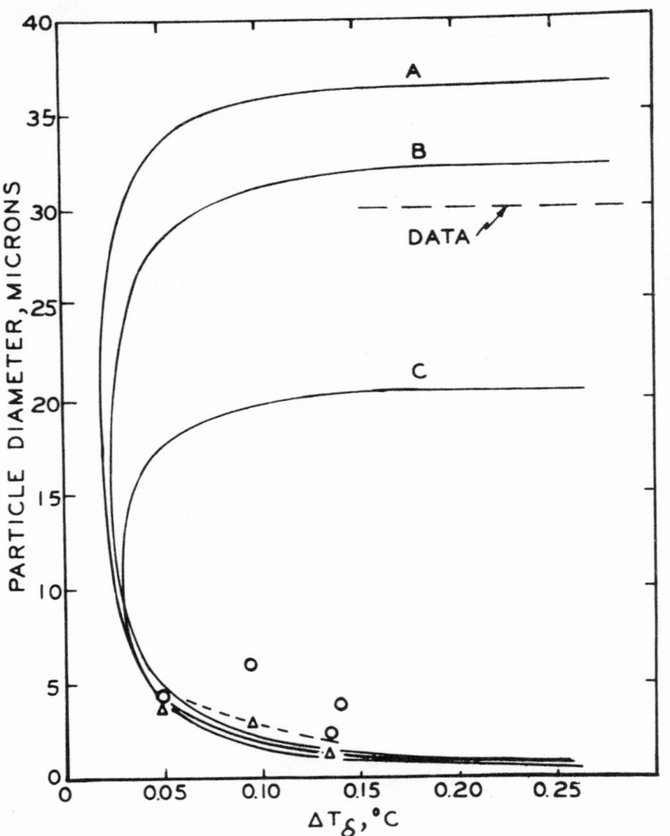

FIGURE 26. EFFECTIVE SIZE RANGE OF PARTICLES WHICH NUCLEATE WATER DROPS AT 23°C. CURVE A ASSUMES ΔT = 0.02°C AND BOUNDARY LAYER THICKNESS OF 200 MICRONS. CURVE B ASSUMES 0.025°C AND 200 MICRONS. CURVE C ASSUMES 0.03°C AND 100 MICRONS. (McCORMICK AND WESTWATER)

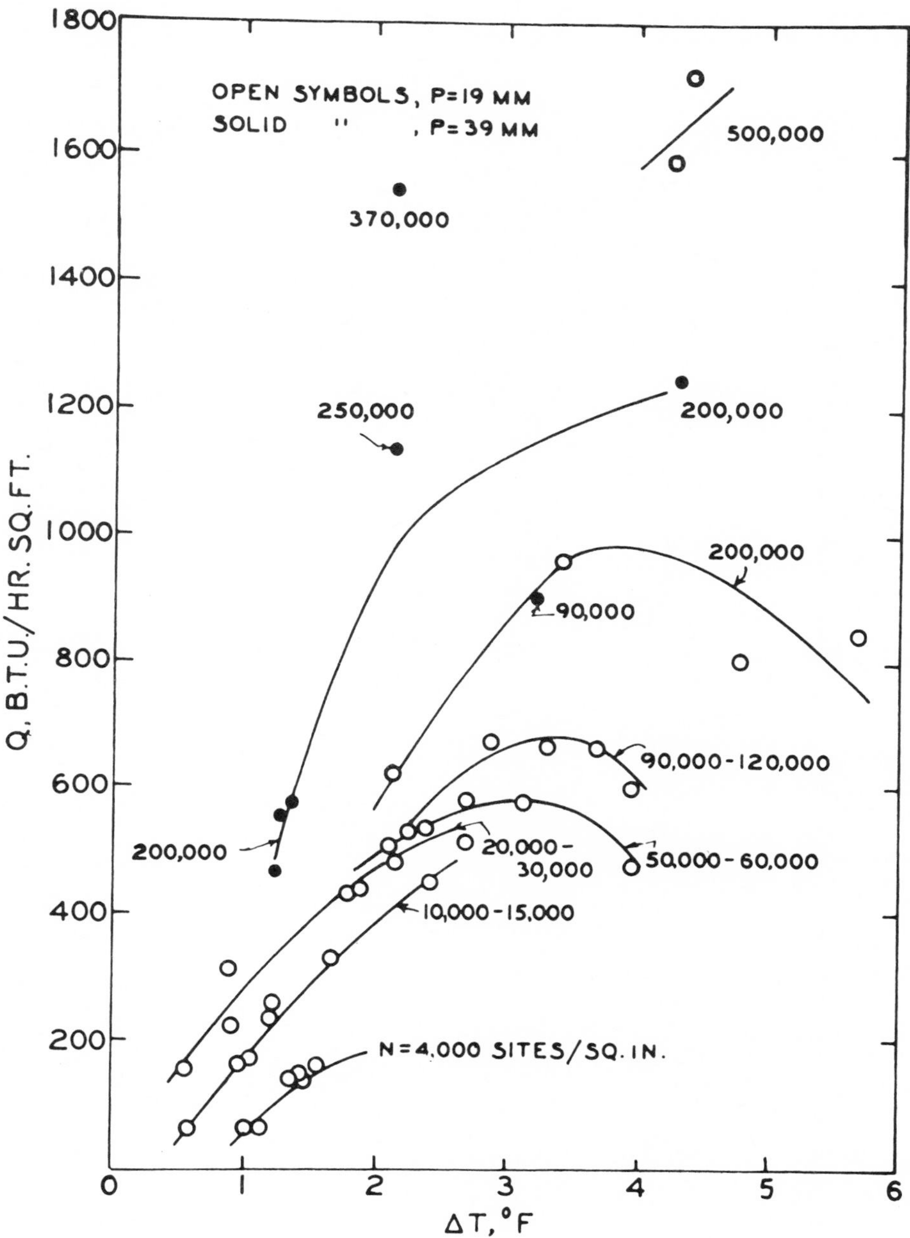

FIGURE 27. HEAT FLUX AS A FUNCTION OF THE POPULATION OF ACTIVE NUCLEATION SITES, FOR WATER ON COPPER PROMOTED WITH BENZYL MERCAPTAN. (McCORMICK AND WESTWATER)

FIGURE 28. COMPARISON OF DATA FOR STEAM FROM 13 WORKERS. (TANNER, POTTER, POPE, AND WEST)

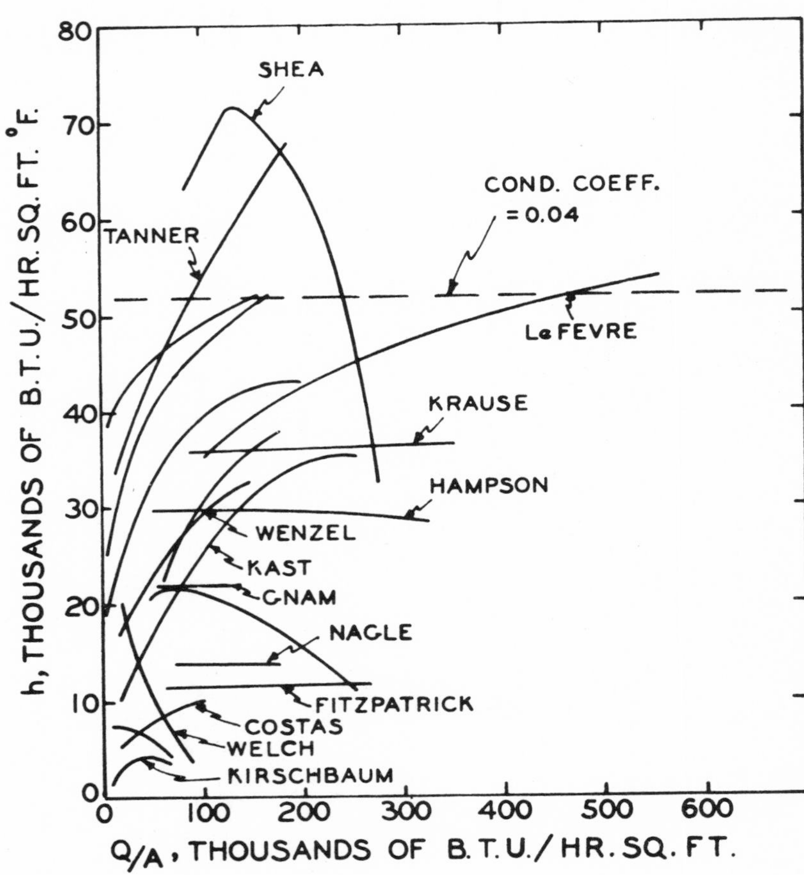

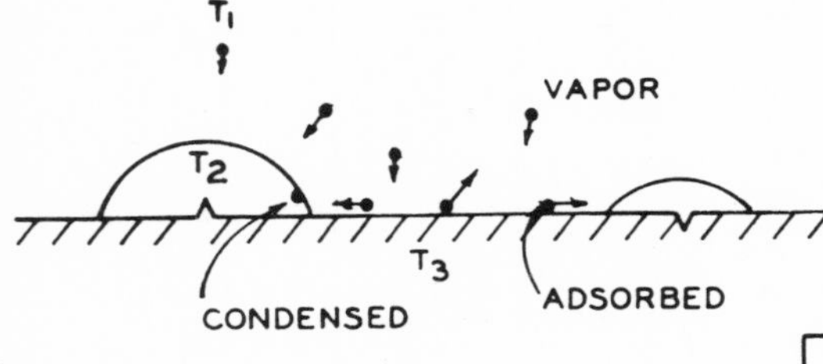

FIGURE 29. MODEL WHICH INCLUDES SURFACE DIFFUSION.

FIGURE 30. EFFECT OF NEARBY DROPS ON DROP GROWTH RATE. TESTS WITH WATER VAPOR AT A TOTAL PRESSURE OF 19 MM. (McCORMICK AND WESTWATER)

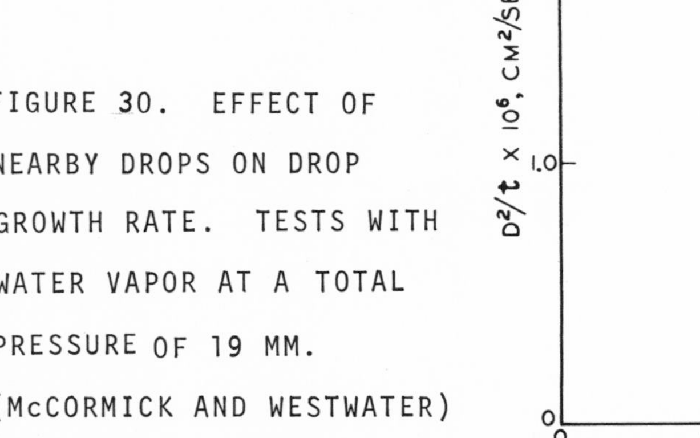

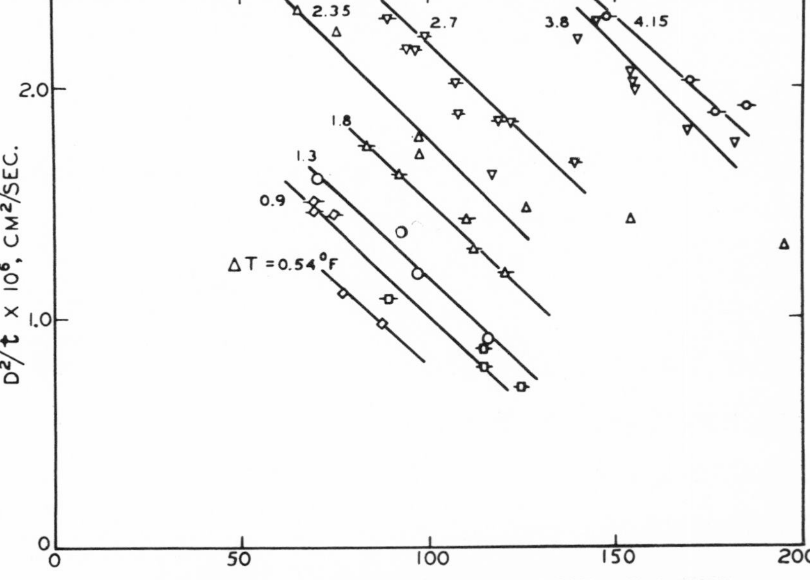

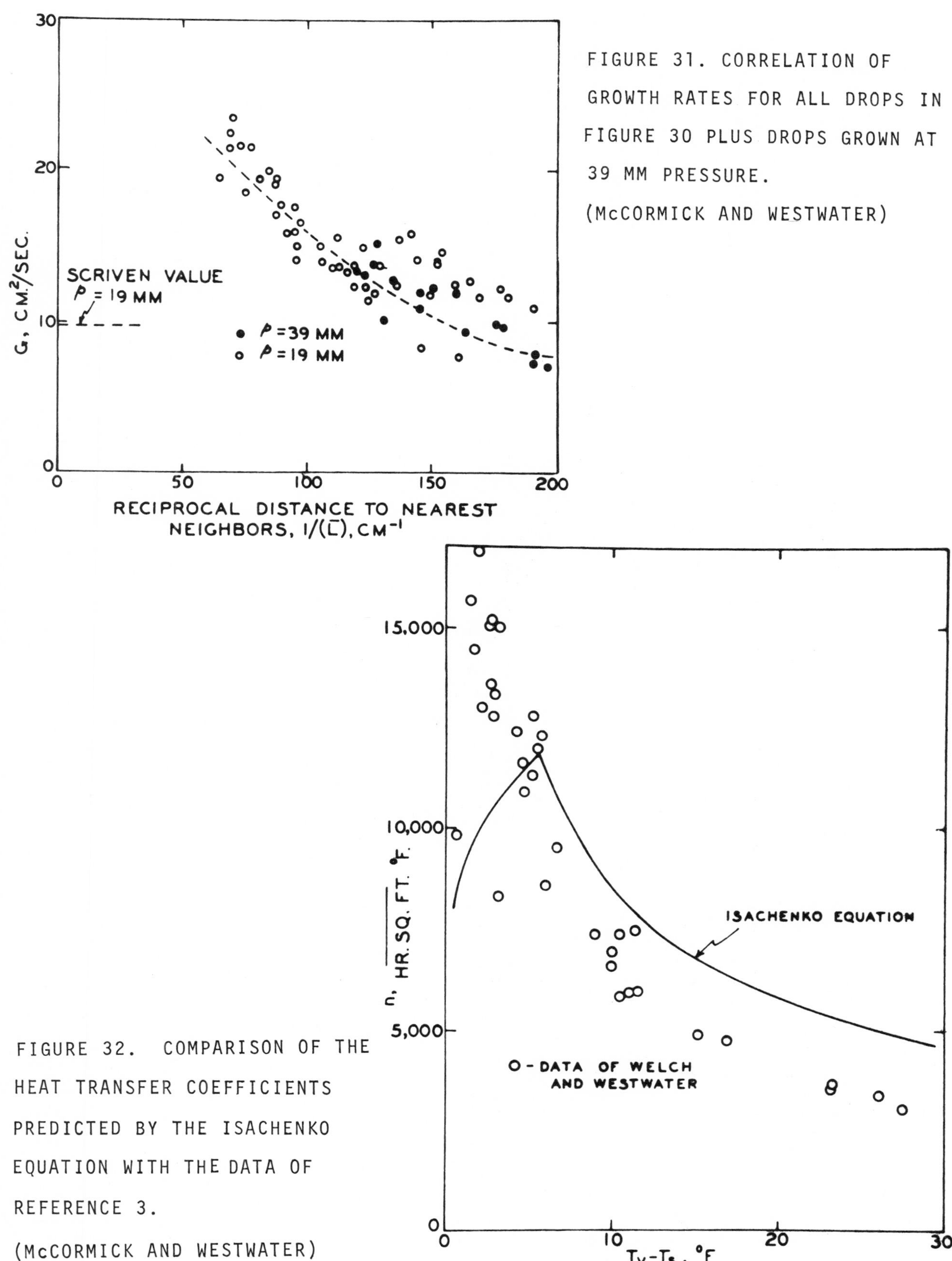

FIGURE 31. CORRELATION OF GROWTH RATES FOR ALL DROPS IN FIGURE 30 PLUS DROPS GROWN AT 39 MM PRESSURE.
(McCORMICK AND WESTWATER)

FIGURE 32. COMPARISON OF THE HEAT TRANSFER COEFFICIENTS PREDICTED BY THE ISACHENKO EQUATION WITH THE DATA OF REFERENCE 3.
(McCORMICK AND WESTWATER)

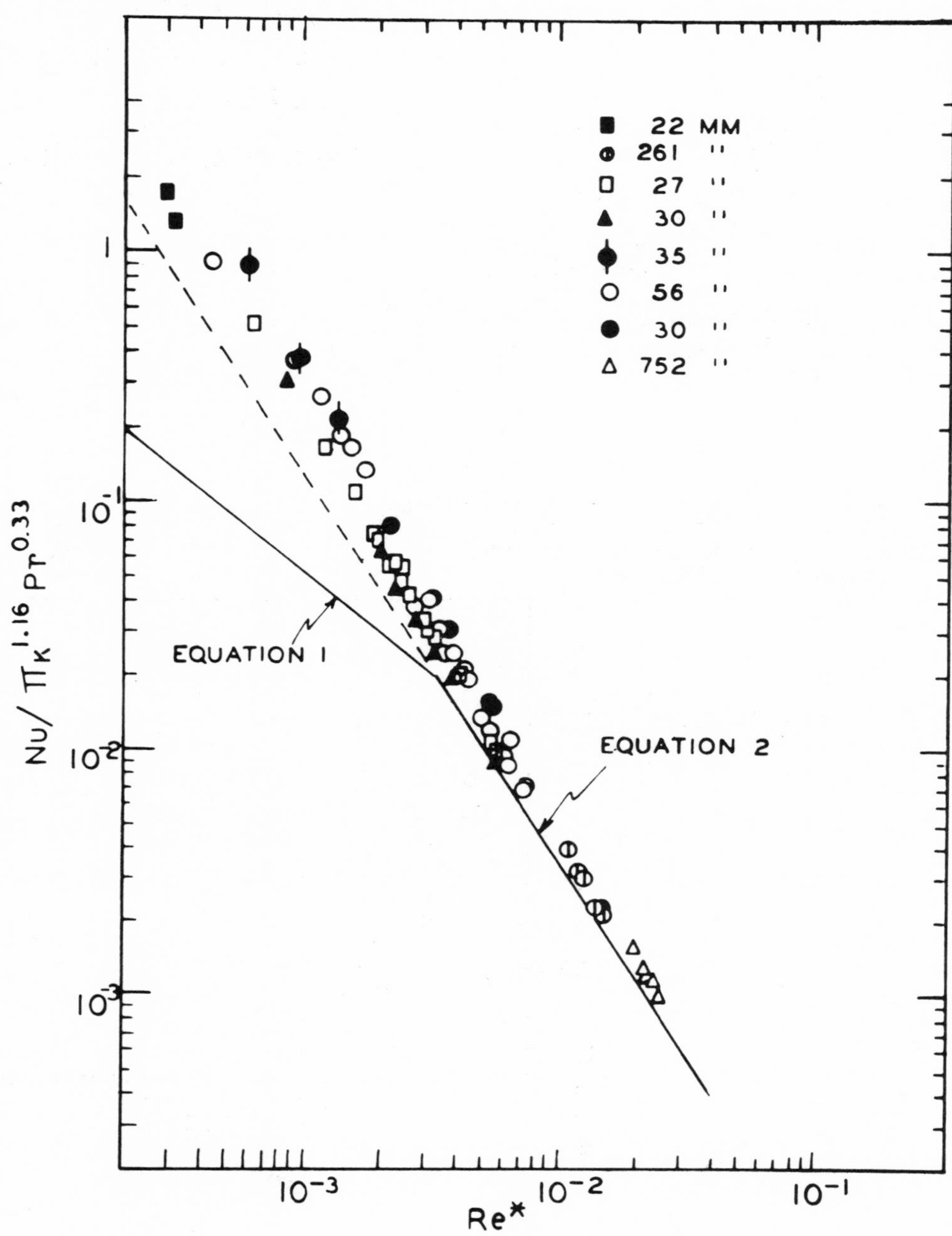

FIGURE 33. COMPARISON OF ETHYLENE GLYCOL DATA WITH THE ISACHENKO EQUATIONS. (PETERSON AND WESTWATER. EQUATIONS (1) AND (2) IN THE GRAPH ARE EQUATIONS (8) AND (9) OF THIS ARTICLE.)

CRITICAL TWO-PHASE FLOW *

H. K. Fauske**

ABSTRACT

Published analytical methods pertaining to gas-liquid critical flows are reviewed. It is shown that, based on recent experimental observations, several assumptions used in these analyses are questionable.

The necessity to include the flow regime dependency in analyses of the two-phase compressibility phenomena is pointed out. Solutions for the case when the flow regime is liquid droplet flow (high quality mixtures) are presented. Considerations include the rate of momentum, mass and heat transfer between the vapor, and the liquid droplets in the approach region to critical flow. Excellent agreement is obtained with steam-water data in the high quality region.

The occurrence of large deviations from equilibrium flow is demonstrated in the low quality region and experimental research currently in progress is briefly discussed.

INTRODUCTION

Recent years have seen considerable effort devoted to the problems associated with two-phase gas-liquid flow mechanics;[(1)***] included in these is critical flow, which is also referred to as maximum, choking, or sonic flow. The main reason for the intensified efforts emanates from the increasing necessity for advanced concepts and technologies in the nuclear and space industries. Critical two-phase flows may take place in a large number of flow systems with working fluids of widely dissimilar properties ranging from cryogenics to liquid metals.[(2-5)] The purpose of this paper is to focus attention on the current status of understanding two-phase critical flow. The first part of the paper summarizes and evaluates previous experimental and analytical work, while the second portion describes work currently in progress at Argonne National Laboratory.

PREVIOUS WORK

Summarizing previous work, this paper will primarily deal with flow systems in which the critical flow phenomenon is not significantly influenced by the shape or

*Work performed under the auspices of the U.S. Atomic Energy Commission.

**Associate Chemical Engineer, Reactor Engineering Division, Argonne National Laboratory, Argonne, Illinois.

***Superscript numbers in parentheses refer to References at the end of this paper.

TABLE 1

RECENT CRITICAL FLOW INVESTIGATIONS

References	Critical Pressure psia	Exit Quality or Stagnation Enthalpy Btu/lb	Testing Fluid	Flow System
Isbin, *et al.*[6]	4-43	0.01-1.0	Water	Annuli and pipes, D = 0.37-1.0 in.
Faletti and Moulton[7]	26-106	0.001-1.0	Water	Annuli, D = 0.2-0.39 in.
Fauske[8]	40-360	0.01-0.7	Water	Pipes, D = 0.27-0.125 in.
Zaloudek[9]	40-110	0.004-0.99	Water	Pipes, D = 0.52 in.
James[10]	14-64	230-1200	Water	Pipes, D = 3-6 in.
Fauske[11]	100-1100	300-670	Water	Pipes, D = 0.249 in.
Zaloudek[12]	23-87	500-1000	Water	Pipe elbow and tee, D = 0.625 in.
Klingebiel[13]	25-60	0.01-0.99	Water	Pipes, D = 0.504 in.
Fauske[14]	17	0.0006-0.1	Air-Water	Pipes, D = 0.25-0.5 in.

size of the flow passage; recent data sources are summarized in Table 1. This excludes typical "metastable flow systems," i.e., discharges from short tubes ($L/D < 12$) and apertures (nozzles and orifices). (See References 11 and 15 through 20.)

Several models have been proposed to predict the data in Table 1. The first analytical model to appear in the literature was the Homogeneous Equilibrium Model (HEM), implying equal velocities and thermal equilibrium between the liquid and vapor phase. The critical flow rate is obtained from the same equation as in single-phase flow:[21]

$$G^2 = \frac{-g_c}{(dv/dp)_s} , \tag{1}$$

where the specific volume for homogeneous flow is

$$v_H = v_g x + v_\ell (1 - x) . \tag{2}$$

For detailed calculation procedures, consult Reference 22. The predicted values, G_{HEM}, are generally much lower than the measured flow rates, G_{exp}, but have been used frequently by researchers as a useful tool for correlating the data[6,7,9] as illustrated in Figure 1. The failure of the HEM to predict the data resulted first in introduction in the literature of such phenomena as "slip" (different vapor and liquid velocity), and "metastability" (deviation from stable thermal equilibrium). These qualitative effects increased the flow rates and brought them in closer agreement with experimental values.

Later, a number of empirical and analytical models have appeared in the literature. All of these analyses assumed thermal equilibrium, separated flow (annular flow pattern), and with each phase represented by a single average velocity. The combination of either a momentum balance (negligible friction and heat losses),

TABLE 2

RECENT CRITICAL FLOW MODELS

References	Equations Used	Void or Velocity Ratio Relationship Used
Isbin, *et al.*[6]	3 and 4	Modified Martinelli-Nelson
Massena[23]	3 and 4	Modified Armand
Fauske[8]	3 and 4	$k = (v_g/v_\ell)^{1/2}$
Levy[24]	3 and 4	Momentum Exchange Model
Ryley,[25] Moody,[26] Zivi[27]	5 and 6	$k = (v_g/v_\ell)^{1/3}$
Cruver and Moulton[28]	3 and 4, 6 and 7	$k = (v_g/v_\ell)^{1/3}$

$$G^2 = \frac{-g_c}{(dv^*/dp)_s} , \qquad (3)$$

where

$$v^* = \frac{v_g x^2}{\alpha} + \frac{v_\ell (1 - x)^2}{1 - \alpha} , \qquad (4)$$

a total energy balance,

$$G^2 = \frac{-2g_c J}{(dv'^2/dh)_s} , \qquad (5)$$

where

$$v'^2 = \frac{x^3 v_g^2}{\alpha^2} + \frac{v_\ell^2 (1 - x)^3}{(1 - \alpha)^2} , \qquad (6)$$

or a mechanical energy balance,

$$G^2 = \frac{-2g_c v_H}{dv'^2/dp} , \qquad (7)$$

with an empirical or analytical relationship for void fraction, α, or velocity ratio, $k = u_g/u_\ell$, resulted in several flow models; these are summarized in Table 2.

The models listed in Table 2 illustrate fairly good agreement with the steam-water data in Table 1, with the exception of the analyses using Martinelli-Nelson and Armand void fractions; these predictions are lower than the measured flow rates at the low qualities. Figures 2, 3, and 4 display comparisons between experimental data and predictions from one of the models. (See Reference 8.) Little or no difference is obtained in predicted values using the other analytical models listed in Table 2. For detailed discussions of the models, see References 2, 3, and 13. Critical flow rate predictions of steam-water mixtures covering a wide range of pressures are displayed in Figure 5. Analytical predictions for fluids other than water are available, (see References 2 and 29) and for suggested pressure drop calculation techniques in the approach region to critical flow, the reader is referred to References 2 and 30 through 33.

Despite the relatively good agreement demonstrated between the annular flow models and data, the models are open to serious criticism: (1) their inability to clearly explain why two-phase choking occurs,[14] and (2) their physical formulism deviates significantly from recent experimental observations.[13,14,34] Measured velocity ratios are considerably smaller than the ones used in the analytical models[14] as illustrated in Figure 6, and the assumption of thermal equilibrium applied in the models appears to be invalid[14] (discussed in the second half of the paper). Furthermore, observation of a homogeneous type flow pattern occurring in the approach region to critical flow invalidates the assumption of annular flow. (See Reference 34.) Apparently, the good agreement between predictions and data results from the high velocity ratios employed compensating for metastability effects which have been neglected in the models. Consequently, applying these models to other fluids than tested could, therefore, give erroneous results and must await experimental verification.

CURRENT WORK

Because of the inherent nonequilibrium associated with two-phase critical flow (velocity and temperature lags between the phases), the most promising approach for improving understanding and prediction methods appears to lie in separate analyses of each flow regime (vapor bubble flow, liquid droplet flow, etc.). Only then can accurate descriptions of transport processes such as momentum, heat, and mass transfer between the phases be carried out.

For one-dimensional steady flow, with each phase represented by a single average velocity, the equations of continuity and momentum, written over a differential length, dy, at the point of choking, in the absence of frictional and gravitational forces are:

$$\frac{d}{dy}\left(\bar{\rho}_g u_g + \bar{\rho}_\ell u_\ell\right) = 0 , \tag{8}$$

$$\frac{d}{dy}\left(\bar{\rho}_g u_g^2 + \bar{\rho}_\ell u_\ell^2\right) + \frac{dP}{dy} = 0 . \tag{9}$$

The actual vapor concentration, $\bar{\rho}_g$, can be formulated in terms of the vapor quality, x, and the velocity ratio, k, by considering the continuity equation for each constituent, resulting in Equation (10),

$$\bar{\rho}_g = \left(\frac{1 - x}{x}\frac{k}{\rho_\ell} + \frac{1}{\rho_g}\right)^{-1} , \tag{10}$$

where $k = u_g/u_\ell$.

The above set of equations holds for all flow regimes. However, a solution of these equations cannot be obtained without specifying x, k, ρ_g, and ρ_ℓ, as well as their derivatives (with respect to pressure), at the point of choking. If thermodynamic equilibrium is not maintained in the accelerating two-phase mixture, the deviations of the above quantities from the equilibrium state must be evaluated. This requires a knowledge of the mode of momentum, mass, and heat transport, as well as the geometrical configuration of the interfaces. Therefore, the flow regime must be specified. An analysis of the case, where the liquid phase appears as droplets in the continuous vapor phase, has been examined (see Reference 35) and is briefly illustrated below.

High Quality Mixtures

At high vapor content by mass and high velocities, droplet dispersed flow is believed to be the predominant flow regime.[36] An evaluation of the transport processes leads to the following conditions prevailing in the approach region to critical flow:

Condition (1):

$$\frac{\partial}{\partial P}\left(\frac{\bar{\rho}_\ell}{\rho_\ell}\right) = 0 \quad \text{and} \quad \frac{\partial}{\partial P}\left(\frac{\bar{\rho}_g}{\rho_g}\right) = 0$$

Condition (2):

$$\frac{\partial x}{\partial P} = 0 \quad \text{and} \quad \frac{\partial \rho_g}{\partial P} = \frac{\rho_g}{KP}$$

Condition (1) above results from a consideration of the transport process of momentum between the phases and, as illustrated in Figure 7, indicates a constant average droplet velocity in the approach region to critical flow. It is assumed that the mass motion of the droplets is due only to the drag force exerted by the vapor as the vapor is accelerated by expansion in the flow duct. (At low pressure, the effect of pressure gradient is negligible.) If droplet size and Reynolds number are known, the average time constant of momentum transfer can be determined. According to Reference 37 the minimum droplet size likely to be present in a steam-water flowing mixture is approximately 10 μ. With this value as the minimum droplet size (maximum drag), and including Reynolds numbers inside as well as outside the Stokesian range, results in time lags of the vapor-droplet motion (steam-water, P = 40 psia) of approximately 10^{-5} sec. Comparing this value with the extremely rapid depressurization taking place in the approach region to critical flow (time rate of change of pressure seen by the continuous fluid phase), it is concluded that the droplet velocity is unable to follow the rapid changes in the vapor phase in this region and remains essentially constant.

Similar calculations carried out to obtain the time constant of heat transport lead to condition (2), characterizing the absence of heat transfer between the phases in the approach region to critical flow, i.e., the droplet temperature remains essentially constant and the vapor expands adiabatically.

Combining Equations (8) and (9) satisfying conditions (1) and (2) results in a solution for the critical flow rate, G:

$$G^2 = \frac{KP\bar{\rho}_g}{x^2} . \qquad (11)$$

At low pressures and high qualities, Equation (10), the volume occupied by the droplets may be neglected, i.e.,

$$\bar{\rho}_g \approx \rho_g .$$

In these cases, no information is required about the slip between the liquid and vapor phase, as shown in Table 3. Measured velocity ratios in the high quality region (50 to 100 per cent and at low pressures ($P < 100$ psia) range from 1 to 2.[13] Furthermore, the exit quality can be approximated by the thermal equilibrium quality since the latter value differs only slightly

TABLE 3

EFFECT OF k on $\bar{\rho}_g$

$P = 40$ psia, $\rho_g = 0.0952$, $x = 0.50$

k	$\bar{\rho}_g$
1	0.0950
2	0.0949
3	0.0947
4	0.0946

from the corresponding stagnation value for qualities greater than 50 per cent.

Calculated values from Equation (11) using equilibrium qualities are displayed in Figure 8 and illustrate good agreement with existing steam-water data in the high quality region[9] (dispersed droplet flow). Also depicted in Figure 8 is the range of predicted values from previous published models (Table 2 in the text).

In Table 4 calculated values from Equation (11) are compared with the most recent data in the literature,[13] illustrating that the theory predicts measured values within 7 per cent.

According to gas-dynamic analyses, the critical velocity is equivalent to the sonic velocity of the fluid,[21] hence explaining the occurrence of choked flow. The velocity of sound in high quality one-component mixtures is, therefore, examined next to see if the same analogy applies to two-phase critical flow. The following assumptions are made in the formulation of the propagation velocity model:

(1) The flow is one-dimensional.

(2) The flow regime is droplet flow, initially at zero velocity in a duct of uniform cross section (see Figure 9).

(3) The effect of mass transfer is negligible.

(4) The volume occupied by the liquid droplets is negligible.

(5) The liquid phase is incompressible.

For the above assumptions, the continuity and the momentum equation written across the wave leads to the following expression for the sonic velocity:

$$a^2 = \frac{\rho_g}{d\rho_g/dP\left[\bar{\rho}_g + \bar{\rho}_\ell(du_\ell/du_g)\right]} . \qquad (12)$$

In single-phase flow, the sonic velocity is normally a thermodynamic property, but in a two-phase mixture, as seen from Equation (12), this quantity is transport-process dependent if nonequilibrium effects are present. Two extreme cases are investigated:

(1) A system wherein disturbances occur in much shorter times than the time constants for momentum and heat transport (high frequencies), i.e.,

$$\frac{du_\ell}{du_g} = 0$$

and

TABLE 4

COMPARISON BETWEEN EXPERIMENTAL AND CALCULATED CRITICAL FLOW RATES

Experimental Data(13)			Calculated	
P	x	G	G	Dev %
52.67	0.6222	307.68	320.0	+ 4.0
32.11	0.5255	227.68	234.7	+ 3.1
41.93	0.5193	296.51	307.5	+ 3.7
52.43	0.5103	369.31	388.0	+ 5.1
33.62	0.9690	143.16	133.2	- 6.9
30.73	0.6160	196.25	192.0	- 2.2
50.82	0.7128	281.77	270.2	- 4.1
30.17	0.7170	170.82	162.3	- 5.0
49.68	0.8089	249.68	233.0	- 6.7
29.95	0.8145	151.67	142.0	- 6.4
48.0	0.9625	198.76	189.5	- 4.7
30.9	0.7230	167.03	164.5	- 1.5

$$\frac{d\rho_g}{dP} = \frac{\rho_g}{KP} ,$$

resulting in

$$a^2_{NE} = \frac{KP}{\bar{\rho}_g} . \tag{13}$$

(2) A system which remains in equilibrium throughout the disturbance (low frequencies),

$$\frac{du_\ell}{du_g} = 1$$

and

$$\frac{d\rho_g}{dP} = \frac{1 - P}{h_{fg}\rho_g} ,$$

where the latter is obtained by assuming ideal gas conditions and applying the well-known Clausius-Clapeyron relationship to the vapor phase, resulting in

$$a^2_{TDE} = \frac{P}{\left[1 - \left(P/h_{fg}\rho_g\right)\right] (\bar{\rho}_g + \bar{\rho}_\ell)} . \tag{14}$$

Calculated values from Equations (13) and (14) are displayed in Figure 10. Case 1 gives calculated propagation velocities independent of the mixture quality and equal to the velocity of sound in saturated vapor, which is in good agreement with experimental data.[38] On the other hand, case 2 indicates that the moisture content can affect greatly the propagation velocity if equilibrium is maintained between the phases.

If the critical vapor velocity is calculated from Equation (13), it results in an expression for the critical flow rate identical to Equation (11), suggesting that the occurrence of choked flow takes place when rarefaction waves of small wave growth times no longer can propagate upstream against the flow.

Low Quality Mixtures

Similar analytical studies as described above are currently under way for mixtures at low vapor content by mass. However, at low qualities, the analysis becomes considerably more complex. While for high quality mixtures exact values of the velocity ratio were not required, Equation (10) illustrates that at low qualities these values must be known. As discussed earlier, available analytic predictions of the velocity ratio appear to be in gross error (see Figure 6). Furthermore, if metastability occurs as the fluid expands, the actual quality cannot be approximated by the stable equilibrium value as was the case for high quality mixtures.

In previous analyses of two-phase critical flow (see Table 2), the quality has been calculated by assuming that the fluid properties vary according to stable equilibrium conditions. That this assumption may lead to erroneous results is illustrated next.

At small qualities ($x < 0.01$), the momentum flux of the vapor phase is negligible compared to that of the liquid, and Equations (8) and (9) can be integrated from pressure $P_1 (x = 0)$ to $P_2 (x = x_E)$ to yield the following expression for the vapor velocity:

$$u_{g_2} = x_E V_{g_2} G \left[\frac{G^2 V_{\ell_2}}{g_c (P_1 - P_2)} + 1 \right] . \qquad (15)$$

Equation (15) represents the minimum vapor velocity if stable thermal equilibrium is maintained between the liquid and the vapor phase. If viscous dissipation is included, larger values of the vapor velocity will result. Thermodynamic considerations dictate that the vapor phase velocity obtained from Equation (15) must be less than that velocity calculated, if saturated vapor alone is assumed to expand isentropically from pressure $P_1 (x = 1.0)$ to pressure $P_2 (x = x_S)$. The latter value can be calculated from the following equation:

$$u_{g,s} = \sqrt{2 g_c J \Delta H} , \qquad (16)$$

where

$$H = h_{g_1} - \left(h_{f_2} + x_s h_{fg_2} \right)$$

and

$$x_s = \frac{s_{g_1} - s_{f_2}}{s_{fg_2}} .$$

Steam-water critical flow data from the current research have been examined with respect to Equations (15) and (16). Typical results are listed in Table 5, which clearly shows that the assumption of thermal equilibrium is not valid in flashing flow, and that the actual quality to be used in Equation (15) should be considerably less than the equilibrium value, i.e., the liquid portion in the two-phase stream is in an unstable state (superheated).

In view of the extreme complexity of the two-phase flow phenomena in the low quality region, it is believed that analytical efforts alone will not improve substantially upon the present state of the art without concurrently performing detailed and basic experiments to guide the theoretical model formulation. Such experiments are presently in progress, including measurements of critical flow rates, critical pressures, and exit void fractions, as well as axial void and pressure profiles, velocity ratio, and observations of flow regimes in low quality mixtures using working fluids ranging from air-water to liquid metals. Furthermore, due to lack of sonic velocity data in low quality one-component mixtures,[39] the program includes an extensive investigation to measure the propagation velocity of pressure waves in two-phase mixtures in order to check analytical models.

TABLE 5

ANALYSIS OF TYPICAL STEAM-WATER CRITICAL FLOWS

G	P_1	P_2	x_E	u_{g_2}	u_{g_s}
lb_m/sec-ft^2	psia	psia		ft/sec	ft/sec
3672	55.1	49.4	0.0046	1450	793
3917	39.8	37.3	0.0032	3333	555
3503	35.2	32.5	0.0039	3107	555

SUMMARY

(1) Published experimental and analytical data on two-phase critical flow are reviewed. While existing analytical models predict two-phase critical flow rates (steam-water) fairly well, several assumptions used in the analysis appear to be not valid according to recent experimental observations. Application of these models to other fluids could give erroneous results and, therefore, must await experimental verification.

(2) The importance of considering each individual flow regime when attempting to understand the two-phase critical flow phenomenon is pointed out. Solutions for the case when the flow regime is liquid droplet flow (high quality mixtures) are presented. Predicted critical flow rates are shown to be in excellent agreement with measured flow rates.

(3) The complexity of two-phase flow phenomena in the low quality region is discussed, and experimental research currently in progress is briefly described.

NOTATION

a = two-phase sonic velocity

D = equivalent diameter

g_c = Newton's constant

G = mass flow rate

h = enthalpy

J = mechanical equivalent of heat

k = velocity ratio

K = isentropic coefficient of expansion for vapor

L/D = length to diameter ratio

P = pressure

P_1 = pressure at the initial point of flashing

P_2 = pressure at the point of choking

u = velocity

v = specific volume

x = quality, ratio of weight of vapor to total weight of mixture flowing

Greek Symbols

α = vapor void fraction

ρ_ℓ = liquid-phase density

$\bar{\rho}_\ell$ = liquid-phase concentration per unit volume of mixture

ρ_g = vapor-phase density

$\bar{\rho}_g$ = vapor-phase concentration per unit volume of mixture

Subscripts

1 refers to the initial point of flashing

2 refers to the point of choking

E refers to thermal equilibrium

f refers to liquid phase

fg refers to change from vapor to liquid phase

g refers to vapor phase

H refers to homogeneous flow

HEM refers to Homogeneous Equilibrium Model

ℓ refers to liquid phase

NE refers to nonequilibrium

s refers to isentropic expansion

TDE refers to thermodynamic equilibrium

REFERENCES

1. Fauske, H. K., "What's New in Two-Phase Flow," Power Reactor Technology and Reactor Fuel Processing, 9:1 (Winter, 1965-1966).

2. Smith, R. V., "Choking Two-Phase Flow Literature Summary and Idealized Design Solutions for Hydrogen, Nitrogen, Oxygen, and Refrigerants 12 and 11," National Bureau of Standards Technical Note 179 (Superintendent of Documents), Washington, D.C.: Government Printing Office, 1963.

3. Isbin, H. S., H. K. Fauske, M. Petrick, C. H. Robbins, R. V. Smith, S. A. Szawlewicz, and F. R. Zaloudek, "Critical Flow Phenomena in Two-Phase Mixtures and Their Relationships to Nuclear Safety," A/Conf. 28/P/232, Proceedings, Third United Nations International Conference on the Peaceful Uses of Atomic Energy, Geneva, 1964.

4. Fauske, H. K., "Compressibility Affects Flow Instability and Burnout," Power Reactor Technology and Reactor Fuel Processing, 9:2 (Spring, 1966).

5. ————, "Two-Phase Compressibility Phenomena and How They Affect Reactor Safety," Power Reactor Technology and Reactor Fuel Processing, 10:1 (1966-1967).

6. Isbin, H. S., J. E. Moy, and A. J. R. Da Cruz, "Two-Phase, Steam-Water Critical Flow," AIChE Journal, 3:3 (1957), p. 361.

7. Faletti, D. W. and R. W. Moulton, "Two-Phase Critical Flow of Steam-Water Mixtures," AIChE Journal, 9 (1963), p. 247.

8. Fauske, H. K., "Critical Two-Phase, Steam-Water Flows," Proceedings, Heat Transfer and Fluid Mechanics Institute. Stanford, Calif.: Stanford University Press, 1961, p. 79.

9. Zaloudek, F. R., "The Low Pressure Critical Discharge of Steam-Water Mixtures from Pipes," Report No. HW-68934 Rev, General Electric Co. (1961).

10. James, R., "Steam-Water Critical Flow Through Pipes," Proceedings, Institute of Mechanical Engineering, 176 (1962), p. 741.

11. Fauske, H. K., "The Discharge of Saturated Water Through Tubes," AIChE Preprint 30, Proceedings, Seventh National Heat Transfer Conference, Cleveland, Ohio, 1964.

12. Zaloudek, F. R., "The Low Pressure Critical Discharge of Steam-Water Mixtures from Pipe Elbows and Tees," Report No. BNWL-34, Pacific Northwest Laboratory (1965).

13. Klingebiel, W. J., "Critical Flow Slip Ratios of Steam-Water Mixtures," Ph.D. thesis, Department of Chemical Engineering, University of Washington, Seattle, Washington, 1964.

14. Fauske, H. K., "Two-Phase Two- and One-Component Critical Flow," Proceedings, Symposium on Two-Phase Flow, University of Exeter, Devon, England, 1 (1965), p. SG101.

15. Friedrich, H. and G. Vetter, "Einfluss der Düsenform auf das Durchflussverhalten von Düsen für Wasser bei verschiedenen thermodynamischen Zuständen," Energie, 14:1 (1962).

16. Fauske, H. K. and T. C. Min, "A Study of the Flow of Saturated Freon-11 Through Apertures and Short Tubes," Report No. ANL-6667, Argonne National Laboratory (1963).

17. Zaloudek, F. R., "The Critical Flow of Hot Water Through Short Tubes," Report No. HW-77594, General Electric Co. (1963).

18. ————, "Steam-Water Critical Flow from High Pressure Systems," Report No. HW-80535, General Electric Co. (1964).

19. Chen, P. C., "Two-Phase Flow Through Apertures," Ph.D. thesis, Department of Chemical Engineering, University of Minnesota, Minneapolis, Minnesota, 1965.

20. Bryers, R. W. and S. C. Hsieh, "Metastable Two-Phase Flow of Saturated Water Through Short Tubes," No. 21d, Proceedings, Sixtieth National Meeting of the American Institute of Chemical Engineers, Atlantic City, New Jersey, September, 1966.

21. Shapiro, A. H. The Dynamics and Thermodynamics of Compressible Fluid Flow. New York: The Ronald Press, 1953.

22. Fauske, H. K., "Contribution to the Theory of Two-Phase, One-Component Critical Flow," Report No. ANL-6633, Argonne National Laboratory (1962).

23. Massena, W. A., "Steam-Water Critical Flow Using the Separated Flow Model," Report No. HW-65739, General Electric Co. (1960).

24. Levy, S., "Prediction of Two-Phase Critical Flow Rate," Journal of Heat Transfer, Transactions, ASME, C, 87 (1965), p. 363.

25. Ryley, D. J., "The Flow of Wet Steam," Engineer, 193 (1952), p. 363.

26. Moody, F. J., "Maximum Flow Rate of a Single Component, Two-Phase Mixture," Journal of Heat Transfer, Transactions, ASME, C, 87 (1965), p. 134.

27. Zivi, S. M., "Estimation of Steady-State Steam Void Fraction by Means of the Principal of Minimum Entropy Production," Journal of Heat Transfer, Transactions, ASME, C, 86 (1964), p. 247.

28. Cruver, J. E. and R. W. Moulton, "Critical Flow of Liquid-Vapor Mixtures," Preprint 29e, Proceedings, AIChE Fifty-fifth National Meeting, Houston, Texas, 1965.

29. Fauske, H. K., "Two-Phase Critical Flow with Application to Liquid-Metal Systems (Mercury, Cesium, Rubidium, Potassium, Sodium, and Lithium)," Report No. ANL-6779, Argonne National Laboratory (1963).

30. Isbin, H. S., H. K. Fauske, T. Grace, and I. Garcia, "Two-Phase Steam-Water Pressure Drops for Critical Flows," Proceedings, Symposium of Two-Phase Fluid Flow, Institute of Mechanical Engineering, London, 1962.

31. Fauske, H. K., "A Theory for Predicting Pressure Gradients for Two-Phase Critical Flow," Nuclear Science and Engineering, 17 (1963), p. 1.

32. Nahavandi, A. N. and R. F. von Hollen, "Two-Phase Pressure Gradients in the Approach Region to Critical Flow," Report No. WCAP-2662, Westinghouse Corp. (1964).

33. Moody, F. J., "Maximum Two-Phase Vessel Blowdown from Pipes," Journal of Heat Transfer, Transactions, ASME, C, 88:3 (1966), p. 285.

34. Uchida, H. and H. Nariai, "Discharge of Saturated Water Through Pipes and Orifices," Proceedings, Third International Heat Transfer Conference, Chicago, Illinois, 1966.

35. Fauske, H. K., "High Quality Two-Phase Choked Flow," Transactions, ANS, 10:1 (June, 1967).

36. Baker, O., "Simultaneous Flow of Oil and Gas," The Oil and Gas Journal, 53 (1954), pp. 185-195.

37. Ryley, D. J., "The Dispersion of Water Globules in Steam," Engineer (1961).

38. Collingham, R. E. and J. C. Firey, "Velocity of Sound Measurements in Wet Steam," I & EC Process Design and Development, 2:3 (1963).

39. Fauske, H. K., "Propagation of Pressure Disturbances in Two-Phase Flow," Proceedings, Symposium on the Dynamics of Two-Phase Flow, Eindhoven, Netherlands, 1967 (in preparation).

FIGURE 1. COMPARISON BETWEEN MEASURED AND CALCULATED FLOW RATES FROM THE HOMOGENEOUS EQUILIBRIUM MODEL.

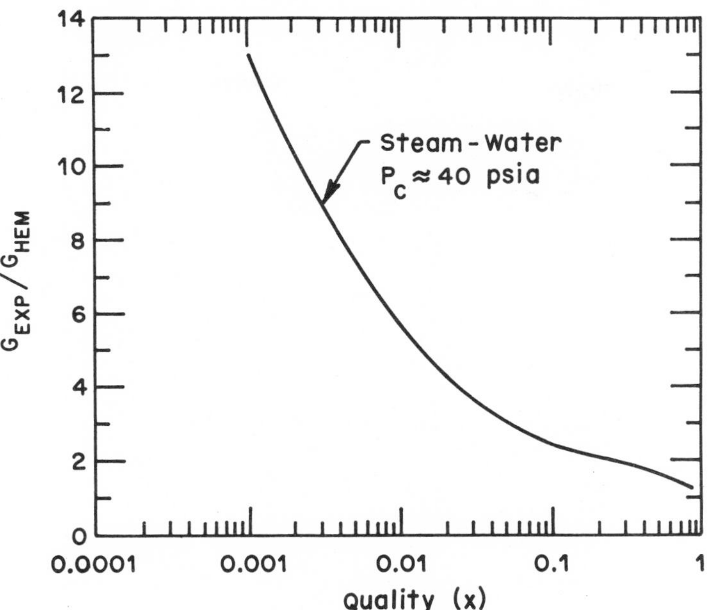

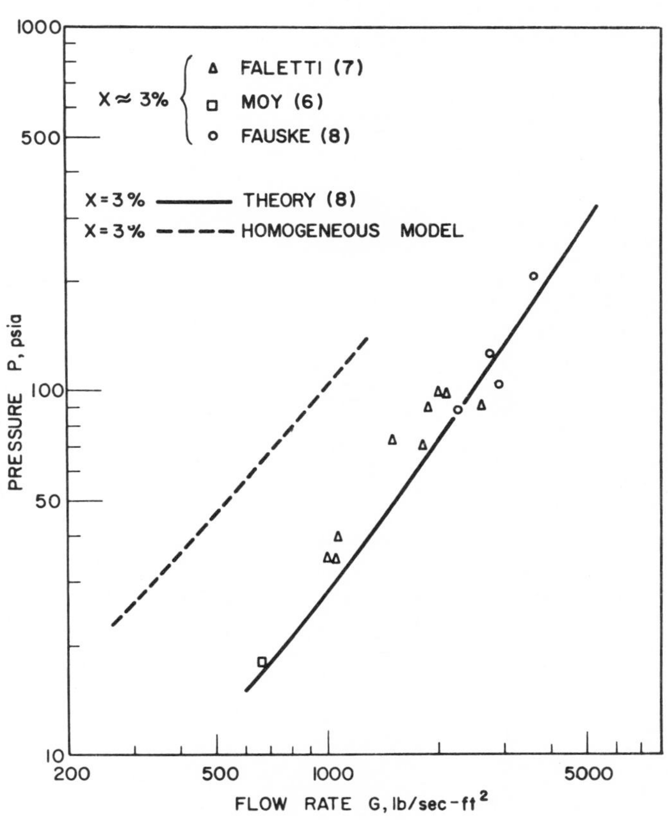

FIGURE 2. COMPARISON BETWEEN MEASURED CRITICAL FLOW RATES AND ANALYTICAL PREDICTIONS.(8) QUALITY EQUAL TO 3 PER CENT.

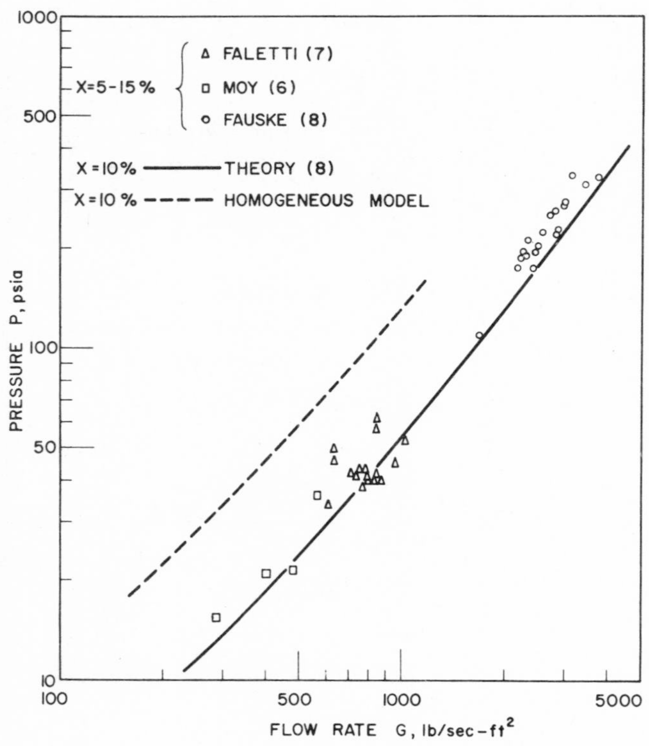

FIGURE 3. COMPARISON BETWEEN MEASURED CRITICAL FLOW RATES AND ANALYTICAL PREDICTIONS.(8) QUALITY EQUAL TO 10 PER CENT.

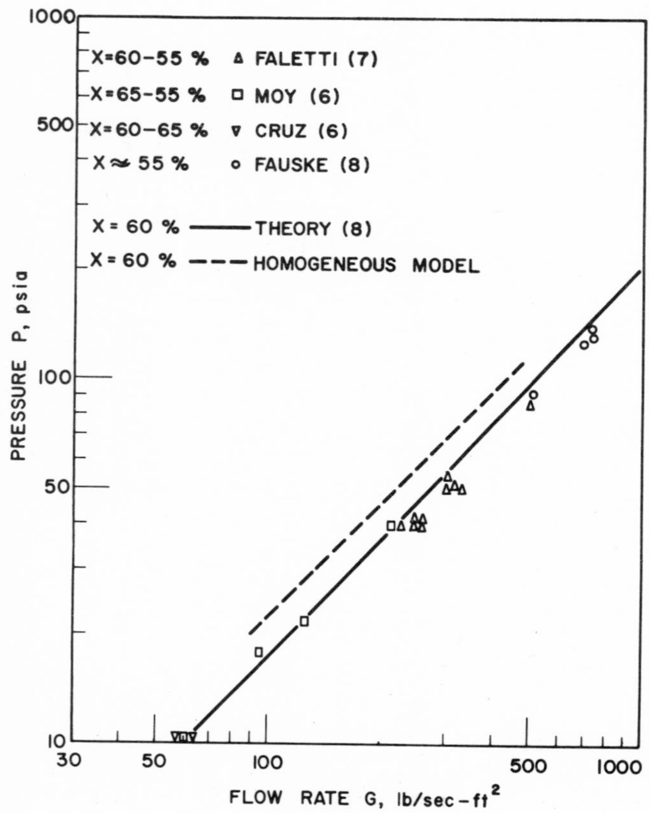

FIGURE 4. COMPARISON BETWEEN MEASURED CRITICAL FLOW RATES AND ANALYTICAL PREDICTIONS.(8) QUALITY EQUAL TO 60 PER CENT.

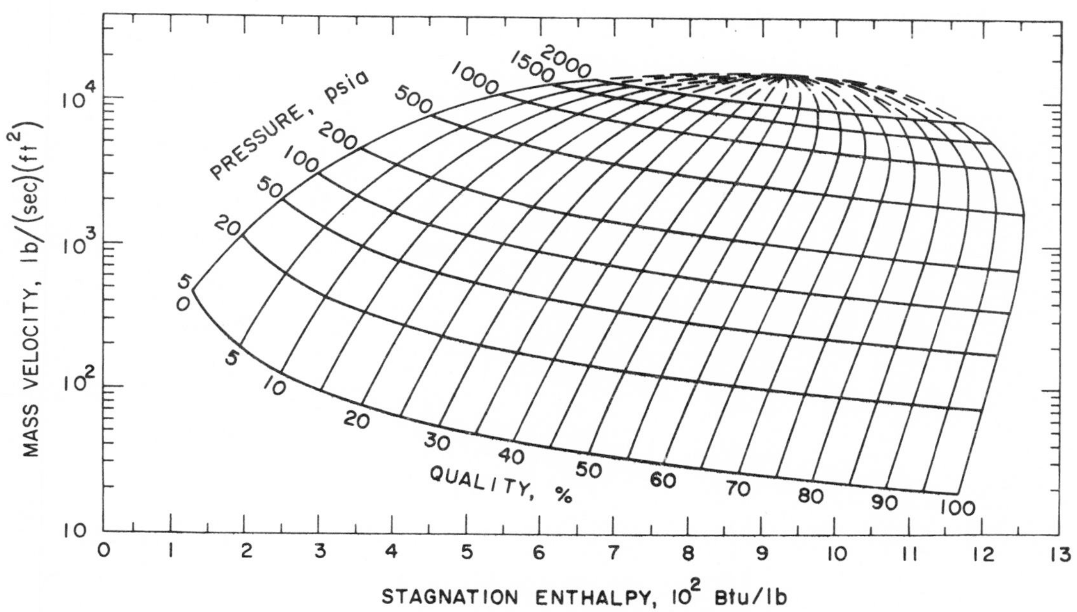

FIGURE 5. ANALYTICAL PREDICTIONS(8) OF CRITICAL FLOW RATES FOR STEAM-WATER MIXTURES.

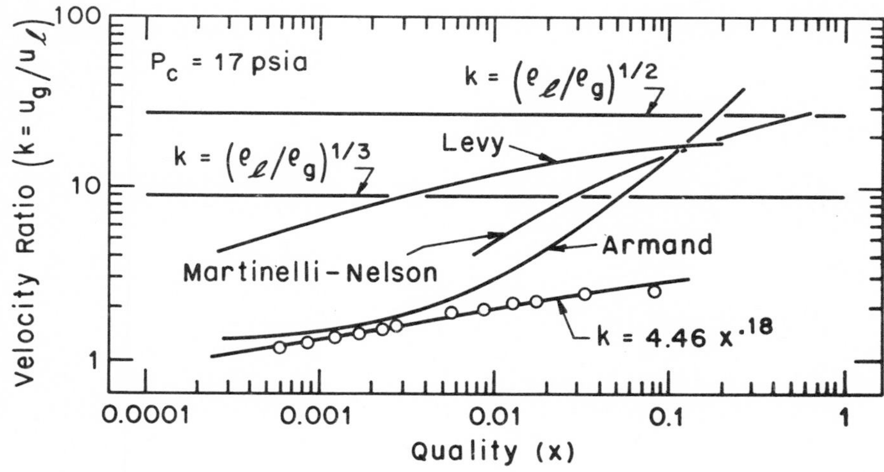

FIGURE 6. EXPERIMENTAL VERSUS CALCULATED VELOCITY RATIOS OF AIR-WATER MIXTURES.

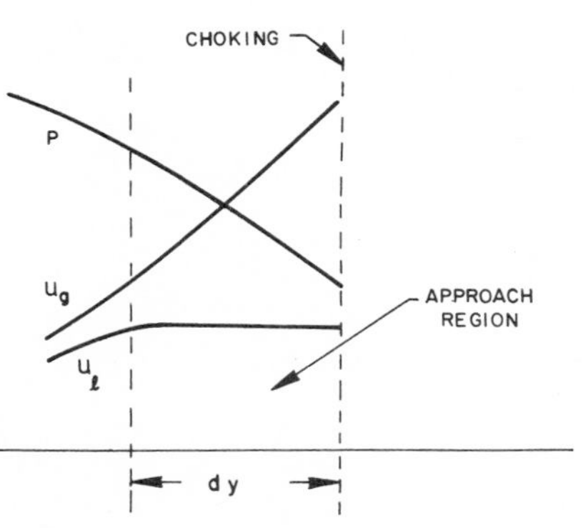

FIGURE 7. PRESSURE AND VELOCITY PROFILES IN THE APPROACH REGION TO CRITICAL FLOW OF A DROPLET MIXTURE.

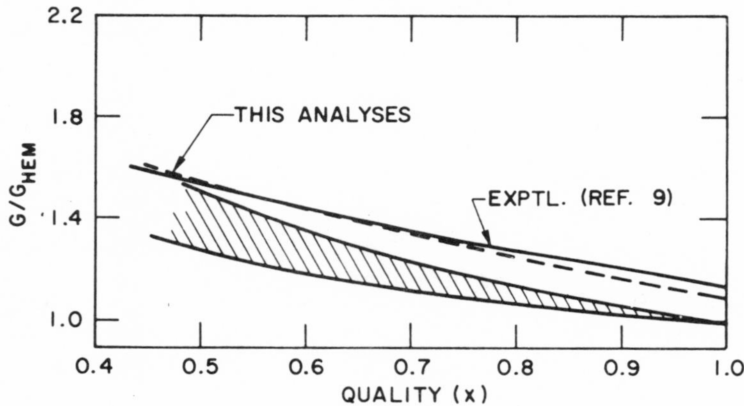

FIGURE 8. COMPARISON BETWEEN CALCULATED AND EXPERIMENTAL CRITICAL FLOW RATES FOR STEAM-WATER MIXTURES AT 40 PSIA (SHADED BAND REPRESENTS EARLIER MODEL PREDICTIONS LISTED IN TABLE 2).

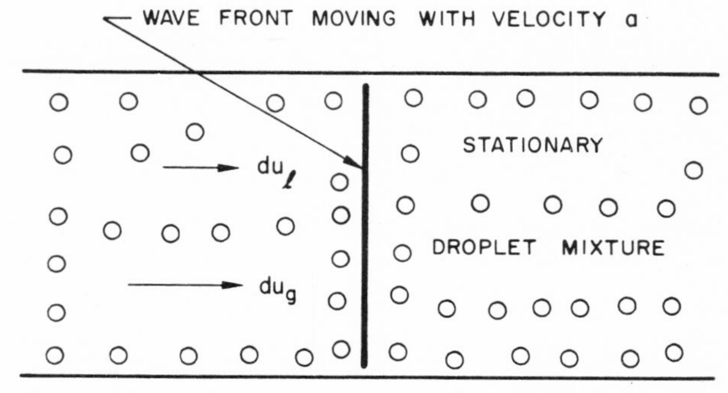

FIGURE 9. ILLUSTRATION OF ONE-DIMENSIONAL PROPOGATION VELOCITY MODEL.

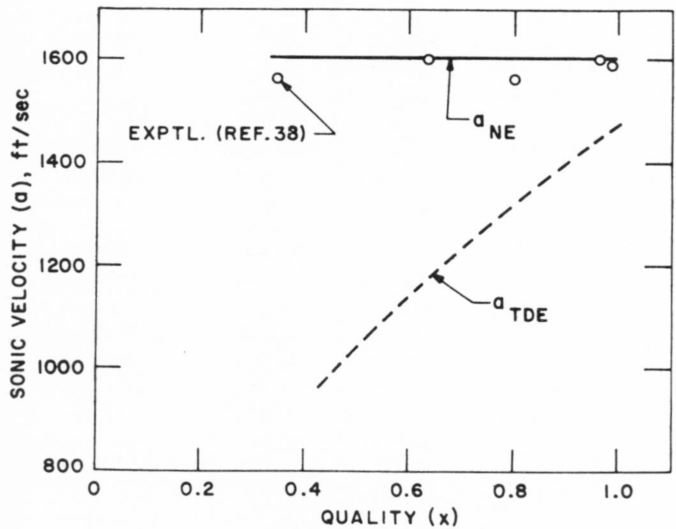

FIGURE 10. COMPARISON BETWEEN CALCULATED AND EXPERIMENTAL VALUES FOR VELOCITY OF SOUND IN STEAM-WATER MIXTURES AT 45 PSIA.

HEAT AND MASS TRANSFER IN SEPARATED FLOWS

R. H. Page*

ABSTRACT

Separated flows are encountered in many heat transfer problems. The classical convective heat transfer approaches are not sufficient for analyzing heat transfer in separated flows, because conventional attached boundary layers do not exist in the separation regions. Separation regions are characterized by free shear layers. The calculation of the heat and mass transfer across free turbulent shear layers is presented in this paper. This is followed by a treatment of the heat and mass transfer pertaining to flow configurations involving both separation and reattachment. Examples of the agreement between theory and experiment are presented.

INTRODUCTION

Separated flows are flows which do not follow the guiding walls. Flow patterns in which the fluid breaks away from a guiding surface are often encountered by engineers working in the heat transfer area. Usually, these flows are turbulent. The Reynolds number (based on free shear layer length) at which transition from laminar to turbulent flow occurs is of the order of 100.(1)** This paper will present methods of analysis that may be used for the calculation of heat and mass transfer in a turbulent separated flow. In order to bring forth the basic ideas in a clear form, our attention will be focused on two-dimensional flow fields.

When one considers a separated flow configuration, two physical observations can be made which are essential to an understanding of the phenomenon. They are: (1) Downstream of the separation point there is always a recirculation pattern. (2) The separated flow always reattaches to a solid boundary (or undergoes a realignment when interacting with a fluid boundary). Thus all separated flows are characterized by a separation point and a reattachment point (realignment point***) with fluid recirculation between them. These two general observations apply to all separated flow configurations regardless of the geometry of the walls or guiding surfaces. In contrast, one notes that the classical separation criteria for attached boundary layers

*Professor and Chairman, Department of Mechanical and Aerospace Engineering, Rutgers, The State University, New Brunswick, New Jersey.

**Superscript numbers in parentheses refer to References at the end of this paper.

***Some authors refer to this as a free rear stagnation point.

are limited to cases of "smooth wall" separation, i.e., $(\delta/r)_S \ll 1.0$. It is well known that separation due to abrupt changes in wall contour cannot be judged by the classical methods used for attached boundary layers; however, the general observations noted above are applicable.

Taking into account the above, a physically perceptive model of separated flow was formally proposed by Dr. H. H. Korst in 1953.(2) This model utilized detailed studies of the various flow components of a separated flow (i.e., attached boundary layer before separation, free shear layer, reattachment, reversed flow, and adjacent flow) in a merged overall analysis in order to account for the interaction between the components. This powerful approach has been used successfully for many problems involving separated flows.(3,4,5)

At present there are two main avenues of approach for dealing with separated flows as a problem of interaction between inviscid and viscous flow mechanisms. In addition to the one mentioned above, which is based on synthesis of individual flow components and often referred to as the Korst-Chapman(6) model, another powerful method which is based on a critical point approach and often referred to as the Crocco-Lees-Reeves model has germinated from Crocco and Lees concepts which were formally proposed in 1952.(7) The Crocco and Lees theory required that the integral conservation equations produce a solution which passed through a critical point of the recompression, which acts much like the throat of a nozzle in determining the solution. Their original model has been extended by the use of multimoment integral methods and has been applied to many separated flow situations.(8,9,10)

The concluding section of this lecture will deal with the overall analyses of separated flow configurations and will follow the Korst approach. One of the components of the separated flow model will be the free shear layer. Since there is a considerable interest in the details of heat and mass transfer across shear layers, these topics will be discussed separately in the next section. They will then be utilized in the overall analyses.

TURBULENT SHEAR LAYER ANALYSIS

Theories for various mixing situations have been published in two well-known books.(11,12) Space is not available here to review the numerous journal articles and reports on mixing. Therefore the author has chosen to select and present the analyses which he has found to be the most accurate and reliable. In this section we shall be concerned with mixing analyses which can be used to calculate turbulent mass and energy transfers across a shear layer. Two situations are considered. The first, single stream mixing, is of importance when one is concerned with the heat transfer between a stream and a stagnant gas. The second, two stream mixing, is important when one wishes to calculate the heat transfer in the shear layer between two streams of different velocity.

Single Stream Mixing

The two-dimensional mixing is shown in Figure 1. The upper gas, a half-infinite jet, has a stagnation enthalpy,

$$h_{o_\infty},$$

while the stagnant gas, of different composition, has an enthalpy,

$$h_{o_{-\infty}}.$$

The turbulent mixing is assumed to be time averaged and the pressure in the shear layer is assumed constant.* It is further assumed that the gases may be treated as nonreacting perfect gases** with an effective turbulent Prandtl number and turbulent Schmidt number of unity.

The velocity and concentration profiles in the shear layer are:(14)

$$\varphi = \frac{1 + \operatorname{erf} \eta}{2}, \tag{1}$$

$$N_a = \frac{1 + \operatorname{erf} \eta}{2}, \tag{2}$$

and

$$N_b = \frac{1 - \operatorname{erf} \eta}{2}, \tag{3}$$

*Because of the two-dimensional nature of the problem, this does not limit the jet itself to a constant pressure field. Only the boundary of the jet is assumed to be at constant pressure.

**See Reference 13 for extensions to cases in which chemical reactions occur.

where

$$\varphi = \frac{u}{u_a} ,$$

$$N = \frac{\text{mass of component}}{\text{mass of mixture}} ,$$

$$\eta = \frac{\sigma y}{x} ,$$

and

$$\sigma = \left[12.0 + 6.16\left(\frac{Ca^2}{1 - Ca^2}\right)\right]^{1/2} .$$

The velocity profile given in Equation (1) is for the fully developed shear layer, which is obtained when the effects of the initial boundary layer are negligible. Equation (1) is an asymptotic solution for the complete solution which includes initial boundary layer effects.[15]

The density and temperature profiles are[14]

$$\frac{\rho}{\rho_\infty} = \frac{T_\infty}{T} \frac{1}{N_a + N_b W_\infty/W_{-\infty}} , \tag{4}$$

$$\frac{T_\infty}{T} = \frac{(1 - C_\infty^2)\, c_p/c_{p\infty}}{h_{o_{-\infty}}/h_{o_\infty} + \left[1 - (h_{o_{-\infty}}/h_{o_\infty})\right] u/u_\infty - C_\infty^2 (u/u_\infty)^2} , \tag{5}$$

where

$$\frac{c_p}{c_{p_\infty}} = N_a + N_b \frac{[k/(k-1)]_{-\infty} W_\infty}{[k/(k-1)]_\infty W_{-\infty}} .$$

Equations (1) through (5) give the local values of the velocity, concentration, and thermodynamic properties as functions of the homogeneous coordinate η and the boundary conditions. These equations are used in a momentum equation in integral form in order to precisely locate them in physical space.[15] It is found that the necessary location of the intrinsic coordinate system (x,y) with respect to the physical (or reference) coordinate system (X,Y) is

$$\eta_m = \eta_R - \int_{-\infty}^{\eta_R} \left(\frac{u}{u_\infty}\right)\left(\frac{\rho}{\rho_\infty}\right) d\eta , \tag{6}$$

in which η_R is a location in the uniform jet and

$$\eta_m = \frac{\sigma y_m}{x} ,$$

where

$$y_m = y - Y,$$

with

$$y_m(0) = 0 .$$

The jet boundary streamline is defined[15] as the streamline which separates the entrained mass from the mass originally in the jet. Its location, η_j, is therefore obtained by application of the principle of conservation of mass to a control volume encompassing the mixing region.[15,16] The following result is obtained:

$$\int_{\eta_j}^{\eta_R} \left(\frac{u}{u_\infty}\right)\left(\frac{\rho}{\rho_\infty}\right) d\eta = \int_{-\infty}^{\eta_R} \left(\frac{u}{u_\infty}\right)^2 \left(\frac{\rho}{\rho_\infty}\right) d\eta . \tag{7}$$

The mass entrained into the shear layer appears in the jet mixing region below the jet boundary streamline. Thus

$$m = \frac{\int_{-\infty}^{j} \rho u dy}{x} , \tag{8}$$

where m is the mass added per unit area. The mass transfer coefficient, β, can be calculated for the given velocity and density profiles.

$$\beta = \frac{m\sigma}{\rho_\infty u_\infty} (1 - C_\infty^2)$$

$$= (1 - C_\infty^2)^{-1} \int_{-\infty}^{\eta_j} \left(\frac{u}{u_\infty}\right)\left(\frac{\rho}{\rho_\infty}\right) d\eta . \tag{9}$$

The energy transfer across the mixing region is found by application of the first law of thermodynamics to a control volume bounded by the jet boundary streamline on the top and a large negative value on the bottom. The energy exchange will include heat transfer and shear work and is positive if the energy transfers from the jet to the surroundings. The dimensionless net energy exchange across the jet boundary streamline[17] is

$$\frac{\Omega_C \sigma}{x \rho_\infty u_\infty h_{o_\infty}} = \int_{\eta_j}^{\eta_R} \frac{\rho}{\rho_\infty} \frac{u}{u_\infty} \left(1 - \frac{h_o}{h_{o_\infty}}\right) d\eta, \quad (10)$$

where Ω_C is the net rate of energy transfer from the gas jet to the stagnant gas per unit of width. The Stanton number for the shear layer is defined in the customary way:

$$St = \frac{\Omega_C}{\rho_\infty u_\infty x \left(h_{o_\infty} - h_{o_{-\infty}}\right)} \quad (11)$$

With some simplification[14] it can be shown that

$$\sigma St = \int_{-\infty}^{\eta_j} \left(\frac{u}{u_\infty}\right)^2 \left(\frac{\rho}{\rho_\infty}\right) d\eta \ . \quad (12)$$

For convenience, an energy transfer (shear work and heat) coefficient can be calculated from the definition:

$$Q = \frac{\sigma St}{1 - C_\infty^2} \ . \quad (13)$$

Some typical results for the transfer coefficients are given in Figures 2 through 7. The parameters α_1 and α_2 which appear on the figures are defined in the notation. Typical values of β are shown in Figures 2, 3, and 4. Figure 2 is for a jet with a molecular weight ten times that of the stagnant gas. Figure 3 is for a jet with a molecular weight identical to that of the stagnant gas. Figure 4 is for a jet with a molecular weight which is one-tenth of that of the stagnant gas. With all other considerations constant, the general trends which can be observed are that: decreasing the molecular weight of the jet increases the mass transfer coefficient; increasing the Crocco number of the jet increases the mass transfer coefficient; and decreasing the temperature ratio of the stagnant gas to the stagnation temperature of the jet increases the mass transfer coefficient.

Figures 5, 6, and 7 show the variation of the energy transfer coefficient in the same manner that Figures 2, 3, and 4 show the variation of the mass transfer coefficient. The energy transfer coefficient undergoes trends which are the same as those described for the mass transfer coefficient. This is to be expected because of the Reynolds' analogy for constant pressure flow with a Prandtl number of unity.

Two Stream Mixing

Two stream mixing plays an important part in separated flow fields in which the recirculating flow serves as one of the mixing streams. Because of this fact, the discussion of the two stream case shall be restricted to two streams of identical composition. Figure 8 shows the typical turbulent two-dimensional configuration. Both streams are assumed to be isobaric, nonreacting perfect gases with an effective turbulent Prandtl number of unity. The solution[18] is again based on a momentum integral method. The velocity profile is

$$\varphi = \frac{\left(1 + \varphi_b\right) + \left(1 - \varphi_b\right) \ \mathrm{erf}\ \eta}{2} , \quad (14)$$

where

$$\varphi_b = \frac{u_b}{u_a} ,$$

$$\eta = \sigma_{II} \frac{y}{x} ,$$

and σ_{II} is the empirical spreading parameter for two stream mixing.[19] The stagnation temperature and density profile are

$$\frac{T_o}{T_{o_a}} = \frac{T_{o_b}}{T_{o_a}} \frac{1 - \varphi}{1 - \varphi_b} + \frac{\varphi - \varphi_b}{1 - \varphi_b} , \quad (15)$$

$$\frac{\rho}{\rho_a} = \frac{1 - C_a^2}{T_o/T_{o_a} - C_a^2 \varphi^2} \ . \quad (16)$$

Using a procedure similar to that followed above for the single stream case, one finds:[18]

$$\eta_m = \eta_{R_a} - \frac{I_2\left(\eta_{R_a}\right) - \varphi_b I_1\left(\eta_{R_a}\right)}{1 - \varphi_b} , \tag{17}$$

$$I_1(\eta_j) = \frac{I_1\left(\eta_{R_a}\right) - I_2\left(\eta_{R_a}\right)}{1 - \varphi_b} , \tag{18}$$

$$\sigma St = \frac{I_2(\eta_j) - \varphi_b I_1(\eta_j)}{1 - \varphi_b} , \tag{19}$$

where

$$I_1 = I_1\left(\eta, C_a^2, \frac{T_{o_b}}{T_{o_a}}, \varphi_b\right)$$

$$= \frac{\varphi_b\left(1 - C_a^2\right)\eta_{R_b}}{\left(T_{o_b}/T_{o_a}\right) - C_a^2 \varphi_b^2} + \int_{\eta_{R_b}}^{\eta} \frac{\left(1 - C_a^2\right)\varphi}{\left(T_o/T_{o_a}\right) - C_a^2 \varphi^2} d\eta , \tag{20}$$

and

$$I_2 = I_2\left(\eta, C_a^2, \frac{T_{o_b}}{T_{o_a}}, \varphi_b\right)$$

$$= \frac{\varphi_b^2\left(1 - C_a^2\right)\eta_{R_b}}{\left(T_{o_b}/T_{o_a}\right) - C_a^2 \varphi_b^2} + \int_{\eta_{R_b}}^{\eta} \frac{\left(1 - C_a^2\right)\varphi}{\left(T_o/T_{o_a}\right) - C_a^2 \varphi^2} d\eta . \tag{21}$$

Reference 18 also includes formulation for other useful properties of the jet mixing region between two streams.

Influence of Finite Initial Boundary Layer

The computations for the fully developed shear layer for the two cases above are relatively simple. When the effects of the initial boundary layer are predominant, the complete solution requires more extensive computations and empirical information on the initial variation of the eddy viscosity. Therefore approximate methods have been developed which utilize the fully developed velocity profile with a correction to account for initial boundary layer effects. Two such methods are worth noting here. An "equivalent mass concept"[18] utilizes a lateral displacement of the fully developed velocity profile. An "origin shift"[20] utilizes an upstream displacement of the origin of the fully developed velocity profiles. These methods are convenient to use when the initial boundary layer effects must be taken into account.

OVERALL ANALYSES

In this section we shall consider convective heat transfer for separated flow configurations. As pointed out in the introduction, this implies that we shall consider flows with separation and reattachment points and consider the interaction of the various components. Thus a system (or control volume) approach will be used. We will consider the steady flow of perfect gases and exclude chemical reactions and radiation heat transfer from the analyses in order to present the methods in a straightforward manner without extraneous complications.

Backstep Model

Figure 9 illustrates the separation region produced by supersonic flow over a backstep. As shown, the possibility of bleeding mass into the separated region is present. The components of such a flow configuration are subject to direct analysis, and an overall solution to the problem is available.(4,16,21,22) The effect on the base pressure of heated mass addition to the separated region ($G_B > 0$) is shown in Figure 10; the effect of heat transfer with no mass addition is shown in Figure 11. Both figures were calculated for fully developed shear layers (that is, thin approaching boundary layers). In each case the approach Mach number has been arbitrarily selected as 2.0 and the specific heat ratio corresponds to the value for air. In Figure 10,

$$\frac{T_{o_B}}{T_{o_a}}$$

refers to the ratio of bleed gas stagnation temperature to free stream temperature while $\mathcal{H}$ is the dimensionless bleed number defined as

$$\mathcal{H} = \left[\frac{G_B\sqrt{T_{o_a}}\left(\mathbb{R}/kg_c\right)^{1/2}}{H\,P_{o1_a}}\right], \tag{22}$$

where G_B is the mass bleed rate per unit width at which gas is added externally to the wake, H is the step height, $\mathbb{R}$ is the gas constant, g_c is the gravitational constant,

$$T_{o_a}$$

is the free stream stagnation temperature, and

$$P_{o1_a}$$

is the stagnation pressure of the free stream before the base.

For a fixed value of mass bleed, G_B, an increase in

$$\frac{T_{o_B}}{T_{o_a}}$$

results in an increase in the base pressure ratio. The direct influence of heat transfer (without mass bleed) on the base pressure ratio is illustrated in Figure 11. Λ is the dimensionless heat transfer number which is defined as

$$\Lambda = \frac{Q_b}{Hc_p P_{o1_a}\sqrt{T_{o_a}}}\left(\frac{\mathbb{R}}{kg_c}\right)^{1/2}\left(\frac{k+1}{2}\right)^{(k+1)/2(k-1)}, \tag{23}$$

where c_p is the specific heat at constant pressure, and Q_b is the heat transfer rate per unit width from the adjacent walls to the wake.

Thus, for a closed wake (i.e., one without mass bleed), heat addition to the gas increases the pressure in the separated region. The converse is also true. The relationships between base pressure and heat transfer for values of Mach numbers between 1 and 10 are shown in Figure 12.(23) Lines for constant values of the heat transfer parameter, $D(D = \sigma\Lambda)$, are plotted. Positive values of D correspond to heat addition to the wake from the wall; negative values of D indicate the direction of the heat transfer is from the gas to the wall.

Figure 12 accounts for the effective heat transfer without specifying the wall temperature -- free stream temperature ratio. In order to specify the overall temperature ratio, one must consider the recirculating flow.

A sketch of the flow model without external mass bleed is shown in Figure 13. Only one-half of the flow is shown since the supersonic flow field will be symmetrical. The influence of unsymmetrical trailing edge conditions can be accounted for by extensions of the basic method.(4) The flow model is composed of the following components: (1) the expansion around the corner 1 to 2, (2) the constant pressure mixing region 2 to 3, (3) the recompression to the centerline direction 3 to 4, and (4) the recirculation region.

Solutions of the model[24] are available for the following conditions: (1) The mixing zone and the recirculating flow are turbulent, (2) the gas is a perfect gas, (3) the influence of the wall boundary layer at 1 is neglected, and (4) the base wall temperature, T_W, is constant.

The relationship between the bulk temperature, T_b, and the wall temperature, T_W, is determined by considering the recirculating flow shown in Figure 14. There are two resistances to heat transfer (the mixing region and the recirculating flow) in series. Page and Dixon[24] used a reflected image of a portion of the free mixing region to estimate the heat transfer through the recirculating flow. They found the theoretical results for air shown in Figure 15. The results are presented and compared with experimental data[25,26,27,28] in terms of a base heating parameter,

$$\frac{q}{P_{o_1}\sqrt{T_{o_a}}},$$

where q is the heat transfer per unit area per unit time,

$$P_{o_1}$$

is the stagnation pressure of the gas flow at 1, and

$$T_{o_a}$$

is the stagnation temperature of the gas flow. The same theoretical and experimental results are shown in Figure 16 as a function of wall temperature, in order to emphasize the fact that the heat transfer is not a linear function of wall temperature. Good agreement is obtained between the available experimental data and the theory. The base heating rate for other gases may be calculated from the theory.[24] This method has been extended in order to calculate the heat transfer in more complex separated flows.[29,30]

Cavity Flow Model

A cavity flow involves a separation point, a reattachment point, and a complex system of interrelated flow regions.[19,31] The analysis requires two steps. First, the dissipative flow model will be used to account for the transfer of mechanical energy to the cavity and its dissipation within the system. The second step will involve a thermodynamic system analysis in order to calculate the convective heat transfer and the cavity Stanton number.

Dissipative Flow Model

The cavity flow model is shown in Figure 17. The separation point, S, and the reattachment, R, are the leading and trailing edges (points) of the cavity. The jet boundary streamline, j, and the discriminating streamline, d, are identified. Since the discriminating streamline is defined as the streamline in the shear layer which discriminates between the flow which recirculates and the flow which passes on downstream, it reattaches at R. The jet boundary streamline originates at S and will be identical to the d-streamline for the case of no mass bleed into the cavity (i.e., $G_B = 0$). Obviously,

$$G_B + \int_{Y_j}^{Y_d} \rho u \, dY = 0.$$

Considering the system boundary shown in Figure 17 (the walls forming the cavity, the j-streamline, and the cross section near R between the j- and d-streamlines) the net amount of mechanical energy transferred to the system must equal the mechanical energy dissipated within the system. Thus[19]

$$\int_o^{\ell_m} \tau_j u_j dx - \int_{Y_d}^{Y_j} \left(\frac{\rho u^3}{2}\right) dY = \int_V e_D \, dV. \quad (24)$$

The first term represents the rate at which shear work is transferred from the external flow to the system. The second term represents the outflow of kinetic energy from the system and is present when $G_B \neq 0$. The dissipation integral can be represented[19] as

$$\int e_D \, dV = E_{D_m} + E_{D_R} + E_{D_{BL}} + E_{D_c}, \quad (25)$$

where dissipation in the jet mixing region (m), the recompression zone (R), the cavity wall boundary layer (BL), and the cavity core (c) are acknowledged as isolated dissipative mechanisms. In some separated flow configurations, the interaction between the dissipative mechanisms may be so strong as to prevent the possibility of isolating the dissipative mechanisms into separate components as done in Equation (25). We shall consider a case where it is possible to isolate the dissipative mechanisms.

We consider a nearly circular cavity in which the core region will exhibit rotation similar to solid body rotation

$$(\therefore E_{D_c} = 0)$$

and the sharp edges at R and S will produce a well-defined mixing region. For such a geometry the dissipation of mechanical energy in the recompression zone may be neglected for incompressible flow,

$$E_{D_R} = 0 ,$$

but may have to be accounted for when shock waves are present in the vicinity of R for supersonic flows.[32] We shall assume

$$E_{D_R} = 0 .$$

Thus, for a nearly circular cavity with a central core which produces a nearly constant flow velocity along the edges of the jet mixing region and the cavity wall boundary layer, Equation (25) becomes

$$\int_V e_D \, dV = E_{D_m} + E_{D_{BL}} . \tag{26}$$

Substituting Equation (26) into Equation (24) yields

$$E_{D_{BL}} = \int_0^{\ell_m} \tau_j u_j \, dx - \int_{Y_d}^{Y_j} \left(\frac{\rho u^3}{2}\right) dY - E_{D_m} . \tag{27}$$

The dissipation in the cavity wall boundary layer has been solved for the incompressible case[31] by assuming that the boundary layer has a 1/n-th power law profile, that the external shear flow retains its velocity distribution, and that the wall shear stress is locally controlled by the thickness of the wall boundary layer and by the velocity at its edge where it joins the external profile. It is found[31] that

$$\frac{E_{D_{BL}}}{\rho_a u_a^3 \, (\ell_m/2)} = f\left(Re_m, \varphi_b, \frac{\ell_w}{\ell_m}, \Delta\right) , \tag{28}$$

where

$$Re_m = \frac{u_a \ell_m \rho_a}{\mu_a} ,$$

$$\varphi_b = \frac{u_b}{u_a} ,$$

ℓ_W is the length of cavity wall, and Δ is a displacement parameter[32] defined as follows:

$$\Delta = \frac{\theta_a}{\ell_m} + \frac{\rho_b \theta_b \varphi_b^2}{\rho_a \ell_m} + \frac{(1 - \varphi_b)\, G_B}{u_a \ell_m \rho_a} , \tag{29}$$

in which θ_a and θ_b are the momentum thicknesses of the approaching boundary layers of the two stream mixing component (Figure 18).

The net transfer of mechanical energy to the wake minus the dissipation in the mixing region, i.e., the right side of Equation (27), can be evaluated for two stream mixing using standard methods with an "equivalent mass bleed concept" and tabulated integrals[18] as

$$\frac{I_4(\eta_d)}{\sigma_{II}} ,$$

where

$$I_4 = \left(1 - C_a^2\right) \int_{-\infty}^{\eta} \left[\varphi \frac{\varphi^2 - \varphi_b^2}{\left(T_o/T_{o_a}\right) - C_a^2 \varphi^2} \right] d\eta . \tag{30}$$

Thus the right side of Equation (27) is a function of φ_b, C_a^2,

$$\frac{T_{o_b}}{T_{o_a}},$$

and Δ. For an incompressible constant pressure shear layer which is nearly isothermal, the right side of Equation (27) will just depend upon φ_b and Δ. Since the functional relationship for the left side of Equation (27) has been presented in Equation (28), it is apparent that the wake reference velocity, φ_b can be determined for an incompressible cavity flow in the following functional form.

$$\varphi_b = f\left(Re_m, \Delta, \frac{\ell_W}{\ell_m}\right) . \qquad (31)$$

Numerical results[31] for Equation (31) are presented in Figure 19. The wake reference velocity, φ_b, is shown as a function of wall-to-mixing length ratio for three values of the length Reynolds number and three values of the displacement parameter. The values of φ_b are significant. It is apparent that cavity flows are characterized by finite wake velocities.

Thermodynamic Flow Model

Recognizing the importance of the finite wake velocities, we will proceed to a thermodynamic system analysis. The system boundary that was used for the dissipative flow model will also be used for the thermodynamic analysis. Figure 20 shows the energy transfers which must be considered. Standard thermodynamic sign conventions are used and thus the heat added to the system and work delivered by the system are positive. Positive directions are indicated by the arrows in Figure 20. The energy balance is

$$\Omega_d + c_p G_B \left(T_{o_B} - T_{o_a}\right) + Q_W + Q_d = 0, \qquad (32)$$

where Ω_d is the energy transport rate in the mixing region[18] (this becomes Ω_j if $G_B = 0$),

$$c_p G_B T_{o_B}$$

is the energy transport rate into cavity due to mass bleed, Q_W is the heat transfer by convection from the cavity walls, and Q_d is the diffusion of thermal energy into the flow above the cavity. (This is due to the introduction of thermal energy at S from the cavity wall boundary layer.)

The changes in density due to the temperature field resulting from Q_d may be neglected if the temperature differences do not become too large. Under those conditions the solution for Q_d may be superimposed on the mixing solution and the velocity and density profiles of the mixing solution [Equations (14) and (16)] may be used in the formulation of the line source heat diffusion problem. The solution has been given[33] in terms of a partition function, Π, in which

$$\Pi_d = -\frac{Q_d}{Q_W} . \qquad (33)$$

Values of the partition function for $G_B = 0$ are given in Figure 21.

Considering the energy transfers which are specified for the heat transfer model, it is apparent that three Stanton numbers may be defined. The cavity Stanton number, St_c, the mixing region Stanton number, St_m, and the average cavity wall Stanton number, $\overline{St}_W$, are:

$$St_c = \frac{Q_W}{\ell_m u_a \rho_a c_p \left(T_W - T_{o_a}\right)} , \qquad (34)$$

$$St_m = \frac{\Omega_j}{\ell_m u_a \rho_a c_p \left(T_{o_a} - T_{o_b}\right)} , \qquad (35)$$

$$\overline{St}_W = \frac{Q_W}{\ell_W u_b \rho_b c_p \left(T_W - T_{o_b}\right)} . \qquad (36)$$

These Stanton numbers have been evaluated[31] in terms of tabulated integrals.[18] A cavity with no mass bleed ($G_B = 0$) is a special case of the above for which

$$\frac{St_c}{St_m} = \frac{1 - T_{o_b}/T_{o_a}}{\left(1 - \Pi_j\right)\left(1 - T_W/T_{o_a}\right)} , \qquad (37)$$

$$\frac{St_c}{St_m} = \left[1 - \Pi_j + \left(\frac{St_m \rho_a \ell_m}{\overline{St}_W \varphi_b \rho_b \ell_W}\right)\right]^{-1} . \qquad (38)$$

Theoretical cavity Stanton numbers are shown in Figure 22 for $G_B = 0$ and small temperature differences. The comparison which is made in Figure 22 between convective heat transfer for attached flow and for separated flow is of special interest. The regions for which the convective heat transfer rates are greater than

$$\frac{St_c}{St_{att}} > 1$$

or less than

$$\frac{St_c}{St_{att}} < 1$$

the equivalent attached flat plate flow are indicated. The primary influence in determining whether the turbulent convective heat transfer is greater or less than the corresponding flat plate flow is the wall length ratio, ℓ_w/ℓ_m.

Experimental confirmation of the theory[19,31] is illustrated in Figures 23 and 24. Figure 23 shows a comparison of the mixing region Stanton numbers obtained from the theory and those experimentally determined for an adiabatic cavity with heat supplied to the core region by means of an electric heating element.[34] Figure 24 illustrates a comparison between the theoretically obtained cavity Stanton number and those experimentally determined in the Department of Mechanical Engineering of the University of Illinois and in the Ames Research Center[35] of the National Aeronautics and Space Administration. Since NASA tests were conducted with an axially symmetric model, the ratio of ℓ_w/ℓ_m was interpreted as the wall cavity surface area to jet mixing surface area and produced a value of ℓ_w/ℓ_m less than unity. The theory is shown to be in good agreement with the experimental data presented in Figures 23 and 24.

Other Geometries and Models

The treatments of the base gas temperature problem[36] and the blunt base heat transfer problem[37] illustrate approaches which rely upon the basic Korst theory.[4] The calculation of the convective heat and mass transfer in separated flows with configurations other than a backstep or nearly circular cavities may be logically approached as extensions of the methods described above. Yet, it must be emphasized that the wall geometry (especially that portion which the recirculating flow encounters) has a major influence on the flow pattern and often other factors, if present, must be considered. For example, three-dimensional cavity flow cells of the Benard type, cavity resonance[34] and cavity pulsation[38] have been observed in separated flow regions.

It is also interesting to note the subsonic experimental data obtained by the Berkeley group (Reference 39), the numerical correlations obtained for rectangular cavities[40] and steps,[41] and other correlations obtained in reviews[42] and recent theories.[43]

SUMMARY

The recirculating flow plays an important role in the determination of the heat transfer to a separated flow. Although one can calculate the heat and mass transfer across a shear layer that has separated from a wall in terms of its local boundary conditions, the calculation of the wall heat transfer must involve an overall analysis with all flow components considered.

Methods of calculating heat and mass transfer for two-dimensional turbulent free shear flows have been reviewed. Also, methods for calculating heat transfer and the effect of mass transfer on two-dimensional turbulent separated flows with reattachment have been reviewed.

The techniques reviewed here have been used successfully for many separated flow heat transfer problems. Some examples of the good agreement between theory and experiment have been presented. The careful extension of this approach to separated flow heat transfer problems with different flow conditions -- or different geometries -- or other changes in new applications should be feasible.

NOTATION

c_p	= specific heat
C	= Crocco number, $u/[2c_pT_o]^{1/2}$
e_D	= dissipation of mechanical energy per unit volume
E_D	= dissipation of mechanical energy, Equation (25)
g_c	= gravitational constant
G_B	= mass bleed rate
h	= enthalpy
H	= step height

I_1, I_2, I_4 = defined integrals, Equations (20), (21), (30)

k = ratio of specific heats

ℓ_m = length of mixing region

ℓ_W = length of cavity wall

m = mass entrainment per unit area

N_a = mass fraction of a (mass of component a per unit volume/density of mixture)

N_b = mass fraction of component b

n = boundary layer velocity power law number

p = pressure

q = heat transfer per unit area per unit time

Q = energy transfer coefficient

Q_b, Q_W, Q_d = heat transfers, Equations (23), (32)

r = local value of radius of curvature of wall

Re_m = Reynolds number $(u_a \ell_m \rho_a / \mu_a)$

St = Stanton number

T = temperature

u = velocity

V = volume

W = molecular weight

x, y = intrinsic coordinate system for velocity profile

X, Y = reference (physical) coordinate system

α_1 = molecular weight parameter $(W_{-\infty}/W_\infty)$

α_2 = specific heat ratio parameter $([k/(k-1)]_{-\infty}/[k/(k-1)]_\infty)$

β = mass transfer coefficient

δ = boundary layer thickness

Δ = displacement parameter, Equation (29)

η = homogeneous coordinate for profiles

$\mathcal{H}$ = bleed number, Equation (22)

θ = boundary layer momentum thickness

μ = viscosity

Π = partition function

ρ = density

σ = similarity parameter for single stream mixing

σ_{II} = similarity parameter for two stream mixing

τ = shear stress

φ = dimensionless velocity (u/u_a)

φ_b = u_b/u_a

Ω = energy transport rate in jet mixing region

$\Omega_c, \Omega_j, \Omega_d$ = energy transports, Equations (10), (32)

Subscripts

a = adjacent potential flow above shear layer

b = flow below shear layer (entrainment stream)

B = bleed flow condition

d = discriminating streamline

j = jet boundary streamline

m = momentum shift

o = stagnation condition

R = reference location in potential flow

R = reattachment point

$\mathbb{R}$ = gas constant

S = separation point

∞ = above shear layer

$-\infty$ = below shear layer

REFERENCES

1. Page, R. H. and W. G. Hill, Jr., "Location of Transition in a Free Jet Region," AIAA Journal, 4:5 (1966), p. 944.

2. Korst, H. H., "Research on Transonic and Supersonic Flow of a Real Fluid at Abrupt Increases in Cross Section," Technical Status Report A-PR-4, AF18 (600)-392, Department of Mechanical Engineering, University of Illinois, Urbana, Illinois, July, 1953.

3. ————, "A Theory for Base Pressures in Transonic and Supersonic Flow," Journal of Applied Mechanics, 23(1956), pp. 593-600.

4. Korst, H. H., W. L. Chow, and G. W. Zumwalt, "Research on Transonic and Supersonic Flow of a Real Fluid at Abrupt Increases in Cross Section (with Special Consideration of Base Drag Problems)," Technical Report ME-TR-392-5, OSR-TR-60-74, Contract AF18 (600)-392. Urbana, Ill.: Department of Mechanical Engineering, University of Illinois, December, 1959.

5. Korst, H. H., "Turbulent Separated Flows," Short Course Proceedings, Von Karman Institute for Fluid Dynamics, Brussels, April, 1967.

6. Chapman, D. R., D. M. Kuehn, and H. K. Larson, "Investigation of Separated Flows in Supersonic and Subsonic Streams with Emphasis on the Effect of Transition," NACA Report 1356, 1958.

7. Crocco, L. and L. Lees, "A Mixing Theory for the Interaction Between Dissipative Flows and Nearly Isentropic Streams," Journal of the Aerospace Sciences, 19:10 (October, 1952), pp. 649-676.

8. Lees, L. and B. L. Reeves, "Supersonic Separated and Reattaching Laminar Flows: I. General Theory and Application to Adiabatic Boundary Layer-Shock Wave Interactions," AIAA Journal, 2:11 (1964), pp. 1907-1920.

9. Reeves, B. L. and L. Lees, "Theory of the Laminar Near-Wake of Blunt Bodies in Hypersonic Flow," AIAA Journal, 3 (1965), pp. 2061-2074.

10. Golik, R. J., W. H. Webb, and L. Lees, "Further Results of Viscous Interaction Theory for the Laminar Supersonic Near Wake," AIAA Journal, preprint (1967), pp. 67-61.

11. Pai, S. I. Fluid Dynamics of Jets. New York: D. Van Nostrand Co., 1954.

12. Abramovich, G. N. The Theory of Turbulent Jets [translation]. Cambridge, Mass.: MIT Press, 1963.

13. Davis, L. R., "The Effect of Chemical Reactions in the Turbulent Mixing Component on the Dynamics and Thermodynamics of Wake Flow Fields," Ph.D. thesis, Department of Mechanical Engineering, University of Illinois, Urbana, Illinois, 1964.

14. Page, R. H. and R. J. Dixon, "Computer Evaluation of an Integral Treatment of Gas Mixing," Third Conference on Performance of High Temperature Systems, December, 1964, Pasadena, California (proceedings in press).

15. Korst, H. H., R. H. Page, and M. E. Childs, "Compressible Two-dimensional Jet Mixing at Constant Pressure," Technical Report ME-TN-392-1. Urbana, Ill.: Department of Mechanical Engineering, University of Illinois, April, 1954.

16. Page, R. H. and H. H. Korst, "Nonisoenergetic Turbulent Compressible Jet Mixing with Consideration of Its Influence on the Base Pressure Problem," Proceedings, Fourth Midwestern Conference on Fluid Mechanics, Purdue University, Lafayette, Indiana, September, 1955, pp. 45-68.

17. Korst, H. H. and W. L. Chow, "Compressible Nonisoenergetic Two-dimensional Turbulent (Prandtl Number Equals Unity) Jet Mixing at Constant Pressure-Auxiliary Integrals-Heat Transfer and Friction Coefficients for Fully Developed Mixing Profiles," Technical Report ME-TN-392-4. Urbana, Ill.: Department of Mechanical Engineering, University of Illinois, January, 1959.

18. ————, "Nonisoenergetic Turbulent ($Pr_t = 1$) Jet Mixing Between Two Compressible Streams at Constant Pressure," NASA-CR-419, April, 1966.

19. Korst, H. H., "Dynamics and Thermodynamics of Separated Flows," AGARD Conference Proceedings No. 4, May, 1966, pp. 701-746.

20. Kessler, T. J., "Two-Stream Mixing with Finite Initial Boundary Layers," AIAA Journal, 5:2 (February, 1967), pp. 363-364.

21. Page, R. H., "On Turbulent Supersonic Diabatic Wakes," American Rocket Society Journal, 29:6 (June, 1959), pp. 443-445.

22. Korst, H. H. and W. L. Chow, "Influence on Base Pressures by Heat and Mass Additions," American Rocket Society Journal, 32 (1962), pp. 1094-1095.

23. Dixon, R. J. and R. H. Page, "The Interdependence of Base Pressure and Base Heat Transfer," American Rocket Society Journal, 31:12 (December, 1961), pp. 1785-1786.

24. Page, R. H. and R. J. Dixon, "Base Heat Transfer in a Turbulent Separated Flow," Proceedings, Fifth International Symposium on Space Technology and Science, Tokyo, September, 1963, AGNE Corp., Tokyo, Japan, pp. 295-308.

25. Swartz, R. J., "Heat Transfer and Pressure Measurements on the Base of a Series of Blunted-Cone-Cylinder Flare Configurations," D2-12874, Aero-Space Division, The Boeing Co., 1962.

26. Naysmith, A., "Heat Transfer and Boundary Layer Measurements in a Region of Supersonic Flow Separation and Reattachment," RAE, TN Aero. 2558, May, 1958.

27. Bloom, M. H. and A. Pallone, "Shroud Tests of Pressure and Heat Transfer Over Short Afterbodies with Separated Wakes," Journal of the Aerospace Sciences, 26:10 (1959), pp. 626-636.

28. Rabinowicz, J., "Measurement of Turbulent Heat Transfer Rates on the Aft Portion and Blunt Base of a Hemisphere-Cylinder in the Shock Tube," GALCIT Memo, No. 41, November, 1957.

29. Page, R. H. and R. J. Dixon, "Base Heating on a Multiple Propulsion Nozzle Missile," Paper No. 63-179, AIAA Summer Meeting, Los Angeles, California, June, 1963.

30. Dixon, R. J. and R. H. Page, "Theoretical Analysis of Launch Vehicle Base Flow," Proceedings, AGARD Conference, No. 4, Brussels, May, 1966, pp. 911-940.

31. Korst, H. H., "Dynamics and Thermodynamics of Separated Flows," Proceedings, Symposium on Single- and Multi-Component Flow Processes, Rutgers, The State University, Engineering Research Publication No. 45 (1965), pp. 69-107.

32. Golik, R. J., "On Dissipative Mechanisms Within Separated Flow Regions," Ph.D. thesis, Department of Mechanical Engineering, University of Illinois, Urbana, Illinois, 1962.

33. Miles, J. B., "Heat Diffusion from a Line Source Into Mixing Region of Two Parallel Streams," AIAA Journal, 2 (1964), pp. 2038-2040.

34. ———, "Stanton Number for Separated Turbulent Flow Past Relatively Deep Cavities," Ph.D. thesis, Department of Mechanical Engineering, University of Illinois, Urbana, Illinois, 1963.

35. Larson, H. R., "Heat Transfer in Separated Flows," IAS Report No. 59-37, January, 1959; also Journal of the Aerospace Sciences, 26:11 (1959), pp. 731-737.

36. Beheim, M. A., J. L. Klann, and R. A. Yeager, "Jet Effects on Annular Base Pressure and Temperature in a Supersonic Stream," NASA Report R-125, 1962.

37. Larson, R. E., A. R. Hanson, F. R. Krause, and W. K. Dahm, "Heat Transfer Below Reattaching Turbulent Flows," AIAA Paper No. 65-825, 1965.

38. Charwat, A. F., C. F. Dewey, Jr., J. N. Ross, and J. A. Hitz, "An Investigation of Separated Flows -- Part II: Flow in the Cavity and Heat Transfer," Journal of the Aerospace Sciences, 28:7 (1961), pp. 513-527.

39. Seban, R. A., "The Effect of Suction and Injection on the Heat Transfer and Flow in a Turbulent Separated Airflow," ASME Paper No. 65-WA/HT-2, 1965.

40. Dhanak, A. M., "Heat Transfer and Drag Due to Cavities on a Re-Entry Surface," Seventeenth International Astronautical Congress, Madrid, 1966 (proceedings in press).

41. Filetti, E. G. and W. M. Kays, "Heat Transfer in Separated Reattached and Redevelopment Regions Behind a Double Step at Entrance to a Flat Duct," ASME Paper No. 66-WA/HT-18, 1966.

42. Hanson, F. B. and P. D. Richardson, "Mechanics of Turbulent Separated Flows as Indicated by Heat Transfer: A Review," Proceedings, ASME Symposium on Fully Developed Separated Flows, ASME, New York, 1964, pp. 27-32.

43. Spalding, D. B., "Heat Transfer from Turbulent Separated Flows," Journal of Fluid Mechanics, 27:1 (1967), pp. 97-109.

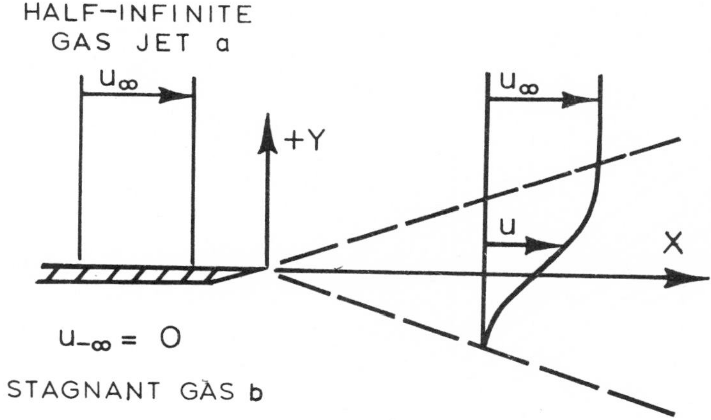

FIGURE 1. SINGLE STREAM MIXING.

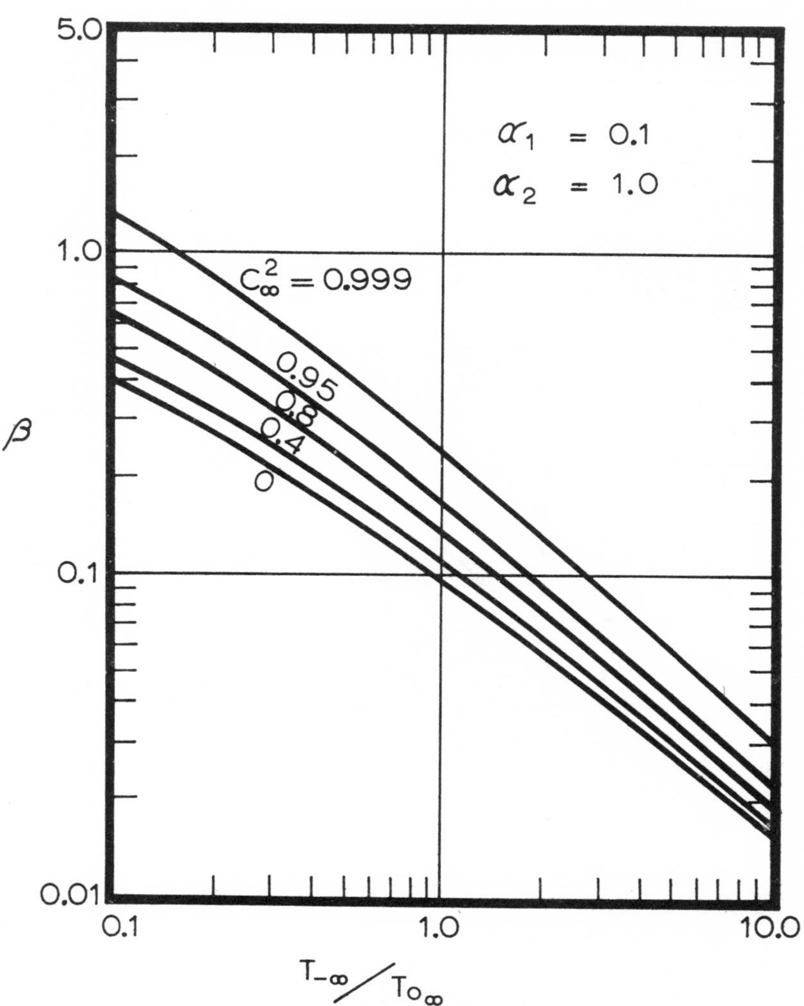

FIGURE 2. MASS TRANSFER COEFFICIENT (HIGH MOLECULAR WEIGHT JET).
[REFERENCE 14]

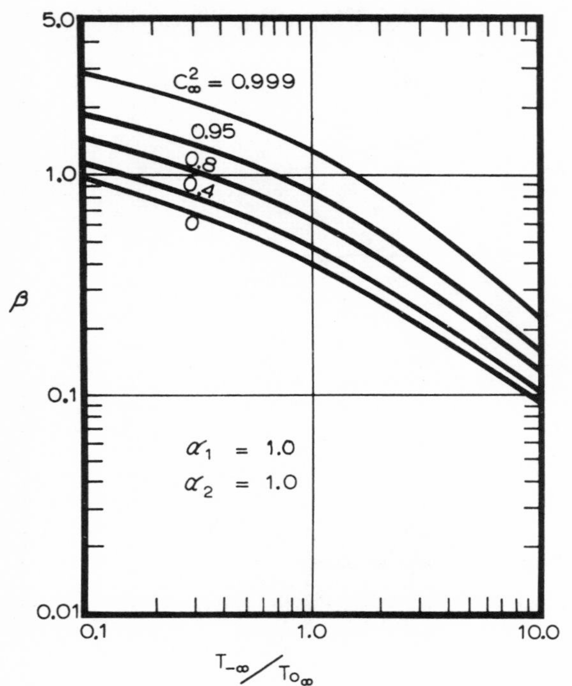

FIGURE 3. MASS TRANSFER COEFFICIENT (EQUAL MOLECULAR WEIGHT JET). [REFERENCE 14]

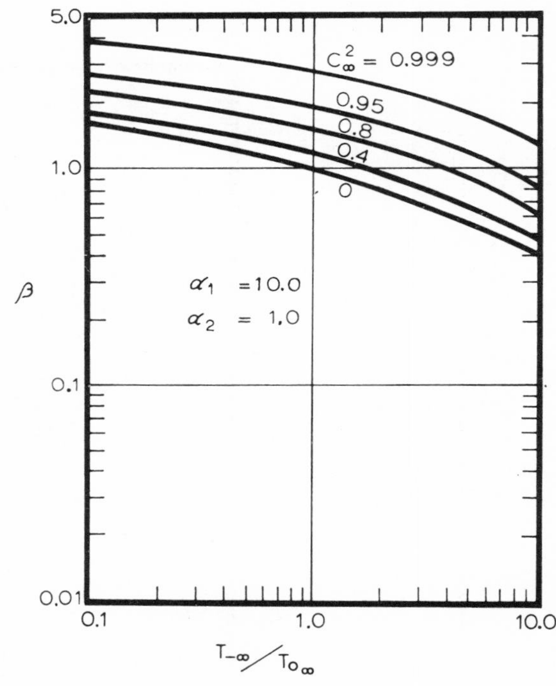

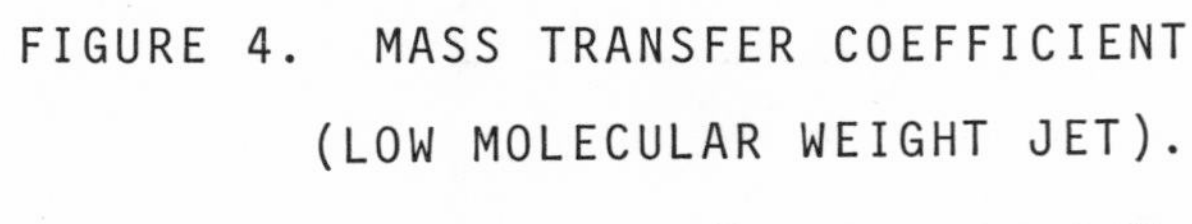
FIGURE 4. MASS TRANSFER COEFFICIENT (LOW MOLECULAR WEIGHT JET). [REFERENCE 14]

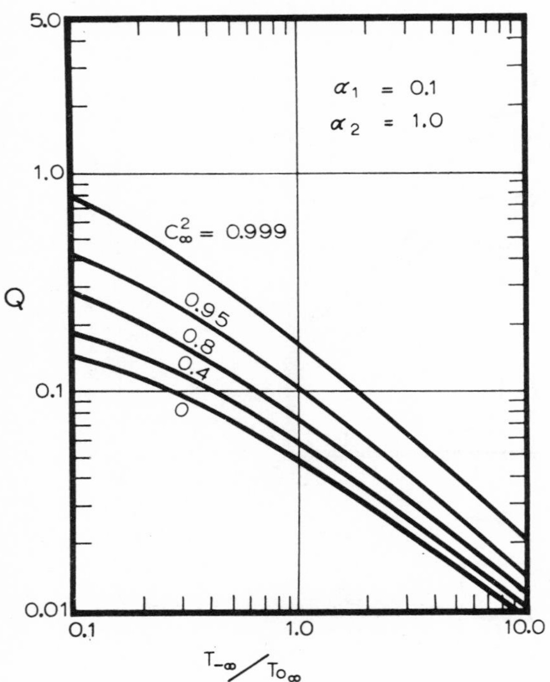

FIGURE 5. ENERGY TRANSFER COEFFICIENT (HIGH MOLECULAR WEIGHT JET). [REFERENCE 14]

FIGURE 6. ENERGY TRANSFER COEFFICIENT (EQUAL MOLECULAR WEIGHT JET). [REFERENCE 14]

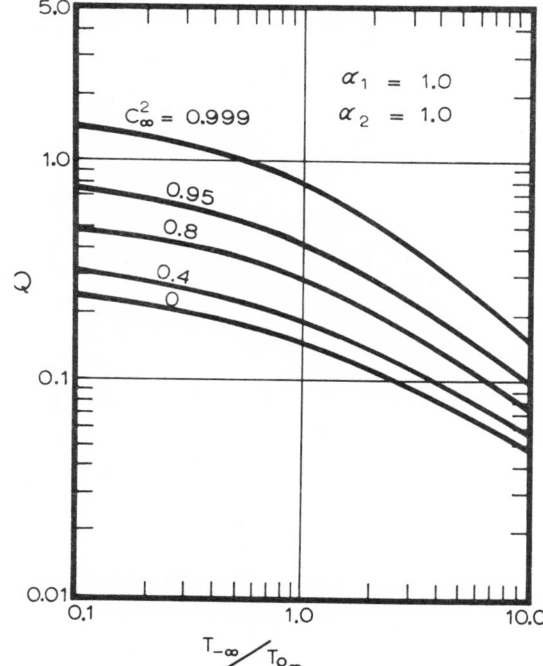

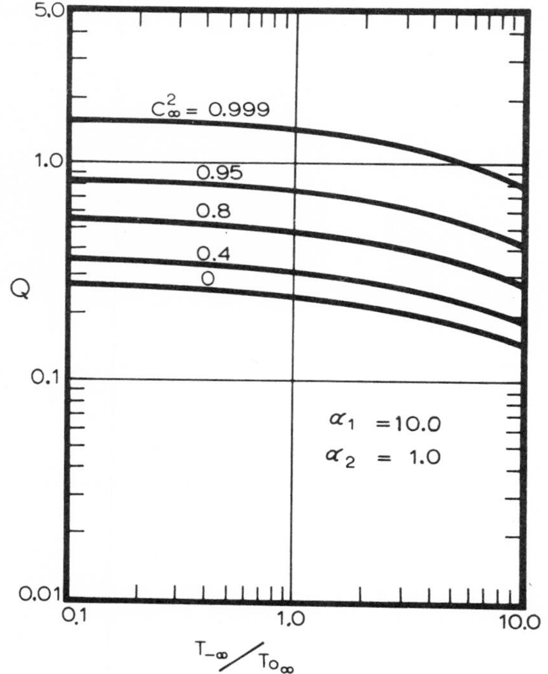

FIGURE 7. ENERGY TRANSFER COEFFICIENT (LOW MOLECULAR WEIGHT JET). [REFERENCE 14]

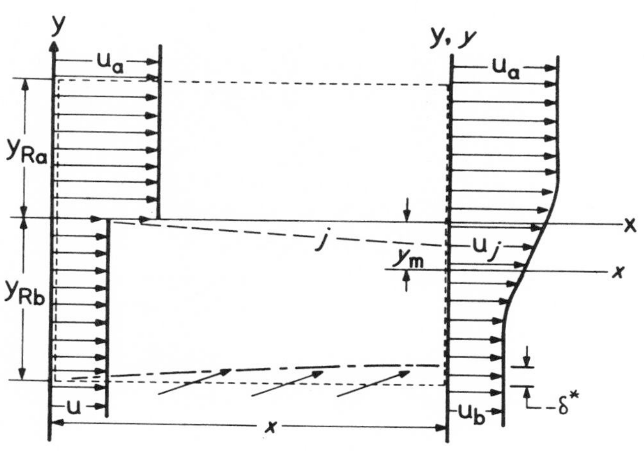

FIGURE 8. TWO STREAM MIXING.

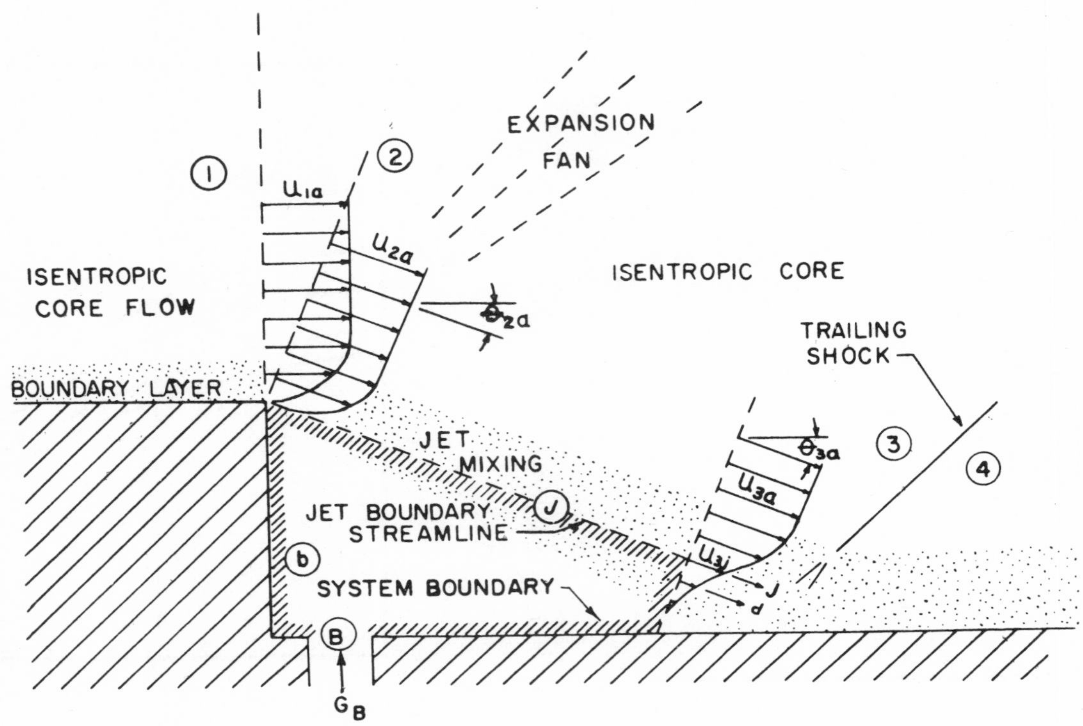

FIGURE 9. SUPERSONIC BACKSTEP. [REFERENCE 16]

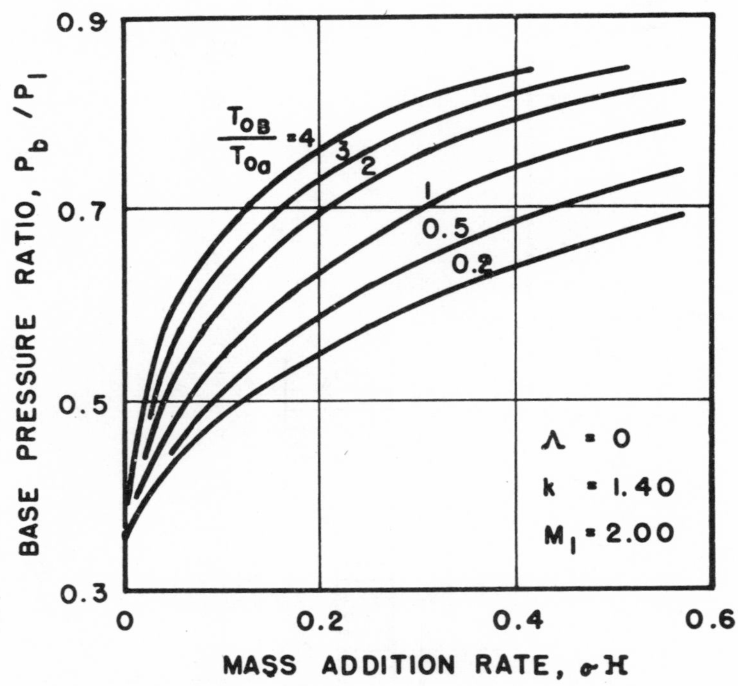

FIGURE 10. INFLUENCE OF BLEED GAS. [REFERENCE 21]

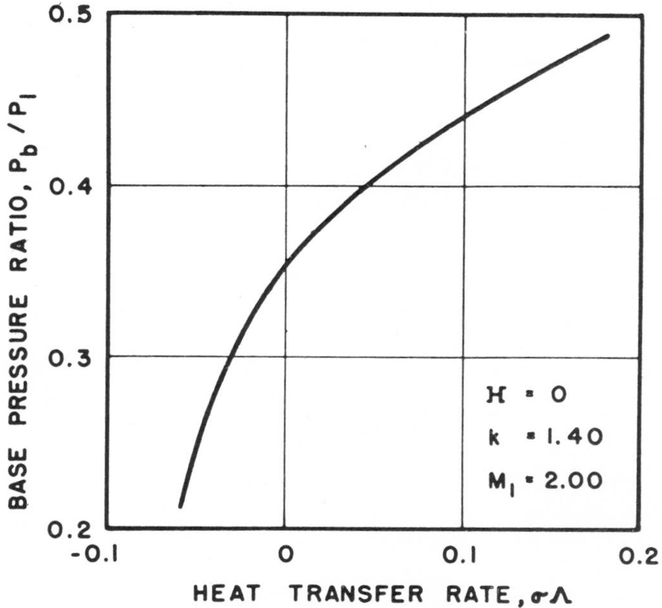

FIGURE 11. INFLUENCE OF HEAT TRANSFER. [REFERENCE 21]

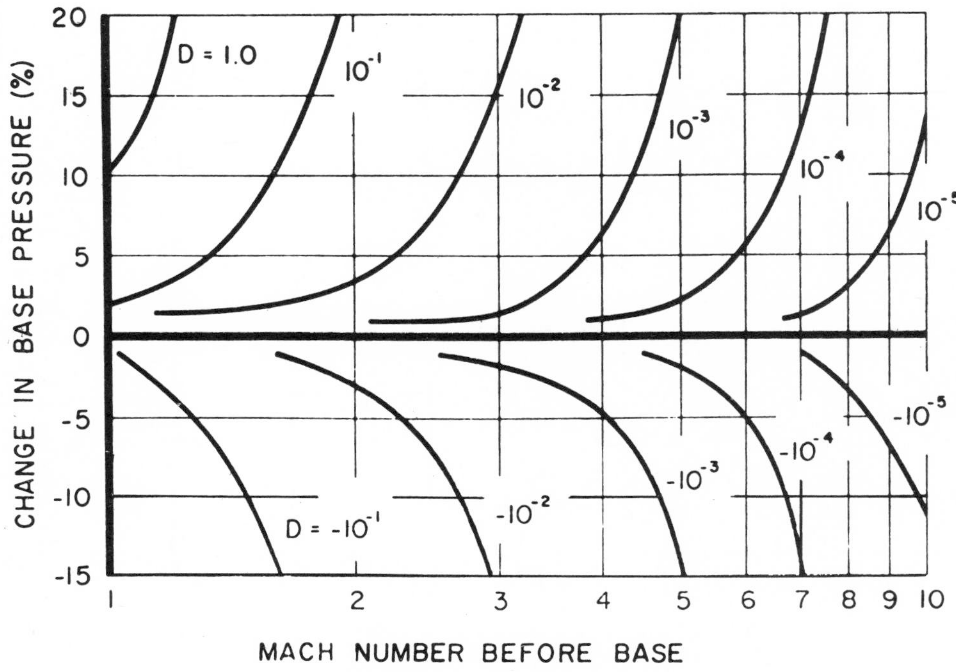

FIGURE 12. INFLUENCE OF HEAT TRANSFER PARAMETER. [REFERENCE 23]

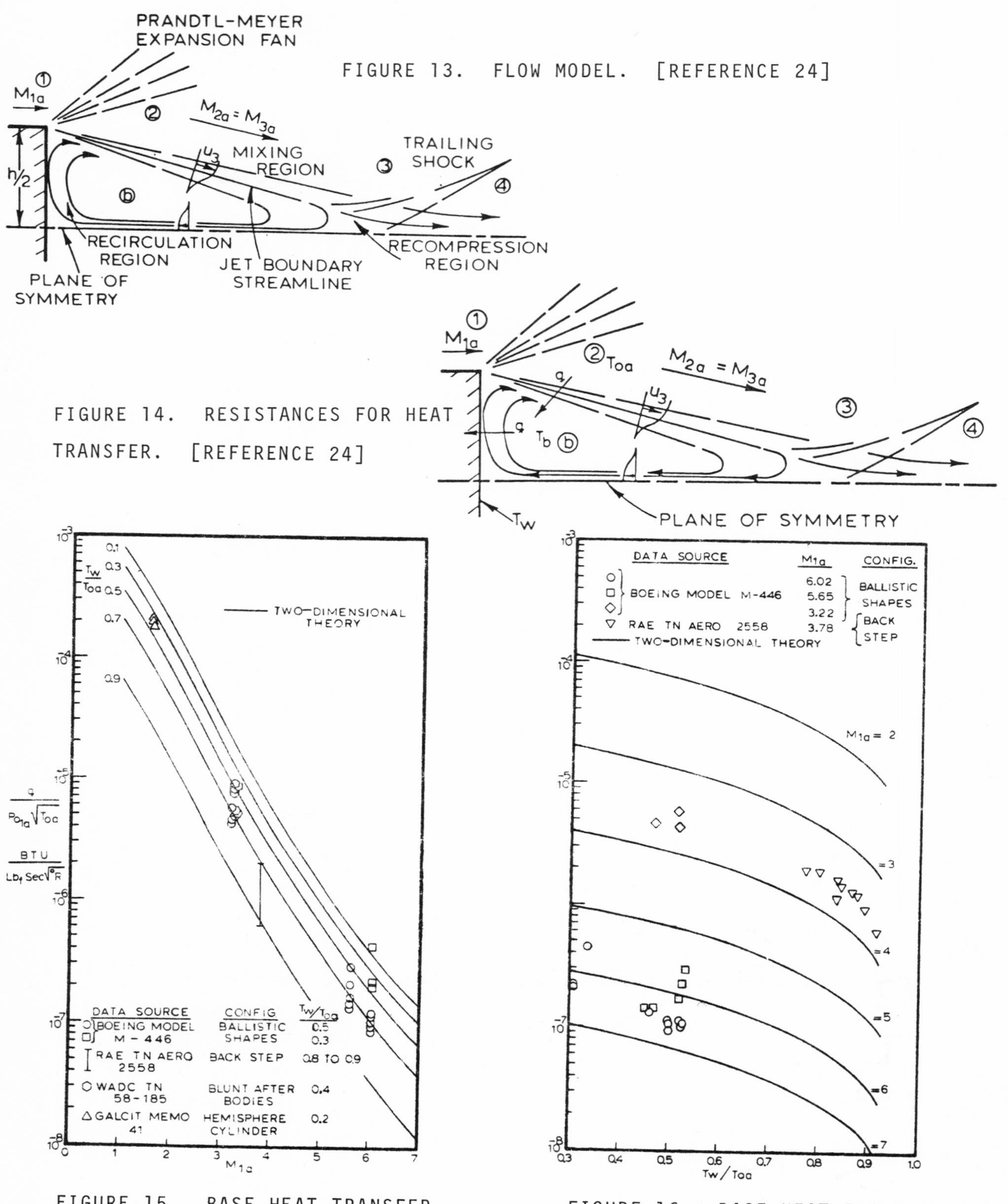

FIGURE 13. FLOW MODEL. [REFERENCE 24]

FIGURE 14. RESISTANCES FOR HEAT TRANSFER. [REFERENCE 24]

FIGURE 15. BASE HEAT TRANSFER CORRELATION (VERSUS MACH NUMBER). [REFERENCE 24]

FIGURE 16. BASE HEAT TRANSFER CORRELATION (VERSUS WALL TEMPERATURE). [REFERENCE 24]

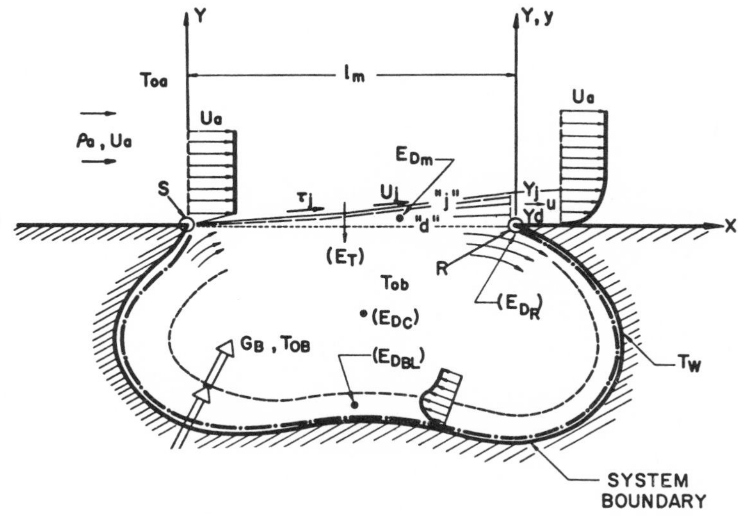

FIGURE 17. CAVITY FLOW MODEL. [REFERENCE 19]

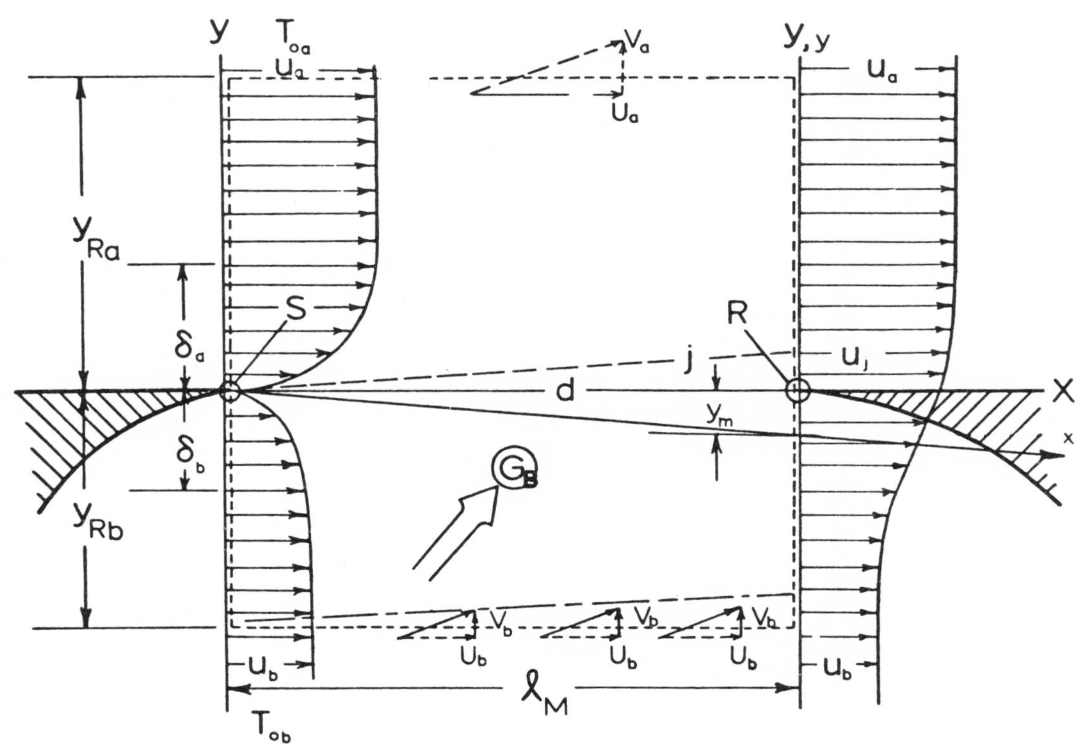

FIGURE 18. TWO STREAM MIXING WITH INITIAL BOUNDARY LAYERS. [REFERENCE 19]

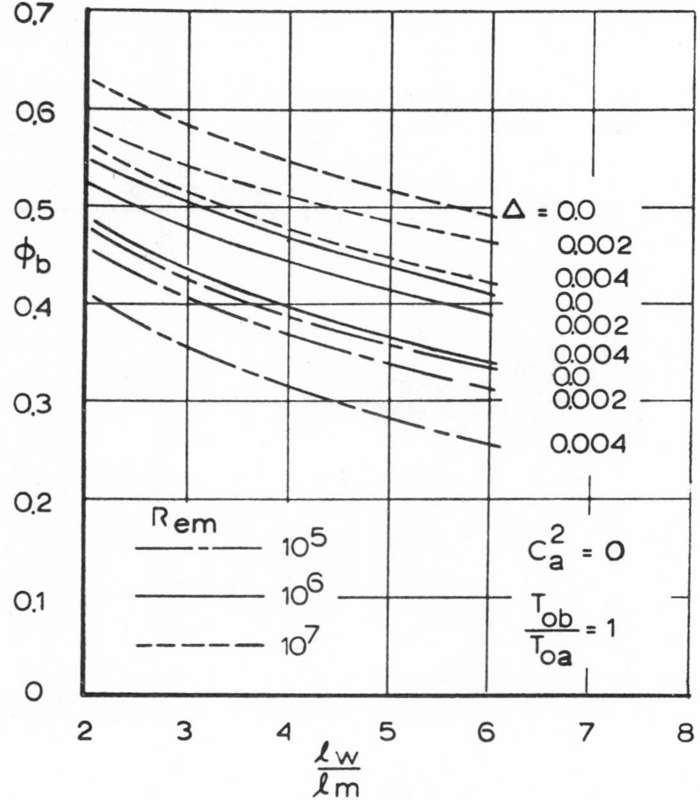

FIGURE 19. WAKE VELOCITY. [REFERENCE 19]

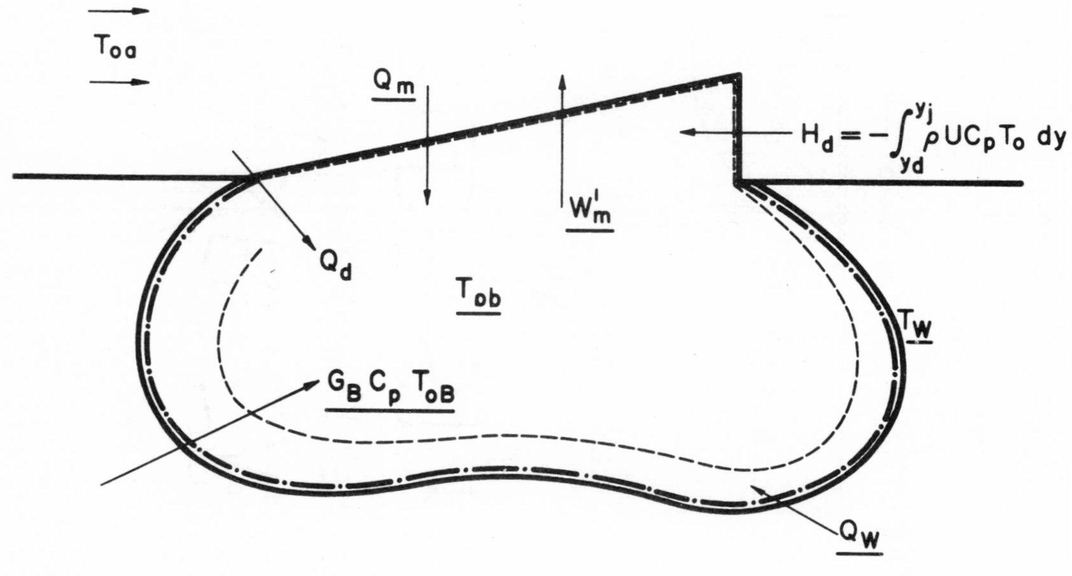

FIGURE 20. ENERGY TRANSFERS. [REFERENCE 19]

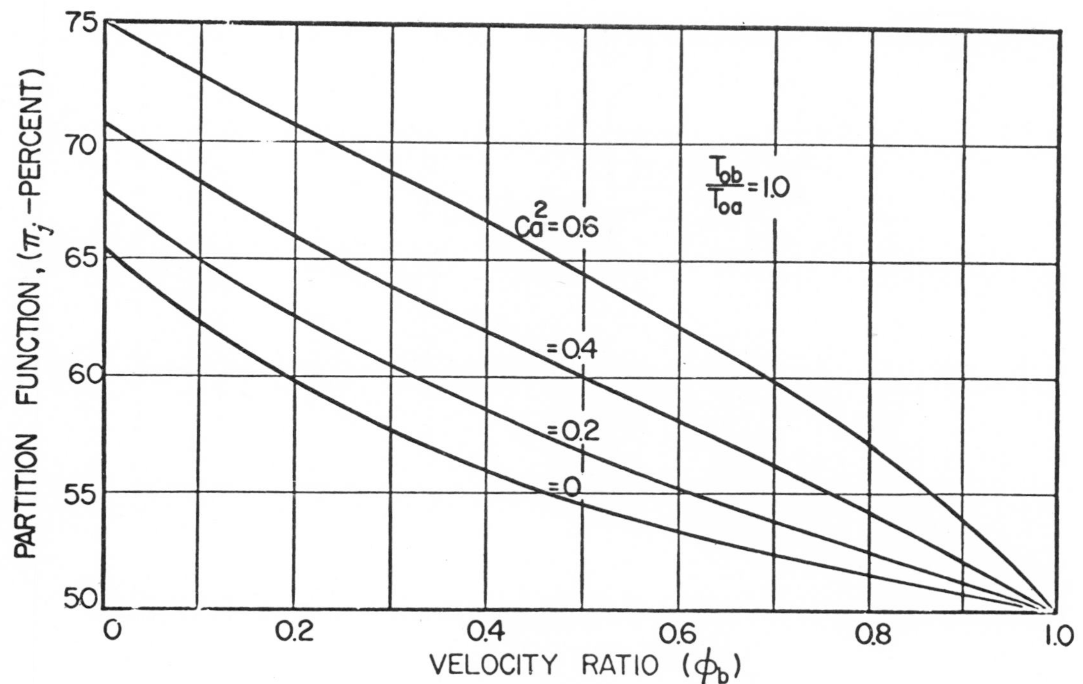

FIGURE 21. PARTITION FUNCTION. [REFERENCE 19]

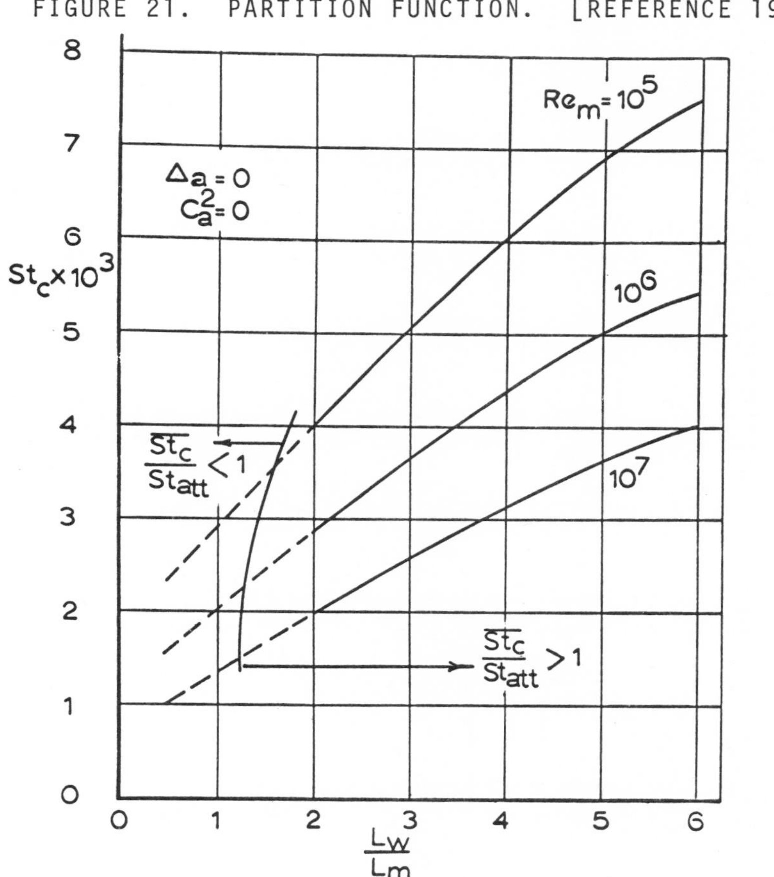

FIGURE 22. STANTON NUMBERS FOR NEARLY CIRCULAR CAVITIES. [REFERENCE 19]

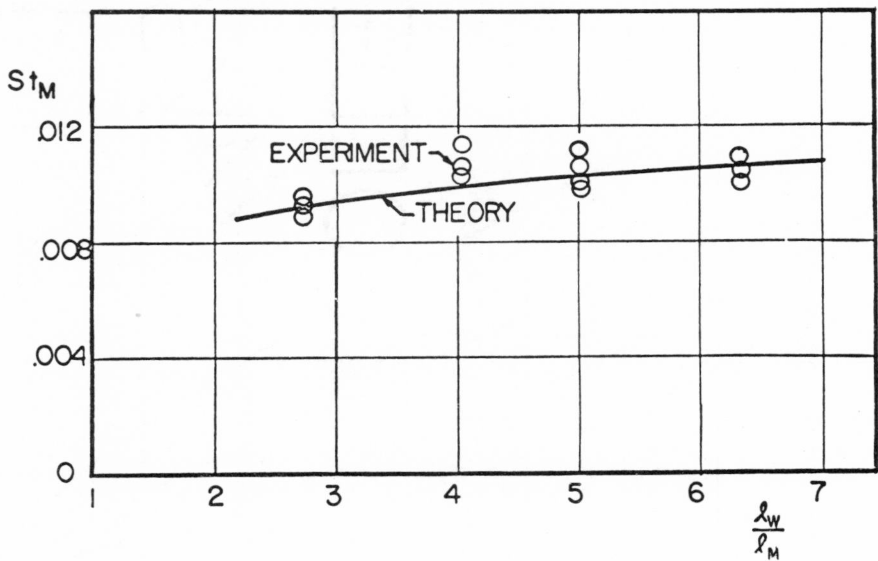

FIGURE 23. MIXING REGION STANTON NUMBER. [REFERENCE 19]

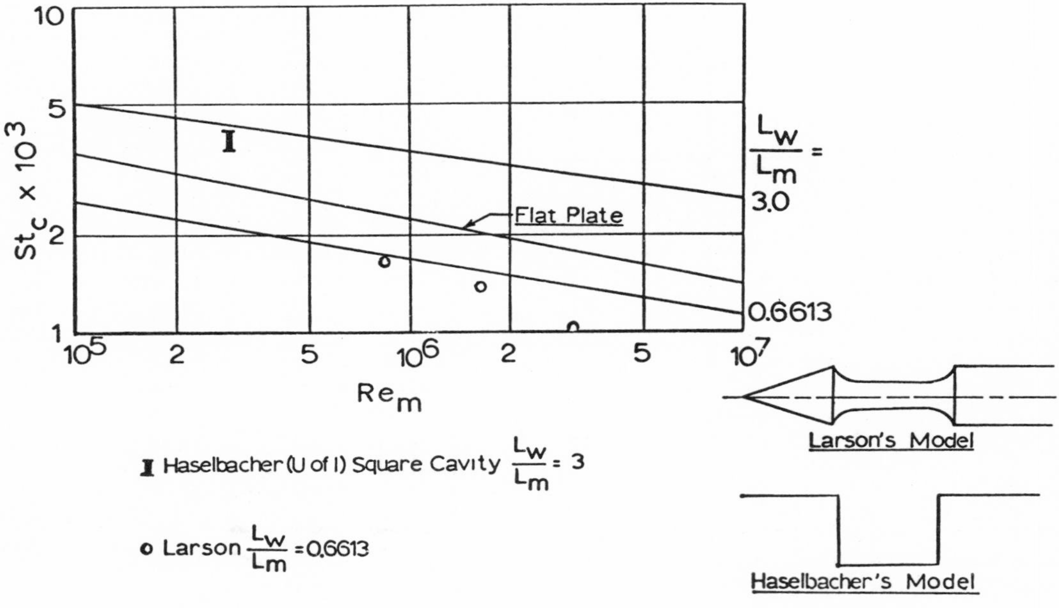

FIGURE 24. CAVITY STANTON NUMBERS. [REFERENCE 19]

MASS TRANSFER COOLING IN LAMINAR BOUNDARY LAYERS

J. P. Hartnett*

ABSTRACT

The available analytical and experimental studies of the mass transfer cooling of surfaces moving at high velocities through an air environment are discussed. Newer results for the mass transfer cooling of surfaces moving through other environments such as nitrogen and carbon dioxide (i.e., a Venus or Mars type of atmosphere), are presented and compared with those for air. Using these results, a proposed design procedure, having applicability to any free stream and to any injectant gas, is presented.

INTRODUCTION

In response to the critical problem of thermally protecting the surfaces of hypersonic vehicles, a new cooling method called mass transfer cooling has been developed. The need for such special cooling of hypersonic vehicles is demonstrated in Figure 1 which shows the equilibrium surface temperature (neglecting dissociation) as a function of Mach number for a fixed free stream temperature of 218°K. This equilibrium value represents the temperature at which there is no heat transfer to or from the plate surface (neglecting radiation and conduction effects), and the surface will approach this temperature level at a rate which is dependent on the properties of the surrounding fluid stream. Noting that the equilibrium temperature for a surface moving at a Mach number of 20 through an air atmosphere is over 10,000°K, a temperature considerably in excess of our present metallurgical capability, it is clear that we must find a way to reduce the equilibrium temperature or alternatively to reduce the rate of approach to this equilibrium value to such a low value that the vehicle can complete its mission before the surface temperature has come up to its design limit. Most of the mass transfer cooling systems accomplish this latter task. A number of mass transfer cooling schemes are diagrammatically shown in Figure 2,(1)** including:

(1) Transpiration cooling

(2) Film cooling
   (a) Using liquid as a coolant
   (b) Using gas as a coolant

(3) Ablation
   (a) Sublimation
   (b) Other ablation phenomena such as melting, erosion, fusion, etc.

---

*Professor and Head, Department of Energy Engineering, University of Illinois at Chicago Circle, Chicago, Illinois.

**Superscript numbers in parentheses refer to References at the end of this paper.

With the exception of film cooling with a gas, these methods all involve the same mechanism in the gaseous phase of the boundary layer. However, there are differences in the conditions at the wall which distinguish the three methods from a thermodynamic as well as a mechanical viewpoint. Transpiration cooling involves the introduction of a coolant gas through a porous surface. Consequently, the rate of fluid injection through the surface and into the boundary layer may be arbitrarily adjusted by purely mechanical means, and the temperature at the surface may thereby be regulated depending upon the injection rate and storage temperature of the coolant gas. It should be noted that a transpiration-cooling system requires pumps, storage tanks, pressure regulators, and accessory plumbing. In addition, the fabrication and maintenance of porous surfaces represents a difficult engineering problem.

A film-cooling system involves pumping a liquid or gas onto the surface through an inlet slot configuration such that a thin film of the material covers the surface. This acts as an insulating coating and, in the case of a liquid, absorbs heat by vaporization. These systems are usually limited by such characteristics as the stability of the liquid film and the pumping power available. Film-cooling systems require essentially all the plumbing and control equipment of a transpiration-cooled operation; however, the surface construction is mechanically simpler.

A sublimation-cooling system is self-controlled, adjusting itself to the appropriate temperature level such that the correct thermodynamic relation (i.e., the Clausius-Clapeyron equation) between the vapor pressure and the surface temperature is satisfied. Heat is absorbed by the material as it sublimes. Thus the heat transfer into the interior is reduced in two ways: (1) direct absorption in the form of heat of sublimation, and (2) reduction of heat transfer because of the movement of the sublimated mass away from the surface. The mass release at the surface depends upon the heat of sublimation and the temperature of the surface. Furthermore, the surface temperature can no longer be arbitrarily controlled and, in fact, will always find "its own level" depending upon the heat load, heat sublimation, and the external flow situation. It should be noted that film-cooling with a liquid is essentially a sublimation process, provided that the surface is completely covered by a liquid layer.

In addition to sublimation, more complex ablation cooling schemes may be visualized. Depending upon the surface material and the flight conditions, it is possible to have such phenomena as fusion of the surface material, mechanical erosion, dissociation of both air and surface material, ionization, and chemical reactions between the components in the boundary layer. The obvious complications involved in an analysis of these complex ablating systems have prevented any really accurate description of the mechanism.

BINARY LAMINAR BOUNDARY LAYER THEORY

The problem of mass transfer cooling has been investigated theoretically for several geometries of practical importance, namely, the flat plate,[2-10] the plane stagnation point,[3,11] and the axisymmetric stagnation point.[11,12] With few exceptions[9,10] the studies have been carried out for a free stream of air and the major emphasis has been placed on the flat plate geometry. We will also restrict our attention to the flat plate analyses, for these bring out clearly the behavior of the boundary layer in the presence of mass addition.

For a binary boundary layer which allows for the injection of a "foreign" gas into a free stream gas of different composition (viz., hydrogen injection into an air boundary layer or nitrogen injection into a carbon dioxide boundary layer), the governing boundary layer conservation equations are:[13]

Continuity:

$$\frac{\partial}{\partial x}(\rho u) + \frac{\partial}{\partial y}(\rho v) = 0 \tag{1}$$

Momentum:

$$\rho u\frac{\partial u}{\partial x} + \rho v\frac{\partial u}{\partial y} = \frac{\partial}{\partial y}\left(\mu\frac{\partial u}{\partial y}\right) \tag{2}$$

Energy:

$$\rho c_p u\frac{\partial t}{\partial x} + \rho c_p v\frac{\partial t}{\partial y}$$

$$= \frac{\partial}{\partial y}\left(k\frac{\partial t}{\partial y}\right) + \mu\left(\frac{\partial u}{\partial y}\right)^2$$

$$+ \rho D_{12}\left(c_{p1} - c_{p2}\right)\frac{\partial t}{\partial y}\frac{\partial W}{\partial y} \tag{3}$$

Species:

$$\rho u\frac{\partial W}{\partial y} + \rho v\frac{\partial W}{\partial y} = \frac{\partial}{\partial y}\left(\rho D_{12}\frac{\partial W}{\partial y}\right). \qquad (4)$$

As written these equations neglect the effects of thermal diffusion and diffusion thermo; such effects are generally small under re-entry conditions but may be significant in laboratory experiments. Excellent reviews of thermal diffusion and diffusion thermo effects are given in References 14 through 16.

The boundary conditions to be imposed on the above system of equations realistically specify the velocity and temperature to be a constant and known value in the free stream and assume the mass fraction of the injected species to be zero in the free stream; at the wall the no-slip condition is imposed on the tangential velocity, and the wall temperature and the mass fraction are specified as constant values. In addition, solutions are also obtained for the adiabatic case where the temperature gradient is taken to be zero at the surface. Clearly we have restricted our analyses to specific thermal conditions, but these particular boundary conditions allow the transformation of the system of partial differential equations into a system of ordinary differential equations which are more amenable to solution. Furthermore, the isothermal case in which the entire surface is held at a constant design temperature is of practical engineering importance.

The boundary conditions take the following form:

At $y = 0$:

$$u = 0,$$

$$t = t_w = \text{constant or } \left(\frac{\partial t}{\partial y}\right)_w = 0, \qquad (5)$$

$$W = W_w = \text{constant.}$$

At $y \to \infty$:

$$u = u_e,$$

$$t = t_e, \qquad (6)$$

$$W = 0.$$

Since we are concerned with a seventh-order system of equations, another boundary condition must be specified. This condition may be obtained by noting that the mass velocity of the free stream gas molecules disappears at the surface of the plate; that is to say, there is no net mass transfer of the boundary layer free stream gas into the plate surface. Therefore, the mass flow of free stream gas by convection away from the surface must be equal and opposite to the diffusive flow of free stream gas toward the surface. This consideration yields the following boundary condition:

$$v = -\frac{\rho D_{12}}{1 - W}\frac{\partial W}{\partial y} \quad \text{at } y = 0. \qquad (7)$$

This system of equations (1-7) forms the starting point for the various investigations cited above. In all cases, a transformation of the coordinates is next introduced with the result that the system of partial differential equations is changed into a new set of interdependent ordinary differential equations. This system of equations is still difficult to solve since, in general, all of the physical properties are functions of the local temperature and concentration of the particular gas mixtures being investigated. To obtain a representative number of solutions in a reasonable time requires the use of high-speed electronic computers.

CONSTANT PROPERTY ANALYSIS

Considerable insight into the mass transfer process may be obtained if it is assumed that the injected coolant has physical properties not markedly different from those of the main stream, thereby permitting the assumption of constant physical properties. The advantage of such a constant property solution is that the dimensionless heat transfer and skin friction behavior are dependent on a minimum number of parameters. This is indicated in Figures 3 and 4 which compare the dimensionless quantities that are of importance for the solid-wall, heat transfer case and those arising in the presence of mass transfer. For the solid wall, it has been demonstrated[17,18] that the constant property solutions for the dimensionless skin friction, Nusselt number, and recovery factor have additional value in that they may be used even when large variations in physical properties are encountered (including dissociation), provided the properties are evaluated at a so-called reference temperature, t^*, which may be given explicitly in terms of the surface, the free stream, and the recovery temperatures:

$$t^* = t_e + 0.5(t_w - t_e) + 0.22(t_{ro} - t_e). \tag{8}$$

It has been established that this so-called Eckert reference formulation applies not only to an air free stream but also to free streams of nitrogen, carbon dioxide, or hydrogen.(10,19) For design purposes it is recommended that this method be applied irrespective of the free stream gas. For future reference we set down the relations for the heat transfer, skin friction coefficient, and recovery factor which are applicable in the absence of mass transfer:

$$c_{fo} = 0.664\sqrt{\frac{u_e x}{\nu}}^{\,*}, \tag{9}$$

$$St_o = \frac{c_{fo}}{2}\,(Pr^*)^{-2/3}, \tag{10}$$

$$r_o = \sqrt{Pr^*}, \tag{11}$$

$$t_{ro} = t_e + r_o\left(\frac{u_e^2}{2c_p^*}\right). \tag{11a}$$

It will be demonstrated that the reference temperature, Equation (8), and Equations (9) through (11) will prove of value in correlating the mass transfer results.

Returning to the mass transfer case, it will be noted from Figure 4 that the heat transfer, q, is defined in terms of the temperature gradient at the wall. This definition is convenient in that this quantity, q, represents the convective heat transferred to the surface from the boundary layer, and in this sense the boundary-layer heat transfer is considered separately from the enthalpy carried across the surface by the coolant. To be consistent with this point of view, the recovery temperature, t_r, is defined as that temperature where the wall-temperature gradient vanishes; in this case there still exists a transport of enthalpy across the surface, by virtue of the coolant flow, but there is no convective heat transferred to the wall from the boundary layer.

Turning to the solutions of Equations (1) through (7) it is obvious that the energy equation, from which the last term is eliminated for the constant property case, is linear and consequently, the general solution of the complete equation may be obtained from the addition of (1) a general solution of the homogeneous equation (that is, neglecting the dissipation term) and (2) a particular solution of the complete equation. The particular solution results in the specification of the recovery factor, a direct measure of the temperature assumed by convection alone. We need, therefore, only to direct our attention to the general solution of the homogeneous equation. To accomplish this solution, a stream function, ψ, is first introduced to satisfy the continuity equation, and a new independent variable, η, and the new dependent variables defined below are substituted into the original equations:

Stream function:

$$u = \frac{\partial\psi}{\partial y}, \quad v = -\frac{\partial\psi}{\partial x} \tag{12}$$

$$f = \frac{\psi}{\sqrt{(\nu u_e x)}}, \quad \eta = \frac{y}{x}\sqrt{\frac{u_e x}{\nu}}, \quad \theta = \frac{t - t_w}{t_e - t_w}, \quad \varphi = \frac{W_w - W}{W_w}. \tag{13}$$

The following equations result:

Momentum:

$$\frac{d^3 f}{d\eta^3} + \frac{1}{2}\frac{d^2 f}{d\eta^2} = 0 \tag{14}$$

Energy:

$$\frac{d^2\theta}{d\eta^2} + \frac{1}{2}\,Pr\,f\frac{d\theta}{d\eta} = 0 \tag{15}$$

Diffusion:

$$\frac{d^2\varphi}{d\eta^2} + \frac{1}{2}\,Sc\,f\frac{d\varphi}{d\eta} = 0 \tag{16}$$

Boundary conditions:

$$\left.\begin{aligned} \frac{df}{d\eta} &= 0 \\ \theta &= 0 \\ \varphi &= 0 \\ f_w &= -2\left(\frac{v_w}{u_e}\right)\sqrt{\frac{u_e x}{\nu}} \end{aligned}\right\} \text{at } \eta = 0 \qquad (17)$$

$$\left.\begin{aligned} \frac{df}{d\eta} &= 1 \\ \theta &= 1 \\ \varphi &= 1 \end{aligned}\right\} \text{at } \eta \to \infty \qquad (18)$$

An important observation common to all flat-plate binary laminar boundary-layer solutions is that the mass transfer into the boundary layer must vary as $1/\sqrt{x}$ if we are to arrive at a system of ordinary differential equations. Further, we have assumed an isothermal surface and it may be shown that this is compatible with the imposed mass transfer distribution.

The velocity profiles for the constant property mass transfer system for several different injection rates are shown in Figure 5. Inspection of these profiles leads to the following conclusions:

(1) The effect of mass addition is to thicken the velocity boundary layer.

(2) The velocity profile becomes S-shaped with mass addition, and since this is known to be an unstable type of profile, it may be concluded that mass transfer is destabilizing.*

(3) The boundary layer "lifts off" the wall at a relatively low value of mass transfer, i.e., at

$$\frac{\rho_w v_w}{\rho_e u_e}\sqrt{\frac{u_e x}{\nu}} = 0.619.$$

Apparently the boundary-layer equations fail to describe the flow field at this mass transfer condition.*

The skin friction coefficient and Nusselt number,[7] presented in Figures 6 and 7, are seen to decrease with increasing mass transfer, both going to zero at the limiting value where the boundary layer "leaves" the wall. The recovery factor, shown in Figure 8, is somewhat reduced by mass transfer but not as markedly as the Nusselt values. At hypersonic velocities the actual heat transfer to the surface is proportional to the product of the Nusselt number and the recovery temperature, and we conclude that a considerable reduction in heat transfer is obtainable with modest amounts of coolant. This is the feature which has drawn attention to this cooling scheme.

VARIABLE PROPERTY ANALYSIS

The binary boundary layer with variable physical properties has been extensively studied for the flat plate geometry. The governing partial differential equations (1-4) are transformed into a set of ordinary differential equations through the introduction of a new independent similarity variable, η, and a corresponding dependent stream function, f. A typical transformation procedure is given below:

Stream function, ψ:

$$u = \frac{\rho_e}{\rho}\frac{\partial\psi}{\partial y}, \quad v = -\frac{\rho_e}{\rho}\frac{\partial\psi}{\partial x} \qquad (19)$$

Dimensionless stream function, f:

$$\psi = \sqrt{u_e \nu_e x}\, f(\eta) \qquad (20)$$

Similarity variable, η:

$$\eta = \frac{1}{2}\sqrt{\frac{u_e}{\nu_e x}} \int_0^y \frac{dy}{\varphi_\mu}\ . \qquad (21)$$

The transformed equations now take the form:

Momentum:

$$\left[\left(\frac{f'}{\varphi_\rho \varphi_\mu}\right)'\right]' + f\left(\frac{f'}{\varphi_\rho \varphi_\mu}\right)' = 0 \qquad (22)$$

Diffusion:

$$\left(\frac{W'}{Sc}\right)' + Sc\, f\left(\frac{W'}{Sc}\right) = 0 \qquad (23)$$

*This behavior is unique to the flat plate and does not occur in the presence of a pressure gradient.

Energy:

$$\left(\frac{\varphi_k T'}{\varphi_\mu}\right)' + Pr_e \frac{\varphi_\mu}{\varphi_k}\left(\varphi_c f + \varphi_{c12} \frac{W'}{Sc}\right)\left(\frac{\varphi_k T'}{\varphi_\mu}\right)$$

$$+ \frac{Pr_e}{4} (\gamma_e - 1) M_e^2 \left[\left(\frac{f'}{\varphi_\rho \varphi_\mu}\right)'\right]^2 = 0. \quad (24)$$

The boundary conditions become:

For $\eta = 0$:

$f' = 0$

$$f = f_w = -\frac{2\rho_w v_w}{\rho_e u_e}\left(Re_{x,e}\right)^{1/2} = \text{constant},$$

$$(25)$$

$W = W_w =$ constant,

$T = T_w =$ constant or $\left(\frac{\partial T}{\partial \eta}\right)_w = 0.$

For $\eta \to \infty$:

$f' = 2,$

$W = 0, \quad (26)$

$T = 1.$

The physical properties of the binary gas mixture are assumed to be known functions of the local temperature and concentration. Generally the density and specific heat of the mixture are assumed to be given by the ideal gas laws, while the transport terms (viscosity, thermal conductivity, and diffusion coefficient) are calculated by mixture laws (for example, by the methods of Reference 20). The governing equations (22-24) taking into account the boundary conditions are transformed into three integral equations for the unknown velocity, temperature, and mass fraction profiles. The resulting equations in conjunction with the physical property information are solved by successive approximation. Initial velocity, temperature, and mass fraction profiles are assumed (allowing the evaluation of the various properties at the local conditions) and are used to obtain new profiles; the procedure is repeated until the equations are satisfied.

This system of equations has been extensively studied for an air boundary layer on a flat plate for the injection of a number of foreign gases including hydrogen,[5,6,7,8] helium,[4,5,6,8] water vapor,[8] air,[2,3] and carbon dioxide.[8] The general features of the heat transfer and skin friction behavior revealed by these solutions is qualitatively similar to the constant property result. In view of the complexity of these calculations a simple design presentation, based on these exact solutions, is clearly desirable for engineering purposes. This has been found possible [21,22] for the air boundary layer by presenting normalized skin friction coefficients and Stanton numbers in the form of c_f/c_{f_o} and St/St_o as functions of the dimensionless blowing parameter

$$\frac{\rho_w v_w}{\rho_e u_e}\left(\frac{Re_x}{C^*}\right)^{1/2}.$$

Here C^* is the Chapman-Rubesin parameter for the free stream gas evaluated at the Eckert reference temperature, Equation (8). An example of the success of this representation is given by Figure 9[21] which demonstrates that a single curve very accurately represents the heat transfer results for a number of exact solutions for the hydrogen-air boundary layer covering a wide range of wall temperature conditions and free stream Mach numbers. Similar behavior is found for the skin friction coefficient.

This same representation was successful for the other coolant gases, and for each coolant gas-air combination a single line gave excellent agreement with the results of the exact calculations. The resulting normalized Stanton numbers and skin friction coefficient are shown on Figures 10 and 11. It is apparent from these figures that the light gases are much more effective than are the heavier gases in reducing the heat transfer and skin friction. Inspection of these figures indicates that the normalized Stanton number, St/St_o, and skin friction coefficient, c_f/c_{fo}, vary linearly with the dimensionless mass transfer parameter for all the gases shown. In particular, the air-into-air results may be expressed by the following two equations:

$$\frac{St}{St_o} = 1 - 1.82\left[\left(\frac{\rho_w v_w}{\rho_e u_e}\right)\left(\frac{Re_x}{C^*}\right)^{1/2}\right], \quad (27)$$

$$\frac{c_f}{c_{fo}} = 1 - 2.08\left[\left(\frac{\rho_w v_w}{\rho_e u_e}\right)\left(\frac{Re_x}{C^*}\right)^{1/2}\right]. \quad (28)$$

It has been demonstrated[21,22] that the results for the other free stream gases can be made to agree with these equations by the simple expedient of multiplying the dimensionless mass transfer parameter by the molecular weight ratio (M_2/M_1) raised to an exponent of (1/3), resulting in the following final design equations:

$$\frac{St}{St_o} = 1 - 1.82\left[\left(\frac{M_2}{M_1}\right)^{1/3}\left(\frac{\rho_w v_w}{\rho_e u_e}\right)\left(\frac{Re_x}{C^*}\right)^{1/2}\right], \quad (29)$$

$$\frac{c_f}{c_{fo}} = 1 - 2.08\left[\left(\frac{M_2}{M_1}\right)^{1/3}\left(\frac{\rho_w v_w}{\rho_e u_e}\right)\left(\frac{Re_x}{c^*}\right)^{1/2}\right]. \quad (30)$$

EXTENSION TO NITROGEN AND CARBON DIOXIDE FREE STREAMS

The governing differential equations, Equations (22) through (24), have been solved for the variable property binary boundary layer on a flat plate for the case where the free stream gas is nitrogen or carbon dioxide while the coolant is hydrogen.[9,23] The mixture properties were calculated by the procedures of Reference 20 and are available.[24] Typical boundary layer temperature, mass fraction, and velocity profiles are shown in Figures 12 and 13 for a free stream Mach number of 12, with the free stream temperature of 218°K, while the wall temperature is specified to be 1308°K. These figures reveal that the maximum temperature within the boundary layer is reduced and moved away from the wall with increasing wall mass fraction (i.e., increasing blowing rate) so that less and less heat is flowing to the wall from the fluid. It may also be noted that the maximum temperature within the boundary layer of the hydrogen-nitrogen mixture is always higher than that of the hydrogen-carbon dioxide mixture when compared at the same blowing rate. Further, the boundary layer is thicker for the hydrogen-carbon dioxide mixture than for the hydrogen-nitrogen layer.

The calculated recovery temperatures are given in Figures 14 and 15, where it may be observed that the influence of mass transfer is to decrease the recovery temperature. It is also clear that the recovery temperature is much higher in the hydrogen-nitrogen boundary layer than in the hydrogen-carbon dioxide case. This can be explained by noting that the free stream total temperature is given by the following formulation:

$$t_t = t_e\left[1 + \frac{\gamma_e - 1}{2} M_e^2\right]. \quad (31)$$

The value of the specific heat ratio, γ_e, for the low molecular weight gas nitrogen is higher than that of carbon dioxide and, accordingly, the free stream total temperature is higher for nitrogen. We therefore expect the recovery temperature to be higher for the hydrogen-nitrogen system than for the hydrogen-carbon dioxide case.

It was possible to represent the analytical results for the normalized Stanton numbers, St/St_o, as a function of the generalized blowing parameter,

$$\frac{\rho_w v_w}{\rho_e u_e}\left(\frac{Re_x}{C^*}\right)^{1/2},$$

as demonstrated in Figures 16 and 17. In each figure all of the analytical results fall within the crosshatched area. Inspection of these two figures reveals that the Stanton number ratio is higher for the hydrogen-carbon dioxide system at low blowing rates, whereas at high blowing rates the hydrogen-nitrogen normalized Stanton numbers are higher than those for hydrogen-carbon dioxide. This of course means that the results are not monotonic in molecular weight and, accordingly, the design equation (27) suggested earlier does not give an accurate prediction for the carbon dioxide free stream. It is clear that the application of Equation (27) to determine heat transfer and skin friction coefficients for situations in which the free stream is other than air is open to considerable question.

GENERALIZED PRESENTATION FOR FOREIGN ATMOSPHERES

Notwithstanding the failure of the proposed design equation when applied to the hydrogen-carbon dioxide system, it was decided to carry out additional analytical studies with the nitrogen and carbon dioxide free streams. In addition, since the transport properties were already calculated, the studies were extended to include a free stream of hydrogen. The range of Mach numbers, free stream and wall temperature values, is the same as noted in Figures 16 and 17. A summary of all the

Stanton number predictions for these foreign gas atmospheres is given in Figure 18, using the presentation which proved successful for the binary system with air as the free stream gas. With the above-noted exception of the hydrogen-carbon dioxide results, the Stanton numbers for all other gas combinations can be successfully correlated by the design equation proposed earlier, Equation (27).

A calculation of the heat transfer requires the knowledge of the recovery temperature (or, equivalently, the recovery factor) as well as the Stanton number. It has been found possible to represent all of the recovery factor calculations for the above-mentioned gas combinations by a single correlation curve. As shown on Figure 19, the normalized recovery factor r/r_o is plotted against the modified blowing parameter:

$$\left[\left(\frac{M_1 + M_2}{2M_1}\right)^{5/4} \left(\frac{M_1}{M_2}\right)^{1/5} \left(\frac{\rho_w v_w}{\rho_e u_e}\right)\left(\frac{Re_x}{C^*}\right)^{1/2}\right].$$

Some caution should be exercised in the use of these recovery factor predictions, for they may be in error if the free stream temperature is very low. However, the departure from unity of the normalized value r/r_o is not great at low and moderate blowing values and, accordingly, the use of the solid wall recovery factor may be acceptable in this range.

RECOMMENDED PROCEDURE FOR CALCULATING HEAT TRANSFER

The following procedure is recommended for calculating heat transfer in the presence of mass transfer. It is assumed that the free stream and coolant gases are specified and that the free stream conditions are known. Further, it is assumed that the coolant mass flow and wall temperature level are specified.

(1) Using Equations (8) through (11a), determine the Eckert reference temperature, t^*, the skin friction coefficient, c_{f_o}, the Stanton number, St_o, and the recovery factor, r_o, for a solid wall exposed to the same free stream conditions and held at the same temperature as the mass transfer cooled wall.

(2) For the known blowing rate and the specific gas combination, calculate the dimensionless parameters

$$\left[\left(\frac{M_2}{M_1}\right)^{1/3} \left(\frac{\rho_w v_w}{\rho_e u_e}\right)\left(\frac{Re_x}{C^*}\right)^{1/2}\right]$$

and

$$\left[\left(\frac{M_1 + M_2}{2M_1}\right)^{5/4} \left(\frac{M_1}{M_2}\right)^{1/5} \left(\frac{\rho_w v_w}{\rho_e u_e}\right)\left(\frac{Re_x}{C^*}\right)^{1/2}\right].$$

Enter Figures 18 and 19 at these values and read off the normalized values St/St_o and r/r_o.

(3) Calculate the Stanton number, St, and the recovery factor, r, and from the known free stream condition determine the heat transfer coefficient, h, and the recovery temperature, t_r:

$$h = (St)\, \rho_e u_e c_{pe}$$

$$t_r = t_e + r \frac{u_e^2}{2c^*_{p2}} ,$$

where c^*_{p2} is the specific heat of the free stream gas evaluated at the reference temperature, t^*.

(4) Calculate the heat transfer by the equation:

$$q = h(t_w - t_r).$$

ACKNOWLEDGMENTS

The author wishes to acknowledge the debt that is owed to his colleagues, in particular to his co-authors of References 21 and 22, and to Professor H. A. Simon and Dr. C. S. Liu, for it is primarily their work that has set the stage for the present review paper.

NOTATION

C = Chapman-Rubesin parameter, $(\rho_2\mu_2)/(\rho_e\mu_e)$

c_f = local skin friction coefficient

c_{f_o} = local skin friction evaluated for solid wall exposed to some free stream conditions and held at same temperature as the actual wall

c_p = specific heat at constant pressure

D_{12} = ordinary diffusion coefficient

f = dimensionless stream function

h = heat transfer coefficient

k = thermal conductivity

M_1 = molecule weight of coolant gas

M_2 = molecule weight of free stream gas

q = local heat transfer rate per unit area, $k_w(\partial t/\partial y)_w$

q_o = local heat transfer rate per unit area evaluated for solid wall exposed to same free stream conditions and held at same temperature as actual wall

r = recovery factor

St = local heat transfer Stanton number, $q/\rho_e u_e c_{p_e}(t_r - t_w)$

St_o = local Stanton number evaluated for solid wall exposed to same free stream conditions and held at same temperature as the actual wall

t = temperature

t_r = recovery temperature, defined as the wall temperature where $k_w(\partial t/\partial y)_w = 0$

t^* = reference temperature defined in Equation (9)

T = dimensionless temperature, t/t_e

u = component of velocity parallel to surface

v = component of velocity normal to surface

w = mass fraction of coolant gas

x = coordinates along the body

y = coordinates normal to the body

Greek Symbols

γ = ratio of specific heats

η = transformed coordinate defined in text

μ = dynamic viscosity

ν = kinematic viscosity

ρ = density

ψ = stream function

φ_c = dimensionless specific heat, c_{pm}/c_{pe}

φ_{c12} = dimensionless specific heat difference, $(c_{p1} - c_{p2})/c_{pe}$

φ_u = dimensionless viscosity, μ_m/μ_e

φ_ρ = dimensionless density, ρ_m/ρ_e

φ_k = dimensionless conductivity, k_m/k_e

Dimensionless Numbers

M = Mach number, ratio of local speed to local speed of sound

Pr = Prandtl number

Re_x = local Reynolds number, $u_e x/\nu_e$

Sc = Schmidt number of the mixture, ν/D_{12}

Subscripts

1 refers to pure coolant

2 refers to pure free stream gas

e evaluated at outer edge of the boundary layer

r refers to recovery conditions, i.e., where $k_w(\partial t/\partial y)_w = 0$

w evaluated at wall conditions

x refers to local conditions

m mixture

Superscripts

$*$ evaluated at the reference temperature, Equation (9)

$''$ differentiation with respect to η

REFERENCES

1. Eckert, E. R. G., "Mass Transfer Cooling, a Means to Protect High-Speed Aircraft," presented at the First International Congress of Aeronautical Sciences, Madrid, Spain, September, 1958.

2. Livingood, J. N. B. and P. L. Donoughe. Summary of Laminar-Boundary-Layer Solutions for Wedge-Type Flow Over Convection- and Transpiration-Cooled Surfaces (NACA TN 3588), December, 1955.

3. Hartnett, J. P. and E. R. G. Eckert, "Mass Transfer Cooling in a Laminar Boundary Layer with Constant Fluid Properties," Transactions, ASME, 79 (February, 1957), p. 247.

4. Baron, J. R. The Binary-Mixture Boundary Layer Associated with Mass Transfer Cooling at High Speeds (Technical Report 160). Cambridge, Mass.: Massachusetts Institute of Technology, Naval Supersonic Laboratory, May, 1956.

5. Sziklas, E. A. An Analysis of the Compressible Laminar Boundary Layer with Foreign Gas Injection (Research Department Report SR-0539-8). United Aircraft Corporation, 1956.

6. Sziklas, E. A. and C. M. Banas, "Mass Transfer Cooling in Compressible Laminar Flow," RAND Symposium on Mass Transfer Cooling for Hypersonic Flight, Santa Monica, California, June, 1957.

7. Eckert, E. R. G., P. J. Schneider, A. A. Hayday, and R. M. Larson, "Mass-Transfer Cooling of a Laminar Boundary Layer by Injection of a Light-Weight Foreign Gas," Jet Propulsion (January, 1958).

8. Sparrow, E. M., W. J. Minkowycz, and E. R. G. Eckert, "Mass-Transfer Cooling of a Flat Plate with Various Transpiring Gases," AIAA Journal, 3: 7 (1965), pp. 1342-1343.

9. Liu, C. S., J. P. Hartnett, and H. A. Simon, "Mass Transfer Cooling in Laminar Boundary Layers with Hydrogen Injected into Nitrogen and Carbon Dioxide Streams," Proceedings, Third International Heat Transfer Conference, Chicago, August, 1966, pp. 15-22.

10. Simon, H. A., C. S. Liu, and J. P. Hartnett. Mass Transfer Cooling of a Flat Plate Exposed to Several Different Free Stream Gases (Department of Energy Engineering Technical Report now in preparation). Chicago: University of Illinois, Chicago Circle.

11. Hayday, A. A. Mass Transfer Cooling in a Steady Laminar Boundary Layer Near the Stagnation Point. 1959 Heat Transfer and Fluid Mechanics Institute. Stanford, Calif.: Stanford University Press, June, 1959.

12. Hoshizaki, H. and H. J. Smith, "The Effect of Helium Injection at an Axially Symmetric Stagnation Point," Journal of the Aerospace Sciences, 26: 6 (June, 1959).

13. Hall, N. A. Flow Equations for Multicomponent Fluid Systems (Technical Report No. 2). Minneapolis, Minn.: University of Minnesota Heat Transfer Laboratory, August, 1955.

14. Eckert, E. R. G., "Diffusion Thermo Effects in Mass Transfer Cooling," Proceedings, Fifth U.S. National Congress of Applied Mechanics. New York: The American Society of Mechanical Engineers, 1966, pp. 639-650.

15. Sparrow, E. M., "Recent Studies Relating to Mass Transfer Cooling," Proceedings, 1964 Heat Transfer and Fluid Mechanics Institute. Stanford, Calif.: Stanford University Press, June, 1964, pp. 1-18.

16. Pun, W. M. and D. B. Spalding, "Influence of Thermal Diffusion on Transpiration Cooling at an Axisymmetrical Stagnation Point," Proceedings, Third International Heat Transfer Conference, 3, August, 1966, pp. 7-14.

17. Eckert, E. R. G., "Engineering Relations for Heat Transfer and Friction in High-Velocity Laminar and Turbulent Boundary Layer Flow Over Surfaces with Constant Pressure and Temperature," Transactions, ASME, 78: 6 (August, 1956), p. 1273.

18. Romig, M., "Stagnation Point Heat Transfer for Hypersonic Flow," Jet Propulsion, 26 (1956), p. 1098.

19. Simon, H. A., C. S. Liu, and J. P. Hartnett. The Eckert Reference Formulation Applied to High-Speed Laminar Boundary Layer of Nitrogen and Carbon Dioxide (NASA CR-420). Washington, D.C.: National Aeronautics and Space Administration, April, 1966.

20. Hirschfelder, J. P., C. F. Curtis, and R. B. Bird. Molecular Theory of Gases and Liquids. New York: John Wiley & Sons, 1954.

21. Gross, J. F., J. P. Hartnett, D. J. Masson, and C. Gazley, Jr. A Review of Binary Boundary Layer Characteristics (U.S. Air Force Project Rand Research Memorandum RM-2516). Santa Monica, Calif.: Rand Corporation, June, 1959.

22. ————, "A Review of Binary Laminar Boundary Layer Characteristics," International Journal of Heat and Mass Transfer, 3:3 (October, 1961), p. 198.

23. Liu, C. S. Mass Transfer Cooling in Laminar Boundary Layers with Hydrogen Injected into Nitrogen and Carbon Dioxide Streams. M.S. thesis, Department of Mechanical Engineering, University of Delaware, Newark, Delaware, June, 1965.

24. Simon, H. A., C. S. Liu, and J. P. Hartnett. Properties of Hydrogen: Nitrogen, Hydrogen:Carbon-Dioxide, and Carbon-Dioxide:Nitrogen Mixtures (NASA CR-387). Washington, D.C.: National Aeronautics and Space Administration, February, 1966.

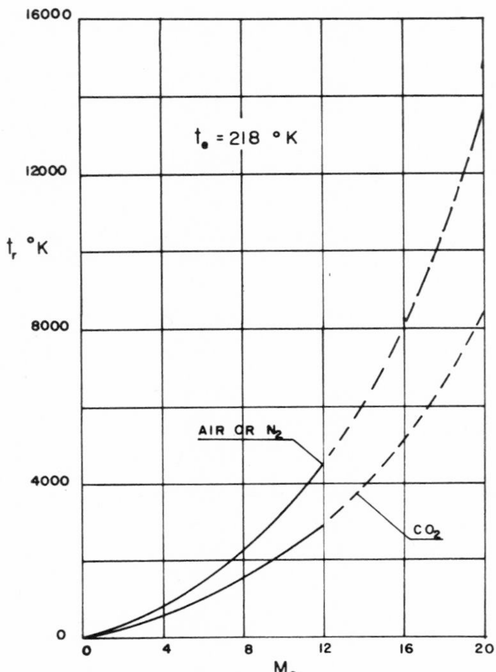

FIGURE 1. RECOVERY TEMPERATURE AS A FUNCTION OF MACH NUMBER FOR t_e = 218°K FREE STREAM OF AIR OR CARBON DIOXIDE, FLAT PLATE.

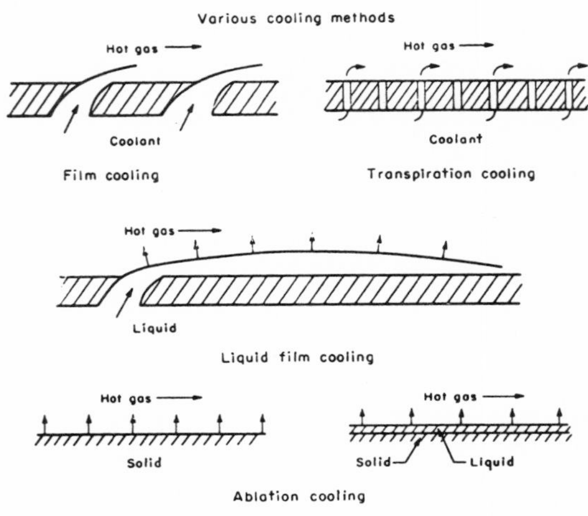

FIGURE 2. VARIOUS MASS TRANSFER COOLING METHODS.

$$q = -k_w \left(\frac{\partial t}{\partial y}\right)_w$$

FOR GIVEN FREE STREAM CONDITIONS

$$t_w = t_r \ \text{WHEN} \ -k_w \left(\frac{\partial t}{\partial y}\right)_w = 0$$

$$t_r = t_r\,(M_e, t_e)$$

DEFINITION OF HEAT-TRANSFER COEFFICIENT, h

$$q = h\ (t_w - t_r)$$

$$Nu = \frac{hx}{k}$$

$$\frac{Nu_x}{\sqrt{Re_x}} = f\,(M_e, t_e, t_w)$$

$$\text{CONSTANT PROPERTIES,}\ \frac{Nu_x}{\sqrt{Re_x}} = f\,(Pr)$$

$$t_r = t_r(Pr)$$

FIGURE 3. DEFINITIONS AND FUNCTIONAL RELATIONS FOR SOLID FLAT PLATE.

$$q = -k_w \left(\frac{\partial t}{\partial y}\right)_w$$

FOR GIVEN FREE STREAM CONDITIONS

$$t_w = t_r \ \text{WHEN} \ -k_w \left(\frac{\partial t}{\partial y}\right)_w = 0$$

$$t_r = t_r\left(M_e, t_e, \frac{\rho_w v_w}{\rho_e u_e}\sqrt{Re_x}, \text{binary gas combination}\right)$$

DEFINITION OF HEAT-TRANSFER COEFFICIENT h

$$q = h\ (t_w - t_r)$$

$$Nu = \frac{hx}{k}$$

$$\frac{Nu_x}{\sqrt{Re_x}} = f\left(M_e, t_e, t_w, \frac{\rho_w v_w}{\rho_e u_e}\sqrt{Re_x}, \text{binary gas combination}\right)$$

$$\text{CONSTANT PROPERTIES,}\ \frac{Nu}{\sqrt{Re_x}} = f\left(\frac{\rho_w v_w}{\rho_e u_e}\ \sqrt{Re_x}, Pr\right)$$

$$t_r = t_r\left(\frac{\rho_w v_w}{\rho_e u_e}\ \sqrt{Re_x}, Pr\right)$$

FIGURE 4. DEFINITIONS AND FUNCTIONAL RELATIONS FOR FLAT PLATE WITH MASS TRANSFER.

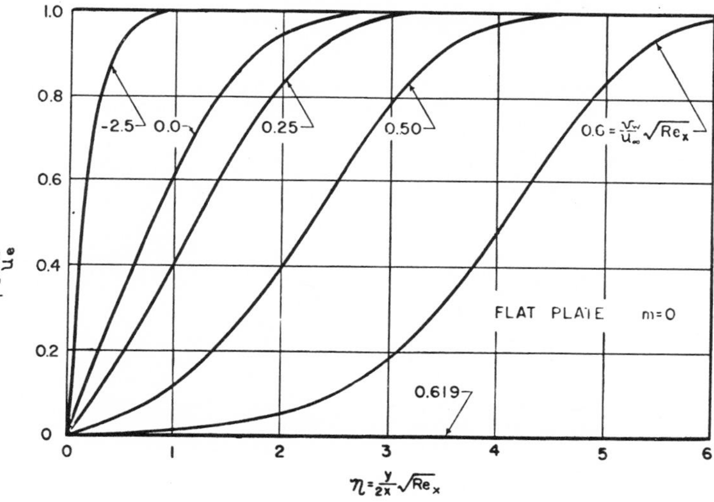

FIGURE 5. VELOCITY PROFILES FOR FLAT PLATE WITH MASS TRANSFER, CONSTANT PROPERTIES.

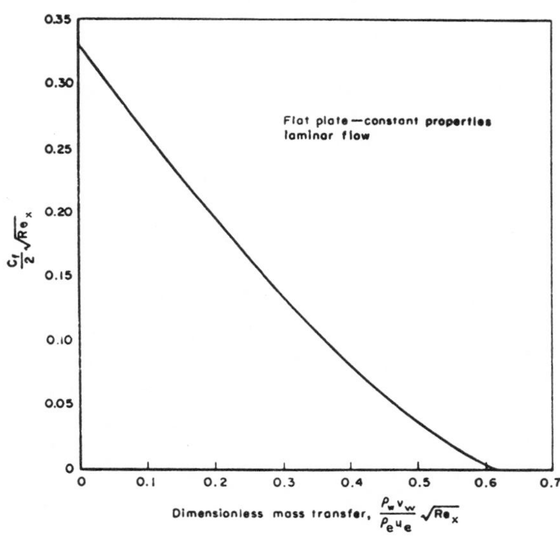

FIGURE 6. SKIN FRICTION COEFFICIENT FOR FLAT PLATE AS A FUNCTION OF THE DIMENSIONLESS MASS TRANSFER, CONSTANT PROPERTIES.

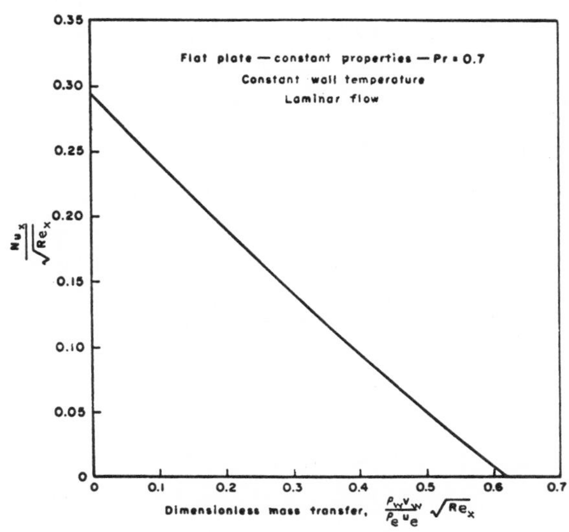

FIGURE 7. NUSSELT NUMBER FOR THE FLAT PLATE AS A FUNCTION OF THE DIMENSIONLESS MASS TRANSFER, CONSTANT PROPERTIES.

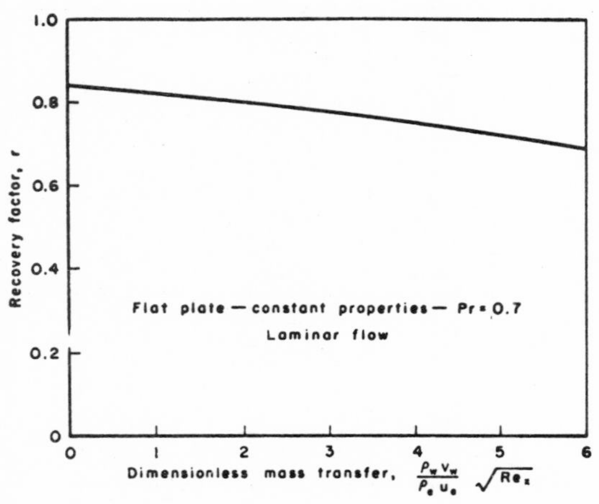

FIGURE 8. RECOVERY FACTOR AS A FUNCTION OF THE DIMENSIONLESS MASS TRANSFER, CONSTANT PROPERTIES.

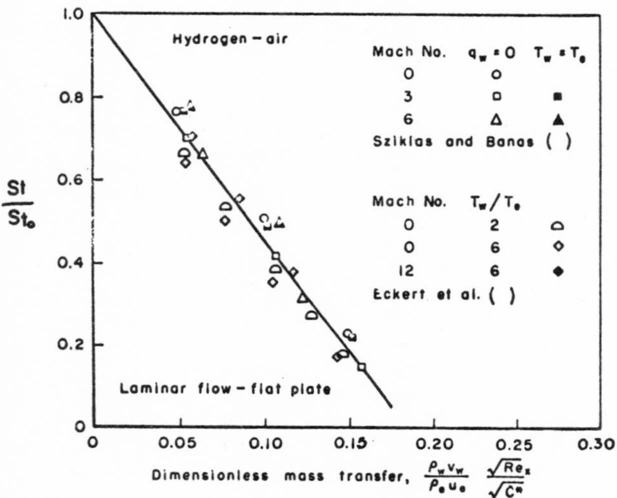

FIGURE 9. NORMALIZED STANTON NUMBERS, St/St_o, AS A FUNCTION OF THE GENERALIZED BLOWING PARAMETER $(\rho_w v_w/\rho_e u_e) \sqrt{R_{ex}/C^*}$ FOR HYDROGEN INJECTION INTO AN AIR BOUNDARY LAYER ON A FLAT PLATE.

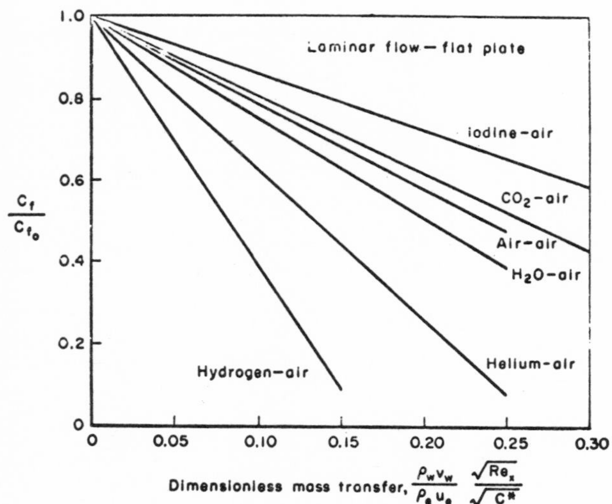

FIGURE 10. SUMMARY OF NORMALIZED SKIN FRICTION COEFFICIENTS, c_f/c_{fo} GENERALIZED BLOWING PARAMETER $(\rho_w v_w/\rho_e u_e) \sqrt{R_{ex}/C^*}$ FOR VARIOUS GASES INJECTED INTO AN AIR BOUNDARY LAYER ON A FLAT PLATE.

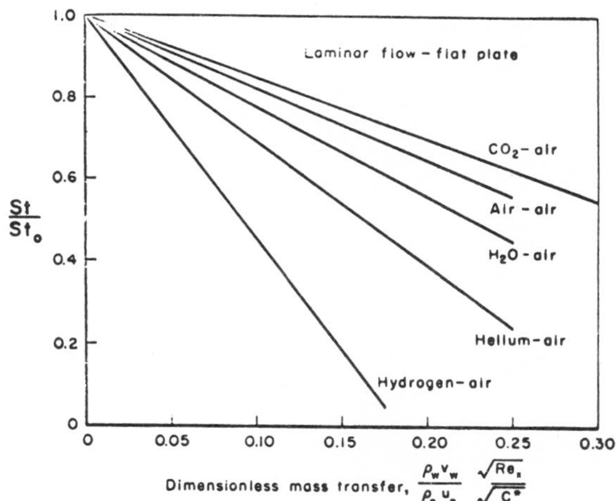

FIGURE 11. SUMMARY OF NORMALIZED STANTON NUMBERS, St/St_0, AS A FUNCTION OF THE GENERALIZED BLOWING PARAMETER $(\rho_w v_w/\rho_e u_e)\sqrt{R_{ex}/C^*}$ FOR VARIOUS GASES INJECTED INTO AN AIR BOUNDARY LAYER ON A FLAT PLATE.

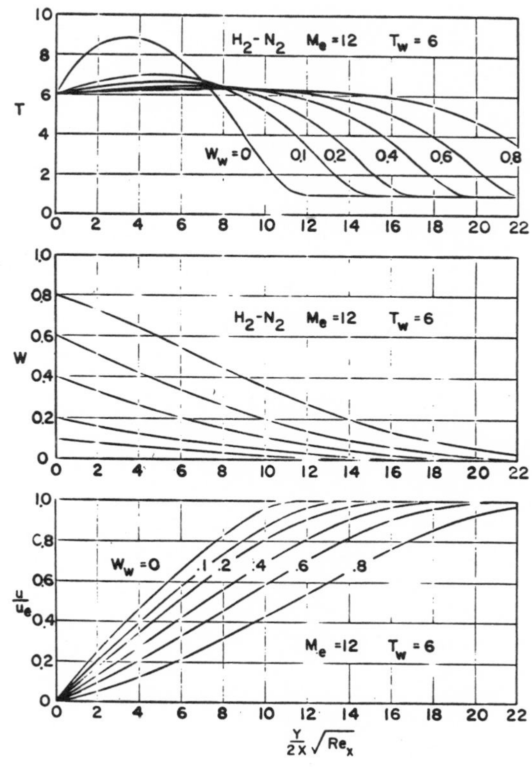

FIGURE 12. TYPICAL VELOCITY, TEMPERATURE, AND MASS FRACTION PROFILES FOR HYDROGEN INJECTED INTO A NITROGEN BOUNDARY LAYER ON A FLAT PLATE.

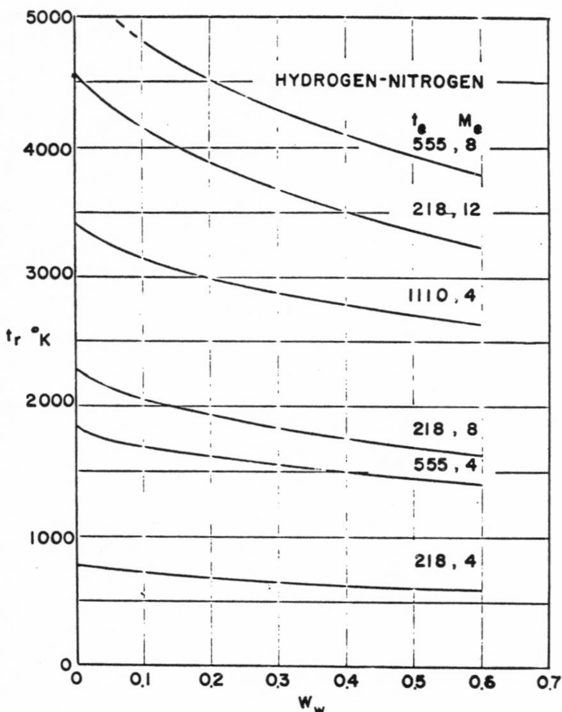

FIGURE 13. TYPICAL VELOCITY TEMPERATURE AND MASS FRACTION PROFILES FOR HYDROGEN INJECTED INTO A CARBON DIOXIDE BOUNDARY LAYER ON A FLAT PLATE.

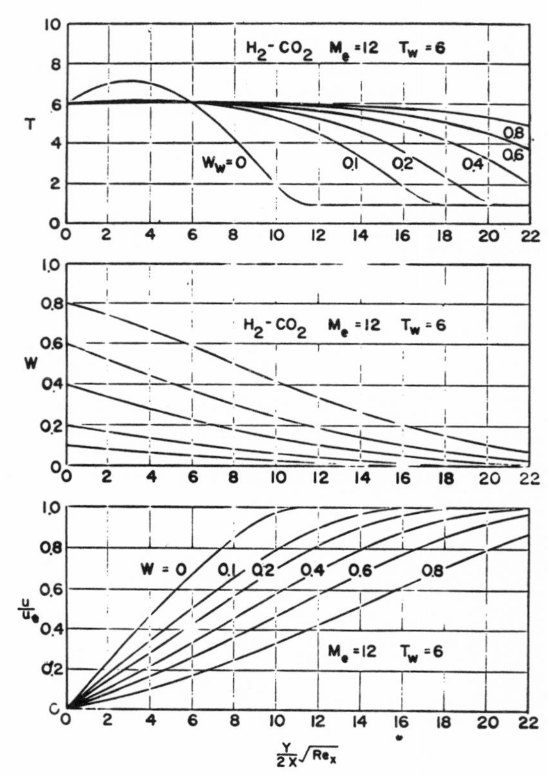

FIGURE 14. RECOVERY TEMPERATURES AS A FUNCTION OF WALL MASS FRACTION FOR HYDROGEN INJECTED INTO A NITROGEN BOUNDARY LAYER ON A FLAT PLATE.

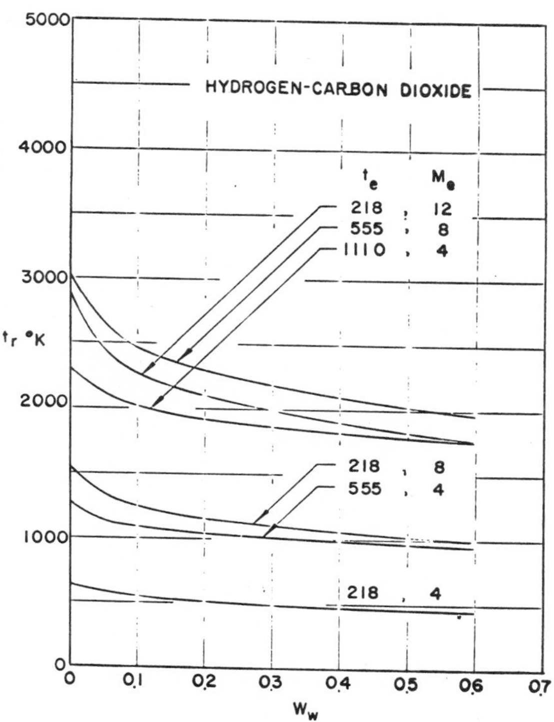

FIGURE 15. RECOVERY TEMPERATURE AS A FUNCTION OF WALL MASS FRACTION FOR HYDROGEN INJECTED INTO A CARBON DIOXIDE BOUNDARY LAYER ON A FLAT PLATE.

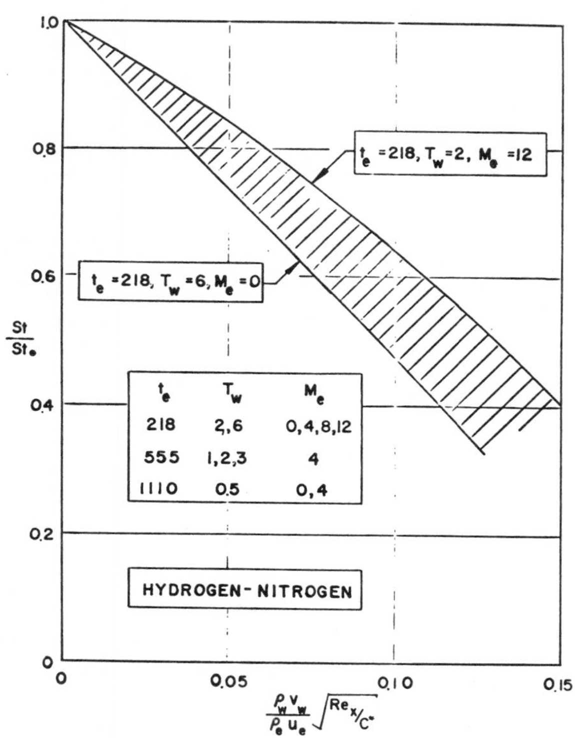

FIGURE 16. NORMALIZED STANTON NUMBERS, St/St_0, AS A FUNCTION OF THE GENERALIZED BLOWING PARAMETER $(\rho_w v_w/\rho_e u_e)\ \sqrt{R_{ex}/C^*}$ FOR HYDROGEN INJECTED INTO A NITROGEN BOUNDARY LAYER ON A FLAT PLATE.

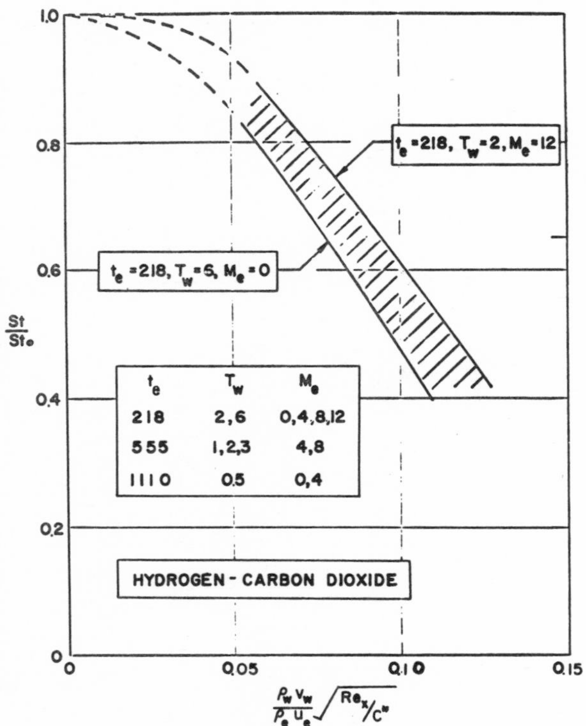

FIGURE 17. NORMALIZED STANTON NUMBERS, St/St_0, AS A FUNCTION OF THE GENERALIZED BLOWING PARAMETER $(\rho_w v_w/\rho_e u_e)\sqrt{R_{ex}/C^*}$ FOR HYDROGEN INJECTED INTO A CARBON DIOXIDE BOUNDARY LAYER ON A FLAT PLATE.

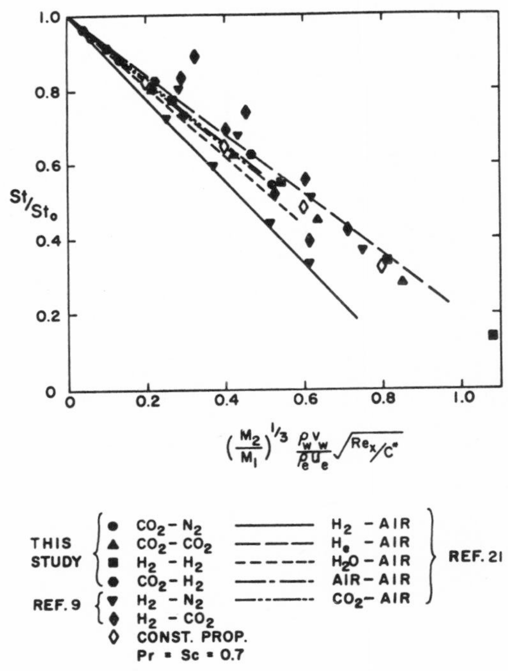

FIGURE 18. NORMALIZED STANTON NUMBERS, St/St_0, AS A FUNCTION OF THE GENERALIZED BLOWING PARAMETER $(M_2/M_1)^{1/3}$ $(\rho_w v_w/\rho_e u_e)\sqrt{R_{ex}/C^*}$ FOR A NUMBER OF DIFFERENT COOLANT GASES AND FREE STREAM GASES, FLAT PLATE GEOMETRY.

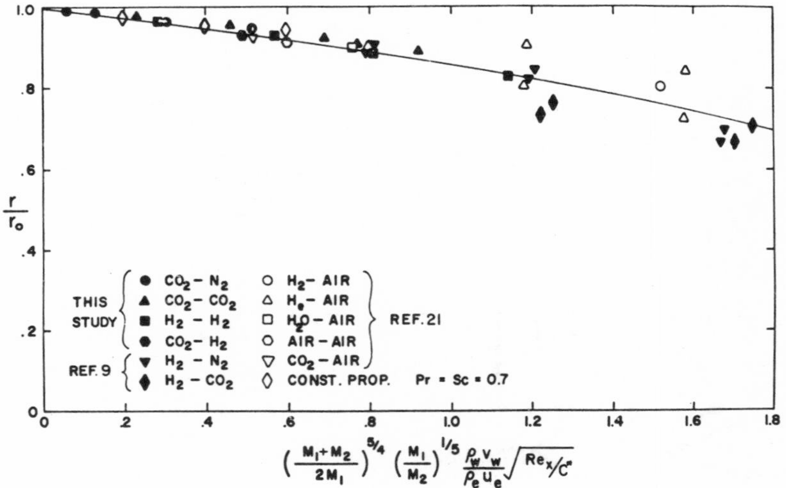

FIGURE 19. NORMALIZED RECOVERY FACTORS, r/r_0, AS A FUNCTION OF THE MODIFIED BLOWING PARAMETER FOR A NUMBER OF DIFFERENT COOLANT GASES, FLAT PLATE GEOMETRY.

HEAT TRANSFER IN CHEMICALLY REACTING GASES

P. M. Chung*

ABSTRACT

The fundamental concepts and the mathematical description of the interaction between the chemically reacting flow field and the solid surfaces are given. The various fundamental governing equations of the chemically reacting, nonequilibrium laminar boundary layers have been formulated for two-dimensional and axisymmetric flows along with their boundary conditions. The general behavior of these equations and the boundary conditions are discussed, and the general effect of the various parameters on the heat transfer are established. It is shown that the appropriately defined total enthalpy difference across the boundary layer largely determines the conduction and diffusion components of the total surface heat transfer. The importance of relating the total enthalpy difference to the particular two components of the heat transfer is presented. By appropriately normalizing the governing equations and the boundary conditions, it is shown that the Damkohler number, defined as the ratio of the characteristic residence time of the species to the chemical reaction time, plays the most important role in determining the overall chemical state of the boundary layers. A few of the typical solutions of the equations are given, and the relationships between the various governing parameters of the equations and the solutions are discussed.

INTRODUCTION

The flow of chemically reacting gases has been the subject of interest to engineers for ages. The real surge of interest in the problem, however, has occurred only during the past decade or so, with the advent of the hypervelocity vehicles in the aerospace technology.

As the chemically reactive flow problems associated with the hypersonic vehicles became increasingly complicated, the more sophisticated treatments of the problems became necessary. In these treatments of the complex problems, the rigorous analyses of the various phenomena occurring in chemically reacting flows, which could have been neglected in the classical problems such as the simple combustion problem, had to be included. As one of the consequences of such analyses, a better understanding of the classical problems resulted.

The most important knowledge of chemically reacting flows has been acquired during the past decade; the more fundamental aspects of this knowledge, upon which the modern understanding of the problems of

*Professor of Fluid Mechanics, Department of Energy Engineering, University of Illinois at Chicago Circle, Chicago, Illinois.

chemically reacting viscous flows is built, will be discussed herein.

In accordance with the nature of the lecture to be delivered, most of this report will be devoted to the general and fundamental aspects of the problem. In the first three sections, the general mathematical as well as the physical concepts of the chemically reacting flows are set forth. The general relationships between the chemical reaction and the heat transfer, and those between the viscous flow and the nonequilibrium reactions are then discussed.

These discussions are based on a recent monograph[(1)*] written by this author.

In the last section, we shall discuss the four typical problems of chemically reacting flows. These are the gas phase and the surface recombinations of dissociated atoms in the stagnation boundary layer flow, the surface catalytic reactions over a flat plate, the development of diffusion flames, and the stagnation merged shock layer of nonequilibrium, chemically reacting air. The first two problems were discussed in the monograph[(1)] whereas the last two problems have been studied since the monograph.

The present lecture note only covers the laminar flows.

GOVERNING EQUATIONS FOR LAMINAR BOUNDARY LAYER

We only consider here the two-dimensional and axisymmetric, steady state flows. The overall continuity and momentum equations are the same as those for the chemically inert flows.[**]

Continuity:

$$\frac{\partial \rho u r^{\epsilon}}{\partial x} + \frac{\partial \rho v r^{\epsilon}}{\partial y} = 0 \tag{1}$$

Momentum:

$$\rho\left(u\frac{\partial u}{\partial x} + v\frac{\partial u}{\partial y}\right) = -\frac{dp}{dx} + \frac{\partial}{\partial y}\left(\mu\frac{\partial u}{\partial y}\right) \tag{2}$$

With the suitable justification[(1)] of the pseudo-binary concentration diffusion applicable for many engineering problems, the conservation of the given chemical species can be expressed by the following equation.

Conservation of i-th species:

$$\rho\left(u\frac{\partial c_i}{\partial x} + v\frac{\partial c_i}{\partial y}\right) = \frac{\partial}{\partial y}\left(\rho D_i\frac{\partial c_i}{\partial y}\right) + W_i, \tag{3}$$

where W_i is the volumic production rate of the i-th species by the gas phase chemical reaction, which is considered here as a known function of C_i, T, and p.

The energy equation for the mixture of the reactive species can be written in various forms. We shall employ the energy equation based on the total enthalpy, h_t, of the mixture defined in the following. This, perhaps, is the most fundamental form of the energy equation because, as it will be seen subsequently, the heat transfer driving potential is essentially the total enthalpy in a chemically reacting flow. The total enthalpy is defined for the present purpose as

$$h_t = \sum_i c_i h_i + \frac{u^2}{2}, \tag{4}$$

where

$$h_i = \int_0^T c_{pi}\,dT + h_i^o. \tag{5}$$

h_i^o is the heat of formation of the i-th species at zero temperature in the gaseous phase. The energy equation becomes,

*Superscript numbers in parentheses refer to References at the end of this paper.

**All symbols are defined in the Notation at the end of this paper.

Energy:

$$\rho\left(u\frac{\partial h_t}{\partial x}+v\frac{\partial h_t}{\partial y}\right)=\frac{\partial}{\partial y}\left\{\frac{\mu}{Pr}\left[\frac{\partial h_t}{\partial y}+(Pr-1)\frac{\partial(u^2/2)}{\partial y}+\sum_i (Le_i-1)h_i\frac{\partial C_i}{\partial y}\right]\right\} \quad (6)$$

Finally, we have the equation of state of the gas mixture.

State:

$$p=\left(\sum_i \frac{C_i}{\hat{M}_i}\right)\rho RT \quad (7)$$

Equations (4) and (5) are employed to relate h_t of the energy equation with the temperature appearing in W_i and in Equation (7). Thus we have five plus i unknown functions: u, v, h_t, T, and ρ, and C_i's and the same number of equations, Equations (1), (2), (3), (4), (6), and (7). A priori knowledge of the pressure, p, is usually assumed.

The boundary conditions for the governing equations are as follows:

At y = 0,

$$u = 0 \quad (8)$$

$$v = v_w \quad (9)$$

$$-\left(\rho D_i\frac{\partial C_i}{\partial y}\right)_w = J_{iw-}-\rho_w v_w(C_{iw}-C_{is}) \quad (10)$$

$$h_t = h_{tw} = \sum_i C_{iw}\left(\int_0^{T_w} c_{pi}\,dT+h_i^o\right)$$

$$\simeq c_pT_w+\sum_i C_{iw}h_i^o\;; \quad (11)$$

and at $y\to\infty$,

$$u = u_e \quad (12)$$

$$C_i = C_{ie} \quad (13)$$

$$h_t = h_{te}\,. \quad (14)$$

The boundary condition (10) is derived by establishing a mass balance for the i-th species between (w) and (s) shown in Figure 2. For the present purpose, we only consider the simple cases of either the solid ablating directly into the gaseous phase at (w-), or the gaseous coolant being passed through a porous solid. In the former case, the J_{iw-} denotes the rate of production of the i-th species by the surface reaction taking place at the surface, whereas in the latter case, J_{iw-} is that which takes place between (s) and (w-).

Often in a problem with surface chemical reaction such as the surface combustion, the surface mass transfer rate is controlled by the net surface chemical reaction rate. For such cases, the boundary condition (9) is not known a priori, but it is coupled with the boundary condition (10). Also, except in the cases in which it is known a priori that the surface is in chemical equilibrium, the surface distribution of the reacting species, C_{iw}, is not known. The boundary condition (11), therefore, is not usually known, and it is coupled with the boundary conditions (9) and (10). We shall see these couplings more explicitly later.

The boundary conditions (12) through (14) are considered to be given from the inviscid solutions as functions of x.

SURFACE HEAT TRANSFER

The heat transfer expressions given in the Section II, Part E of the monograph[1] by the present author were originally derived for the ablating solids, and, hence, are directly applicable only when the solid surface is ablating into the gaseous phase. When a gaseous coolant is being passed through a porous solid, however, certain terms of the expressions must be interpreted a bit differently compared to the case of the ablating solid. The explanation of this fact was not presented with sufficient clarity in that monograph. As the substitute of that section, therefore, a rather detailed discussion of the heat transfer relationship for chemically reacting boundary layers is given in the following, which applies to both the case of the ablating solid and the case of blowing through the porous solid.

The mass balance of the i-th species between (s) and (w) is given by Equation (10). Similarly, we establish an energy balance between (s) and (w) as

$$(\rho v)_w \left(\sum_i c_{is} h_{is} \right)^* = (\rho v)_w \left(\sum_i c_{iw} h_{iw} \right) - \left[\lambda \frac{\partial T}{\partial y} + \sum_i \rho_i D_i h_i \left(\frac{\partial c_i}{\partial y} \right) \right]_w . \tag{15}$$

The left side of Equation (15) is the net energy flowing into the control volume, between (s) and (w), at the boundary (s), whereas the right side is that flowing out of the control volume at (w).

It is important at this point to agree among ourselves as to precisely what we mean by "surface heat transfer." As we shall see later, there are two surface heat transfer quantities that are of interest. They are the heat transfer at the gas-solid interface, (w), due to conduction and diffusion, which will be denoted by $(q_{cd})_w$, and that into the solid interior at (w-) due to conduction, and diffusion (when the solid is porous) which will be denoted by $(q_{cd})_{w-}$.

The expressions for these heat transfer quantities will be derived in the following equations. By the use of these heat transfer relations, the general effect of chemical reactions on heat transfer will be clearly shown.

Heat Transfer at the Gas-Solid Interface Due to Conduction and Diffusion, $(q_{cd})_w$

The last two terms of Equation (15) constitute $(q_{cd})_w$. Hence, by rewriting Equation (15), we have

$$(q_{cd})_w = \left(\lambda \frac{\partial T}{\partial y} + \rho \sum_i D_i h_i \frac{\partial c_i}{\partial y} \right)_w$$

$$= \rho_w v_w \sum_i \left(c_{iw} h_{iw} - c_{is} h_{is} \right). \tag{16}$$

Recognizing from Equations (4) and (5) that

$$\frac{\partial T}{\partial y} = \frac{1}{c_p} \left(\frac{\partial h_t}{\partial y} - \sum_i h_i \frac{\partial c_i}{\partial y} - u \frac{\partial u}{\partial y} \right), \tag{17}$$

we derive from Equation (16),

$$(q_{cd})_w = \frac{\mu_w}{Pr} \left[\frac{\partial h_t}{\partial y} + \sum_i (Le_i - 1) h_i \frac{\partial c_i}{\partial y} \right]_w . \tag{18}$$

The general effect of chemical reaction on $(q_{cd})_w$ can be readily seen from the energy equation, Equation (6), and the heat transfer expression, Equation (18).

We consider that the Lewis numbers, Le_i, are sufficiently close to one such that the last terms of Equations (6) and (18) are negligible. Also, we assume for the moment that h_{tw} is uniform along the surface, for it will simplify our manipulation without changing the basic argument. Then, by defining

$$H_t = \frac{h_t - h_{tw}}{h_{te} - h_{tw}} , \tag{19}$$

Equations (6) and (18), respectively, become

$$\rho \left(u \frac{\partial H_t}{\partial x} + v \frac{\partial H_t}{\partial y} \right)$$

$$= \frac{\partial}{\partial y} \left\{ \frac{\mu}{Pr} \left[\frac{\partial H_t}{\partial y} + (Pr - 1) \frac{\partial (\mu^2/2)}{\partial y} \right] \right\} \tag{20}$$

$$(q_{cd})_w = \frac{\mu_w}{Pr} \left(\frac{\partial H_t}{\partial y} \right)_w (h_{te} - h_{tw}) . \tag{21}$$

Equation (20) shows that the chemical reaction affects the solution only through its effect on ρ, u, v, and μ. These

*When the i-th species are in solid phase, h_i^o may be considered as a negative quantity. Also, note that $(\rho v)_s = (\rho v)_{w-} = (\rho v)_w$.

effects usually are not too strong, and the term

$$\frac{\mu_w}{Pr}\left(\frac{\partial H_t}{\partial y}\right)_w$$

in Equation (21) is a very insensitive function of the chemical reaction compared to the enthalpy difference $(h_{te} - h_{tw})$. The surface heat transfer $(q_{cd})_w$ is hence proportional to the enthalpy difference. The total enthalpy difference, $(h_{te} - h_{tw})$, is therefore basically the driving potential of heat transfer, $(q_{cd})_w$, in a chemically reacting flow as it is with the conventional chemically inert flow, provided that h_t is defined properly.

The quantitative effect of the chemical reaction on heat transfer can be elucidated by decomposing the enthalpy difference, with the aid of Equations (4) and (5), as

$$(q_{cd})_w = \frac{\mu_w}{Pr}\left(\frac{\partial H_t}{\partial y}\right)_w \left(\sum_i C_{ie}\int_o^{Te} c_{pi}dT - \sum_i C_{iw}\int_o^{Tw} c_{pi}dT + \frac{u_e^2}{2} + \sum_i C_{ie}h_i^o - \sum_i C_{iw}h_i^o\right)$$

$$\simeq \frac{\mu_w}{Pr}\left(\frac{\partial H_t}{\partial y}\right)_w \left[c_p\,(T_e - T_w) + \frac{u_e^2}{2} + \left(\sum_i C_{ie}h_i^o - \sum_i C_{iw}h_i^o\right)\right]. \qquad (22)$$

The effect of the chemical reaction on heat transfer, $(q_{cd})_w$, is therefore determined by the term

$$\sum_i C_{ie}h_i^o - \sum_i C_{iw}h_i^o \quad .$$

By noticing the fact that the quantity

$$\sum_i C_{ie}h_i^o - \sum_i C_{iw}h_i^o$$

is zero for chemically inert flow, the ratio of the heat transfer in a chemically reacting flow to that in a chemically inert flow is

$$\frac{(q_{cd})_w \text{ for chemically reacting}}{(q_{cd})_w \text{ for chemically inert}} \simeq 1 + \frac{\sum_i C_{ie}h_i^o - \sum_i C_{iw}h_i^o}{c_p(T_e - T_w) + u_e^2/2} \quad . \qquad (23)$$

It should be noted that the general relationships discussed in the present section are true whether the chemical reaction is a gas phase reaction or a surface reaction. Also, it is true whether or not there exists a surface mass transfer.

In the general reacting cases in which the chemical state at (w) is not in chemical equilibrium, h_{tw} is not known *a priori* because the C_{iw}'s are not known. Hence the relationships such as Equations (22) and (23) are not yet quantitatively complete. It is, however, shown here that the total enthalpy difference is basically the driving potential for the heat transfer, $(q_{cd})_w$, in a chemically reacting flow. It is also equally important to remember that the total enthalpy difference is the driving potential of heat, provided we are referring to $(q_{cd})_w$, which is comprised of the conductive and diffusive energy transfers only at the gas-solid interface.

Returning to Equation (16), we see that one may experimentally determine $(q_{cd})_w$ in a laboratory by measuring the quantity

$$\rho_w v_w \sum_i \left(C_{iw}h_{iw} - C_{is}h_{is}\right)$$

between the solid interior and the surface, if the C_{iw}'s are known. One should remember that it is important to compute correctly the term

$$\sum_i \left(c_{iw}h_{iw} - c_{is}h_{is} \right)$$

in determining $(q_{cd})_w$ in a laboratory experiment. Rosner (see Reference 2), for instance, commenting on a recent experimental study of chemically reacting flows, clearly pointed out the importance of this fact.

Heat Transfer into the Solid Interior by Conduction and Diffusion, $(q_{cd})_{w-}$

First we establish the mass balance of the i-th species and the heat balance between (w) and (w-) as follows.

$$(\rho v)_w c_{iw-} + J_{iw-} = -\rho_w D_{iw} \left(\frac{\partial c_i}{\partial y} \right)_w + \rho_w v_w c_{iw} + J_{idw-} , \tag{24}$$

$$(\rho v)_w \left(\sum_i c_i h_i \right)_{w-} - (q_{cd})_{w-} = -\lambda_w \left(\frac{\partial T}{\partial y} \right)_w - \rho_w \sum_i D_{iw} h_{iw} \left(\frac{\partial c_i}{\partial y} \right)_w + \rho_w v_w \sum_i h_{iw} c_{iw} , \tag{25}$$

where J_{iw-} and J_{idw-} are the rate of production by surface chemical reaction and the rate of diffusion away from (w-) into the porous solid interior of the i-th species, respectively. In each of the above equations, the left side represents the net addition into the control volume, between (w-) and (w), and the right side denotes the net flow out of the control volume.

Equation (25) is rewritten as

$$(q_{cd})_{w-} = \lambda_w \left(\frac{\partial T}{\partial y} \right)_w + \rho_w \sum_i D_{iw} h_{iw} \left(\frac{\partial c_i}{\partial y} \right)_w - \rho_w v_w \sum_i \left(c_{iw} h_{iw} - c_{iw-} h_{iw-} \right) \tag{26}$$

$$= (q_{cd})_w - \rho_w v_w \sum_i \left(c_{iw} h_{iw} - c_{iw-} h_{iw-} \right) . \tag{27}$$

In Equations (26) and (27),

$$\sum_i \left(c_{iw} h_{iw} - c_{iw-} h_{iw-} \right) = \sum_i c_{iw} h_{iw} - \sum_i c_{iw-} \left(h_{iw} - h^o_{iL} \right)$$

$$= \sum_i h_{iw} \left(c_{iw} - c_{iw-} \right) + \sum_i c_{iw} h^o_{iL}$$

$$= \sum_i h_{iw} \left(c_{iw} - c_{iw-} \right) + h^o_L , \tag{28}$$

where

$$h^o_{iL} > 0 \text{ and } h^o_L > 0.$$

We next multiply Equation (24) through by h_{iw} and sum up each term over all the species. Then we combine the resulting equation with Equation (26) with the aid of Equation (28) and produce the equation

$$(q_{cd})_{w-} = \lambda_w \left(\frac{\partial T}{\partial y}\right)_w - \sum_i h_{iw} J_{iw-} - (\rho v)_w h_L^o + \sum_i h_{iw} J_{idw-} . \tag{29}$$

The last term is zero when a solid is ablating. This is the equation, along with Equations (26) and (27), given in Reference 1 in which the symbol q_w was employed for $(q_{cd})_{w-}$. For this case, Equation (29) shows explicitly that the heat transfer rate into the interior of the ablating solid is determined by the net balance of the conduction at the gas-solid interface, heat generation due to the surface chemical reaction, and the heat absorbed by the ablating material. The effect of the gas phase reaction on $(q_{cd})_{w-}$ expresses itself through $(\partial T/\partial y)_w$ in the relationship of Equation (29).

CHEMICAL STATE OF BOUNDARY LAYER

The boundary layer is a dynamic system in that each differential volume fixed with respect to the body is continuously occupied by a different fluid element. Therefore, the actual rate of chemical reaction taking place in a given differential volume in the boundary layer depends not only on the chemical kinetics of the reaction, but also on the length of the time each fluid element resides in that differential volume. The local chemical state of the boundary layer, hence, is not determined by the absolute rate of the chemical reaction, but rather by the reaction rate relative to the speed at which the fluid element is continuously replaced. More explicitly, the local chemical state is determined by the local ratio of the residence time to the reaction time, which is generally defined as the Damkohler number.

With this physical background, let us return to the governing equations and boundary conditions.

The particular form of the energy equation, Equation (6), shows that the effect of the chemical reaction explicitly enters into that equation only through the surface boundary condition, Equation (11), via C_{iw}. The effects of the gas phase and the surface chemical reactions on the boundary layer are most directly expressed by the species conservation equation, Equation (3), and the surface boundary condition, Equation (10), respectively. The previously stated physical fact that the local chemical state of the boundary layer is determined by the reaction rate relative to the speed at which the fluid element is continuously replaced, is equivalent to the statement that the chemical states are determined by the magnitudes of the terms W_i and J_{iw-} relative to those of the rest of the terms in Equations (3) and (10), respectively. In order to properly compare the magnitudes of the various terms, we first nondimensionalize each term to the order of one. This can be most conveniently accomplished by employing the standard boundary layer transformation in the following manner.

Transformation of Species Conservation Equation

We define

$$\eta = \frac{r^\varepsilon u_e}{\sqrt{2s}} \int_0^y \rho \, dy , \tag{30}$$

$$s = \int_0^x \rho_e u_e \mu_e r^{2\varepsilon} dx , \tag{31}$$

and

$$f(\eta, s) = \frac{\psi}{\sqrt{2s}} , \tag{32}$$

where ψ is the usual stream function. Equations (3) and (10) then become, after some manipulation (see Reference 1),

$$\frac{1}{Sc_i}\frac{\partial}{\partial \eta}\left(\ell \frac{\partial m_i}{\partial \eta}\right) + f\frac{\partial m_i}{\partial \eta} + \frac{2}{C_{ie}}\frac{L}{u_\infty}\frac{\rho_e \mu_e}{(\rho_e \mu_e)_o}\frac{1}{[F(X)]^2}\left(\frac{w_i}{\rho}\right) = 2s\left(\frac{\partial f}{\partial \eta}\frac{\partial m_i}{\partial s} - \frac{\partial f}{\partial s}\frac{\partial m_i}{\partial \eta}\right), \tag{33}$$

and at the surface,

$$\left(\frac{\partial m_i}{\partial \eta}\right)_w + \left(\frac{Sc_i}{\ell}\right)_w \left(f + 2s\frac{\partial f}{\partial s}\right)_w (m_{iw} - m_{is}) = -\left[\left(\frac{\ell w}{Sc_i}\right) c_{ie}\sqrt{\frac{(\rho_e\mu_e)_o u_\infty}{2L}}\, F(X)\right]^{-1} J_{iw-}, \quad (34)$$

$$F(X) = \frac{\left[\rho_e\mu_e/(\rho_e\mu_e)_o\right](u_e/u_\infty)(r/L)^{\epsilon}}{\left\{\int_o^X \left[\rho_e\mu_e/(\rho_e\mu_e)_o\right](u_e/u_\infty)(r/L)^{2\epsilon}\, dX\right\}^{1/2}}. \quad (35)$$

At $\eta \to \infty$, $m_i(\infty) = 1$.

Equations (33) and (34) can be made more explicit by elaborating on the w_i/ρ and J_{iw-}.

The net production rate of the i-th species by gas phase reaction, w_i/ρ, can be written in general from the law of mass action[1] as

$$\frac{W_i}{\rho} = k_f(T)\varphi_f(C_j,T,p) - k_r(T)\varphi_r(C_j,T,p), \quad (36)^*$$

where $k_f\varphi_f$ and $k_r\varphi_r$ are the forward and the reverse reaction rates, respectively, and k_f and k_r are the specific forward and reverse rate coefficients, respectively. Equation (36) can be rewritten as

$$\frac{W_i}{\rho} = \alpha\left(C_{ie},T_e,p\right) k_f(T)\left[\hat{\varphi}_f\left(C_j,T,p\right) - K_E(T)\hat{\varphi}_r\left(C_j,T,p\right)\right], \quad (37)$$

where K_E is the equilibrium constant for the gas phase reaction. $\alpha(C_{ie},T_e,p)$ is the normalizing function for the terms within the bracket which are now dimensionless.

Similarly, the net production rate of the i-th species by surface reaction, J_{iw-}, can be written as

$$J_{iw-} = \alpha_J\left(C_{ie},T_e,p\right) K_f\left[\hat{\varphi}_f\left(C_{jw},T_w,p\right) - K_{JE}\hat{\varphi}_r\left(C_{jw},T_w,p\right)\right]. \quad (38)$$

Equations (33) and (34) are now rewritten with the use of Equations (37) and (38) as

$$\frac{1}{Sc_i}\frac{\partial}{\partial \eta}\left(\ell\frac{\partial m_i}{\partial \eta}\right) + f\frac{\partial m_i}{\partial \eta} - 2s\left(\frac{\partial f}{\partial \eta}\frac{\partial m_i}{\partial s} - \frac{\partial f}{\partial s}\frac{\partial m_i}{\partial \eta}\right) = -\zeta_i\left[\hat{\varphi}_f\left(C_j,T,p\right) - K_E(T)\hat{\varphi}_r\left(C_j,T,p\right)\right] \quad (39)$$

and

$$\frac{\ell w}{Sc_i}\left(\frac{\partial m_i}{\partial \eta}\right)_w + \left(f + 2s\frac{\partial f}{\partial s}\right)\left(m_{iw} - m_{is}\right) = -\zeta_{Ji}\left[\hat{\varphi}_f\left(C_{jw},T_w,p\right) - K_{JE}\hat{\varphi}_r\left(C_{jw},T_w,p\right)\right], \quad (40)$$

---

*C_j in φ shows that φ may be a function of other chemical species present in addition to the i-th species.

where

$$\zeta_i = \frac{\alpha k_f}{(c_{ie}/2)(u_\infty/L)\left[\left(\rho_e\mu_e\right)_o/\left(\rho_e\mu_e\right)\right]\;[F(X)]^2} \tag{41}$$

and

$$\zeta_{Ji} = \frac{\alpha_J K_f}{c_{ie}\left[u_\infty\left(\rho_e\mu_e\right)_o/2L\right]^{1/2}\;F(X)}\;. \tag{42}$$

We have now completed the transformation of the species conservation equation and its surface boundary condition which show the effects of the gas phase and the surface reactions, respectively, on the chemical state of the boundary layer. We shall now discuss the chemical behavior of the boundary layer by the use of the transformed equations, Equations (39) through (42).

Role of Damkohler Numbers and the Chemical Behavior

First, we shall discuss the physical significance of the functions ζ_i and ζ_{Ji} defined by Equations (41) and (42). We shall then study the effect of these functions on the behaviors of Equations (39) and (40), hence, on the chemical behavior of the boundary layer.

For

$$Sc_i = 0(1),$$

each term of Equations (39) and (40) is of the order one.* This means that for

$$Sc_i = 0(1)$$

the residence time (or, inversely, the transport rate of the i-th species) due to the diffusion is of the same order of magnitude as that due to the convection, and, hence, either of the residence times (or the transport rates) may be compared to the chemical reaction time (or the reaction rate) in studying the chemical state of the boundary layer. When

$$Sc_i \neq 0(1),$$

the shorter of the two times must be employed for the purpose.

The diffusive flux of the i-th species is

$$\rho D_i\frac{\partial c_i}{\partial y} = \left(\frac{\partial m_i}{\partial \eta}\,\frac{\ell}{Sc_i}\right)\left[c_{ie}\sqrt{\frac{\left(\rho_e\mu_e\right)_o u_\infty}{2L}}\;F(X)\right]. \tag{43}$$

Since

$$\frac{\partial m_i}{\partial \eta}\,\frac{\ell}{Sc_i}$$

is of order one for

$$\frac{1}{Sc_i} = 0(1),$$

Equation (43) shows that the characteristic order of magnitude of the diffusive flux is given by the quantity in the bracket on the right side of that equation.

Considering Equation (38) for the surface reaction, the quantity in the bracket expresses the normalized difference between

*In a near-equilibrium flow, there may often exist the isolated singular regions in which $(\partial^2 m_i)/\partial y^2$ may increase without limit (see Reference 1).

the forward and reverse reaction rates which are equal in the limit of the surface chemical equilibrium. The term $\alpha_J K_f$, on the other hand, represents the characteristic reaction rate at which the forward and the reverse reactions are being equilibrated.

We now see from the above arguments concerning Equation (38) and Equation (43) that the function ζ_{Ji} given in Equation (42) is the ratio of the characteristic surface reaction rate to the characteristic diffusion rate, or, inversely, it is the ratio of the characteristic residence time to the characteristic surface reaction time. The function ζ_{Ji}, therefore, is the surface Damkohler number.

For the gas phase reaction given by Equation (37), the characteristic reaction rate αk_f has the dimension of the reciprocal of time. $1/(\alpha k_f)$ is, therefore, the characteristic gas phase reaction time. The order of magnitude of the diffusive residence time is readily derived in the following manner. The order of magnitude of the boundary layer thickness is first obtained from Equations (30) and (31), by considering the fact that the boundary layer thickness, δ, corresponds to $\eta = 0(1)$, as

$$\delta \simeq \sqrt{\frac{2(\rho_e\mu_e)_o L}{u_\infty}} \, \frac{1}{F(X)} \, \frac{\rho_e\mu_e}{(\rho_e\mu_e)_o} \, \frac{1}{\rho} \, . \tag{44}$$

The characteristic diffusive residence time is then obtained by dividing the boundary layer thickness, Equation (44), by the diffusion velocity, $D_i(\partial C_i/\partial y)$, from Equation (43). We then see that the function ζ_i given by Equation (41) is the ratio of the characteristic residence time to the characteristic gas phase reaction time. The function ζ_i is, therefore, the gas phase Damkohler number.

Having seen the physical meaning of the Damkohler numbers ζ_i and ζ_{Ji}, we shall now discuss their effects.

First, it is obvious that in the limits of $\zeta_i \to 0$ and $\zeta_{Ji} \to 0$, Equations (39) and (40) degenerate, respectively, to the species conservation equation and its surface boundary condition for the conventional chemically inert flows. Physically, however, $\zeta_i \to 0$ and $\zeta_{Ji} \to 0$ do not necessarily mean that the fluid is chemically inert; that is, $\alpha k_f = 0$ and $\alpha_J K_{Jf} = 0$, but rather mean that the reaction rates are extremely small compared to the transport rates. These limits of vanishingly small Damkohler numbers are, therefore, referred to as chemically "frozen" flows.

On the other hand, when ζ_i for the gas phase reaction becomes much greater than one, we see from Equation (39) that the quantity in the bracket of that equation must approach zero, since all other terms of Equation (39) remain as order one.* The vanishing of the quantity in the bracket means that the forward and the reverse reaction rates are equal and, hence it means that the gas phase chemical state of the boundary layer is in chemical equilibrium. Usually for such "equilibrium" flows, the algebraic equation

$$\frac{\hat{\Phi}_f(C_j, T, P)}{\hat{\Phi}_r(C_j, T, P)} = K_E \tag{45}$$

replaces the differential equation, Equation (39), and the analysis is greatly simplified. A similar argument holds for the surface chemical state when $\zeta_{Ji} \gg 1$.

For all other values of the Damkohler numbers, the chemical state at the boundary layer is in a general nonequilibrium state and the full equations, Equations (39) and (40), must be analyzed.

Before we conclude the present section on the general behavior of the chemically reacting boundary layers, let us elaborate further on the surface heat transfer relationship developed earlier.

Surface Heat Transfer

Equations (22) and (27) show that for $Le_i \simeq 1$ the heat transfers $(q_{cd})_w$ and $(q_{cd})_{w-}$ can be readily obtained from the solution of the energy equation alone, provided that the C_{iw}'s are known. The energy equation, Equation (20), is identical in form to that for the chemically inert boundary layers, and solution of the equation is readily obtainable in the conventional manner in terms of H_t. The determination of C_{iw}'s, however, necessitates the solution of Equations (33) and (34), which usually is quite a painstaking task. Solution of Equations (33) and (34) is further complicated by the fact that these equations are coupled with the energy equation, because w_i/ρ is usually a

---

*Except for the isolated singular regions in which the first term has a singularity. Discussions of these singular regions are given in References 1 and 3, and will not be included here.

function of the temperature among other variables.

When $\zeta_i \to 0$ and $\zeta_{Ji} \to 0$, that is, when all the reactions are frozen, the C_{iw}'s are known *a priori* when $v_w = 0$, or they are readily obtainable when $v_w \neq 0$ by integrating Equations (33) and (34) in the standard manner without the chemical reaction terms. The determination of the C_{iw}'s is also rather simple when either $\zeta_i \to \infty$ or $\zeta_{Ji} \to \infty$. For these cases, the C_{iw}'s can be readily obtained by setting the appropriate coefficient of ζ_i or ζ_{Ji}, given in Equation (39) or Equation (40), to zero. For most cases, the C_{iw}'s given by the gas phase equilibrium relationship and those given by the surface equilibrium relationship are practically equal. They are exactly equal when the surface enters into the surface reaction only as a catalyst. Therefore, when $Le_i = 1$, the surface heat transfer is practically the same whether $\zeta_i \to \infty$, $\zeta_{Ji} \to \infty$, or $\zeta_i \to \infty$ and $\zeta_{Ji} \to \infty$ simultaneously, since the heat transfer depends only on the C_{iw}'s.

We close this portion of the discussion with the following example. Let us consider the flow of a partially dissociated diatomic gas such as oxygen or nitrogen over a cooled body. We consider that C_e is the mass fraction of the dissociated atoms in the inviscid region. We readily see that $C_w = C_e$ when $\zeta = \zeta_J = 0$. On the other hand, when the wall is cooled, the gas phase and the surface equilibrium relationships show that $C_w \to 0$ for $\zeta_i \to \infty$ or $\zeta_{Ji} \to \infty$. The maximum variation of heat transfer due to the atom recombination, between the frozen and the equilibrium limits, can be computed from Equations (22) and (27), or Equation (23). The latter shows for this case that

$$\frac{(q_{cd})_{w,\ equil.}}{(q_{cd})_{w,\ frozen}} = 1 + \frac{C_e \Delta h^o}{c_p(T_e - T_w) + u_e^2/2} , \qquad (46)$$

where

$$\Delta h^o = h^o_{atom} - h^o_{molecule}$$

$$= \text{dissociation energy} > 0.$$

TYPICAL SOLUTIONS

Thus far we have discussed in detail the various fundamental aspects of the governing equations, and the general chemical behavior of the boundary layers both on the mathematical and the physical grounds. In particular, the basic effects of the chemical reactions on the heat transfer have been elaborated.

In this section, we shall show a few of the typical solutions of the chemically reacting viscous flow problems. The detailed mathematics involved in the solution of these problems are beyond the scope of the present lecture note. We shall only discuss some of the salient physical features of the solutions. For the more detailed analyses of these and other reacting flow problems, the reader is referred to the referenced literature.

Recombination of Dissociated Atoms at the Stagnation Region

In general, when ζ_i and ζ_{Ji} are finite and nonvanishing, the functions such as h_t and m_i depend explicitly on both variables η and s; hence solution of the partial differential equations such as Equation (39) is required. However, because of the geometric symmetry, the functions depend only on η, and therefore, a similarity transformation exists at the stagnation region of two-dimensional or axisymmetric blunt bodies.

Various exact and approximate studies of the governing equations have been carried out in the past which analyzed the gas phase and catalytic surface recombinations of dissociated air for stagnation boundary layer flows. The approximate but rather general solutions were obtained by Chung and Liu[4] for this problem. The general heat transfer results given in that reference are reproduced here in Figure 3. In the figure, Γ_g and Γ_w are, respectively, the gas phase and the surface Damkohler numbers derivable from the general expressions, Equations (41) and (42). The surface reaction considered is the catalytic recombination which is known to be of first order reaction. By first-order surface reaction, we mean that the reaction is linearly proportional to the concentration of the reactant (atoms) at the surface.

The results shown in Figure 3 are for Le = 1. The curves in the figure are the results of the approximate analysis of Reference 4, whereas the circles are those of more exact analysis of Reference 5.

For the particular recombination processes of air atoms, the equilibrium concentration of atoms at the cooled surface temperature of less than about 2000°K is practically zero, whether the equilibrium at the surface is due to $\Gamma_g \to \infty$ or $\Gamma_w \to \infty$. Hence Figure 3 shows that the same maximum heat transfer is obtained as either Γ_g or Γ_w increases without limit.

Catalytic Surface Reaction Along Flat Plate

Both the gas phase and the surface Damkohler numbers increase with x if T_w, T_e, and the pressure remain constant, because the residence time increases with x as the boundary layer thickness does. This effect of increasing boundary layer thickness on the Damkohler numbers appears in Equations (41) and (42) through the function F(X) defined by Equation (35). Also, the variation of the chemical state of the boundary layer with respect to the Damkohler number depends, among other things, on the order of the chemical reaction.

The above two effects, the effects of the increasing boundary layer thickness and of the various reaction orders, are very clearly seen in the analysis of the catalytic surface reaction along a flat plate, with the frozen gas phase reaction, carried out by Rosner.[6] Figure 4 shows the surface distribution of the reactant for the various reaction order, n. The surface reaction kinetics are expressed by

$$- J_{iw-} = K_r \left(\rho_w C_w\right)^n , \tag{47}$$

where C is the reactant mass fraction. The surface Damkohler number, derivable from Equation (42), is given by the following equation for this problem.

$$\zeta_f = \zeta_J = \frac{Sc^{2/3}}{.332} \frac{\left(\rho_w^n K_r\right) C_e^{n-1}}{\left[\ell u_e \left(\rho_e \mu_e\right)/2\right]^{1/2}} \sqrt{x} . \tag{48}$$

The Damkohler number is zero at the leading edge since there the residence time is zero due to the fact that the boundary layer thickness is zero.* The reaction is therefore frozen at x = 0. ζ_J continuously increases with x for a given finite K_r; hence the surface approaches the equilibrium state as x is continuously increased. We see in Figure 4 that the higher the order of the chemical reaction, the longer it takes (it takes a larger Damkohler number) to reach the equilibrium state.

Development of Diffusion Flames

There has been a considerable amount of work done recently on the diffusion flames in connection with supersonic combustion. (See, for instance, Reference 7.) The ignition and the combustion are more complicated when they occur on the hypersonic vehicles than those which take place in the conventional surroundings. Two of the major causes of the complication are that due to the reduced Damkohler number and that due to the high stagnation temperature which may prevent the complete release of the heat of combustion. These two aspects have been analyzed in detail recently by Chung, et al.[8] The major findings of this analysis[8] are briefly discussed in the following.

For convenience of analysis, Reference 8 considered the stagnation flow configuration shown in Figure 5, in which a jet of gaseous fuel is injected into the oncoming stream of oxidant (such as air), and a mixing layer is formed. The rate of fuel injection is considered to be such that the thickness of the total fuel layer is much greater than that of the mixing layer. This particular flow is chosen so that (1) the similarity transformation of the governing equations is possible, and (2) the surface condition will not directly influence the energetics of the gaseous combustion zone within the mixing layer. The chemical kinetics of the combustion are assumed to be given by the simple equation,

*Note that the Damkohler number is finite at the stagnation point of a blunt body, since there the boundary layer thickness is finite.

$$\underline{a} \text{ molecules of oxidant} + \underline{b} \text{ molecules of fuel} \underset{k_r}{\overset{k_f}{\rightleftarrows}} \underline{d} \text{ molecules of product} . \tag{49}$$

The most important finding of the analysis is that for a given surrounding under which a combustion takes place there may be one or more possible combustion rates, depending on the fuel characteristics in relation to the surrounding conditions. This means that there may be one or more physically possible solution of the boundary value problem posed by Equations (1) through (14) for the combustion of initially unmixed reactants. Thus, as the Damkohler number is increased, the chemical state of the mixing layer (or boundary layer) will go through the transition from a frozen to an equilibrium state, which may be of the simple type shown in Figure 6 or the multiple type shown in Figure 7.

In Figures 6 and 7,

$$H = \frac{c_p}{\Delta h^o} T ,$$

where Δh^o is the heat of combustion. The subscript M denotes the maximum temperature point which exists within the combustion zone. The value of H_M is minimum when no combustion takes place (chemically frozen), whereas it is maximum when the combustion rate is maximum (chemical equilibrium).

The subscripts 1, 2, and 3 denote the oxidant, fuel, and the combustion product, respectively. Γ_f is the gaseous Damkohler number derivable from Equation (41).

The simple transition shown in Figure 6 is analogous to that observed for the atom recombination in Figure 3. The multiple transition shown in Figure 7 implies certain combustion anomalies. For the particular case of Figure 7, for instance, no combustion will take place for

$$\Gamma_f \lesssim 10^2 ,$$

whereas the combustion will proceed at a maximum rate when

$$\Gamma_f \gtrsim 10^{11} .$$

For

$$10^2 \lesssim \Gamma_f \lesssim 10^{11} ,$$

however, we may have any one of the three possible chemical states of the mixing layer. Hence this region may imply certain instabilities as to the ignition and extinction.

It was found[8] that the transition of the chemical state with respect to Γ_f will be of the simple category of Figure 6 if

$$\left(\frac{T_{M,E}}{T_{M,F}}\right)^{a+b-1} \exp\left[-\frac{E_a}{R}\left(\frac{1}{T_{M,F}} - \frac{1}{T_{M,E}}\right)\right] \cdot 10^4 \gg 1 , \tag{50}$$

whereas it will be of the multiple category of Figure 7 if

$$\left(\frac{T_{M,E}}{T_{M,F}}\right)^{a+b-1} \exp\left[-\frac{E_a}{R}\left(\frac{1}{T_{M,F}} - \frac{1}{T_{M,E}}\right)\right] \cdot 10^4 \ll 1 . \tag{51}$$

In deriving the above criteria, it was assumed that

$$k_f = k_{fo} \exp\left(\frac{-E_a}{RT}\right) , \tag{53}$$

where k_{fo} is a constant.

In Figures 6 and 7, $\Gamma_E = 0$ implies that the reverse reaction of Equation (49) is negligible compared with the forward reaction. As the stagnation temperature increases with the u_∞, Γ_E becomes non-negligible. As was expected, it was found in References 3 and 8 that the increasing

Γ_E reduces the chemical heat release, and hence the maximum temperature, for a given Γ_f. With the continuous increase of Γ_E, the energy addition by combustion becomes impractical.

Nonequilibrium Merged Shock Layer

This last example to be considered is somewhat different from those preceding in that it is not a boundary layer problem. We shall, in the present note, limit our discussion to the problem of the nonequilibrium layer formed at the axisymmetric stagnation region of a hypersonic vehicle. The nose radius is considered to be small (of the order of half an inch), and only the flight through the atmospheric air is considered. Our interest is with the flight Mach numbers of the order of 20 to 25.

At the low altitudes, for example, below about 150,000 ft, the flow near the stagnation region can be readily divided into the infinitesimally thin shock, the thick inviscid shock layer, and the thin boundary layer. The gas phase Damkohler number for the dissociation in the inviscid shock layer is sufficiently large to ensure the equilibrium state at the boundary layer edge. The gas phase and the surface Damkohler number for the recombination in the boundary layer, on the other hand, may be quite small, and the analysis of the nonequilibrium boundary layer is important. One of such analyses has been discussed earlier.

As the altitude is increased to above about 150,000 ft, the dissociative Damkohler number behind the shock is no longer sufficiently large and the shock layer is in a chemically nonequilibrium state. Furthermore, at these altitudes, the boundary layer is no longer thin; in fact, the entire shock layer is viscous and no distinction can be made between the shock layer and the boundary layer. The problem between the altitudes of about 150,000 ft and 180,000 ft is then to analyze the nonequilibrium viscous shock layer. Such analysis has been carried out, for instance, in References 9 through 11.

With the increasing altitude beyond about 180,000 ft, the shock itself becomes thick and diffuse, and the entire flow region between the free stream and the surface becomes a viscous merged shock layer. In this regime, the gas phase chemical relaxation and the relaxations of the momentum and the translational energy of the flow themselves become coupled; these relaxations take place simultaneously. The residence time in this regime is not only insufficient for the chemical reaction to reach equilibrium, but also it is not sufficient for the complete conversions of the kinetic energy of the free stream into the pressure and the static energy. The problem above the altitude of about 180,000 ft is then to analyze the nonequilibrium merged shock layer. Such analysis was carried out recently by Chung, *et al.*,[11] and subsequently it has been extended to include the ionization by Lee, *et al.*[12]

The typical temperature and atom concentration profiles across the merged layer are shown in Figure 8. The above-mentioned coupling effects of the chemical reaction and the merging (incompletion of the shock transition) are apparent.

The governing equations of the merged shock layer are of course somewhat different from those given for the boundary layer in the second section of this paper. We shall terminate our discussion here, because the details of the problem are beyond the scope of this lecture.

NOTATION

C = mass fraction

c_p = constant pressure specific heat of gas mixture

c_{pi} = constant pressure specific heat of i-th species

D = binary diffusion coefficient

E_a = activation energy

$F(X)$ = function of body shape defined by Equation (35)

f = stream function defined by Equation (32)

H_t = normalized total enthalpy defined by Equation (19)

h_i = enthalpy of i-th species defined by Equation (5)

h_i^o = heat of formation of i-th species at zero temperature in gas phase

h_{iL}^o = heat of sublimation of i-th species

h_L^o = $\sum_i C_i h_{iL}^o$

h_t = total enthalpy of gas mixture defined by Equation (4)

J_i = rate of production of i-th species by surface chemical reaction

J_{id} = rate of diffusion of i-th species into the solid interior

K_E = gas phase equilibrium constant

K_f = forward surface reaction rate coefficient

K_r = reverse surface reaction rate coefficient

K_{JE} = surface equilibrium constant

k_f = forward gas phase reaction rate coefficient

k_r = reverse gas phase reaction rate coefficient

L = characteristic length such as nose radius

Le_i = Lewis number for i-th species

ℓ = $(\rho\mu)/(\rho\mu)_e$

$\hat{M}$ = molecular weight

m = C/C_e

Pr = Prandtl number of gas mixture, $\mu c_p/\lambda$

p = pressure

q_{cd} = heat transfer by conduction and convection considered positive in (-y) direction

R = universal gas constant

r = distance defined in Figure 1 for axisymmetric bodies

Sc_i = Schmidt number for i-th species, $\mu/(\rho D_i)$

s = function defined by Equation (31)

T = absolute temperature

u = x-component of velocity

v = y-component of velocity

W_i = production rate of i-th species

X = x/L

x = direction and distance along the surface of body

y = direction and distance normal to surface

α, α_J = normalizing functions for chemical reaction terms

ζ_i = gas phase Damkohler number defined by Equation (41)

ζ_{Ji} = surface Damkohler number defined by Equation (42)

Γ = particular Damkohler numbers defined in the text

Γ_E = nondimensionalized gas phase equilibrium constant

ϵ = number which is zero or one depending on whether the flow is two dimensional or axisymmetric

η = similarity variable defined by Equation (30)

λ = thermal conductivity of gas mixture

μ = dynamic viscosity of gas mixture

ν = kinematic viscosity of gas mixture

ρ = density of gas mixture

ψ = stream function defined in standard manner

Subscripts

E = equilibrium

e = boundary layer edge

F = frozen

i = i-th species

o = reference point such as the boundary layer edge at the stagnation point

s = solid interior (see Figure 2)

w = gas-solid interface (see Figure 2)

w- = surface (see Figure 2)

∞ = free stream

REFERENCES

1. Chung, P. M., "Chemically Reacting Nonequilibrium Boundary Layers," in Advances in Heat Transfer, Vol. II. New York: Academic Press, 1965, pp. 110-270.

2. Rosner, D. E., "Mass Injection Effects on Convective Heat Transfer from a Partially Dissociated Gas" (Invited discussion of ASME Paper No. 66-WA/HT-24). AeroChem Corp., Princeton, New Jersey. TP-148, January, 1967.

3. Chung, P. M. and V. Blankenship, "Equilibrium Structure of Thin Diffusion Flame Zone," The Physics of Fluids, 9 (August, 1966), pp. 1569-1577.

4. Chung, P. M. and S. W. Liu, "Simultaneous Gas-Phase and Surface Atom Recombination for Stagnation Boundary Layer," AIAA Journal, 1 (April, 1963), pp. 929-931.

5. Goodwin, G. and P. M. Chung., "Effects of Nonequilibrium Flows on Aerodynamic Heating During Entry into the Earth's Atmosphere from Parabolic Orbits," in Advances in Aeronautical Sciences, Vol. IV. New York: Pergamon Press, 1961, pp. 997-1018.

6. Rosner, D. E., "The Apparent Chemical Kinetics of Surface Reactions in External Flow Systems: Diffusional Falsification of Activation Energy and Reaction Order," AIChE Journal, 9 (May, 1963), pp. 321-331.

7. Ferri, A., "Supersonic Combustion Progress," Astronautics and Aeronautics (August, 1964), pp. 32-37.

8. Chung, P. M., E. Fendell, and J. Holt, "Nonequilibrium Anomalies in the Development of Diffusion Flames," AIAA Journal, 4 (June, 1966), pp. 1020-1026.

9. Chung, P. M., "Hypersonic Viscous Shock Layer of Nonequilibrium Dissociating Gas," NASA TR, R-109 (1961).

10. Tong, H., "Effects of Dissociation Energy and Vibrational Relaxation on Heat Transfer," AIAA Journal, 4 (January, 1966), pp. 14-18.

11. Chung, P. M., J. Holt, and S. W. Liu, "Merged Stagnation Shock Layer of Nonequilibrium Dissociating Gas," Aerospace Corporation Technical Report, TR-0158(3240-10)-8, January, 1968.

12. Lee, R. H. C. and T. Zierten, "Merged Layer Ionization in the Stagnation Region of a Blunt Body," Proceedings, 1967 Heat Transfer and Fluid Mechanics Institute, LaJolla, California, 1967.

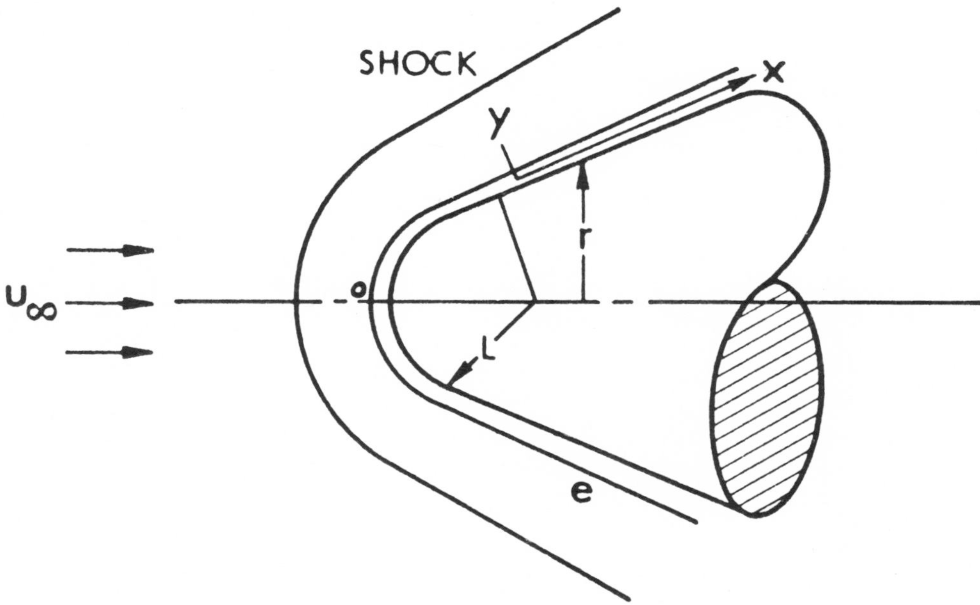

FIGURE 1. BOUNDARY LAYER COORDINATES.

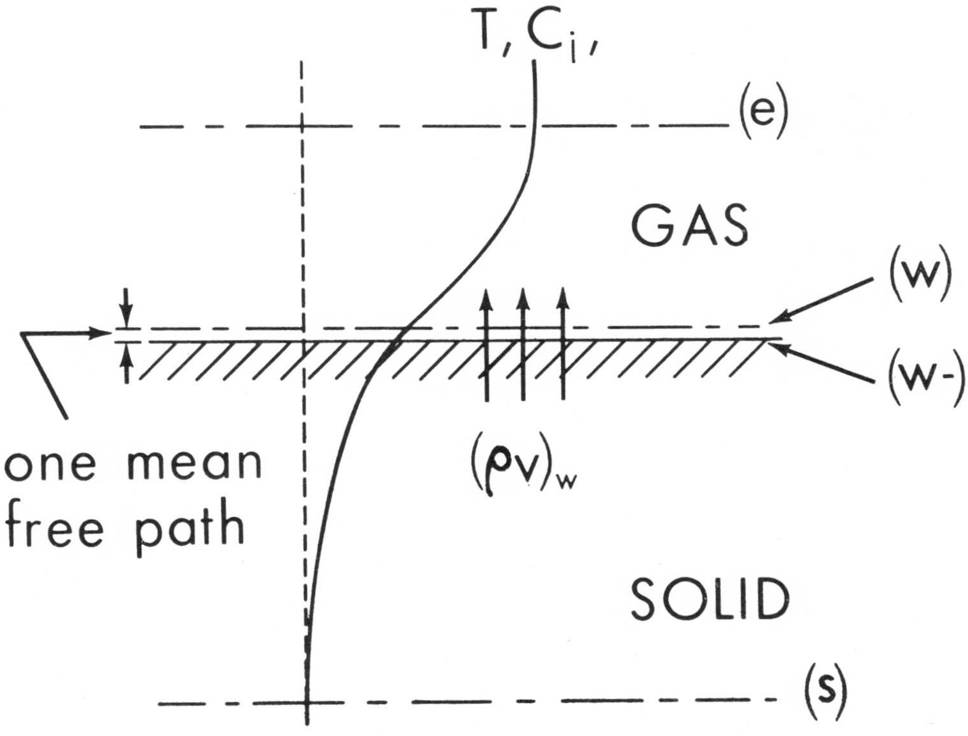

FIGURE 2. HEAT AND MASS BALANCES AT THE SURFACE.

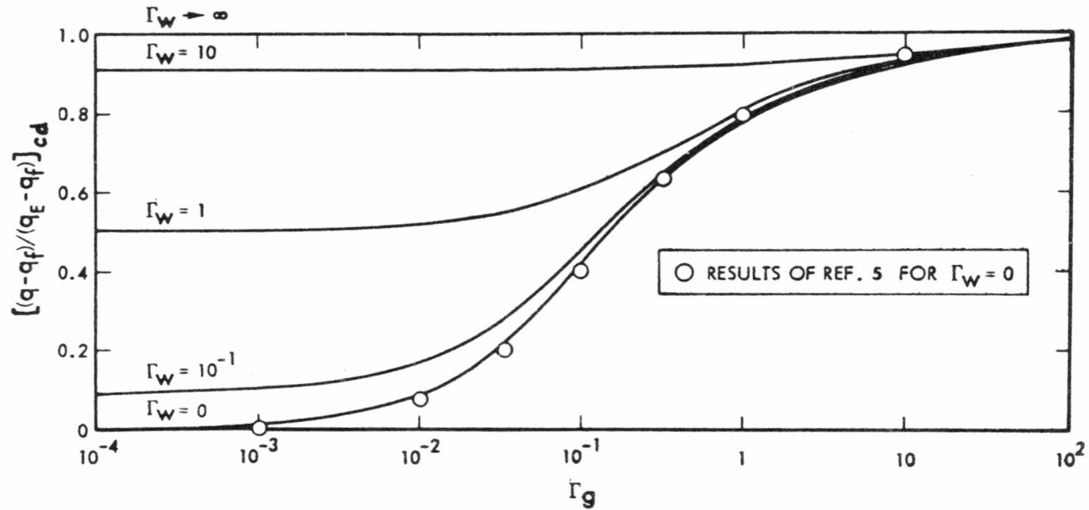

FIGURE 3. VARIATION OF STAGNATION HEAT TRANSFER DUE TO NONEQUILIBRIUM GAS PHASE AND SURFACE REACTIONS.

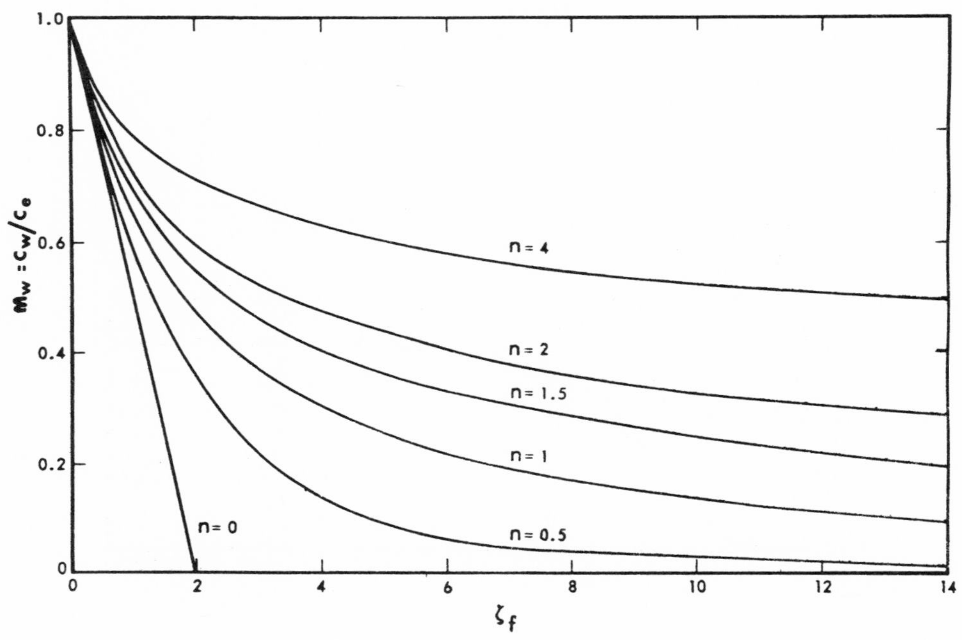

FIGURE 4. SURFACE DISTRIBUTION OF REACTANT ALONG A CATALYTIC FLAT PLATE.

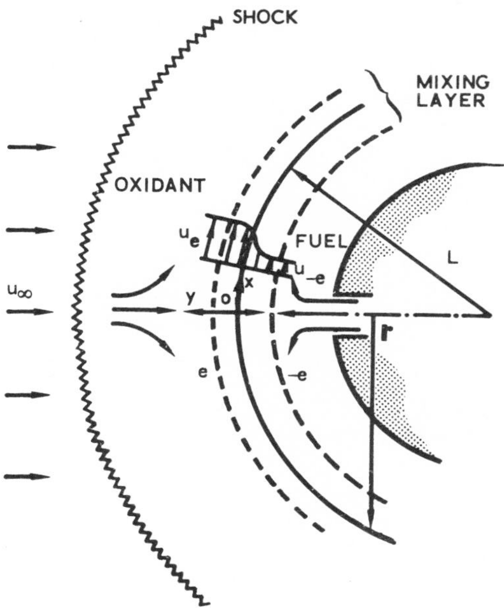

FIGURE 5. STAGNATION MIXING LAYER ($\Lambda = u_{-e}/u_e$).

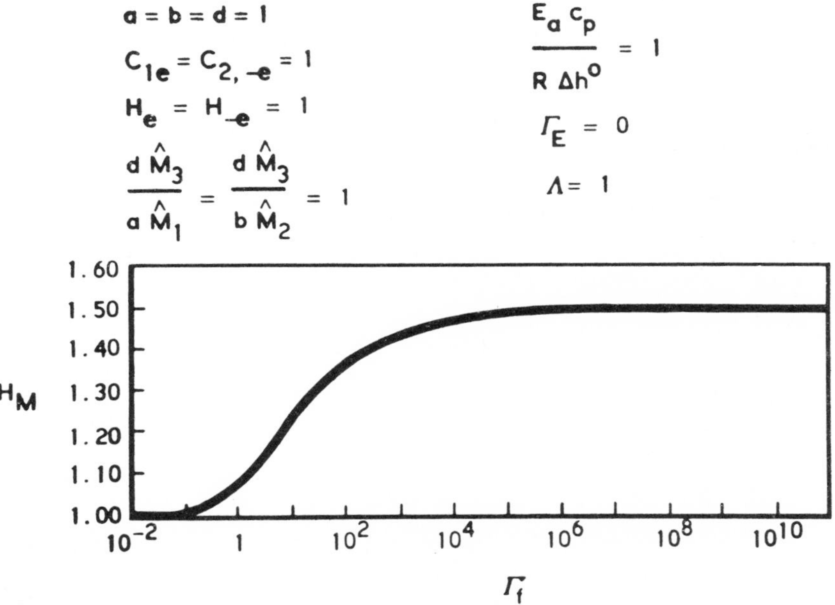

FIGURE 6. SIMPLE TRANSITION BETWEEN FROZEN AND EQUILIBRIUM LIMITS FOR DIFFUSION FLAMES.

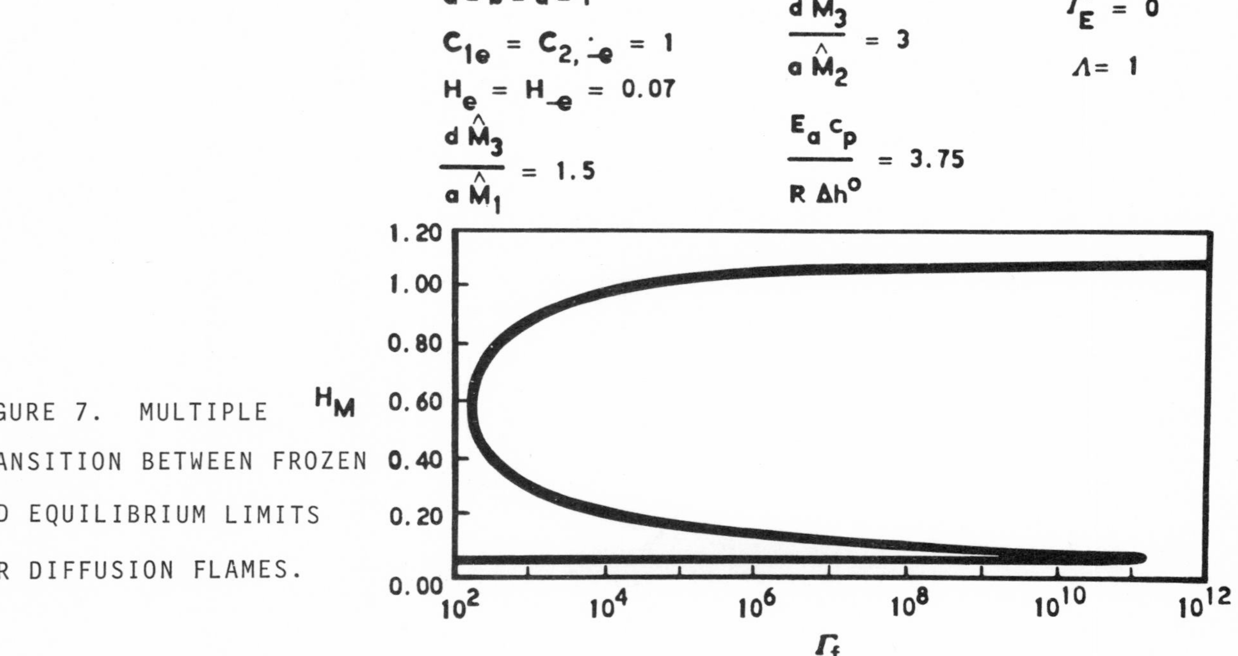

FIGURE 7. MULTIPLE TRANSITION BETWEEN FROZEN AND EQUILIBRIUM LIMITS FOR DIFFUSION FLAMES.

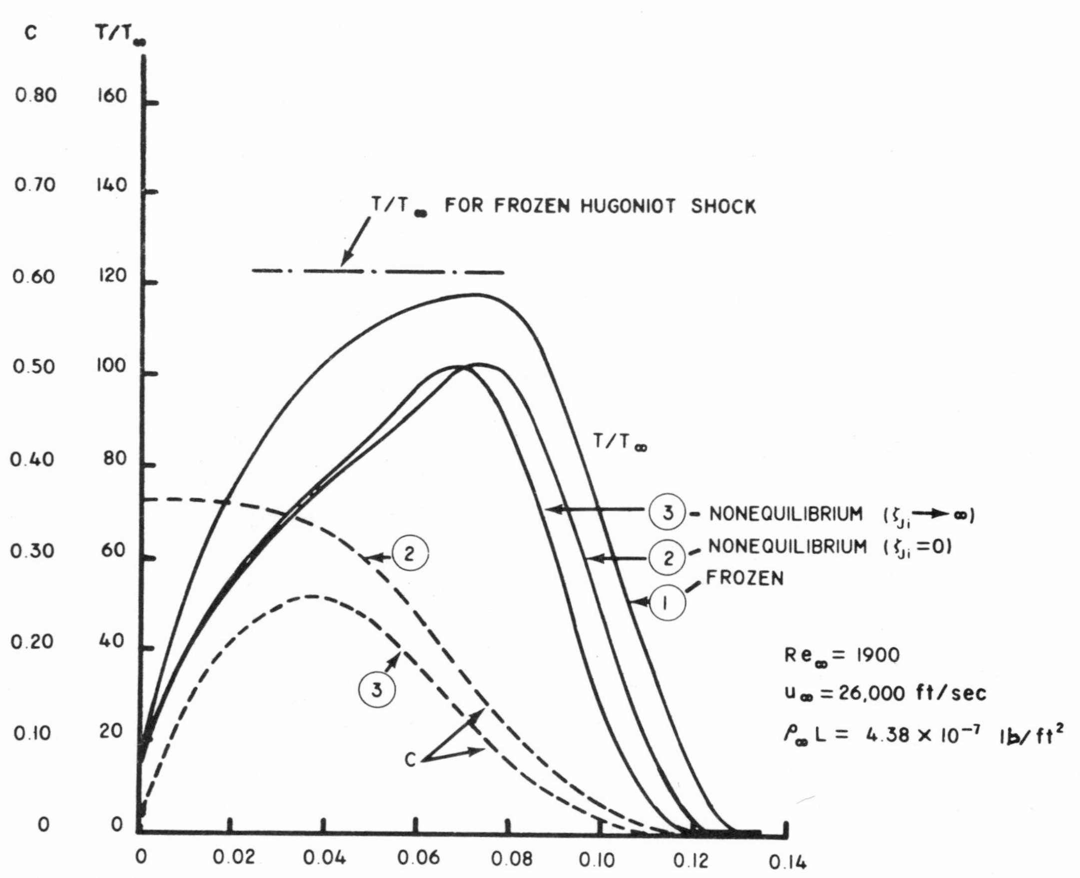

FIGURE 8. NONEQUILIBRIUM MERGED SHOCK LAYER.

RECENT DEVELOPMENTS IN TURBULENCE MEASURING AND ANALYZING TECHNIQUES

B. G. Jones*

INTRODUCTION

Turbulence measuring equipment has improved markedly over the past twenty years chiefly as a result of rapid advances in the development of necessary electronic components. These include not only the turbulence sensing instrumentation, but also recording and analyzing systems which provide the capacity, speed, and versatility so necessary in studying and interpreting the statistical behavior of turbulent phenomena. The purpose of this paper is to discuss, in general, the measuring and analyzing techniques which are currently available and to describe, in detail, several of these techniques which are being used in research activities here at the University of Illinois.

Perhaps it is interesting to reflect on the reasons why turbulence measurements are important in fluid mechanics, in general, and in convective heat transfer, in particular. In a broad sense we may argue that the majority of practical dynamic fluid flow systems are turbulent and that the associated turbulent transport mechanisms dominate those of molecular transport. Thus studies of systems in which the flow is turbulent are important. However, what quantities must be measured so that the transport mechanisms can be adequately examined and described? To answer this it is necessary to examine whether the detailed structure or the macroscopic behavior is desired. In the latter, a detailed structure is usually assumed and the resultant analytical or empirical predictions checked against locally measured mean values or overall system balances (i.e., integral techniques). Although these techniques have proven extremely useful, they do not provide the fundamental information which is required to check and to construct analytical models of the transport mechanisms. It is, therefore, believed necessary to make detailed local microscopic measurements of the randomly fluctuating turbulence parameters in order to study turbulent transport mechanisms successfully.

The measured values which are specifically required vary, not only with the transported quantity (i.e., mass, momentum, energy) and the media (i.e., liquid, gas: compressible, incompressible), but also with the geometry and dynamic flow conditions. However, as will be described in detail later, the important parameters include local mean flow characteristics, local macroscopic turbulent parameters, local microscopic turbulent distributions, and local correlation values which couple the transport processes. These observations may include instantaneous measurements of mass velocity, temperature, pressure, concentration, etc. Normally, not all of these would be important simultaneously to understanding many engineering oriented problems, but there are currently some turbulent fields in which all of the above, and more, are present and interacting, e.g.,

*Associate Professor of Mechanical and of Nuclear Engineering. University of Illinois, Urbana, Illinois.

combustion, ionized wakes of re-entry vehicles, etc. In this discussion such complex interactions will not be considered, but rather the fundamentals of utilizing available instrumentation and the application of modern analyzing techniques will be discussed, i.e., anemometer velocity, pressure, and temperature sensing equipment associated with turbulent convective heat transfer. In addition to the hot-wire and hot-film anemometer sensors, the cooled-film sensor and the aspirating probe will be discussed.

With anemometer sensor studies, the measured results are based on a reference frame which does not move in the same manner as the fluid elements, namely the Eulerian frame. However, it is often useful to examine turbulent transport phenomena with respect to a reference frame moving with the fluid elements, namely the Lagrangian frame. The turbulent trajectories of solid particles suspended in a fluid turbulent field provide directly Lagrangian turbulence data for the particle. A system in which data in both reference frames was obtained will be discussed later.

In studying the detailed turbulent structure, Lagrangian characteristics potentially can reveal more insight into the mechanisms of turbulent transport than Eulerian measurements. By observing the trajectory of many fluid particles the three-dimensional structure can be found from their ensemble. In contrast, the Eulerian observations cannot yield directly the three-dimensional turbulent structure of the fluid particle motion, since interpretation is related to the mean fluid velocity. It is only under restrictive conditions, which will be discussed later, that this interpretation can be validly employed.

In order to interpret the measured signals as meaningful characteristics of turbulence, they must be reduced and analyzed. Two approaches are open: first, the direct or indirect analysis of the dynamic analog output from the sensing instruments and second, the conversion of the analog data to digital form with subsequent reduction and analysis by a digital computer. Detailed discussion of these analyzing techniques is given later.

TURBULENCE MEASUREMENTS

Eulerian Turbulence Measurements

Although several sensing techniques are actively employed to measure fluid turbulence, anemometer systems have been the most widely used. Innovations have been continually applied to the standard anemometry methods to improve their sensitivity and to broaden the variety of media in which they can be used. In this section a variety of sensing techniques will be briefly mentioned and the cooled-film sensor, used with an anemometer, will be discussed in detail later.

Hot-Sensor Anemometer Principles

Hot-sensor anemometry techniques have developed from the original analysis of King[(1)*] which predicted that the rate of convective heat transfer of a small heated circular cylinder held perpendicular to the mean flow direction was given by

$$Nu = A + B\, Re^{1/2} , \tag{1}$$

where Nu is the instantaneous Nusselt number of the cylinder, Re is the instantaneous Reynolds number based on the cylinder diameter and the instantaneous fluid velocity, and A and B are coefficients which vary with fluid characteristics and flow conditions but are reasonably constant for each application. This relation, although known to have inadequacies for extreme flow conditions, is satisfactory for most practical anemometry applications. With the aid of the relation the instantaneous cooling rate of the small heated sensor is interpreted from the output of an anemometer operating in one of two modes, constant current or constant temperature.

The constant current mode of operation is electronically simple and as a result was employed first. A constant selected heating current is supplied to the sensor and a change in cooling coefficient alters the sensor resistance which appears as a voltage change across it. This output signal can then be related through [Equation (1)] to the mass flow rate and to the fluctuating ambient temperature by writing a heat balance for the sensor. This type of system has wide frequency response and relatively low noise level, provided thermal inertia for the sensor is not significant.

---

*Superscript numbers in parentheses refer to references at the end of this paper.

In contrast, the constant temperature system, through the use of a feedback amplifier, maintains the sensor at a chosen temperature above ambient and monitors the current required to do this. Again a heat balance on the sensor provides the relation between the film coefficient, the temperature difference, and the mass flow velocity. This type of system has the advantage of not being sensitive to the thermal capacity of the sensor and is a necessary property for operation of large wire or surface film sensors.

Since the rate of energy transfer from a heated surface to its surroundings is dependent principally upon the relative temperature difference and the convective heat transfer coefficient, changes in either of these appear as changes in the anemometer output. For constant temperature anemometer operation, with relatively large mean temperature differences compared to the turbulent temperature fluctuations, the cooling rate is principally dependent on mass flux variations. For incompressible fluid flows this reduces to a dependence on the instantaneous velocity variations. However, in many turbulent heat transfer studies the temperature fluctuations are of major importance. Until recently means for monitoring these temperature fluctuations adequately were not directly available with constant temperature anemometers. This is no longer the case, as several manufacturers of this equipment now provide a suitable resistance bridge and output amplifier which operates hot-wire and hot-film sensors as resistance thermometer elements. The sensitivity available is up to 0.2 volts per ^{o}C with frequency response of more than 1 kHz. This enables small fluctuations in temperature to be monitored.

For further details of anemometry principles the reader is referred specifically to References 2 through 5 on the fundamentals of hot-sensor anemometry, and to technical literature provided by suppliers of anemometer equipment.

Hot-Wire and Hot-Film Sensor Characteristics

Anemometry was developed initially using the constant current mode of operation. For this mode of operation, the sensing element is required to have very small thermal inertia, as it must essentially behave as a pseudo resistance thermometer element. The practical form of sensor for this operation is very small diameter (0.0038 to 0.005 mm) and relatively long (1.0 to 2.0 mm) metal wire. The most common materials are tungsten, platinum, and a platinum-iridium alloy. The choice of sensor for specific applications strongly depends on the environment, flow conditions, sensitivity requirements, and the temperature of operation; all of which influence the corrosive resistance and the strength requirements.

The successful development of the constant temperature anemometer has made the use of hot-film, in addition to hot-wire, sensors feasible. In contrast to the delicate wires required for reasonable sensitivity from hot-wire sensors, the hot-film provides a rugged and sensitive configuration which is ideally suited for measurements in liquids.

This sensor is usually made of a thin (less than 100 Å) platinum film which is fused to the surface of a ceramic (usually quartz) supporting structure. The film shape is quite arbitrary but is commonly placed on a wedge, cone, or cylinder. Flush mounted wall sensors are also used in some studies.

Measurements in a variety of liquids have required either that the fluid system be of very low ion content (i.e., nonconducting) or that the sensor be electrically insulated from the fluid. A spattered quartz or silicon dioxide insulation layer has been widely used for this purpose.

Because the physical strength and thermal diffusivity of the hot-film sensor are mainly determined by the substrate material, most films are platinum, which provides high oxidation resistance and therefore good stability. Film sensors also provide wider flexibility in configuration, relatively lower rate of heat loss to supports, and less fouling by accumulation of foreign material than comparable wire sensors. However, wire sensors provide better characteristics in some applications, particularly when a small sensor dimension is important. Since wires typically have diameters an order of magnitude smaller than cylindrical film sensors, a much closer approach to solid surfaces is possible with wires. In addition, the higher temperature resistance of tungsten wires makes them superior in noise level. The combination of smaller size and higher temperature resistance enables tungsten wire sensors to provide higher maximum frequency response than platinum film sensors.

Because of the relatively low-frequency characteristics of turbulence in liquids and the potentially serious requirement for electrically insulating the sensor from the media, hot-film sensors with quartz coatings are ideally suited for monitoring in liquids. However, certain precautions must be taken to insure stable operation of these sensors in liquids. Runstadler,[6] Grant,[7] and Widell[8] have discussed the requirements for stable operation of hot-films in water and other liquids. The particularly serious problem of air bubbles and boiling on hot-film sensors in low-pressure water flow is discussed in detail by Rasmussen.[9] He discusses the two major sources of bubble generation, namely, thermal and electrolytic. This generation may be reduced or eliminated by using the appropriate combination of a distilled water media, quartz coated sensors, AC heating currents, and low film temperatures. The conditions under which the bubble problem is most serious for hot-sensor use in water are at low velocities and high thermal load on the sensor.

Turbulence characteristics in water at atmospheric pressure using a bare platinum wedge geometry hot-film sensor are presented later in this paper. By following the suggestions of Ling,[10] to remove all lint from the system, and Bankoff and Rosler,[11] to use highly demineralized and de-airated water, stable operation of a linearized constant temperature anemometer sensor was readily obtained.

Other investigators have successfully applied hot-film sensors to isothermal turbulence measurements in liquid organic solvents[12] (i.e., benzene, toluene, and cyclohexane) and in liquid metals[13] (i.e., mercury).

Cooled-Film Sensor Characteristics

Cooled-film sensors were primarily designed for measurements in high-temperature environments (i.e., chemical combustion flames) which exceed the limitations of normal hot-wire and hot-film sensor operation. A non-cooled sensor must operate with sensor temperatures higher than that of its environment. The cooled sensor, since it uses an internal thermal sink, can operate below, as well as above, the temperature of the environment. By accurately controlling the internal cooling rate and operating the sensor in the constant temperature mode, the cooled-film sensor responds in a similar manner to the standard hot-film sensor, except that the effects of temperature and velocity fluctuations in the environment invert the anemometer response when the film temperature is less than its environment.

The upper limit of the environmental temperature in which the sensor may be used is restricted only by the maximum internal cooling rate that is available for the sensor. It is desirable to use various fluids for internal cooling so that stable flow patterns and internal coolant temperatures may be employed for widely varying applications. When liquid coolants are used boiling must be carefully avoided, since this introduces serious noise in the anemometer output. For lowest noise effects laminar single phase internal coolant flow, with good temperature and mass flow rate regulation, must be used.

Cooled sensors are commercially available from T.S.I.* in cylindrical geometry with 0.15 mm outside diameter and active film sensor length of 1.5 mm. The hollow quartz cylinder with 0.10 mm inside diameter provides the internal coolant passage. The sensor extension tubes are sealed to coolant supply tubes, which also provide the electrical connections between the anemometer bridge and the sensor.

In addition to their ability to operate in high-temperature environments, these sensors provide higher frequency response than the same size of hot-film sensor. They also provide the ability to emphasize temperature or mass flux sensitivity by controlling the film and coolant temperatures with respect to the environmental temperature conditions. By judiciously controlling the relative sensitivity, this sensor enables temperature fluctuations to be monitored directly and at high frequencies. The higher frequency response results from increasing the electrical power input to the sensor, which increases the closed loop gain of the constant temperature anemometer. Therefore, in low-velocity or low-density environments, the cooled-film sensor performance exceeds that of similar hot-film and hot-wire sensors.

A detailed thermal analysis of the cooled film sensor and its sensitivity characteristics are presented in the Appendix. Experimental verification of the analytical predictions are currently in progress here.

Aspirating Probes: Temperature and Composition Sensors

A constant temperature anemometer sensor mounted just upstream of a sonic

*Thermo-Systems Incorporated, St. Paul, Minnesota.

orifice of an aspirating device is sensitive to the velocity, temperature, and composition of the bleed fluid. Since sonic velocity is maintained through the orifice, the velocity past the sensor is nearly independent of the environment velocity for incompressible flow fields. Therefore the velocity past the sensor depends only on the temperature and composition of the bleed from the flow field being studied. For many flows of practical interest, either temperature or composition is constant, and the sensor provides the instantaneous variations in the other parameter. In two-component gas mixtures where the heat transfer characteristics of the two gases differ sufficiently (e.g., helium and air) the relative percentage variations of the two components can be readily determined.

The frequency response of this device is directly related to that of the anemometer sensor being used and to the aspirating passage geometry from the turbulent fluid media to the sensor. The sensitivity is also related to the bleed fluid transit time, which should be kept short.

Other Sensing Devices

Local instantaneous temperature, pressure, density, composition, and velocity characteristics are essential for the study of the microscopic structure of turbulent flow fields. The anemometer methods which have been described represent but one of the means available to monitor these values.

The major instrumentation requirement is to obtain sufficiently high frequency response and sensitivity to observe the details of the fluctuating behavior. Recent advances in the manufacture of rapid response thermocouples make fluctuating temperatures in liquids and gases directly available. By using vacuum-deposited plating of thermocouple junctions, probe surface temperatures are measured with response times as little as one microsecond.*

Pressure sensing probes, with high-frequency response behavior, are also becoming available from T.S.I. A recently developed pressure transducer uses a heated filament in a small capillary tube to measure the flow between a reference pressure and the fluctuating pressure. By suitably choosing the mean pressure differential, the fluid, and the tube dimensions, the sensitivity to pressure differences is made equivalent to a hot-wire's sensitivity to velocity changes in low-speed bleed flow. A configuration which bleeds a controlled fluid into the flow being studied insures a clean and constant temperature fluid in contact with the sensor. Thus composition and thermal effects do not perturb the pressure fluctuations. In contrast, bleeding from the studied flow may be used to examine temperature and composition variations in a manner similar to the aspirating probe.

Other methods have utilized measurements of local chemical concentrations to infer the associated fluid turbulence. Hanratty[(14)] has developed this "electrochemical" method for the study of turbulent characteristics very near the wall.

The above tabulation is by no means complete, but is intended to emphasize those devices which are specifically applicable to both isothermal and nonisothermal subsonic turbulent flow fields.

Because of the high fidelity of the solid state, integrated circuits used in modern anemometer electronics, the component which most often limits the system performance is currently the sensor. This explains the current major effort by anemometry supply companies to improve their sensors and to expand the configuration and type of sensors which they supply from stock.

Lagrangian Turbulence Measurements

The tagging of small fluid elements and the subsequent observing of their detailed turbulent behavior is a difficult experimental problem. Not only must the tagged elements be small, but ideally their behavior should not differ from that of the surrounding untagged fluid elements. In attempts to accomplish this experimentally, concentration and energy have been used to tag fluid elements.[(15)] However, these measurements have provided only macroscopic transport characteristics and the detailed turbulent structure is not directly measured, but can only be inferred from mean dispersion results.

In contrast to tagging fluid elements, small solid particles have been used to simulate tagged fluid elements.[(16,17)] By choosing very small, neutrally buoyant particles and by analyzing their detailed trajectories, both microscopic and macroscopic Lagrangian turbulence characteristics of the particles are obtained. In the limit, when the particle size is reduced to a

*Heat Technology Laboratory, Inc., Huntsville, Alabama.

small fraction of the smallest turbulence scale of the fluid, the measured characteristics of the solid particle behavior should provide a good approximation to the Lagrangian turbulence structure of the fluid.

In past studies here the smallest particle size which could be successfully monitored was approximately 0.6 cm in diameter. The fluid turbulence field was the central core region of the fully developed flow of water in a 18.4 cm diameter tube having a Reynolds number of 50,000. The associated Eulerian integral scale was 0.070 seconds (2.1 cm) and the dissipative scale was 0.026 seconds (0.80 cm) at the tube centerline. Since the particle size was not small compared to these scales, it was not appropriate to infer that its motion represented the Lagrangian fluid turbulence. However, recent experimental improvements now allow the particle diameter to be reduced by a factor of five and the experimental program is currently using these particles. Discussion of the particle trajectory data reduction is given in a later section.

Measures of Fluid Turbulence

Many models and methods have been extensively employed to characterize the behavior and transport of turbulent fluid flow fields. These may be broadly divided into two categories. The first, based on phenomenological considerations, includes global or integral quantities which do not provide detailed information on the structure of the turbulence and which are usually insensitive to the structure. In general, these models are strongly oriented to overall system behavior and are often used to correlate experimental data of this behavior. However, they have had limited success in predicting the behavior of new flow fields. The second, based on statistical considerations, includes quantities which are directly measures of the local structure of the random fluctuations in the turbulent field. Until recently adequate means by which the local structure could be monitored were not available. However, improvements in monitoring equipment now enable many of the fluctuating quantities to be monitored directly and provide experimental data to check the validity of proposed models which include the local structure. Since all descriptions of the flow field must be compatible, it is expected that the detailed understanding of the turbulent structure will enable improvements to be made in the phenomenological descriptions.

The phenomenological approaches normally use long time averaged turbulent quantities such as mean temperature and velocities. In contrast, the statistical approaches emphasize the details of the fluctuating structure and require that statistical data analysis be used. Details of the statistical methods are given in the section Statistical Analysis of Turbulence Data.

It has been customary in many analytical models for turbulent transport to consider that the molecular diffusion principle could be extended to accommodate turbulent, as well as molecular, transport. Undoubtedly such models are appealing since the methods for analyzing diffusion processes are well developed and the mean driving potentials in turbulent systems can be readily measured. On the other hand, the statistical models for all but the most ideal fluid turbulence are analytically difficult and are not easily implemented to solve practical flow fields.

For diffusional processes to be applicable, several criteria must be satisfied. The most critical one for turbulent transport in confined regions is that the diffusion length be small compared to the characteristic dimension of the region. Except for turbulence in the atmosphere and in the ocean, this criterion is usually not adequately satisfied.

Consider, as an illustration, the turbulent transport of momentum in an incompressible, two-dimensional, turbulent flow. Hinze[18] has formulated and discussed the diffusion based models which are used to describe this process. In these the concepts of mixing length and eddy diffusivity are employed to characterize the turbulent flow field. In conjunction with the local mean gradient of velocity, these quantities are interpreted as the local turbulent shear stresses. However, little, if any, of the strongly correlated structure of these shear stresses can be accommodated in such models. Indeed, many models do not provide for variation of either mixing length or eddy diffusivity across the flow region. Experimental evaluation of these quantities shows, however, that there is significant variation across the confined region and that the mixing lengths are a significant fraction of the channel width. Such models have obvious inconsistencies near solid boundaries. Thus the applicability of diffusion based models becomes questionable.

If the diffusion model is questionable for confined turbulent shear flows, are there alternative descriptions which can predict turbulent transport? Although the answer is not definite, the statistical formulation of turbulent transport indicates the physical parameters of the flow field which must be examined. It would seem reasonable to suggest that detailed experimental studies of the statistical behavior of the flow field should be fruitful in providing improved phenomenological models. For instance, what characteristic transport length should be used? From the analysis of the statistical behavior we find that the turbulent integral scale should be considered for main transport effects and that the turbulent dissipative scale should characterize local, small scale effects. Examination of the power spectral density and correlation distributions provides additional insight into the turbulent structure which should improve understanding of the transport mechanisms.

In conjunction with the physical parameters suggested from the statistical formulations of turbulent transport, it is interesting to speculate on the possibility of replacing some of the common flow parameters, which are based on molecular fluid properties, mean flow characteristics, and flow configuration, by their turbulent counterparts. For example, a pseudo Reynolds number could be based on the integral scale, the root mean square of the fluctuating velocities, and the viscous transport related to the turbulent shear stresses. Studies currently in progress here are examining these statistical parameters, but sufficient data has not been collected and compared to verify the utility of such parameters in correlating and predicting turbulent transport. It is expected, however, in the combined transport of momentum and energy, that the strong coupling of the turbulent transport mechanisms could be usefully examined in this manner.

HANDLING AND STORAGE OF TURBULENCE DATA

The turbulence sensing devices, reviewed in the previous section, provide fluctuating signal outputs from which the characteristics of the turbulent phenomena may be interpreted. Since it is often inconvenient to subject the signals directly to on-line analyzing systems, the use of recording equipment is increasing rapidly. By storing data in dynamic form, it is readily retrieved for repeated analyses without the necessity of rerunning the experiment. These procedures are particularly advantageous for time varying or short duration experiments in which long-time steady operation cannot be attained or when the costs of repeated operation of the experiment are prohibitive. For turbulence data it is readily seen that, unless dynamic records are made, human conversion of static records to dynamic form would effectively eliminate processing of the recorded data. In fact, the lack of suitable recording devices prohibited early investigators from studying much of the detail which can be routinely reprocessed from records today.

Two forms of magnetic tape records are commonly utilized to dynamically store the data, namely, analog (continuous) and digital (discrete). The former is typically recorded in one of two modes -- frequency-modulated (FM) or direct (DR) and usually with multichannel capabilities. The latter normally provides a digital tape which may include multichannel input data and which may be used directly as input to a digital computer.

Use of the multichannel capability of most analog recorders enables many phenomena to be simultaneously stored on magnetic tape. This allows repeated simulation of complete experimental systems. It also provides a means by which a variety of analyses may be applied to the same data without relying on the repeatibility of the experiment.

To analyze transient turbulence phenomena, often many experimental observations are used to form an ensemble average. Such procedures are readily accommodated, since tape markers enable the appropriate segment to be accurately selected from the record of each experiment included in the ensemble. This technique was used to advantage in analyzing turbulence in the wake of a simulated re-entry vehicle in a ballistic range.[19]

In many instances the ability to change the time base of the record, from real time to either an expanded or contracted time base, is highly desirable. Time base expansion is particularly useful for analyzing devices whose frequency range would normally be insufficient to examine them.[19] On the other hand, time base contraction may provide reduced handling and analyzing times. Multichannel analog recorders are well suited for these procedures.

The utilization of the tape looping capability of analog recorders provides a convenient and rapid method of subjecting short data records to repetitive analyses. It is also frequently used to provide a continuous signal for detailed analyses of turbulent characteristics from relatively short data records.

Application of digital magnetic tape storage is undergoing rapid expansion since digital analyzing techniques for statistical phenomena are being widely employed. Two complimentary procedures for obtaining digital tape records are commonly used in conjunction with analog-to-digital conversion (ADC): The first uses direct, on-line coupling to the sensing instruments; the second employs analog recording systems which are replayed into the conversion equipment, off-line. Both make use of a multiplexing unit to sequence the incoming data channels in any desired order to the ADC and then to the digital tape transport. The use of the portable analog tape recorders offers added advantages in data acquisition, since these recorders are rugged and may readily be employed with a variety of experimental systems with data reduction and analysis being completed conveniently in central laboratory settings.

In most data handling procedures signal preconditioning is necessary to maintain good signal-to-noise characteristics and often to provide suitable signal amplitude and power for subsequent components in the handling system. This is regularly accomplished with low noise level amplifiers and suitable analog filters which can be adjusted to retain signals of desired bandwidth.

Calibration information for each channel of recorded data may be conveniently handled by frequently interspersing known signals through the complete handling system. This not only provides reliable checks, but also removes the necessity of calibrating each component in the handling system. Similar procedures may be used to establish accurately experimental time when changes in time base are involved.

Successful application of FM recorders and ADC systems to anemometer data in both water and air, as well as to several interdependent channels of particle trajectory data, is readily demonstrated by the results shown in the following section of this paper. Figure 1 shows the data acquisition system employed for one of these studies, i.e., for a turbulence signal from a single channel anemometer. Standard analog analyzing devices were used to evaluate directly the turbulence characteristics from the anemometer signal. This immediately provided information on the validity of the experimental data and also permanent results for checking against subsequently analyzed digital results. The raw digitized data was submitted to the computer for interpretation and reduction with calibration information to give refined data to be submitted for statistical analysis. Similar procedures were used for handling the other data.

It should be realized that, by using digital equipment to refine the data, sophisticated calibration concepts are readily applied. This has been adequately demonstrated by Jones[16] and Shirazi[17] in connection with instantaneous particle trajectory interpretation. By using the digital computer to invert the static calibration of signal voltages for a three-dimensional set of positions, the instantaneous position was interpreted from the digitized set of signal voltages. The time derivative of these signals provided the instantaneous velocities which were submitted to statistical analyses.

STATISTICAL ANALYSIS OF TURBULENCE DATA

As indicated in the previous section, two approaches to statistically analyze randomly fluctuating signals are available: the use of analog devices, into which continuously varying signals are fed; and the use of digital computers, into which discrete values from ADC of the continuous signals are fed. In this section both analyzing procedures will be discussed; however, greater emphasis will be placed on the digital techniques and on results from these analyses.

Definitions of Statistical Quantities for Fluid Turbulence

The definitions of statistical quantities which are commonly used to describe fluid turbulence were given by Taylor.[20] These are similar to those used in the fields of time series in statistics and of communication engineering. However, the normalizing quantities and the asymmetry of the Fourier integral representation used in fluid turbulence are chosen differently to provide appropriate physical interpretation of the statistical quantities.

Analytical descriptions of the statistical characteristics of random fluctuations are more completely developed for phenomena with Gaussian behavior. In addition, the assumption of stationarity is usually employed. It is useful for data reduction to provide measures of departure from Gaussian behavior. This is most conveniently done by examining central moments of the turbulence signals. Other measures which describe the detailed structure of turbulent phenomena are correlations and power spectral densities, both of which can be estimated from the fluctuating signals.

From practical considerations, since only samples of finite length can be analyzed, it is desirable to limit examination to turbulence data whose statistical characteristics satisfy the ergodic condition; that ensembles of estimates are equivalent to long-time-average estimates. However, estimates of variability are necessary to specify the accuracy of the statistical characteristics obtained from finite samples. But variability is only exactly specified for Gaussian behavior. Fortunately, it is often reasonable to utilize variability estimates, defined for Gaussian behavior, to approximate those for phenomena which are somewhat perturbed from satisfying Gaussian conditions. Frenkiel and Klebanoff[21,27] have examined methods and measures for determining the degree of departure from Gaussian characteristics. In these studies they employed very long samples (i.e., 12.5 seconds with 160,000 equally spaced data values) from grid generated turbulence in 50 ft/sec air with a constant current anemometer system. Using non-Gaussian probability distributions of the Gram-Charlier type, they obtained good agreement with measured values.

For the purposes of this discussion, only single channel turbulence data is considered. Multichannel data may be treated in a parallel manner, and the appropriate definitions are well known. In addition, the concept of interrelating turbulence characteristics defined with respect to space to those defined with respect to time (Taylor's hypothesis) is well known. However, since it is strictly appropriate only for homogeneous flow fields with very low turbulence levels, discretion must be exercised when interpreting turbulent parameters in flow fields with high intensity, nonhomogeneous, or nonisotropic turbulence (e.g., jets, shear flows, etc.).

Since turbulence data may be treated as a randomly varying phenomena with respect to time, its steady mean component may be extracted leaving only the fluctuating component on which to perform the statistical calculations. Letting the random continuous signal be $X(t)$, where t is time and X may represent velocity, the temporal mean central moments may be evaluated from the general expression

$$\overline{x^n} = \lim_{T\to\infty} \frac{1}{T} \int_0^T \left[X(t) - \overline{X}\right]^n dt \ , \qquad (2)$$

where

$$x(t) = X(t) - \overline{X} \ ,$$

n is any nonzero positive integer, and $\overline{X}$ is the temporal mean of the random signal. Over a finite interval T the estimate of Equation (2) may be approximated by

$$\overline{x^n} \simeq \frac{1}{T} \int_0^T x^n(t) \, dt \ , \qquad (3)$$

for phenomena which are stationary.

The physical interpretations of these quantities are:

(1) n = 1: gives $\overline{x}$, the central moment mean.

(2) n = 2: gives $\overline{x^2}$, the variance, which is a measure of the turbulent kinetic energy of the velocity fluctuations in fluid turbulence. The intensity of the velocity fluctuations is given by

$$\frac{\left(\overline{x^2}\right)^{1/2}}{\overline{X}} \ .$$

(3) n = 3: gives $\overline{x^3}$, which when normalized with the variance gives a measure of the skewness of the probability distribution; namely,

$$\frac{\overline{x^3}}{\left(\overline{x^2}\right)^{3/2}} \ .$$

(4) n = 4: gives $\overline{x^4}$, which gives a measure of the flatness of the probability distribution when normalized with the variance; namely,

$$\frac{\overline{x^4}}{\left(\overline{x^2}\right)^{2}} \ .$$

(5) n > 4: gives higher order central moments which are less directly interpreted in terms of their physical significance, but from which other characteristics of the probability distribution are determined. Frenkiel and Klebanoff[21] have utilized values of n of 1 through 6 to interpret departure from normal distributions in grid generated turbulence.

The autocovariance is defined as

$$Q(\tau) = \lim_{T\to\infty} \frac{1}{T} \int_0^T x(t)x(t+\tau)\,dt , \quad (4)$$

which may be approximated for finite sample length, T, as

$$Q(\tau) \simeq \frac{1}{T} \int_0^T x(t)x(t+\tau)\,dt . \quad (5)$$

The normalized autocovariance, the autocorrelation, is defined, for a finite sample length, as

$$R(\tau) = \frac{Q(\tau)}{\text{variance}} \simeq \frac{\int_0^T x(t)x(t+\tau)\,dt}{\int_0^T x^2(t)\,dt} . \quad (6)$$

The power spectral density (spectrum) may be evaluated from the autocorrelation through the integral cosine transform pair

$$F(f) = 4 \int_0^\infty R(\tau) \cos 2\pi f\tau \; d\tau \quad (7)$$

and

$$R(\tau) = \int_0^\infty F(f) \cos 2\pi f\tau \; df , \quad (8)$$

where f is the cyclic frequency. In practice the maximum value of the lag time, τ, is finite but is sufficiently large such that $R(\tau)$ may be considered as zero for larger values of τ. Likewise, the maximum value of frequency is finite but sufficiently large such that F(f) may be considered as zero for greater values of f. Thus the upper limits of the integrals in Equations (7) and (8) become the maximum τ and f values, respectively. The choice of these values is strongly dependent on the specific characteristics of the experiment in which the turbulence is being monitored. For example, in comparable studies of turbulent duct flows with air and water, these values normally differ by two or three orders of magnitude.

The spectrum represents the contribution to the total variance by frequencies between f and f + df. Summing over all frequencies, the area of the spectrum curve is unity, thus

$$\int_0^\infty F(f)\,df = 1 . \quad (9)$$

Other commonly defined quantities are the scales of turbulence. They provide physically meaningful parameters by which turbulent flow fields can be compared. The two most used scales are:

(1) The Integral Time Scale: $\mathcal{T}$. It is defined as

$$\mathcal{T} = \int_0^\infty R(\tau)\,d\tau \simeq \int_0^{\tau_{max}} R(\tau)\,d\tau , \quad (10)$$

which is equivalent to

$$\mathcal{T} = \frac{F(0)}{4} , \quad (11)$$

from Equation (7), where τ_{max} is the maximum lag time and F(0) is the zero frequency intercept of the spectrum. Taylor suggests that this is representative of the average characteristic dimension of strongly correlated regions of fluid.

(2) The Dissipative Time Scale: τ. It is defined as

$$\frac{1}{\tau^2} = -\frac{1}{2}\left[\frac{\partial^2 R(t)}{\partial t^2}\right]_{t=0} , \quad (12)$$

which implies that the acceleration is finite and thus agrees with the physical behavior of continuum fluid mechanics. It is suggested to be indicative of the most rapid changes that occur in the velocity fluctuations and which are responsible for the majority of the dissipation in turbulent flows. An alternative definition is obtained from the use of Equation (8) in Equation (12); namely,

$$\frac{1}{\tau^2} = 2\pi^2 \int_0^\infty f^2 F(f)\,df . \quad (13)$$

These turbulence characteristics may be conveniently cataloged into two groups:

macroscopic, those which describe overall behavior as long-time-averaged quantities, and microscopic, those which characterize the detailed fine structure. The single quantities of intensity, variance, integral scale, dissipative scale, etc., represent macroscopic characteristics. The distributed quantities of spectrum and correlation are microscopic properties.

These definitions are certainly not an exhaustive set but represent those most commonly employed to study and compare experimental turbulence data. In the following discussion the methods by which these are extracted from data will be described. Some typical data are also included to illustrate the methods.

Analog Analyzing Methods

The ability to directly analyze data with analog devices is attractive since in many cases discrepancies in system performance may be recognized and corrected immediately. This is particularly true for much of the detailed statistical structure which requires a series of time-averaged quantities and additional algebraic manipulation to obtain the properly normalized quantities. It has become common practice to combine direct analyzing methods with analog magnetic tape record/replay techniques to cover a wider range of turbulence quantities which may be evaluated with electric analog devices. In addition, the magnetic tape records provide permanent dynamic data for repeated analyses without relying on the exact repeatability of the experimental test conditions.

To perform analog analyses on continuous signals for the particular turbulence characteristics defined in the previous section requires a separate set of electronic components for each quantity. For example, to evaluate the mean flow velocity requires an averaging voltmeter and the turbulence intensity requires a true root-mean-square (RMS) voltmeter. Both of these components are readily available and are the most commonly employed direct monitoring devices. However, to proceed to higher order velocity moment calculations requires increasingly more complex circuitry, including multipliers as well as a ratiometer, if direct readout is desired. The dissipative scale requires the additional complication of a differentiating operation which is strongly sensitive to high-frequency noise in the system. All of the above operations provide only gross, long-time-average measures of the turbulent behavior. To determine by analog methods distributed characteristics (e.g., spectrum and autocorrelation) of turbulent fluid flow requires considerably more sophisticated electronics. As an example, the evaluation of the autocorrelation requires that a variable delay be applied to the signal, that the direct and delayed signals be multiplied together, and that the product be long-time-averaged. By discretely or continuously varying the delay time the autocovariance of Equation (5) is obtained. The autocorrelation requires that this be normalized with the variance.

No mention of the requirements for stability and frequency response have been specified for the analog equipment in the discussion above. In general, most solid state components provide adequate electronic stability. However, the required frequency response varies widely depending on the media and flow conditions in which the turbulence is being studied. For water flows in channels and ducts the frequency range varies over approximately 0.1 to 100 Hz. For low-speed air flows in similar configurations the range is approximately 10 to 10,000 Hz. No general analog analyzing instruments are available which operate over this wide frequency range, and very few are capable of accurate frequency response below 10 Hz. In fact, only recently have RMS voltmeters had adequate low frequency for use with water data.

The above discussion indicates that an adequate analog analyzing system requires many electronically different components. This is not only expensive, but also is a complex system to operate and maintain. In contrast, digital analyzing procedures can perform each function by simply adding subroutines to general analysis programs. However, by judiciously using specific analog analyzing procedures, it is believed that analysis of continuous random signals can compliment digital evaluation methods.

Digital Analyzing Methods

Application of digital analyzing procedures to randomly fluctuating signals to obtain estimates of their statistical characteristics has been extensively employed in communication engineering.[22,23,24] Brief discussions of their applications to the study of fluid turbulence have been given by Pasquill[25] and Panofsky and McCormick.[26] Recent applications of these techniques are reported by Jones,[16] Shirazi,[17] Frenkiel and Klebanoff,[21,27] and Raichlen.[28] For the convenience of the reader and to establish the validity of employing digital techniques to analyze the fluid turbulence characteristics defined earlier, a brief derivation of the equivalent quantities for discrete data is presented. Only equal time increments will be considered, but this is not restrictive since in practice most ADC systems use this mode.

Consider a continuous signal which is interrogated sequentially each increment in time, Δt, the averaging time. For a given sample time, T, there will be N pieces of digital data, where

$$N = 1 + \frac{T}{\Delta t} . \tag{14}$$

It is convenient to show the process of digitizing a typical turbulence signal by the diagram of Figure 2. The continuous signal $X(t)$ has a temporal mean $\overline{X}$ and a central difference fluctuating component $x(t)$. One first computes the temporal mean from

$$\overline{X} = \frac{1}{N} \sum_{k=1}^{N} X(k) , \tag{15}$$

which is the discrete representation of the finite sample time integral

$$\overline{X} = \frac{1}{T} \int_{0}^{T} X(t)\,dt \tag{16}$$

of Equation (3). By subtracting $\overline{X}$ from each $X(k)$, the discrete central differences $x(k)$ are obtained for the sample. The general central moments of Equation (3) are readily formed since the arguments $x^n(k)$ may be evaluated at each index k of 1 through N and summed over all N. The general relation is

$$\overline{x^n} = \frac{1}{N} \sum_{k=1}^{N} x^n(k) \quad ; \quad n = 1,2, \ldots . \tag{17}$$

The autocovariance approximation of Equation (5) for finite sample size and stationary behavior is given in discrete form by

$$Q(k\Delta t) = \frac{1}{(N - k)\Delta t} \sum_{i=1}^{N-k} x(i\Delta t) \times \Big((i + k)\Delta t\Big) \Delta t , \tag{18}$$

where all values of lag number $k = 0,1,2, \ldots, N - 1$ are possible. However, in order to obtain reasonable statistical estimates Southworth[29] recommends that the lag number should be $\leq N/10$. For digital computer evaluations it is convenient to consider $\Delta t = 1$. We rewrite Equation (18) as

$$QP(p) = \frac{1}{N - p + 1} \sum_{i=1}^{N-p+1} x(i)\; x(i + p - 1) , \tag{19}$$

where p is the indexing parameter with range $1 \leq p \leq 0.1N$. For nonstationary data, Equations (18) and (19) must be replaced by expressions which include local means of the lagged segments. Details of the derivation are given by Southworth.[29]

The autocorrelation is obtained by dividing Equation (19) by the variance from Equation (17) with $n = 2$ to obtain the form

$$RP(p) = \frac{N}{N - p + 1} \cdot \frac{\sum_{i=1}^{N-p+1} x(i)\; x(i + p - 1)}{\sum_{i=1}^{N} \left[x(i)\right]^2} . \tag{20}$$

The spectrum, defined by Equation (7), may be approximated by using trapezoidal integration as

$$F(f) = 4\Delta t \left[\frac{1}{2} RP(1) + \sum_{k=1}^{M-1} RP(k+1) \cos 2\pi f k \Delta t + \frac{1}{2} RP(M+1) \cos 2\pi f M \Delta t \right], \quad (21)$$

where M is the maximum lag number. The frequency is represented by

$$f = \frac{q}{2M\Delta t}, \quad (Hz); \quad q = 0,1,2,\ldots,M. \quad (22)$$

Equation (21) may be rewritten as

$$FP(q) = 2\Delta t \left[RP(1) + 2 \sum_{k=1}^{M-1} RP(k+1) \cos \frac{\pi q k}{M} + RP(M+1) \cos \pi q \right], \quad (23)$$

which represents the discrete value of the spectrum over the frequency bandwidth of $1/(2M\Delta t)$ on all internal points and $1/(4M\Delta t)$ for the end points. To index this for computer evaluation the indices q must be replaced by $j = q + 1$ and k by $p = k + 1$ to obtain

$$FP(j) = 2\Delta t \left[RP(1) + 2 \sum_{p=2}^{M} RP(p) \cos \pi(p-1)(j-1) + RP(M+1) \cos \pi(j-1) \right], \quad (24)$$

where $j, p = 1,2,3, \ldots, M+1$.

To evaluate the integral scale it is most convenient to use Equation (11), since it is directly obtained during the calculation of the spectrum. However, the integral form of the dissipative scale, Equation (13), is preferred, since in digital calculations differentiation tends to emphasize uncertainties whereas integration tends to smooth them.

The flow diagram for the digital computer program which has been extensively employed here to study single channel anemometer data is shown in Figure 3. The program employs the discrete forms of the turbulence parameters as derived in this section.

Ideally, one would like either to analyze a very long turbulence record which exhibits a stationary nature or to form an ensemble of a large number of short records from stationary data. From practical considerations the first approach is severely limited by computational time and computer storage, while the second approach is limited by the large number of data sets required for a proper ensemble. Thus compromises become necessary.

Digital calculations typically exhibit different optimum sampling and averaging times for evaluating macroscopic and microscopic characteristics. For example, consider the variance and the spectrum. The variance is evaluated as the second moment of the fluctuating velocity component averaged over the sampling time. This is a simple, rapid procedure and, being a measure of the turbulent kinetic energy of the fluid, may use relatively long averaging times since most of this energy is contained in the low-frequency components. Also, to insure small variability limits long sampling times should be used when the data is stationary. On the other hand, the spectrum is a much more detailed quantity as it is a measure of the distribution of kinetic energy over the full frequency range. Thus it must include sufficiently short averaging times to include all high-frequency effects. The use of both long samples and short averaging times requires unacceptably long computational times. In recent studies here these effects have been examined in detail and will be reported in a separate paper, currently in preparation. The general conclusions show that: the macroscopic moment parameters are relatively insensitive to averaging time but strongly affected by sampling time (i.e., long sampling time required); the autocorrelation and spectrum are strongly affected at short time and high frequency, respectively, by the averaging time, whereas their long-time and low-frequency behavior depend upon the sampling time and to some extent the lag time; the integral scale is strongly affected by the sampling time; and the dissipative scale is most strongly affected by the averaging time (i.e., short ones required).

Mention has been made that averaging times must be sufficiently short to insure inclusion of all high-frequency characteristics. Of equal importance is the consideration given to the low-frequency behavior. If there should exist a nonturbulent slow variation of the mean signal strength, then the fluctuating velocities would appear significantly larger than they actually are. This would result in inaccuracies of such quantities as the variance and the zero frequency intercept of the spectrum. It is suggested that "running mean" average velocities should be applied over sufficiently long sub-samples to remove the nonturbulent variations from the fluctuation velocities. This can be accomplished by applying either analog or digital high pass filtering to the raw data prior to analysis.

Examination of the total program running time shows strong dependence on the total number of data points, N, and the maximum lag time, M, used to evaluate the autocorrelation and spectrum. For example, the complete program shown in the flow diagram of Figure 3 was run on an IBM 7094 computer using the same set of data with varying averaging, sampling, and maximum lag times. The running times were:

Δt	N	M	Total Run Time
0.002 sec.	6000	500	4 min. - 24 sec.
0.002 sec.	12000	500	6 min. - 57 sec.
0.002 sec.	18000	500	10 min. - 0 sec.
0.004 sec.	9000	500	5 min. - 56 sec.
0.004 sec.	9000	250	2 min. - 45 sec.
0.006 sec.	6000	166	1 min. - 13 sec.
0.010 sec.	3600	100	0 min. - 29 sec.

By suitably choosing sampling and averaging times, the complete structure of the turbulence data may be obtained. Selecting the optimum times to evaluate each specific property provides one way to affect economy of computing time. However, recent advances in computing techniques now provide an alternative procedure for evaluating the power spectrum. It is the fast Fourier transform, which is both a direct and an efficient technique. It has been thoroughly documented in a special issue of the IEEE Transactions on Audio and Electroacoustics.(30) The major advantages of the fast Fourier transform are speed and economy in computation of spectra from large samples. It requires N log N operations, whereas the method outlined earlier requires N^2 operations. For samples of $N = 10^3$ or larger the time saved in this part of the turbulence structure evaluation is large and is a significant fraction of the total operation time required. The comparison of these new techniques with those of the direct approach detailed in this section is also being studied here and will be reported shortly.

Digitally Analyzed Experimental Results from Anemometer Data

Turbulence data was obtained from the isothermal flow of water in a circular tube at two Reynolds numbers (50,000 and 100,000) using a single anemometer channel with a wedge shaped hot-film sensor. Digital computer analyses, employing the definitions from the previous section and the flow diagram of Figure 3, were performed on the digitized anemometer data. These analyses were optimized and performed with a 4-millisecond averaging time, a sampling time of 36 seconds, and a lag number of 250 with 9000 total data points per set.

Turbulence intensities were evaluated and compared with the corresponding analog RMS analyzer results. Agreement was within ± 8 per cent over all radial traverse positions at three axial stations shown in Figure 4; namely, stations 0, 1, and 3. Other macroscopic turbulence characteristics such as skewness, flatness, integral scale, and dissipative scale were obtained for all runs. In general, the skewness and flatness values indicated that only minor deviations from Gaussianity existed for most experimental runs. This indicated that the turbulence in the central core region was reasonably Gaussian and homogeneous. Some values are shown in the legends of Figures 5 and 7.

Microscopic turbulence characteristics were also evaluated for all data records. Figure 5 shows the radial variation of the averaged power spectral density at exit from the test section for 50,300 Reynolds number. Hamming smoothing procedures were applied in these evaluations. Averages were performed over all available data records at each radial position. Little variation was found in the evaluated results from individual data records. The corresponding, non-smoothed, velocity autocorrelations are shown in Figure 6.

The detailed turbulence characteristics of the developing flow were examined. It was found that the microscopic structure developed somewhat faster than the macroscopic structure, particularly the mean velocity. This is readily observed in comparing the mean velocity profiles in Figure 4 and the macroscopic results, tabulated in Figure 7, with the power spectral density and velocity autocorrelation distributions shown in Figures 7 and 8, respectively. The power spectral density reported by Ippen and Raichlen[31] is presented in Figure 7 for comparison. Their result was obtained from turbulent water flow in an open channel and is similar to the fully developed pipe flow result. Many other quantities were measured, analyzed, and displayed in Reference 16, to which the reader is directed for details.

Digitally Analyzed Experimental Results from Particle Trajectory Data

Particle trajectory data was obtained using individual solid particles of 0.57-cm diameter. The particles were approximately neutrally buoyant. Each emitted light of constant and uniform intensity. These particles were inserted into the Borda mouth entrance of the water loop shown in Figure 4. As each particle moved with the turbulent water flow along the core region of the test section, its position was monitored by a group of photomultiplier tubes. These tubes were mounted on a carriage which moved at the same mean axial velocity as the test particle. The phototube output voltage signals were recorded on FM magnetic tape, and replayed through ADC to obtain digital magnetic tape records. Six simultaneous channels of data were recorded which enabled the instantaneous particle position to be computed with the aid of static calibration information. Details of these procedures are given in Reference 16.

The averaging time employed to digitize the data was 4 milliseconds and the sampling time was 12 seconds. This provided the maximum useful signal length for a Reynolds number of 50,000 in the 15 feet of carriage travel.

In computing the instantaneous particle velocities, time derivatives were required. This procedure tends to emphasize noise contributions at high frequencies and in some cases can obscure the turbulence characteristics at high frequency. Considerable difficulty was experienced from this aspect; however, meaningful results were obtained by employing both analog and digital low pass filtering and by reducing all noise contributions to a minimum in the phototube monitoring system.

Typical results from several particle trajectories for the spectrum and autocorrelation are shown in Figures 9 and 10, respectively. For comparison purposes the corresponding properties of the Eulerian fluid turbulence are displayed.

Additional discussion of the coupling between the Lagrangian particle and the Eulerian fluid parameters is presented in References 16, 17, 32, and 33.

SUMMARY

This article has reviewed many of the available turbulence sensing and measuring devices as well as the procedures by which the data may be analyzed. Two specific observations, which focus attention on the current limitations in experimentally observing and analyzing turbulence, may be drawn from the discussion and from studies of the current literature. The first is that unless significant improvements are made in turbulence sensors, only minor advances are to be expected in data acquisition techniques, since the recent advances in the associated electronic components have removed the majority of difficulties previously associated with them. The second is that the recent trends of handling data by digital techniques have opened new avenues of analyzing and interpreting turbulence information and are expected to advance our understanding of turbulent phenomena, particularly with respect to its detailed structure.

REFERENCES

1. King, L. V., "On the Convection of Heat from Small Cylinders in a Stream of Fluid," Philosophical Transactions Royal Society Series A, 214 (November 13, 1914), p. 373.

2. Grant, H. P. and R. E. Kronauer, "Fundamentals of Hot-Wire Anemometry," Symposium on Measurements in Unsteady Flow, ASME, May, 1962, p. 44.

3. Rasmussen, C. G. and B. B. Madsen, "Hot-Wire and Hot-Film Anemometry," DISA-S and B Inc. Technical Information Bulletin, 1967.

4. Thermo-Systems Inc., "Hot Film and Hot Wire Anemometry: Theory and Application," Thermo-Systems Inc. Bulletin TB5, 1968.

5. Flow Corporation, "Comparison of Hot Wire Circuits: Constant Current, Constant Temperature and Constant Resistance Ratio," Flow Corporation Bulletin 40, 1962.

6. Runstadler, P. W., Jr., "Stable Operation of Hot Film Probes in Water," Symposium on Measurements in Unsteady Flow, ASME, May, 1962, p. 83.

7. Grant, H. P., "Hot-Wire in Liquid Flow Measurement," Flow Corporation, Technical Memorandum Bulletin 89B, 1963.

8. Widell, K. E., "Coated Hot-Film Probes for Measurements in Liquids," Note on New Products, DISA Information No. 3 (January, 1966).

9. Rasmussen, C. G., "The Air Bubble Problem in Water Flow Hot-Film Anemometry," DISA Information No. 5 (June, 1967).

10. Ling, S. C., "Heat Transfer Characteristics of Hot Film Sensing Element Used in Flow Measurement," Transactions, ASME, Series D, Journal of Basic Engineering (September, 1960), p. 629.

11. Bankoff, S. G. and R. S. Rosler, "Constant-Temperature Hot-Film Anemometer as a Tool in Liquid Turbulence Measurements," The Review of Scientific Instruments, 33:11 (November, 1962), p. 1209.

12. Patterson, G. K. and J. L. Zakin, "Hot-Film Anemometer Measurements of Turbulence in Pipe Flow: Organic Solvents," AIChE Journal, 13:3 (May, 1967), p. 513.

13. Easley, D. C., "Characteristics of a Hot-Film Anemometer in a Liquid Mercury System," M.S. thesis, Department of Chemical Engineering, Purdue University, Lafayette, Indiana, 1966.

14. Hanratty, T. J., "Use of Electrochemical Techniques to Study Turbulence Close to a Wall," Fluid Dynamics Research Review, University of Illinois, Urbana, Illinois (March 6-7, 1967), p. 7.

15. Baldwin, L. V. and T. J. Walsh, "Turbulent Diffusion in the Core of a Fully Developed Pipe Flow," AIChE Journal, 7:1 (March, 1961), p. 53.

16. Jones, B. G., "An Experimental Study of the Motion of Small Particles in a Turbulent Fluid Field Using Digital Techniques for Statistical Data Processing," Ph.D. thesis, Nuclear Engineering Program, University of Illinois, Urbana, Illinois, June, 1966.

17. Shirazi, M. A., "On the Motion of Small Particles in Turbulent Flow," Ph.D. thesis, Department of Mechanical Engineering, University of Illinois, Urbana, Illinois, February, 1967.

18. Hinze, J. O. Turbulence. New York: McGraw Hill Book Company, 1959.

19. Fox, J., W. H. Webb, B. G. Jones, and A. G. Hammitt, "Hot-Wire Measurements of Wake Turbulence in a Ballistic Range," AIAA Journal, 5:1 (1967), p. 99.

20. Taylor, G. I., "Statistical Theory of Turbulence," Parts 1-4, Proceedings, Royal Society, London, A, 151 (1935), p. 421.

21. Frenkiel, F. N. and P. S. Klebanoff, "Correlation Measurement in Turbulent Flow Using High-Speed Computing Methods," The Physics of Fluids, 10:8 (1967), p. 1737.

22. Blackman, R. B. and J. W. Tukey. The Measurement of Power Spectra. New York: Dover Publications, Inc., 1958.

23. Blackman, R. B. Data Smoothing and Prediction. Reading, Mass.: Addison-Wesley Publishing Co., Inc., 1965.

24. Bendat, J. S. and A. G. Piersol. Measurement and Analysis of Random Data. New York: John Wiley & Sons, 1966.

25. Pasquill, F. Atmospheric Diffusion. London: D. Van Nostrand Co., 1962.

26. Panofsky, H. A. and R. A. McCormick, "Properties of Spectra of Atmospheric Turbulence at 100 Meters," Quarterly Journal Royal Meteorological Society, 80 (1964), p. 546.

27. Frankiel, F. N. and P. S. Klebanoff, "Higher Order Correlations in a Turbulent Field," The Physics of Fluids, 10:3 (1967), p. 507.

28. Raichlen, F., "Some Turbulence Measurements in Water," Journal of the Engineering Mechanics Division, ASCE, 93: EM 2, Proceedings Paper 5195 (April, 1967), p. 73.

29. Southworth, R. W., "Autocorrelation and Spectral Analysis," Article 19 in Mathematical Methods of Digital Computers. A. Ralston and H. S. Wilf, eds. New York: John Wiley & Sons, 1960.

30. "Special Issue on Fast Fourier Transforms," IEEE Transactions on Audio and Electroacoustics, AU-15:2 (June, 1967).

31. Ippen, A. T. and F. Raichlen, "Turbulence in Civil Engineering: Measurements in Free Surface Streams," Proceedings, ASCE, 83: HY5 (1957), p. 1392.

32. Jones, B. G., B. T. Chao, and M. A. Shirazi, "An Experimental Study of the Motion of Small Particles in a Turbulent Fluid Field Using Digital Techniques for Statistical Data Processing," Sermak and Goodman, eds., Proceedings, Tenth Midwestern Mechanics Conference, Fort Collins, Colorado, Vol. 4 (August 21-23, 1967), p. 1249.

33. Shirazi, M. A., B. T. Chao, and B. G. Jones, "On the Motion of Small Particles in a Turbulent Fluid," Sermak and Goodman, eds., Proceedings, Tenth Midwestern Mechanics Conference, Fort Collins, Colorado, Vol. 4 (August 21-23, 1967), p. 1179.

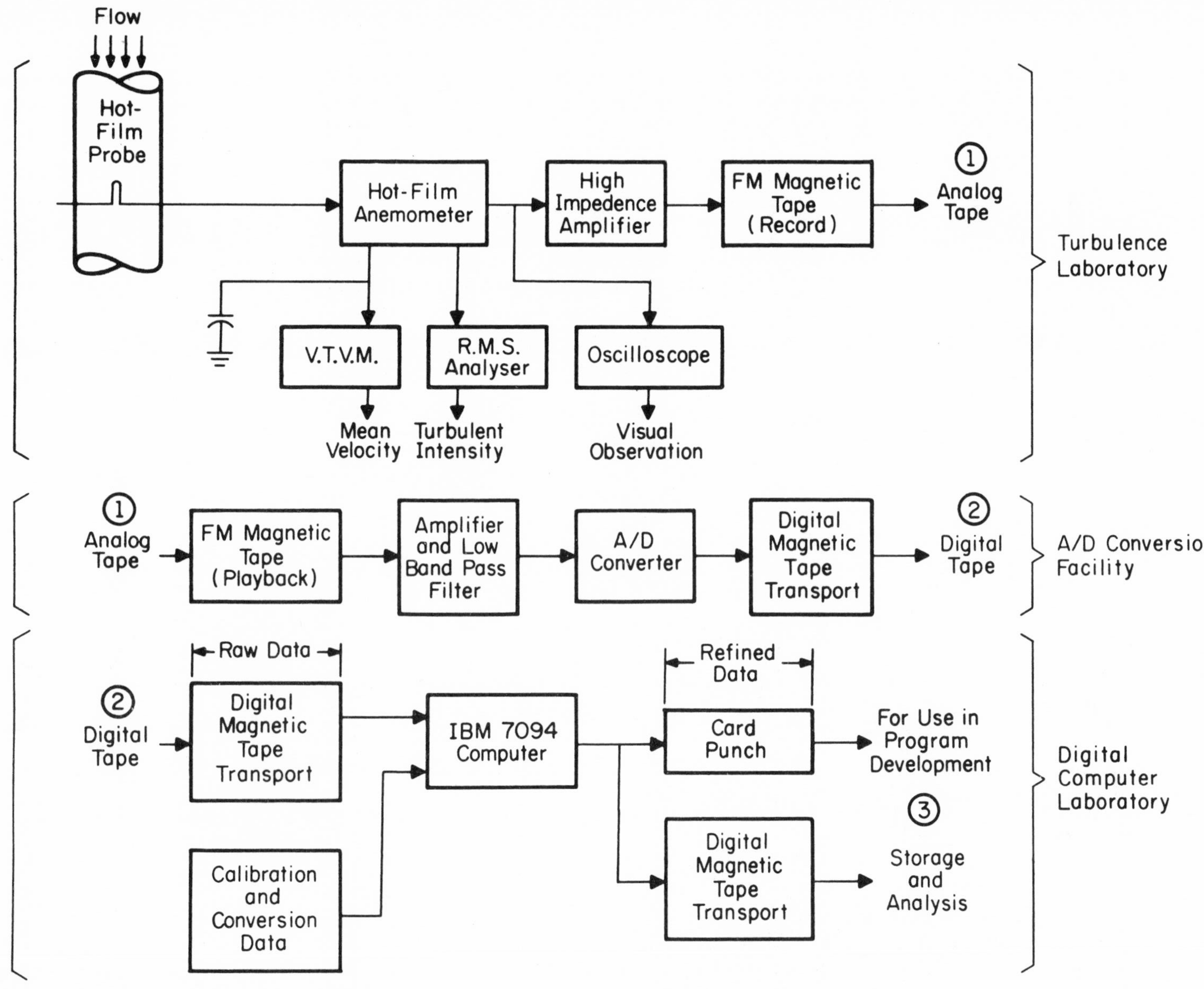

FIGURE 1. ANEMOMETER DATA ACQUISITION SYSTEM.

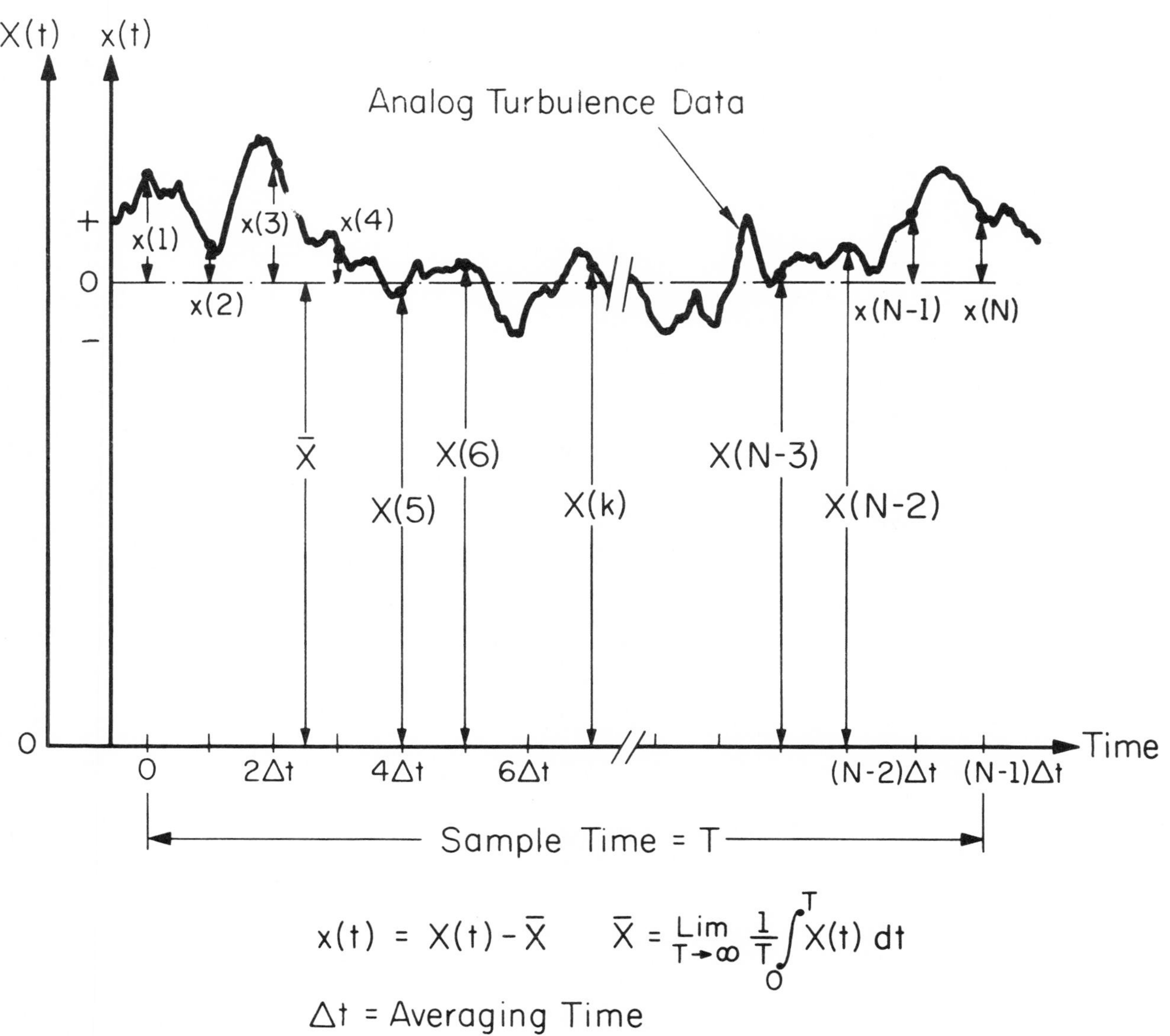

FIGURE 2. DIGITAL REPRESENTATION OF ANALOG TURBULENCE DATA.

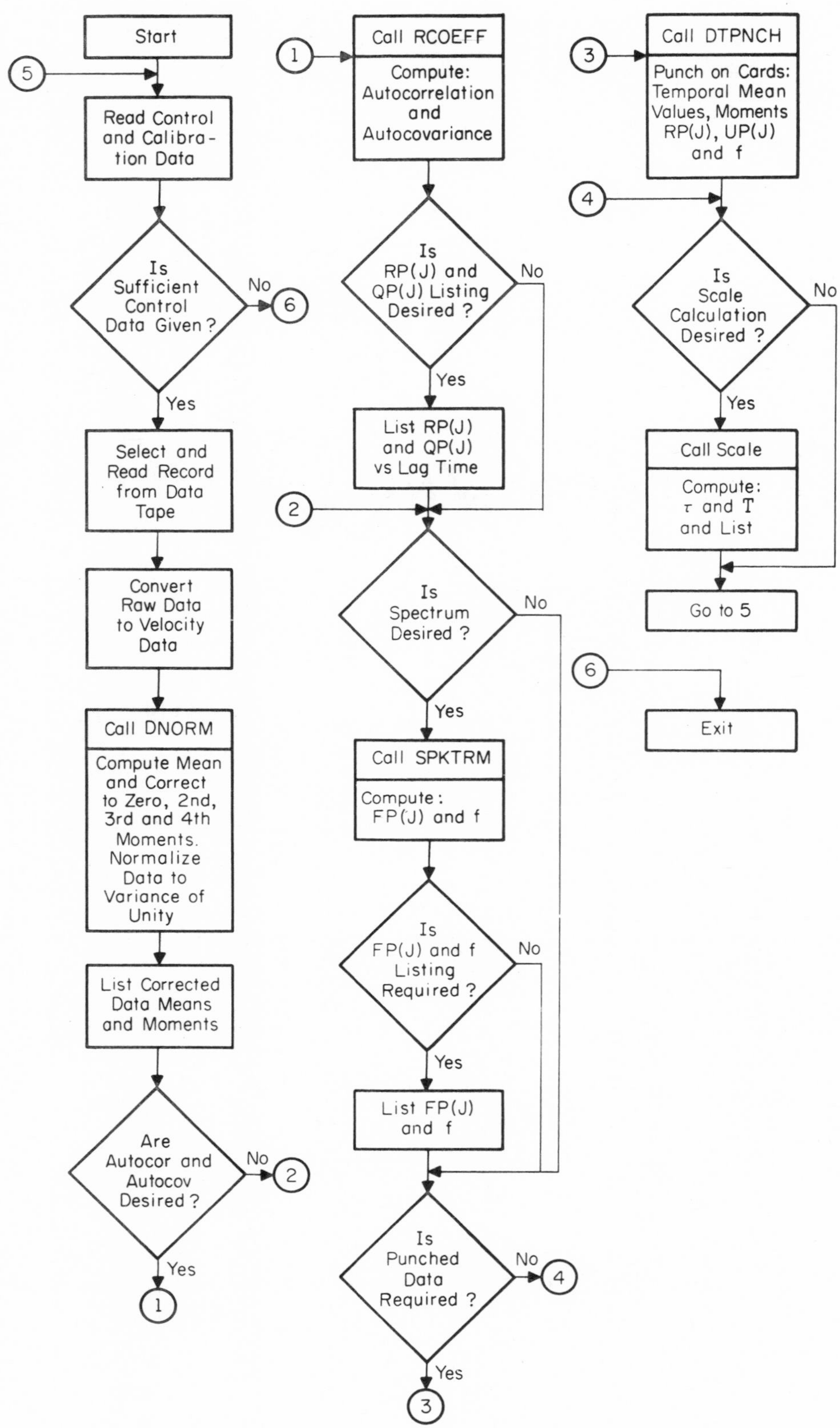

FIGURE 3. FLOW DIAGRAM FOR STATISTICAL ANALYSIS PROGRAM.

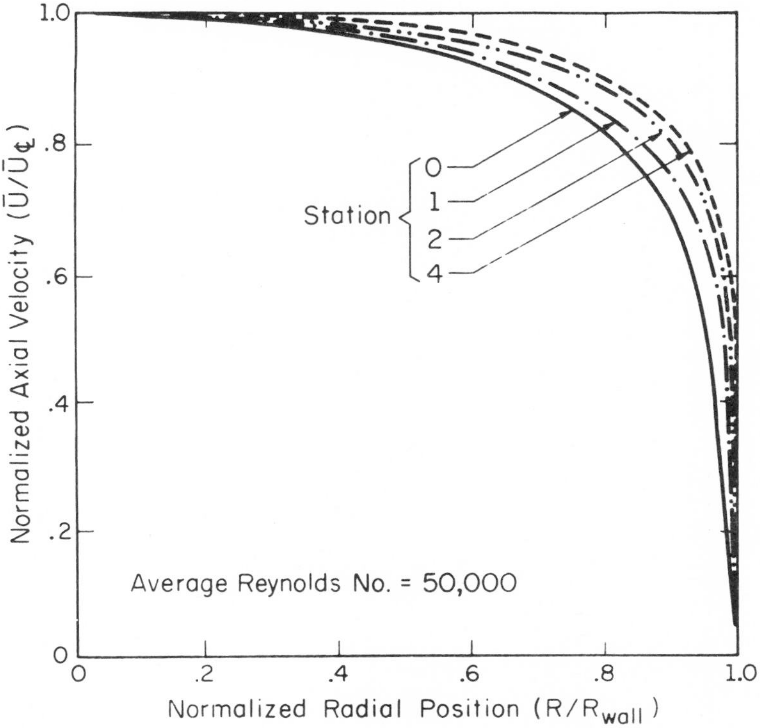

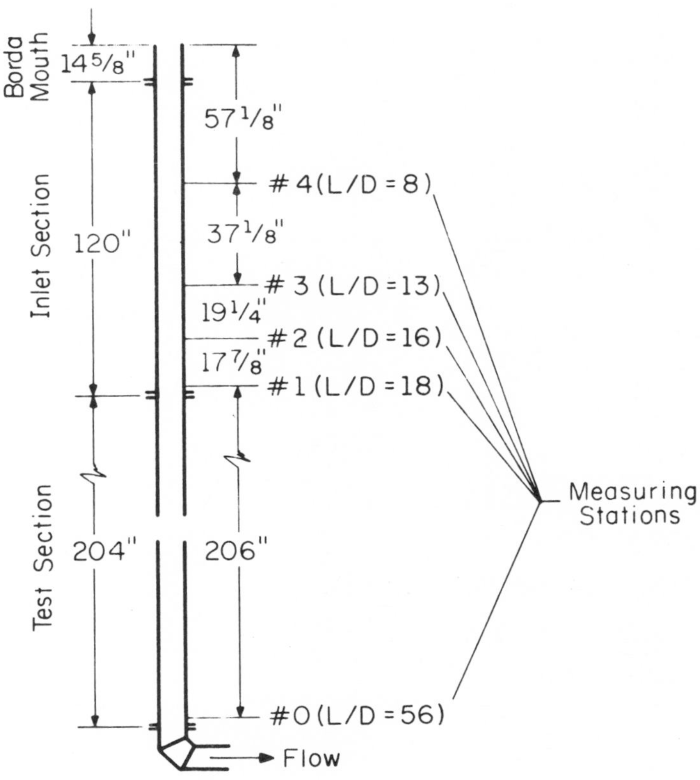

FIGURE 4. RADIAL VELOCITY PROFILES IN THE INLET AND TEST SECTIONS.

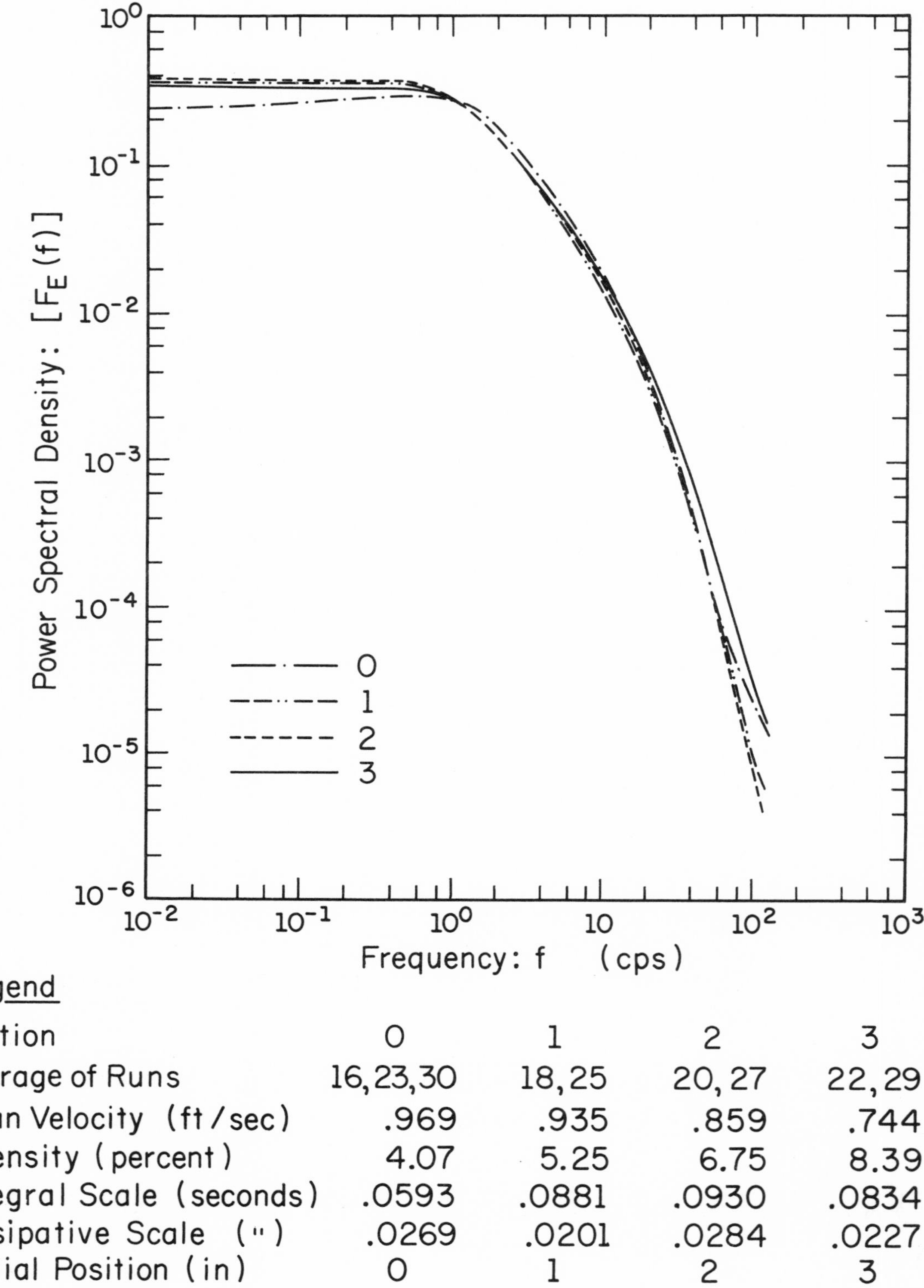

Legend

Station	0	1	2	3
Average of Runs	16,23,30	18,25	20,27	22,29
Mean Velocity (ft/sec)	.969	.935	.859	.744
Intensity (percent)	4.07	5.25	6.75	8.39
Integral Scale (seconds)	.0593	.0881	.0930	.0834
Dissipative Scale (")	.0269	.0201	.0284	.0227
Radial Position (in)	0	1	2	3

FIGURE 5. RADIAL VARIATION OF THE EULERIAN POWER SPECTRUM AT EXIT FROM THE TEST SECTION FOR Re = 50,300.

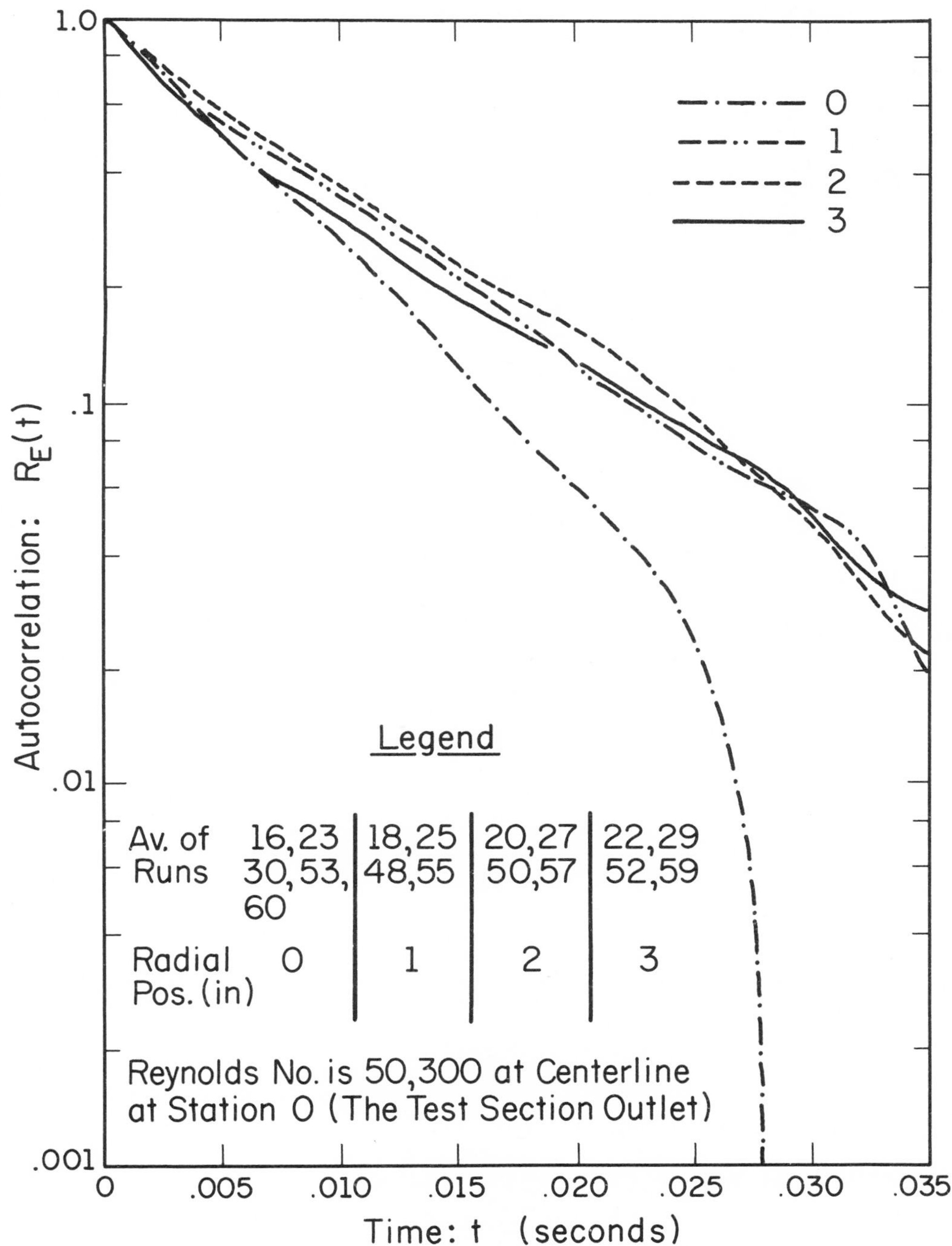

FIGURE 6. RADIAL VARIATION OF THE EULERIAN AUTOCORRELATION AT EXIT FROM THE TEST SECTION FOR Re = 50,300.

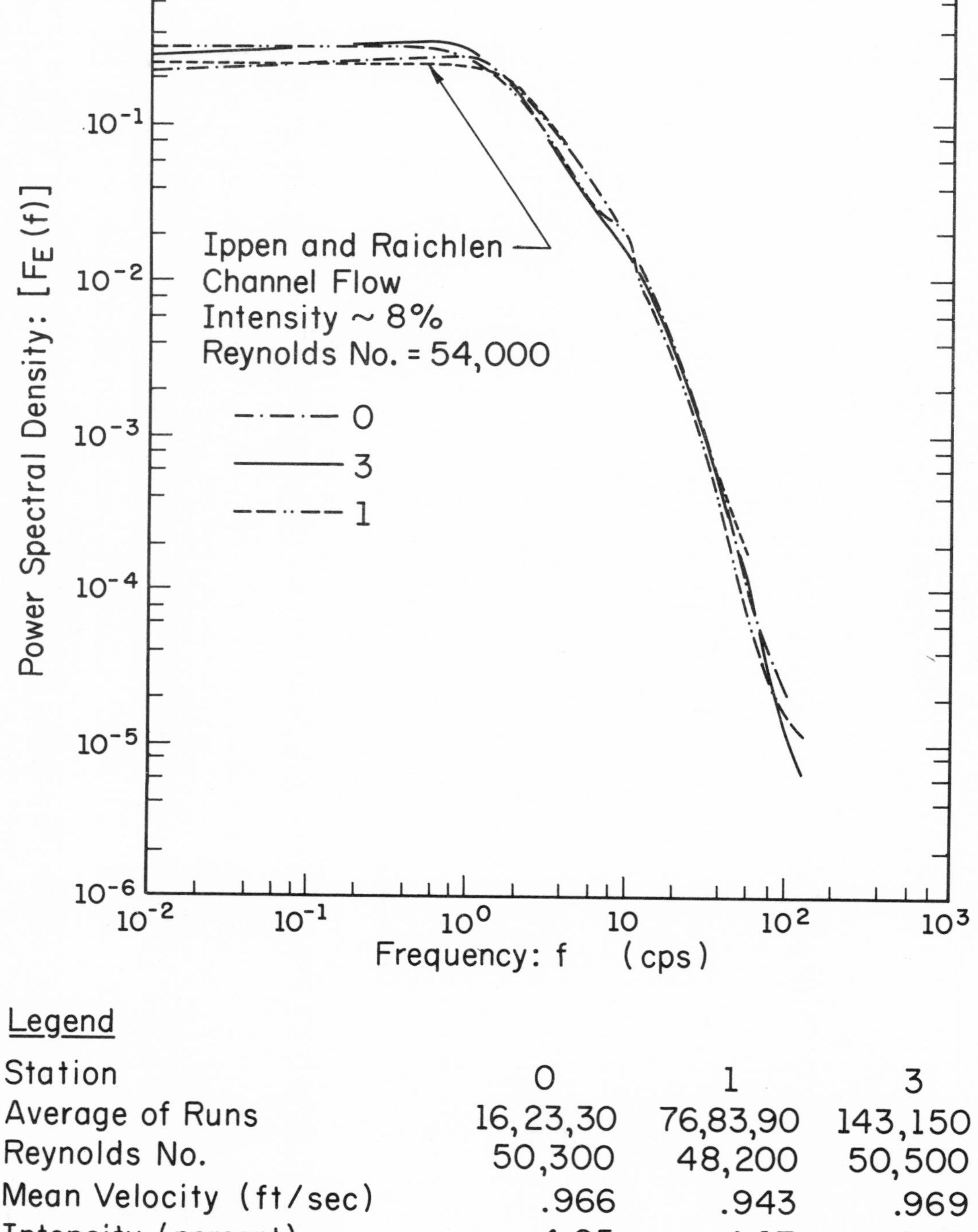

Legend

Station	0	1	3
Average of Runs	16,23,30	76,83,90	143,150
Reynolds No.	50,300	48,200	50,500
Mean Velocity (ft/sec)	.966	.943	.969
Intensity (percent)	4.05	4.17	6.48
Integral Scale (seconds)	.0554	.0790	.0708
Dissipative Scale (")	.0261	.0257	.0252

FIGURE 7. AXIAL VARIATION OF THE EULERIAN POWER SPECTRUM AT THE TUBE CENTERLINE FOR Re NEAR 50,000.

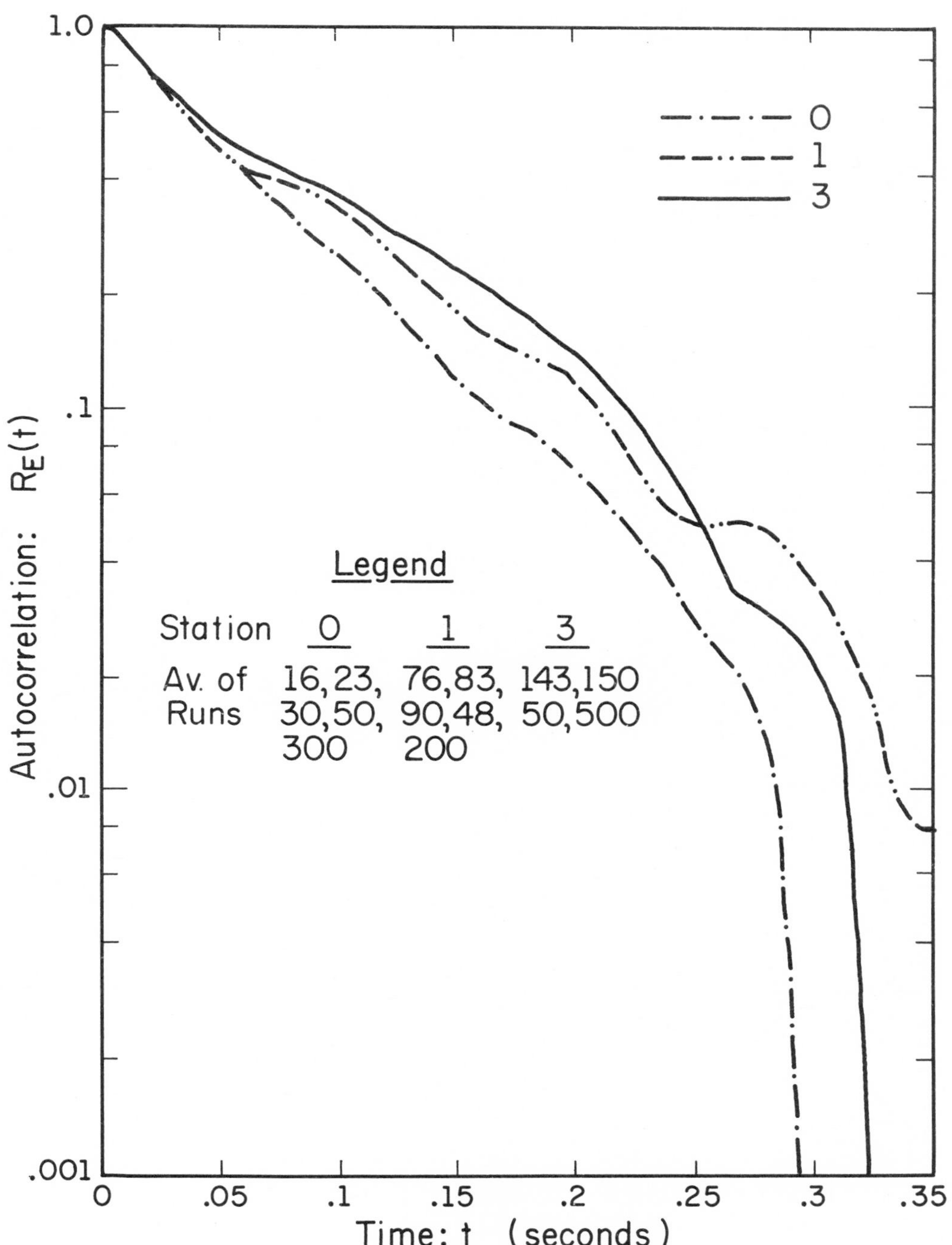

FIGURE 8. AXIAL VARIATION OF THE EULERIAN AUTOCORRELATION AT THE TUBE CENTERLINE FOR Re NEAR 50,000.

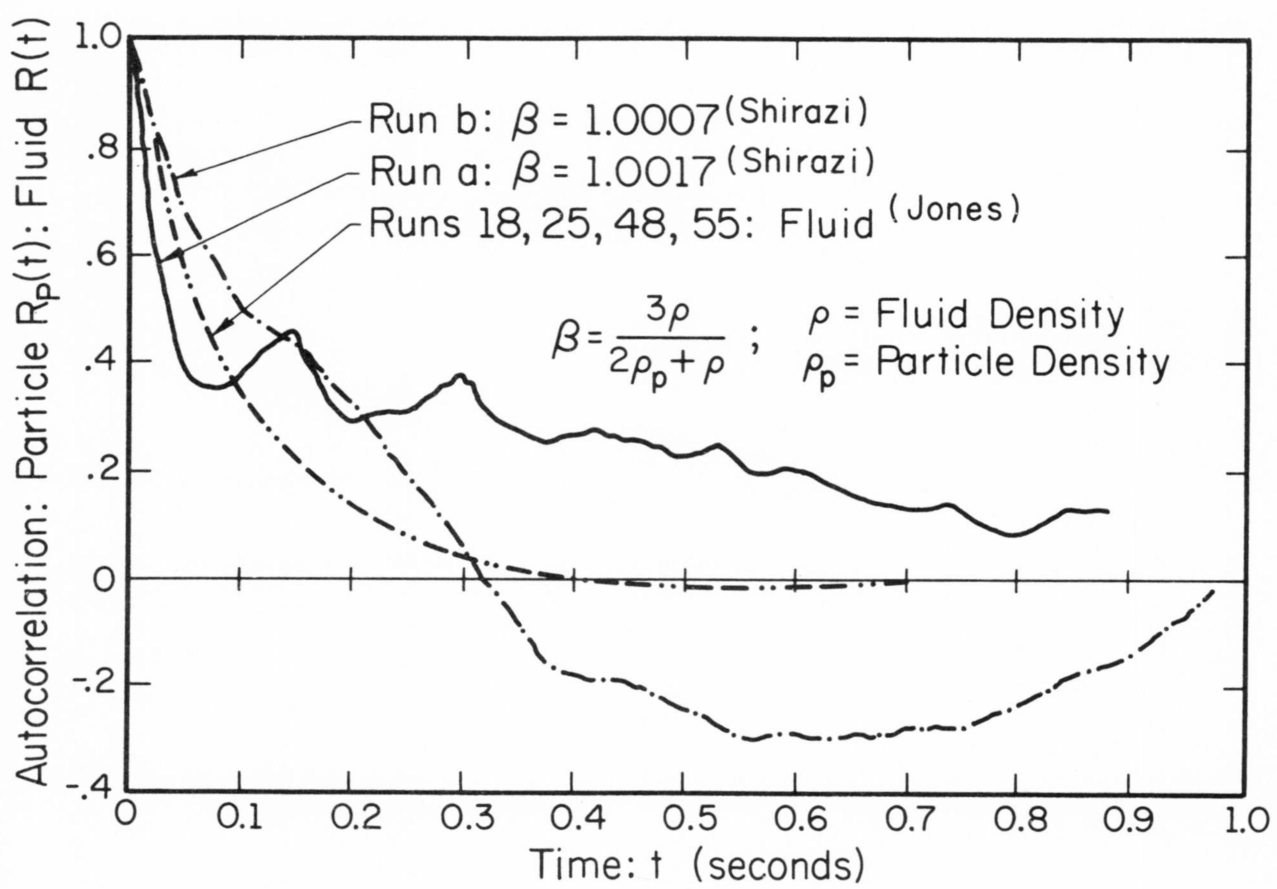

FIGURE 9. AUTOCORRELATION ESTIMATES FROM PARTICLE TRAJECTORY AND HOT-FILM ANEMOMETER EXPERIMENTAL DATA.

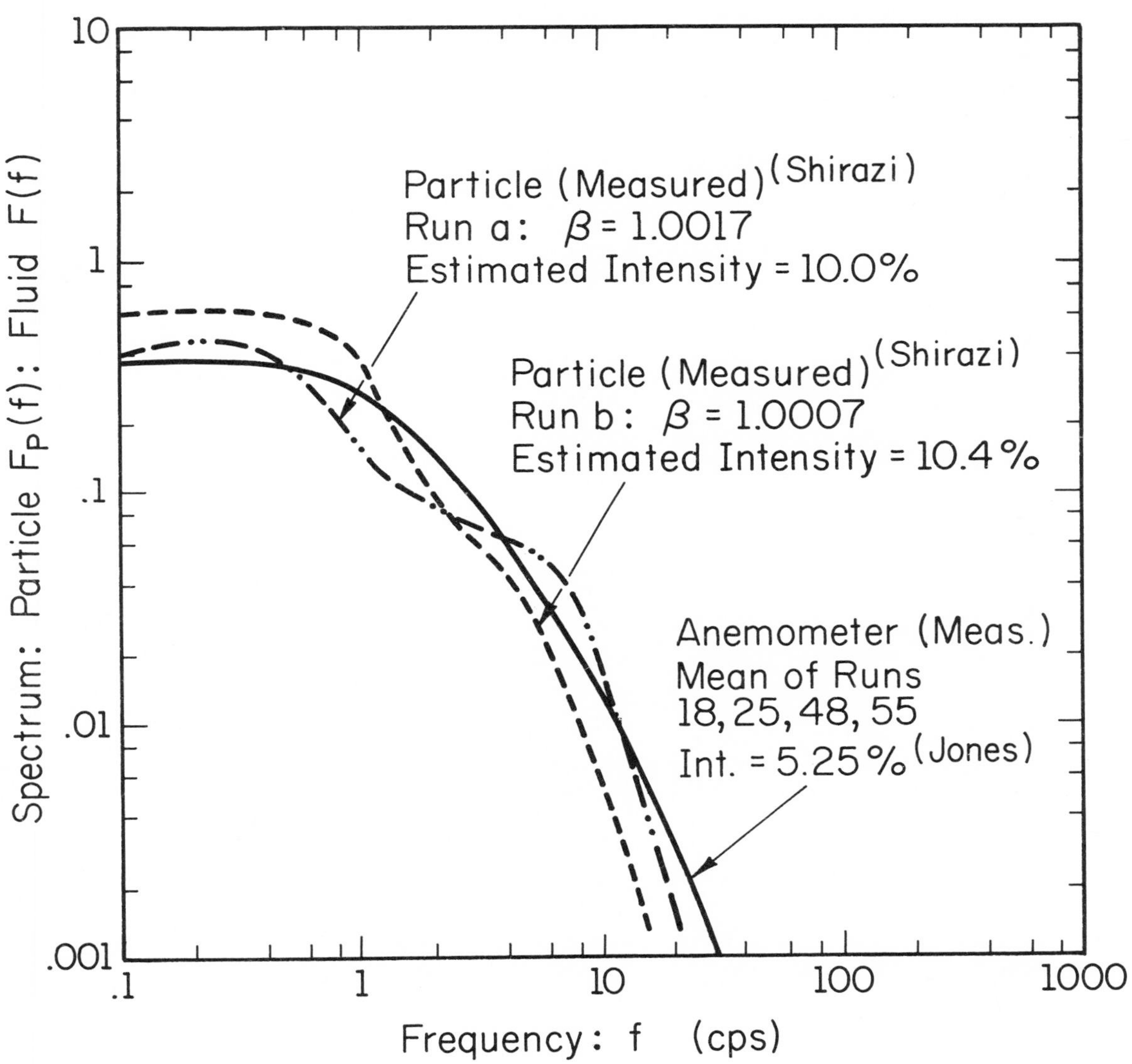

FIGURE 10. SPECTRUM ESTIMATES FROM PARTICLE TRAJECTORY AND HOT-FILM ANEMOMETER EXPERIMENTAL DATA.

APPENDIX: COOLED-FILM SENSOR OPERATION IN SUBSONIC TURBULENT FLOWS

INTRODUCTION

The brief discussion of the cooled-film sensor presented earlier in the section Cooled-film Sensor Characteristics gives only general applications and does not analyze its behavior. In this section analytical relations will be developed from which the detailed response characteristics are obtained. By choosing a small temperature difference between the film temperature and the fluid mean temperature and also using relatively low velocity fluid flow, the direct application of the sensor to distinguish between temperature and velocity (mass flux) fluctuations is examined. The extended variety and range of applications are not examined in detail, since relations describing the energy transfer by small cylinders in compressible flow, rarified flow, or supersonic flow are not sufficiently well correlated to warrant detailed study now. However, as these relations become available, operation and interpretation of the sensor behavior in a wide variety of flows could be examined in much the same manner that is used below. At least one study has been performed[1] in which the ionized wake of a simulated reentry vehicle has been probed with a cooled-film sensor.

ENERGY TRANSFER RELATIONS AND CONSIDERATIONS

The cooled-film sensor was developed by Fingerson[2] for use with constant temperature anemometers. It is currently available as a 1-micron-thick by 0.080-in.-long layer of platinum on a 0.006-in. diameter hollow pyrex glass tube. Figure A1 shows a typical probe shape with a cutaway of the active sensor portion which indicates the energy transfer paths with the associated temperatures in the sensor and surrounding fluids. The explicit relations governing the energy flow, for a one-dimensional radially symmetric model, are:

Ohmic Heating (positive to film):

$$q_o = I^2R = I^2 \frac{S_c \ell}{\pi d t_f} \left[1 + \alpha_f T_f \left(\frac{T_f - T_c}{T_f} \right) \right] \quad \text{(A1)}$$

Convective Heating (positive to film):

$$q_A = h_A A_f \Delta T_f = \frac{Nu\ k_A}{d} \pi d \ell\ T_f \left[\frac{T_r - T_f}{T_f} \right] \quad \text{(A2)}$$

Conduction Through Glass (positive to coolant):

$$q_T = \frac{2\pi\ k_g \ell\ T_f}{\ln\left[d/(d - 2t_g)\right]} \left[\frac{T_f - T_g}{T_f} \right] \quad \text{(A3)}$$

Convection to Sensor Coolant (positive to coolant):

$$q_c = C\ \pi\ \ell\ k_c\ T_f \left[\frac{T_g - T_c}{T_f} \right] W^{1/3} \quad \text{(A4)}$$

where C and W are given by McAdams[3] for $W > 10$ as $C = 1.62$ and $W = 4\dot{\omega}\ Cp_c/\pi \ell k_c$.

To a first approximation, the average Nusselt number for low-speed flow may be represented in Equation (A2) by

$$Nu = A + B\ Re^n , \quad \text{(A5)}$$

with $A = 0.3$, $B = 0.5$, and $n = 0.5$.

Combining Equations (A1) through (A5) an energy balance may be obtained as

$$q_c = q_T = q_o + q_A , \quad \text{(A6)}$$

which has the physical restriction requiring the ohmic heating to be positive so that electronic feedback may be used to control the film at the selected temperature. However, the balance relation accommodates $T_f > T_r$ and $T_f < T_r$, both of which can be realized within practical limits of operation. Examination of these relations, within the cited physical restrictions, provides a means by which control of the sensor's relative response to temperature and velocity fluctuations in the fluid can be obtained.

OPERATIONAL CONSIDERATIONS

Two specific operating conditions may be employed, namely: $T_r = T_f$ and $T_r \neq T_f$. The first implies that $q_o = q_T$ and that q_A will be zero in the mean and relatively small unless the fluctuating components are large deviations from the mean condition. The second allows a mean temperature differential between the fluid and the sensor which can provide relatively high rates of energy transfer, q_A, as a reference condition. By examining these two modes of operation, conclusions can be drawn regarding the choice of operating levels for a series of measurements. This is particularly pertinent when the mean ambient temperature, $\overline{T}_r$, varies throughout the observation.

Close control of the internal coolant flow and temperature are necessary to provide stable and calibrated operation. In subsonic, low-temperature flows adequate control of T_c is readily obtained with water or compressed air in laminar flow. However, in more severe ambient conditions such as combustion flames and highly ionized re-entry wakes the coolant flow rate must be increased. Often this results in transition or turbulent flow existing in the sensor which necessarily results in marked increases in sensor noise level. This condition can be avoided in some cases by choosing an appropriate coolant to delay transition or to prevent boiling. Additional study is required in this area to clarify which combinations of fluid, flow rate, sensor geometry, and operating temperatures should be employed in a given application.

SENSOR SENSITIVITY RELATIONS

The two operating conditions indicated in the preceding paragraphs are treated separately below. In the interest of brevity, detailed development is given only for the case where $T_r = T_f$.

Case (1): $T_r = T_f$

The condition of no average driving potential, i.e., $T_r = T_f$, makes the average $q_A = 0$. Thus Equation (A6) reduces to a simple equality between q_c, q_T, and q_o from which the following identities result. From Equation (A1) and (A4),

$$I_o^2 S_c \frac{\ell}{\pi d t_f}\left[1 + \alpha_e T_f \left(\frac{T_f - T_c}{T_f}\right)\right] = C \pi \ell k_c T_f \left[\frac{T_g - T_c}{T_f}\right] W^{1/3}, \qquad \text{(A7)}$$

while from Equations (A3) and (A4),

$$\frac{2\pi k_t \ell T_f}{\ln\left[d/(d - 2t_g)\right]}\left[\frac{T_f - T_g}{T_f}\right] = C \pi \ell k_c T_f \left[\frac{T_g - T_c}{T_f}\right] W^{1/3}. \qquad \text{(A8)}$$

By introducing the identity

$$\frac{T_g - T_c}{T_f} = \frac{T_f - T_c}{T_f} - \frac{T_f - T_g}{T_f} \qquad \text{(A9)}$$

into Equation (A8) and after simplifying it becomes

$$C k_c \left[\frac{T_f - T_c}{T_f}\right] = \left[C k_c + \frac{2k_g}{W^{1/3} \ln\left[d/(d - 2t_g)\right]}\right]\left(\frac{T_f - T_g}{T_f}\right)$$

or

$$\frac{T_g - T_c}{T_f} = \frac{T_f - T_c}{T_f}\left[1 - \left\{1 + \frac{2k_g/k_c}{C W^{1/3} \ln\left[d/(d - 2t_g)\right]}\right\}^{-1}\right]. \qquad \text{(A10)}$$

It is convenient to introduce the following dimensionless parameters:

$$\left.\begin{aligned} &\tau_c = \frac{T_f - T_c}{T_f} \quad : \quad \tau = \frac{T_r - T_f}{T_f}, \\ &\text{and} \\ &X = 1 - \left\{1 + \frac{2k_g/k_c}{C W^{1/3} \ln\left[d/(d - 2t_g)\right]}\right\}^{-1}. \end{aligned}\right\} \qquad \text{(A11)}$$

Introducing Equations (A11) and (A10) into Equation (A7) gives the reference condition in terms of measurable parameters and system constants as

$$I_o^2 S_c \frac{\ell}{\pi d t_f} \left[1 + \alpha_c T_f \tau_c\right] = C \pi \ell k_c T_f \tau_c X W^{1/3} . \qquad \text{(A12)}$$

For the operating condition, q_A represents the instantaneous heat exchange between the surrounding fluid and the sensor. Thus Equation (A6) becomes

$$I^2 S_c \frac{\ell}{\pi d t_f} \left[1 + \alpha_c T_f \tau_c\right] + \pi \ell k_A T_f \tau \, Nu = I_o^2 S_c \frac{\ell}{\pi d t_f} \left[1 + \alpha_c T_f \tau_c\right]$$

$$= C \pi \ell k_c T_f \tau_c X W^{1/3} ,$$

which reduces to

$$\frac{I^2}{I_o^2} = 1 - \frac{Z \, Nu \, \tau}{\tau_c} , \qquad \text{(A13)}$$

where

$$Z = \frac{k_A}{k_c} \cdot \frac{1}{C X W^{1/3}} ,$$

a constant for each experimental test. Examination of Equation (A13) shows that $\tau = 0$ requires that $I^2 = I_o^2$. The limit of operation is set by the physical restriction that $I^2 \geq 0$. Thus, when $I^2 = 0$,

$$\tau = \frac{\tau_c}{Z \, Nu} . \qquad \text{(A14)}$$

The fluctuating sensitivity, for the assumption of constant material properties, is obtained from the general relation

$$\delta I^2 = \frac{\partial I^2}{\partial T_r} \delta T_r + \frac{\partial I^2}{\partial(\rho u)} \delta(\rho u) , \qquad \text{(A15)}$$

which may be normalized by I^2 to give

$$\frac{\delta I^2}{I^2} = S_{T_r} \frac{dT_r}{T_r} + S_{\rho u} \frac{d(\rho u)}{\rho u} , \qquad \text{(A16)}$$

where the temperature and mass flux sensitivities are given, respectively, by

$$S_{T_r} = \frac{T_r}{I^2} \frac{\partial I^2}{\partial T_r}$$

and

$$S_{\rho u} = \frac{\rho u}{I^2} \frac{\partial I^2}{\partial(\rho u)} .$$

Evaluation of S_{T_r} and $S_{\rho u}$ from Equation (A13) gives

$$S_{T_r} = - \frac{I_o^2}{I^2} \cdot \frac{T_r}{T_f} \cdot \frac{Z \, Nu}{\tau_c} \qquad \text{(A17)}$$

and

$$S_{\rho u} = - \frac{1}{2} \frac{I_o^2}{I^2} \cdot \frac{\tau}{\tau_c} \cdot Z(Nu - A) . \qquad \text{(A18)}$$

Combining Equations (A17) and (A18) to form the sensitivity ratio and observing that

$$\frac{T_r}{T_f} = \tau + 1$$

gives

$$\frac{S_{\rho u}}{S_{T_r}} = \frac{1}{2} \cdot \frac{Nu - A}{Nu} \cdot \frac{\tau}{\tau + 1} . \qquad \text{(A19)}$$

For the reference condition, since $q_A = 0$, there is no driving potential available and there is no direct means of eliminating A, the natural convection term, in Equation (A5). However, it is normally small compared to Nu and may be ignored in many applications.

Case (2): $T_r \neq T_f$

The reference condition for this case is chosen to have no forced flow over the sensor, i.e., Re = 0, but where A is finite as a result of natural convection. However, operating with or without forced cooling the total energy flow to the sensor coolant is constant. Thus the power to the sensor, normalized to no forced cooling, may be written as

$$\frac{I^2}{I_o^2} = \frac{1 - ZNu(\tau/\tau_c)}{1 - ZNu_o(\tau/\tau_c)} , \qquad (A20)$$

which reduces to Equation (A13) for $Nu_o = 0$.

The fluctuating sensitivities, defined as in Case (1), become

$$S_{T_r} = -\frac{I_o^2}{I^2} \cdot \frac{\tau + 1}{\tau_c} \cdot \frac{Z(Nu - A)}{\left[1 - ZNu_o(\tau/\tau_c)\right]^2} \qquad (A21)$$

and

$$S_{\rho u} = -\frac{1}{2} \frac{I_o^2}{I^2} \cdot \left(\frac{Nu}{Nu_o} - 1\right) \cdot \frac{ZNu_o(\tau/\tau_c)}{1 - ZNu_o(\tau/\tau_c)} . \qquad (A22)$$

These combine to give

$$\frac{S_{\rho u}}{S_{T_r}} = \frac{1}{2} \cdot \frac{Nu - A}{Nu - Nu_o} \cdot \frac{\tau}{\tau + 1} \cdot \left(1 - ZNu_o \frac{\tau}{\tau_c}\right) , \qquad (A23)$$

which reduces to Equation (A19) for $Nu_o = 0$.

PREDICTED COOLED-FILM PERFORMANCE

Only results for Case (1) are presented, since it represents the simpler one while retaining the main parametric characteristics. As an example, Figure A2 shows the evaluated results of Equation (A13) for fixed τ_c and Z, but with varying Re and τ. These results indicate that I^2/I_o^2 is relatively sensitive to the cooling mass flux or, for the incompressible flow, the velocity. The relative sensitivity to mass flux and recovery temperature variations is shown in Figure A3 for the same sensor conditions.

It is noted that both graphical results indicate nonlinear response characteristics. However, for many flows of interest the relative fluctuations are below 10 per cent and the assumption of a linear response is adequate.

By suitably choosing τ and Nu, in the general case, the relative sensitivity ratio may be controlled to accentuate temperature sensitivity or mass flux sensitivity. An upper limit of one half may be approached with large thermal overheat ($\tau \gg 1$) and with significant forced convection ($Nu \gg A$). The lower limit of zero is approached as $\tau \to 0$ and represents complete dominance of sensitivity by temperature fluctuations. The former represents operation as a hot-wire (or a hot-film), while the latter is a unique characteristic of the cooled-film sensor.

Applications of the cooled-film sensor enable much larger variation in sensitivity ratio than the hot-wire. By suitably choosing the operating levels it becomes feasible to differentiate between velocity and temperature fluctuations in a more definite manner than with a hot-wire sensor.

ACKNOWLEDGMENTS

The author wishes to thank Professor B. T. Chao, Mr. W. E. Burchill, and Mr. B. W. Spencer of the University of Illinois for their suggestions and discussion during the preparation of this paper. Thanks are also extended to Mr. G. Morris who prepared all figures for the article.

NOTATION

- A = empirical constant in Equation (A5)
- A_f = area of film's cooling surface
- B = empirical constant in Equation (A5)
- C = empirical constant in Equation (A4)
- C_{pc} = specific heat of coolant
- d = outside diameter of glass tube
- h_A = convective heat transfer coefficient of film
- I = electric current to the film
- I_o = reference electric current to the film
- k_c = thermal conductivity of coolant
- k_g = thermal conductivity of glass tube
- ℓ = film length
- Nu = Nusselt number
- q_A = convective heat flux to film
- q_c = convective heat flux to internal coolant
- q_o = ohmic heating flux to film
- q_T = conductive heat flux through glass
- R = electrical resistance of film
- Re = Reynolds number

S_c = electrical resistivity of film at T_c

$S_{\rho u}$ = sensitivity of the sensor to mass flux

S_{T_r} = sensitivity of the sensor to surrounding temperature

t_f = wall thickness of film

t_g = wall thickness of glass tube

T = temperature in glass tube

T_c = average coolant temperature

T_f = average film temperature

T_g = temperature at glass tube -- coolant interface

T_r = recovery temperature in the fluid surrounding the sensor

u = velocity of the fluid surrounding the film

W = parameter defined after Equation (A4)

X = parameter defined in Equation (A11)

Z = parameter defined after Equation (A13)

α_f = temperature coefficient of resistivity of the film

ν = kinematic viscosity of the surrounding fluid

ρ = density of the surrounding fluid

$\tau;\tau_c$ = dimensionless temperatures defined by Equation (A11)

$\dot{\omega}$ = mass flow rate of the coolant

REFERENCES FOR THE APPENDIX

1. Trottier, D. C., A. M. Ahmed, and D. Ellington, "Cooled-Film Anemometer Measurements in the Hypersonic Wake," CARDE Technical Note, No. 1720, 1966.

2. Fingerson, L. M., "A Heat Flux Probe for Measurement in High Temperature Gases," Ph.D. thesis, Department of Mechanical Engineering, University of Minnesota, Minneapolis, Minnesota, 1961.

3. McAdams, W. H., Heat Transmission. 3rd ed. New York: McGraw-Hill Book Co., 1954, p. 259.

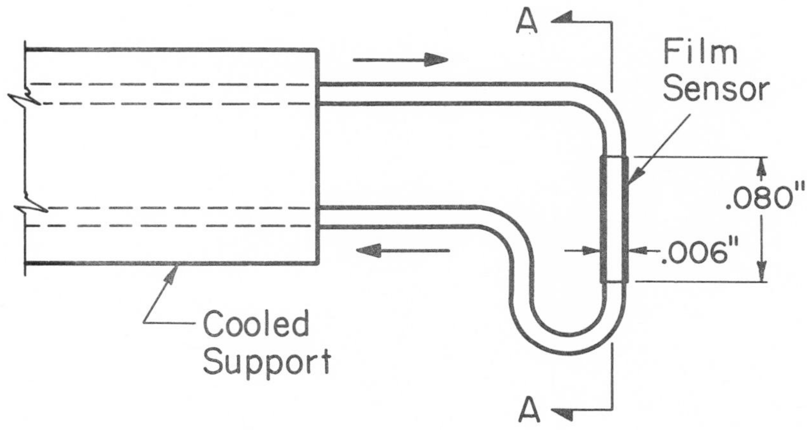

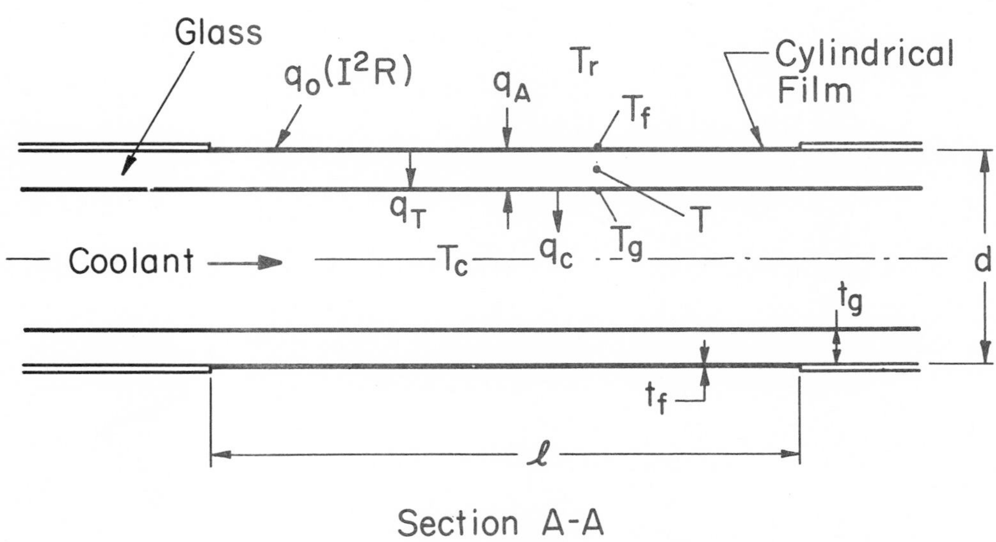

FIGURE A1. COOLED-FILM SENSOR SCHEMATIC.

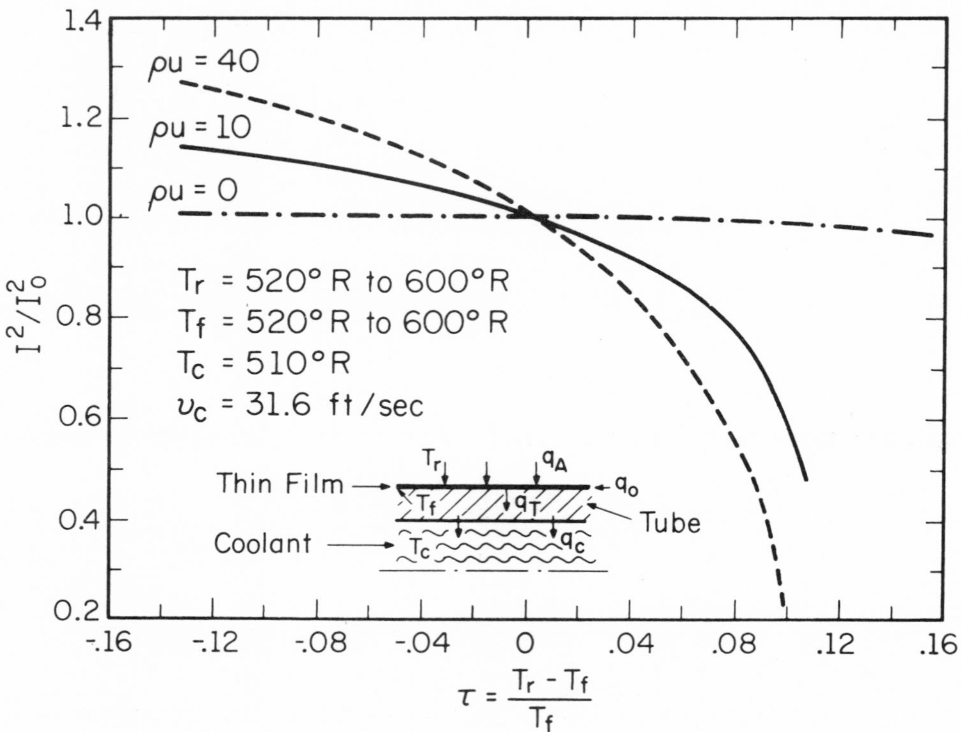

FIGURE A2. STEADY STATE HEAT LOSS CHARACTERISTICS FOR A COOLED-FILM SENSOR.

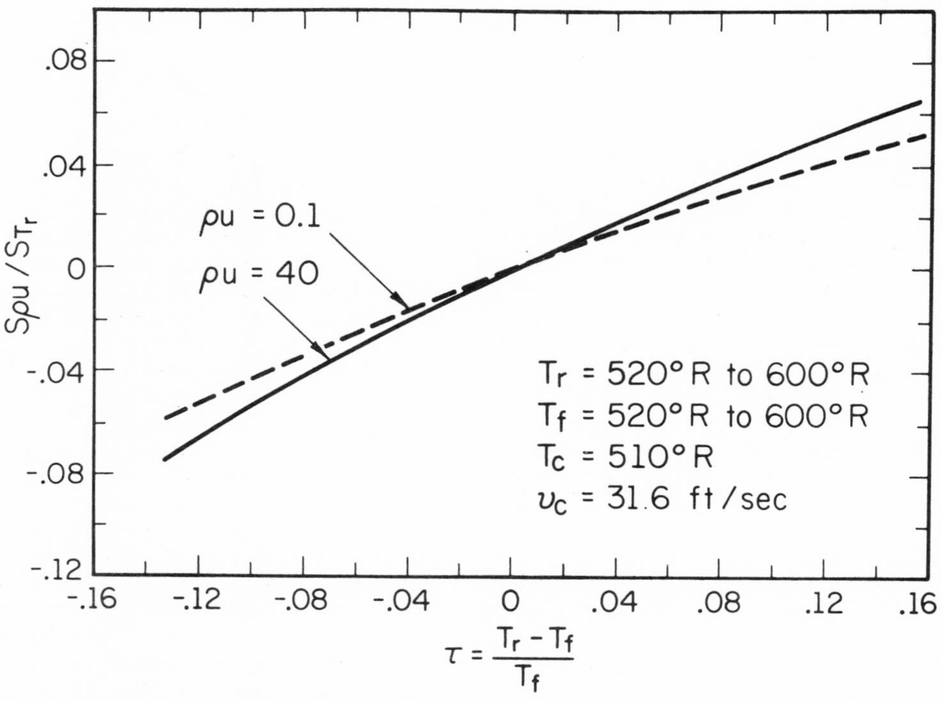

FIGURE A3. MASS FLUX SENSITIVITY RATIO FOR A COOLED-FILM SENSOR.

RADIATIVE TRANSFER BETWEEN TWO CONCENTRIC GRAY SPHERES SEPARATED BY RADIATING GAS

R. Viskanta*

ABSTRACT

The problem of energy transfer through a spherical shell of gray absorbing-emitting and uniformly heat generating medium confined between two concentric spheres is studied. The spheres are considered to be opaque, diffuse as well as gray, and are maintained at different but uniform temperatures. Radiative transfer theory is used to formulate the problem rigorously. A superposition method is employed to transform the energy equation into two singular Fredholm integral equations of the second kind for two universal functions. The numerical solution of these equations is obtained by the method of successive approximations. Results are reported graphically as well as in a tabular form and are compared with available solutions.

INTRODUCTION

This paper analyzes steady state radiative transfer between two concentric, gray, opaque spheres separated by a gray absorbing and emitting medium which is generating heat uniformly. The primary objective is the precise determination of the local emissive power distribution and the local radiation flux. The secondary objective is the determination of the optical conditions under which radiation transfer through an absorbing and emitting medium contained between two gray opaque spheres can be approximated by that through a plane layer contained between two heated, opaque parallel walls.

The basic theory of radiative transfer is developed in books by Chandrasekhar,[1] Kourganoff,[2] and Sobolev[3] and deals primarily with plane layers. A number of specific adaptations of the theory and results to radiative transfer problems of engineering nature have recently been reported, and the book by Sparrow and Cess[4] furnishes an up-to-date survey of the applications and calculations.

Radiative transfer in curvilinear geometries has received comparatively little attention. Certain aspects of such problems are discussed by Chandrasekhar[1] and Kuznetsov[5] in connection with radiative transfer in spherical atmospheres. The transfer of radiation in a homogeneous spherical medium has been considered and exact solutions have been reported by Heaslet and Warming.[6,7] Heat transfer by radiation between two concentric black spheres maintained at the same temperatures and containing an absorbing, emitting, and heat generating gas have been predicted numerically by Sparrow, et al.[8] The transfer of radiant energy between two

*Professor of Mechanical Engineering, Purdue University, Lafayette, Indiana.

black spheres kept at different temperatures and separated by an absorbing emitting gas has been studied by Rhyming.[9] Radiant heat transfer in a spherical shell has been analyzed by Chisnell,[10] and unsteady heating and cooling of a spherical mass of gas by radiation has been considered by Viskanta and Lall.[11,12] In the above references, except 1 and 5, the analyses are restricted to constant absorption. The transformation from physical to optical radial distance in a spherically symmetric problem lacks utility and the variation of the absorption coefficient with distance introduces extreme computational difficulties; therefore results have been reported only for constant absorption.

The analysis of radiative transfer is made difficult by the basic non-localized character of the phenomenon. Therefore, for many practical problems in which complicated curvilinear geometries arise, the physical model has to be idealized for the sake of simplicity rather than reality. For example, the shocked gas region in front of a blunt body has been approximated by a plane layer.[13,14] The validity of approximations of this type has not been established, and the effects of curvature on radiant heat transfer have not been estimated. Furthermore, some of the approximate techniques which have been developed and used with success for analyzing radiation transfer for one-dimensional plane layers do not appear to be as good for the situations in which the flux is not constant.[10,15,16] This study has been undertaken with the intention of obtaining precise solutions for radiation transfer in a spherically symmetric geometry so that the results would be used to estimate the effects of curvature and to check the validity of the various approximate methods of solution. The present study extends and supplements the work of references 6 through 10.

ANALYSIS

Description of the System

The physical system considered is shown schematically in Figure 1. It consists of two concentric spheres. The radial distance is denoted by r and the inner and the outer radii by r_1 and r_2, respectively. The space between the two spheres is filled with an absorbing and emitting medium which is considered capable of generating heat at a uniform rate per unit of volume, S. The medium is characterized by a constant solumetric absorption coefficient, κ, which is independent of frequency, i.e., the medium is taken as gray. The index of refraction of the medium is considered to by unity; however, the analysis and results would also be valid for an arbitrary index of refraction as long as it is independent of temperature.

The problem is completely defined by prescribing the physical conditions at the inner and outer surfaces (denoted by subscripts 1 and 2, respectively). The opaque walls are assumed to be gray and the temperatures, T, and emissivities, ϵ, of the bounding surfaces are taken to be constant but different. The problem is to predict the temperature distribution and the radiative flux.

Conservation of Energy Equation

For a problem possessing spherical symmetry, the steady state conservation and radiant energy equations can be expressed as

$$\kappa(4\pi\, I_b - \mathcal{G}) = S \tag{1}$$

and

$$\frac{1}{r^2}\frac{d}{dr}(r^2\, \mathcal{F}) = \kappa(4\pi\, I_b - \mathcal{G}), \tag{2}$$

where I_b is Planck's function. The local radiation flux is defined as

$$\mathcal{F}(r) = \int_{\Omega=4\pi} I(r,\mu)\, \mu d\Omega$$

$$= 2\pi \int_{-1}^{1} I(r,\mu)\, \mu d\mu \tag{3}$$

and the incident radiation as

$$\mathcal{G}(r) = \int_{\Omega=4\pi} I(r,\mu)\, d\Omega$$

$$= 2\pi \int_{-1}^{1} I(r,\mu)\, d\mu, \tag{4}$$

where $I(r,\mu)$ is the intensity of radiation. The physical meaning of Equation (2) can be clarified by noting that the term $4\pi\kappa I_b$ represents the local rate of emission of radiant energy per unit of volume, and $\kappa\mathcal{G}$ represents the local rate of absorption of energy per unit of volume. The intensity of radiation $I(r,\mu)$ at point r and making an angle θ ($\mu = \cos\theta$) between the pencil of radiation and the extension of the radius, r, is governed by the equation of transfer:[1]

$$\mu \frac{\partial I}{\partial r} + \frac{(1-\mu^2)}{r}\frac{\partial I}{\partial \mu} = \kappa\left[I_b(r) - I(r,\mu)\right]. \quad (5)$$

The boundary conditions on the intensity at the two opaque and diffuse walls can be approximated by

$$I(r,\mu) = I(r_1,\mu) = I_1 \text{ for all } \mu\ , \quad (6)$$

$$I(r,\mu) = I(r_2,\mu) = I_2 \text{ for all } \mu\ . \quad (7)$$

The intensities of radiation, I_1 and I_2, leaving the inner and outer spheres, respectively, are considered to be independent of direction and represent the contributions both due to emission and reflection of radiation.

The solution of the equation of transfer (not restricted to a constant absorption coefficient) with the boundary conditions (6) and (7) has been obtained by Kuznetsov,[5] and the lengthy mathematical details need not be repeated here. For the case of constant absorption coefficient, the incident radiation, $\mathcal{G}$, can be shown to be[5]

$$\mathcal{G}(\tau) = \frac{2}{\tau}\left[F_1 g_1(\tau) + F_2 g_2(\tau) + \int_{\tau_1}^{\tau_2} K(\tau,t)\, E_b(t)\, dt\right], \quad (8)$$

where the optical radial distance, τ, is defined as

$$\tau = \int_0^r \kappa dr = \kappa r\ , \quad (9)$$

and the inner and outer optical radii as

$$\tau_1 = \kappa r_1 \quad \text{and} \quad \tau_2 = \kappa r_2\ .$$

In Equation (8),

$$F_1 = \pi I_1$$

and

$$F_2 = \pi I_2$$

represent the radiative fluxes leaving inner and outer surfaces, respectively, and

$$E_b = \sigma T^4$$

is the black body emissive power, where σ is the Stefan-Boltzmann constant and T is the absolute temperature. The functions $g_1(\tau)$, $g_2(\tau)$, and the kernel, $K(\tau,t)$, are defined as

$$g_1(\tau) = \tau_1 E_2(\tau - \tau_1) - E_3(\tau - \tau_1) + E_3\left(\sqrt{\tau^2 - \tau_1^2}\right), \quad (10)$$

$$g_2(\tau) = \tau_2 E_2(\tau_2 - \tau) + E_3(\tau_2 - \tau) - \sqrt{\tau_2^2 - \tau_1^2}\, E_2\left(\sqrt{\tau_2^2 - \tau_1^2} + \sqrt{\tau^2 - \tau_1^2}\right) - E_3\left(\sqrt{\tau_2^2 - \tau_1^2} + \sqrt{\tau^2 - \tau_1^2}\right), \quad (11)$$

and

$$K(\tau,t) = \left[E_1\left(|\tau - t|\right) - E_1\left(\sqrt{\tau^2 - \tau_1^2} + \sqrt{t^2 - \tau_1^2}\right)\right] t, \quad (12)$$

where the exponential integral function $E_n(\tau)$ is defined as

$$E_n(\tau) = \int_0^1 \mu^{n-2} e^{-\tau/\mu} d\mu = \int_1^\infty e^{-\tau u} u^{-n} du. \quad (13)$$

Substitution of Equation (8) into Equation (2), and integration with respect to τ from $\tau = \tau_1$ to $\tau = \tau_2$ yields, after interchanging the order of integration,

$$\tau^2 \mathcal{G}(\tau) = 2\left[F_1 h_1(\tau) + F_2 h_2(\tau) + \int_{\tau_1}^{\tau_2} H(\tau,t)\, E_b(t)\,dt\right], \tag{14}$$

where

$$h_1(\tau) = \tau\tau_1 E_3(\tau-\tau_1) - (\tau-\tau_1)\, E_4(\tau-\tau_1) - E_5(\tau-\tau_1)$$
$$+\sqrt{\tau^2-\tau_1^2}\; E_4\left(\sqrt{\tau^2-\tau_1^2}\right) + E_5\left(\sqrt{\tau^2-\tau_1^2}\right), \tag{15}$$

$$h_2(\tau) = -\tau\tau_2 E_3(\tau_2-\tau) + (\tau_2-\tau)\, E_4(\tau_2-\tau) + E_5(\tau_2-\tau)$$
$$-\sqrt{\tau_2^2-\tau_1^2}\,\sqrt{\tau^2-\tau_1^2}\; E_3\left(\sqrt{\tau_2^2-\tau_1^2}+\sqrt{\tau^2-\tau_1^2}\right)$$
$$-\left(\sqrt{\tau_2^2-\tau_1^2}+\sqrt{\tau^2-\tau_1^2}\right) E_4\left(\sqrt{\tau_2^2-\tau_1^2}+\sqrt{\tau^2-\tau_1^2}\right)$$
$$-\,E_5\left(\sqrt{\tau_2^2-\tau_1^2}+\sqrt{\tau^2-\tau_1^2}\right), \tag{16}$$

and

$$H(\tau,t) = \Bigg[\tau\,\mathrm{sign}(\tau-t)\, E_2\left(|\tau-t|\right) + E_3\left(|\tau-t|\right)$$
$$-\sqrt{\tau^2-\tau_1^2}\; E_2\left(\sqrt{t^2-\tau_1^2}+\sqrt{\tau^2-\tau_1^2}\right) - E_3\Big(\sqrt{t^2-\tau_1^2}$$
$$+\sqrt{\tau^2-\tau_1^2}\Big)\Bigg]\, t. \tag{17}$$

We note that even though Equations (8) and (14) are expressed in terms of different functions, they are identical to those derived by Rhyming.[9]

As a result of the previous formulas, the conservation of energy equation, Equation (1), can be expressed as

$$E_b(\tau) = \left(\frac{S}{4\kappa}\right) + \frac{1}{2\tau}\left[F_1 g_1(\tau) + F_2 g_2(\tau) + \int_{\tau_1}^{\tau_2} K(\tau,t)\, E_b(t)\,dt\right]. \tag{18}$$

The above integral equation fixes the temperature distribution, i.e., $E_b(\tau)$. The solution of this equation is quite a formidable task. Additional mathematical complexities result from the fact that the radiative fluxes, F_1 and F_2, leaving the inner and outer walls must be explicitly written in terms of temperatures and emissivities. Expressions for these fluxes will be given later.

Solution of the Energy Equation

Inspection of Equation (18) reveals that the energy equation has six independent parameters: τ_1, τ_2, F_1, F_2, S, and κ. In addition, the radiative fluxes, F_1 and F_2, leaving surfaces 1 and 2, respectively, cannot in general be evaluated before the temperature distribution has been completely determined. In view of this, a very large number of individual solutions are required to cover the complete range of parameters of physical interest. Since the energy equation is linear in E_b, this difficulty can be partly overcome by subdividing the general problem into the following subproblems:[17,18]

Problem (1):
Inner sphere at 0
Outer sphere at $F_2 - F_1$
No internal heat generation

Problem (2):
Inner sphere at F_1
Outer sphere at F_1
Internal heat generation S

The transformation

$$E_b(\tau) = F_1 + (F_2 - F_1)\,\varphi(\tau) + \left(\frac{S}{\kappa}\right)\varphi_s(\tau)\ , \tag{19}$$

where

$$\varphi(\tau) = \frac{E_b(\tau) - F_1}{F_2 - F_1} \tag{20}$$

and

$$\varphi_s(\tau) = \frac{E_b(\tau) - F_1}{S/\kappa} \tag{21}$$

permits one to express the solution of the linear integral equation in terms of the two independent singular Fredholm integral equations of the second kind:

$$\varphi(\tau) = \frac{1}{2\tau}\left[g_2(\tau) + \int_{\tau_1}^{\tau_2} K(\tau,t)\,\varphi(t)\,dt\right] \tag{22}$$

and

$$\varphi_s(\tau) = \frac{1}{2\tau}\left[\frac{\tau}{2} + \int_{\tau_1}^{\tau_2} K(\tau,t)\,\varphi_s(t)\,dt\right]. \tag{23}$$

Substitution of Equation (19) into Equation (14) results in the following expression for the local radiation flux:

$$\tau^2 \mathcal{F}(\tau) = (F_2 - F_1)\,Q(\tau) + \left(\frac{S}{\kappa}\right) Q_s(\tau)\ , \tag{24}$$

where

$$Q(\tau) = 2\left[h_2(\tau) + \int_{\tau_1}^{\tau_2} H(\tau,t)\,\varphi(t)\,dt\right] \tag{25}$$

and

$$Q_s(\tau) = 2\int_{\tau_1}^{\tau_2} H(\tau,t)\,\varphi_s(t)\,dt\ . \tag{26}$$

Equations (22) and (23) represent the dimensionless energy equations for subproblems 1 and 2, respectively. The functions $\varphi(\tau)$ and $\varphi_s(\tau)$ are universal functions of the optical radial distance, τ, and the optical radii, τ_1 and τ_2, only. Equations (22) and (23) are independent of the parameters (except τ_1 and τ_2) affecting the particular conditions to be specified at the two walls and may be considered as universal functions in terms of black wall conditions, i.e., $\varepsilon_1 = \varepsilon_2 = 1$, as has been done by Heaslet and Warming for the planar medium.[17]

An integral equation of the same form as Equation (22) has been obtained by Rhyming[9] who has considered radiation transfer between two concentric black spheres. Unfortunately, he has presented his results in terms of temperature rather than the more general function $\varphi(\tau)$. Rhyming has solved the integral equation by the method of undetermined parameters and has reported results for three values of inner to outer wall temperature ratios, T_1/T_2 = 2, 5, and 25. The numerical solution of the integral Equation (23) has been obtained by Sparrow, et al.,[8] who have considered radiation transfer between two concentric black spheres maintained at the same temperatures and containing a radiating and heat generating gas. Closed form analytical solutions of Equations (22) and (23) do not appear possible; however, exact solutions can be obtained numerically. Before proceeding to the numerical method of solution it is advantageous to discuss limiting cases.

Limiting Solutions

Exact closed form solutions can be obtained for the optically thin ($\tau_2 \rightarrow 0$) and thick ($\tau_2 \rightarrow \infty$) conditions. These limiting cases are well known and the details need not be given.

According to Kourganoff,(2) $\tau E_1(\tau) \to 0$ as $\tau \to 0$ and the $E_2(\tau)$ and $E_3(\tau)$ functions have the following asymptotic expansions as $\tau \to 0$:

$$E_2(\tau) = 1 + 0(\tau)$$

and

$$E_3(\tau) = \frac{1}{2} - \tau + 0(\tau^2) \ .$$

Thus, with aid of these expressions, the solutions of Equations (22) and (23) in the optically thin limit ($\tau_2 \to 0$) become

$$\tau\varphi(\tau) = \frac{1}{2}\left(\tau + \sqrt{\tau^2 - \tau_1^2}\right) \tag{27}$$

and

$$\tau\varphi_s(\tau) = \frac{\tau}{4} \ , \tag{28}$$

respectively. According to Equation (27) there is a slip (jump) of temperature at the two boundaries. This is consistent with the results for the similar limiting case of an optically thin medium confined between two parallel plates.

For the optically thick limiting case, ($\tau_2 \to \infty$), the solutions of Equations (22) and (23) are

$$\tau\varphi(\tau) = \tau_2\left(\frac{\tau - \tau_1}{\tau_2 - \tau_1}\right) \tag{29}$$

and

$$\tau\varphi_s(\tau) = \frac{1}{8}\Big[\tau(\tau_2^2 - \tau^2) - \tau_1(\tau_2 + \tau_1)(\tau_2 - \tau)\Big] \ , \tag{30}$$

respectively. These solutions can be verified by direct substitution of Equations (29) and (30) into Equations (22) and (23), respectively, or more simply by use of the Rosseland diffusion approximation as is done, for example, in References 9 or 17.

As expected, in this limiting case the temperature distribution is continuous.

Radiative Flux and Temperature Distribution

With the functions $\varphi(\tau)$ and $\varphi_s(\tau)$ determined, it remains only to give explicit expressions for the temperature distributions and radiative fluxes in terms of them and the surface temperatures and emissivities.

Energy balances at gray, diffuse walls 1 and 2 yield the following expressions for the radiative fluxes, F_1 and F_2, respectively,

$$F_1 = E_{b_1} - \left(\frac{1}{\epsilon_1} - 1\right)\mathcal{F}_1 \tag{31}$$

and

$$F_2 = E_{b_2} + \left(\frac{1}{\epsilon_2} - 1\right)\mathcal{F}_2 \ , \tag{32}$$

where

$$\mathcal{F}_1 = \mathcal{F}(\tau_1)$$

and

$$\mathcal{F}_2 = \mathcal{F}(\tau_2) \ .$$

Combining Equation (1) with Equation (2) and integrating, we get

$$\tau^2\mathcal{F}(\tau) = \left(\frac{S}{3\kappa}\right)\tau^3 + \text{const.} \tag{33}$$

Thus, in the absence of internal heat generation, the local radiation flux, $\mathcal{F}(\tau)$, is inversely proportional to τ^2. It follows from Equation (33) that

$$\mathcal{F}_2 = \left(\frac{\tau_1}{\tau_2}\right)^2\mathcal{F}_1 + \left(S\frac{\tau_2}{3\kappa}\right)\left[1 - \left(\frac{\tau_1}{\tau_2}\right)^3\right] . \tag{34}$$

Subtracting Equation (31) from Equation (32) and making use of Equation (34) to eliminate $\mathcal{F}_2$ in favor of $\mathcal{F}_1$, we obtain:

$$F_2 - F_1 = E_{b_2} - E_{b_1} + \left[\left(\frac{1}{\epsilon_1} - 1\right) + \left(\frac{1}{\epsilon_2} - 1\right)\left(\frac{\tau_1}{\tau_2}\right)^2\right]\mathcal{F}_1 + \left(\frac{S\,\tau_2}{3\kappa}\right)\left(\frac{1}{\epsilon_1} - 1\right)\left[1 - \left(\frac{\tau_1}{\tau_2}\right)^3\right] . \tag{35}$$

To calculate $\mathcal{F}_1$ we specialize Equation (24) by letting $\tau = \tau_1$, and obtain

$$\tau_1^2 \mathcal{F}(\tau_1) = \tau_1^2 \mathcal{F}_1 = (F_2 - F_1)\, Q(\tau_1) + \left(\frac{S}{\kappa}\right) Q_s(\tau_1) \, . \tag{36}$$

Eliminating $F_2 - F_1$ between Equations (35) and (36) yields for the radiation flux at the inner sphere $\mathcal{F}_1$.

$$\mathcal{F}_1 = \frac{\left(E_{b_1} - E_{b_2}\right) Q(\tau_1) + (S/\kappa)\left\{(\tau_2/3)(1/\varepsilon_1 - 1)\left[1 - (\tau_1/\tau_2)^3\right] Q(\tau_1) + Q_s(\tau_1)\right\}}{\tau_1^2 - \left[(1/\varepsilon_1 - 1) + (1/\varepsilon_2 - 1)(\tau_1/\tau_2)^2\right] Q(\tau_1)} \, . \tag{37}$$

The desired temperature distribution is obtained by combining Equations (19), (31), and (35). Finally, one obtains

$$\begin{aligned} E_b(\tau) = {} & E_{b_1} + \left(E_{b_2} - E_{b_1}\right) \varphi(\tau) - \left(\frac{1}{\varepsilon_1} - 1\right) \mathcal{F}_1 \\ & + \left[\left(\frac{1}{\varepsilon_1} - 1\right) + \left(\frac{1}{\varepsilon_2} - 1\right)\left(\frac{\tau_1}{\tau_2}\right)^2\right] \mathcal{F}_1\, \varphi(\tau) \\ & + \left(\frac{S}{\kappa}\right)\left\{\varphi_s(\tau) + \left(\frac{\tau_2}{3}\right)\left(\frac{1}{\varepsilon_2} - 1\right)\left[1 - \left(\frac{\tau_1}{\tau_2}\right)^3\right] \varphi(\tau)\right\} . \end{aligned} \tag{38}$$

The radiation flux at the inner wall, $\mathcal{F}_1$, and the temperature distribution, $E_b(\tau)$, have been expressed in terms of the physical parameters of the problem

$$T_1 \left(E_{b_1}\right), \quad T_2 \left(E_{b_2}\right),$$

ε_1, ε_2, τ_1, τ_2, S/κ, τ and the universal functions $\varphi(\tau)$, $\varphi_s(\tau)$, $Q(\tau_1)$, and $Q_s(\tau_1)$. The local radiation flux, $\mathcal{F}(\tau)$, can be simply evaluated from Equation (33). The constant is obtained by letting $\tau = \tau_1$ and there results

$$\tau^2 \mathcal{F}(\tau) = \tau_1^2 \mathcal{F}_1 + \left(\frac{S}{3\kappa}\right)(\tau^3 - \tau_1^3). \tag{39}$$

Method of Solution

The successive approximation method was used to solve integral Equations (22) and (23). When applied to integral Equation (22), it yields

$$\varphi_n(\tau) = \frac{1}{2\tau}\left[g_2(\tau) + \int_{\tau_1}^{\tau_2} K(\tau, t)\, \varphi_{n-1}(t)\, dt\right] \quad n = 1, 2, 3, \ldots , \tag{40}$$

where $\varphi_o(\tau)$ is the zeroth approximation. While this procedure can be carried out analytically, the process is very tedious; and, therefore, the method is performed numerically on a digital computer.

Unfortunately, there are difficulties associated with the numerical integration, since the exponential integral function $E_1(\tau)$ has a logarithmic singularity at $\tau = 0$. Thus the kernel, $K(\tau,t)$, becomes singular at $t = \tau$. In order to eliminate the numerical difficulty, the integral is rewritten in the form

$$\int_{\tau_1}^{\tau_2} f(t)\, E_1\left(|\tau - t|\right) dt = \int_0^1 \zeta_1(\tau,t)\, dt + \int_0^1 \zeta_2(\tau,t)(-\ell n\, t)\, dt \ , \qquad (41)$$

where $f(t)$, $\zeta_1(\tau,t)$, and $\zeta_2(\tau,t)$ are finite over the range of interest. The first integral does not possess any singularities and is evaluated using the standard Gaussian quadrature.[19] The singularity in the second integral is avoided by use of the special Gaussian quadrature,[19]

$$\int_0^1 \zeta_2(\tau,t)(-\ell n\, t)\, dt = \sum_{i=1}^{m} A_i\, \zeta_2(\tau,t_i) \ , \qquad (42)$$

where A_i and t_i are the Gaussian weights and points, respectively.

For a uniform homogeneous heat generating medium ($\tau_1 = 0$) the integral Equation (23) possesses another numerical difficulty. In the form given in Equation (23), $\varphi_s(0)$ cannot be found numerically. Since $\varphi_s(\tau)$ is finite for all $\tau(0 \le \tau \le \tau_2)$, the limit of

$$\int_0^{\tau_2} K(\tau,t)\, \varphi_s(t)\, dt$$

as $t \to 0$ is zero. Applying L'Hospital's rule, the expression for $\varphi_s(0)$ is found to be

$$\varphi_s(0) = \frac{1}{4} + \int_0^{\tau_2} e^{-t}\, \varphi_s(t)\, dt \ . \qquad (43)$$

Obviously, when the method of successive approximations is carried out numerically, the continuous function, $\varphi(\tau)$, must be replaced by discrete values. Instead of dividing the interval ($\tau_1 \le \tau \le \tau_2$) into equal subdivisions, the size of the subdivisions near the boundaries is decreased.

Lastly, the question of when to terminate the procedure arises. The error criterion

$$\left| \frac{\varphi_n(\tau) - \varphi_{n-1}(\tau)}{\varphi_n(\tau)} \right| < \epsilon$$

must be used with caution. If the sequence $\varphi_n(\tau)$, ($n = 0, 1, 2 \ldots$) convergences slowly to $\varphi(\tau)$, this error check will terminate the method too quickly. As an additional check the flux integrals $Q(\tau)$ and $Q_s(\tau) - \tau^3/3$ are evaluated throughout the interval $\langle \tau_1, \tau_2 \rangle$ to ensure that they remain constant for subproblems (1) and (2), respectively. This readily follows from Equations (24) and (39).

RESULTS AND DISCUSSION

The "exact" numerical solutions to Equations (22) and (23) have been obtained for a wide range of parameters of physical interest and are presented graphically in Figures 2 through 8 and in Tables 1 through 4. As remarked previously, some comparable results have already been reported. The

TABLE 1

VALUES OF THE FUNCTION $\phi(\tau)$ AT THE TWO WALLS AND THE FLUX INTEGRAL $Q(\tau)$ FOR $\tau_2 = 1.0$

τ_1	$\varphi(\tau_1)$	$\varphi(\tau_2)$	$-Q(\tau)$
0.01	0.4986	1.0000	$0.996(10)^{-4}$
0.05	0.4906	0.9992	$0.246(10)^{-2}$
0.10	0.4824	0.9971	$0.9680(10)^{-2}$
0.20	0.4674	0.9887	$0.3764(10)^{-1}$
0.30	0.4557	0.9750	$0.8282(10)^{-1}$
0.40	0.4472	0.9558	0.1449
0.50	0.4420	0.9306	0.2244
0.60	0.4403	0.8982	0.3226
0.70	0.4425	0.8566	0.4422
0.80	0.4497	0.8012	0.5874
0.90	0.4645	0.7208	0.7662
0.95	0.4770	0.6591	0.8732
0.99	0.4929	0.5722	0.9720

TABLE 2

VALUES OF FUNCTION $\phi(\tau)$ AT THE TWO WALLS AND FLUX INTEGRAL $Q(\tau)$ FOR DIFFERENT RATIOS OF INNER TO OUTER OPTICAL RADII τ_1/τ_2

(a) $\tau_1/\tau_2 = 0.1$

τ_2	$\varphi(\tau_1)$	$\varphi(\tau_2)$	$-Q(\tau)$
0.1	0.4983	0.9974	$0.9970(10)^{-4}$
0.5	0.4913	0.9973	$0.2461(10)^{-2}$
1.0	0.4824	0.9971	$0.9680(10)^{-2}$
5.0	0.4097	0.9968	0.2079
10.0	0.3329	0.9972	0.6839

(b) $\tau_1/\tau_2 = 0.5$

τ_2	$\varphi(\tau_1)$	$\varphi(\tau_2)$	$-Q(\tau)$
0.1	0.4941	0.9325	$0.2475(10)^{-2}$
0.5	0.4707	0.9311	$0.5930(10)^{-1}$
1.0	0.4420	0.9306	0.2244
2.0	0.3890	0.9325	0.8006
5.0	0.2737	0.9449	3.623
10.0	0.1763	0.9613	9.585
20.0	0.1011	0.9767	22.50

(c) $\tau_1/\tau_2 = 0.9$

τ_2	$\varphi(\tau_1)$	$\varphi(\tau_2)$	$-Q(\tau)$
0.1	0.4963	0.7180	$0.8056(10)^{-2}$
0.5	0.4818	0.7187	0.1970
1.0	0.4645	0.7208	0.7662
2.0	0.4324	0.7275	2.898
5.0	0.3563	0.7554	15.44
10.0	0.2755	0.7985	49.22
20.0	0.1912	0.8537	139.7

(d) $\tau_1/\tau_2 = 0.95$

τ_2	$\varphi(\tau_1)$	$\varphi(\tau_2)$	$-Q(\tau)$
0.1	0.4976	0.6563	$0.8995(10)^{-2}$
0.5	0.4883	0.6573	0.2219
1.0	0.4770	0.6591	0.8732
2.0	0.4557	0.6643	3.379
5.0	0.4017	0.6856	19.18
10.0	0.3372	0.7226	66.42
20.0	0.2589	0.7785	209.7
40.0	0.1799	0.8423	592.2

functions $\varphi(\tau)$ and $\varphi_s(\tau)$ were calculated to a degree of accuracy in excess of three significant figures. The values of the function $\varphi_s(\tau)$ checked with the solutions given in References 6 and 8 and were found to agree with the graphical results with no discernible differences.

Emissive Power Distributions

The dimensionless emissive power distributions, $\varphi(\tau)$, for subproblem (1) are given in Figures 2 through 6. The effect on $\varphi(\tau)$ of decreasing the inner optical radius, τ_1, with the outer optical radius fixed at unity, is illustrated in Figure 2. The influence of radius ratio τ_1/τ_2 on $\varphi(\tau)$ is shown in Figures 3 through 6 and corresponds to ratios of 0.1, 0.5, 0.9, and 0.95, respectively. In each figure the "exact" numerical solutions are displayed as unbroken lines, and the thin and thick approximations according to Equations (27) and (29) are shown by broken lines. As expected, the effect of curvature increases with decreasing τ_1/τ_2. It is seen from Figure 2 that even for $\tau_1 = 0.99\,\tau_2$ the effect of curvature is still evident. The optically thin approximation appears to be most useful for small values of τ_2 and for both small and large ratios of τ_1/τ_2. On the other hand, the validity of the optically thick approximation appears to be much more limited. In examining the results presented in Figures 2 and 3, we note that for small values of τ_1/τ_2 the emissive power distribution and therefore the temperature changes rapidly within a narrow region close to the inner sphere resembling the laminar boundary layer effect. Concurrently, the energy jump at the inner sphere is much larger than that at the outer sphere. Examination of the results presented in Table 1 reveals that while $\varphi(\tau_2)$ increases with decreasing τ_1, $\varphi(\tau_1)$ has a minimum value of about 0.440 at $\tau_1 \simeq 0.6$. Thus it appears that for a fixed τ_2 there is a radius τ_1 at which the energy jump at the inner sphere is minimum. As τ_1/τ_2 increases the emissive power varies much more gradually between the two spheres, but the energy jump at the outer sphere increases. When τ_1 approaches τ_2 the variation of $\varphi(\tau)$ becomes similar to that of an analogous function governing radiative transfer through a plane layer of a radiating medium confined between two infinitely large parallel plates.[17] However, inspection of Figure 6 shows that even for

$$\frac{\tau_1}{\tau_2} = 0.95$$

the curvature effect is appreciable, and there is considerable difference between the results for the two geometries.

Equation (23) has also been solved for a range of optical radii, τ_2, and several ratios of τ_1/τ_2. The solutions were primarily obtained for comparison with the previously reported results and for estimation of the accuracy of the method. The function $\varphi_s(\tau)$ was compared with the exact solutions given by Heaslet and Warming[6] for the limiting case of

$$\frac{\tau_1}{\tau_2} = 0$$

and with the numerical results of Sparrow, *et al.*,[8] who have reported graphically the distribution of $\varphi_s(\tau)$ for several values of τ_2 between 0.1 and 2.0. The functions $\varphi_s(\tau)$ obtained agreed with the results presented graphically in the above references and therefore are not reproduced here. Some of the values of $\varphi_s(\tau)$ at the two walls and the flux integral $Q_s(\tau) - \tau^3/3$ obtained in this study are given in Tables 3 and 4. Inspection of Table 3 reveals good agreement between the values for $\varphi_s(\tau_2)$ given by Heaslet and Warming[6] and those of this paper. Since the prediction of $\varphi_s(\tau)$ at the walls and therefore of the slip is quite critical because the function $\varphi_s(\tau)$ is logarithmically singular at the boundaries, the small discrepancy between the exact and the numerical solutions is very encouraging. The results for the flux integral given in Table 4 also show good agreement. These comparisons establish the validity and further substantiate the accuracy of the solutions.

In a recent report Chou[20] has studied radiative transfer between concentric spheres and coaxial cylinders by what he termed the "method of regional averaging." He has reported values of 0.847×10^{-4} and 0.830 of the flux integrals

$$-Q_s(\tau) + \frac{\tau^3}{3}$$

for $\tau_2 = 0.1$ and 2.0, respectively, at

$$\frac{\tau_1}{\tau_2} = 0.5.$$

TABLE 3

VALUES OF THE FUNCTION $\phi_s(\tau)$ AT THE CENTER AND THE BOUNDARY FOR UNIFORM HEAT GENERATION AND $\tau_1 = 0$

τ_2	$\varphi_s(0)$	$\varphi(\tau_2)$ Present	$\varphi(\tau_2)$ Reference 6
0.1	0.2759	0.2626	
0.25	0.3184	0.2816	
0.50	0.3989	0.3138	0.314
0.75	0.4922	0.3466	
1.00	0.5989	0.3798	0.380
1.50	0.8537	0.4471	0.447
2.00	1.167	0.5158	0.516
3.00	1.972	0.6551	0.656
5.00	4.323	0.9379	0.939

TABLE 4

VALUES OF THE FUNCTION $\phi_s(\tau)$ AT THE WALLS AND THE FLUX INTEGRAL FOR UNIFORM HEAT GENERATION AND $\tau_1/\tau_2 = 0.5$

τ_2	$\varphi_s(\tau_1)$	$\varphi_s(\tau_2)$	$-Q_s(\tau) + \tau^3/3$ Present	$-Q_s(\tau) + \tau^3/3$ Reference 8
0.1	0.2585	0.2604	$0.8020(10)^{-4}$	$0.802(10)^{-4}$
0.25	0.2718	0.2759	$0.1276(10)^{-2}$	
0.50	0.2957	0.3017	$0.1049(10)^{-1}$	$0.104(10)^{-1}$
0.75	0.3214	0.3273	$0.3630(10)^{-1}$	
1.00	0.3490	0.3526	$0.8812(10)^{-1}$	$0.880(10)^{-1}$
1.50	0.4081	0.4026	0.3098	
2.00	0.4723	0.4517	0.7619	0.765
3.00	0.6082	0.5472	2.719	
5.00	0.9010	0.7332	13.47	

These results are a few per cent lower than those given in Table 4; however, it is not possible to judge the general validity of the method on the basis of the limited results available.

Radiative Fluxes

Once $\varphi(\tau)$ and $\varphi_s(\tau)$ have been determined, Equations (37), (38), and (39) permit ready calculation of the radiative fluxes at the two walls and the temperature distributions for arbitrary physical conditions. The temperature slip at the inner and outer spheres also follow immediately from Equation (38). For the sake of brevity only the radiative fluxes for the case of black walls are presented graphically.

From a practical point of view, the most interesting results are given in Figure 7, in which the radiative fluxes at the inner and outer spheres are compared with the radiative flux across a plane layer confined between two infinitely large parallel plates. The radiative fluxes across the planar medium fall between the fluxes at the two walls. Even for a very small spacing, $\tau_2 - \tau_1 = 0.01$ and $\tau_1/\tau_2 = 0.95$, the difference between the flux across a plane layer and the flux at the outer sphere of the spherical shell is still greater than 10 per cent. The radiative fluxes at the inner and outer spheres approach the radiative flux for a planar medium as $\tau_1/\tau_2 \to 1$ and $\tau_2 - \tau_1$ becomes large.

The radiative flux based on the Rosseland diffusion approximation can be obtained by accounting for the temperature slip at the walls as suggested by Deissler (see Reference 21). Following the procedure outlined by him, one can obtain the following expression for the radiative flux at the inner wall:

$$-\frac{Q(\tau_1)}{\tau_1^2} = \left[\left(\frac{1}{\varepsilon_1} - \frac{1}{2}\right) + \left(\frac{1}{\varepsilon_2} - \frac{1}{2}\right)\left(\frac{\tau_1}{\tau_2}\right)^2 + \frac{3}{8}\tau_1\left[1 - \left(\frac{\tau_1}{\tau_2}\right)^3\right] + \frac{3}{4}\tau_1\left(1 - \frac{\tau_1}{\tau_2}\right)\right]^{-1} . \quad (44)$$

The predictions of Equation (44) are compared with the exact results in Figure 8. It is seen from the figure that the validity of Equation (44) is much more limited than of a similar expression for the radiative flux across a plane layer.[21] The present findings are in agreement with the results reported by Perlmutter and Howell[15] for radiative transfer through a cylindrical shell of radiating medium confined between two coaxial cylinders. Calculations not included here have shown that as the emissivity of the walls is decreased the agreement between the exact results and those based on Equation (44) becomes better.

For small optical radii, τ_2, very simple approximations to the Equations (22) and (23) can be obtained by expanding $\varphi(t)$ and $\varphi_s(t)$ in a Taylor series around $t = \tau$ and retaining only the first term. The resulting expressions for $\varphi(\tau)$ and $\varphi_s(\tau)$ are used to evaluate the flux integrals at $\tau = \tau_1$. The results based on these approximations are given in Tables 5 and 6. Comparison of the flux integrals with the exact results given in Tables 2, 3, and 4 shows a surprisingly good agreement even for intermediate values of τ_2.

CONCLUSIONS

Based on the solutions of the rigorously formulated problem the following conclusions may be drawn:

(1) The results obtained by the present method of successive approximations agree very well with the existing exact and numerical solutions. The validity of the optically thin and thick approximations is very limited.

(2) The effect of curvature on the dimensionless emissive power distributions and the radiative fluxes is appreciable even when the inner optical radius is close to the outer optical radius, i.e., $\tau_1 = 0.95\ \tau_2$.

(3) The range of validity of the expression for the radiative flux based on the Rosseland diffusion approximation with the jump boundary condition is much more restricted than for the case when the local flux is constant.

(4) A number of assumptions have been introduced in the formulation of the problem, and it is recognized that some of them may be unrealistic under certain conditions. The analysis is, however, based on exact

TABLE 5

VALUES OF FUNCTION $\phi(\tau)$ AT THE WALLS AND FLUX INTEGRAL $Q(\tau)$ BASED ON THE TAYLOR SERIES APPROXIMATION

(a) $\tau_1/\tau_2 = 0.5$

τ_2	$\varphi(\tau_1)$	$\varphi(\tau_2)$	$-Q(\tau_1)$
0.1	0.4835	0.9344	$0.2474(10)^{-2}$
0.5	0.4183	0.9410	$0.5913(10)^{-1}$
1.0	0.3415	0.9506	0.2216
2.0	0.2138	0.9684	0.7555
5.0	0.03994	0.9937	2.400

(b) $\tau_1/\tau_2 = 0.9$

τ_2	$\varphi(\tau_1)$	$\varphi(\tau_2)$	$-Q(\tau_1)$
0.1	0.4949	0.7187	$0.8056(10)^{-2}$
0.5	0.4746	0.7228	0.1970
1.0	0.4498	0.7299	0.7660
2.0	0.4026	0.7481	2.895
5.0	0.2818	0.8119	15.27
10.0	0.1496	0.8952	46.29

TABLE 6

VALUES OF FUNCTION $\phi(\tau)$ AND THE FLUX INTEGRAL $Q_s(\tau_1)$ BASED ON THE TAYLOR SERIES APPROXIMATION

(a) $\tau_1/\tau_2 = 0$

τ_2	$\varphi_s(0)$	$\varphi_s(\tau_2)$
0.10	0.2763	0.2623
0.25	0.3210	0.2798
0.50	0.4122	0.3063
0.75	0.5293	0.3294
1.00	0.6796	0.3491
1.50	1.120	0.3797
2.00	1.847	0.4015

(b) $\tau_1/\tau_2 = 0.5$

τ_2	$\varphi_s(\tau_1)$	$\varphi_s(\tau_2)$	$-Q_s(\tau_1) + \tau_1^3/3$
0.10	0.2583	0.2602	$0.8020(10)^{-4}$
0.25	0.2706	0.2750	$0.1275(10)^{-2}$
0.50	0.2908	0.2979	$0.1049(10)^{-1}$
0.75	0.3105	0.3185	$0.3633(10)^{-1}$
1.00	0.3292	0.3367	$0.8824(10)^{-1}$
1.50	0.3636	0.3668	0.3114
2.00	0.3931	0.3898	0.7681
3.00	0.4370	0.4208	2.770
5.00	0.4800	0.4517	14.04

radiative transfer theory and the results presented in the paper may prove useful as a standard in evaluating approximate methods. It is in this spirit that the paper was written.

ACKNOWLEDGMENTS

The author would like to acknowledge the contributions of Mr. Alfred L. Crosbie who assisted with the analysis and numerical computations.

REFERENCES

1. Chandrasekhar, S. Radiative Transfer. London: Oxford University Press, 1950.

2. Kourganoff, V. Basic Methods in Transfer Problems. London: Oxford University Press, 1952.

3. Sobolev, V. V. A Treatise on Radiative Transfer. New York: D. Van Nostrand Co., 1963.

4. Sparrow, E. M. and R. D. Cess. Radiation Heat Transfer. Belmont, Calif.: Brooks/Cole Publishing Co., 1966.

5. Kuznetsov, E. S., "Radiative Transfer in a Gaseous Envelope Surrounding an Absolutely Black Sphere," Izvestiia Akademiia Nauk SSSR, Seriya Geofizicheskaya 15, No. 3 (1951), p. 69 (In Russian).

6. Heaslet, M. A. and R. F. Warming, "Application of Invariance Principles to a Radiative Transfer Problem in a Homogeneous Spherical Medium," Journal of Quantitative Spectroscopy and Radiative Transfer, 5 (1965), p. 669.

7. ————, "Radiation Flux from a Slab or a Sphere," Journal of Mathematical Analysis and Applications, 14 (1966), p. 359.

8. Sparrow, E. M., C. M. Usiskin, and H. A. Hubbard, "Radiation Heat Transfer in a Spherical Enclosure Containing a Participating, Heat Generating Gas," Transactions, ASME, Journal of Heat Transfer, C83 (1961), p. 199.

9. Rhyming, I. L., "Radiative Transfer Between Concentric Spheres Separated by an Absorbing and Emitting Gas," International Journal of Heat and Mass Transfer, 9 (1966), p. 315.

10. Chisnell, R. F., "Radiant Heat Transfer in a Spherically Symmetric Medium," LMSC 6-76-66-20, Lockheed Missiles and Space Co., Sunnyvale, California (1966).

11. Viskanta, R. and P. S. Lall, "Transient Cooling of a Spherical Mass of High-Temperature Gas by Thermal Radiation," Transactions, ASME, Journal of Applied Mechanics, E32 (1965), p. 740.

12. ————, "Transient Heating and Cooling of a Spherical Mass of Gray Gas by Thermal Radiation," Proceedings, 1966 Heat Transfer and Fluid Mechanics Institute, M. A. Saad and J. A. Miller, eds. Stanford, Calif.: Stanford University Press, 1966, pp. 181-197.

13. Pai, S. I. Radiation Gas Dynamics. New York: Springer-Verlag New York, Inc., 1966.

14. Hoshizaki, H. and K. H. Wilson, "Convective and Radiative Heat Transfer During Superorbital Entry," AIAA Journal, 5 (1967), p. 25.

15. Perlmutter, M. and J. R. Howell, "Radiant Heat Transfer Through a Gray Gas Between Concentric Cylinders Using Monte Carlo," Transactions, ASME, Journal of Heat Transfer, C86 (1964), p. 169.

16. Cess, R. D., "On the Differential Approximation in Radiative Transfer," ZAMP, 17 (1966), p. 776.

17. Heaslet, M. A. and R. F. Warming, "Radiative Transport and Wall Temperature Slip in an Absorbing Planar Medium," International Journal of Heat and Mass Transfer, 8 (1965), p. 979.

18. Viskanta, R., "Effectiveness of a Layer of an Absorbing-Scattering Gas in Shielding a Surface from Incident Thermal Radiation," J. Franklin Institute 280, 483 (1965).

19. Stroud, A. H. and D. Secret. Gaussian Quadrature Formulas. Englewood Cliffs, N.J.: Prentice-Hall, Inc., 1966.

20. Chou, Y. S., "Study of Radiant Heat Transfer by the Method of Regional Averaging," LMSC 6-77-67-6, Lockheed Missiles and Space Co., Sunnyvale, California (1967).

21. Deissler, R. G., "Diffusion Approximation for Thermal Radiation in Gases with Jump Boundary Condition," Transactions, ASME, Journal of Heat Transfer, C86 (1964), p. 240.

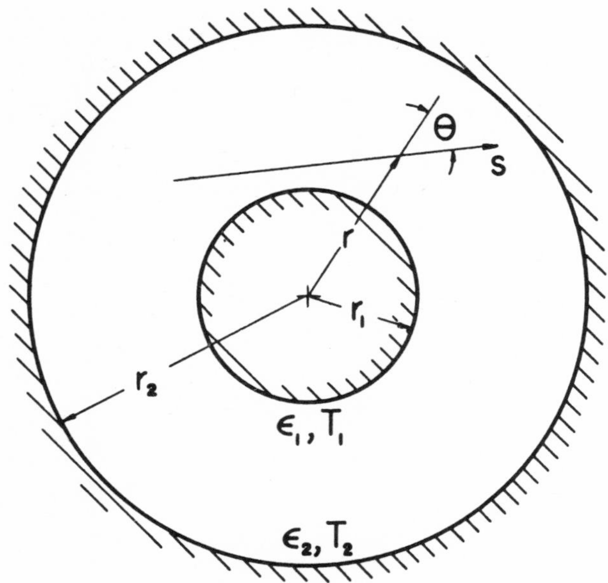

FIGURE 1. PHYSICAL MODEL AND COORDINATE SYSTEM.

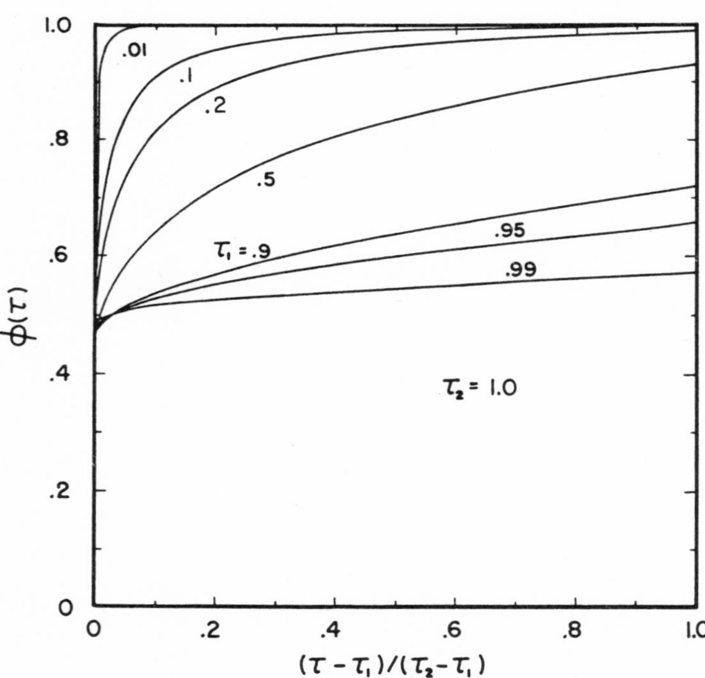

FIGURE 2. EFFECT OF OPTICAL RADIUS ON DIMENSIONLESS EMISSIVE POWER DISTRIBUTION WITH NO HEAT GENERATION FOR $\tau_2 = 1.0$.

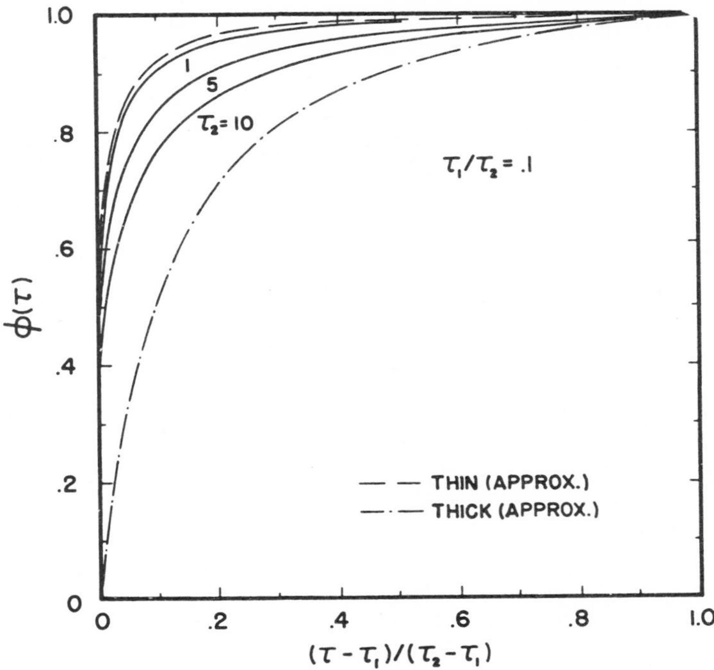

FIGURE 3. DIMENSIONLESS EMISSIVE POWER DISTRIBUTION WITH NO HEAT GENERATION FOR $\tau_1/\tau_2 = 0.1$.

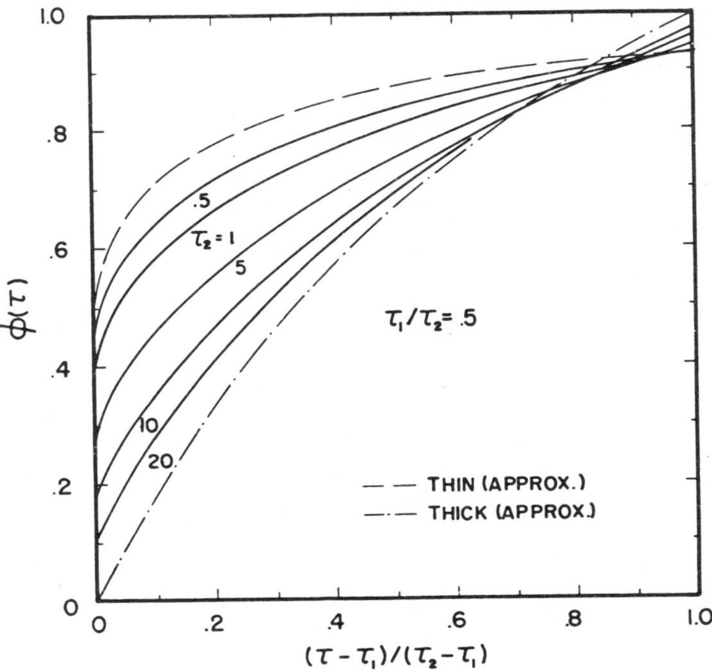

FIGURE 4. DIMENSIONLESS EMISSIVE POWER DISTRIBUTION WITH NO HEAT GENERATION FOR $\tau_1/\tau_2 = 0.5$.

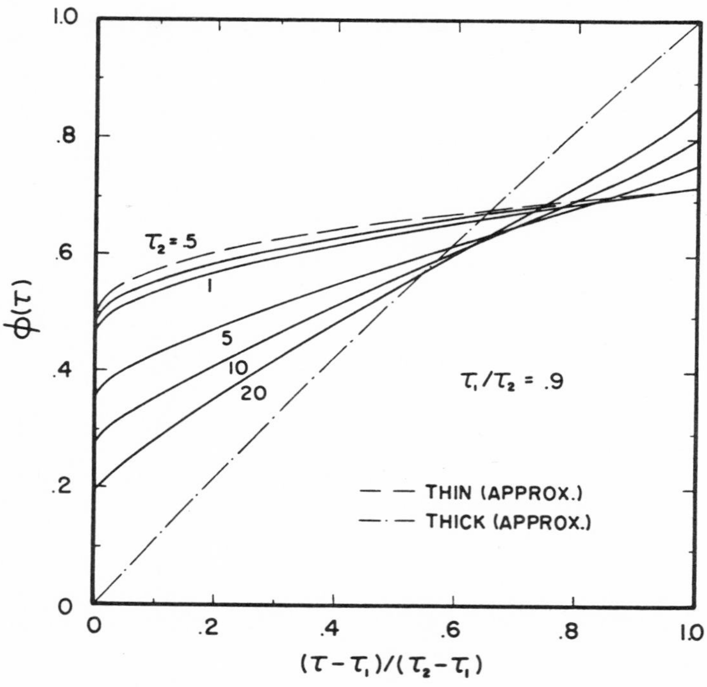

FIGURE 5. DIMENSIONLESS EMISSIVE POWER DISTRIBUTION WITH NO HEAT GENERATION FOR τ_1/τ_2 = 0.9.

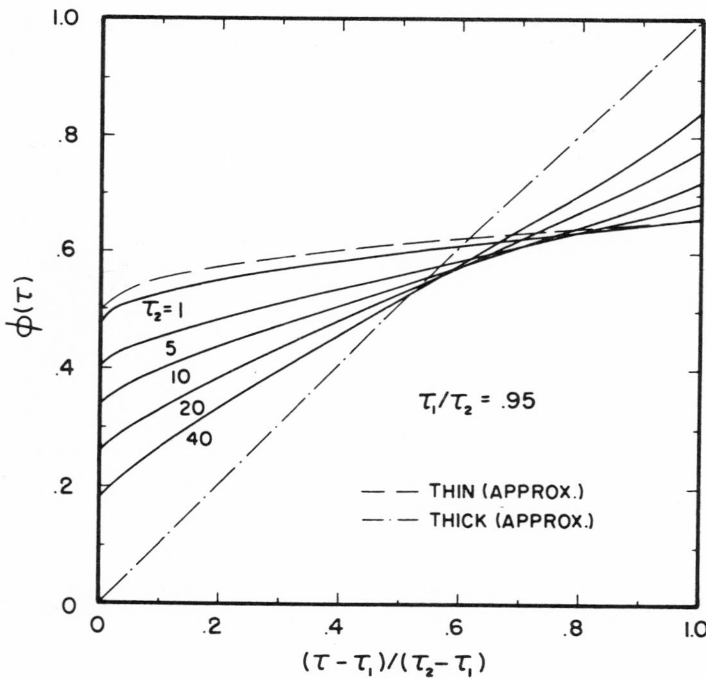

FIGURE 6. DIMENSIONLESS EMISSIVE POWER DISTRIBUTION WITH NO HEAT GENERATION FOR τ_1/τ_2 = 0.95.

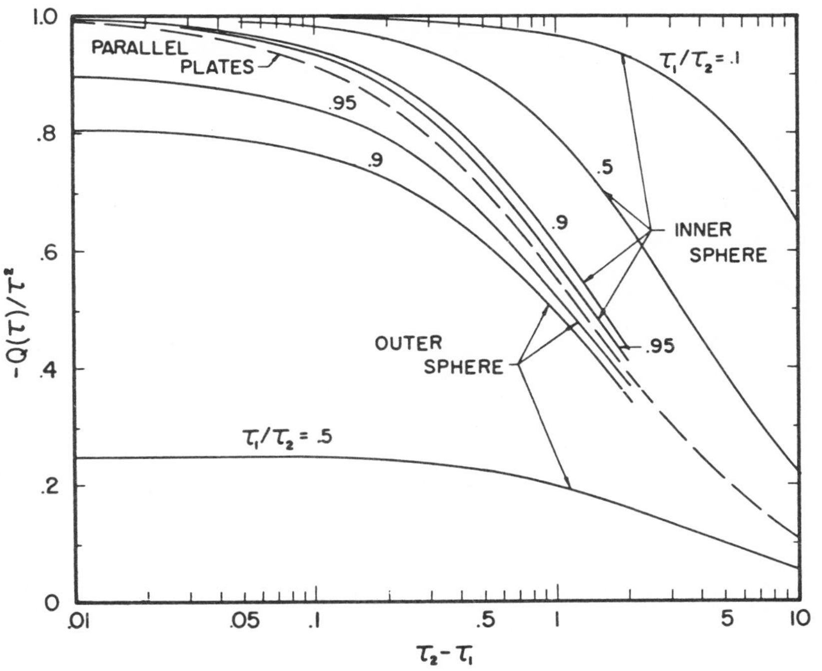

FIGURE 7. EFFECT OF τ_1/τ_2 ON THE RADIATIVE FLUXES AT THE WALLS WITH NO HEAT GENERATION FOR $\varepsilon_1 = \varepsilon_2 = 1.0$.

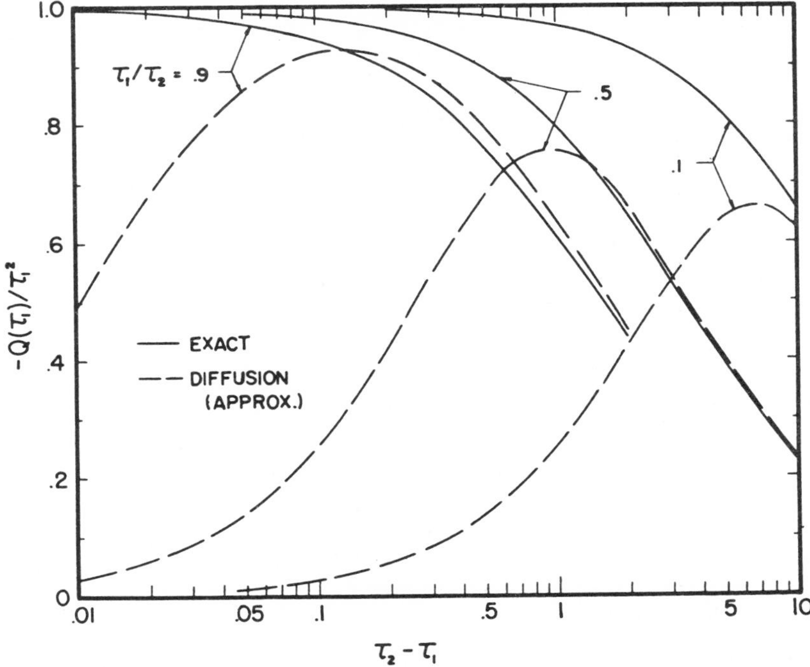

FIGURE 8. COMPARISON BETWEEN EXACT AND APPROXIMATE PREDICTIONS OF RADIATIVE FLUX AT INNER SPHERE WITH NO HEAT GENERATION FOR $\varepsilon_1 = \varepsilon_2 = 1.0$.

HEAT TRANSFER IN BIOENGINEERING

J. C. Chato*

ABSTRACT

Several aspects of heat transfer problems related to biology are explored primarily from the engineering viewpoint. The discussion is divided into four sections: (1) instrumentation and equipment, (2) internal heat transfer, (3) external heat transfer, and (4) miscellaneous heat transfer problems. A listing of thermal conductivity and diffusivity data on biological materials is given in the Appendix.

INTRODUCTION

Bioengineering, or biomedical engineering, is a relatively new and only partially organized area of activity which is still in the developmental stage. At present, it is questionable whether the term can be defined exactly; it means different things to different people working inside and outside the field. Perhaps the real problem in definition arises from the fact that it is difficult to determine the boundaries between biology, physiology, or medicine on the one hand, and engineering on the other. Certainly there was and is research and even developmental work done by biologists and medical doctors which could be classified as at least partly engineering. It is equally clear, however, that engineers are performing work today which can be claimed to be essentially biology. Because of the complicated "systems" involved in biology, in many cases the really meaningful advances are made as a result of cooperation between competent people on both sides of this diffuse boundary.

For our present purposes, let us define bioengineering as an engineering activity which originated in the realm of biology or in its related fields. Thus heat transfer in bioengineering deals with engineering heat transfer problems which originated in biology or medicine. Since this field is rather broad, the discussion will be confined to specific topics within the sphere of the author's interest and activities.

INSTRUMENTATION AND EQUIPMENT

For several years the author has been working with cryosurgical devices.[1,2] These provide localized cooling or freezing of the tissue. The cooling effect in the temperature range of 0° to 27° C can be used most effectively for the exploration of the central nervous system, particularly the interior of the brain. Cooling the nervous

*Associate Professor of Mechanical Engineering, University of Illinois, Urbana, Illinois.

system in this temperature range effectively eliminates the normal nerve functions which, however, return to normal as soon as the temperature is allowed to rise to the usual level. These so-called reversible lesions can show clearly the roles that various regions of the nervous system play in a living system. Below 0° C the tissue can be destroyed by the freezing-thawing process. Thus surgery -- cryosurgery -- can be performed by cold probes, in many cases with distinct advantages over other methods. One of these advantages stems from the previously described nerve deadening effect of cold: the cooling provides its own anesthesia. Another advantage is that since the only physical cutting occurs where the cryoprobe is introduced into the tissue, whereas the volume of the tissue to be destroyed is much larger than that of the probe, usually very few blood vessels are ruptured and little bleeding occurs. In special applications other advantages may occur. For example, if a cataract is to be removed from the eye, a cryoprobe cannot only freeze the tissue but will also freeze to it, thus providing two desirable effects simultaneously. First, the hard frozen tissue is much easier to cut than the normal, jelly-like substance; and second, the probe acts as a holder of the tissue and tremendously facilitates its handling and final removal.

The previous paragraph described the medical side of the technique. The engineering aspects have to do with the development of an appropriate device and with the determination of the temperature distribution around the cold tip inside the tissue.

The problem of cooling the tip of a long and narrow needle of less than 2 mm in diameter was solved by the author employing the common vapor expansion refrigeration principle.[2] Warm refrigerant liquid is forced by its own pressure to flow down to the tip of the probe along the outer annulus formed by two concentric tubes. A tiny valve formed between the two tubes at the tip creates the pressure drop which causes the liquid to evaporate and cool its environment. The cold vapor is bled off along the inner tube at a rate controlled by a needle valve. The bleed-off rate determines the regrigerant pressure at the tip and, consequently, also its temperature. Since the warm refrigerant in the outer annulus acts as an insulator for the cold vapor, very little additional insulation is necessary. The temperature at the tip is monitored by a thermocouple. A schematic diagram of a simple setup is shown in Figure 1. The equipment can be made more sophisticated by adding automatic controls and a compressor-condenser system to provide for recirculation of the refrigerant. Neither one is essential, however, and the latter would actually make the device more cumbersome, expensive, and dangerous in the operating room. There are other solutions to this problem such as using liquid nitrogen, or employing the Joule-Thompson effect in a tiny high-pressure counter-current heat exchanger. For temperatures down to about -50° C, the refrigerant expansion device is simpler, and probably also safer, since either too little or too much refrigerant flow will warm the tip. To obtain temperatures below -50° C, the outlet hose has to be connected to a simple vacuum system. At these low temperatures insulation along the stem of the probe is needed for most applications.

Another instrument currently under development by the author is for the measurement of thermal conductivity and diffusivity of biological materials both in vivo and in vitro. The device under development is small enough to be introduced into the body tissues with no more difficulty than a hypodermic needle. It measures local conditions in a volume small enough to be considered in most cases as homogeneous tissue. This latter consideration is important, since various organs or tissues located near each other may have considerably different properties. The instrument is related to others used previously for the measurement of blood flow.[3-10] The heart of the device is a small, spherical, resistance type of temperature sensitive element such as a thermistor bead. It is completely imbedded in the material to be tested and acts as a heat source to its surroundings. A controlled power source drives the bead in either of two modes of operation. One is to suddenly increase the bead temperature above that of its surroundings, and keep it at this constant elevated level by varying the power input; the other is to suddenly apply a fixed heat input and monitor the resulting temperature variation. From a practical standpoint the first method is easier to achieve and, except for a very brief initial warmup time, does not have an inherent error due to the thermal capacitance of the bead. The theoretical analysis of the system is based on the model of a spherical heat source located in an infinite conducting medium.

It will be shown later that for tissues without blood flow a plot of the heating power, q, against $(\text{time})^{-1/2}$ should be a straight line with an intercept proportional to k alone and a slope proportional to $k/\sqrt{\alpha}$ or $\sqrt{k\rho c_p}$. Thus both k and α can be obtained simultaneously. Figure 2 shows typical curves for a 1.75 per cent Agar solution. Preliminary experiments yielded the following typical thermal conductivity values in watts/cm-°C:

Lean neck beef, fresh	$3.42\text{-}5.10 \times 10^{-3}$,
Beef fat, fresh	$2.25\text{-}2.28 \times 10^{-3}$.

Additional data on thermal conductivities and diffusivities obtained by the method described are given in the Appendix.

Equipment for the maintenance of specific environments for living systems could also be included in this discussion. The problem here is to provide appropriate heat and mass transfer rates at the surface of the "system" such as the human body. Thus it is related primarily to external transport phenomena to be discussed later. The range of applications is extremely broad and it could include such diverse processes as freeze-preservation of biological media, the growing of cell cultures, conventional air conditioning, and the maintenance of acceptable "microclimates" in protective suits worn in any hostile environment. The author is currently working on methods of freeze-preservation; and on protective suits primarily for space use, but secondarily for industrial and other applications. The latter work requires not only an engineering approach, but also work in physiology to determine the basic "design parameters" required for a good engineering system. For example, the amount, type, and location of cooling applied to a human body has to be determined in great part from physiological data, some of which is not yet available. The ultimate aim of the work is the development of integrated man-in-suit systems which can cope with the particular tasks set for them.

INTERNAL HEAT TRANSFER

The temperature and heat flow distributions inside a living "system" become important factors under various circumstances. The effectiveness of the previously described instruments depends on the heat transfer processes occurring in their vicinity. Cold-blooded animals have to control the heat flow inside and at the surface of their bodies to keep their temperatures within relatively narrow operating limits. In warm-blooded animals, the temperature distribution inside the body will determine whether they are comfortable, begin to sweat, or begin to shiver. From the engineering standpoint the modeling difficulties are manifold. First of these is the nonhomogeneous and nonisotropic nature of tissues. Although this is easily observable under a microscope, the corresponding thermophysical properties are very difficult to measure quantitatively. It seems, however, that with proper sample size selection, both homogeneity and isotropy can be assumed for most tissues. On the one hand, "proper size" means that it is sufficiently large to contain a great enough number of cells and fibers to average their individual behavior. On the other hand, the sample to be considered must be small enough to contain only one type of tissue. The exact limitations depend on the particular application and the judgment of the investigator. For example, the thermophysical property measurements should be done on as small homogeneous samples as possible. For comfort or thermoregulatory considerations, however, a much more overall approach is usually quite adequate.

The second difficulty in adequate modeling arises from the fact that living tissues or organisms are systems, and very active and closely controlled systems at that. They react not only to influences exerted directly on them, but also to influences affecting other parts of the body. For instance, sweating occurs not only on heated body surfaces, but in other areas as well. It occurs in response to presently not entirely understood temperature signals from the hypothalamus in the brain and from other thermoreceptors in the body. It is important, therefore, to know not only the thermophysical properties of the various tissues, but also their behavior under the conditions considered. In many instances the behavioral aspects such as changes in blood supply or contraction of muscles are far more important than the basic heat transfer mechanisms themselves.

Three major phenomena influence heat transfer within biological systems: conduction, possible internal heat generation, and blood flow. The first two of these can be analyzed by mathematical methods which are quite well developed. Blood flow, however, represents an unusual condition from the standpoint of engineering practice because it represents a diffuse and generally pulsatile flow inside a heat conducting solid medium. An approximate analog of this phenomenon is the diffusion of a fluid

*1 watt/cm - °C = 57.79 Btu/hr-ft-°F.

through a porous medium. An important difference is, however, that while in the porous solid the direction of flow is usually quite well defined, as in the case of injection- or sweat-cooling of an aerodynamically heated body. In a biological system the blood vessels are not uniform and a well-defined flow direction may not exist. In addition, the blood in the larger arteries and veins may be at a different temperature from that of the surrounding tissue. There are two relatively simple ways of modeling the heat transfer effects of blood flow. One is to include a quantity of heat equivalent to the rate of heat given off or absorbed by the blood in a unit volume, as an internal heat generation or heat absorption term in the describing equation. The other, more realistic way, is to assume that the blood reaches a region through relatively large vessels at some arterial temperature, T_a, and is cooled or heated in the capillaries to the existing local temperature level.

The general energy balance equation can be written as

$$\rho c_p \frac{\partial (T - T_o)}{\partial t} = \text{div}\left[k \text{ grad } (T - T_o)\right] - c_b\left(w_v T_v - w_a T_a\right) + Q. \quad (1a)$$

For the second model described above, $T_v = T$. If k is assumed to be constant and pooling of the blood is neglected, i.e., $w_v = w_a = w$, then the equation can be changed into a more convenient form for calculation purposes.

$$\frac{\partial (T - T_o)}{\partial t} = \alpha \nabla^2 (T - T_o) - \frac{wc_b}{\rho c_p}(T - T_o) + \frac{Q'}{\rho c_p}, \quad (1b)$$

where

$$Q' = Q + wc_b(T_a - T_o) . \quad (2)$$

Q' and w are usually considered to be constants.

This equation can be solved analytically for certain geometries and boundary conditions.[6,7,12] For instance, the substitution

$$T - T_o = T' e^{-(wc_b/\rho c_p)t} \quad (3)$$

reduces Equations (1) to

$$\frac{\partial T'}{\partial t} = \alpha \nabla^2 T' + \frac{Q'}{\rho c_p} e^{(wc_b/\rho c_p)t} . \quad (4)$$

This equation can be solved by the methods described in Reference 12.

If spherical symmetry can be assumed, Equations (1) can be simplified by the substitution

$$T'' = \left(T - T_o - \frac{Q'}{wc_b}\right) r \quad (5)$$

to the form

$$\frac{\partial T''}{\partial t} = \alpha \frac{\partial^2 T''}{\partial r^2} - \frac{wc_b}{\rho c_p} T'' . \quad (6)$$

The solution to Equation (6) is for the case of constant $wc_b/\rho c_p$, zero initial T'', and boundary conditions of either constant temperature or convective heat transfer at the surface

$$T'' = \frac{wc_b}{\rho c_p} \int_o^t e^{-(wc_b/\rho c_p)t'} U(t')dt' + U e^{-(wc_b/\rho c_p)t} , \quad (7)$$

where U is the solution with identical boundary conditions but with $wc_b = 0$. Other solutions for biological conditions are given by References 6 and 7.

As an example, let us apply the theory to the technique described before of measuring thermophysical properties of tissues using a small, spherical heat source. First, it will be assumed that the blood flow has been occluded in the living tissue or that the sample is tested in vitro. Internal heat generation will be ignored. The governing equation then becomes

$$\frac{\partial (T - T_o)}{\partial t} = \alpha\left[\frac{\partial^2 (T - T_o)}{\partial r^2} + \frac{2}{r} \cdot \frac{\partial (T - T_o)}{\partial r}\right] . \quad (8)$$

Consider the boundary conditions when the heated bead is suddenly raised to a constant temperature level, T_R, at $t = 0$. The tissue will be considered infinitely large in comparison to the bead.

$$T = T_R \quad \text{for} \quad r = R \quad \text{and} \quad t > 0, \quad (9)$$

$$T = T_o \tag{10}$$

for $t = 0$ and $r > R$,

$$q = -4\pi R^2 k \frac{\partial T}{\partial r} \tag{11}$$

for $r = R$ and $t > 0$.

By the change of variables indicated in Equation (5), the problem can be transformed into one for a semi-infinite solid. The resulting temperature distribution and heat flow from the bead can be expressed by

$$\frac{T - T_o}{T_R - T_o} = \frac{R}{r} \operatorname{erfc} \frac{r - R}{2\sqrt{\alpha t}} \tag{12}$$

and

$$q = 4\pi R k (T_R - T_o) + \frac{4\sqrt{\pi} R^2 k (T_R - T_o)}{\sqrt{\alpha t}} . \tag{13}$$

Thus, as it was stated previously, a plot of q against $t^{-1/2}$ should be a straight line which can be established with a test run of a few seconds.

If blood flow exists, the mathematical solution is more complicated. If again spherical symmetry is assumed, the governing relation is given by Equation (6). The boundary conditions can be assumed as

$$T'' = \left(T_R - T_o - \frac{Q'}{wc_b}\right) R \tag{14}$$

for $r = R$ and $t > 0$,

$$T'' = -\frac{Q'}{wc_b} r \tag{15}$$

for $t = 0$ and $r > R$ or 0,

$$q = -4\pi R^2 k \left(\frac{1}{r}\frac{\partial T''}{\partial r} - \frac{T''}{r^2}\right) \tag{16}$$

for $r = R$ and $t = 0$.

Similar to Equation (3), let

$$T'' = T' e^{-(wc_b/\rho c_p) t} \tag{17}$$

Substituting into Equation (6) yields

$$\frac{\partial T'}{\partial t} = \alpha \frac{\partial^2 T'}{\partial r^2} , \tag{18}$$

with the boundary conditions

$$T' = \left(T_R - T_o - \frac{Q'}{wc_b}\right) R\, e^{(wc_b/\rho c_p) t} , \tag{19}$$

for $r = R$ and $t > 0$,

$$T' = -\frac{Q'}{wc_b} r \tag{20}$$

for $t = 0$ and $r > R$ or 0,

$$q = -4\pi R^2 k \left(\frac{1}{r}\frac{\partial T'}{\partial r} - \frac{T'}{r^2}\right) e^{-(wc_b/\rho c_p) t} \tag{21}$$

for $r = R$ and $t > 0$.

The solution of Equation (18), subject to these boundary conditions, is in terms of the temperature distribution and heat supplied as follows:

$$\frac{T - T_o - Q'/wc_b}{T_R - T_o - Q'/wc_b} = \frac{R}{2r} \left\{ \exp\left[-(r - R)\sqrt{\frac{wc_b}{k}}\right] \operatorname{erfc}\left[\frac{r - R}{2\sqrt{\alpha t}} - \sqrt{\frac{wc_b}{\rho c_p} t}\right] + \exp\left[(r - R)\sqrt{\frac{wc_b}{k}}\right] \operatorname{erfc}\left[\frac{r - R}{2\sqrt{\alpha t}} + \sqrt{\frac{wc_b}{\rho c_p} t}\right] \right\} - \frac{Q'/wc_b}{T_R - T_o - Q'/wc_b} \exp\left[-\frac{wc_b}{\rho c_p} t\right] \tag{22}$$

$$q = \left(T_R - T_o - \frac{Q'}{wc_b}\right)\left[4\pi Rk + \frac{4\sqrt{\pi}R^2k}{\sqrt{\alpha t}} \exp\left(-\frac{wc_b}{\rho c_p} t\right) + 4\pi R^2 \sqrt{k\, wc_b}\, \operatorname{erf}\sqrt{\frac{wc_b}{\rho c_p} t}\right]. \quad (23)$$

When w = 0 and Q' = 0 these equations reduce to Equations (12) and (13), respectively, as it is to be expected. As $t \to \infty$, the final, steady state heat flow becomes

$$q_{t\to\infty} = \left(T_R - T_o - \frac{Q'}{wc_p}\right)\left[4\pi RK\left(1 + R\sqrt{\frac{wc_b}{k}}\right)\right]. \quad (24)$$

Thus the apparent thermal conductivity increases by a factor of $(1 + R\sqrt{wc_b/k})$ as a result of the blood flow. It is important to note that this increase is a function of the geometry, R, as well as the blood flow and tissue conductivity.

If the thermal properties of the tissue are known, the blood flow can be calculated by fitting Equation (22) or (23) to appropriate experimental data. Levy, et al.,[10] used the heat loss from a heated bead kept at a constant temperature level above that of an unheated one to monitor changes in tissue blood flow. An alternate thermal method for measuring blood flow was used by Stow, et al.,[5,8] and Pearl, et al.,[6,9] which consisted of changing the blood flow by occlusion or release of the main supply artery and recording the temperature variation of the tissue. If it is assumed that during a rapid occlusion or release the temperature field remains essentially constant, the blood flow can be related to the change of the time derivative of tissue temperature, $\partial T/\partial t$, and the difference between arterial blood and tissue temperatures, $T_a - T$. From Equation (1a)

$$w_2 - w_1 = \frac{(\partial T/\partial t)_1 - (\partial T/\partial t)_2}{(c_b/\rho c_p)(T - T_a)}, \quad (25)$$

where the subscripts 1 and 2 refer to conditions just before and just after the change of flow. In order to eliminate the necessity of determining T_a, two sets of experiments can be performed with different steady temperature fields. One can be with the normal, uniform temperature distribution, the other with the steady addition of heat from a heater. Applying Equation (25) to the two cases yields two equations which allow the elimination of T_a, if it is assumed that $w_2 - w_1$ is the same for both. With complete occlusion, $w_2 - w_1 = -w$; with complete release, $w_2 - w_1 = w$.

The blood supply temperature, T_a, can vary considerably in the body. Within the torso it is essentially deep body temperature, but in the limbs it can be considerably lower. The larger arteries and veins are located parallel and near each other in the extremeties. These blood vessels form heat exchangers, sometimes with regulated bypass arrangements, which conserve body heat in a very ingenious fashion. In a cold environment a considerable portion of the heat lost from the arteries in the limbs is transferred to the adjacent veins and is carried back to the torso. Thus steep temperature gradients can be established along the limbs with relatively little heat loss to the surroundings. In a warmer environment the bypass veins located near the skin carry a greater portion of the return blood and provide for higher heat losses to the surroundings. In order to investigate the heat transfer characteristics of arteries and veins, a study of heat transfer to pulsating flows is required. Although there is quite a large body of literature on the fluid mechanics of pulsating blood flows, virtually no reference was discovered by the author relating to the corresponding heat transfer problem.[14,15] Work is now in progress to remedy the situation. The current model consists of a rigid tube in which combined free and forced convection laminar flows can be studied in detail. Thus the work at present is essentially engineering in nature. Some of the recent results on steady flows are discussed by the author in another paper in this book. The physiological complications such as elastic tube walls will have to follow after the simpler model is better understood. One advantage of this approach is that useful engineering data are obtained on the preliminary models, while the complications are introduced only gradually. For example, the rigid wall model is directly applicable to the design of heat exchangers with pulsatile flows such as the extracorporeal blood cooling devices used in surgery.

The preservation of biological media by freezing poses a number of bioengineering problems. There are two different major purposes for freezing biological media: first, preservation for survival over extended periods; second, preservation of the cell structure for microscopic examination. The first requires carefully controlled cooling and warming rates usually with a substance such as PVP (polyvinylpyrrolidone) added to prevent cell damage. In such problems the influence of cooling rates, storage temperatures, and warming rates has to be determined in order to find the optimum process which produces the greatest survival rate over the longest period of time. The second purpose requires extremely rapid cooling rates and the warming is of no consequence, since the slide preparation and examination is done with the frozen specimen.

EXTERNAL HEAT TRANSFER

Some aspects of this topic are well established, such as those relating to human comfort conditions under relatively normal circumstances. The general problem includes the physiological responses of humans under various environmental and working conditions, and the combined heat and mass transfer occurring between the human and his environment. An excellent review of this area is given by Stoll.[16] The specific problem on which the author is working was mentioned before; it deals with the development of protective suits for space, and for industrial and other possible applications. Such effort is necessarily divided between the determination of the necessary physiological requirements and the engineering development of the various concepts for the suit. In a space environment the human has to be able to perform satisfactorily for various lengths of time under widely varying conditions. Consequently, a space suit must provide a completely closed microclimate which can be adequately controlled in any possible external environment and under any possible internal heat load due to the activities of the occupant. Industrial environments are hardly ever as severe as those in space, but protective suits for deep-sea diving may entail almost as many difficulties as those encountered in space. The physiological work involves the determination of the necessary environments for the various parts of the body, as well as for the entire body. For example, the minimum ventilation requirements for the hands and feet have to be established. The optimal distribution of cooled body surfaces and surface temperatures will have to be known in order to be able to judge the feasibility of a partially cooled suit. Heat transfer between the various parts of the body and their environment can be studied by well-established engineering methods, as well as newer techniques. Infrared scanning devices are available today which can map the temperature distribution on the skin. Figure 3 shows an interferometer picture of the palm of the hand with the surface temperatures given at two locations. In spite of the non-uniformity of the hand, the interference fringes are rather clear.

Birkebak, et al.,[17-19] studied heat transfer through animal integuments in both free and forced convection. Reference 18 is, in addition, a discussion of heat transfer in biological systems.

Last, but not least, the external heat and mass transfer phenomena occurring around the body are closely connected with the thermoregulatory behavior of the "biological system." Cold-blooded animals have to control their body temperature by moving in and out of warm and cold environments. Warm-blooded animals, however, have an involuntary control system which can react in several ways to the various heat and work stresses imposed upon the body: the blood distribution in the various parts of the body can change, sweating or shivering can begin, or the metabolic heat production rate can change. In order for a protective suit to function properly, it has to cope with all possible changes occurring both in the environment and within its occupant. Thus the problem of external heat transfer includes the design of thermoregulatory systems which match the thermoregulatory behavior of the "biological system" considered.

MISCELLANEOUS "BIOENGINEERING" HEAT TRANSFER PHENOMENA

Only a few examples will be described here. These will illustrate heat transfer phenomena occurring in nature which may be related to bioengineering in varying degrees.

Parkhurst, et al.,[20] and Curran and Nottage[21] studied heat transfer to leaves both in the quiescent and in the oscillating state. They found that the data could be correlated if appropriately defined Nusselt and Reynolds numbers were used. The correlation was very similar to those found in engineering practice. Oscillations of the leaves seemed to have negligible effects.

Heat transfer through animal integuments was already mentioned.[17-19] Heath[22] is studying the thermoregulatory behavior of cold-blooded animals and insects. His "bioengineering problems" include convective heat transfer at the skin, radiative properties of insect wings which are used by their owners as parasols, and body temperature variations together with their behavioral control.

It may be appropriate to close on a curious note. A sphinx moth, by the name of *Pholus achemon*, controls its body temperature by the use of a built-in evaporative air conditioner: he sticks his long, wet tongue out and after it cools, due to evaporation, he pulls it back again to reduce his body temperature from the inside out. The study of this phenomenon actually belongs to the realm of human activity called bionics, the study of biological systems for adaptation to human needs, rather than bioengineering; but it is an interesting use of heat and mass transfer in nature.

NOTATION*

c_b = specific heat of blood ($\theta/M - T$)

c_p = specific heat of tissue ($\theta/M - T$)

k = thermal conductivity ($\theta/t - L - T$)

q = heat flow rate (θ/t)

Q = heat generation rate per unit volume ($\theta/t - L^3$)

r = radial location (L)

R = radius of measuring sphere (L)

t = time

T = temperature (T)

T', T'' = transforms defined by Equations (3) and (5) (T)

w = blood flow rate per unit volume ($M/t - L^3$)

Greek Symbols

α = thermal diffusivity (L^2/t)

ρ = density of tissue (M/L^3)

Subscripts

a = arterial conditions

v = venous conditions

o = initial and/or uniform conditions in the tissue

*Units in parentheses are: M, mass; F, force; L, length; T, temperature; θ, heat (F - L); t, time.

REFERENCES

1. Mark, V. H., J. C. Chato, et al., "Localized Cooling in the Brain," *Science*, 134 (1961), pp. 1521-1522.

2. Chato, J. C., U.S. Patent No. 3,272,203, Surgical Probe, 1966.

3. Gibbs, F. A., "A Thermoelectric Blood Flow Recorder in the Form of a Needle," *Proceedings, Society of Experimental Biology and Medicine*, 31 (1933), pp. 141-146.

4. Grayson, J., "Internal Calorimetry in the Determination of Thermal Conductivity and Blood Flow," *Journal of Physiology*, 118 (1952), p. 54.

5. Stow, R. W. and J. F. Schieve, "Measurement of Blood Flow in Minute Volumes of Specific Tissues in Man," *Journal of Applied Physiology*, 14 (1959), pp. 215-224.

6. Perl, W., "Heat and Matter Distribution in Body Tissues and the Determination of Tissue Blood Flow by Local Clearance Methods," *Journal of Theoretical Biology*, 2 (1962), pp. 201-235.

7. ————, "An Extension of the Diffusion Equation to Include Clearance by Capillary Blood Flow," *Annals of the New York Academy of Science*, 108 (1963), p. 92.

8. Stow, R. W., "Thermal Measurement of Tissue Blood Flow," *Transactions, New York Academy of Science*, 27 (1965), pp. 748-758.

9. Perl, W. and S. A. Cucinell, "Local Blood Flow in Human Leg Muscle Measured by a Transient Response Thermoelectric Method," Biophysical Journal, 5 (1965), pp. 211-230.

10. Levy, L., H. Graichen, J. A. J. Stolwijk, and M. Calabresi, "Evaluation of Local Tissue Blood Flow by Continuous Direct Measurement of Thermal Conductivity," Journal of Applied Physiology, 22 (1967), pp. 1026-1029.

11. Chato, J. C., "A Survey of Thermal Conductivity and Diffusivity Data on Biological Materials," ASME Paper No. 66-WA/HT-37, 1966.

12. Carslaw, H. S. and J. C. Jaeger. Conduction of Heat in Solids. 2nd ed. Oxford: Clarendon Press, 1959.

13. ————. Conduction of Heat in Solids. 2nd ed. Oxford: Clarendon Press, 1959, p. 135.

14. Lighthill, M. J., "The Response of Laminar Skin Friction and Heat Transfer to Fluctuations in the Stream Velocity," Proceedings, Royal Society of London, 224 (1954), pp. 1-23.

15. Ostrach, S., "Compressible Laminar Boundary Layer and Heat Transfer for Unsteady Motions of a Flat Plate," NACA TN 3569, 1955.

16. Stoll, A. M., "Heat Transfer in Biotechnology," in Advances in Heat Transfer, Vol. IV, J. P. Hartnett and T. F. Irvine, Jr., eds. New York: Academic Press, 1967.

17. Birkebak, R. C., C. J. Cremers, and E. A. LeFebvre, "Heat Transfer Applied to Animal Systems," Journal of Heat Transfer, Transactions, ASME, C, 88 (1966), pp. 125-130.

18. Birkebak, R. C., "Heat Transfer in Biological Systems," International Review of General and Experimental Zoology, Vol. II, J. L. Felts and R. J. Harrison, eds. New York: Academic Press, 1966, pp. 268-344.

19. Birkebak, R. C. and C. J. Cremers, "Forced Convection Heat Transfer from Biological Surfaces," ASME Paper No. 66-WA/HT-55, 1966.

20. Parkhurst, D. F., P. R. Duncan, D. M. Gates, and F. Kreith, "Convection Heat Transfer from Broad Leaves of Plants," Journal of Heat Transfer, Transactions, ASME, C, 90:1 (1968), pp. 71-76.

21. Curran, D. G. T. and H. B. Nottage, "Heating of a Leaf," ASME Paper No. 67-WA/HT-41, 1967.

22. Heath, J. E., Department of Physiology and Biophysics, University of Illinois, Urbana, Illinois, personal communication.

APPENDIX: THERMOPHYSICAL PROPERTIES OF BIOLOGICAL MEDIA

The following table is a collection of thermophysical property data on biological media based partly on the author's previous survey (Reference 11 in the main part of this paper), but extended to include new data known to the author.

Although specific heat data are not tabulated, some of the references such as 12 in this Appendix, list some values. The range of specific heats is from 0.185 gcal/gm°C (Btu/lb°F) for materials frozen at -185°C to about 1.0 gcal/gm°C (Btu/lb°F) for materials with very high water content at room temperature.

Reference	Material	Thermal conductivity, k w/(cm°C) x 10^3	Thermal diffusivity, α cm^2/sec x 10^3	Thermal inertia, $k\rho c_p$ w^2 sec/(cm^4 $°C^2$) x 10^3
	Internal Tissues, in Vitro			
1	Animal muscle	4.06-4.18		
2	Beef muscle	1.98		
3	Muscle	4.60		
4	Beef muscle	2.80		
5	Beef muscle	5.27		
6	Beef muscle	5.28		
7	Lean neck beef, fresh	3.42-5.10		
6	Calf muscle (veal)	5.45		
7	Lean neck pork, fresh	6.13-6.83		
8	Dog muscle	21.8-24.3		
4	Human muscle	3.85		
5	Human muscle	4.40		
9	Muscle (dry)			9.80
9	Muscle (moist)			20.0
8	Dog liver	12.6 -14.5		
8	Dog lung	2.19- 2.64		
8	Dog brain	7.12- 7.95		
10	Rat and rabbit livers	4.95		
10	Rabbit kidney	5.02		
11	Rabbit liver	11.7		
6	Beef kidney	5.25		
6	Beef brain	4.97 and 4.15		
6	Beef liver	4.88		
12	Bovine liver	5.06 and 5.53		
6	Beef lung	2.82		
6	Beef bone marrow	2.20		
9	Bone			8.75
1	Animal fat	1.32- 1.56		
2	Beef fat	2.04		
5	Beef fat	2.22		
13	Beef fat, 2-30 per cent water, 20°C	0.94- 3.46		
6	Bovine fat (fresh)	2.30		
6	Bovine fat (melted)	1.90		
7	Beef fat, fresh	2.25-2.28		
7	Pork fat, fresh	3.59-3.71		
3	Subcutaneous fat	1.60		
4	Human and beef fat	2.04		
5	Human fat	2.00		
9	Human fat			4.55 ± 0.5

Reference	Material	Thermal conductivity, k w/(cm°C) x 10^3	Thermal diffusivity, α cm^2/sec x 10^3	Thermal inertia, $k\rho c_p$ w^2 sec/(cm^4 °C^2) x 10^3
14	Dead, dry bone	6.06		
15	Bovine bones:			
	Midshaft of metacarpus 1/2 month dead	5.13-5.45		
	Midshaft of metatarsus 1/2 month dead	4.55-6.35		
	Midshaft of tibia 2 months dead	6.54-8.67		
	Midshaft of tibia 7 months dead	5.95-7.63		
	Neck of metacarpus 1/2 month dead	4.12-5.29		
	Neck of metatarsus 1/2 month dead	4.40-5.61		
	Neck of tibia 2 and 7 months dead	5.95-8.73		
16	Cat brain, 1 1/2 hours after expiration		0.63-0.64	
16	Cat brain, 2 1/2 hours after expiration		0.44-0.50	
	(see also Frozen and Cold Materials)			
	<u>Skin Tissues</u>			
17	Skin (calc. by 23)	1.88		
18	Skin	2.09		
3	Epidermis	2.09		
3	Dermis	3.70		
19	Skin	2.09		
20	Skin	3.35		
21	0-2 mm skin	3.76		
21	Cool living skin	5.45		
21	Very warm living skin	28.0		
22	Cold living hand	3.35		
22	Normal living hand	9.60		
6	Chicken skin	3.56		
23	0-0.45 mm living underarm skin		0.60	
23	0-0.9 mm living underarm skin		1.00	
23	0-1.3 mm living underarm skin		1.30	
23	0-1.0 mm living thigh skin		5.45	
23	1-2 mm living thigh skin		9.62	
23	Mucous membrane of the cat's tongue		1.30	
24	0-0.26 mm skin		0.40	
24	0-0.45 mm skin		0.60	
24	0-0.90 mm skin		0.85-1.20	
24	0-1.32 mm skin		0.90-1.60	

Reference	Material	Thermal conductivity, k w/(cm°C) x 10^3	Thermal diffusivity, α cm^2/sec x 10^3	Thermal inertia, $k\rho c_p$ w^2 sec/(cm^4 °C^2) x 10^3
9	Excised skin (dry)			9.62
9	Excised skin (moist)			13.1
9	Living skin (no blood flow)			15.8
9	Living skin (with blood flow)			15.8-70.0
25	Living forearm skin			22.8
25	Living average skin			17.5
25	Living fingertip skin			40.0
26	Living skin (calc. by 9)			21.9
27	Living skin (calc. by 9)			21.0
28	Living inner forearm skin			24.5 ± 3.5
29	Irradiated skin			13.1-31.7
30	Porcine living skin	4.14	0.82-0.86	
	Internal Tissues, in Vivo			
16	Cat brain		1.12-1.24	
14	Living bone	22.8		
15	Bovine bones:			
	Midshaft of tibia	8.88-30.8		
	Rib	4.03-11.1		
	Frontal bone	4.87-21.5		
15	Caprine bones			
	Midshaft of tibia	4.63-5.04		
	Neck of tibia	3.30-8.19		
	Rib	1.58-2.84		
	Frontal bone	4.72-6.50		
30	Periosteum of the tibia			
	Bovine	19.0-27.6		
	Caprine	8.10-22.5		
30	Cartilage of the scapula			
	Bovine	17.9-27.6		
	Caprine	13.7-19.0		
	Fluids			
8	Dog blood (anticoagulant EDTA)	6.70-7.54		
8	Dog plasma (anticoagulant EDTA)	6.28-7.12		
6	Blood (42 and 43 per cent hematocrit)	5.09 and 5.30		
6	Human blood	$k \times 10^3 \simeq 5.7-(1.21 \times 10^{-2})$ (% hematocrit)		
6	Plasma	5.71		
46	Dog blood (1% heparin)	$k \times 10^3 = 6.0[1-0.2$ (volume fraction of red blood cells)] $- 0.0226\,(37-T°C)$		
32	Rat blood	5.27		
32	Human blood	5.06		
32	Human plasma	5.82		
32	Human blood corpuscles	4.81		
32	Cow's milk	5.31		
32	Skimmed milk	5.74		
32	Top of milk	4.23		
32	Cream (double Devon)	3.11		
32	Egg white	5.56		
32	Egg yolk	3.38		

Reference	Material	Thermal conductivity, k w/(cm $^{\circ}$C) x 10^3	Thermal diffusivity, α cm^2/sec x 10^3	Thermal inertia, $k\rho c_p$ w^2 sec/(cm^4 $^{\circ}$C^2) x 10^3
6	Egg yolk	4.2		
32	Cod-liver oil	1.70		
6	Beef vitreous humor	5.94		
6	Beef aqueous humor	5.78		
6	Urine	5.61		
6	Gastric juice	4.45		
6	PVP additive	4.74		
	<u>Frozen Then Thawed Materials</u>			
6	Slow freezing and thawing just below 0°C increased the thermal conductivity by 12-17 per cent. Quick freezing and thawing to -195°C increased the thermal conductivity by only 5 per cent.			
6	Slow freeze-thaw:			
	Bovine liver, fresh, ground	4.41		
	Bovine liver ground, frozen <3 days	4.96		
	Bovine liver ground, frozen >5 days	5.60		
	Bovine liver fresh, whole	4.98		
	Bovine kidney ground, frozen 5 days	5.76		
	Bovine kidney fresh, whole	5.12		
	Bovine brain ground, frozen 4 days	5.99		
	Bovine brain fresh, whole	4.96		
6	Quick freeze-thaw Bovine liver	5.27		
	<u>Frozen and Cold Materials</u>			
33	Eye of loin beef, U.S. good grade, aged 69.5 per cent water			
	at 0°C	2.94		
	frozen at - 3°C	9.90		
	-10°C	10.4		
	-17°C	10.7		
34	Beef at -130°C	15.5		
35	Frozen beef	22.5		
35	Dehydrated beef	0.35		
7	Lean neck beef -7°C	4.15-4.73		
7	Beef fat, -9°C	2.25-2.30		

Reference	Material	Thermal conductivity, k w/(cm °C) x 10^3	Thermal diffusivity, α cm²/sec x 10^3	Thermal inertia, $k\rho c_p$ w² sec/(cm⁴ °C²) x 10^3
36	Lean beef, 78.5 per cent water, perpendicular to fiber at 0° C	4.80		
	- 5° C	10.6		
	-10° C	13.5		
	-20° C	15.7		
36	Fat beef, 74.5 per cent water, perpendicular to fiber at 0° C	4.78		
	- 5° C	9.30		
	-10° C	12.0		
	-20° C	14.3		
36	Beef fat, 7 per cent water at 0° C	2.04		
	- 5° C	2.12		
	-10° C	2.27		
	-20° C	2.54		
37	Beef fat (udder), 89 per cent fat, 9 per cent water	2.89		
37	Meat at 0° C	4.80		
37	Lean sirloin beef, 0.9 per cent fat, 75 per cent water, -10 to -25° C, parallel to fiber	13.6-15.4		
37	Lean flank beef, 3.4 per cent fat, 74 per cent water, -10 to -25° C, perpendicular to fiber	10.7-12.1		
12	Bovine liver slowly frozen, at -32° C	1.92 and 1.99		
12	Bovine liver quickly frozen at -195°C	0.99		
11	Rabbit liver	10.9		
37	Lean pork leg, 6.1 per cent fat, 72 per cent water, -10 to -25° C, parallel to fiber,	14.3-16.1		
	perpendicular to fiber	12.3-13.8		
37	Pork fat, 93 per cent fat 6 per cent water	2.10		
36	Pork, 76.8 per cent water, perpendicular to fiber at 0° C	4.78		
	- 5° C	7.67		
	-10° C	9.90		
	-20° C	12.9		

Reference	Material		Thermal conductivity, k $w/(cm^{o}C) \times 10^3$	Thermal diffusivity, α $cm^2/sec \times 10^3$	Thermal inertia, $k\rho c_p$ $w^2 sec/(cm^4\,{}^oC^2) \times 10^3$
36	Pork fat, 3.1 per cent water at	0° C	1.86		
		-5° C	2.27		
		-10° C	2.54		
		-20° C	2.91		
7	Lean neck pork,	-8° C	6.99-7.40		
7	Pork fat,	-9° C	3.90-4.08		
38	Plasma at	-10° C	20.3	9.7	
		-20° C	21.0	11.4	
		-40° C	23.1	15.1	
		-60° C	25.8	18.8	
		-80° C	28.6	22.9	
		-100° C	31.9	26.9	
38	Whole blood at	-10° C	16.4	8.7	
		-20° C	17.4	10.4	
		-40° C	19.2	13.6	
		-60° C	21.4	16.9	
		-80° C	23.8	20.4	
		-100° C	26.6	23.7	
38	Packed cells at	-10° C	12.4	6.8	
		-20° C	13.1	8.2	
		-40° C	15.1	11.0	
		-60° C	17.3	14.1	
		-80° C	19.8	17.2	
		-100° C	22.6	20.4	
	Animal Integuments				
39-45	Fur and feathers (considerable scatter)		$k \times 10^3 = 0.252 + 0.0478$(thickness in cm)		

REFERENCES FOR THE APPENDIX

1. Breuer, H., "The Thermal Conductivity of Muscle and Fat," Archiv für die gesamte Physiologie, 204 (1924), p. 442.

2. Hardy, J. D. and G. F. Soderstrom, "Heat Loss from the Nude Body and Peripheral Blood Flow at 22-35° C," Journal of Nutrition, 16 (1938), p. 493.

3. Henriques, F. C., Jr. and A. R. Moritz, "Studies in Thermal Injury, I," American Journal of Pathology, 23 (1947), p. 531.

4. Hatfield, H. S. and L. G. C. Pugh, "Thermal Conductivity of Human Fat and Muscle," Nature, 168 (1951), p. 918.

5. Hatfield, H. S., "Measurement of the Thermal Conductivity of Animal Tissue," Proceedings, Physiological Society, March 20-21, 1953, p. 35P.

6. Poppendiek, H. F., "Thermal and Electrical Conductivity of Biological Fluids and Tissues," DDC AD Nos. 608 768, 613 560, 624 897, 630 303, and 630 712, 1964-66.

7. Author's preliminary data.

8. Ponder, E., "The Coefficient of Thermal Conductivity of Blood and Various Tissues," Journal of General Physiology, 45 (1962), p. 545.

9. Lipkin, M. and J. D. Hardy, "Measurement of Some Thermal Properties of Human Tissues," Journal of Applied Physiology, 7 (1954), p. 212.

10. Grayson, J., "Internal Calorimetry in the Determination of Thermal Conductivity and Blood Flow," Journal of Physiology, 118 (1952), p. 54.

11. H. T. Meryman, quoted in Reference 12 of Appendix.

12. Poppendiek, H. F., et al., "Whole Organ Freezing and Thawing Heat Transfer and Thermal Properties," DDC AD Nos. 638 648, 641 900, 646 333, 653 101, 1966-67.

13. Lapshin, A., "Heat Penetration and Conduction in Raw and Melting Fat," Myasnaya Industriya S.S.S.R., 25:2 (1954), p. 55.

14. Vachon, R. I., F. J. Walker, D. F. Walker, and G. H. Nix, "'In Vivo' Determination of Thermal Conductivity of Bone Using the Thermal Comparator Technique," Proceedings, Seventh International Conference on Medical and Biological Engineering, Stockholm, 1967, p. 502 (preliminary data).

15. Kirkland, R. W., "In Vivo Thermal Conductivity Values for Bovine and Caprine Osseous Tissue," Proceedings, Annual Conference on Engineering in Medicine and Biology, 9, Boston, Massachusetts, 1967, p. 204 (preliminary data).

16. Trezek, G. J., D. L. Jewett, and T. E. Cooper, "Measurements of In Vivo Thermal Diffusivity of Cat Brain," Proceedings, NBS Seventh Conference on Thermal Conductivity, Gaithersburg, Maryland, November, 1967.

17. Klug, F., "Research on the Conduction of Heat in the Skin," Zeitschrift für Biologie, 10 (1874), p. 73.

18. Lefevre, J., Chaleur Animale et Bioenergetique. Paris: Maison et Cie, 1911; also Vol. VIII (1929).

19 Lomholt, S., "Investigations on the Temperature Distribution in the Skin During Irradiation with Visible Light," Strahlentherapie, 35 (1930), p. 324.

20. Roeder, F., "Basic Research on the Thermal Regulation of Humans," Zeitschrift für Biologie, 95 (1934), p. 164.

21. Büttner, R., "The Influence of Blood Circulation on the Transport of Heat in the Skin," Strahlentherapie, 55 (1936), p. 333.

22. Aschoff, J. and F. Kaempffer, "Heat Transfer Through the Skin and Its Change During Vasco-constriction," Archiv für die gesamte Physiologie, 249 (1948), p. 112.

23. Hensel, H., "The Physiology of Thermoreception," Ergebnisse der Physiologie, 47 (1952), pp. 180, 182.

24. ————,"The Intracutaneous Variation of Temperature Due to External Temperature Stimuli" and "Intracutaneous Temperature Sensation and Heat Transfer," Archiv für die gesamte Physiologie, 252 (1950), p. 146.

25. Buettner, K., "Effects of Extreme Heat and Cold on Human Skin, I. Temperature Changes Caused by Different Kinds of Heat Applications; II. Surface Temperature, Pain, and Heat Conductivity in Experiments with Radiant Heat," Journal of Applied Physiology, 3 (1951), p. 691.

26. Hardy, J. D., H. Goodell, and H. G. Wolff, "The Influence of Skin Temperature upon the Pain Threshold as Evoked by Thermal Radiation," Science, 114 (1951), p. 149.

27. Hardy, J. D. and T. W. Oppel, "Studies in Temperature Sensation. IV. The Stimulation of Cold Sensation by Radiation," Journal of Clinical Investigation, 17 (1938), p. 771.

28. Vendrik, A. J. H. and J. J. Vos, "A Method for the Measurement of the Thermal Conductivity of Human Skin," Journal of Applied Physiology, 11 (1957), p. 211.

29. Stoll, A. M. and L. C. Greene, "Relationship Between Pain and Tissue Damage Due to Thermal Radiation," Journal of Applied Physiology, 14 (1959), p. 373.

30. Davis, T. P., "The Heating of Skin by Radiant Energy," in Temperature, Its Measurement and Control in Science and Industry, Vol. III, Part 3, "Biology and Medicine." New York: Reinhold Publishing Corporation, 1963, pp. 149-169.

31. Kirkland, R. W. and R. I. Vachon, Auburn University, Auburn, Alabama, preliminary data, personal communication, 1967.

32. Spells, K. E., "The Thermal Conductivities of Some Biological Fluids," Physics in Medicine and Biology, 5:2 (1960), p. 139; also DDC AD No. 229 167, 1959.

33. Miller, H. L. and J. E. Sunderland, "Thermal Conductivity of Beef," Food Technology, 17:4 (1963), p. 490.

34. Awbery, J. H. and E. Griffiths, "Thermal Properties of Meat," Journal of the Society Chemical Industry (London) 52 (1933), p. 326.

35. Tappel, A. L., A. Conroy, M. R. Emerson, L. W. Regier, and G. F. Stewart, "Freeze Dried Meats, Preparations and Properties," Food Technology, 9 (1955), p. 401.

36. Cherneeva, L. I., "Study of the Thermal Properties of Foodstuff," Report of VNIKHI, Gostorgisdat, Moscow, USSR, 1956.

37. Lentz, C. P., "Thermal Conductivity of Meats, Fats, Gelatin Gels, and Ice," Food Technology, 15:5 (1961), p. 243.

38. Rinfret, A. P., Linde Division, Union Carbide Company, Tonawanda, New York, personal communication.

39. Scholander, P. F., V. Walters, R. Hock, and L. Irving, "Body Insulation of Some Arctic and Tropical Mammals and Birds," Biological Bulletin, 99 (1950), p. 225.

40. Hammel, H. T., "Thermal Properties of Fur," American Journal of Physiology, 182 (1955), p. 369.

41. Hart, J. S., "Seasonal Changes in Insulation of the Fur," Canadian Journal of Zoology, 34 (1956), p. 53.

42. Birkebak, R. C., C. J. Cremers, and E. H. LeFebvre, "Thermal Modeling Applied to Animal Systems," Journal of Heat Transfer, Transactions, ASME, C, 88 (1966), pp. 125-130.

43. Lentz, C. P. and J. S. Hart, "The Effect of Wind and Moisture on Heat Loss Through the Fur of Newborn Caribou," Canadian Journal of Zoology, 38 (1960), p. 679.

44. Moote, I., "The Thermal Insulation of Caribou Pelts," Textile Research Journal, 25 (1955), p. 832.

45. Birkebak, R. C., "Heat Transfer in Biological Systems," International Review of General and Experimental Zoology, Vol. II, J. L. Felts and R. J. Harrison, eds. New York: Academic Press, 1966, pp. 268-344.

46. Singh, A. and P. L. Blackshear, Jr., "The Thermal Conductivity of Stationary and Moving Blood," Proceedings, Seventh International Conference on Medical and Biological Engineering, Stockholm, 1967, p. 400.

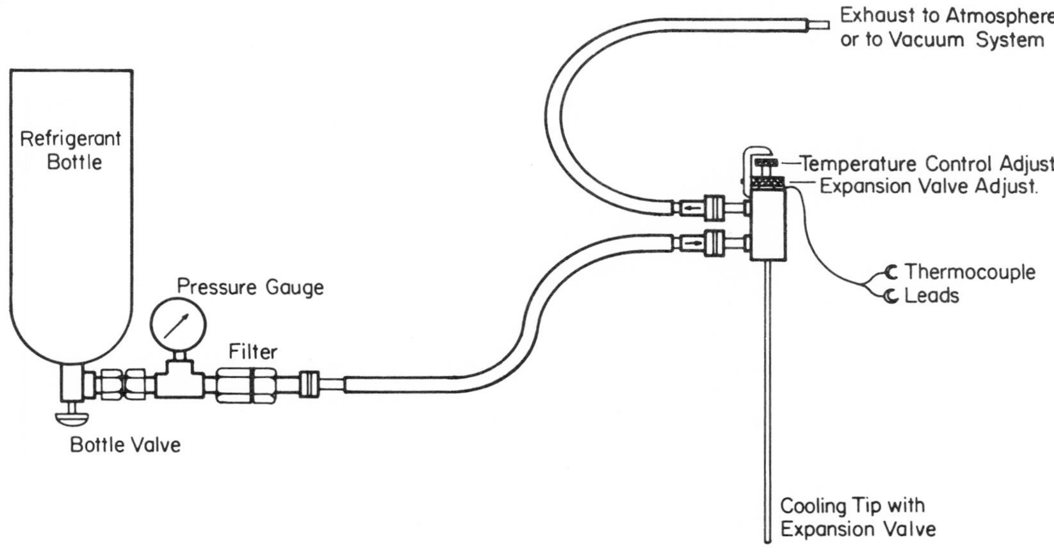

FIGURE 1. SCHEMATIC DIAGRAM OF A SIMPLE SETUP FOR A CRYOPROBE.

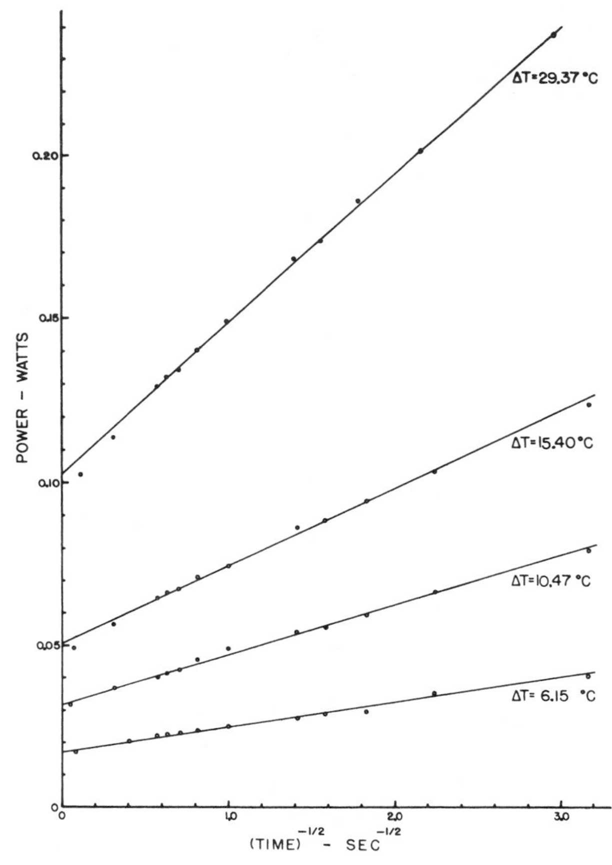

FIGURE 2. THERMISTOR POWER CURVES IN A 1.75 PER CENT AGAR SOLUTION.

FIGURE 3. INTERFEROMETER PICTURE OF THE PALM OF THE HAND. AMBIENT TEMPERATURE = 79.7°F; SURFACE TEMPERATURE AT LOWER ARROW ≃ 85.2°F; SURFACE TEMPERATURE AT UPPER ARROW ≃ 88.0°F

HEAT TRANSFER PROCESSES OF PARTICULATE SUSPENSIONS

S. L. Soo*

ABSTRACT

Fundamental concepts and recent developments relating to the study of heat transfer involving a particulate suspension are discussed. Characteristics of a flowing suspension, effects of inertia, impingement, deposition, radiation, and electric charges play significant parts. These effects give a suspension advantages as well as limitations as a heat transfer fluid. Cases illustrated are pipe flow, channel flow with diffuse radiation, channel flow with electric field, and flow over a cylinder with deposition of liquid droplet and flow of liquid films. Useful data and topics in need of further study are discussed.

INTRODUCTION

Heat transfer in the flow of particulate suspensions may have to be accounted for in physical systems because of either intentionally or unintentionally introduced particulate matter in a working fluid. Addition of particulate matter was found to enhance the heat transfer rate of a gas.(1)** However, increase of heat transfer rate above that of a gas alone due to unintentionally introduced particulate matter such as combustion product, ash, and soot, unless provided for in the design, may cause damage to an apparatus. Accurate data are needed to provide a critical or optimum design.

*Professor of Mechanical and of Nuclear Engineering, University of Illinois, Urbana, Illinois.

**Superscript numbers in parentheses refer to References at the end of this paper.

Systems constituting "particulate" suspensions include: (1) solid particles in a gas or liquid, (2) liquid droplets in a gas, (3) liquid globules in another immiscible liquid, (4) gas bubbles in a liquid, and (5) liquid film bubbles in a gas. A general discussion was given in a recent book(2) in which it is demonstrated that the above list is also the descending order of rigor in which these systems can be treated. Further, when we are concerned with the modification of heat transfer rate, solid particles added to a liquid do not contribute as significantly as when they are added to a gas, although the basic interactions are similar. For these reasons, the present discussion deals with heat transfer in the flow of dilute gaseous suspensions of solid particles and liquid droplets (particulate phases), in the absence of phase change and chemical reactions. (Note that the word "dilute" remains to be defined.) Moreover, since it has been demonstrated that distribution in particle size and multiplicity of species are readily accounted for under the multiphase concept,[2,3] the present treatment is restricted to cases involving one species and size of spherical particulate matter; that is, a monodispersed suspension. With these simplified systems, we shall first outline and identify the basic nature of a particulate suspension. This will be followed by discussions of recent results in the light of basic relations, and an identification of topics in need of further study.

BASIC PARAMETERS AND INTERACTIONS

Study of heterogeneous systems such as suspensions is facilitated by proper choice of parameters. For a suspension of one species and size of particles, we identify the mass of particulate matter per unit volume as density of the particulate cloud, ρ_p, for density of material, $\bar{\rho}_p$, constituting each particle which is itself homogeneous. Thus the volume fraction of the particulate phase, φ, is given by

$$\rho_p = \varphi \bar{\rho}_p . \tag{1}$$

Excluding the volume fraction occupied by the particles, the mass of fluid in a given overall volume is less, and we identify the density of the fluid phase as ρ for density of fluid material $\bar{\rho}$, thus

$$\rho = (1 - \varphi) \bar{\rho}. \tag{2}$$

The density of the mixture is then $\rho + \rho_p$.

For n spherical particles of radius a per unit volume,

$$\rho_p = nm_p = n\left(\frac{4\pi}{3}\right) a^3 \bar{\rho}_p, \tag{3}$$

m_p is the mass of each particle; and the mean interparticle spacing is given by

$$n^{-1/3} = \left(\frac{4\pi}{3\varphi}\right)^{1/3} a. \tag{4}$$

Therefore, for $\varphi = 1/100$, particles are 3 to 4 diameters apart. Due to large density of particulate phase in a gaseous suspension, however, we note that the mass ratio is given by:

$$m^* = \frac{\rho_p}{\rho} = \left(\frac{\bar{\rho}_p}{\bar{\rho}}\right)\left(\frac{\varphi}{1 - \varphi}\right). \tag{5}$$

For a suspension of water droplets in atmospheric air of $\varphi = 1/100$, $m^* = 10$. Note that $\varphi = 1/100$ still does not mean that the suspension is dilute.

At this point, it might appear that we can begin to formulate all the thermodynamic and transport properties of such a mixture. Many authors have, including a formulation for specific heat of a suspension by Pfeffer, Rossetti, and Lieblein,[1] ratio of specific heats by Kliegel,[4] its thermal conductivity by Gorring and Churchill,[5] and its viscosity by Happel[6] through extending the Einstein's equation for a colloid.[7] It is seen that these results are only valid for the case of identical velocities and temperatures of the phases, and the latter condition is not to be expected when a flowing suspension undergoes heat transfer with a wall. Obviously, difficulties arise in the basic definitions of these properties when we attempt to speak of specific heat and thermal conductivity of a suspension when the temperatures and temperature gradients of the phases are different, and also when we attempt to speak of viscosity of a suspension when the velocity gradients of the phases are different. Such effects have been analyzed and identified.[2] It should surprise no one that the measured apparent viscosity of a suspension of air and talc powder by Sproull[8] should show a decrease with the addition of talc powder rather than an increase as predicted by Einstein.[7] Einstein assumed identical velocities of the phases which are pertinent to a colloid. Therefore any logical treatment of a suspension must account for the following facts:

(1) For particles greater than 1μ in diameter (2a), random Brownian motion of particles is negligible. The particles do not contribute to the static pressure of the system and the particle velocity $\underset{\sim}{U}_p$ arises due to viscous drag force exerted by the fluid interacting with the inertia of the particle. We have, in general, $\underset{\sim}{U}_p \neq \underset{\sim}{U}$, the velocity of the fluid.

(2) The fluid phase and the particle phase may have different families of streamlines. Thus the total time, t, derivatives must be identified by:

$$\frac{d}{dt} = \frac{\partial}{\partial t} + U_j\left(\frac{\partial}{\partial x_j}\right) \tag{6}$$

for the fluid phase, and

$$\frac{d}{dt_p} = \frac{\partial}{\partial t} + U_{pj}\left(\frac{\partial}{\partial x_j}\right) \qquad (7)$$

for the particle phase.

(3) Acceleration of particles has to be carried out at the expense of the energy of the fluid, but deceleration of particles does not necessarily raise the static pressure or mass velocity of the fluid. Especially for a dilute suspension, dissipation in the wakes of particles simply constitutes heat addition to the system. This is a special character of free particles, not just a matter of coordinate transformation.

(4) Even a solid or liquid particle of 1μ in diameter may consist of the order of 10^{10} molecules or atoms. The energy of such a particle is identified by its body temperature, T_p, although the temperature, T, of the gaseous phase represents the kinetic energy of random motion of its molecules. Because of the large heat capacity of a particle when compared to similar volume of the gas, in general, $T_p \neq T$.

(5) Due to small size and high thermal conductivity when compared to a gas, the temperature is uniform throughout a particle at a given instant. This is seen in the case of a magnesia particle whose thermal diffusivity, α_p, is of the order of 5×10^{-7} m^2/sec, the time constant for transient heat conduction is

$$\frac{\alpha_p}{a^2} = 2 \times 10^2 \text{ sec}^{-1} ,$$

even for a 100μ particle.[9]

(6) Although addition of heat to a gas increases its total temperature (due to its thermal energy and kinetic energy of mass motion), addition of heat to a cloud of particles simply increases its static temperature. The kinetic energy of a cloud of particles arises due to fluid drag only.

(7) Because of items (1) and (4) and dissimilar gradients, we must identify viscosity, μ, and thermal conductivity, κ, of the fluid phase in the mixture and those of the particle phase in the mixture, μ_p and κ_p; they are, in general, different from those of materials constituting the phases, $\overline{\mu}$, $\overline{\kappa}$, and $\overline{\mu}_p$, $\overline{\kappa}_p$. It was shown that for a gaseous suspension of $\varphi \ll 1$,

$$\frac{\rho_p}{\rho} < 1,$$

$$\mu_{mixture} \lesssim \overline{\mu},$$

$$\kappa_{mixture} \lesssim \overline{\kappa},$$

$$\mu_p \sim \overline{\mu}\left(\frac{\rho_p}{\rho}\right) ,$$

$$\kappa_p \sim \overline{\kappa}\left(\frac{c_p\rho_p}{c\rho}\right);$$

and for

$$\frac{\rho_p}{\rho} \ll 1,$$

$$\mu \sim \overline{\mu},$$

$$\kappa \sim \overline{\kappa},$$

$$\mu_p \sim \rho_p D_p,$$

$$\kappa_p \sim c_p \rho_p D_p.$$

(See References 2, 10, 11, and 48 at the end of this paper.) D_p is the diffusivity of the particles and c_p is the specific heat of the particles. The situation is such that shear stress of the mixture arises due to viscosity of the fluid phase and the momentum transferred by the diffusion of particles. $\rho_p/\rho \ll 1$ may be considered as the definition for a truly dilute suspension.

(8) Recognizing these basic characters of a suspension, we see that even when our interest is in the heat exchange with a suspension in the free stream, interactions of momentum and energy between the phases must be accounted for. For a dilute suspension ($\rho_p/\rho \ll 1$), the fluid-particle interactions are given by the momentum and energy equation of the particles: (when velocity and temperature gradients are absent)

$$\frac{d\underset{\sim}{U}_p}{dt_p} = F(\underset{\sim}{U} - \underset{\sim}{U}_p) + \underset{\sim}{f}'_p + \underset{\sim}{f}_p, \qquad (8)$$

$$\frac{dT_p}{dt_p} = G(T - T_p) + \frac{\mathcal{G}_p}{c_p} , \qquad (9)$$

where $\underset{\sim}{f}_p$ is the field force per unit mass of the particle phase, and $\underset{\sim}{f}'_p$ is the force due to fluid field: apparent mass, relative acceleration, change in flow field and fluid shear,[2,3] and $\mathcal{G}_p$ is the heat generated per unit mass of the particle phase including radiative input. F is the inverse of relaxation time or time constant for momentum transfer between the fluid and the particle; G is the time constant for convective heat transfer between the phases. For mass, m_p, of a particle, F and G are defined by:

$$F = \frac{\text{drag force}}{m_p \left| \underset{\sim\sim}{U} - \underset{\sim\sim}{U_p} \right|} , \tag{10}$$

$$G = \frac{\text{heat exchanged}}{c_p m_p \left| T - T_p \right|} . \tag{11}$$

It is seen that only for small relative motion, or the particle Reynolds number

$$N_{Re} = \frac{2a\left| \underset{\sim\sim}{U} - \underset{\sim\sim}{U_p} \right| \bar{\rho}}{\bar{\mu}} \lesssim 1,$$

that is, motion in the Stokes' law range, F and G are constants, or

$$F = \frac{9\bar{\mu}}{2a^2\, \bar{\rho}_p} , \tag{12}$$

$$G = \frac{3\bar{\kappa}}{c_p a^2 \bar{\rho}_p} . \tag{13}$$

Modification of these relations for large $(N_{Re})_p$ and high particle concentration $\varphi > 0.08$ has been extensively treated.(2) Recent results given by Khorguani(12) on a cluster of particles show a reduction of drag force from that of a single sphere.

It is noted that in identifying $\underset{\sim\sim}{U}$, $\underset{\sim\sim}{U_p}$; T, T_p; we refer to the mass character of the fluid and the particle cloud; distribution of fluid velocity and fluid temperature around each particle is accounted for in terms of F and G.

(9) As has been noted, $\underset{\sim\sim}{f_p'}$ in Equation (8) includes forces due to fluid shear. The lift force acting on a particle is notable in the boundary layer. This lift force due to Magnus effect was estimated by Saffman.(13) It was demonstrated experimentally that the measured impact rate of particles at the wall in turbulent pipe flow without external electric field is two orders of magnitude smaller than an estimate based on random motion.

(10) In a dilute particulate suspension, the effect of particle-particle interaction is negligible. However, there is an interaction length, L_p, of particles with fluid due to Brownian motion, wall interaction, wake effect or fluid turbulence, or their combinations;

$$L_p \sim \frac{\langle \Delta u^2 \rangle^{1/2}}{F} , \tag{14}$$

where $\langle \Delta u^2 \rangle^{1/2}$ is the RMS of relative velocity between the particles and the fluid. L_p may become large when compared to the characteristic physical dimension of a system. Take the example of flow over a flat plate,(2,48) with the flat plate along the x-direction, x = 0 at the leading edge, $U_i = u$, $U_{pi} = u_p$, $x_i = x$, $x_j = y$, and free stream velocity U, we have the boundary condition of particle velocity at the wall:

$$u_{pw} = U(1 - \xi) + L_p\left(\frac{\partial u_p}{\partial y}\right)_{y=0} , \tag{15}$$

where $\xi = Fx/U$ and $\xi = 1$ for $Fx/U > 1$. We have an interaction Knudsen number,

$$N_{Kp} = \frac{L_p}{x} . \tag{16}$$

Hence not only may the particles have different streamlines from the fluid, but the phases may exist in different flow regimes.(2) For example, in a turbulent fluid, the motion of suspended particles could be in a regime analogous to that of a rarefied gas.

(11) For similar reasons, the interaction length for convection heat transfer between the particle and the fluid is

$$L_T = \frac{\langle \Delta u^2 \rangle^{1/2}}{G} = L_p\left(\frac{F}{G}\right) . \tag{17}$$

For the example in item (10) with temperatures T_w at the wall and T_∞ in the free stream, we have the boundary condition for particle temperature given by:

$$T_{pw} - T_w = (T_\infty - T_w)(1 - \xi_T) + L_T\left(\frac{\partial T_p}{\partial y}\right)_{y=0} , \tag{18}$$

with

$$\xi_T = \frac{Gx}{U} = \frac{\xi F}{G}$$

and for $\xi_T = 1$ at $Gx/U > 1$; T_{pw} is the temperature of the particles at the wall. It is further noted that, since the temperature of the particle phase is that of the body temperature of the particles, heat conduction by surface contact must be accounted for.(2) For elastic collision of particles with a wall, it is readily shown that an accommodation coefficient, α, is given by:

$$\alpha \sim \frac{15(2.94)}{16(2)^{2/5}} \left[\frac{\left(\bar{\kappa}_p \bar{\kappa}_w\right)^{1/2}}{c_p \bar{\rho}_p a(\Delta u_w)} \right] \left(N'_{Im}\right)^{4/5}, \quad (19)$$

where the impact number N'_{Im} is given by:

$$N'_{Im} = \frac{5\pi^2}{2} (\Delta u_w)^2 (1 + r^*) \bar{\rho}_p k_p \left(1 + \frac{k_p}{k_w}\right), \quad (20)$$

r^* is the ratio of the reflected speed to the incoming speed Δu_w, $k = (1 - \nu^2)/\pi E$, ν is the Poisson ratio, E is the modulus of elasticity. The temperatures of an incoming particle, T_{pi}, and a reflected particle, T_{pr}, are related by

$$\frac{T_{pr} - T_{pi}}{T_w - T_{pi}} = \alpha. \quad (21)$$

Since $T_{pw} \sim (T_{pi} + T_{pr})/2$, the heat flux, J_{qp}, due to impaction of particles at the wall is given by:

$$\frac{J_{qp}}{c_p \dot{m}_{pw}} = (T_r - T_i) = \frac{2\alpha}{2 - \alpha} (T_{pw} - T_w), \quad (22)$$

where $\dot{m}_{pw}$ is the mass flux of impact at the wall. Equations (18) and (22) account for both the effect of conduction by surface contact at the wall and the effect of convection heat transfer with the fluid.

(12) Even though the fluid phase may be transparent to radiation at the temperature above, say, 1000°F, heat transfer to a suspension by radiation is usually significant. Further, the smaller the particles in the suspension are, the more significant is the effect of the electric charges it carries. An uncharged micron-sized particle is a special case. Particles collect charges by contact with a surface, from atmospheric electricity, from an ionized gas, or by emission.[2]

BASIC FORMULATION

Based on the recognized behavior of a suspension given in the above section, we now present the simplest non-trivial formulations for the general motion of a dilute suspension ($\rho_p/\rho \ll 1$) where the fluid is incompressible ($\rho \sim \bar{\rho}$ = constant). The continuity equations of the phases are:

$$\frac{\partial U_i}{\partial x_i} = 0, \quad (23)$$

and

$$\frac{\partial \rho_p}{\partial t} + \frac{\partial \rho_p U_{pi}}{\partial x_i} = 0. \quad (24)$$

Equation (23) is simply the continuity condition of an incompressible fluid. Equation (24) speaks of the basic behavior of a suspension that ρ_p does not arise due to state parameter such as temperature and pressure, rather it is varied by the relative motion between the fluid and the particles. When the particle phase is slowed down, ρ_p increases. Whether the fluid is compressible or not is immaterial. It was demonstrated experimentally in steady pipe flow that mass ratio and mass flow ratio of the particle phase are different.[14]

The momentum equations of the phases are:

$$\rho \left(\frac{dU_i}{dt}\right) = -\left(\frac{\partial P}{\partial x_i}\right) + \left(\frac{\partial}{\partial x_j}\right) (\bar{\mu}\Delta_{ji}) + K \left[\rho_p F(U_i - U_{pi}) + \rho_p f'_{pi}\right] + \rho \mathcal{F}_i, \quad (25)$$

and

$$\frac{dU_{pi}}{dt_p} = F(U_i - U_{pi}) + f'_{pi} + f_{pi} + \frac{1}{\rho_p} \frac{\partial}{\partial x_j} \left[\rho_p D_p (\Delta_p)_{ji}\right]. \quad (26)$$

In Equation (25), K is an "effectiveness" of momentum transfer,

$$K = 1 \quad \text{for} \quad |U_i| > |U_{pi}|,$$

and

$$1 > K \geq 0 \quad \text{for} \quad |U_i| < |U_{pi}|.$$

Note that this does not constitute a discontinuity in a natural phenomenon because K changes from 1 to 0 at $U_i = U_{pi}$ and

and

$$f'_{pi} = 0;$$

and $f'_{pi} = 0$; $\mathcal{F}_i$ is the field force per unit mass of the fluid;

$$\Delta_{ji} = \frac{\partial U_i}{\partial x_j} + \frac{\partial U_j}{\partial x_i},$$

the deformation tensor and the dilation

$$\Theta = \frac{1}{2}\Delta_{kk} = 0$$

from Equation (23).

The energy equations of the phases are:

$$\rho \frac{d}{dt}\left(\frac{U^2}{2} + cT\right) + \rho_p \frac{d}{dt_p}\left(\frac{U_p^2}{2}\right) = \frac{\partial P}{\partial t} + \frac{\partial}{\partial x_j}\left(\bar{\mu} U_j \Delta_{ji}\right) + \frac{\partial}{\partial x_j}\left(\bar{\kappa}\frac{\partial T}{\partial x_j}\right)$$
$$- c_p\rho_p G(T - T_p) + \rho F(U_j - U_{pj})^2 + \mathcal{G}, \quad (27)$$

and

$$\frac{dT_p}{dt_p} = G(T - T_p) + \frac{\mathcal{Q}_p}{c_p} + \frac{1}{c_p\rho_p}\frac{\partial}{\partial x_j}\left(c_p\rho_p D_p \frac{\partial T_p}{\partial x_j}\right), \quad (28)$$

where c is the specific heat at constant pressure of the fluid, $\mathcal{G}$ is the rate of heat generated per unit mass of the fluid including radiative input. Equation (27) signifies the fact that U_p arises due to U and Equation (28) speaks of the fact that heat added to a particle cloud raises the body temperature of the particles.

The terms including ρ_p in Equations (25) and (27), of course, may be dropped when $\rho_p \to 0$, that is, a very dilute suspension. In this case, the state of the fluid and its motion are entirely unaffected by the presence of particles, although those of the particle phase still arise due to fluid motion[15] and fluid temperature. It cannot be overemphasized that ρ_p can be taken to be constant only for one-dimensional uniform motion.

A further relation is that of diffusion of particles in a field. For a dilute suspension Equation (24) may be replaced by

$$\frac{d\rho_p}{dt} = -\frac{\partial}{\partial x_j}\left[-D_p\frac{\partial\rho_p}{\partial x_j} + \rho_p(U_{pj} - U_j)\right], \quad (29)$$

where the flux $\rho_p(U_{pj} - U_j)$ is given by Equation (26).

PIPE FLOW

Most of the studies on heat transfer of a suspension were made with the pipe flow system, with determination of local and mean heat transfer coefficients.[1] The reason is obvious; the system is simple to work with and the gain, if any, of heat transfer rate due to addition of particles is readily demonstrated. In addition, since pipes constitute usual heat transfer passages, the results are readily applicable to design. This is, however, the point where anomaly of results and disagreements among various contributors began.[16-25]

All referred to loading, that is, the ratio of the rates of throughput of particles to gas, as a basic parameter. In the strict sense loading is the mean mass flow ratio of particle to gas, or

$$\dot{m}^* = \frac{\rho_p \bar{U}_p}{\rho \bar{U}} = m^* \frac{\bar{U}_p}{\bar{U}}, \quad (30)$$

where $\bar{U}_p$ and $\bar{U}$ are mean velocity of pipe flow. This is at least one source of anomaly besides m^*, because modification of heat transfer rate of a gas by the added particles is influenced by the fluid-particle interaction and electrostatic charge on the particles.[14] Some authors, as a matter of logical choice, based their correlations on a weighted specific heat ratio:

$$c^* = \frac{c_p\rho_p\bar{U}_p}{c\rho\bar{U}} = m^*\left(\frac{c_p}{c}\right)\left(\frac{\bar{U}_p}{\bar{U}}\right). \quad (31)$$

Denoting h_m as the heat transfer coefficient of the suspension and h as that due to gas alone, and $N_{Re} = 2R_o\overline{U\rho}/\overline{\mu}$ as the gas phase Reynolds number, $2R_o$ as the pipe diameter, various correlations were presented.

Farbar and Morley,[16] Depew,[17] and Danziger,[18] all correlated with

$$\frac{h_m}{h} = f(N_{Re}, \dot{m}^*), \tag{32}$$

using silica-alumina catalyst (10 to 210μ), glass particles (30 to 200μ) of $\dot{m}^* = 0$ to 41.6, with fluid temperature from 75 to 1050°F. It was reported by Farbar and Depew[17] that for similar $\dot{m}^*$, large particles contribute less to the increase in h_m, and a decrease in (h_m/h) with increase N_{Re} for given (2a) and $\dot{m}^*$. Danziger's result is based on cooling the suspension for 2a = 50μ.

Schluderberg, Whitelaw, and Carlson (see References 19, 20, and 21) gave

$$\frac{h_m}{h} = 0.78\ (1 + C^*)^{0.45} \tag{33}$$

from experiments with graphite suspension (1 to 5 μ) in various gases at 90° to 1100°F. The lack of reproducibility of their results was discussed by Wachtell, Waggener, and Steigelmann[22] who suggested correlation with

$$\frac{h_m}{h} = 16.9\ (N_{Re})^{-0.3}\ (1 + C^*)^{0.45}. \tag{34}$$

Both studies used tube diameters of 0.313 to 0.875 in., and $\dot{m}^* = 0$ to 90.

Gorbis and Bakhtiozin[23] also studied graphite particles and gave

$$\frac{h_m}{h} = 1 + \left[6.3\ N_{Re}^{-0.3}\ (N_{Re})_p^{-0.33}\right] C^*, \tag{35}$$

where $(N_{Re})_p = 2a\ \overline{U}_p\ \overline{\rho}/\overline{\mu}$, and $\overline{U}_p$ is for a gravity system.

Tien and Quan[24] with 30 to 200 μ glass and lead particles showed that (h_m/h) actually decreased below 1 for $0 < \dot{m}^* < 4$. This was no accident since a decrease in the friction factor below clean air in similar conditions is well known.[2] Explanations were offered basing the damping effect of particles on turbulence;[14,25] but the fact remains that such an effect must be accounted for in applications.

Investigations by Mickley and Trilling[26] were made with a fluidized gas-solid mixture. Their results were in agreement with Farbar and Morley,[16] Danziger,[18] and Schluderberg,[21] and reported an effect of particle size. Study by List[27] also indicated such an effect, but there was no agreement between these two studies.[26,27]

Some of the above results are summarized in Figure 1, in which the anomaly of decrease of h_m below h at certain $\dot{m}^*$ is unaccounted for.

Studies by Soo and Trezek[14] have demonstrated that the electric charge induced on particles due to surface interaction with the wall of the pipe has an important effect on the concentration distribution and deposition, as well as velocity distribution of the particle phase. Therefore both the particle material and the pipe material have an effect on the heat transfer rate. The electrostatic charge effect is represented by the electroviscous number for turbulent pipe flow (electrostatic force to turbulent force):

$$N_{ev} = \left(\frac{\rho_{po}}{4\epsilon_o}\right)^{1/2} \left(\frac{q}{m_p}\right) \frac{R^2}{D_p}, \tag{36}$$

where q is the charge per particle; ρ_{po} and D_p are the density at the center of the pipe and diffusivity of the particle phase, respectively; ϵ_o is the permittivity. The effect of electric charge on velocity and concentration of the particle phase is such that at large N_{ev}, ρ_{pw} at the wall increases rapidly. It is not surprising that results such as obtained by Schluderberg[21] are similar to those of a fluidized bed obtained by Mickley and Trilling.[26]

Recent study by Soo[49] has shown that the previous semi-empirical correlations for concentrations and velocities distributions in pipe flow[14] can be replaced by rigorous computations. This is extended here to include heat transfer by convection in pipe flow. For fully developed pipe flow of a suspension of finite N_{Kp} in an incompressible fluid Equations (25) to (29) can be expressed in dimensionless quantities as follows:

$$-\frac{\partial P^*}{\partial x^*} - 2N_m \frac{\rho_{po}}{\rho} \rho_p^* (u^* - u_p^*) + \frac{1}{r^*}\frac{d}{dr^*}\left(r^*\tau^*\right) = 0 \quad , \tag{37}$$

$$\rho_p^* (u^* - u_p^*) + N_{DF}^2 \frac{1}{r^*} \frac{d}{dr^*} \left(r^* \rho_p^* \frac{du_p^*}{dr^*} \right) = 0, \tag{38}$$

$$u^* \frac{\partial T^*}{\partial x^*} = \frac{1}{r^*} \frac{\partial}{\partial r^*} \left(r^* q^* \right) - \frac{G}{F} N_m \frac{c_p \rho_{po}}{c \rho} \rho^* (T^* - T_p^*), \tag{39}$$

$$N_m^{-1} \rho_p^* u_p^* \frac{\partial T_p^*}{\partial x^*} = \frac{G}{F} \rho_p^* (T^* - T_p^*) - N_{DF}^2 \frac{1}{r^*} \frac{\partial}{\partial r^*} \left(r^* \rho_p^* \frac{\partial T_p^*}{\partial r^*} \right), \tag{40}$$

$$-N_S N_{DF}^2 \rho_p^* (u_p^* - u^* - \Delta u^*) \left| \frac{du^*}{dr^*} \right|^{1/2} r^* + 4N_{ev}^2 N_{DF}^2 \rho_p^* \int_0 \rho_p^* r^* dr^* = r^* \frac{d\rho_p^*}{dr^*}, \tag{41}$$

where

$$u^* = \frac{u}{u_o},$$

$$u_p^* = \frac{u_p}{u_o},$$

$$\rho_p^* = \frac{\rho_p}{\rho_{po}},$$

$$T^* = \frac{(T - T_w)}{(T_1 - T_w)},$$

$$T_p^* = \frac{(T_p - T_w)}{(T_1 - T_w)},$$

$$r^* = \frac{r}{R}$$

$$x^* = \frac{x}{R},$$

$$p^* = \frac{P}{\frac{1}{2}\rho u_o^2},$$

$$\tau^* = \frac{\tau}{\frac{1}{2}\rho u_o^2},$$

$$q^* = \frac{J_q}{\rho c u_o (T_1 - T_w)};$$

u, u_p are velocities in the x-direction, r is the radius, R is the pipe radius, J_q is the heat flux at the wall, T_1 is the inlet temperature of the suspension. Subscript o refers to the condition at the center of the pipe. The dimensionless parameters which have not been explained before are as follows: the momentum transfer number (flow time/relaxation time) is

$$N_m = F \frac{R}{u_o}, \tag{42}$$

the diffusion response number (relaxation time to diffusion time) is

$$N_{DF} = \left(\frac{D_p}{FR^2} \right)^{1/2}, \tag{43}$$

and the shear response number (Magnus force/inertia force) is

$$N_S = \frac{3}{4\pi} (81.2) \frac{\bar{\rho}}{\bar{\rho}_p} (N_{Re})^{-1/2} \frac{R}{a}; \tag{44}$$

the Magnus force is the lift exerted on a particle due to fluid shear,[13] and

$$N_{Kp} = \frac{L_p}{R}$$

from boundary conditions according to Equations (15) and (18). Depending on the flow regime, if the suspension is so dilute that the fluid phase is unaffected by the presence of the particles,[50,51]

$$\tau^* = \tau^*(N_{Re}), \tag{45}$$

$$q^* = q^*(N_{Re}, N_{Pr}), \tag{46}$$

where N_{Pr} is the Prandtl number of the fluid. In such a case u^* and T^* are given and Equations (38), (40), and (41) can be solved to determine u_p^*, T_p^*, and ρ_p^*. The case with negligible heat transfer has been solved for a turbulent gas-solid suspension; the Magnus effect is significant only within the thickness of the laminar sublayer. However, charges induced on the particles by impact of particles with the wall produce a higher density at the wall than at the center of the pipe. Velocity distribution of particles is characterized by a slip velocity at the wall and a lag in velocity in the core from the fluid phase. These results are verified by earlier measurement (Figure 2 with $\beta = N_{DF}^{-2}$, $\alpha = N_{ev}^2 N_{DF}^2$).

In general, τ^* and q^* are influenced by the presence of the particles(2) and simultaneous solution of Equations (37) to (41) is necessary but cannot be carried out yet without further knowledge of τ^* and q^* for turbulent flow with the presence of particles. It is seen, however, that empirical correlations for friction in pipe flow should include:

$$\frac{\rho_{po}}{\rho} \text{ or } m^*,\ N_m,\ N_{DF},\ N_{Kp},$$

$$N_{ev}\ (\text{or } N_S),\ \text{and } N_{Re}\ .$$

(N_{ev} is more significant in turbulent gaseous suspensions than in liquid suspensions; N_S is more significant in liquid suspensions.) Empirical correlations of heat transfer by convection should include:

$$m^*,\ \frac{c_p}{c},\ N_m,\ N_{DF},\ N_{Kp},\ \alpha,$$

$$N_{ev}(\text{or } N_S),\ \frac{G}{F},\ N_{Re},\ \text{and } N_{Pr}\ ,$$

if generality of correlated results is expected. It is noted that for heat transfer correlation, C^* in Equation (31) should logically be replaced by $m^*(c_p/c)$ according to Equation (39).

A useful point of view is that the curves in Figure 1 really represent a rolled surface with N_{ev} as the third coordinate; that is, the curves are at least parametric with N_{ev}.

For the pipe flow system, Tien[28] solved the energy equations for steady pipe flow with radiative heat input to a dilute suspension of small optical depth (every particle sees the wall) given in the following form:

$$\rho u \left(\frac{\partial T}{\partial x}\right) = \left(\frac{\rho}{r}\right) \frac{\partial}{\partial r} \left[r(\epsilon_H + \alpha) \frac{\partial T}{\partial r}\right] - G\rho_p\ (T - T_p), \qquad (47)$$

$$u \frac{\partial T_p}{\partial x} = G(T - T_p) + G_r(T_w^4 - T_p^4), \qquad (48)$$

where x is the axial coordinate, ϵ_H is the eddy diffusivity for heat given by Pai,[29] and Sleicher and Tribus,[30] α is the thermal diffusivity of the fluid, $G_r = 3\sigma_r e_p/ac_p\bar{\rho}_p$ (in $sec^{-1}\ {}^oK^{-3}$), where σ_r is the Stefan-Boltzmann constant, and e_p is the emissivity of the particle. He made the assumption that ρ_p is a constant, which is nearly true for small electric charge per particle, and for steady flow with $u = u(r)$ only given by turbulent flow condition. With the approximation $T_w^4 - T_p^4 \sim 4T_w^3(T_w - T_p)$, following separation of variables, Tien solved the problem by reducing the radial distribution of fluid temperature to a Sturm-Liouville differential equation. For a correlating parameter

$$G^* = \frac{6\varphi(N_{Nu})_p\ (R/a)^2}{1 + \left[\bar{\kappa}(N_{Nu})_p \big/ 4\sigma_r \epsilon_p a T_w^3\right]}\ . \qquad (49)$$

Tien computed for 100μ alumina silica particles in air at $\dot{m}^* = 1.0$, $c_p/c = 1.2$, and $T_w = 2000^oF$. His results of the asymptotic Nusselt number

$$(N_{Nu})_a = \frac{h_{ma}(2R)}{\bar{\kappa}}$$

based on $h_{ma} = h_m$ at far downstream from the thermal entrance of the pipe are shown in Figure 3, together with those without the effect of radiation, for two N_{Re} and at various N_{Pr}, the Prandtl number $c\bar{\mu}/\bar{\kappa}$ Although the effect of radiation remains to be proven experimentally, Tien's result without the effect of radiation is confirmed by the measurement of Farbar and Morley.[16]

The above results show that the knowledge of the effect of concentration distribution or electric charge on particles and the effect of radiation remain to be furthered by future studies. Future reports should also indicate pipe materials, or preferably mean electric charge per particle, used in experiments.

EFFECT OF DIFFUSE RADIATION IN CHANNEL FLOW

When dealing with transmission of radiant energy through and to a cloud of particles of sufficient optical depth, absorption, emission, and scattering must be taken into account. The Milne equation[30,31] gives the transfer of energy along a ray as:

$$\frac{dI(s,\theta,\varphi)}{ds} = -\rho_p\beta_m I(s,\theta,\varphi) + \rho_p\alpha_m I_{bb}(s) + \frac{\rho_p\sigma_m}{\pi}\int_0^{2\pi}\int_0^{\pi} I(s,\theta',\varphi')S(\theta,\varphi;\theta',\varphi')\sin\theta' d\theta' d\varphi', \tag{50}$$

where I is the monochromatic intensity of radiation, I_{bb} is that of the black body, s is the distance along a ray, β_m is the monochromatic mass extinction coefficient, α_m is the monochromatic mass absorption coefficient, θ is the polar angle, φ is the azimuthal angle, and S is a scattering function.

$$\beta_m = \sigma_m + \alpha_m, \tag{51}$$

where σ_m is the monochromatic scattering coefficient. The optical depth τ is given by

$$d\tau = \rho_p\beta_m dx, \tag{52}$$

where x is the geometrical distance from a surface. Love and Grosh[32,33] solved Equation (30) by using the Gaussian quadrature formula,[31] and further simplified the problem by taking an isothermal scattering medium at T_p between two parallel walls at T_1 and T_2. Love[33] gave the radiant heat flux $\dot{q}_r$ in the form

$$\dot{q}_{r_1} = T_1^4\sum_{j=1}^{n} A_jM_j - T_2^2\sum_{j=1}^{n} A_jN_j - T_p^4\sum_{j=1}^{n} A_jQ_j, \tag{53}$$

for the net radiation flux at wall 1, where A_j's are universal coefficients, and M_j, N_j, and Q_j are dimensionless parameters which are functions of T, a, complex refractive index m_r, wall reflectivities, and optical depth τ;

$$\tau = \left(\frac{3}{4}\right)\varphi K_e y_o,$$

y_o is the gap height, K_e is the extinction cross section, φ is the volume fraction of particles.

The method of Love and Grosh can be extended to approximate the case of slug flow of a suspension through a channel with walls at a given temperature T_w. Since the middle plane between the plates is a plane of zero heat flux,

$$\dot{q}_r = T_w^4\sum_{j=1}^{n} A_jM_j - T_p^4\sum_{j=1}^{n} A_jQ_j, \tag{54}$$

with $2y_o$ as the height of the channel. Neglecting radiation in the flow direction and the convection of the fluid phase, the energy equation of the particle phase is given in the form

$$c_p\rho_p u_p y_o \frac{dT_p}{dx} = \dot{q}_r . \tag{55}$$

Results of computation shown in Figure 4 are applicable for the case of y_o = 0.2 ft, ρ_p = 0.01 lb/ft^3 of 2μ iron particles (c_p = 0.118), m_r = 1.25 - 1.25 i, reflectivity of the wall of 0.1, T_w = 2000°F, with flow in the x-direction at mean velocity u_p, and T_p = 1000°F at x = 0. Figure 4 shows variation of particle temperatures and wall heat flux along the channel. It is seen that for u_p = 10 fps, x = 10.6 ft for particle temperature to reach 1900°F.

The above approximation for constant u_p and T_p is valid to a turbulent suspension through a channel. Inclusion of convection and fluid-particle interaction should yield further useful information. Another interesting problem is to extend the above solution to pipe flow.

EFFECT OF ELECTRIC FIELD IN CHANNEL FLOW

The fact that particulate matter of a suspension is usually charged due to surface interaction led to an investigation by Chao and Min[34] on the enhancement of heat transfer rate of a gas-solid suspension by introducing a steady or alternating electric field, E. Substantial particle charges were induced by introducing aluminum baffles which produced charge (q) to mass (m_p) ratio of the order of 3×10^{-4} coulomb/kg (positive) on 30μ glass particles in air. (85μ and 270μ were also tested but we shall use the case of 30μ in our discussion.) Tests were carried out in a passage of 1/4-in. height from heat transfer surface on each side of a plate. Electric fields of 1.6×10^6 v/m were imposed at various frequencies between the plate and each heat transfer surface. Suspension of various mass flow ratios was passed through this passage of 1 in. wide over an instrumented length of 18 in. Their results for the case of Reynolds number of 4380 based on passage height and gas phase properties are shown in Figures 5 and 6, giving the increase in Nusselt numbers (ΔN_{Nu}) and the increase in pressure drops above the case without electric field, which checks with the trend of Farbar and Morley.[16] Measurements were made also on the mass flow distribution.

It is seen that the case with steady electric field is similar to sedimentary flow with turbulence and a gravity of $E(q/m_p)$. The high concentration near the wall is close to the fluidized bed condition,[26] and the turbulence normal to the wall due to electric field alone has an intensity amounting to

$$\langle v_q^2 \rangle^{1/2} \sim \frac{E(q/m_p)}{F} \,.$$

(See Reference 35.) Since $F \sim 134\ \mathrm{sec}^{-1}$ for a 30μ glass particle in air,

$$\langle v_q^2 \rangle^{1/2} \sim 3.7\,\mathrm{m/sec}.$$

The fact that such a steady electric field gives larger increase in Nusselt number and pressure drop is readily understood.

The contribution of the alternating electric field is interesting in that the pressure drop is reduced with increasing frequency, but the increase in the Nusselt number passes through an optimum between 14.5 cps and 60 cps. In order to achieve some understanding of the basic mechanism, we shall first investigate the motion of particles with the presence of fluid turbulence and alternating electric field. Equation (26) for the component of motion v and v_p normal to the wall, takes the form (using the Lagrangian coordinate system)

$$\frac{dv_p}{dt} = F(v - v_p) + E\,\frac{q}{m_p} \,. \tag{56}$$

Since $\bar{\rho}_p \gg \bar{\rho}$,[36] where the fluid turbulence is given by:

$$v = \sum A_n \sin(2\pi nt), \tag{57}$$

where n is the frequency and

$$\frac{1}{2} A_n^2 = \langle v^2 \rangle f(n)\ dn, \tag{58}$$

with spectrum f(n) given by:

$$f(n) \sim 2\pi\lambda \langle v^2 \rangle^{1/2} \exp\left(\frac{-\lambda^2\pi^2 n^2}{\langle v^2 \rangle}\right) \tag{59}$$

for Lagrangian microscale λ of the stream. The electric field is given by, for frequency f, and assuming a simple wave,

$$E = E_o \sin(2\pi ft). \tag{60}$$

Equation (56) is readily integrated to give

$$v_p = F \sum_n \frac{A_n \sin(2\pi nt - \theta_n)}{\left[1 + (2\pi n/F)^2\right]^{1/2}} + E_o\left(\frac{q}{m_p}\right)\frac{\sin(2\pi ft - \theta_f)}{\left[1 + (2\pi f/F)^2\right]^{1/2}}, \tag{61}$$

where

$$\theta_n = \tan^{-1}\left(\frac{2\pi n}{F}\right)$$

and analogously for θ_f. Application of the Parseval's theorem gives the intensity of particle motion as:

$$\langle v_p^2 \rangle = \frac{1}{2}\sum_n \frac{A_n^2}{1 + (2\pi n/F)^2} + \frac{1}{2}\,\frac{\left[E_o(q/m_p)/F\right]^2}{1 + (2\pi f/F)^2} + A_f\,\frac{E_o(q/m_p)/F}{1 + (2\pi f/F)^2}\,, \tag{62}$$

where A_f is A_n at $n = f$, and the last term gives the interference of turbulence and alternating electric field. Integration for $f(n)$ given in the above yields

$$\langle v_p^2\rangle = \langle v^2\rangle \frac{\pi^{1/2}}{K} \exp\left(\frac{1}{K^2}\right) \operatorname{erfc}\left(\frac{1}{K}\right) + \frac{1}{2} \frac{\left[E_o(q/m_p)/F\right]^2}{1 + (2\pi f/F)^2}$$

$$+ \frac{2\pi^{1/4}(2\Delta f D)^{1/2} \quad E(q/m_p)/F}{1 + (2\pi f/F)^2} \exp\left[\frac{-\pi^2\lambda^2 f^2}{2\langle v^2\rangle}\right], \tag{63}$$

where D is the stream diffusivity,

$$D = \sqrt{\frac{\pi}{2}}\, \lambda \langle v^2\rangle^{1/2}$$

and Δf is the width of interference (for instance, the experiment was carried out with a nearly square wave form),

$$K = \frac{2}{9}\left[2a\langle v^2\rangle^{1/2} \frac{\bar{\rho}}{\bar{\mu}}\right]\left(\frac{a}{\lambda}\right)\left(\frac{\bar{\rho}_p}{\bar{\rho}}\right)$$

is the turbulent interaction parameter.[2,35] It is seen that the first term on the right side of Equation (63) is just the intensity of particle motion induced by stream turbulence;[35,36] the second term is that due to electric field which decreases with increasing frequency, f, as shown in Figure 7. The third term is that due to interference over Δf between the turbulent field and electric field at frequency f which is exponentially curtailed at high frequency. Figure 7 shows that the velocity of collision of particles with the wall is less at high frequency.

Integrating Equation (61) once more gives, for

$$v_p = \frac{dy_p}{dt},$$

the amplitude of particle oscillation due to electric field as:

$$\text{amplitude} = \left[\frac{E_o(q/m)}{2\pi f F}\right]\left[1 + \left(\frac{2\pi f}{F}\right)^2\right]^{-1/2}, \tag{64}$$

which for the 30μ particles is given in inches in Figure 8 for various frequencies. This quantity when compared to the passage height is revealing: below 20 cps, switching of the field produces particle motion from the one wall to the other traversing the whole passage, but long dwell at the wall is expected at low frequency; however, above 20 cps, some of the particles will simply oscillate without colliding with the wall. Greater pressure drop occurs at low frequencies because of greater particle concentration at the wall and more particles are slowed down and subsequently accelerated in the core at high fluid velocity.

It is noted that collision with the wall is an effective means of transfer of heat to the particles. Based on the area and duration of surface contact of a sphere with another surface given by Herz and Rayleigh[37] (1881), heat transfer coefficient due to impaction h_{wp} is given by:[3,38]

$$(N_{Nu})_{wp} = h_{wp}\frac{R}{\bar{\kappa}}$$

$$= \eta(2.94)\left[\frac{3(2)^{2/5}}{4}\right]\varphi\left(\frac{\bar{\kappa}_p}{\bar{\kappa}}\right)\frac{R}{a}\left[\left(\frac{k_p}{k_w}\right)^{1/2} + \left(\frac{k_w}{k_p}\right)^{1/2}\right]^{4/5}(N_{Im})^{4/5}, \tag{65}$$

where

$$N_{Im} = 5\pi^2\langle v_p^2\rangle\bar{\rho}_p(k_p k_w)^{1/2}$$

is the impaction number and

$$k_p = \frac{(1 - \nu_p^2)}{\pi E_p},$$

$$k_w = \frac{(1 - \nu_w^2)}{\pi E_w};$$

and ν_p, ν_w, E_p, E_w are the Poisson ratios and moduli of elasticity of the particle and wall material, respectively; η is the fraction of particles actually reaching the surface (to account for loss due to Magnus effect and diffusion). More collisions take place with increasing frequency below 20 cps in the present system, while the intensity of impact is not greatly altered at relatively low frequencies. All these effects contribute to an optimum frequency at a given flow condition. For other flow Reynolds numbers in the above system, at $N_{Re} = 5840$, ΔN_{Nu} is insensitive over the range of 3.5 to 14.5 cps, but drops to a low value at 60 cps; at 2920 and 1460 the optimum frequency appears to be between 7.7 and 14.5 cps, at a greater improvement in ΔN_{Nu} than f = 0, but (ΔN_{Nu}) drops again to low values at 60 cps.

FLOW OF LIQUID-DROPLET SUSPENSION OVER A CIRCULAR CYLINDER

Motivated by the possibility of improving heat transfer coefficient of a gas, studies were made on the utilization of gas-liquid-drop suspension. Acrivos(39) experimented with flow over tubes, Tifford(40) studied the flow over tube bundles; a comprehensive review was given by Collier and Pulling.(42) Goldstein, Yang, and Clark(43) presented an analytical study of laminar flow of a gas-liquid-droplet suspension over a circular cylinder. They treated heat transfer through a moving layer of liquid due to accumulated liquid droplets and this layer is carried along by the suspension in the system as shown in Figure 9. Their basic relations include the equations of the liquid layer, and its interaction with the boundary layer of the suspension. Of particular interest is their correlation with a parameter

$$N_E = \varphi \, N_{Re}, \tag{66}$$

with

$$N_{Re} = \frac{v_o R_o \bar{\rho}}{\bar{\mu}} ,$$

where v_o is the velocity of the free stream of the suspension and R_o is the pipe radius. With x measured along the circumference from the stagnation point, Goldstein(44) computed the particle trajectories and determined that for $N_E \geq 0.1$ the liquid drop trajectory is straight, while $N_E \leq 0.1$ the presence of liquid film does not influence the gas boundary layer for any drop trajectories. The droplet diameter does not enter into the final correlations at all. Their results on heat transfer for

$$N_{Nu} = \frac{h_m x}{\bar{\kappa}}$$

compared favorably with the experimental results of Acrivos(39) as shown in Figure 9, for the case

$$N_{Re} = 8 \times 10^4,$$

$$a = 76\mu,$$

$$R_o = 0.75 \text{ in.},$$

$$N_{Pr}(\text{liquid}) = 10,$$

and

$$E = 0.02.$$

It is noted that the studies by Langmuir and Blodgett,(45) Brun and Mergler,(46) and Ranz(47) gave the fraction impacted η as shown in Figure 10, with the fluid-particle momentum transfer parameter given by:

$$\psi = \frac{v_o}{FR_o} = \frac{1}{9}\left(\frac{\bar{\rho}_p}{\bar{\rho}}\right) N_{Re} \left(\frac{a}{R_o}\right)^2 , \tag{67}$$

with correction for deviation from Stokes' law accounted for by:

$$\Phi = 18 \left(\frac{\bar{\rho}}{\bar{\rho}_p}\right) N_{Re} . \tag{68}$$

It is seen that for the above numerical quantities, $\psi = 150$, even though $\Phi \simeq 1.5 \times 10^3$, the deviation from straight line motion of liquid droplet is still very small. The rate of heat transfer is expected to be significantly smaller than predicted without accounting for particle size if a = 7.6μ in the above $\eta \sim 0.5$. Extending studies into this range would be pertinent to environmental effects such as cooling rate in a fog. However, if a large heat transfer rate is of primary interest, we should keep ψ, say, above 100, then the assumption of straight particle trajectory would be adequate.

DISCUSSION

There is no question that the heat transfer rate by convection of a gas can be increased by the addition of solid particles or liquid droplets forming a dilute suspension. This fact has been illustrated with cases of pipe flow (utilizing increased heat capacity), channel flows with significant effect of radiation (making use of the increased absorptivity), or with an applied electric field (utilizing the charge on the particles), and liquid film cooling using suspension instead of a liquid alone (a saving in the quantity of liquid).

However, use of intentionally added particulate matter in a gas is strongly hampered by the inertia and electric charge effects of the particles. Deposition tends to occur where it should not. For the same reason, the use of liquid droplets requires added equipment to produce liquid droplets when they are needed.

A more productive viewpoint is that great usefulness of results of heat transfer studies will be found in obtaining better design knowledge for cases where particles are not deliberately added for improving the heat transfer rate. This includes solid particle handling systems such as in the manufacture of some of the oxides, and systems where particles are introduced because the principal process produces a solid phase such as in some combustion systems or a liquid phase such as fog in the atmosphere. For instance, to produce heat exchange with thorium oxide in its manufacture, we have to design a system with best available information. Accuracy of data becomes more useful if, for example, electric data of the test system is reported, including pipe material as well as particle material. For high temperature systems handling combustion products, accurate knowledge of heat transfer coefficients including radiation makes possible a critical design so that the temperature limit (such as scaling) will not be exceeded. The heat transfer of fog is definitely interesting to environmental studies.

NOTATION

a = radius of a particle

c = specific heat at constant pressure of the fluid

c_p = specific heat of the particles

C^* = specific heat ratio

D = stream diffusivity

D_p = diffusivity of the particles

e_p = emissivity of the particles

E = modulus of elasticity

f = frequency

$\underset{\sim}{f}_p$ = field force per unit mass of the particle phase

$\underset{\sim}{f}'_p$ = force due to fluid field

F = inverse of relaxation time or time constant for momentum transfer between the fluid and the particles

$\mathcal{F}_i$ = field force per unit mass of the fluid

$\mathcal{G}$ = rate of heat generated per unit mass of the fluid

g_p = heat generated per unit mass of the particle phase including radiative input

G = time constant for convective heat transfer

h = heat transfer coefficient due to gas alone

h_m = heat transfer coefficient of the suspension

I = monochromatic intensity of radiation

I_{bb} = monochromatic intensity of the black body

J_q = heat flux at the wall

J_{qp} = heat flux due to impaction of particles at the wall

K = effectiveness of momentum transfer

K_e = extinction cross section

L_p = interaction length of particles with fluid

L_T = interaction length for convection heat transfer

m_p = mass of a particle

m_r = complex refractive index

m^* = mass ratio

$\dot{m}_{pw}$ = mass flux of impact at the wall

$\dot{m}^*$ = mass flow ratio

N_{DF} = diffusion response number

N_{ev} = electroviscous number

N_{Kp} = Knudsen number of particle-fluid interactions

N_m = momentum transfer number

N_{Nu} = Nusselt number

N_{Pr} = Prandtl number of the fluid

N_{Re} = Reynolds number

N_{Im} = impact number

q = charge per particle

$\dot{q}_r$ = radiant heat flux

r = radius

r^* = ratio of the reflected speed to the incoming speed of a particle

R, R_o = pipe radius

s = distance along a ray

S = scattering function

t = time

T = temperature of the gaseous phase

T_p = body temperature of a particle

T_{pi} = temperature of an incoming particle

T_{pr} = temperature of a reflected particle

T_{pw} = temperature of the particles at the wall

T_w = temperature at the wall

T_∞ = temperature in the free stream

u, u_p = velocities in the x-direction

U = free stream velocity

$\underset{\sim}{U}$ = velocity of the fluid

$\underset{\sim}{U}_p$ = velocity of the particulate phase

$\overline{U}$ = mean velocity of the fluid in pipe flow

$\overline{U}_p$ = mean velocity of the particles in pipe flow

v = velocity of the free stream of the suspension

Greek Symbols

α = accommodation coefficient or thermal diffusivity

α_m = monochromatic mass absorption

α_p = thermal diffusivity of particle material

β_m = monochromatic mass extinction coefficient

Δ = shear tensor

Δf = width of interference of frequency

Δu_w = incoming speed of impaction

ϵ_o = permittivity

η = fraction of particles actually reaching the surface

θ = polar angle

κ = thermal conductivity of the fluid phase in the mixture

$\overline{\kappa}$ = thermal conductivity of the fluid material

κ_p = thermal conductivity of the particle phase in the mixture

λ = Lagrangian microscale

μ = viscosity of the fluid phase in the mixture

$\overline{\mu}$ = viscosity of the fluid material

μ_p = viscosity of the particle phase in the mixture

ν = Poisson ratio

ξ = Fx/U

ρ = density of the fluid phase

$\overline{\rho}$ = density of fluid material

ρ_p = density of the particulate cloud

$\overline{\rho}_p$ = density of material

ρ_{po} = particle density at the center of the pipe

σ_m = monochromatic scattering coefficient

τ = optical length

Φ = parameter defined by Equation (68)

φ = volume fraction of the particulate phase, azimuthal angle only in Equation (50)

Subscripts

f = frequency

m = mass

o = condition at the center of the pipe or reference condition

p = particle

w = wall

* = dimensionless quantity

REFERENCES

1. Pfeffer, R., S. Rossetti, and S. Lieblein, "Analysis and Correlation of Heat-Transfer Coefficient and Friction Factor Data for Dilute Gas-Solid Suspension," NASA TN D-3603 (1966).

2. Soo, S. L. Fluid Dynamics of Multiphase Systems. Waltham, Mass.: Blaisdell Publishing Co., 1967.

3. ————, "Dynamics of Multiphase Flow Systems," Industrial and Engineering Chemistry Fundamentals, 4 (1965), p. 426.

4. Kliegel, J. R. and G. R. Nickerson, "Flow of Gas-Particle Mixtures in Axially Symmetric Nozzles," Space Technology Laboratory Report TR-59-0000-00746, Paper No. 1713-61, American Rocket Society Meeting (1961).

5. Gorring, R. L. and S. W. Churchill, "Thermal Conductivity of Heterogeneous Materials," Chemical Engineering Progress, 57:7 (1961), pp. 53-59.

6. Happel, J., "Viscosity of Suspensions of Uniform Spheres," Journal of Applied Physics, 28:11 (1957), pp. 1288-1292.

7. Einstein, A., "Zur Theorie der Brownschen Bewegung," Annalen der Physik, Leipzig, 19 (1906), pp. 371-381.

8. Sproull, W. T., "Viscosity of Dusty Gases," Nature, 190:4780 (1961), p. 976.

9. Kreith, F. Principles of Heat Transfer. Scranton, Pa.: International Textbook Co., 1958, p. 143.

10. Soo, S. L., "Fluid Dynamics of Multiphase Systems," Paper No. 36E, AIChE Conference, Dallas, Texas, 1966.

11. ————, "Fluid Mechanics of Non-Equilibrium Systems," Proceedings, Seventeenth Congress of the International Astronautical Federation, Madrid, 1967, pp. 3-11.

12. Khorguani, V. G. Studies in Physics of the Atmosphere and Ocean. Washington, D.C.: Joint Publications Research Service, July 14, 1966, pp. 38-44.

13. Saffman, P. G., "The Lift on a Small Sphere in a Slow Shear Flow," Journal of Fluid Mechanics, 22 (1965), p. 385.

14. Soo, S. L. and G. J. Trezek, "Turbulent Pipe Flow of Magnesia Particles in Air," Industrial and Engineering Chemistry Fundamentals, 5 (1966), p. 388.

15. Soo, S. L., "Laminar and Separated Flow of a Particulate Suspension," Astronautica Acta, 11:422 (1965); also Applied Mechanics Review, 19, item 2637 (1966).

16. Farbar, L. and M. J. Morley, "Heat Transfer to Flowing Gas-Solids Mixtures in a Circular Tube," Industrial and Engineering Chemistry, 49(1957), p. 1143.

17. Farbar, L. and C. A. Depew, "Heat Transfer Effects to Gas-Solids Mixtures Using Solid Spherical Particles of Uniform Size," Industrial and Engineering Chemistry Fundamentals, 2:2 (1963), p. 130.

18. Danziger, W. J., "Heat Transfer to Fluidized Gas-Solids Mixtures in Vertical Transport," Industrial and Engineering Chemistry (Process Design and Development) 2:4 (1963), pp. 269-276.

19. Babcock and Wilcox Co., "Gas Suspension Coolant Project," Report No. BAW-1159 to AEC, August, 1959.

20. ————, "Gas Suspension Task II," Report No. BAW-1181, November, 1959.

21. Schluderberg, D. C., R. L. Whitelaw, and R. W. Carlson, "Gaseous Suspensions -- a New Reactor Coolant," Nucleonics, 19:8 (1961), pp. 68-70, 72, 74, 76.

22. Washtell, G. P., J. P. Waggener, and W. H. Steigelmann, "Evaluation of Gas-Graphite Suspensions as Nuclear Reactor Coolants," Report No. NYO-9672, AEC, August, 1961.

23. Gorbis, Z. R. and R. A. Baktiozin, "Investigation of Convection Heat Transfer to a Gas-Graphite Suspension Under Conditions of Internal Flow in Vertical Channels," Soviet Journal of Atomic Energy, 12:5 (1962), pp. 402-409.

24. Tien, C. L. and V. Quan, "Local Heat Transfer Characteristics of Air-Glass and Air-Lead Mixtures in Turbulent Pipe Flow," ASME Paper 62-HT-15 (1962).

25. Peskin, R. L. and H. A. Dwyer, "A Study of the Mechanics of Turbulent Gas-Solid Shear Flows," ASME Paper 65-WA/FE-24 (1965).

26. Mickley, H. S. and C. A. Trilling, "Heat Transfer Characteristics of Fluidized Beds," Industrial and Engineering Chemistry, 43 (1951), p. 1220.

27. List, H. L., "Heat Transfer to Flowing Gas-Solids Mixture," Ph.D. thesis, Polytechnic Institute of Brooklyn, Brooklyn, New York, 1958.

28. Tien, C. L., "Transport Processes in Two-Phase Turbulent Flow," Ph.D. thesis, Princeton University, Princeton, New Jersey, 1959.

29. Pai, S. I., "On Turbulent Flow in Circular Pipe," J. Franklin Institute, 256 (1953), p. 337.

30. Sleicher, C. A., Jr. and M. Tribus, "Heat Transfer in a Pipe with Turbulent Flow and Arbitrary Wall-Temperature Distribution," Transactions, ASME, 79 (1957), p. 789.

31. Chandrasekhar, S. Radiative Transfer. New York: Dover Publications, 1960.

32. Love, T. J., Jr. and R. J. Grosh, "Radiation and Heat Transfer in Absorbing, Emitting and Scattering Media," ASME Transactions, Journal of Heat Transfer, 87C (1964), p. 161.

33. Love, T. J., Jr., "An Investigation of Radiant Heat Transfer in Absorbing, Emitting and Scattering Media," Ph.D. thesis, Purdue University, Lafayette, Indiana; also Report ARL-63-3, Aeronautical Research Laboratory, Office of Aerospace Research, USAF (1963).

34. Min, K. and B. T. Chao, "Particle Transport and Heat Transfer in Gas-Solid Suspension Flow Under the Influence of an Electric Field," Nuclear Science and Engineering, 26 (1966), pp. 534-546.

35. Soo, S. L., "Statistical Properties of Momentum Transfer in Two-Phase Flow," Chemical Engineering Science, 5 (1956), p. 57.

36. Chao, B. T., "Turbulent Transport Behavior of Small Particles in Dilute Suspension," Österreichisches Ingenieur-Archiv, 18, 1/2, 7 (1964).

37. Timoshenko, S. Theory of Elasticity. New York: McGraw-Hill Book Co., 1934, p. 339.

38. Soo, S. L., "Comments on Mechanics of Particle Collision in One-Dimensional Dynamics of Gas-Particle Mixtures," The Physics of Fluids, 8:9 (1965), p. 1751.

39. Acrivos, A., "Research Investigation of Two-Component Heat Transfer," Aerospace Research Laboratory Report ARL 64-116 (1964).

40. Tifford, A. N., "Exploratory Investigation of Laminar Boundary Layer Heat Transfer Characteristics of Gas-Liquid-Spray Systems," Aerospace Research Laboratory Report ARL 64-136 (1964).

41. Elperin, I. T., "Heat Transfer of Two-Phase Flow with a Bundle of Tubes," Inzhenerno-Fizicheski Zurnal, 4:8 (1961), p. 30.

42. Collier, J. G. and D. J. Pulling, "Heat Transfer to Two-Phase Gas-Liquid Systems, Part II," Report of United Kingdom Atomic Energy Authority, AERE-R-3809 (1962).

43. Goldstein, M. E., W. J. Yang, and J. A. Clark, "Momentum and Heat Transfer in Laminar Flow of Gas with Liquid-Droplet Suspension Over a Circular Cylinder," ASME Paper No. 66-WA/HT-33 (1966).

44. Goldstein, M. E., "Boundary Layer Analysis of Two-Phase (Gas Liquid) Flow Over a Circular Cylinder and Oscillating Flat Plate," Ph.D. thesis, University of Michigan, Department of Mechanical Engineering, Ann Arbor, Michigan, 1965.

45. Langmuir, I. and K. Blodgett, "Mathematical Investigation of Water Droplet Trajectories," General Electric Research Laboratory, Report RL-225, 1944-45; also Journal of Meteorology, 5 (1948), p. 175; also AAF-TR-5418, February, 1946.

46. Brun, R. J. and H. H. Mergler, "Impingement of Water Droplets on a Cylinder in an Incompressible Flow Field and Evaluation of Rotating Multicylinder Method for Measurement of Droplet-Size Distribution, Volume-Median Droplet Size, and Liquid-Water Content in Clouds," NACA TN 2904 (1953).

47. Ranz, W. E., "On Sprays and Spraying," Engineering Research Bulletin No. 65, University Park, Pa.: Pennsylvania State University, 1956.

48. Soo, S. L., "Non-Equilibrium Fluid Dynamics -- Laminar Flow Over a Flat Plate," Journal of Applied Mathematics and Physics (ZAMP), (In press)(1968).

49. ————, "Pipe Flow of Suspensions" (to be published).

50. Schlichting, H. Boundary Layer Theory. New York: McGraw-Hill Book Co., 1960, p. 508.

51. Rohsenow, W. M. and H. Y. Choi. Heat, Mass and Momentum Transfer. Englewood Cliffs, N.J.: Prentice-Hall, Inc., 1961, p. 196.

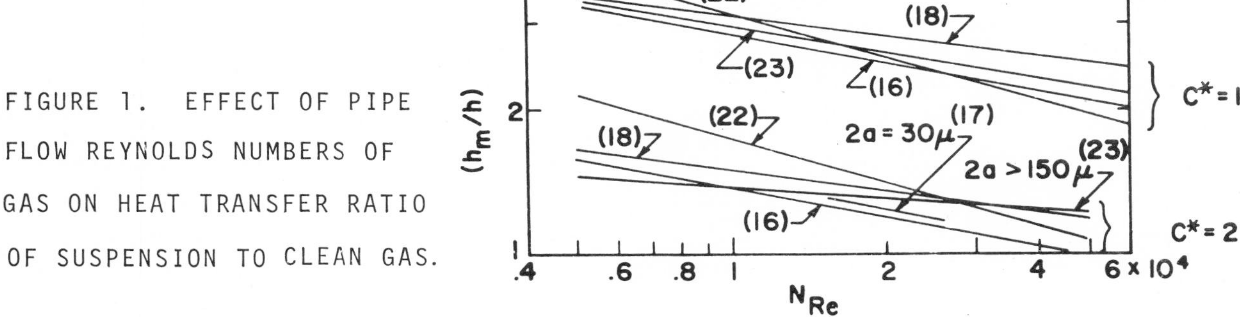

FIGURE 1. EFFECT OF PIPE FLOW REYNOLDS NUMBERS OF GAS ON HEAT TRANSFER RATIO OF SUSPENSION TO CLEAN GAS.

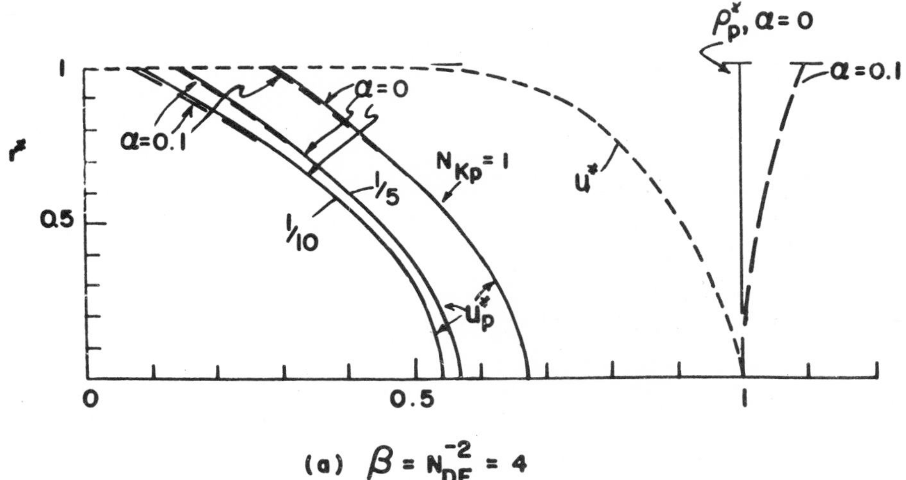

(a) $\beta = N_{DF}^{-2} = 4$

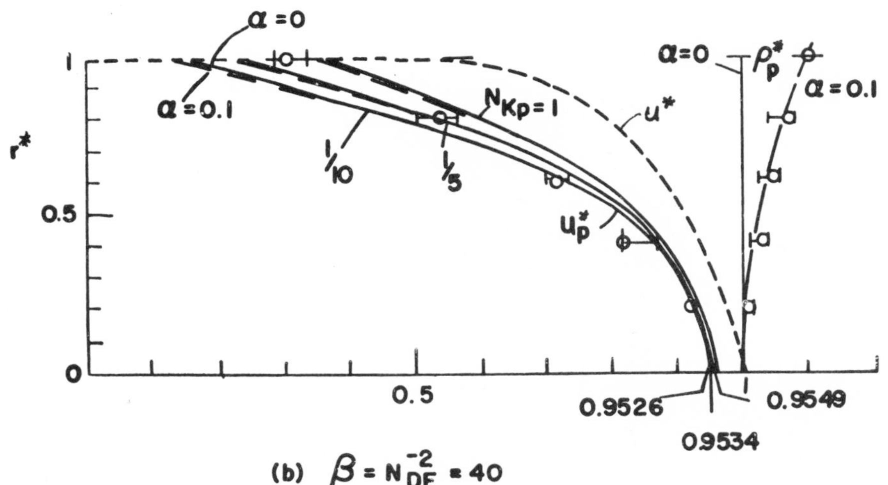

(b) $\beta = N_{DF}^{-2} = 40$

FIGURE 2. VELOCITY AND DENSITY PROFILES IN FULLY DEVELOPED TURBULENT PIPE FLOW OF A SUSPENSION

(O MEASURED VALUE FOR SOLID-GAS MASS RATIO OF 0.45, U_o = 138 fps, 5-IN. PIPE; RANGE FOR OTHER MASS RATIOS.)

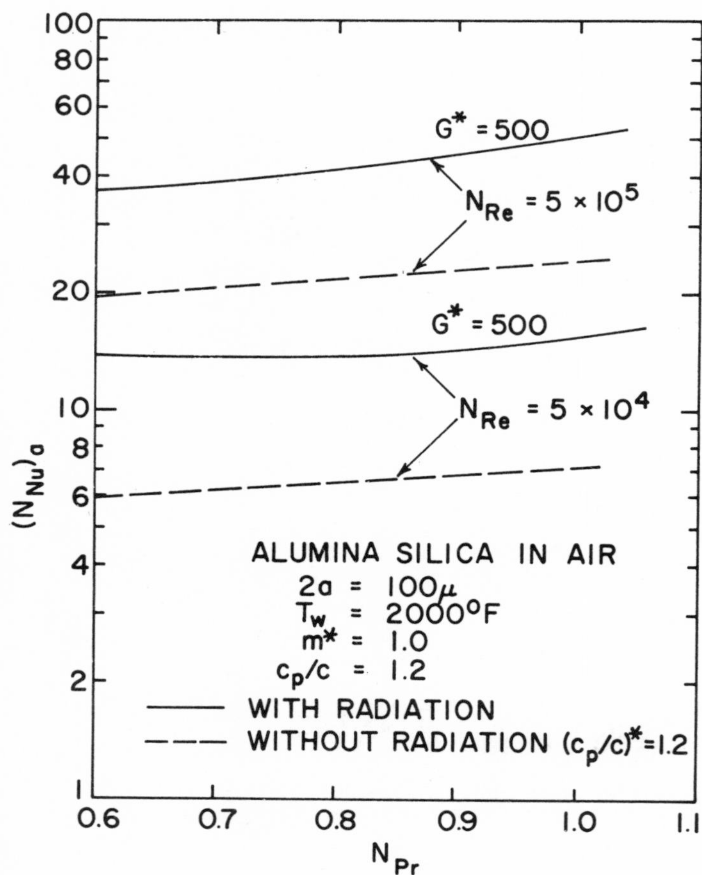

FIGURE 3. ASYMPTOTIC PIPE FLOW NUSSELT NUMBER OF SUSPENSION[28].

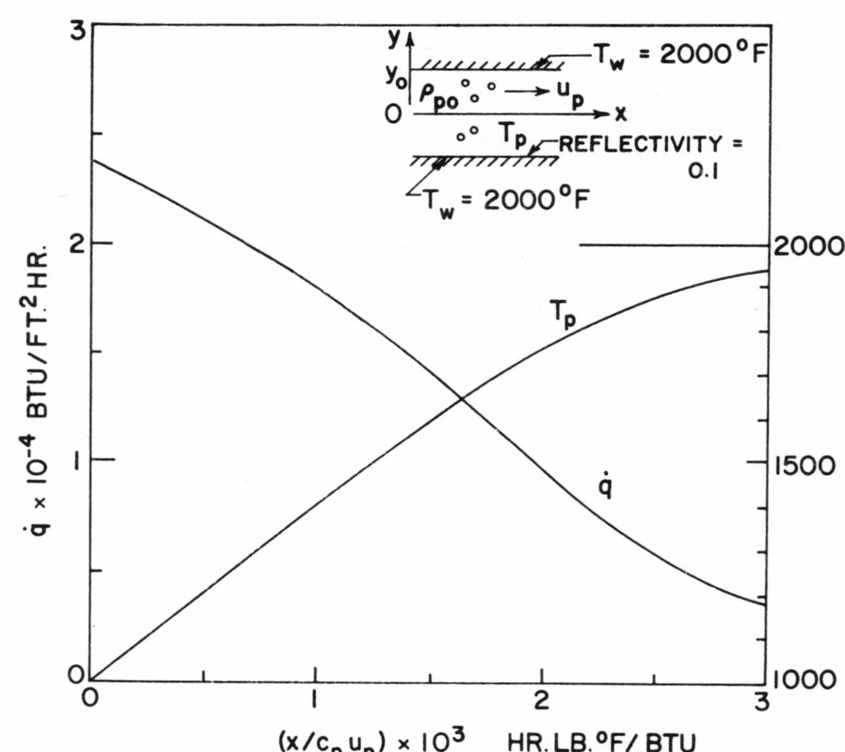

FIGURE 4. RADIATIVE HEAT TRANSFER OF A SUSPENSION.

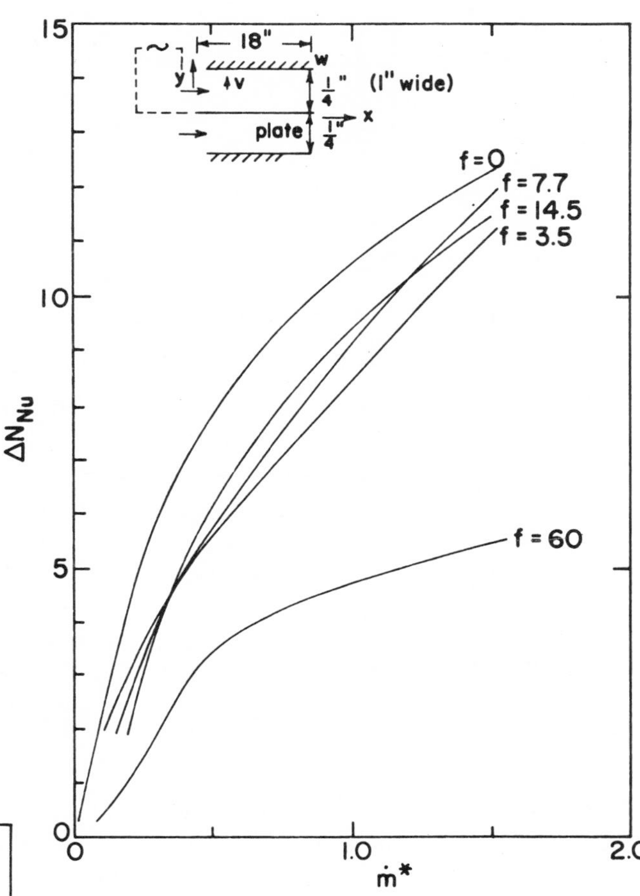

FIGURE 5. INCREASE IN NUSSELT NUMBER AT VARIOUS FREQUENCIES (34).

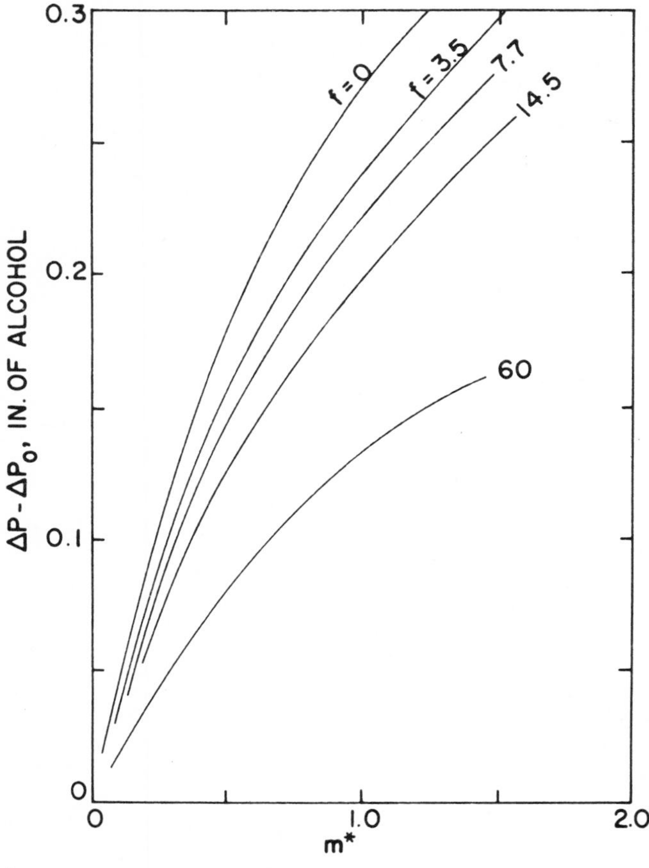

FIGURE 6. INCREASE IN PRESSURE DROP ABOVE THE CASE OF GAS FLOW ALONE (34).

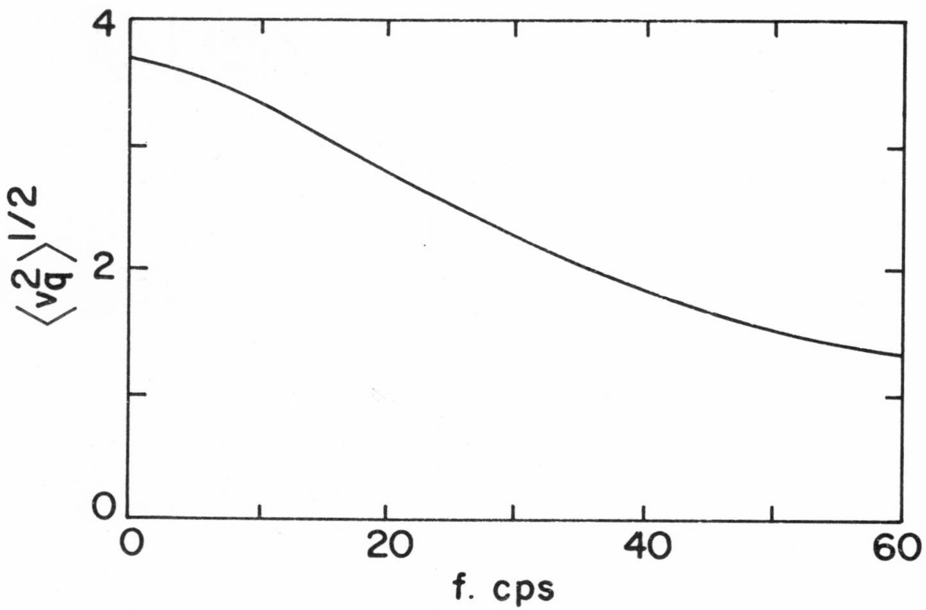

FIGURE 7. INTENSITY OF PARTICLE MOTION DUE TO ELECTRIC FIELD $\langle v_q^2 \rangle^{1/2} = 2^{-1/2}\, [E_0\ (q/m)/F]\, [1 + (2\ \pi f/F)^2]^{-1/2}$.

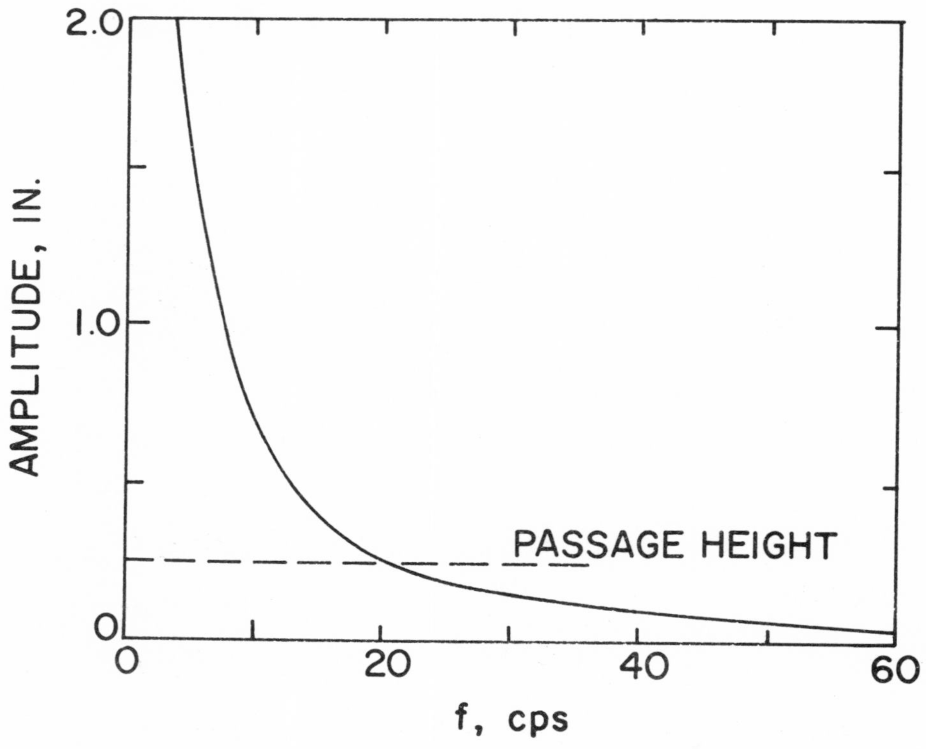

FIGURE 8. AMPLITUDE OF PARTICLE MOTION DUE TO ELECTRIC FIELD.

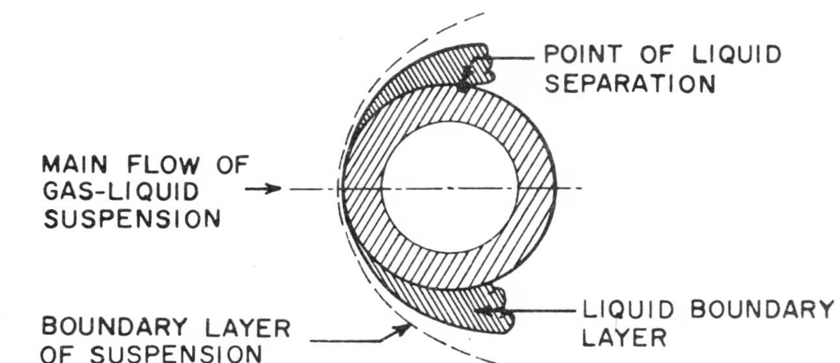

FIGURE 9. FLOW OF GAS LIQUID DROPLET SUSPENSION OVER CYLINDER.

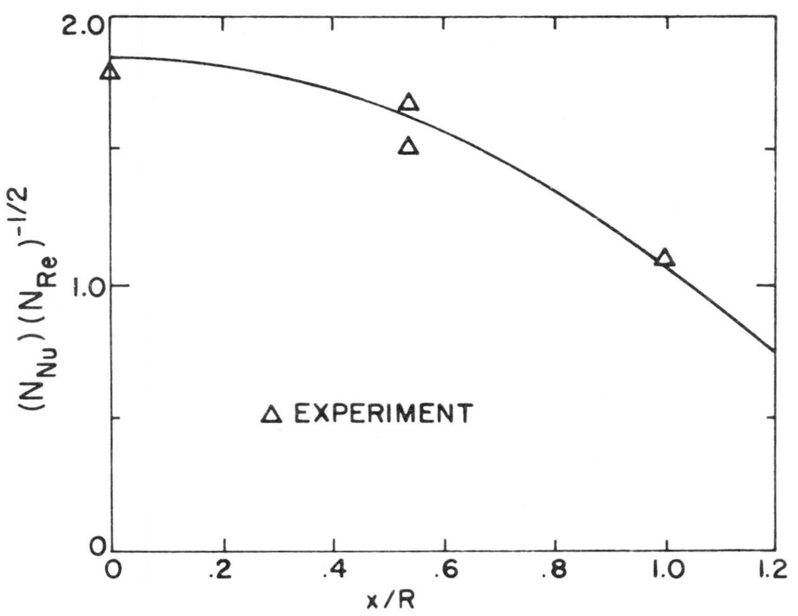

FIGURE 10. COMPARISON OF ANALYTICAL AND EXPERIMENTAL RESULTS OF HEAT TRANSFER FROM CYLINDER BY GAS-LIQUID SUSPENSION, FOR $N_{Re} = 8 \times 10^4$, $a/R = 0.004$, $N_{Pr} = 10$, $N_E = 0.02$, (43) $a = 76\mu$, $R = 0.75$ in., (39) $\psi = 150$.

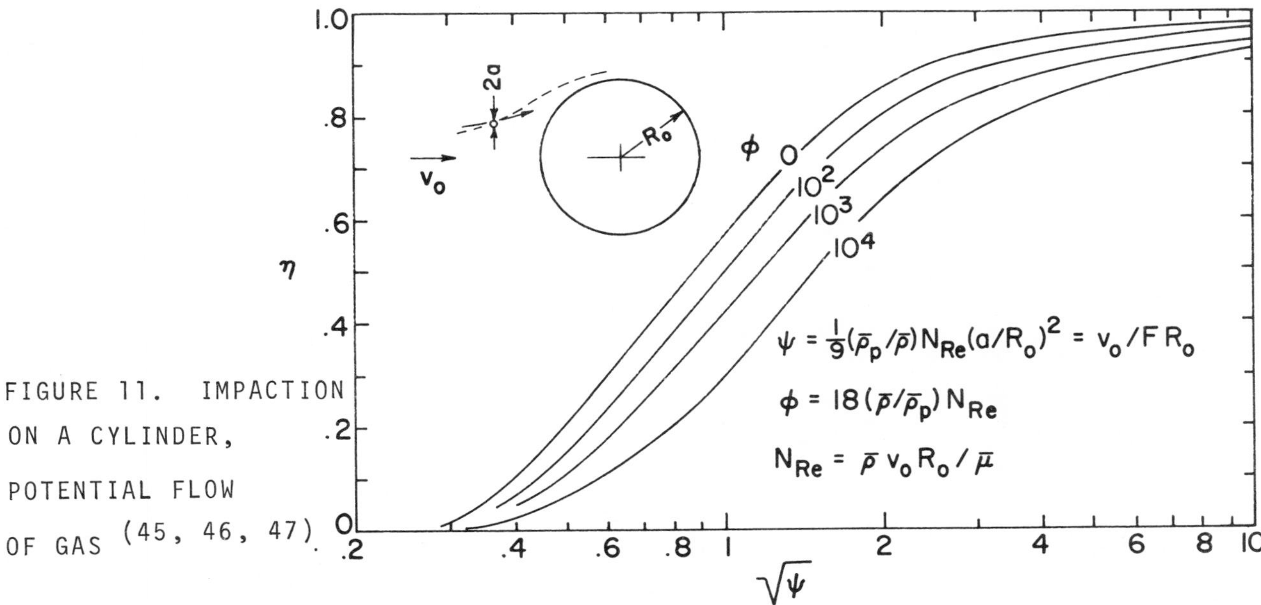

FIGURE 11. IMPACTION ON A CYLINDER, POTENTIAL FLOW OF GAS (45, 46, 47).

COMBINED FREE AND FORCED CONVECTION FLOWS IN CHANNELS

J. C. Chato*

ABSTRACT

The behavior of combined free and forced convection flows in parallel channel systems is examined. The existence of metastable flow regimes is shown both theoretically and experimentally. The effects of the various flow parameters on the magnitude of the metastable regimes are explored.

Since more accurate calculations require more detailed knowledge of the flow in each channel, the second part of this paper describes combined free and forced convection flows in single channels. Emphasis is placed on single-phase, laminar flows and on the transition to turbulence.

INTRODUCTION

Natural circulation or free convection flows occur in every fluid system that has an appreciable amount of heat transfer. In some cases the convection currents set up by density differences serve as means for driving the fluid around its circuit. Examples of application of this principle can be found in boilers, internal combustion engine cooling systems, gas turbine blade cooling schemes, and nuclear reactors. In other cases, natural convection effects are relatively unimportant under normal operating conditions; but can become critically important during transient conditions as in startup, shutdown, or during emergency situations such as a pump failure. Certain atomic reactors are good examples for this situation.

STEADY STATE ANALYSIS OF PARALLEL CHANNEL SYSTEMS

The following analysis is based in part, on the author's previous work.(1-4)**

If within a set of N interconnected channels the heat inputs differ, some sort of natural circulation will tend to establish itself. In order to analyze the flow pattern, the pressure change from the bottom to the top of any channel can be written as follows:

$$\Delta p = - \int_0^h \rho g \, dz \mp \int_0^h 4f \frac{V^2}{2} \rho \frac{dz}{D} \quad \text{(cont.)}$$

*Associate Professor of Mechanical Engineering, University of Illinois, Urbana, Illinois.

**Superscript numbers in parentheses refer to References at the end of this paper.

$$\mp C \frac{\bar{V}_m^2}{2} \rho_m \pm \Delta p_e \ . \tag{1}$$

In this equation the signs on the top of the double signs apply to upward, the signs on the bottom apply to downward flows. The density, ρ, is a cross-sectional average, while ρ_m is a mean value for the entire channel. The terms on the right side of the equation represent the hydrostatic pressure, the loss due to friction, the combined losses due to entrance, exit, and acceleration effects, and the pressure rise due to fluid machinery such as a pump.* The entrance, exit, and acceleration losses could also be represented by separate terms. Their combined effect, however, is usually small in comparison to the other terms and for the purposes of this analysis can be represented satisfactorily by the single expression given in Equation (1).

If the pressure drops in the connecting headers can be neglected, or can be included in the pressure drops of the appropriate channels, then the pressure changes across all channels must be the same.

$$\Delta p_1 = \Delta p_n \tag{2}$$

where n = 2, 3, ..., N.

For a closed system, continuity, i.e., the conservation of mass, requires that

$$\sum_{n=1}^{N} \bar{w}_n = \sum_{n=1}^{N} \rho_n \bar{V}_n A_n$$

$$= \sum_{n=1}^{N} (\rho_n \bar{V}_n A_n)_m = 0 \ . \tag{3}$$

The bar above a flow parameter indicates a directionalized quantity, defined here as positive for upward and negative for downward flow.

The total heat flow into a channel can be expressed in terms of a mean specific heat and the change of the mean fluid temperature from entrance to exit, ΔT_o, if no phase change occurs.

$$Q = wc_{pm} \Delta T_o \ , \tag{4}$$

where the heat input, Q, is positive or negative depending on the sign of ΔT_o.

Assume that the friction factor can be expressed as

$$f = \frac{C_f}{(Re)^s} \ . \tag{5}$$

C_f and s depend on both the Reynolds and Rayleigh (or Grashof) numbers, but can be taken as constants over limited ranges.

In order to solve the equations, the temperature dependence of two quantities has to be established. These are the mean cross-sectional density, ρ, and a combination of the viscosity and density defined as

$$\varphi = \frac{\mu^s}{\rho} \ . \tag{6}$$

Assume that both of these properties vary linearly with temperature, i.e.,

$$\rho = \rho_o(1 - \beta\Delta T) \ , \tag{7}$$

$$\varphi = \varphi_o(1 - \gamma\Delta T) \ . \tag{8}$$

If, in addition, it is assumed that all channels have the same height, h, and that the temperatures in them vary linearly from entrance to exit, then a set of 2N nondimensional working equations can be obtained which relate the 4N variables (X_n, Re_n, q_n, and T_{on} or ρ_{on}).

$$\frac{\rho_{o1}}{\rho_{on}} \left[\left(1 + \tau_1 a_1 Re_1^{1-s_1} \overline{Re_1}\right) X_1 - a_1 Re_1^{1-s_1} \overline{Re_1} (\tau_1 - 1) + a_1 b_1 \overline{Re_1} \frac{Re_1}{X_1} - p_{e1} \frac{\overline{Re_1}}{Re_1} \right]$$

$$= \left(1 + \tau_n a_n Re_n^{1-s_n} \overline{Re_n}\right) X_n - a_n Re_n^{1-s_n} \overline{Re_n} (\tau_n - 1) + a_n b_n \overline{Re_n} \frac{Re_n}{X_n} - p_{en} \frac{\overline{Re_n}}{Re_n} \ , \tag{9}$$

where n = 2, 3, ..., N.

*For a hydraulic motor Δp_e is a negative quantity.

$$\sum_{n=1}^{N} \overline{Re}_n \, k_n D_n \mu_{on} = 0 \,, \tag{10}$$

$$q_n = Re_n \, (1 - X_n) \,, \tag{11}$$

where $n = 1, 2, \ldots, N$.

If, by selecting C_f judiciously, s can be taken as unity, Equation (9) can be simplified.

$$\frac{\rho_{o1}}{\rho_{on}} \left[(1 + \tau_1 a_1 \overline{Re}_1) \, X_1 - a_1 \overline{Re}_1 (\tau_1 - 1) + a_1 b_1 \overline{Re}_1 \frac{Re_1}{X_1} - p_{e1} \frac{\overline{Re}_1}{Re_1} \right]$$

$$= (1 + \tau_n a_n \overline{Re}_n) X_n - a_n \overline{Re}_n \, (\tau_n - 1) + a_n b_n \overline{Re}_n \frac{Re_n}{X_n} - p_{en} \frac{\overline{Re}_n}{Re_n} \,, \tag{9a}$$

where $n = 2, 3, \ldots, N$.

Analysis of a Three-Channel System

The simplest configuration in which several flow patterns can occur with a given set of heat inputs is a three-channel system. Its study can give considerable insight into the behavior of multi-channel flows. In many cases a complicated equipment can be described approximately as a three-channel system by lumping together channels with similar heat inputs.

Assume that the channels are long enough to make $b_n \simeq 0$, and that $s = 1$. Let the first channel have the highest heat input and the third channel have the lowest, i.e., smallest positive or highest negative heat input. Equation (9a) can be written in the following form:

$$(1 + \tau_n a_n \overline{Re}_n) X_n - a_n \overline{Re}_n \, (\tau_n - 1) - p_{en} \frac{\overline{Re}_n}{Re_n}$$

$$= \frac{\rho_{o3}}{\rho_{on}} \left[(1 + \tau_3 a_3 \overline{Re}_3) X_3 - a_3 \overline{Re}_3 (\tau_3 - 1) - p_{e3} \frac{\overline{Re}_3}{Re_3} \right] , \tag{12}$$

where $n = 1, 2$.

The case of an adiabatic third channel and $p_{en} = 0$ was investigated previously. (See References 1 and 2.) The results can be summarized as follows.

The governing equation becomes:

$$q_n = Re_n \left(1 + a_n \overline{Re}_n\right) \frac{1 - \left[\rho_{o3} \left(1 + a_3 \overline{Re}_3\right)\right] / \left[\rho_{on} \left(1 + a_n \overline{Re}_n\right)\right]}{1 + \tau_n a_n \overline{Re}_n} \,, \tag{13}$$

where $n = 1, 2$.

Since $a_n Re_n < 1$ for the applications of interest, the following relations can be established for the heat input to the second channel.

$$q_2 < 0 \text{ for } \frac{\rho_{o3} \, (1 + a_3 \overline{Re}_3)}{\rho_{o2} \, (1 + a_2 \overline{Re}_2)} > 1$$

$$q_2 = 0 \text{ for } \frac{\rho_{o3} \, (1 + a_3 \overline{Re}_3)}{\rho_{o2} \, (1 + a_2 \overline{Re}_2)} = 1 \text{ or } Re_2 = 0$$

$$q_2 > 0 \text{ for } \frac{\rho_{o3} \, (1 + a_3 \overline{Re}_3)}{\rho_{o2} \, (1 + a_2 \overline{Re}_2)} < 1 \,.$$

Figure 1 shows typical characteristics of a system of three identical channels for which all T_o, a, and τ values are the same, and Q_1 is constant. The most significant curve is that for the heat input ratio, Q_2/Q_1, which rises from negative to a positive maximum critical value, Q_{2c}/Q_1, falls to zero at $Re_2 = 0$, then rises again.

For the more general case of the adiabatic third channel with $p_{en}=0$, in the regime

$$\left[\frac{\rho_{o3}}{\rho_{o2}}\left(1 + a_3\overline{Re}_3\right) - 1\right] < a_2\overline{Re}_2 < 0 ,$$

the heat input to the second channel, Q_2, is positive, but the flow in it is negative, i.e., downward. The analysis shows that the heat input range $0 < Q_2 < Q_{2c}$ is metastable where one unstable and two metastable flow patterns exist for a given Q_2. The two metastable regimes are:

$$\left[\frac{\rho_{o3}}{\rho_{o2}}\left(1 + a_3\overline{Re}_3\right) - 1\right] < a_2\overline{Re}_2 < a_2\overline{Re}_{2c}$$

and

$$\overline{Re}_2 > 0 .$$

References 1 and 2 give the critical heat input ratios with identical entrance temperatures in all three channels. If two out of three channels have the same diameter, then increasing the diameter of either the first or the second channel will increase the critical heat input ratio from zero to an asymptotic value of 0.1717; while decreasing the diameter of the third will increase the critical heat input ratio from zero to 1.0. For three identical channels the theoretical critical heat input ratio is 0.0714.

The author and his co-workers have proven experimentally the existence of the metastable regime using water,[1] refrigerant 113,[2,3] and air.[4] The experimental equipment was designed to provide identical or nearly identical entrance temperatures. For these conditions, with the assumption that

$$|\tau_n a_n \overline{Re}_n| \ll 1 ,$$

the previous analysis leads to the following equation for the critical heat input ratio.

$$\frac{Q_{2c}}{Q_1} = \frac{r_c}{|r_c|}\left\{\frac{\left[(a_2/a_1) + (D_2a_3/D_3a_1)\right] r_c^2 + (D_1a_3/D_3a_1)r_c}{1 + (D_1a_3/D_3a_1) + (D_2a_3/D_3a_1)r_c}\right\}\left[\frac{1 + \tau_1 a_1\overline{Re}_1}{1 + \tau_2 a_2\overline{Re}_2}\right]\frac{D_2}{D_1}, \tag{14}$$

where

$$r_c \equiv \frac{\overline{Re}_{2c}}{\overline{Re}_1}$$

can be found from the quadratic equation

$$\left(\frac{D_2a_2}{D_1a_1} + \frac{D_2^2a_3}{D_1D_3a_1}\right)r_c^2 + 2\left(\frac{D_3a_2}{D_1a_3} + \frac{D_2}{D_1} + \frac{a_2}{a_1} + \frac{D_2a_3}{D_3a_1}\right)r_c + 1 + \frac{D_1a_3}{D_3a_1} = 0 . \tag{15}$$

Comparison of the critical heat input rates (using $C_f = 16$, $s = 1$) indicated fair agreement between theory and experiment. The greatest deviations occurred generally with bigger diameters, greater heat inputs, or at very low critical heat input ratios.

Effects of Property Variations

The previous analysis is based on a rather simple modeling of the flow. Although it serves quite well to study approximate behavior of multiple-channel systems, for more accurate results more detailed and accurate descriptions of the flow are needed. The effects of actual property variations in individual channels will be discussed in later sections. Here manifestations of these effects in multiple-channel systems, and particularly on the metastable regime of a three-channel setup will be described.

Figure 2 shows that increasing the magnitude of the heat input rate tends to decrease the critical heat input ratio. One partial explanation for this behavior is the increase of the friction factor with increasing heat input rates, i.e., increasing Rayleigh (or Grashof) numbers. Such an increase in the first, most strongly heated channel increases the pressure drop for a given flow rate. The result is the same as when the channel diameter is reduced, i.e., the critical heat input ratio goes down. The same can be said about the effect of the entrance, exit, and acceleration losses as expressed by the parameter, b. Its increase also results in a decrease of the critical heat input ratio.

Changing the channel entrance temperatures and, consequently, the corresponding properties also has marked effects. Consider the case with an adiabatic third channel as expressed by Equation (13). Since

$$a_n \, Re_n \ll 1$$

and

$$\frac{q_n}{Re_n} \ll 1,$$

the following qualitative deductions can be made if $q_1 > 0$, $q_2 > 0$, and ρ_{o3} are all assumed to be known and fixed:

(1) The density, ρ_{on}, has much stronger influence than the viscosity, μ_{on}.

(2) A decrease of the density at the entrance, ρ_{on}, increases the absolute value of the flow and the corresponding Reynolds number, Re_n. This situation generally corresponds to an increase in the entrance temperature, T_{on}, where n = 1, 2.

(3) A decrease of the density at the entrance of the second channel, ρ_{o2}, corresponding to an increase in the entrance temperature, T_{o2}, reduces the metastable flow regime until it disappears when

$$\frac{\rho_{o2}}{\rho_{o3}} = 1 + a_3 \, \overline{Re}_3 \, .$$

Effects of Boiling

The effects of boiling, obviously, cannot be calculated by the previous method. The author performed a number of experiments with three-channel systems to gain at least qualitative insight into the influence of boiling, particularly in the metastable regime.

In these experiments boiling in the second heated channel produced far more drastic results than boiling in the first. The obvious reason was that, even if there was downward flow in this channel, the vapor bubbles moved upward against the flow and thereby very strongly disturbed it. In the usual case with any appreciable amount of vapor formation, a large bubble grew near the bottom of the second channel. The bubble rose upward and reduced or reversed the downward flow of liquid. After the bubble left the channel at the top, the existing pressure differences forced the liquid from the top header down into the second channel, thereby starting the process all over again. The result was a steady oscillation. This was augmented in many cases by concurrent periodic boiling in the first channel. As the bubble in the second channel temporarily slowed down the downward flow, the upward flow in the first channel also had to slow down. Since the heat inputs remained constant, usually boiling developed during this period at the top of the first channel, causing an even more forceful downflow of the liquid in the second channel after each bubble departure. This type of instability was observed both in water[1] and in refrigerant 113.[3] For example, in a 45 1/4-in.-high three-channel system with two heated 0.4-in. i.d. and one unheated 0.2-in. i.d. tubes, such boiling oscillation occurred with a period of about 8.5 seconds, with refrigerant 113.

The exact effects of boiling in the first channel were more difficult to determine. This was due partly to the fact that in the glass tubes used the boiling was very erratic. The smooth walls did not provide adequate sites for nucleate boiling to develop. In most cases boiling had to be induced by a thin, flat copper ring which was introduced at the top and could be located around the inside surface of the tube at any axial position. Thus the amount and the axial starting point of boiling could be controlled both by the heat input and by the position of the copper ring. The basic question to be answered at least qualitatively was whether the driving force due to the reduction in mean density increased faster than the pressure losses. If it did, then the critical heat input ratio was expected to increase as boiling developed in the first channel. Figure 3 shows the experimental results obtained with refrigerant 113. In all cases the critical heat input ratio increased by about 30 per cent as boiling developed in the first channel.

It should be pointed out that instabilities and periodic oscillations may develop in a multi-channel system in many ways, only one of which was described above. Another, similar periodic flow pattern was observed in the experimental three-channel system just below the critical heat input rate but without boiling: the flow in the second heated channel changed directions in more or less regular periods. A quantitative explanation of this phenomenon, however, would require a much more detailed study than the previous analyses can provide. This phenomenon, oscillations without boiling, seems to be unusual; at least the publications on oscillations in multichannel systems refer overwhelmingly to two-phase flows.

Effects of Pumping

Examination of the governing equations, Equation (9), (9a), or (12), indicates that natural circulation effects certainly become insignificant if

$$|p_e| \gg 1.$$

However, a more detailed study on the previously used specific case of a three-channel system with an adiabatic third channel yields a more stringent restriction. Assume that in the regime of interest the flow in the third channel is always downward, i.e., $\overline{Re}_3 < 0$; and that a pump is located in the most practical position, i.e., next to the third channel. Then

$$p_{e3} \frac{\overline{Re}_3}{Re_3} < 0$$

$$p_{e1} = p_{e2} = 0.$$

Also assume that p_{e3} is a known constant. The governing equation, corresponding to Equation (13), then becomes:

$$q_n = Re_n \left(1 + a_n \overline{Re}_n\right) \frac{1 - \left[\rho_{o3}\left(1 + a_3\overline{Re}_3 + p_{e3}\right)\right] / \left[\rho_{on}\left(1 + a_n\overline{Re}_n\right)\right]}{1 + \tau_n a_n \overline{Re}_n}, \tag{16}$$

where n = 1, 2.

The metastable flow regime occurs in the range:

$$\left[\frac{\rho_{o3}}{\rho_{o2}}\left(1 + a_3\overline{Re}_3 + p_{e3}\right) - 1\right] < a_2\overline{Re}_2 < 0.$$

The relation indicates that pumping in the third channel reduces this metastable flow regime. Since

$$\frac{\rho_{o3}}{\rho_{o2}} \simeq 1 ,$$

the metastable regime reduces to zero when

$$p_{e3} \simeq - a_3\overline{Re}_3 .$$

Since

$$| a_3\overline{Re}_3 | \ll 1 ,$$

this limitation is much more stringent than the one given at the beginning of this section. Further reflection on the characteristics of natural circulation flows leads to the conclusion that, since the driving head without boiling is of the order of

$$(1 - X_n) = \frac{q_n}{Re_n}$$

in dimensionless form, the natural circulation effects will become small if

$$| p_{en} | \gg \left|\frac{q_n}{Re_n}\right| \simeq | a_n\overline{Re}_n | . \tag{17}$$

Equation (17) will be satisfied in any practical forced circulation system under normal operating conditions. However, in case of a pump failure not only will the pumping head vanish, but the pump itself will contribute a considerable pressure loss to the flow. The result will be a rapid, or possibly violent, slowdown in the return line accompanied by strong disturbances in the rest of the system. Since the increased resistance in the return channel increases the metastable flow regime, a pump failure may cause reverse flow in a heated channel.

FREE CONVECTION IN SINGLE CHANNELS

In order to determine correctly the behavior of a multi-channel system, the effects of heating or cooling in each single channel must be known or at least estimated.

Vertical Channels

The behavior patterns of the flow in a vertical, nonadiabatic channel fall into two basic categories according to the axial direction of the density gradient. In the first case the density decreases upward and buoyancy drives the fluid in the direction of the mean flow. In the second case the density increases upward and buoyancy tends to force the fluid opposite to the mean flow. Thus heating in upflow is similar to cooling in downflow, while cooling in upflow is similar to heating in downflow.

Since there is heat transfer, a nonuniform temperature profile develops. Then the buoyancy effects distort the velocity profiles from those found in adiabatic flows. Qualitatively, a temperature field which reduces the density causes the fluid to change velocity in the upward direction. The opposite effect is found if the density is increased. Although the heat flux is greatest at the wall, viscosity keeps the velocity here always at zero. Consequently, velocity changes due to buoyancy effects should first occur near, but not at the wall, then propagate inward as the flow progresses along the channel.*

Other important effects of heat transfer are the development of turbulence at very low Reynolds numbers and the change from symmetric to asymmetric flow configurations.

There have been numerous studies on nonadiabatic vertical pipe flows since Graetz first developed an infinite series solution for the uniform and parabolic velocity profiles with constant wall temperature.[5] Most of these involved a number of simplifications and assumptions in order to make the solutions tractable. The various theoretical approaches can be grouped in the three categories described in the following paragraphs.

(1) The velocity profiles were assumed to be the same as those in the appropriate adiabatic flows, i.e., the fluid properties were considered essentially constant. The Graetz solutions were typical of these. Sellars, et al.,[6] extended the Graetz solutions to include the constant wall heat flux and constant wall temperature gradient cases. Singh[7] also used parabolic and uniform velocity profiles and included axial heat conduction, viscous dissipation, and internal heat generation effects. Kays[8] and Ulrichson and Schmitz[9] used Langhaar's developing velocity profiles[10] to find numerical solutions to the heat transfer in the entrance region. To solve the same problem for high Prandtl numbers, Tien and Pawelek[11] employed local straight line velocity approximations in the thin thermal boundary layer. Roy[12] and Siegel, et al.,[13] used parabolic velocity and fourth order temperature distributions. Petukhov and Tsvetkov[14] and also Grigull and Tratz[15] obtained numerical solutions for the developing temperature distributions, the latter with parabolic velocity profiles.

(2) The flow was assumed to be fully developed, i.e., the axial gradients were either zero or constant and, consequently, the velocity and temperature profiles became independent of the axial distance. Since this approach allowed for the variation of one or more properties, the results tended to be more realistic and hence more widely applicable than those of the previous method, provided the channel was long enough to render entrance effects relatively small.

Ostroumov[16,17] and Hallman[18] solved the differential equations for fully developed flow with linear temperature dependence of the density. The latter also included internal heat generation. Brown and Grassman[19,20] treated mass flux as well as heat flux. Deissler[21] considered variable density, viscosity, and thermal conductivity. Hanratty, et al.,[22] gave results for uniform wall heat flux with heating and cooling in both upflow and downflow. Pigford[23] used an integral method allowing for simultaneous, linear variations of the density and viscosity for high flow rates in short tubes. A similar method was applied by Rosen and Hanratty[24] to obtain solutions to the problem of heating in upflow with an initially fully developed laminar flow. They assumed power series for the velocity and temperature profiles.

Using complex functions, Tao[25] developed solutions for the constant wall temperature gradient case with internal heat generation. Bradley and Entwistle[26] performed numerical calculations for the cases of upward flow with cooling with uniform wall temperature or axial temperature gradient. They included the actual variations of specific heat, density, viscosity, and thermal conductivity for air. The

*Internal heat generation is not considered in this discussion.

two-dimensional case with linear density variation was treated by Vernier.[27]

(3) The governing equations were solved without the assumption of fully developed flow and with variations of the fluid properties. These methods were the most accurate; but, because of the complexity of the equations, the solutions had to be obtained by numerical methods. The author and W. T. Lawrence[28,29] used this approach for water, allowing for nonlinear variations of density and viscosity. Worsøe-Schmidt and Leppert[30,31] developed an implicit finite difference method using boundary-layer type equations for air. Experimental data were compared with theoretical predictions in References 20, 27, 28, and 29. Additional experimental work was reported in References 32 through 37.

Only the second and third approaches described previously can be considered as free or natural convection analyses, since density variations are not included in the first approach. In the following paragraphs the results of these methods will be discussed together with some general conclusions about nonadiabatic vertical channel flows.

Heat transfer affects the velocity and temperature profiles very strongly. Marked deviations occur already at $Gr/Re \simeq 4$. The overall quantities, however, such as the friction factor and Nusselt number, are only slightly affected at this value. The behavior of these two quantities is very similar. For the case of decreasing density in the upward direction, the theoretical friction factor and Nusselt number increase approximately exponentially with increasing Gr/Re. For the case of increasing density in the upward direction, these two quantities decrease very rapidly and the friction factor even becomes negative due to reverse flows occurring at the wall. For example, Brown's[19,20] analysis for fully developed flow yields the following relations:

$$Nu = 4.73 \qquad \text{for } \frac{Gr}{Re} < 12.8, \tag{18a}$$

$$Nu \simeq 2.5 \left(\frac{Gr}{Re}\right)^{0.25} \qquad \text{for } \frac{Gr}{Re} > 12.8$$

$$\text{for } \frac{\partial\rho}{\partial z} < 0 \ , \tag{18b}$$

$$fRe = 16 \qquad \text{for } \frac{Gr}{Re} < 8.4 \ , \tag{19a}$$

$$fRe \simeq 5.8\left(\frac{Gr}{Re}\right)^{0.475} \qquad \text{for } \frac{Gr}{Re} > 8.4$$

$$\text{for } \frac{\partial\rho}{\partial z} < 0 \ . \tag{19b}$$

Measured temperature profiles[20] and Nusselt numbers[20,32] agreed quite well with Equation (18) for moderate heat input rates. Equation (19) lies just above the upper range of the friction factor data observed with oil and water.[33]

For $\partial\rho/\partial z > 0$ both the Nusselt number and the friction factor decrease very rapidly beyond Gr/Re = 12.8 and 8.4, respectively, according to this analysis.

Although the fully developed flow analyses may represent the conditions existing at a given value of Gr/Re, in practice this parameter changes very rapidly with temperature. For example, for water with constant wall heat flux, the value of Gr/Re at 100°F is 2.5 times that at 60°F. Consequently, the profiles may never become fully developed but change continually along the channel. The exact behavior depends on the variations of the properties of the fluid, particularly the density and viscosity, with temperature. If Gr/Re increases with temperature, as in water, the velocity profiles will continually change until the velocity gradients become too steep and turbulence develops. Such transition to turbulence occurs if the channel is long enough. Of course, boiling may also start; but this is not considered here. Figure 4 shows a typical set of developing velocity profiles for upward flow of water with heating, calculated by Lawrence and Chato.[28,29] Figure 5 shows excellent agreement between calculated and measured centerline velocities. Figure 6 shows the development of the temperature profiles. The agreement between theory and experiment is excellent here too. Since the flows were never fully developed, the pressure gradients were not constant but varied continually as shown in Figure 7. The importance of using nonlinear density and viscosity functions is demonstrated in Figures 8 and 9. In the latter, a comparison is shown with the fully developed solution of Hanratty, <u>et al.</u>[22] The disparity between the various approximations is quite striking, and it indicates the importance of considering more exact, nonlinear properties when accurate predictions are needed.

With gases the situation is quite different because the viscosity increases and, consequently, Gr/Re tends to decrease with temperature. An implicit finite difference method was developed by Worsøe-Schmidt[30,31] for "perfect" gases allowing for nonlinear variations of the specific heat, viscosity, and thermal conductivity. Their results show that the velocity profiles distort in the entrance region similarly to the profiles shown in Figure 4, but then they straighten out again and eventually return to their initial, parabolic shape. For the particular properties chosen in Reference 30, this straightening out started at an axial distance of

$$\frac{z/R}{Re_o Pr_o} \simeq 0.015.$$

Because of this smoothening of the distortions, turbulence is not as likely to develop in such gases as in liquids.

For upward flowing gases with heating and a parabolic velocity profile at the entrance, the following local friction factors were recommended:[31]

$$f = \frac{16}{Re_m}\left(\frac{T_w}{T_b}\right)$$

$$\text{for } \left(\frac{T_w}{T_b}\right) < 1.2 \text{ to } 1.5. \qquad (20a)$$

For air and helium:

$$f = \frac{15.5}{Re_m}\left(\frac{T_w}{T_b}\right)^{1.10}$$

$$\text{for } \quad 1.5 < \left(\frac{T_w}{T_b}\right) < 3 . \qquad (20b)$$

For carbon dioxide:

$$f = \frac{15.5}{Re_m}\left(\frac{T_w}{T_b}\right)^{1.25}$$

$$\text{for } \quad 1.2 < \left(\frac{T_w}{T_b}\right) < 2 . \qquad (20c)$$

For upward flowing gases with cooling the following friction factor was suggested:[30]

$$f = \frac{16}{Re_m}\left(\frac{T_w}{T_b}\right)^{0.81}$$

$$\text{for } \quad 0.5 < \left(\frac{T_w}{T_b}\right) < 1 . \qquad (21)$$

The Nusselt numbers were usually correlated in terms of the deviation from the constant property value.[30,31,38] In particular, Reference 31 gives the following values for the upward flow, constant heat flux case.

For air and helium:

$$Nu_z = Nu_p + 0.025(q^o)^{1/2} \quad \frac{(Gz_z - 3)\ (Gz_z - 20)}{Gz_z^{3/2}}$$

$$\text{for } 3 < Gz_z < 1000 \quad \text{and} \quad 0 < q^o < 20 . \qquad (22a)$$

For carbon dioxide:

$$Nu_z = Nu_p + 0.07(q^o)^{1/2} \quad \frac{(Gz_z - 8)}{Gz_z^{1/2}}$$

$$\text{for } 10 < Gz_z < 1000 \quad \text{and} \quad 0 < q^o < 5 . \qquad (22b)$$

The constant property Nusselt number, Nu_p, can be evaluated from:

$$Nu_p = 1.58\, Gz_z^{0.3} \qquad \text{for} \quad Gz_z > 26 , \qquad (23a)$$

$$Nu_p = 4.2 \qquad \text{for} \quad Gz_z < 26 . \qquad (23b)$$

The results of Reference 9 indicate that the constant property Nusselt number is smaller with constant wall temperature and approaches a value of 3.6 at small Graetz numbers. In general the constant wall temperature case always approaches the constant property behavior as the fluid temperature approaches the wall temperature along the tube. Other channel geometries such as annular passages were investigated in References 39 through 48.

Inclined and Horizontal Tubes

Inclined, circular tubes with upward flow were investigated theoretically by Iqbal and Stachiewicz.[49] The assumption of fully developed flow was made, the density was varied only in the buoyancy term, and constant wall heat flux was considered. The results showed that the Nusselt number has a maximum with a tube inclination of between 20 to 60 degrees from the horizontal.

Most recent reports on horizontal tubes include both the constant wall temperature and constant wall heat flux cases. Jackson, et al.,[50] and Oliver[51] studied the constant wall temperature situation with fully developed profiles at the entrance. The latter summarized previous work and reported experimental data on relatively non-Newtonian fluids, including water. His empirical correlation for the mean Nusselt number is:

$$Nu_m = 1.75 \left(\frac{\mu_b}{\mu_w}\right)^{0.14} \left[Gz_m + 5.6 \times 10^{-4} \left(Gr_m Pr_m \frac{h}{D}\right)^{0.7}\right]^{1/3} \quad \text{for} \quad Gz_m > \pi\, Nu_m , \tag{24}$$

where

$$Gz_m \equiv \frac{wc_p}{Kh} ,$$

$$Gr_m \equiv \frac{\beta\,(T_w - T_b)\,D^3 g}{\nu_b^2} ,$$

and

$$Pr_m \equiv \frac{c_p \mu_b}{K} .$$

Casal and Gill[52] examined the fully developed, horizontal flow with constant axial temperature gradient, using a perturbation method. Their correlation for the friction factor is:

$$f = \frac{8}{Re_r}\left[1 + \frac{(Ray_m/Re_r)^2}{2304}\left(34.8\, Re_r Pe_r + 0.21\, Re_r^2\, Pe_r^2\right)\right] , \tag{25}$$

where

$$Ray_m \equiv \frac{\beta(\partial T/\partial z)\, R^4 g}{\nu_o^2} ,$$

$$Re_r \equiv \frac{\rho VR}{\mu} ,$$

and

$$Pe_r \equiv \frac{\rho_o V_o R c_p}{K} .$$

McComas and Eckert,[53] and Shannon and Depew[54] performed experiments with constant wall heat fluxes. They could not establish satisfactory correlations. In general, however, they found that in the entrance region the Nusselt number was the same as in pure forced convection; but further downstream the Nusselt number increased with increasing Grashof numbers to between 10 to 100 per cent above that in forced convection. This result indicates that the secondary flows due to buoyancy effects in horizontal tubes develop much slower than the velocity and temperature profiles at the entrance.

A theoretical treatment of the horizontal annulus was given by Abbott.[55]

Transition to Turbulence

The development of unstable flow is generally explained by some kind of laminar instability theory. For example, Lin[56] and Pai[57] discussed a theory according to which, for any velocity profile, there is a Reynolds number beyond which the profile will be unstable to disturbances within a certain range of wave numbers. For a given velocity profile there is a minimum, critical Reynolds number below which the flow will be laminar for any disturbance. Ryan and Johnson,[58] Hanks,[59] and Scheele and Greene[60] defined related stability criteria, which are based essentially on the idea that the ratio of the local acceleration force to the local viscous force cannot exceed a fixed number. This criterion was expressed in terms of a critical Reynolds number as follows.[60]

$$Re_c = 808 \left(\frac{\left| d^2U/dR^2 + (1/R)(dU/dR) \right|}{\left| U\, dU/dR \right|}\right)_{\text{minimum}} , \tag{26}$$

where $U \equiv V/Vm$, $R \equiv r/R$, and the constant was chosen to yield $Re_c = 2100$ for the adiabatic, parabolic velocity profile. Although the authors'[60] experimental data was in agreement with this criterion, its validity will have to be tested on much more extensive data. Other analyses were performed by Brown,[19,20] Joseph,[61] and Gill.[62]

Transition to turbulence or to unstable flows was observed by many authors. However, not all of them made quantitative evaluations. References 28, 29, 32, 33, 63, 64, and 65 contain experimental data and some correlations. For the case of $\partial\rho/\partial z < 0$ such as heating in upflow, turbulence may occur at Reynolds numbers as low as 30. For the case of $\partial\rho/\partial z > 0$ such as heating in downflow Scheele, et al.,[63,64] reported a drastic shift to asymmetric, but laminar flow patterns before the development of turbulence. This occurred very soon after flow reversal developed at the wall, i.e., when $Gr/Re \geq 25$.

Figure 10, from the work by Lawrence and Chato,[28,29] shows transition data in a heated, circular, vertical channel with upward flow. It is not known whether the curve is asymptotic at the right end. In a nonadiabatic channel with a water-like fluid, Gr/Re increases with temperature, and consequently, transition will occur if the channel is long enough. In these experiments turbulence was detected by the temperature fluctuations. It always occurred after the velocity profiles developed an inflection point. A dimensional correlation has been developed for the transition point in water, which depended on the characteristics of the local velocity profiles. Since these profiles have to be computed numerically, however, for quick estimates Figure 10 can be used. For comparison, Scheele, et al.,[63] found that as the Reynolds number decreased, the critical, local value of Gr/Re decreased asymptotically to about 42.5 for the case of upward flow with constant heat influx. The same reference indicated a decrease of the critical, local value of Gr/Re with increasing Reynolds number for the case of downward flow with constant heat flux. The highest Gr/Re occurred as $Re \rightarrow 0$ and its magnitude was 75 ± 3.

Metais, et al.,[66] attempted to map the various flow regimes for combined free and forced convection. Their plot, however, was based on somewhat meager data and can be considered only a very rough guide for preliminary estimates.

Once turbulence develops, buoyancy effects play a much less important role because the profiles become more uniform. Although there is considerable work done on the low Reynolds number, turbulent regime, somewhat modified forced convection correlations seem to be rather sufficient in predicting the behavior of such flows.

CONCLUSION

Although considerable advances have been made in the understanding of combined free and forced convection laminar flows, there are still gaps in our knowledge. Particularly, transient flow behavior and the transition to turbulence need further studies.

NOTATION*

a	$= 2C_f\left(\nu_o^2/gD^3\right)$
A	= area (L^2)
A_c	= cross-sectional area (L^2)
b	$= (C/4C_f)(D/h)$
c_p	= specific heat (θ/M-T)
C	= constant for evaluating entrance, exit, and acceleration losses
C_f	= constant in friction factor equation
D	= hydraulic diameter of channel (L)
f	= friction factor
g	= gravitational acceleration (L/t^2)
Gr	$= R^4 g\,(q/A)\,\beta/\nu^2 K$, Grashof number
Gz_m	$= wc_p/Kh$, mean Graetz number
Gz_z	$= wc_p/Kz$, local Graetz number
h	= length of channel (L)
h_T	= heat transfer coefficient ($\theta/t\text{-}L^2\text{-}T$)
k	$= A_c/D^2 = \pi/4$ for circle
K	= thermal conductivity (θ/t-L-T)
Nu	$= h_T D/K$, Nusselt number
Δp	= pressure change (F/L^2)
p	$= \Delta p/gh\rho_o$, dimensionless pressure change
Pe_r	$= \rho_o V_o R c_p/K$, Peclet number
q	$= Q\beta/2kD\mu_o c_p$, dimensionless heat input rate

*Units in parentheses are: M, mass; F, force; L, length; T, temperature; θ, heat (F-L); t, time.

q^o = $QD/2K_oT_oA_w$, dimensionless heat input rate

Q = heat input rate (θ/t)

r = radial distance (L)

r_c = $\overline{Re}_{2c}/\overline{Re}_1$, critical flow rate ratio

R = channel radius

Ray = $g\beta\ (\partial T/\partial z)\ R^4/\nu\alpha = 4\ Gr/Re$, Rayleigh number

Ray_m = $g\beta\ (\partial T/\partial z) R^4/\nu^2$, modified Rayleigh number

Re* = VD/ν, Reynolds number

Re_r = VR/ν, Reynolds number based on radius

T = absolute temperature (T)

ΔT_o = change of bulk temperature from entrance to exit (T)

U = V/V_m, dimensionless axial velocity

V* = axial velocity (L/t)

w* = mass flow rate (M/t)

X = $\rho/\rho_m = 1 - \beta\Delta T_o/2$

z = axial coordinate (L)

α = thermal diffusivity (L^2/t)

β = thermal expansion coefficient (T^{-1})

γ = temperature coefficient for φ (T^{-1})

Δ = finite change

μ = dynamic viscosity (F-t/L^2)

ν = kinematic viscosity (L^2/t)

ρ = density (M/L^3)

τ = γ/β

φ = property group defined by Equation (6)

Subscripts

b = bulk

c = critical value

e = fluid machinery

m = mean value

n = n-th channel

o = channel entrance

p = constant property value

w = wall

z = local value

*A bar above these symbols indicates directionalized quantities, positive for upward and negative for downward flow.

REFERENCES

1. Chato, J. C., "Natural Convection Flows in Parallel-Channel Systems," Journal of Heat Transfer, Transactions ASME, C, 85 (1963), pp. 339-345.

2. Chato, J. C. and W. T. Lawrence, "Natural Convection Flows in Single and Multiple Channel Systems," in Developments in Heat Transfer. Cambridge, Mass.: MIT Press, 1964, pp. 371-388.

3. Chato, J. C., "Metastable Natural-Convection Flows in Multiple Channel Systems with and Without Boiling," lecture notes for Two-Phase Gas-Liquid Flow Special Summer Program, MIT, Cambridge, Massachusetts, 1964.

4. Manglis, P. C., "An Approximate Theory of Fluid Flow in Heated Vertical Tubes," M.E. thesis, Department of Mechanical Engineering, Massachusetts Institute of Technology, Cambridge, Massachusetts, 1965.

5. Graetz, L., "Über die Wärmeleitfähigkeit von Flüssigkeiten," Annalen der Physik, 18 (1883), p. 79; 25 (1885), p. 337.

6. Sellars, J. R., M. Tribus, and J. S. Klein, "Heat Transfer to Laminar Flow in a Round Tube or Flat Conduit -- The Graetz Problem Extended," Transactions, ASME, 78 (1956), pp. 441-448.

7. Singh, S. N., "Heat Transfer by Laminar Flow in a Cylindrical Tube," Applied Science Research, A7 (1958), pp. 325-340.

8. Kays, W. M., "Numerical Solutions for Laminar-Flow Heat Transfer in Circular Tubes," Transactions, ASME, 77 (1955), pp. 1265-1274.

9. Ulrichson, D. L. and R. A. Schmitz, "Laminar-Flow Heat Transfer in the Entrance Region of Circular Tubes," International Journal of Heat and Mass Transfer, 8 (1965), pp. 253-258.

10. Langhaar, H. L., "Steady Flow in the Transition of a Straight Tube," Journal of Applied Mechanics, 9 (1942), p. A-55.

11. Tien, C. and R. A. Pawelek, "Laminar Flow Heat Transfer in the Entrance Region of Circular Tubes," Applied Science Research, A.13:4,5(1964), p. 317.

12. Roy, D. N., "Laminar Heat Transfer in the Inlet of a Uniformly Heated Tube," Journal of Heat Transfer, Transactions, ASME, C, 87 (1965), p. 425.

13. Siegel, R., E. M. Sparrow, and T. M. Hallman, "Steady Laminar Heat Transfer in a Circular Tube with Prescribed Wall Heat Flux," Applied Science Research, A7 (1958), p. 386-392.

14. Petukhov, B. S. and F. F. Tsvetkov, "Calculation of Heat Exchange Under Conditions of Laminar Flow of Fluids in Tubes Over a Range of Small Peclet Numbers," Inzhenerno fizicheskii Zhurnal, 4 (1961), p. 10.

15. Grigull, U. and H. Tratz, "Thermischer Einlauf in ausgebildeter laminarer Rohrströmung," International Journal of Heat and Mass Transfer, 8 (1965), pp. 669-678.

16. Ostroumov, G. A., "The Mathematical Theory of Heat Transfer in Circular, Vertical Tubes with Combined Forced and Free Convection," (in Russian), Zhurnal Tekhnicheskoi Fiziki, 20:6 (1950), pp. 750-757.

17. ————, "Free Convection Under the Conditions of the Internal Problem," NACA TM 1407, 1958.

18. Hallman, T. M., "Combined Forced and Free Laminar Heat Transfer in Vertical Tubes with Uniform Internal Heat Generation," Transactions, ASME, 78 (1956), pp. 1831-1841.

19. Brown, W. G. and P. Grassman, "Der Einfluss des Auftriebs auf Wärmeübergang und Druckgefälle bei erzwungener Strömung in lotrechten Rohren," Forschung auf dem Gebiete des Ingenieurwesens, 25:3 (1959), pp. 69-78.

20. Brown, W. G., "The Superposition of Natural and Forced Convection at Low Flow Rates in a Vertical Tube," Forschung auf dem Gebiete des Ingenieurwesens, 26, VDI - Forschungsheft 480 (1960).

21. Deissler, R. G., "Analytical Investigation of Fully Developed Laminar Flow in Tubes with Heat Transfer with Fluid Properties Variable Along the Radius," NACA TN 2410, 1951.

22. Hanratty, T. J., E. M. Rosen, and R. L. Kabel, "Effect of Heat Transfer on Flow Field at Low Reynolds Numbers in Vertical Tubes," Industrial and Engineering Chemistry, 50:5 (1958), pp. 815-820.

23. Pigford, R. L., "Nonisothermal Flow and Heat Transfer Inside Vertical Tubes," Chemical Engineering Progress Symposium Series, 51:17 (1955), pp. 79-92.

24. Rosen, E. M. and T. J. Hanratty, "Use of Boundary Layer Theory to Predict the Effect of Heat Transfer on the Laminar Flow Field in a Vertical Tube with a Constant Temperature Wall," AIChE Journal, 7:1 (1961), p. 112.

25. Tao, L. N., "Heat Transfer of Combined Free and Forced Convection in Circular and Sector Tubes," Applied Science Research, A9 (1960), pp. 357-368.

26. Bradley, D. and A. G. Entwistle, "Developed Laminar Flow Heat Transfer from Air for Variable Physical Properties," International Journal of Heat and Mass Transfer, 8 (1965), pp. 621-638.

27. Vernier, P. H., "Convection naturelle dans un canal vertical de section rectangulair chauffe uniformement," Report No. 2197, Centre d'Etudes Nucleaire de Grenoble (1962).

28. Lawrence, W. T., "Entrance Flow and Transition from Laminar to Turbulent Flow in Vertical Tubes with Combined Free and Forced Convection," Sc.D. thesis, Department of Mechanical Engineering, Massachusetts Institute of Technology, Cambridge, Massachusetts, 1965.

29. Lawrence, W. T. and J. C. Chato, "Heat Transfer Effects on the Developing Laminar Flow Inside Vertical Tubes," Journal of Heat Transfer, Transactions, ASME, C, 88:2 (1966), pp. 214-222.

30. Worsøe-Schmidt, P. M. and G. Leppert, "Heat Transfer and Friction for Laminar Flow of Gas in a Circular Tube at High Heating Rate," International Journal of Heat and Mass Transfer, 8 (1965), pp. 1281-1301.

31. Worsøe-Schmidt, P. M., "Heat Transfer and Friction for Laminar Flow of Helium and Carbon Dioxide in a Circular Tube at High Heating Rate," International Journal of Heat and Mass Transfer, 9 (1966), pp. 1291-1295.

32. Hallman, T. M., "Experimental Study of Combined Forced and Free Laminar Convection in a Vertical Tube," NASA TN D-1104, 1961.

33. Kemeny, G. A. and E. V. Somers, "Combined Free and Forced Convective Flow in Vertical Circular Tubes -- Experiments with Water and Oil," Journal of Heat Transfer, Transactions, ASME, C, 84 (1962), pp. 339-346.

34. Kays, W. M. and W. B. Nicoll, "Laminar Flow Heat Transfer to a Gas with Large Temperature Differences," Journal of Heat Transfer, Transactions, ASME, C, 85 (1963), pp. 329-338.

35. Hudson, J. L. and S. A. Khan, "Experimental Investigation into the Effect of the Thermal Boundary Condition on Heat Transfer in the Entrance Region of a Pipe," Mechanical Engineering Science, 6:3 (1964), p. 250.

36. Brown, C. K. and W. H. Gauvin, "Combined Free and Forced Convection. Part I, Heat Transfer in Aiding Flow; Part II, Heat Transfer in Opposing Flow," Canadian Journal of Chemical Engineering, 43:6 (1965), p. 306.

37. Davenport, M. E. and G. Leppert, "The Effect of Transverse Temperature Gradients on the Heat Transfer and Friction for Laminar Flow of Gases," Journal of Heat Transfer, Transactions, ASME, C, 87:2 (1965), pp. 191-196.

38. McEligot, D. M. and T. B. Swearingen, "Prediction of Wall Temperatures for Internal Laminar Heat Transfer," International Journal of Heat and Mass Transfer, 9 (1966), pp. 1151-1152.

39. Heaton, H. S., W. C. Reynolds, and W. M. Kays, "Heat Transfer in Annular Passages. Simultaneous Development of Velocity and Temperature Fields in Laminar Flow," International Journal of Heat and Mass Transfer, 7:7 (1964), p. 763.

40. Sutherland, W. A. and W. M. Kays, "Heat Transfer in an Annulus with Variable Circumferential Heat Flux," International Journal of Heat and Mass Transfer, 7 (1964), pp. 1187-1194.

41. Viskanta, R., "Heat Transfer with Laminar Flow in a Concentric Annulus with Prescribed Wall Temperature," Applied Science Research, A12: 6 (1964), p. 463.

42. Sparrow, E. M. and S. H. Lin, "The Developing Laminar Flow and Pressure Drop in the Entrance Region of Laminar Ducts," Journal of Basic Engineering, Transactions, ASME, D, 86 (1964), p. 827.

43. Rao, S. K. L., "Heat Transfer of Combined Free and Forced Convection in a Truncated Sectorial Tube," Bulletin, Polska Akademia Nauk, Serie des Sciences Techniques, 12:7 (1964), p. 383.

44. Khlebutin, G. N. and G. F. Shaidurov, "Heat Convection in a Vertical Annular Tube," Inzhenerno fizicheskii Zhurnal, 8:1 (1965), p. 3.

45. Saidikov, I. N., "Laminar Heat Transfer in the Starting Portion of a Rectangular Passage," Inzhenerno fizicheskii Zhurnal, 8:4 (1965), p. 423.

46. Cheng, K. C., "Laminar Heat Transfer in Noncircular Ducts by Moiré Method," Journal of Heat Transfer, Transactions, ASME, C, 87:2 (1965), p. 308.

47. Quarmby, A., "Note on Developing Laminar Flow in Annuli," Journal of Basic Engineering, Transactions, ASME, 88 (1966), pp. 811-812.

48. Sparrow, E. M. and A. Haji-Sheikh, "Flow and Heat Transfer in Ducts of Arbitrary Shape with Arbitrary Thermal Boundary Conditions," Journal of Heat Transfer, Transactions, ASME, C, 88 (1966), pp. 351-358.

49. Iqbal, M. and J. W. Stachiewicz, "Influence of Tube Orientation on Combined Free and Forced Laminar Convection Heat Transfer," Journal of Heat Transfer, Transactions, ASME, C, 88:1 (1966), pp. 109-116.

50. Jackson, T. W., J. M. Spurlock, and K. R. Purdy, "Combined Free and Forced Convection in a Constant Temperature Horizontal Tube," AIChE Journal, 7:1 (1961), pp. 38-41.

51. Oliver, D. R., "The Effect of Natural Convection on Viscous-Flow Heat Transfer in Horizontal Tubes," Chemical Engineering Science, 17 (1962), p. 335.

52. del Casal, E. P. and W. N. Gill, "A Note on Natural Convection Effects in Fully Developed Horizontal Tube Flow," AIChE Journal, 8 (1962), pp. 570-574.

53. McComas, S. T. and E. R. G. Eckert, "Combined Free and Forced Convection in a Horizontal Circular Tube," Journal of Heat Transfer, Transactions, ASME, C, 88:2 (1966), pp. 147-153.

54. Shannon, R. L. and C. A. Depew, "Combined Free and Forced Laminar Convection in a Horizontal Tube with Uniform Heat Flux," ASME Paper No. 67-HT-52, 1967.

55. Abbott, M. R., "A Numerical Method for Solving the Equations of Natural Convection in a Narrow Concentric Cylindrical Annulus with a Horizontal Axis," Quarterly Journal of Mechanics and Applied Mathematics, 17:4 (1964), p. 471.

56. Lin, C. C. Hydrodynamic Stability. London: Cambridge University Press, 1955.

57. Pai, S. Viscous Flow Theory. Vol. I. Princeton, N. J.: D. Van Nostrand Co., 1956, p. 309.

58. Ryan, N. W. and M. M. Johnson, "Transition from Laminar to Turbulent Flow in Pipes," AIChE Journal, 5:4 (1959), pp. 433-435.

59. Hanks, R. W., "The Laminar-Turbulent Transition for Flow in Pipes, Concentric Annuli, and Parallel Plates," AIChE Journal, 9 (1963), pp. 45-48.

60. Scheele, G. F. and H. L. Greene, "Laminar-Turbulent Transition for Nonisothermal Pipe Flow," AIChE Journal, 12 (1966), pp. 737-740.

61. Joseph, D. D., "Variable Viscosity Effects on the Flow and Stability of Flow in Channels and Pipes," The Physics of Fluids, 7:11 (1964), p. 1761.

62. Gill, A. E., "On the Behavior of Small Disturbances to Poisseuille Flow in a Circular Pipe," Journal of Fluid Mechanics, 21, Part 1 (1965), p. 145.

63. Scheele, G. F., E. M. Rosen, and T. J. Hanratty, "Effect of Natural Convection on Transition to Turbulence in Vertical Pipes," Canadian Journal of Chemical Engineering, 38 (1960), pp. 67-73.

64. Scheele, G. F. and T. J. Hanratty, "Effect of Natural Convection on Stability of Flow in a Vertical Pipe," Journal of Fluid Mechanics, 14, Part 6 (1962), pp. 244-256.

65. ———, "Effect of Natural Convection Instabilities on Rates of Heat Transfer at Low Reynolds Numbers," AIChE Journal, 9:2 (1963), pp. 183-185.

66. Metais, B. and E. R. G. Eckert, "Forced, Mixed, and Free Convection Regimes," Journal of Heat Transfer, Transactions, ASME, C, 86:2 (1964), pp. 295-296.

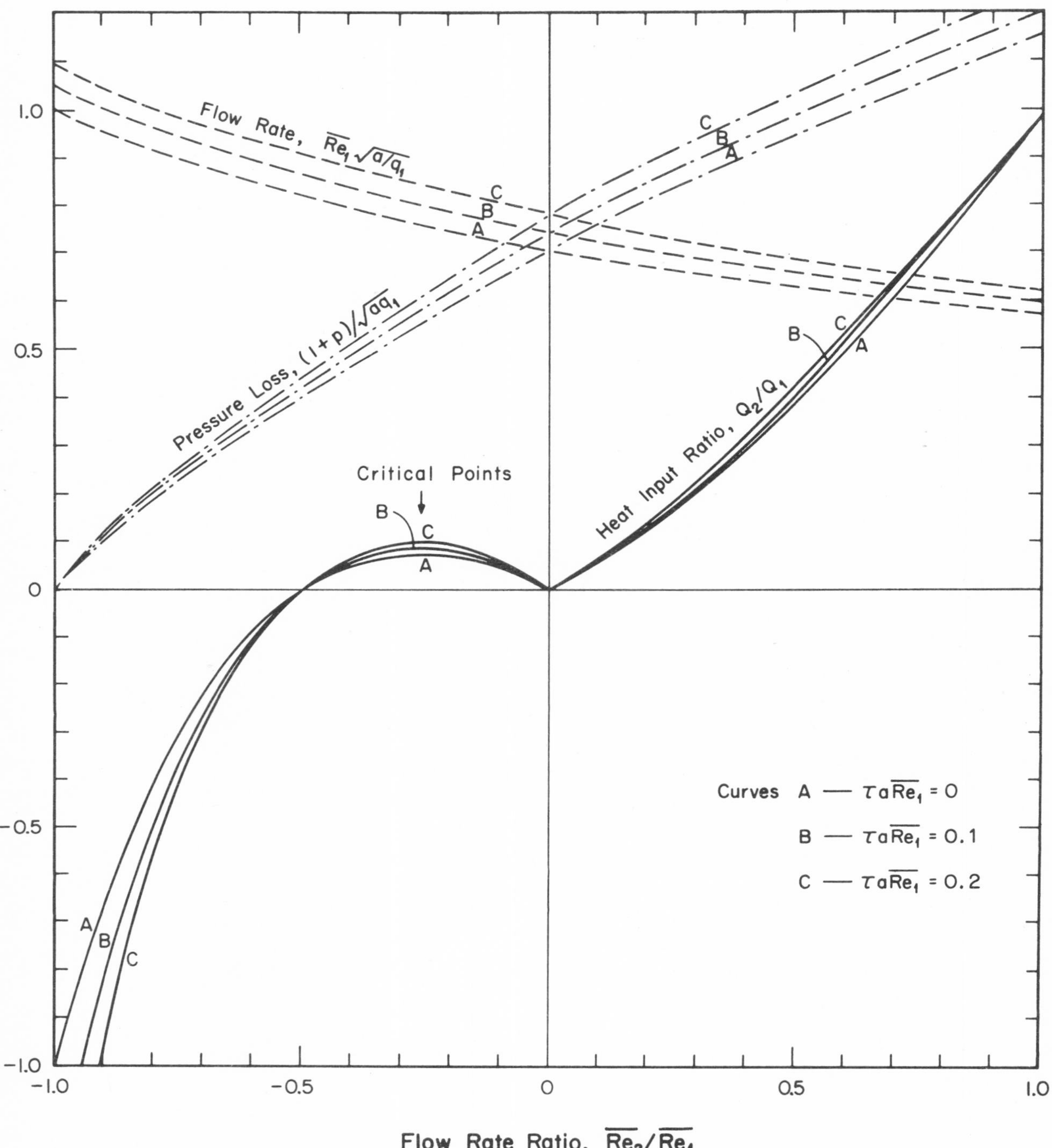

FIGURE 1. PERFORMANCE CHARACTERISTICS OF A SYSTEM OF THREE IDENTICAL CHANNELS WITH IDENTICAL ENTRANCE TEMPERATURES. THE THIRD CHANNEL IS ADIABATIC AND q_1 IS CONSTANT. (AFTER REFERENCE 2)

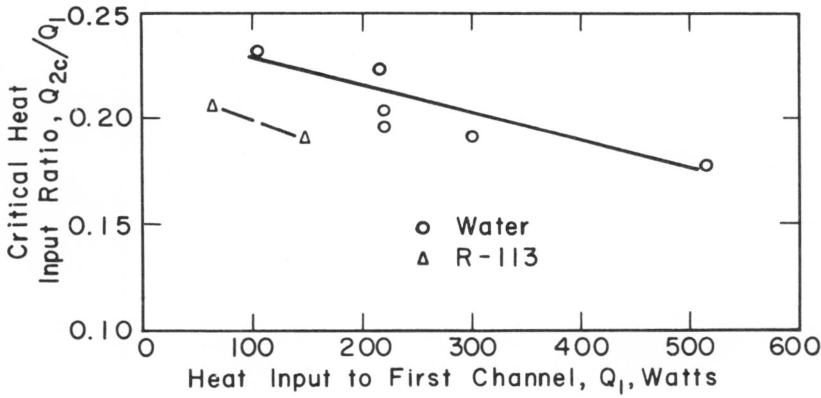

FIGURE 2. EFFECT OF THE HEAT INPUT RATE ON THE CRITICAL HEAT INPUT RATIO IN A THREE-CHANNEL SYSTEM. (h = 45.25 in., D_1 = 0.4 in., D_2 = 0.4 in., D_3 = 0.3 in.) (AFTER REFERENCE 3)

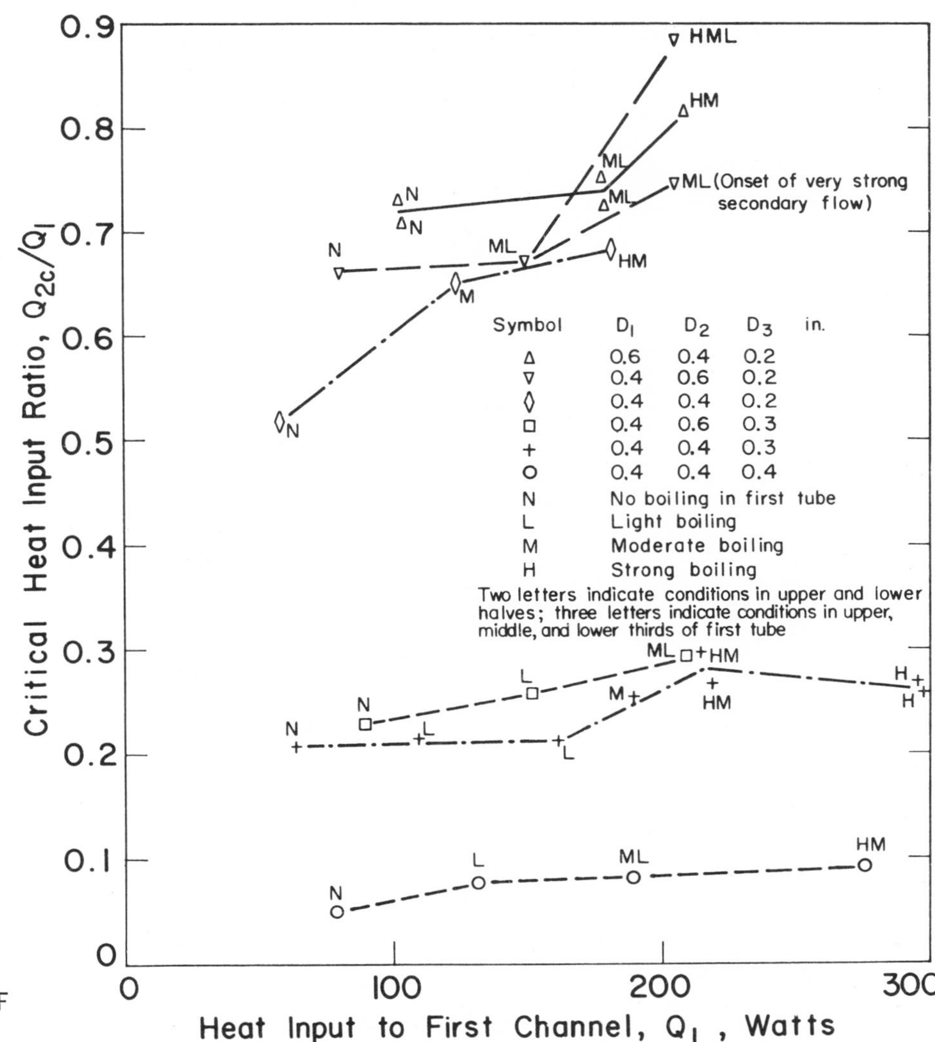

FIGURE 3. EFFECT OF BOILING IN THE FIRST CHANNEL OF A THREE-CHANNEL SYSTEM WITH REFRIGERANT 113. (AFTER REFERENCE 3)

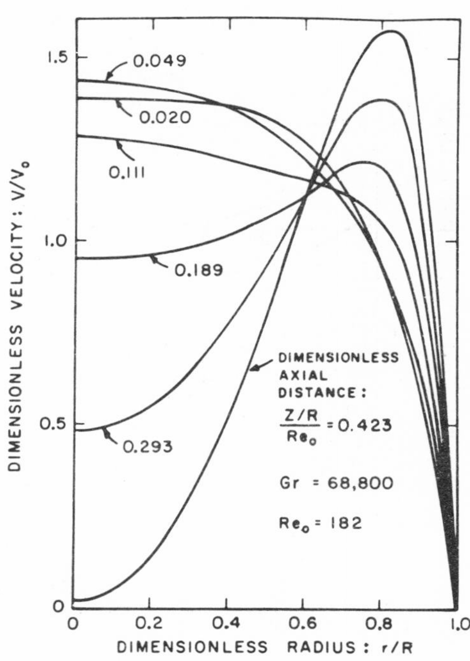

FIGURE 4. THEORETICAL DEVELOPING VELOCITY PROFILES IN A HEATED CIRCULAR CHANNEL WITH UPWARD FLOW. (AFTER REFERENCES 28, 29)

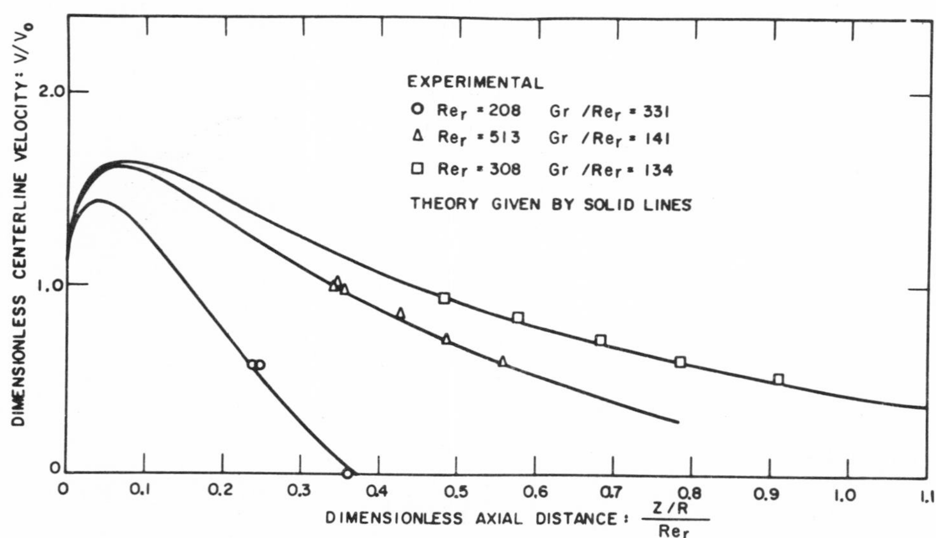

FIGURE 5. EXPERIMENTAL AND THEORETICAL DEVELOPING CENTERLINE VELOCITIES. (AFTER REFERENCES 28, 29)

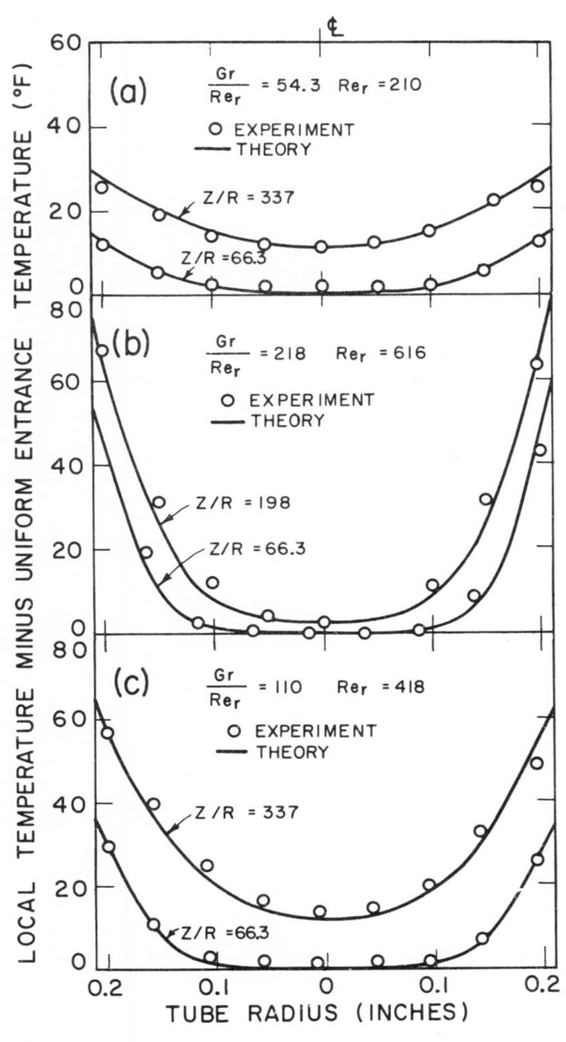

FIGURE 6. EXPERIMENTAL AND THEORETICAL DEVELOPING TEMPERATURE PROFILES. (AFTER REFERENCES 28, 29)

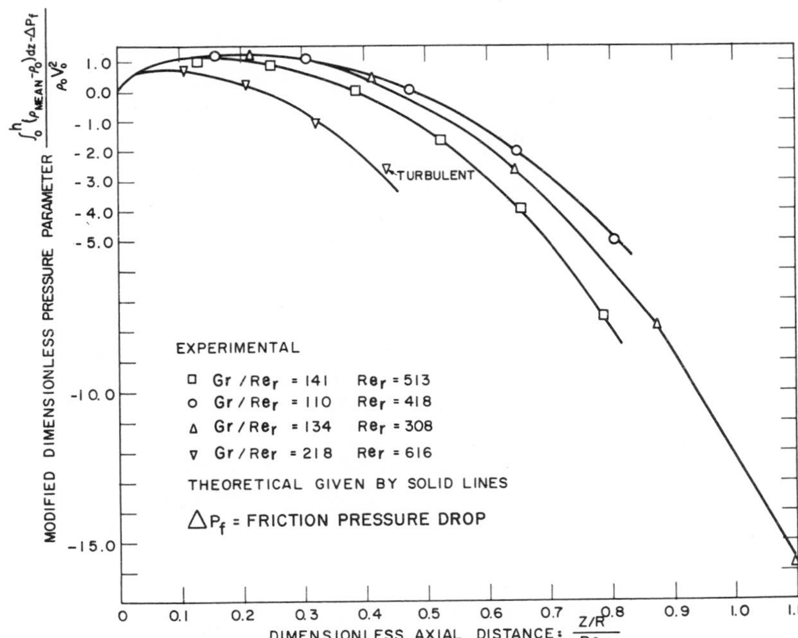

FIGURE 7. EXPERIMENTAL AND THEORETICAL VARIATION OF THE PRESSURE PARAMETER ALONG THE TUBE. (AFTER REFERENCES 28, 29)

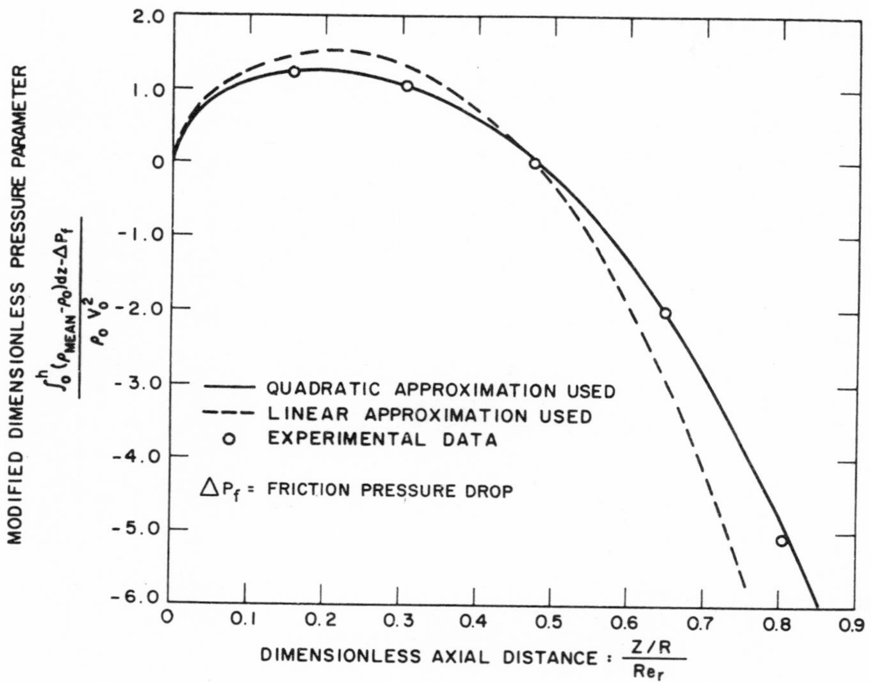

FIGURE 8. COMPARISON OF THE EXPERIMENTAL PRESSURE DROP PARAMETER WITH THE THEORETICALLY CALCULATED VALUES USING LINEAR AND QUADRATIC DENSITY FUNCTIONS. (AFTER REFERENCES 28, 29)

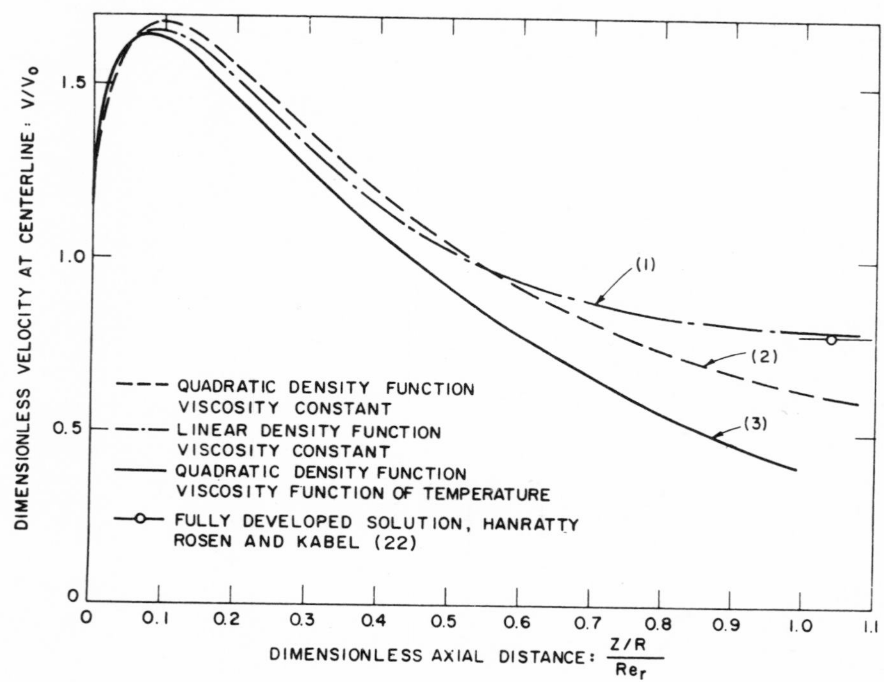

FIGURE 9. COMPARISON OF THE THEORETICALLY CALCULATED CENTERLINE VELOCITIES BASED ON DIFFERENT DENSITY AND VISCOSITY FUNCTIONS. (AFTER REFERENCES 28, 29)

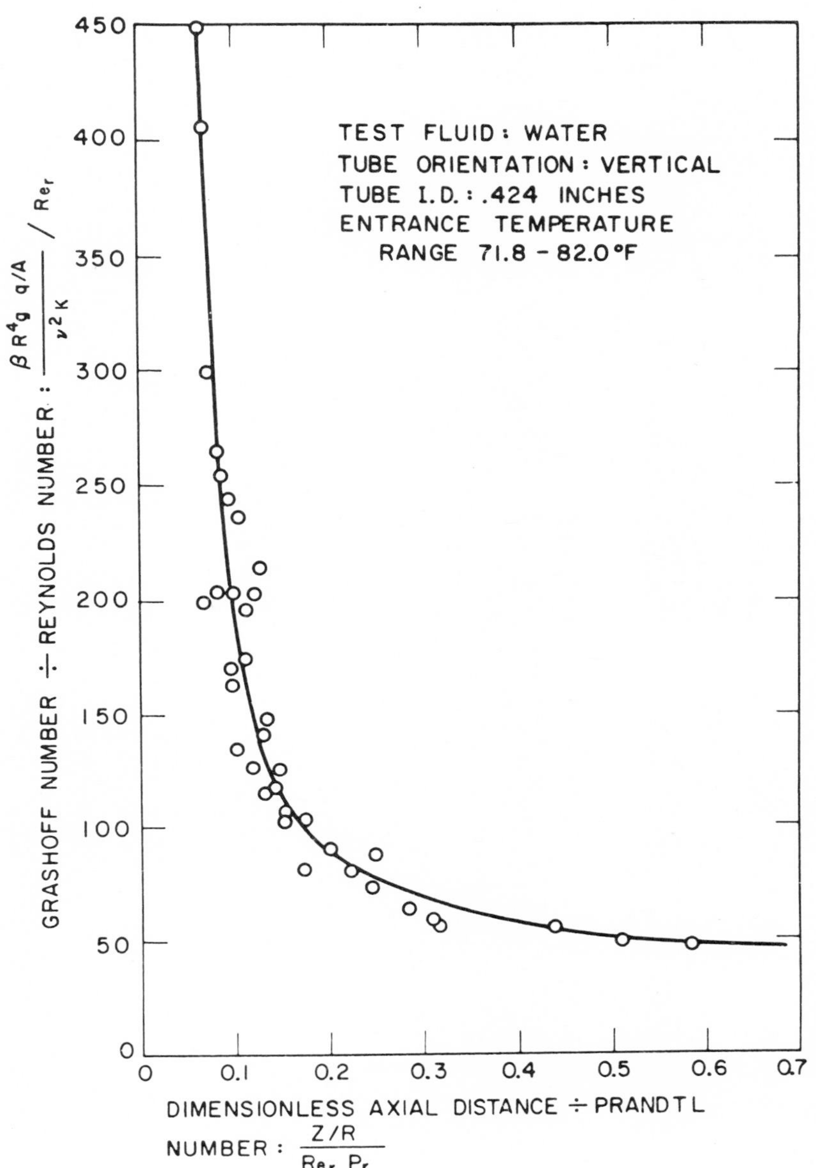

FIGURE 10. TRANSITION TO TURBULENCE IN A HEATED, CIRCULAR CHANNEL WITH UPWARD FLOW. (AFTER REFERENCES 28, 29)